Chemistry in Context

Applying Chemistry to Society

Fifth Edition

Lucy Pryde Eubanks
Clemson University

Catherine H. Middlecamp
University of Wisconsin–Madison

Norbert J. Pienta
University of Iowa

Carl E. Heltzel
Transylvania University

Gabriela C. Weaver
Purdue University

A Project of the American Chemical Society

Boston   Burr Ridge, IL   Dubuque, IA   Madison, WI   New York   San Francisco   St. Louis
Bangkok   Bogotá   Caracas   Kuala Lumpur   Lisbon   London   Madrid   Mexico City
Milan   Montreal   New Delhi   Santiago   Seoul   Singapore   Sydney.   Taipei   Toronto

Higher Education

CHEMISTRY IN CONTEXT: APPLYING CHEMISTRY TO SOCIETY, FIFTH EDITION

Published by McGraw-Hill, a business unit of The McGraw-Hill Companies, Inc., 1221 Avenue of the Americas,
New York, NY 10020. Copyright © 2006, 2003, 2000, 1997, 1994 by American Chemical Society. All rights reserved.
No part of this publication may be reproduced or distributed in any form or by any means, or stored in a database or
retrieval system, without the prior written consent of The McGraw-Hill Companies, Inc., including,
but not limited to, in any network or other electronic storage or transmission, or broadcast for distance learning.

Some ancillaries, including electronic and print components, may not be available to customers outside
the United States.

 This book is printed on recycled, acid-free paper containing 10% postconsumer waste.

1 2 3 4 5 6 7 8 9 0 QPD/QPD 0 9 8 7 6 5

ISBN 0–07–282835–8

Editorial Director: *Kent A. Peterson*
Sponsoring Editor: *Thomas D. Timp*
Managing Developmental Editor: *Shirley R. Oberbroeckling*
Senior Marketing Manager: *Tamara L. Good-Hodge*
Senior Project Manager: *Gloria G. Schiesl*
Senior Production Supervisor: *Kara Kudronowicz*
Lead Media Project Manager: *Judi David*
Senior Media Technology Producer: *Jeffry Schmitt*
Designer: *Rick D. Noel*
Cover/Interior Designer: *Rokusek Design*
(USE) Cover Image: © *Masterfile, Spider Web by R. Ian Lloyd*
Lead Photo Research Coordinator: *Carrie K. Burger*
Photo Research: *Karen Pugliano*
Supplement Producer: *Brenda A. Ernzen*
Compositor: *The GTS Companies/York, PA Campus*
Typeface: *10/12 Times Roman*
Printer: *Quebecor World Dubuque, IA*

The credits section for this book begins on page 603 and is considered an extension of the copyright page.

Library of Congress Cataloging-in-Publication Data

Chemistry in context : applying chemistry to society. — 5th ed. / Lucy Pryde Eubanks ... [et al.].
        p.   cm.
"A Project of the American Chemical Society."
Includes index.
ISBN 0–07–282835–8 (acid-free paper)
    1. Biochemistry. 2. Environmental chemistry. 3. Geochemistry. I. Eubanks, Lucy P. II. American Chemical Society.

QD415.C482   2006
540—dc22   2004029004
                                                            CIP

Brief Contents

Contents

Chapter 3

The Chemistry of Global Warming   114

Chapter 4

Energy, Chemistry, and Society   170

Chapter 5

The Water We Drink   218

Chapter 6

Neutralizing the Threat of Acid Rain   266

Chapter 7

The Fires of Nuclear Fission   308

Chapter 11

Nutrition: Food for Thought 484

Chapter 12

Genetic Engineering and the Chemistry of Heredity 528

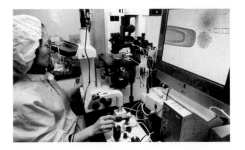

Appendix 1

Measure for Measure: Conversion Factors and Constants 569

Appendix 2

The Power of Exponents 570

Appendix 3

Clearing the Logjam 571

Appendix 4

Answers to Your Turn Questions Not Answered in Text 573

Appendix 5

Answers to Selected End-of-Chapter Questions Indicated in Color in the Text 582

Preface

Following in the tradition of its first four editions, the goal of *Chemistry in Context,* fifth edition, is to establish chemical principles on a need-to-know basis within a contextual framework of significant social, political, economic, and ethical issues. We believe that by using this approach, students not majoring in a science develop critical thinking ability, the chemical knowledge and competence to better assess risks and benefits, and the skills that lead them to be able to make informed and reasonable decisions about technology-based issues. The word *context* derives from the Latin word meaning "to weave." Thus, the spider web motif on the cover, used for the first four editions, continues with this edition because a web exemplifies the complex connections between chemistry and society.

Chemistry in Context is not a traditional chemistry book for nonscience majors. In this book, chemistry is woven into the web of life. The chapter titles of *Chemistry in Context* reflect today's technological issues and the chemistry principles imbedded within them. Global warming, acid rain, alternative fuels, nutrition, and genetic engineering are examples of such issues. To understand and respond thoughtfully in an informed manner to these vitally important issues, students must know the chemical principles that underlie the sociotechnological issues. This book presents those principles as needed, in a manner intended to better prepare students to be well-informed citizens.

Organization

The basic organization and premise remain the same as in previous editions. The focal point of each chapter is a real-world societal issue with significant chemical context. The first six chapters are core chapters in which basic chemical principles are introduced and expanded upon on the need-to-know basis. These six chapters provide a coherent strand of issues focusing on a single theme—the environment. Within them, a foundation of necessary chemical concepts is developed from which other chemical principles are derived in subsequent chapters. Chapters 7 and 8 consider alternative (nonfossil fuel) energy sources— nuclear power, batteries, fuel cells, and the hydrogen economy. The emphases in the remaining chapters are carbon-based issues and chemical principles related to polymers, drugs, nutrition, and genetic engineering. Thus, a third of the text has an organic/biochemistry focus. These latter chapters provide students with the opportunity to concentrate on additional interests beyond the core topics, as time permits. Most users teach seven to nine chapters in a typical one-semester course.

What's New

Art Program

The art program for the fifth edition has been totally updated for consistency and accuracy. Chemical structures emphasize the important details of bonding and reactive sites. Details of chemical processes are emphasized in many figures, and real-world data have been updated and their presentation clarified to help students understand the information.

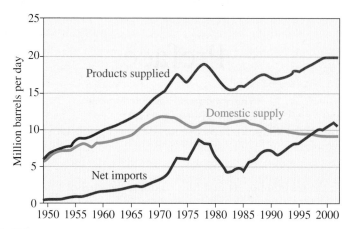

Figure 4.12

U.S. petroleum product use, domestic production and imports.

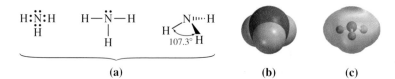

Figure 4.13

Sources of crude oil and petroleum products imported by the U.S. (August 2003).

Spartan Charge–Density Diagrams

Charge–density diagrams show charge distribution in molecules and have been added together with Spartan space-filling models to provide other representations of a molecule. This type of diagram will help the student understand solubility, acidic and basic properties, and the reasons behind many reactions.

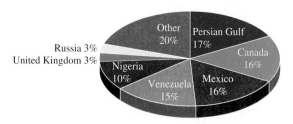

Figure 3.10

Representations of NH_3

Chapter 0, "Why the Spiderweb?"

Chapter 0 walks students through use of all of the resources available to them in *Chemistry in Context*. Much of the information found in this preface in previous editions is now placed in the new introductory chapter. Although written for students, Chapter 0 will also serve to explain the pedagogy, problem-solving opportunities, and many of the media resources of the fifth edition to instructors.

New and Updated Content

The major focus of a new edition is to update topical content. All information is as up-to-date as possible using a printed format. The resources of the World Wide Web allow students to acquire real-time data, seek out current information, and make their own risk–benefit analysis about topics at the science–society interface.

This edition introduces many new or expanded topics while keeping the same chapter organization. The discussion of air quality now includes more information on ozone as a secondary pollutant. The role of volatile organic compounds and free radicals in changing air quality, as well as information about how to interpret the Air Quality Index, are included. Computer modeling as the basis for making decisions receives more attention, particularly in discussing and updating issues of stratospheric ozone depletion and global warming. Energy-related topics now include biodiesel fuels and expanded coverage of fuel cells, hybrid cars, and the hydrogen economy. Nanotechnology is introduced early in this edition, and followed with examples of its potential uses in storing hydrogen and developing new materials. The comparison of nuclear power with fossil-fuel power is explored more explicitly. The role of reactive forms of nitrogen in forming acid rain is explained in greater detail than in previous editions. Organic functional groups have been formally introduced in the context of polymers, enabling a smooth transition into the rest of the organic and biochemical chapters that follow. The coverage of drugs in the news has been expanded to include designer steroids, drugs of abuse, and herbal remedies. Nutrition topics have been reorganized to enable students to better understand and make decisions about popular diets. Stem cell research has been introduced in the final chapter of the text, along with increased information about DNA fingerprinting, cloning, transgenic foods, and the Human Genome Project.

New and Updated Resources on the Online Learning Center (www.mhhe.com/cic)

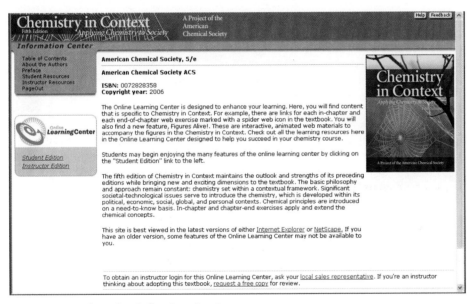

Opening Screen from the *Online Learning Center*

The *Online Learning Center* (OLC) is a comprehensive, book-specific Web site offering excellent tools for both the instructor and the student. Instructors can create an interactive course with the integration of this site, and a secured Instructor Center stores your essential course materials to save you prep time before class. This Instructor Center offers the Instructors Resource Guide and a *brand new* test bank of questions. The Student Center offers Web Exercises, Figures Alive Interactives and quiz questions for each chapter. The *Online Learning Center* content has been created for use in PageOut,

WebCT, and Blackboard course management systems. Questions developed by Devon Latimer from the University of Winnipeg are included in the WebCT content.

Figures Alive Interactives, marked by this icon near the figure in the text, lead the student through the discovery of various layers of knowledge inherent in the figure and enables them to develop their own understanding. Each chapter now has an interactive learning experience tied to a specific figure in the chapter. The self-testing segments built into Figures Alive! are based on the same categories as the chapter-end problems—*Emphasizing Essentials, Concentrating on Concepts,* and in many cases, *Exploring Extensions.*

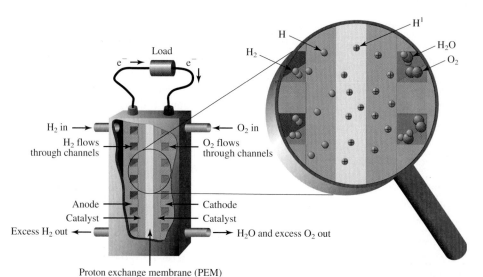

A screen shot from Chapter 8: "Energy from Electron Transfer"

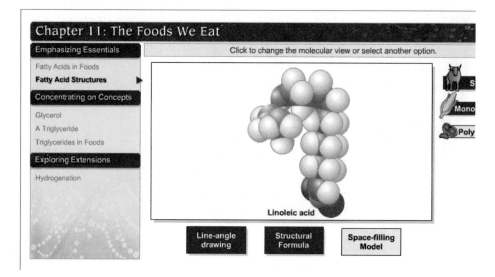

A screen shot from Chapter 11: "Nutrition: Food for Thought"

The *Instructors Resource Guide,* edited by Carl Heltzel (Transylvania University), can be found on the *Online Learning Center* under Instructor Resources and on the Instructor's Testing and Resource CD-ROM. The guide contains:

- A chemical topic matrix provides listing of chemical principles commonly covered in a general chemistry course.

- Course syllabi give some indication about the scope, pace and scheduling of the course.

- Topical essays give a variety of background material and pragmatic suggestions for teaching strategies and student development goals.

- Answers for suggested responses to many of the open-ended questions in the Consider This sections and the solutions to the in-chapter and chapter-end exercises and questions.

- The instructors guide for the laboratory experiments.

The *Digital Content Manager* is a multimedia collection of visual resources allowing instructors to utilize artwork from the text in multiple formats to create customized classroom presentations, visually based tests and quizzes, dynamic course Web site content, or attractive printer support materials. The Digital Content Manager is a cross-platform CD containing an image library, photo library, and a table library.

A screen shot from the Digital Content Manager

The *Instructor's Testing and Resource CD* contains the electronic file of the Instructor's Resource Guide. It also contains a *new* feature designed to help provide testing options for the instructor. For the first time, *Chemistry in Context* has a test bank specific to the text. These individuals wrote and edited questions for the test bank:

Eric Bosch	*Southwest Missouri State University*
Mark Freilich	*University of Memphis*
Penny J. Gilmer	*Florida State University*
Amy J. Phelps	*Middle Tennessee State University*
Julie Smist	*Springfield College*
Thomas Zona	*Illinois State University*

The test bank was accuracy checked by:

Marcia L. Gillette	*University of Indiana–Kokomo*

Art for the test bank was prepared by:

I. Dwaine Eubanks	LATEst IDEas, Inc.

This resource contains approximately 50 multiple-choice questions for every chapter. The questions are comparable to the problems in the text. The test bank is formatted for easy integration into the following course management systems: WebCT, and Blackboard. You may also choose to use these questions as models for writing your own classroom-specific test questions.

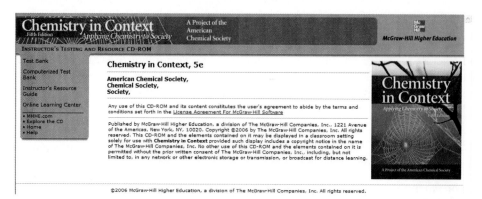

Opening Screen from the Instructor's Testing and Resource CD-ROM

Other New and Updated Resources

The *transparency set* contains selected four-color illustrations from the text reproduced on acetate for overhead projection. Because of the cost associated with producing this set, the transparency set will not automatically be sent to users. Please request transparencies if you use this mode of course delivery.

For those whose course includes a laboratory component, a *Laboratory Manual*, compiled and edited by Gail Steehler (Roanoke College), is available for the fifth edition. The experiments use microscale equipment (wellplates and Beral-type pipets) and common materials. Project-type and cooperative/collaborative laboratory experiments are included. Additional experiments are available on the *Online Learning Center*, as is the instructor's guide.

Special Acknowledgments

It is always a pleasure to bring a new textbook or new edition to fruition. But the work is not done by just one individual. It is a team effort, one comprised of the work of many talented individuals. The fifth edition builds on the proud tradition of prior author teams, led by A. Truman Schwartz (Macalester College) for the first and second editions, and by Conrad L. Stanitski (University of Central Arkansas) for the third and fourth editions. We have been fortunate to have the continuing, unstinting support of Sylvia A. Ware, Director of the ACS Division of Education, who helped to create the first edition of *Chemistry in Context*. We also recognize the able assistance of Jerry A. Bell and Marta Gmurczyk of the ACS Education Division office during preparation of both the fourth and fifth editions.

The McGraw-Hill team has been superb in all aspects of this project. Kent Peterson (Director of Editorial), Thomas Timp (Sponsoring Editor), and Shirley Oberbroeckling (Managing Developmental Editor) led this outstanding team. Tami Hodge serves as the Senior Marketing Manger. The Senior Project Manager is Gloria Schiesl, who coordinates the production team of Rick Noel (Designer), Carrie Burger (Lead Photo Researcher), Kara Kudronowicz (Production Supervisor), and Brenda Ernzen (Supplemental Production Coordinator). The Senior Media Producer is Jeffry Schmitt and Judi David serves as Lead Media Project Manager. The team also benefited from the knowledgeable editing of Linda Davoli and from the persistent work of Karen Pugliano in tracking down elusive photos. Dwaine Eubanks brought both his chemical knowledge and computer-based artistic skills together to establish a new high standard for the art in this edition.

The fifth edition is the product of a collaborative effort among writing team members—Lucy Pryde Eubanks, Catherine Middlecamp, Norbert Pienta, Carl Heltzel and Gabriela Weaver. This is the maiden voyage in this realm for Carl Heltzel and Gabriela

Weaver, new coauthors and colleagues. We welcome them to the team and have benefited from their diverse expertise.

We are very excited by the new features of this fifth edition, which exemplify how we continue to "press the envelope" to bring chemistry in creative, appropriate ways to non-science majors, while being honest to the science. We look forward to your comments.

Lucy Pryde Eubanks
Senior Author and Editor-in-Chief
October 2004

Further Acknowledgments

We would like to thank these individuals, whose comments were of great help to us in preparing this revision.

John Allen	*Southeastern Louisiana State University*
Joseph Bariyanga	*University of Wisconsin–Milwaukee*
Edward J. Baum	*Grand Valley State University*
David R. Bjorkman	*East Carolina University*
Bruce S. Burnham	*Rider University*
Joseph Chaiken	*Syracuse University*
Theo Clark	*Truman State University*
Cynthia Coleman	*State University of New York–Potsdam*
Kimberley R. Cousins	*California State University–San Bernardino*
Dru L. DeLaet	*Southern Utah University*
Jeannine Eddleton	*Virginia Tech University*
Rosemary Effiong	*University of Tennessee at Martin*
John Galiotos	*Houston Community College*
Stephen J. Glueckert	*University of Southern Indiana*
Tammy Jahnke	*Southwest Missouri State University*
Cindy Kepler	*Bloomsburg University*
Kevin Kittredge	*Miami University–Middletown*
Sara-Kaye Madsen	*Truman State University*
Eric Miller	*San Juan College*
Maria Pacheco	*Buffalo State College*
Linda Pallack	*Washington & Jefferson College*
Kutty Pariyadath	*University of South Carolina–Aiken*
Holly Phaneuf	*Salt Lake Community College*
Brian Polk	*Rollins College*
Bert Ramsey	*Eastern Michigan University*
Kresimir Rupnik	*Louisiana State University*
Benjamin E. Rusiloski	*Delaware Valley College*
Anne Marie Sokol	*New York State University–Buffalo*
John Todd	*Bowling Green State University*
Chris Truitt	*Texas Tech University*
John Vincent	*University of Alabama*
Marcy Whitney	*University of Alabama*
Thomas Zona	*Illinois State University*
Martin Zysmilich	*George Washington University*

Chemistry
in Context

chapter

0

Why the Spiderweb?

Dear Students,

Have you wondered why the image of a spiderweb appears on the cover of your chemistry text? Perhaps you connected it with the World Wide Web or with the incredible influence the Web has wielded on all aspects of our lives. However, when the first edition of *Chemistry in Context* was published in 1994, few college courses used the resources of the Web. Therefore, this was not the origin of the spiderweb motif. Rather, our title, *Chemistry in Context,* provides the clue to the choice of the spiderweb. The word *context* derives from the Latin word meaning "to weave." The spiderweb reminds you that this text emphasizes the strong and complex connections that exist among chemistry, societal, and personal concerns. Therefore, we continue the tradition of using a spiderweb in this fifth edition.

The spiderweb motif carries another and more subtle message, however. Spiders are industrious little creatures, often rebuilding their webs each day. Their most difficult task always is to establish the first anchoring thread for the web. Spiders do this by releasing a long sticky silken thread that blows with the wind until it attaches and becomes securely fastened. In *Chemistry in Context,* the authors have established anchoring threads for you by choosing several of today's real-world issues that have significant chemical context. You will need to work every day to build your own network of connecting strands, forming a strong, resilient web of knowledge, attitudes, and skills that will help you in *Applying Chemistry to Society,* the subtitle of this text.

Because we want your web to be well formed, we have, through four previous editions, designed ways to enhance your weaving. You will soon discover that the chemistry is presented when you need to apply it. This approach will help you to become a well-informed citizen no matter what career path you choose. For example, if you are following the thread of learning about air quality, you need to know what substances are found in air and why the concentrations of such substances matter, even at very low levels. If you are considering the issues surrounding global warming, it is fundamental to understand how evidence is gathered and evaluated and why the shapes of molecules make a difference in the story. Energy is a thread that reappears throughout the text—in the context of combustion of fossil fuels, nuclear power, or alternative means of providing energy for transportation. Understanding what makes drinking water safe requires some specific chemical knowledge, enough to make informed choices that have both health and financial consequences. Many of the later chapters deal with the chemistry involved in personal issues such as nutrition, drugs, and genetic engineering. In every case, chemistry can help to inform decision making about important societal and personal choices.

As is the case with our industrious spiders, consistent practice in refining your web of knowledge will be beneficial. Do not just "read" this text, much as you might read a novel. Rather, stop and participate in the activities embedded in the text and on the text-specific *Online Learning Center.* Become an active learner, for this is one of the best ways for you to check your growing understanding. Here is a sampling of the different opportunities for learning you will find within *Chemistry in Context.*

Your Turn exercises give you a chance to practice a skill or calculation that has just been introduced in the text. Answers are often given following the exercise, or are found in Appendix 4. Plan to complete all the Your Turns as you proceed through a chapter in *Chemistry in Context.* Here is an example.

Your Turn 1.5   Using Parts Per Million (ppm)

a.   The EPA sets the permissible limit for the average concentration of carbon monoxide in an 8-hr period at 9 ppm. Express this concentration as a percentage.
b.   Exhaled air typically contains about 75% nitrogen gas. Express this concentration in parts per million.

Consider This activities give you a chance to use what you are learning to make informed decisions. They often require risk–benefit analysis, consideration of opposing viewpoints, speculation on the consequences of a particular action, or formulation and defense of a personal decision. These activities may require additional research, often from Web sources, as is the case with this example.

Consider This 5.30   Purifying Water Away From Home

How can you purify water when you are hiking? Use the resources of the Web to explore some of the possibilities. What are the relative costs and effectiveness of these alternatives? Are any of the methods similar to those used to purify municipal water supplies? Why or why not?

Sceptical Chymist activities require you to marshal your analytical skills to respond to various statements and assertions. The unusual spelling comes from an influential book written in 1661 by Robert Boyle, an early investigator studying the properties of air. His experimentation challenged some of the "conventional wisdom" of the time, which is what you will do in these activities. Here is an example.

Sceptical Chymist 11.29   Low-Fat Cheese

A popular brand of low-fat shredded cheddar cheese advertises that it provides 1.5 g of fat with 15 Calories from total fat per serving. There are 50 Calories per serving. Of the total fat, 1.0 g is saturated fat. A serving is defined as $\frac{1}{4}$ cup or 28 g. Is this a "low-fat" cheese? Defend your decision with some calculations. Remember that the dietary recommendation is that no more than 30% of Calories should come from fat.

A Chapter Summary is found at the end of each of *Chemistry in Context's* 12 threads of study. The summary is keyed to specific sections in each chapter, providing you with an efficient way to review topics. Here is a partial list from Chapter 9, The World of Plastics and Polymers, as an example.

Chapter Summary

Having studied this chapter, you should be able to:

- Understand the nature of plastics and polymers, their typical properties and molecular structures (9.1)
- Describe typical uses for the Big Six polymers (9.2)
- Be able to name and draw several functional groups (9.2)
- Understand the molecular mechanism of addition polymerization (9.3)

End-of-chapter questions are of three types.

- **Emphasizing Essentials** These questions give you the opportunity to practice the fundamental skills developed in the chapter. This set of questions relates most closely to the Your Turn exercises within the chapter. Answers are provided in Appendix 5 for questions with numbers printed in blue.

- **Concentrating on Concepts** Questions in this category require you to focus on the chemical concepts developed in the chapter and their relationships to the topics under discussion. Such questions allow you to integrate and apply chemical concepts. This set of questions most closely resembles Consider This activities within the chapter. Answers are provided in Appendix 5 for questions whose numbers printed in blue.

- **Exploring Extensions** These questions challenge you to go beyond the information presented in the text. They provide an opportunity for extending and integrating the skills, concepts, and communication abilities practiced in the chapter.

End-of-text material Yes, there is more! You will want to check out these resources as well. At the back of the book is an appendix containing conversion factors, and another one that will help you review operations with exponents. You will find the appendix on logarithms particularly useful for understanding the concept of pH in Chapter 6. There are answers to Your Turns that have not been already answered in the text and answers to selected end-of-chapter questions. A very useful resource is the **Glossary,** a handy place to do a quick check of vocabulary. Each term is keyed to a page number where the term is first defined and explained in context. The **Index** will lead you efficiently to information in the text, figures, and tables.

Figures Alive! This icon in the text alerts you to visit McGraw-Hill's *Online Learning Center* at www.mhhe.com/cic. Each chapter has a figure that "comes alive," guiding you through practicing chapter essentials, developing concepts, and, in many cases, exploring extensions based on a figure in the text. Return visits to this resource are highly encouraged! Here is an example of a figure from Chapter 1 that is used as the basis for interactive learning.

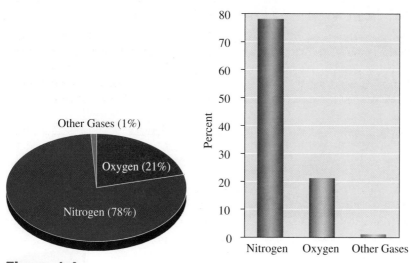

Figure 1.4
The composition of dry air, by volume.

Figures Alive! Visit the *Online Learning Center* to learn more about these graphs as well as about the atoms and molecules that make up air. Look for the Figures Alive! icon throughout this text.

Other media resources are found at McGraw-Hill's *Online Learning Center.*

- **Quiz questions** Each chapter has two sets of quiz questions. Check your own understanding and receive immediate feedback on your success.

- **Web links** The *Online Learning Center* provides quick access to many of the Web sites useful for those Consider This and Sceptical Chymist activities marked with this Web icon.

- **Other features** Look for Green Chemistry links when you see this icon in the text. You will also find chapter overviews, information about *Chemistry in Context* and its authors, and other useful features. This site is under constant development, so check for new resources not listed here.

As you journey through *Chemistry in Context*, remember the image of that industrious spider and its beautiful web. It illustrates that many interconnections are necessary to create a harmonious whole, and so it is with the topics in this text. Remember to think about how the particular thread under discussion connects to others. Said another way, any individual risk–benefit analysis must be considered in light of its potential connections with other societal and personal concerns. Looking at the "big picture" of how strands are connected is what some call a life cycle analysis.

And now, dear students, it is time for you to start weaving your own webs. We hope this orientation will prove useful. Our best wishes for a successful and most enjoyable experience with *Chemistry in Context*, fifth edition.

Most sincerely yours,

The Author Team

Chemistry in Context, fifth edition

Left to right:
Carl E. Heltzel, Transylvania University
Lucy Pryde Eubanks, Clemson University
Catherine H. Middlecamp, University of Wisconsin–Madison
Gabriela C. Weaver, Purdue University
Norbert J. Pienta, University of Iowa

The Air We Breathe

The "blue marble," our Earth, as seen from outer space.

"The first day or so, we all pointed to our countries. The third or fourth day, we were pointing to our continents. By the fifth day, we were aware of only one Earth."

**Prince Sultan Bin Salmon Al-Saud,
Saudi Arabian astronaut**

Individually and collectively, we take the air we breathe for granted. Yet our atmosphere is a fragile, thin veil of essential gases interspersed with pollutants in differing amounts. It surrounds the third planet from the Sun, helping to make habitable the place we call home. The striking words of astronaut James Irwin compel us to consider the awesome spectacle of our home planet: "Finally it shrank to the size of a marble, the most beautiful marble anyone can imagine." Only a few men and women have actually observed what James Irwin saw in July, 1971, but most of us have seen the spectacular photographs of the Earth taken from outer space. From that vantage point, our planet looks magnificent—a blue and white ball composed of water, earth, air, and fire. It is where thousands upon thousands of species of plants and animals live in a global community. More than 6 billion of us belong to one particular species with special responsibilities for the protection of our beautiful "blue marble."

As we move in from outer space, an aerial view of Earth from a satellite reveals more detail about the blue marble. The landforms visible in the computer-enhanced photograph of Figure 1.1 include rivers, lakes, islands, mountains, forests, and prairies. Truly, our planet has great geological diversity, and many biological species inhabit these varied environments. Although the gases of our atmosphere are invisible in this photo, we can see white areas of condensed water vapor, better known as clouds. These clouds that both shade us and give us rain, as well as the invisible air that surrounds them, are resources beyond price.

At ground level we arrive at the communities that we know best: the cities, towns, ranches, and farms where we live, study, play, work, and sleep. Our families, friends, and neighbors can be found here (Figure 1.2). We are shaped by the people, customs, and laws, as well as by the climate, natural resources, and air quality in these regional environments.

As individuals, we simultaneously inhabit these concentric communities. Our personal lives are embedded not only in our immediate surroundings, but also in our countries and the entire globe. Changes in any of these environments affect us and we, in turn, have obligations at each level of community, from personal to global. This book is about some of those responsibilities and the ways in which a knowledge of chemistry can help us meet them with intelligence, understanding, and wisdom.

Wherever you live, to be an informed (and healthy) member of your community, you should know about the air you breathe. In air are the chemical substances that are essential for your existence, as well as a few that can endanger it. To understand the

Figure 1.1

The Great Lakes, imaged by SeaWiFs (Sea-viewing Wide Field-of-view Sensor) aboard a satellite launched in 1997.

Figure 1.2
A festival in Los Angeles.

chemical complexities of air, you will need to become familiar with certain chemical facts and concepts. Therefore, this chapter begins by considering the composition of air, its major and minor constituents (including pollutants), and how the concentration of each can be expressed.

1.1 Everyday Breathing

We begin by asking you to do something you do automatically and unconsciously thousands of times each day—to take a breath. You certainly do not need textbook authors to tell you to breathe! A doctor or nurse may have encouraged your first breath, but from then on nature took over. Even when you hold your breath in a moment of fear or suspense, you soon involuntarily gasp a lungful of that invisible stuff we call air. Indeed, you could not survive more than 5–10 minutes without a fresh supply of air.

Consider This 1.1 **Take a Breath**

What total volume of air do you exhale in a typical day? Although you could simply guess, a simple experiment can enable you to come up with a reasonably accurate answer. You will need to determine how much air you exhale in a single "normal" breath and how many breaths you "normally" take per minute. Once you establish this information, determine how much air you exhale in a day (24 hours). Describe the experiment you performed, provide the data you obtained, and list any factors you believe may have affected the accuracy of your answer.

The previous exercise addresses how much air you breathe, but not the equally important topic of *what* you breathe and whether it might be harmful. For a commentary on air quality we turn to a statement from a Shakespearean play about a troubled young man named Hamlet: ". . . this most excellent canopy, the air, look you, this brave o'erhanging firmament, this majestical roof fretted with golden fire, why, it appears no other thing to me but a foul and pestilent congregation of vapors." To be sure, the speaker

Table 1.1	Changes in Air Quality	
Criteria Air Pollutant	**1982–2001 (%)**	**1992–2001 (%)**
Carbon monoxide	−62	−38
Nitrogen dioxide	−24	−11
Ozone		
1 hr	−18	− 3
8 hr	−11	0
Sulfur dioxide	−52	−35
PM_{10}	...	−14
$PM_{2.5}$	...	...
Lead	−94	−25

Source: EPA, *Latest Findings on National Air Quality: 2001 Status and Trends.*
http://www.epa.gov/airtrends/reports.html

... Trend data not available

Note: These percents represent overall changes. Changes for a particular urban area may differ.

> At ground level, ozone is an air pollutant. At high altitudes, ozone is beneficial, as you will find out in Chapter 2.

had a lot on his mind, especially the allegation that his uncle killed Hamlet's father before marrying his mother. Hamlet spends the rest of the play trying to decide what to do about his dysfunctional family, and no further reference to air pollution is made, except perhaps in the observation that "Something is rotten in the state of Denmark."

Actually, the air is probably worse in Los Angeles, Mexico City, or Bangkok than in Elsinore (Hamlet's home) or nearby Copenhagen. But, wherever you live, there is a good chance that the lungful of air you just inhaled contains some substances that, depending on their amount, can be harmful. The health threat can be so serious that laws are passed to limit pollution by curtailing some of the ways we normally do things. Although it is impossible to completely remove all pollutants from the air, the air quality has improved in the United States over the past three decades. This improvement has occurred through a combination of governmental actions, chemical ingenuity, and allowing the atmosphere to regenerate naturally.

In 1970, the passage of the Clean Air Act set into motion a federal mandate to reduce air pollution and improve air quality, establishing national air quality standards. The dramatic results, shown in Table 1.1, demonstrate the percentage decrease in six major pollutants for the period 1982–2001. The U.S. Environmental Protection Agency, better known as the EPA, calls these **criteria air pollutants** or more simply **criteria pollutants.** For each, the EPA has set permissible levels in the air based on their effects on human health and on the environment. Four of the pollutants are atmospheric gases: carbon monoxide, nitrogen oxides, ozone, and sulfur dioxide. The other two, particulate matter (PM) and lead, are minuscule suspended particles on the order of a millionth of a meter. PM includes dust, soot, dirt, and even droplets of liquid, bacteria, or viruses. As you can see in Table 1.1, particulate matter is classified by size in micrometers. A **micrometer** (μm) is 10^{-6} of a meter (m) and is sometimes simply referred to as a micron. PM_{10} has an average diameter of 10 μm or less, which is on the order of 0.0004 inches or about 1/6 the diameter of a human hair. $PM_{2.5}$, also called fine particles, has an average diameter less than 2.5 μm, which is closer to 1/20 of a human hair. From these definitions, you can see that $PM_{2.5}$ is a subset of PM_{10}. In Section 1.3, we will revisit air quality and the effects of these different pollutants.

Appendix 1 units and conversions

> The EPA sets standards both for $PM_{2.5}$ and for PM_{10}, as they have different health effects. PM_{10} standards were first set in 1990.

Consider This 1.2   A Visit to the EPA

The U.S. Environmental Protection Agency (EPA) maintains an extensive Web site. Go to the *Online Learning Center* or directly to EPA's Web site, where you will find many consumer-friendly documents on air quality. Select a document that relates to the air you breathe. Report its title, URL, and two or three interesting things that you learned.

Figure 1.3
Clouds consist of condensed water vapor. As we will see in Chapter 3, clouds are one of several factors that influence Earth's temperature.

1.2 What's in a Breath? The Composition of Air

The air we breathe is a **mixture,** that is, a physical combination of two or more substances present in variable amounts. For the moment we will focus on only five components of air: oxygen, nitrogen, argon, carbon dioxide, and water. The first four normally exist as gases. Although we usually think of water as a liquid, it can also be a gas, in which case we may refer to it as "water vapor." Just like oxygen and nitrogen, water vapor is a colorless gas that you cannot see. Although you *can* see steam and clouds, these are not water vapor as such. Rather, they are condensed water vapor, that is, tiny droplets of liquid water (Figure 1.3).

> Mixtures do not need to be gases. Soil is a mixture of solids and liquids. We will revisit mixtures in Section 1.6.

The concentration of water vapor in air varies widely. It can be close to 0% in dry desert air or as much as 5% in a tropical rain forest. Because of this variability, reference tables typically list the composition of air with no humidity. The normal composition of dry air is 78% nitrogen, 21% oxygen, and 1% other gases by volume. **Percent** means "parts per hundred" and is sometimes abbreviated as pph. In this case, the parts are molecules (or, in a few cases, atoms).

> If you had 100 liters of air, you would have
> • 78 liters of nitrogen.
> • 21 liters of oxygen.
> • 1 liter of other gases.

Figure 1.4 displays the composition of air in the form of a pie chart and a bar graph. Both of these are important, widely used methods for displaying numerical information, and we will use both in this text. The pie chart emphasizes the fractions of the total, whereas the bar graph uses height to emphasize the relative sizes of each. Regardless of how we present the data, notice that 99% of dry air is made up of only two substances: nitrogen and oxygen.

Life on Earth bears the stamp of oxygen. Indeed, it is difficult to conceive of life on any planet without this remarkable chemical. Oxygen is absorbed into our blood via the lungs and reacts with the foods we eat to release the energy needed for all life processes within our bodies (Chapter 11). Oxygen is also a participant in burning, rusting, and other corrosion reactions. As a constituent of water and of many rocks, oxygen

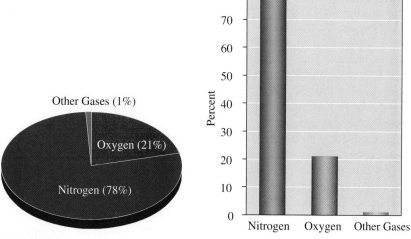

Figure 1.4

The composition of dry air, by volume.

Figures Alive! Visit the *Online Learning Center* to learn more about these graphs as well as about the atoms and molecules that make up air. Look for the Figures Alive! icon throughout this text.

The terms *molecule* and *atom* are defined in Section 1.7.

is the most abundant element by mass in the Earth's crust and in the human body. Given this broad distribution and high reactivity, it is somewhat surprising that oxygen was not isolated as a pure substance until the 1770s. But once isolated, oxygen proved to be of great significance in establishing the principles of the young science of chemistry.

Consider This 1.3　　More Oxygen?

Humans are accustomed to living in an atmosphere of 21% oxygen where a paper match burns completely in less than a minute, a fireplace consumes a small pine log in about 20 minutes, and you exhale a certain number of times a minute. Burning, rusting, and the rate of most metabolic processes in plants and animals depend on the concentration of oxygen. How would life on Earth be different if the oxygen content in the atmosphere were doubled? List at least four effects.

Nitrogen is the most abundant substance in the air and constitutes over three fourths of the air we inhale. However, it is much less reactive than oxygen, and it is exhaled from our lungs unchanged (Table 1.2). Although nitrogen is essential for life and is a part of all living things, most plants and animals obtain the nitrogen they require from other sources, not directly from the atmosphere.

Section 6.12 describes the route by which atmospheric nitrogen becomes part of living plants and animals.

Table 1.2	Typical Composition of Inhaled and Exhaled Air	
Substance	**Inhaled air (%)**	**Exhaled air (%)**
Nitrogen	78.0	75.0
Oxygen	21.0	16.0
Argon	0.9	0.9
Carbon dioxide	0.04	4.0
Water	0.0	4.0

The remaining 1% of air is mostly argon, a substance so unreactive that it is said to be "chemically inert." This inertness is recognized in the name *argon*, which means "lazy." As you can see from Table 1.2, argon is exhaled from our lungs unchanged.

The percentages we have been using to describe the composition of the atmosphere are based on volume. Thus, we could closely approximate 100 liters (L) of dry air by mixing 78 L of nitrogen, 21 L oxygen, and 1 L argon (78% nitrogen, 21% oxygen, and 1% argon). But because the volume of a gas sample increases with increasing temperature and decreases with increasing pressure, all gas volumes must be measured at the same temperature and pressure.

An alternative way to represent the composition of air is in terms of the molecules and atoms present in the mixture. This works because equal volumes of gases contain equal numbers of molecules, providing the gases are at the same temperature and pressure. Thus, if you took a sample of 100 air molecules and atoms (an unrealistically small amount of air), 78 would be nitrogen molecules, 21 would oxygen molecules, and 1 would be an argon atom. In other words, when we say that air is 21% oxygen, we mean that there are 21 molecules of oxygen per 100 molecules and atoms in the air. We will soon explain why nitrogen and oxygen are made up of molecules and argon of atoms.

Some atmospheric components are present at less than one part per hundred or 1%. Such is the case with carbon dioxide, which has a concentration of about 0.0375%. Although we could express this as 0.0375 molecules of carbon dioxide per 100 molecules and atoms in the air, it does not make sense to talk about a fraction of a molecule. Accordingly, we scale up the measurement from parts per hundred to **parts per million (ppm),** which means one part out of a million and is 10,000 times less concentrated than 1 part per hundred (pph). Having 0.0375 pph is equivalent to having 375 ppm. Through a series of relationships, we can show that the difference between pph (= percent) and ppm is a factor of 10,000, that is, moving the decimal four places:

0.0375% means 0.0375 parts per hundred

means 0.375 parts per thousand

means 3.75 parts per ten thousand

means 37.5 parts per hundred thousand

means 375 parts per million

Thus, out of a sample consisting of 1,000,000 air molecules and atoms, 375 of them will be molecules of carbon dioxide. Hence, the carbon dioxide concentration is 375 ppm, or 0.0375%.

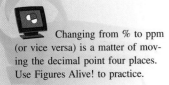
Changing from % to ppm (or vice versa) is a matter of moving the decimal point four places. Use Figures Alive! to practice.

1 liter = 1.06 quart
Conversion factors are found in Appendix 1.

One part per million corresponds to:
• 1 second in nearly 12 days
• 1 step in a 568-mile journey
• 1 penny out of $10,000
• a pinch of salt on 20 pounds of potato chips

Sceptical Chymist 1.4   **Really One Part in a Million?**

It has been said that a part per million is the same as one second in nearly 12 days. Is this a correct analogy? How about for one step in a 568-mile journey?

Your Turn 1.5   **Using Parts Per Million**

a.   The EPA sets the permissible limit for the average concentration of carbon monoxide in an 8-hr period at 9 ppm. Express this concentration as a percentage.

b.   Exhaled air typically contains about 75% nitrogen gas. Express this concentration in parts per million.

Answers
a. 0.0009%   b. 750,000 ppm

Every time we exhale, we (together with our fellow members of the animal kingdom) add carbon dioxide to the atmosphere. Table 1.2 indicates the difference between inhaled dry air and exhaled air. Clearly some changes have taken place that use up oxygen and give off both carbon dioxide and water. Not surprisingly, chemistry is involved. In the biological process of metabolism, oxygen reacts with foods to yield carbon dioxide and water. However, most of the water in exhaled air is simply the result of evaporation from the moist surfaces within the lungs. Note that even exhaled air still contains 16% oxygen. Some people mistakenly think that in respiration most of the oxygen is replaced with carbon dioxide. But if this were true, mouth-to-mouth resuscitation would not work.

Look for more about carbon dioxide in Chapter 3.

1.3   What Else Is in a Breath?

Our noses tell us that the air is obviously different in a pine forest, a bakery, an Italian restaurant, a locker room, and a barnyard. Even blindfolded, we can *smell* where we are. Pine needles, fresh bread, garlic, sweat, and manure all have distinctive odors that are carried by molecules. Hence, air must contain trace quantities of substances not included among the five substances listed in Table 1.1. Although the major components of air are odorless (at least to our noses), many of the other airborne substances have pronounced odors. In fact, the human nose is an extremely sensitive odor detector. In some cases, only a minute trace of a substance is needed to trigger the olfactory receptors responsible for detecting odors. Thus, tiny amounts of substances can have a powerful effect on our noses, as well as on our emotions.

Our noses also warn us to avoid certain places. But some of the more dangerous air pollutants have no odor, and others are dangerous at concentrations low enough that we are unable to detect them by smell. As a result, it may be necessary to rely on specialized scientific equipment to monitor the presence of such substances in the air. It is rather surprising that the gases that cause serious air pollution are present in relatively small amounts, generally in the parts per million to parts per billion range. Yet even at such low concentrations, they can do significant harm.

In this chapter, we focus on four gases that contribute to air pollution at the surface of the Earth. One of these gases, carbon monoxide, is odorless; the other three—ozone, sulfur dioxide, and nitrogen oxides—have characteristic (and unpleasant) odors. With sufficient exposure, each of these is hazardous to health, even at concentrations well below 1 ppm. Together with particulate matter (PM), they represent the most serious air pollutants at the Earth's surface. Let's now look at the health effects of each.

PM comes from many sources, including trucks, cars, and coal-burning power plants. Forest fires and blowing dust are natural sources of particulate matter.

Carbon monoxide is known as the "silent killer," because your senses cannot detect it. Once in the lungs, it enters the bloodstream and disrupts the delivery of oxygen throughout the body. In extreme cases, such as breathing auto exhaust or furnace emissions in a confined space, carbon monoxide inhalation can be fatal. But charcoal grills, kerosene heaters, and propane stoves also generate carbon monoxide (Figure 1.5) and must be used outdoors or vented if used indoors. With low-level exposure to carbon monoxide from improper venting, victims first experience symptoms such as dizziness, headache, and nausea, all easily mistaken for the onset of a respiratory or sinus infection. To add to the difficulties in diagnosing carbon monoxide poisoning, people exposed at the same time may have different symptoms. Carbon monoxide inhalation can be quite serious for individuals suffering from cardiovascular disease. With severe symptoms, emergency medical care is necessary.

Ozone is a special form of oxygen. It has a characteristic sharp odor that is frequently detected around photocopiers, electric motors, transformers, and welding torches. Unlike normal oxygen, ozone is toxic. It affects the respiratory system, and even at very low concentrations it will reduce lung function in normal, healthy people during periods of exercise. Symptoms include chest pain, coughing, sneezing, and pulmonary congestion. At the Earth's surface, ozone is definitely a bad actor, but as you will see in Chapter 2, it plays an essential role at high altitudes (10–30 miles).

Sulfur oxides and *nitrogen oxides* are respiratory irritants that can affect breathing and lower resistance to respiratory infections. People most susceptible include the elderly,

Figure 1.5

Propane camping stoves need venting to avoid a build-up of carbon monoxide.

young children, and individuals with lung diseases, such as emphysema or asthma. A particularly severe example of these effects occurred during the London fog of December 1952. The fog lasted five days and led to approximately 4000 deaths. The oxides of sulfur and nitrogen also contribute to acid precipitation, a subject explored in Chapter 6.

Particulate matter or PM is not just one chemical; rather the term includes a mixture of tiny solid particles and liquid droplets that are emitted into the air, form in the air from other pollutants, or are caught up into the air by the wind. Some particulate matter is visible, and you may recognize it as soot or smoke. Of concern here, though, are the particles too tiny for you to see. Once airborne, particles with a diameter of 10 μm or less can get deep into the lungs and cause a variety of mischief. Because particles smaller than 2.5 μm are even more worrisome as a health hazard, the standards for these are stricter (as we soon will see in Table 1.5).

The elderly and children are at highest risk, as are those with existing breathing ailments such as asthma or bronchitis. Exposure to particulate matter is linked to heart disease. As reported in *Circulation,* a journal of the American Heart Association, $PM_{2.5}$ particles in urban air are linked with an increased risk of heart attacks. In the study, the risk for heart attack in Boston peaked both 2 hr and 24 hr after patients were exposed to increased levels of the particles. A 2003 article in *Circulation* examines pollution data from over 150 cities and reveals that the risk of heart disease increases as the amount of $PM_{2.5}$ increases. Thus particulate matter may affect more than just the lungs.

To better understand the effects of *what* you are breathing *when* you are likely to be breathing it, the EPA has developed the color-coded Air Quality Index (AQI) shown in Table 1.3, with its distinctive logo (Figure 1.6). If you live in a metropolitan area, you may find the daily AQI forecast listed in your daily newspaper. National newspapers carry the information as well. For example, *USA Today* currently reports the daily AQI for 36 cities. If you had checked their listing for a hot summer day (August 15, 2003, to be exact), you would have found that 13 of these were green ("good") and none were maroon ("hazardous"). But nine cities, including Dallas, Detroit, Baltimore, and the Washington, DC, metropolitan area, were orange ("unhealthy for sensitive groups"). People who fall in the "sensitive group" vary with the pollutant, but include those with respiratory diseases such as asthma, and those who are either quite young or old.

To see the cumulative effects of air quality, consult Table 1.4. If you were one of the 2 million people living in the Houston metropolitan area during the past 7 years, the air was of good or moderate quality 90–95% of the time. Alternatively, the air was unhealthy for you for up to three weeks out of a year. By looking at the data over a period of years, you can see the fluctuation in the values, depending in large part on the weather during that particular year. Sporadic events such as volcanic eruptions and forest fires also influence air quality.

Figure 1.6
EPA air quality logo.

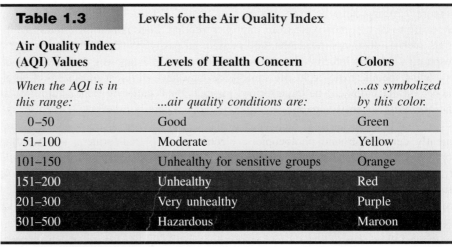

Table 1.3	Levels for the Air Quality Index	
Air Quality Index (AQI) Values	**Levels of Health Concern**	**Colors**
When the AQI is in this range:	*...air quality conditions are:*	*...as symbolized by this color.*
0–50	Good	Green
51–100	Moderate	Yellow
101–150	Unhealthy for sensitive groups	Orange
151–200	Unhealthy	Red
201–300	Very unhealthy	Purple
301–500	Hazardous	Maroon

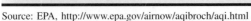
Source: EPA, http://www.epa.gov/airnow/aqibroch/aqi.html

Table 1.4	Air Quality Index Values for Houston			
Year	Good (0–50)	Moderate (51–100)	Unhealthy for Sensitive Groups (101–150)	Unhealthy (>150)
1997	258	54	33	20
1998	253	70	23	19
1999	223	82	34	26
2000	166	147	37	16
2001	180	144	25	16
2002	196	136	24	9
2003	169	154	26	16

Values reported are the number of days per year. "Unhealthy" includes all three of the unhealthy categories (red, purple, and maroon). The total number of days does not always add up to 365 (=1 year), as on some days, no data were reported.

Consider This 1.6   AQI Where You Live

You can use the EPA Web site's AirData section to construct an Air Quality Index Summary Report for a city of your choice (other than Houston). Further directions are found on the *Online Learning Center*. Select the most recent year, and generate the report. How does the air quality in the city you selected compare to that for Houston for the same year?

Air quality may also be listed by individual pollutant. For example, Figure 1.7 shows the newspaper AQI forecast for carbon monoxide, ozone, and particulates for March 31, 2004, in Phoenix. Although the color coding is not employed, the criteria are the same.

Consider This 1.7   Orange Alert

Unlike the numerical scale (see Figure 1.7) used by the *Arizona Republic* to forecast pollution, some newspapers predict air quality using color codes.

a. What are the advantages of using qualitative color codes (green, yellow, orange, red) rather than a numerical scale for a given air pollutant? The disadvantages?

b. A "red alert" is forecast for today. List five or more actions you could take to help reduce air pollution, particularly if everyone were to follow your suggestions.

Pollution

Good: 0-50 Moderate: 51-100
Unhealthful: More than 100

Carbon monoxide	**6**
Ozone	**34**
Particulates	**60**

Figure 1.7

Air quality forecast for central Phoenix, March 31, 2004. These values are not in ppm; rather, a relative scale is given where 100 is considered unhealthy.

Air pollution is primarily an urban problem, and more than 50% of all Americans live in cities with populations over 500,000. Many of these cities (as we saw with Houston) fail to meet the national air quality standards during certain periods of time. In spite of recent improvement in air quality, we still have difficulties, especially with nitrogen oxides and ozone. Furthermore, our present air quality standards may provide only a small margin of safety in protecting public health. The American Lung Association estimates that $50 billion in health benefits could be realized annually in the United States if air quality standards were met throughout the country.

We face difficult political and economic choices. Are we willing to spend the money that would be needed to truly clean up the air we breathe? To stimulate the economy, what regulations can we afford to drop or relax? Would economic gains compensate for the hidden health costs? In considering the risks with the benefits, a tightening of regulations could mean a boon to health and a significant reduction in health care costs associated primarily with respiratory conditions. The improved air quality that we have enjoyed could be short-lived. How do we assess the risks and benefits? We now turn to questions of risk.

1.4 Taking and Assessing Risks

Air quality provides an opportunity for our first look at the subject of risk, one to which we will return repeatedly throughout this text. Indeed, it is an issue that is central to life itself, because everything we do carries a certain level of risk. Certain activities that carry high risks are labeled as such. For example, by law, cigarette packages are required to carry a warning. Similarly, on a bottle of wine you will find the words "GOVERNMENT WARNING" followed by a statement about the risks of birth defects for pregnant women and about the risks of driving a car or operating machinery under the influence of alcohol. Some practices have been declared illegal because the level of risk is judged unacceptable to society. However, many activities carry no warning. In these cases, presumably the risk is quite low, the risk is obvious or unavoidable, or the benefits of the activity far outweigh the risk.

One feature of such warnings is a characteristic of risk itself. The warnings do not say that a specific individual *will* be affected by a particular activity. They only indicate the statistical probability or chance that an individual will be affected. For example, if the odds of dying from an accident while traveling 300 miles in a car are approximately one in a million, this means that, on average, one person out of every million people traveling 300 miles by car would be killed in an accident. Such predictions are not simply guesses, but are the result of **risk assessment:** evaluating scientific data and making predictions in an organized manner about the probabilities of an occurrence.

Consider This 1.8   Cell Phones

Driving down the highway? If you are using a cell phone while driving, your ability to handle your car has been called into question. As this book went to press, talking while driving was controversial. Using cell phones as an example, state the risks and the benefits of answering a phone call while driving. Are the risks obvious or unavoidable? Do the benefits of the activity far outweigh the risk? Do other factors apply?

For air pollutants, the assessment of risk requires knowing two factors: the **toxicity,** the intrinsic health hazard of a substance, and the **exposure,** the amount of the substance encountered. Exposure is the easier factor to evaluate because it depends simply on the concentration of the substance in the air, the length of time a person is exposed, and the amount of air inhaled into the lungs in a given period. As you saw earlier in this chapter, the last factor depends on lung capacity and breathing rate.

One cubic meter (m^3) is a volume 1 meter $\times$ 1 meter $\times$ 1 meter. 1 m^3 = 1000 liters (about 250 gallons)

1 μg is about the mass of a period printed on a page.

Concentrations of pollutants in air are usually expressed either as parts per million (ppm) or as micrograms per cubic meter ($\mu g/m^3$). Earlier, when talking about particulate matter, we encountered the prefix micro in micrometers (μm), meaning 10^{-6}, or a millionth of a meter. Thus, one **microgram** (μg) is 10^{-6}, or a millionth of a gram (g).

Let us consider the risk related to carbon monoxide. Millions of tons of carbon monoxide are spread throughout the atmosphere. Yet, by itself, this prodigious amount does not tell the true story of risk associated with this substance. Carbon monoxide is not evenly distributed in the atmosphere. In some places the concentration is so low that no health concerns arise. In others, such as with a faulty appliance with poor ventilation, the carbon monoxide concentration may be hazardous. To assess our risk from this pollutant, we need to consider our exposure to it as well as its toxicity.

Moderate quality air can contain carbon monoxide (CO) at a concentration of about 5000 μg CO per cubic meter of air. How much carbon monoxide are we breathing? To answer this, we need to return to how much air we breathe in a day. As you saw earlier, each of us breathes 10,000–20,000 L of air daily, depending on our lung capacity and other factors. Assuming the lower value, 10,000 L, first we need to express this in terms of cubic meters. This way, we can compare the value with others, such as those set by the National Ambient Air Quality Standards (NAAQS). Noting that 1 m^3 = 1000 L, the 10,000 L that we breathe each day is equivalent to 10 cubic meters of air:

Ambient refers to the outside air, that is, the air surrounding or encircling us. Indoor air is the topic of Section 1.13.

$$10,000 \text{ L air} \times \frac{1 \text{ m}^3}{1000 \text{ L}} = 10 \text{ m}^3 \text{ air}$$

Using this value together with the amount of carbon monoxide in a cubic meter of moderate quality air,

$$10 \text{ m}^3 \text{ air} \times \frac{5,000 \text{ μg CO}}{1 \text{ m}^3 \text{ air}} = 50,000 \text{ μg CO in a day}$$

we would be exposed to 50,000 μg of carbon monoxide in a day, or 0.05 g. Someone who breathes twice as much of this air in a day obviously would be exposed to twice as much carbon monoxide, or 100,000 μg (a tenth of a gram). Thus, exercise or other heavy exertion increases your exposure to pollutants.

With values such as 50,000 μg CO per day, it makes sense to use **scientific notation,** that is, a system for writing number as the product of a number and 10 raised to the appropriate power. Using scientific notation avoids turning the text into strings of zeros, either before or after the decimal point. This particular number, 50,000, can be written in scientific notation as 5×10^4. Here, the easy way to make this conversion is to simply count the number of zeroes to the right of the initial 5. There are four of them. The number 5 is then multiplied by 10^4 to obtain 5×10^4 μg of CO breathed per day.

Scientific notation becomes even more useful when considering much larger numbers, such as the number of molecules in a typical breath. The value is more than 20,000,000,000,000,000,000,000 molecules, a number large enough to take your breath away! In scientific notation this particular number is written as 2×10^{22} molecules. Why 20,000,000,000,000,000,000,000 equals 2×10^{22} takes a bit more explaining. This value is simply another instance in the following series:

The air you breathe contains mainly molecules, but there are a few atoms as well. We will explain shortly.

$1 \times 10^1 = 10$
$1 \times 10^2 = 10 \times 10 = 100$
$1 \times 10^3 = 10 \times 10 \times 10 = 1000$

Note that 10^1 is 10, that is, 1 followed by one zero. Next, 10^2 is 100, or 1 followed by two zeros. And next 10^3 is 1000, or 1 followed by three zeros. Continuing this pattern, 10^{22} is 1 followed by 22 zeros. Therefore, 2×10^{22} is 2 followed by 22 zeros or 20,000,000,000,000,000,000,000. If exponents and scientific notation are new or confusing to you, consult Appendix 2.

Being exposed to 5×10^4 μg CO over the course of a day might seem like a worrisome quantity. But keep in mind that this amount, when inhaled over a 24-hr period, is less than the amount determined as hazardous. To understand why, we consult the National Ambient Air Quality Standards set by the EPA (Table 1.5). The values for each pollutant are expressed in a concentration over a time period. In the case of CO, this period is either 1 hr or 8 hr. If you examine these values, you can see that higher concentrations of CO can be tolerated for shorter periods.

The Clean Air Act requires that the EPA set standards for pollutants. The most recent amendments to the Clean Air Act were in 1990.

Returning to our initial value for CO in air of moderate quality, 5000 μg CO/m^3 of air, we now express this as 5×10^3 μg CO/m^3. Clearly, this value is less than the concentrations for either period. In the case of an 8-hr exposure, 5×10^3, or 5000, is less than 1×10^4, or 10,000. Similarly, for a 1-hr period, 5×10^3 is also less than 4×10^4 (40,000). Thus, when evaluating toxicity, compare the exposure level to the

Table 1.5	National Ambient Air Quality Standards, 1999	
Pollutant	**Standard (ppm)**	**Approximate Equivalent Concentration of Standard ($\mu g/m^3$)**
Carbon monoxide		
8-hr average	9	1×10^4
1-hr average	35	4×10^4
Nitrogen dioxide		
Annual average	0.053	100
Ozone		
1-hr average	0.12	235
8-hr average	0.08	157
Lead		
Quarterly average	...	1.5
Particulates*		
PM_{10}, 24-hr average	...	150
PM_{10}, annual average	...	50
$PM_{2.5}$, 24-hr average	...	65
$PM_{2.5}$, annual average	...	15
Sulfur dioxide		
Annual average	0.03	80
24-hr average	0.14	365
3-hr average	0.50	1300

more serious health consequences. [

... Data not available

* PM_{10} refers to airborne particles 10 μm in diameter or less, and $PM_{2.5}$ to those less than 2.5 μm. These two categories are monitored separately, and the standards for $PM_{2.5}$ are still controversial.

minimum amount required for the pollutant to become a public health risk or to be present at toxic levels. Although a pollutant may be present, it is not considered hazardous unless it exceeds the amount that causes harmful effects.

Your Turn 1.9   **Time Matters**

For carbon monoxide, we just saw that the limits of exposure were higher for shorter periods. Do the limits for the other pollutants follow this pattern as well? Consult Table 1.5 to find out.

Let's examine the information about other pollutants found in Table 1.5. Again, these values, based on scientific studies, are the maximum concentrations considered to be safe for the general population. If you did Your Turn 1.9, you found that air quality standards differ in terms of length of exposure. For example, it is hazardous to breathe carbon monoxide at 9 ppm for a period of 8 hours. In contrast, it is hazardous to breathe ozone at a concentration of 0.08 ppm for the same period. These values give us a basis on which to evaluate the relative amounts that are hazardous. Comparing 9 ppm with 0.08 ppm, we can see that ozone is about 100 times more hazardous to breathe than carbon monoxide. However, the actual danger may not be a factor of 100. In the case of breathing air polluted with ozone, your senses can detect it and you are likely to move to less polluted air (perhaps indoors) if you can. In contrast, you may not know that you are inhaling carbon monoxide at a hazardous level, because it is odorless (and because after breathing it for a while your judgment may be affected).

Your Turn 1.10   Particle Size Matters Too

We mentioned earlier that "fine" particulate matter (<2.5 μm in diameter) had more serious health consequences than "coarse" particulate matter (<10 μm in diameter). Does Table 1.5 bear this out?

Although the standards for the pollutant gases are expressed as parts per million, the concentrations of sulfur oxides and nitrogen oxides are sufficiently low that they also could conveniently be reported in **parts per billion (ppb),** which means one part out of one billion, or 1000 times less concentrated than 1 part per million.

sulfur oxides	0.030 ppm = 30 ppb
nitrogen oxides	0.053 ppm = 53 ppb

As you can see from these values, to convert from parts per million to parts per billion, you need to move the decimal point three places to the right.

Whether in parts per million or parts per billion, toxicities are difficult to accurately assess, because it is unethical to experiment on humans with substances such as carbon monoxide or sulfur dioxide. This leaves scientists with three choices: human population studies, animal studies, and bacterial studies. Population studies involve collecting data on affected groups of people. For example, a researcher may determine the percentage of people who get lung cancer after smoking one pack of cigarettes per day. Such studies are necessarily limited and may require years of observation to obtain results that are statistically significant and reflect accurately the long-term risk. For this reason, animal studies have been a widely used substitute. Aside from the ethical questions of animal rights, the problem remains that animals may not respond to toxic substances the same as humans. As a result, scientists must interpret animal studies with great caution. A more recent field of toxicity measurements relies on studies with bacteria. An advantage of using bacteria is that they grow and reproduce very rapidly, allowing many studies to be done quickly and inexpensively.

Even if data were available to calculate the risks from a given pollutant, we still would have to ask what level of risk was acceptable and for what groups of people. Various government agencies are charged with establishing safe limits of exposure for the major air pollutants. Table 1.5 gives current outdoor air quality standards established by the EPA for the pollutants discussed in this chapter. Some states, including California and Oregon, have their own, stricter, standards.

Your Turn 1.11   Living Downwind

Copper metal can be recovered from copper ore by smelting, a refining process that may release sulfur dioxide (SO_2) into the atmosphere. A woman living downwind of a smelter typically inhales 1050 μg of SO_2 in a 24-hr period.

a.   If her lungs handled about 15,000 liters (15 m^3) of air per day, would she exceed the 24-hr average for the National Ambient Air Quality Standards for SO_2? Show calculations to support your answer.

b.   Would she exceed the annual average?

Finally, an important factor in dealing with risks is not only the actual risk, but people's perception of a particular risk. For example, the risks of driving are far higher than the risks of flying. Each day in the United States alone, more than 100 people die in automobile accidents. Yet many people avoid taking a flight because of their fear of falling out of the sky. Similarly, many fear living near a nuclear power plant. Yet living on a flood plain or near a small dam is a far riskier proposition. In both of these cases, other factors may be at work as well. For example, the media's portrayal of an unfortunate airline accident may heighten people's fears about flying. After an accident, there often is a large ripple effect of public concern.

Consider This 1.12   **Risk Analysis**

A publication of the American Chemical Society, *Chemical Risk: A Primer*, states: "The general public is uncomfortable with uncertainties. Too often we think in terms of absolutes and demand that scientists and decision makers be held accountable for their risk decisions." Do you agree or disagree with these statements? Support your opinion with reasonable arguments, giving a specific example from your personal experience in considering a risk of importance to you.

1.5   The Atmosphere: Our Blanket of Air

The most familiar kinds of air pollution occur in the **troposphere,** the region of the atmosphere that lies directly above the surface of the Earth. Figure 1.8 provides the names of regions of the atmosphere and some reference points in relation to altitude. As one rises in the troposphere, the temperature decreases until it reaches about −40 °C (also −40 °F). That temperature roughly marks the beginning of the **stratosphere,** the region of the atmosphere above the troposphere that includes the ozone layer. The temperature of the stratosphere increases from about −40 °C at 20 kilometers (km) to 0 °C (32 °F) at 50 km. Above that altitude, the temperature of the atmosphere again begins to decrease on passing through the **mesosphere,** the region of the atmosphere above an altitude of 50 km. The issues we will study in the first three chapters of this book will take us to these various regions, which differ in atmospheric properties and phenomena (see Figure 1.8). Bear in mind that no sharp physical boundaries separate these layers. The atmosphere is a continuum with gradually changing composition, concentrations, pressure, and temperature. In fact, temperature changes account for the organization of the atmosphere.

Look for more about the ozone layer in Chapter 2.

The relative concentrations of the major components of the atmosphere are nearly constant at all altitudes. In other words, the concentration of oxygen remains about 21% and that of nitrogen is 78%. However, you might know from the experience of hiking in high mountains or from flying in a plane (Figure 1.9) that the air gets "thinner" with increasing altitude. As you climb higher up into the atmosphere, there is less air, that is, fewer molecules in a given volume. Moreover, as you climb higher, the mass of air above you decreases.

Air has a lower density at higher elevations. The concept of density is introduced in Chapter 5.

Your Turn 1.13   **Altitude and Pollutants**

Pollutants are minor components of the atmosphere. Their concentrations usually are not constant at different altitudes. In general, do you think the concentrations are higher or lower near the surface of the Earth? In the special case of pollutants emitted from a tall smokestack, what would you expect?

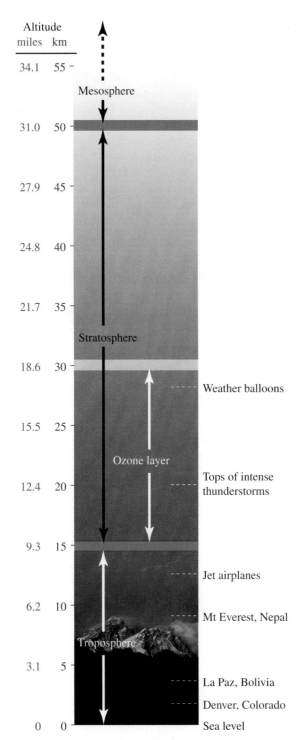

Figure 1.8
The regions of the lower atmosphere.

The **atmospheric pressure,** the force with which the atmosphere presses down on a given area, decreases with increasing altitude as shown in Figure 1.10. Atmospheric pressure is measured with a device called a barometer. At sea level, such as in Boston or Los Angeles, the barometric pressure is 14.7 pounds per square inch (lb/in.^{2}). A pressure of this magnitude is defined as 1 atmosphere (1 atm). In Denver, the "mile-high city," the pressure is 12.0 lb/in.^{2}, or about 0.8 atm. You will note from Figure 1.10 that the plot of pressure versus altitude is not a straight line. Above about 20 km, the

As altitude ↑, atmospheric pressure ↓.

Figure 1.9

Flight across the Rockies. A view out of the aircraft window shows the upper troposphere at about 35,000 feet.

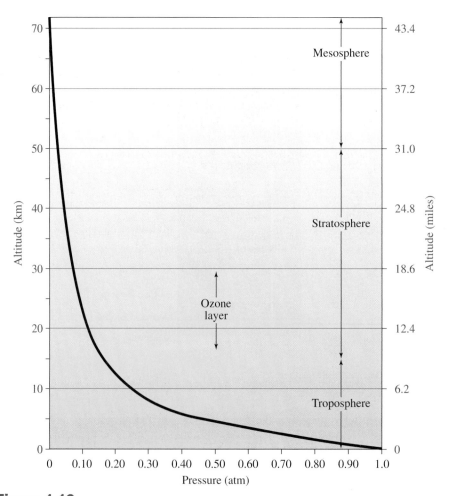

Figure 1.10

The pressure of the atmosphere changes with altitude. An increase in altitude is accompanied by a drop in air pressure.

pressure drops very sharply with increasing altitude. In this region, the pressure decreases by about 50% for every 5-km increase in altitude (1 km = 0.62 mile, 5 km = 3.1 mi). At higher altitudes, the pressure decreases more gradually. Somewhere above 100 km, the atmosphere simply fades into the almost perfect vacuum of outer space.

1.6 Classifying Matter: Mixtures, Elements, and Compounds

As we described the atmosphere and air quality, we used a bit of chemical terminology. Before proceeding, then, some clarification is in order. To begin with, we will examine the way in which chemists describe the composition of different types of matter. Matter can be classified either as a single pure substance or as a mixture of two or more pure substances, as seen in Figure 1.11.

In Section 1.2, we defined a mixture as a physical combination of two or more substances in variable amounts.

Much of the matter we encounter in everyday life is in the form of mixtures. A breath of air is a mixture of gases. Polluted air is also a mixture that, depending on the pollutants, has different compositions. Exhaled air is a different mixture from inhaled air. As composition of a mixture varies, so do many of its properties. For example, gasoline is a mixture of several different compounds, and, as we will see in Chapter 4, as the composition of the mixture is changed, the properties of the gasoline change.

Elements and compounds are the two pure substances of most interest to us (see Figure 1.11). The two most plentiful components of air are nitrogen and oxygen. These both are examples of **elements,** substances that cannot be broken down into simpler ones by any *chemical* means. There are over 100 elements, and all common forms of matter are composed of them. About 90 elements occur naturally on planet Earth and, as far as we know, in the universe. The other 10 or so elements have been created from existing elements through artificially induced nuclear reactions. Plutonium is probably the best known of the artificially produced elements, although it does occur in trace concentrations in nature. In some cases, the total amount of a newly created form of matter is so small that there is some uncertainty (perhaps even some controversy) over just how many elements have been identified.

An alphabetical list of the elements and their **chemical symbols,** one- or two-letter abbreviations for the elements, appears in the inside back cover of the text. These symbols, established by international agreement, are used throughout the world. Some of the symbols are quite obvious to those who speak English. For example, oxygen is O, nitrogen is N, carbon is C, and sulfur is S. Most symbols, however, consist of two letters: Ni for nickel, Cl for chlorine, Ca for calcium, and so on. Other symbols appear to have little relationship to their English names. Thus, Fe is iron, Pb is lead, Au is gold, Ag is silver, Sn is tin, Cu is copper, and Hg is mercury. All these metals were

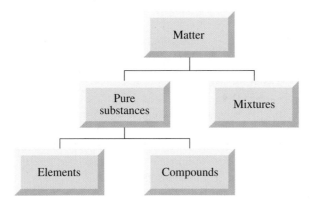

Figure 1.11
The classification of matter.

known to the ancients and hence were given names long ago. Their symbols reflect their Latin names, for example, *ferrum* for iron, *plumbum* for lead, and *hydrargyrum* for mercury.

Elements have been named for properties, planets, places, and people. Hydrogen (H) means "water former," a name that reflects the fact that this flammable gas burns in oxygen to form water. Neptunium (Np) and plutonium (Pu) were named after the two most recently discovered members of our solar system. Berkelium (Bk) and californium (Cf) honor the Berkeley lab where a team of researchers first produced these two elements. Albert Einstein, Dmitri Mendeleev, and Lise Meitner (codiscoverer of nuclear fission) have attained elementary immortality in einsteinium (Es), mendelevium (Md), and meitnerium (Mt), respectively. The most recently discovered element, darmstadtium (Ds), was first detected in 1994 in the city of Darmstadt, Germany. This is the heaviest element currently known, and only a few atoms of darmstadtium have been produced.

It is particularly appropriate that Mendeleev should have his own element, because the most common way of arranging the elements reflects the periodic system developed by this 19th-century Russian chemist. Figure 1.12 is the **periodic table,** an orderly arrangement of all the elements based on similarities in their properties. We will explain how it is ordered and the significance of all the numbers in Chapter 2. For the moment, it is sufficient to note that about the time of the American Civil War, Mendeleev arranged the 66 elements that were then known into vertical columns according to the properties they had in common. These vertical columns are called **groups** and are given numbers. The members of Group 1A include lithium (Li), sodium (Na), potassium (K), and three other very reactive metals. Similarly, Group 7A consists of very reactive nonmetals, including fluorine (F), chlorine (Cl), bromine (Br), and iodine (I). Nitrogen and oxygen, the two most common elements in the atmosphere, are side by side in Groups 5A and 6A.

Some helpful generalizations come from examination of the elements that make up the periodic table. For example, the vast majority of elements are solids; some are gases; and only two, bromine and mercury, are liquids at room temperature and pressure. Most elements are metals, as you can see by the shading on the periodic table in Figure 1.12. **Metals** are elements that are shiny and conduct electricity and heat well; they include familiar substances such as iron, gold, and copper. Far fewer are **nonmetals,** elements that have varied appearances and don't conduct well, such as sulfur, chlorine, and oxygen. The Group 8A elements are known as the **noble gases,** elements that are inert and do not readily undergo chemical reactions. In fact, some of the noble gases (helium, neon, and argon) do not combine chemically with *any* other elements. Radon, however, is a noble gas that is radioactive, as are over a dozen other naturally occurring elements. As we will see in Section 1.13 of this chapter, radon affects the quality of indoor air. Thus, the periodic table is a very handy database, an amazingly useful way of organizing the building blocks of the universe. We will refer to it often throughout the text.

Lothar Meyer, a German chemist, also developed a periodic table at the same time as Mendeleev.

Metals and nonmetals (and the ions they form) are discussed in Section 5.7

Radioactivity is an important topic in Chapter 7.

Consider This 1.14   Adopt an Element

Periodic tables on the Web list the properties of elements, their date of discovery, their naturally occurring isotopes, and much more. Thus, the Web can give you quick access to information that it might take you hours to find using reference books. Use the periodic table links at the *Online Learning Center* to learn more about an element of your choice. Find out what year your element was discovered; whether it occurs naturally as a solid, liquid, or gas; its appearance; where it is found; and any two other facts, such as toxicity, cost, uses, and so on. Following the directions given by your instructor, get together with other students to see the larger trends.

The periodic table of the elements

Key:
24
Cr
52.00

24 — Atomic number
52.00 — Atomic mass

1 1A	2 2A		3 3B	4 4B	5 5B	6 6B	7 7B	8	9 8B	10	11 1B	12 2B	13 3A	14 4A	15 5A	16 6A	17 7A	18 8A
1 **H** 1.008																		2 **He** 4.003
3 **Li** 6.941	4 **Be** 9.012												5 **B** 10.81	6 **C** 12.01	7 **N** 14.01	8 **O** 16.00	9 **F** 19.00	10 **Ne** 20.18
11 **Na** 22.99	12 **Mg** 24.31												13 **Al** 26.98	14 **Si** 28.09	15 **P** 30.97	16 **S** 32.07	17 **Cl** 35.45	18 **Ar** 39.95
19 **K** 39.10	20 **Ca** 40.08		21 **Sc** 44.96	22 **Ti** 47.88	23 **V** 50.94	24 **Cr** 52.00	25 **Mn** 54.94	26 **Fe** 55.85	27 **Co** 58.93	28 **Ni** 58.69	29 **Cu** 63.55	30 **Zn** 65.39	31 **Ga** 69.72	32 **Ge** 72.61	33 **As** 74.92	34 **Se** 78.96	35 **Br** 79.90	36 **Kr** 83.80
37 **Rb** 85.47	38 **Sr** 87.62		39 **Y** 88.91	40 **Zr** 91.22	41 **Nb** 92.91	42 **Mo** 95.94	43 **Tc** (98)	44 **Ru** 101.1	45 **Rh** 102.9	46 **Pd** 106.4	47 **Ag** 107.9	48 **Cd** 112.4	49 **In** 114.8	50 **Sn** 118.7	51 **Sb** 121.8	52 **Te** 127.6	53 **I** 126.9	54 **Xe** 131.3
55 **Cs** 132.9	56 **Ba** 137.3		57 **La** 138.9	72 **Hf** 178.5	73 **Ta** 180.9	74 **W** 183.9	75 **Re** 186.2	76 **Os** 190.2	77 **Ir** 192.2	78 **Pt** 195.1	79 **Au** 197.0	80 **Hg** 200.6	81 **Tl** 204.4	82 **Pb** 207.2	83 **Bi** 209.0	84 **Po** (210)	85 **At** (210)	86 **Rn** (222)
87 **Fr** (223)	88 **Ra** (226)		89 **Ac** (227)	104 **Rf** (261)	105 **Db** (262)	106 **Sg** (266)	107 **Bh** (264)	108 **Hs** (269)	109 **Mt** (268)	110 **Ds** (271)	111	112	113	114	115	(116)	(117)	(118)

58 **Ce** 140.1	59 **Pr** 140.9	60 **Nd** 144.2	61 **Pm** (145)	62 **Sm** 150.4	63 **Eu** 152.0	64 **Gd** 157.3	65 **Tb** 158.9	66 **Dy** 162.5	67 **Ho** 164.9	68 **Er** 167.3	69 **Tm** 168.9	70 **Yb** 173.0	71 **Lu** 175.0
90 **Th** 232.0	91 **Pa** 231.0	92 **U** 238.0	93 **Np** (237)	94 **Pu** (244)	95 **Am** (243)	96 **Cm** (247)	97 **Bk** (247)	98 **Cf** (251)	99 **Es** (252)	100 **Fm** (257)	101 **Md** (258)	102 **No** (259)	103 **Lr** (262)

Metals
Metalloids
Nonmetals

The 1–18 group designation has been recommended by the International Union of Pure and Applied Chemistry (IUPAC) but is not yet in wide use. In this text we use the standard U.S. notation for group numbers (1A–8A and 1B–8B). No names have been assigned for elements 111–115. Elements 116–118 have not yet been synthesized.

Figure 1.12
The periodic table of the elements.

Two other components of the atmosphere, water and carbon dioxide, are examples of **compounds,** pure substances made up of two or more elements in a fixed, characteristic chemical combination. For example, water is a compound of the elements oxygen and hydrogen. Similarly, carbon dioxide is a compound of the elements oxygen and carbon. There is no residual uncombined oxygen or carbon in carbon dioxide. In CO_2, the two elements are chemically combined and are no longer in their elemental forms.

As its name implies, carbon dioxide is made up of chemically combined carbon and oxygen in a fixed composition. All pure samples of carbon dioxide contain 27% carbon and 73% oxygen by weight (or mass). Thus, a 100-g sample of carbon dioxide will always consist of 27 g of carbon and 73 g of oxygen, chemically combined to form this particular compound. These values never vary, no matter the source of the carbon dioxide. This illustrates the fact that every compound exhibits a constant characteristic chemical composition.

In contrast, carbon monoxide is also a compound of carbon and oxygen. However, pure samples of carbon monoxide contain 43% carbon and 57% oxygen by weight. Thus, 100 g of carbon monoxide contain 43 g of carbon and 57 g of oxygen, a much different composition from that of carbon dioxide. This is not surprising, because carbon monoxide and carbon dioxide are two different compounds.

The physical properties (such as boiling point) and chemical reactivity of compounds are also constant. Consider water, a compound for which a 100-g sample consists of 11 g of hydrogen and 89 g of oxygen, (11% hydrogen and 89% oxygen by weight). At room temperature, pure water is a colorless, tasteless liquid. It boils at 100 °C and it freezes at 0 °C. All samples of pure water have these same properties.

Although only about 100 elements exist, over 20 million compounds have been isolated, identified, and characterized. Some are very familiar naturally occurring substances such as water, salt, and sugar. But most known compounds do not exist in nature; rather, they were chemically synthesized by men and women across our planet. The motivation for making new compounds is almost as varied as the compounds themselves. The reasons may be to make synthetic fibers and plastics, to find drugs to cure AIDS or cancer, to create materials with a memory, or just for the creativity and intellectual fun of it. In Chapters 9 and 10, we will look at some examples of new compounds synthesized by chemists.

> A gram is approximately the mass of a peanut or a small paper clip.

Your Turn 1.15   **Classifying Pure Substances**

Using your everyday knowledge of materials, classify each of these as an element, a compound, or as a mixture.

a.	water	**b.**	nickel	**c.**	U.S. nickel coin
d.	diamond	**e.**	sulfur dioxide	**f.**	lemonade

Answers

a.	compound	**b.**	element	**c.**	mixture

1.7 **Atoms and Molecules**

The definitions of elements and compounds given earlier are valid, in spite of the fact that no assumptions were made about the physical structure of matter. But modern insights into the organization of matter help us better understand matter, the "stuff" of the universe. It is now well established that elements are made up of **atoms,** the smallest unit of an element that can exist as a stable, independent entity. The word *atom* comes from the Greek for "uncuttable." Today we know that atoms consist of smaller particles, and that atoms can be "split" by high-energy processes. However, atoms remain indivisible by chemical or mechanical means.

Atoms are extremely small, many billions of times smaller than anything we can directly see. Because of this small size, huge numbers of atoms must be in any sample

> Atomic structure is further discussed in Section 2.2, and atom "splitting" (nuclear fission) is a topic of Chapter 7.

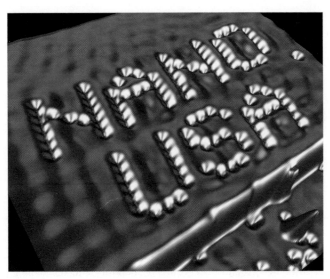

Figure 1.13

Carbon monoxide molecules lined up on a copper surface, as seen with a scanning tunneling microscope.

of matter that we can see or touch or weigh by conventional means. For example, in a single drop of water you might find 5×10^{21} atoms. This is about a trillion times greater than the approximately 6 billion people on Earth, enough to give each person a trillion atoms from that water drop.

As Figure 1.13 reveals, the invisible have recently become visible. Using a scanning tunneling microscope, scientists at the IBM Almaden Research Center lined up 112 carbon monoxide molecules on a copper surface to spell "NANO USA." **Nanotechnology** refers to work at the atomic and molecular (nanometer) scale: 1 nanometer (nm) $= 1 \times 10^{-9}$ m. Each letter is 4 nm high by 3 nm wide. At this size, about 250 million nanoletters could fit on a cross section of a human hair, corresponding to about three hundred 300-page books.

The existence of atoms provides a means of refining our earlier definitions of elements and compounds. Elements are made up of atoms of one type. For example, the element carbon is made up of carbon atoms only. By contrast, compounds are made up of the atoms of two or more elements. In the compound carbon dioxide, two types of atoms are present: carbon and oxygen. Similarly, water is made up of both hydrogen and oxygen atoms.

But we need to be careful with our language. The carbon and oxygen atoms in carbon dioxide are not present as such. Rather, the carbon and oxygen atoms are chemically combined to form a carbon dioxide **molecule,** a combination of a fixed number of atoms held together in a certain spatial arrangement. More specifically, two oxygen atoms are combined with one carbon atom to form a carbon dioxide molecule. We represent this molecule with its chemical formula, CO_2. Similarly, in the water molecule, H_2O, three atoms are bonded together: two hydrogens and one oxygen. Millions of compounds other than carbon dioxide and water exist as molecules.

A **chemical formula** is a symbolic way to represent the elementary composition of a substance. For CO_2, the elements are present in a ratio of one carbon atom for every two oxygen atoms. In a similar manner, the chemical formula for water, H_2O, indicates two hydrogen atoms for each oxygen atom. Note that when an atom is used once in a formula, such as oxygen in H_2O or carbon in CO_2, the subscript of "1" is omitted. We will say more about the chemical formulas of compounds in the section that follows.

Elements have chemical formulas as well. Some elements exist as single atoms, such as helium or radon. We represent these as He and Rn, respectively. Other elements exist as molecules. For example, nitrogen and oxygen are found in our atmosphere as N_2 and O_2 molecules. We call these **diatomic molecules,** meaning that they contain two atoms per molecule. In contrast, helium consists of individual He atoms.

Table 1.6	**Classification of Matter**	
Substance	**Observable Properties**	**Microscopic Level**
Element	Cannot be broken down into simpler substances	Only one kind of atom
Compound	Fixed composition, but capable of being broken down into elements	Two or more atoms in fixed combination
Mixture	Variable composition of elements, compounds, or both	Variable assortment of atoms, molecules, or both

Table 1.6 summarizes how we can describe elements, compounds, and mixtures. It lists both the behavior we can observe experimentally and the theory that scientists use to explain what is happening at the incredibly small atomic level. Both ways are correct and complement each other.

Your Turn 1.16   Elements and Compounds

Name the element(s) present in each of these substances. Identify each substance as an element or as a compound.

a. sulfur dioxide, SO_2 **b.** carbon tetrachloride, CCl_4
c. hydrogen peroxide, H_2O_2 **d.** sucrose, $C_{12}H_{22}O_{11}$
e. chlorine, Cl_2 **f.** nitrogen monoxide, NO

Answers
a. sulfur, oxygen (compound)   **e.** chlorine (element)

We now can apply these concepts to the atmosphere. Air is a mixture, which means that its composition can vary with time and place. Some of its components, such as nitrogen, oxygen, and argon, are elements; others, notably carbon dioxide and water, are compounds. All of the compounds and some of the elements are present as molecules (for example, CO_2 and O_2). The other elements exist as uncombined atoms (for example, Ar).

Furthermore, we can now explain what we said earlier about air being mostly molecules (see Section 1.2), but with some atoms as well. Dry air is composed mainly of the elements nitrogen and oxygen, that is, N_2 molecules and O_2 molecules. If it is humid, add in some water vapor in the form of H_2O molecules. Remember also the 375 ppm of carbon dioxide, which means that there will be 375 CO_2 molecules in the 1×10^6 molecules and atoms that make up the air. Which atoms are these? Air contains Ar atoms (just under 1%), as well as tiny amounts of He (helium) and Xe (xenon) atoms and even tinier amounts of Rn (radon) atoms.

Sceptical Chymist 1.17   The Chemistry of Lawn Care

News reports and advertisements should be viewed with a critical eye for scientific accuracy, bias, and timeliness, among other criteria. For example, a lawn care service advertisement identifies the fertilizers it uses as "a balanced blend of nitrogen, phosphorus, and potassium. They have an organic nature, made up of carbon molecules. These fertilizers are biodegradable and turn into water." Comment on the chemical correctness of this information. What changes would you suggest?

1.8 Formulas and Names: The Vocabulary of Chemistry

If chemical symbols are the alphabet of chemistry, then chemical formulas are the words. And the language of chemistry, like any other language, has rules of spelling and syntax. The chemical symbols must be combined in ways that correctly correspond to the composition of the compounds in question. Although this system of chemical symbolism and nomenclature is logical, precise, and extremely useful (at least to chemists), it presents somewhat of a barrier to others studying the discipline. This, of course, is true of any specialized vocabulary, whether in chemistry, accounting, music, sports, or any other area. This book does not seek to make its readers expert in all aspects of chemical nomenclature; however, some familiarity with the rules of writing formulas and naming simple chemical compounds will be helpful. In this chapter we will consider only compounds consisting of two elements.

The chemical formula of a compound reveals the elements that combine to make it (by chemical symbols) and the atom ratio of those elements (by the subscripts). The name usually conveys similar information. For example, the name *magnesium oxide* indicates a compound consisting of magnesium (Mg) and oxygen (O). Because the magnesium and oxygen combine in a one-to-one atomic ratio, the formula of magnesium oxide is MgO. Similarly, *sodium chloride* indicates a compound composed of the elements sodium (Na) and chlorine (Cl) and has the formula NaCl.

The rule for naming such two-element compounds is simple: The name of the more metallic element comes first, followed by the name of the less metallic element, modified to end in the suffix "ide." For example, let's name the compound composed of potassium (K) and iodine (I). First we need to determine which of these two elements is more metallic, and for this we turn to the periodic table. The metallic elements are on the left side of the periodic table (green in the periodic table) and the nonmetallic elements on the right (blue). Thus, potassium is more metallic than iodine. Applying the rule, potassium is first and the compound is named potassium iod*ide*. The formula turns out to be KI, as we will explain later.

Your Turn 1.18   Naming Simple Compounds

Name the compound that contains each pair of elements.

a. bromine and magnesium   **b.** oxygen and barium
c. hydrogen and chlorine   **d.** sodium and sulfur

Answers
b. barium oxide   **c.** hydrogen chloride

Your Turn 1.19   More Naming Exercises

Name the compound that has each of these formulas.

a. $ZnCl_2$ **b.** Al_2O_3
c. CaS   **d.** Li_3N

Answers
a. zinc chloride   **b.** aluminum oxide

check on this.

Your Turn 1.20   Even More Naming Exercises

Go to the *Online Learning Center* and find the "naming compounds" practice exercises. Complete the exercises given there as assigned by your instructor.

Table 1.7		**Prefixes Used in Naming Compounds**	
Prefix	**Meaning**	**Prefix**	**Meaning**
Mono	One	Hexa-	Six
Di- or Bi-	Two	Hepta-	Seven
Tri-	Three	Octa-	Eight
Tetra-	Four	Nona-	Nine
Penta-	Five	Deca-	Ten

Why calcium chloride has the formula $CaCl_2$ (and not CaCl) is explained in Section 5.7.

Unfortunately, writing chemical formulas is a bit more complicated than we just suggested. Not all compounds exhibit one-to-one atomic ratios such as KI and MgO. For example, calcium chloride has two atoms of chlorine for each atom of calcium and the chemical formula is $CaCl_2$. Someone familiar with the periodic table and atomic structure should be able to predict the atomic ratio and the formula of just about any two-element compound and to name it. But we have not yet provided you with enough information to develop that skill. At present it is probably sufficient if you can name a two-element compound when presented with its formula. Thus, the formula H_2S represents a compound called hydrogen sulfide, a gas with the unmistakable smell of rotten eggs.

One of the complications associated with the rules we have been applying is that the name of the compound does not always unambiguously reveal its formula. A way of eliminating the ambiguity is to use prefixes that indicate the number of atoms of an element specified by the formula. A good example is one of the atmospheric compounds we have already introduced, carbon dioxide. *Di-* means "two" and thus the name carbon *di*oxide implies that each molecule of the compound includes two oxygen atoms. The corresponding formula, CO_2, indicates the two oxygen atoms with a subscript 2 on the symbol O.

The use of the prefixes listed in Table 1.7 makes it possible to distinguish between two or more compounds consisting of the same elements, but in different atomic ratios. Thus, carbon *mon*oxide also consists of carbon and oxygen, but the prefix *mon-* or *mono-* indicates that only one oxygen atom is in each molecule of this compound. It follows that the formula of carbon monoxide is CO. Using the same logic and the same set of prefixes, the chemical name for SO_3 is sulfur trioxide. Note that in most compounds in which a formula contains only one atom of an element, the *mono-* prefix is omitted. Carbon monoxide is an exception; the *mono-* prefix is used to avoid possible confusion with carbon dioxide, CO_2.

Your Turn 1.21   Chemical Formulas

What information does each chemical formula convey? Each substance is found in the atmosphere in very small amounts.

a. NO_2 **b.** SO_2 **c.** N_2O_4 **d.** O_3

Answer

a. A molecule of the compound represented by the formula NO_2 consists of one atom of the element nitrogen combined with two atoms of the element oxygen.

Your Turn 1.22   Names from Chemical Formulas

Use appropriate prefixes from Table 1.7 to name the first three compounds given in Your Turn 1.21. The last is ozone, a form of the element oxygen.

Answers

a. nitrogen dioxide   **c.** dinitrogen tetraoxide

1.9 Chemical Change: Oxygen's Role in Burning

The first pollutant listed in Table 1.5 is carbon monoxide, CO; however, all air, polluted or not, contains carbon dioxide, CO_2. Carbon monoxide and carbon dioxide can both arise from the same source: combustion. **Combustion** is the rapid combination of oxygen with a substance. When elemental carbon or carbon-containing compounds burn in air, oxygen combines with the carbon to form CO_2 or CO (or both). Similarly, combustion reactions produce water and sulfur dioxide by the burning of hydrogen and sulfur, respectively.

Combustion is a major type of **chemical reaction,** a process whereby substances described as **reactants** are transformed into different substances called **products.** Chemical reactions can be represented by a **chemical equation,** a representation of a chemical reaction using chemical formulas. To students, chemical equations are probably better known as "the thing with the arrow in it." Chemical equations are the sentences in the language of chemistry. They are made up of chemical symbols (corresponding to letters) that often are combined in the formulas of compounds (the "words" of chemistry). Like a sentence, a chemical equation conveys information, in this case about the chemical change taking place. But, as we now will see, a chemical equation must also obey some of the same constraints that apply to a mathematical equation.

At its most fundamental level, a chemical equation is very simple indeed. It is a qualitative description of the reaction:

$$\text{Reactant(s)} \longrightarrow \text{Product(s)}$$

By convention, the reactants are always written on the left and the products on the right. The arrow represents a chemical transformation and is read as "is converted to" or "yields." Thus, reactants are converted to products in the sense that the reaction gives products whose properties are different from those of the reactants.

The combustion of carbon to produce carbon dioxide (for example, the burning of charcoal in air, Figure 1.14) can be represented in several ways. One way is by a "word equation":

$$\text{carbon} + \text{oxygen} \longrightarrow \text{carbon dioxide}$$

It is more common to use chemical formulas to represent the elements and compounds involved:

$$C + O_2 \longrightarrow CO_2 \qquad [1.1]$$

This compact symbolic statement conveys a good deal of information. A translation of equation 1.1 might sound something like this: "One atom of the element carbon reacts with one molecule of the element oxygen to yield one molecule of the compound carbon dioxide."

If we use a black sphere to represent a carbon atom and a red sphere to represent an oxygen atom, the rearrangement of atoms by this reaction can be represented like this:

Likewise, the burning of sulfur produces the air pollutant sulfur dioxide.

$$S + O_2 \longrightarrow SO_2 \qquad [1.2]$$

Now using yellow spheres for sulfur atoms, we can represent the formation of sulfur dioxide.

Figure 1.14
The burning of charcoal in air.

Oxygen is a diatomic molecule, O_2.

In Chapter 3, you will find out why the CO_2 molecule is linear, and the SO_2 molecule is bent.

Equation 1.2 is balanced because an equal number of sulfur atoms are in the reactants and products, and the numbers of oxygen atoms in the reactants and products are also equal.

It is possible to pack even more information into a chemical equation by specifying the physical states of the reactants and products. A solid is designated by *(s)*, a liquid by *(l)*, and a gas by *(g)*. Because carbon and sulfur are solids, and oxygen, carbon dioxide, and sulfur dioxide are gases at ordinary temperatures and pressures, equations 1.1 and 1.2 become:

$$C(s) + O_2(g) \longrightarrow CO_2(g)$$

$$S(s) + O_2(g) \longrightarrow SO_2(g)$$

We will designate the physical states when this information is particularly important, but in most cases we will omit it for simplicity.

You will note that equation 1.1 has some of the characteristics of a mathematical equation; in this case, the number and kinds of atoms on the left equal those on the right:

Left side: 1 C, 2 O   Right side: 1 C, 2 O

> Consider this analogy in terms of the reorganization of reactants to form products in a chemical reaction. The building materials used to construct an apartment building (reactants) can be disassembled and rearranged to build three houses and a garage (products).

This is the test for a correctly balanced equation. Atoms are neither created nor destroyed in a chemical reaction, and the elements present do not change when converted from reactants to products. This relationship is called the **law of conservation of matter and mass:** in a chemical reaction, matter and mass are conserved. The mass of the reactants consumed equals the mass of the products formed. The total mass does not change, as matter is neither created nor destroyed.

Atoms are rearranged during a chemical reaction. That is what a chemical change is all about: The atoms in the products are in a different arrangement than they were as reactants. Therefore, there is no requirement that the number of *molecules* must be the same on both sides of the arrow. In equation 1.1, one atom of carbon plus one molecule of oxygen yields one molecule of carbon dioxide. This looks suspiciously like $1 + 1 = 1$. This is no cause for alarm; a chemical equation is not exactly the same as a mathematical equation. Remember, a chemical equation represents a transformation, not a simple equality. In a correctly balanced chemical equation, some things must be equal, others need not be. Table 1.8 summarizes.

Equation 1.1 describes the combustion of pure carbon in an ample supply of oxygen. However, if the oxygen supply were limited, the products would include CO. For the sake of argument, let us say that pure carbon monoxide is formed. First we write the chemical formulas for the reactants and product:

$$C + O_2 \longrightarrow CO \text{ (unbalanced equation)}$$

Is this chemical equation balanced? No. There are two oxygen atoms on the left but only one on the right. We cannot balance the equation by simply adding an additional oxygen atom to the product side. Once we write the *correct* chemical formulas for the reactants and products, we cannot change them. To do so would imply a different reaction. All we can do is to use whole-number coefficients (or occasionally fractional ones) in front of the various chemical formulas. In simple cases like this, the coefficients can

Table 1.8	**Characteristics of Chemical Equations**

Always Conserved

Identity of atoms in reactants = Identity of atoms in products
Number of atoms in reactants = Number of atoms in products
Mass of all reactants = Mass of all products

May Change

Number of molecules in reactants may differ from number of molecules in products
Physical states (*s, l,* or *g*) of reactants may differ from physical states of products

be found quite easily by simple trial and error. If we place a 2 to the left of the symbol CO, it signifies two molecules of carbon monoxide. This corresponds to a total of two carbon atoms and two oxygen atoms. Two oxygen atoms are also on the left side of the arrow, so the oxygen atoms have been balanced.

$$C + O_2 \longrightarrow 2\,CO \text{ (still not balanced)}$$

But now the carbon atoms do not balance. Fortunately, this is easily corrected by placing a 2 in front of the C.

$$2\,C + O_2 \longrightarrow 2\,CO \text{ (balanced equation)} \qquad [1.3]$$

The balanced equation also can be represented by colored spheres, using black for carbon and red for oxygen atoms.

> A **subscript** follows a chemical symbol, as in O_2 or CO_2.
>
> A **coefficient** precedes a symbol or a formula, for example 2 C or 2 CO.

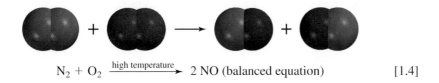

It is evident from comparing equations 1.1 and 1.3 that, relatively speaking, more O_2 is required to form CO_2 from carbon than is needed to form CO. This matches the conditions we stated for the formation of carbon monoxide, namely, that the supply of oxygen was limited.

Another air pollutant, nitrogen monoxide (commonly called nitric oxide), is produced from nitrogen and oxygen. In the presence of something hot, such as an automobile engine or a forest fire, these two atmospheric gases will combine. Again, we begin by writing the chemical formulas for the reactants and product.

$$N_2 + O_2 \xrightarrow{\text{high temperature}} NO \text{ (unbalanced equation)}$$

The equation is not balanced: there are two oxygen atoms on the left, but only one on the right. The same is true for nitrogen atoms. Placing a 2 to the left of the formula NO supplies two nitrogen *and* two oxygen atoms, and the equation is now balanced.

> Nitrogen and oxygen both are diatomic molecules.

$$N_2 + O_2 \xrightarrow{\text{high temperature}} 2\,NO \text{ (balanced equation)} \qquad [1.4]$$

Your Turn 1.23   Chemical Equations

Balance these equations and draw a representation of each using spheres. For the latter, both H_2O and NO_2 are bent molecules, with O and N, respectively, as the middle atom. The next chapter explains why.

a. $H_2 + O_2 \longrightarrow H_2O$

b. $N_2 + O_2 \longrightarrow NO_2$

Answer

a. Balanced equation: $2\,H_2 + O_2 \longrightarrow 2\,H_2O$

Your Turn 1.24   More Chemical Equations

Go to the *Online Learning Center* and find the "balancing equations" practice exercises. Complete the exercises given there as assigned by your instructor.

Consider This 1.25   Advice from Grandma

A grandmother offered this advice to rid the garden of pesky caterpillars. "Hammer some iron nails about a foot up from the base of your trees, spacing them every four to five inches." According to this grandmother, the iron nails convert the tree sap (a sugary substance containing carbon, hydrogen, and oxygen atoms) into ammonia (NH_3), a substance the caterpillars cannot stand. Comment on the accuracy of grandma's chemistry (allowing that the nails may still work, regardless of her explanation).

1.10 Fire and Fuel: Air Quality and Burning Hydrocarbons

Look for other examples of burning fuels in Chapter 4.

In the previous section, we saw that the combustion of carbon, nitrogen, and sulfur can produce the air pollutants CO, NO, and SO_2. Another way to produce both CO and CO_2 is by the combustion of **hydrocarbons,** compounds of hydrogen and carbon. Hydrocarbons have many natural sources, but we typically obtain them from petroleum. Two common mixtures of hydrocarbons are gasoline and kerosene. Methane (CH_4), the simplest hydrocarbon, is the primary component of natural gas.

Given an ample supply of oxygen, a hydrocarbon will burn completely: All the carbon will combine with oxygen to form carbon dioxide, and all the hydrogen will combine with oxygen to form water. We can use this reaction to illustrate how to balance more complex chemical equations. Again, we begin by writing the chemical formulas for the reactants and products:

$$CH_4 + O_2 \longrightarrow CO_2 + H_2O \text{ (unbalanced equation)}$$

One reason why this equation is not as easy to balance as the previous ones is because O appears in *both* of the products: CO_2 + H_2O. When balancing equations like this one, it is easiest to start with the element(s) that are present only in *one substance* on each side of the chemical equation, in this case, C and H. If we start with C, we note that the expression is already balanced with respect to it. Now, to bring the H atoms into balance, we must place the number 2 in front of H_2O.

$$CH_4 + O_2 \longrightarrow CO_2 + 2\,H_2O \text{ (still not balanced)}$$

The coefficient 2 multiplies all the parts of the substance to the right of it, in this case water. Thus, it means that 4 H atoms and 2 O atoms are present. Because a CO_2 molecule contains 2 O atoms, there are now a total of 4 O atoms on the right of the equation and 2 O atoms on the left. We equalize the number of O atoms by placing a 2 before O_2.

$$CH_4 + 2\,O_2 \longrightarrow CO_2 + 2\,H_2O \text{ (balanced equation)} \qquad [1.5]$$

A nice feature of chemical equations is that you can always tell if they are balanced by counting atoms on both sides of the arrow.

Carbon: Left: $1\ \text{CH}_4\ \text{molecule} \times \dfrac{1\ \text{C atom}}{1\ \text{CH}_4\ \text{molecule}} = 1\ \text{C atom}$

Right: $1\ \text{CO}_2\ \text{molecule} \times \dfrac{1\ \text{C atom}}{1\ \text{CO}_2\ \text{molecule}} = 1\ \text{C atom}$

Hydrogen: Left: $1\ \text{CH}_4\ \text{molecule} \times \dfrac{4\ \text{H atoms}}{1\ \text{CH}_4\ \text{molecule}} = 4\ \text{H atoms}$

Right: $2\ \text{H}_2\text{O molecules} \times \dfrac{2\ \text{H atoms}}{1\ \text{H}_2\text{O molecule}} = 4\ \text{H atoms}$

Oxygen: Left: $2\ \cancel{O_2\ molecules} \times \dfrac{2\ O\ atoms}{1\ \cancel{O_2\ molecule}} = 4\ O\ atoms$

Right: $\left(1\ \cancel{CO_2\ molecule} \times \dfrac{2\ O\ atoms}{1\ \cancel{CO_2\ molecule}} \right)$

$+ \left(2\ \cancel{H_2O\ molecules} \times \dfrac{1\ O\ atom}{1\ \cancel{H_2O\ molecule}} \right) = 4\ O\ atoms$

This tabulation confirms that the equation is indeed balanced.

One of the most widely used fuels in automobiles is gasoline, a mixture of dozens of hydrocarbons. One is octane, C_8H_{18}. If a sufficient supply of oxygen is delivered to the vehicle's engine when the octane burns, only carbon dioxide and water are formed.

$$2\ C_8H_{18} + 25\ O_2 \longrightarrow 16\ CO_2 + 18\ H_2O \qquad [1.6]$$

In reality, however, not all of the carbon is converted to carbon dioxide. The amount of oxygen present and the amount of time available for reaction (before the materials are ejected in the exhaust) are insufficient for the reaction represented by equation 1.6 to occur as written. Instead, some CO is formed. An extreme situation is represented by equation 1.7, in which all of the carbon in the octane is converted to carbon monoxide.

$$2\ C_8H_{18} + 17\ O_2 \longrightarrow 16\ CO + 18\ H_2O \qquad [1.7]$$

Note that the coefficient of O_2 in equation 1.5 is 25, whereas the corresponding coefficient in equation 1.7 is 17. Less oxygen is used in the reaction where CO is formed.

What really happens in a car's engine is a combination of the two reactions. Most of the carbon released in automobile exhaust is in the form of CO_2, although some CO is also produced. The relative amounts of these two gases indicate how efficiently the car burns the fuel, which in turn is evidence of how well tuned the engine is. States that monitor auto emissions check for this by sampling exhaust emissions using a probe that detects CO. The measured CO concentrations are compared with established standards, for example, 1.20% in the state of Minnesota. A car whose CO emissions exceed the standard must be serviced so that it complies (Figure 1.15).

Your Turn 1.26   **Combustion Reactions**

a. "Bottled gas" or "liquid petroleum gas" (LPG) is mostly propane, C_3H_8. Balance this equation for the burning of propane.

$$C_3H_8 + O_2 \longrightarrow CO_2 + H_2O$$

b. Cigarette lighters burn butane, C_4H_{10}. Write a balanced equation, assuming plenty of oxygen.

Answer

a. $C_3H_8 + 5\ O_2 \longrightarrow 3\ CO_2 + 4\ H_2O$

Your Turn 1.27   **Balancing Equations**

Demonstrate that equations 1.6 and 1.7 are balanced by counting the number of atoms of each element on either side of the arrow.

Answer

Equation 1.6 contains 16 C, 36 H, and 50 O on each side.

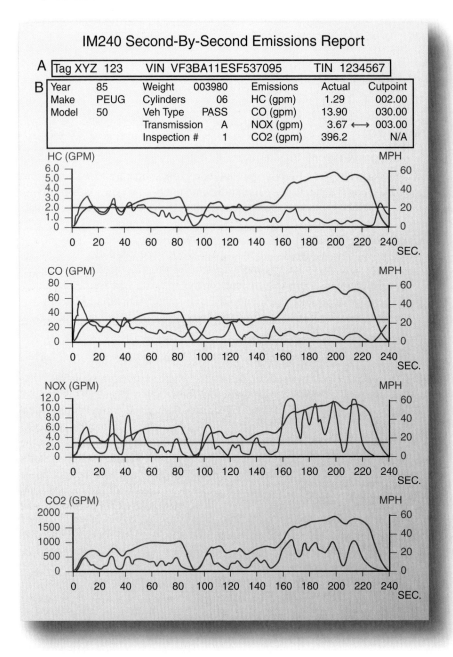

Figure 1.15

An auto emission report.

GPM is the abbreviation for grams per mile.

HC is the abbreviation for hydro-carbons.

1.11 Air Pollutants: Direct Sources

By now you should recognize that each lungful of air you inhale contains mostly nitrogen and oxygen. In contrast, air pollutants such as carbon monoxide, the oxides of sulfur and nitrogen, particulate matter and ozone are present in only tiny amounts. But even at the part per million or part per billion level, these minor components compromise the quality of the air we breathe. To better understand how to lower the levels of pollutants, we need to know their chemical properties and how they are produced. In this section, we will examine two major sources of pollutants: the coal-fired plants that generate electricity, and the tailpipes of cars, trucks, and other vehicles.

See Chapter 4 for more about coal and its chemical composition.

Your Turn 1.28   Tailpipe Gases

What comes out of the tailpipe of an automobile? Start your list now and build on it as you work through this section.

Hint: Some of the air that enters the engine also comes out in the exhaust. Also remember to include the products of combustion.

In the United States, burning coal is the major source of electric power. It also is the major source of SO_2. Coal is mostly carbon and hydrogen, and thus the major products of its combustion are carbon dioxide and water. But coal is a complex mixture of variable composition and contains other elements as well. Relevant here is that most coals contain 1–3% sulfur and rock-like minerals. When coal is burned, this sulfur is converted into gaseous sulfur dioxide, and the minerals are converted into fine ash particles. If not removed, the sulfur dioxide and these particles go right up the smokestack. Since hundreds of millions of tons of coal are burned in this country, millions of tons of sulfur dioxide and ash are emitted.

Once emitted, sulfur dioxide can react further with oxygen to form sulfur trioxide, SO_3.

$$2\,SO_2 + O_2 \longrightarrow 2\,SO_3 \qquad\qquad [1.8]$$

Although normally quite slow, this reaction is much faster in the presence of small ash particles. The ash particles also aid another process. If the humidity is high enough, they promote the conversion of water vapor into an aerosol of tiny water droplets that we call fog. **Aerosols** consist of particles, both liquid and solid, that stay suspended in the air rather than settle out. Smoke is a familiar aerosol made up of tiny particles of solids and liquids.

In Chapter 6, we will see how sulfuric acid aerosols contribute to haze.

Here, the aerosol of interest is one made up of tiny droplets of sulfuric acid, H_2SO_4. It forms because sulfur trioxide dissolves readily in water droplets to produce sulfuric acid.

$$H_2O + SO_3 \longrightarrow H_2SO_4 \qquad\qquad [1.9]$$

If inhaled, the sulfuric acid aerosol droplets are small enough to become trapped in the lung tissue where they can cause severe damage.

The good news? Sulfur dioxide emissions in the United States are declining. For example, in 1985, approximately 20 million tons of SO_2 was emitted from the burning of coal. Today the value is closer to 15 million tons. This decrease can be credited to the Clean Air Act of 1970 that mandated reductions in emissions from coal-fired electric power plants. More stringent regulations were established in the Clean Air Act Amendments of 1990. But continued progress will not come cheaply. Look for more information about the strategies and technologies available to reduce atmospheric SO_2 and their economic and political costs in Chapter 6.

Your Turn 1.29   Sulfur Dioxide from the Mining Industry

Burning coal is not the only source of sulfur dioxide. As you saw earlier in Your Turn 1.11, the smelting of ores to produce metals is another source. For example, silver and copper metal can be produced from their sulfide ores. Write balanced chemical equations.

(continued on p. 40)

Your Turn 1.29   Sulfur Dioxide from the Mining Industry (*continued*)

a. Silver sulfide (Ag_2S) is heated with oxygen gas to produce silver and sulfur dioxide.
b. Copper sulfide (CuS) is heated with oxygen gas to produce copper and sulfur dioxide.

Answer

a. $Ag_2S + O_2 \longrightarrow 2Ag + SO_2$

With more than 220 million vehicles (more than one for every two Americans), the United States has more automobiles per capita than any other nation. Would you expect sulfur dioxide to be coming out all these tailpipes? Happily, the answer is no, because most cars have internal combustion engines fueled by gasoline. We already discussed the combustion of octane in gasoline to form carbon dioxide and water vapor (equation 1.6), both of which come out the tail pipe. Because gasoline contains little to no sulfur, burning it is not a significant source of sulfur dioxide. Nonetheless, the tailpipe puffs out its share of air pollutants. The ubiquitous motor car, as well as light trucks and SUVs, add to the atmospheric concentrations of carbon monoxide, nitrogen oxides, and ozone, as well as a number of other unhealthful substances.

Let's first examine the tailpipe as a source of carbon monoxide. Cars account for about 60% of carbon monoxide emissions nationwide. But remember to think in terms of *all* the tailpipes out there, not just those attached to an automobile. Nonroad vehicles such as farm and construction equipment, boats, snowmobiles, and gasoline-powered lawn mowers also emit carbon monoxide. Gasp! So do heavy trucks, SUVs, and the three "m's": motorcycles, minibikes, and mopeds.

Your Turn 1.30   Other Tailpipes

What about the exhausts of lawn mowers and boats? On the EPA Web page entitled *Nonroad Vehicles and Equipment,* look up any tailpipe of interest and summarize your findings in a brief report. Be sure to cite the EPA documents you consulted.

Severe wildfires in recent years have resulted in small increases in total CO emissions.

A dramatic reduction in CO emissions has occurred even though the number of cars has doubled in the past 25 years. According to a report by the EPA entitled *Latest Findings on National Air Quality,* the "2001 ambient average CO concentration is almost 62% lower than that for 1982 and is the lowest level recorded during the past 20 years." This decrease is due to several factors, including better engine design, computerized sensors that better adjust the fuel/oxygen mixture, and most importantly, that all new cars since the mid-1970s must have **catalytic converters** (Figure 1.16), devices installed in the exhaust stream to reduce emissions. In general, a **catalyst** is a chemical substance that participates in a chemical reaction and influences its speed without undergoing permanent change. Catalytic converters in vehicles have two functions. The first is to lower carbon monoxide emissions using metals such as platinum and rhodium to catalyze the combustion of CO to CO_2. Other catalysts convert nitrogen oxides back to N_2 and O_2, the atmospheric gases that formed them.

A modern high-performance automobile, capable of operating at high speeds and with fast acceleration, not only emits carbon in the form of carbon monoxide, but also in the form of unburned hydrocarbons. These are often referred to as VOCs, or volatile organic compounds. This term requires some explaining. A substance is **volatile** if it readily passes into the vapor phase. Gasoline and nail polish remover are both volatile; when you spill a few drops, these drops quickly evaporate. A substance is an **organic**

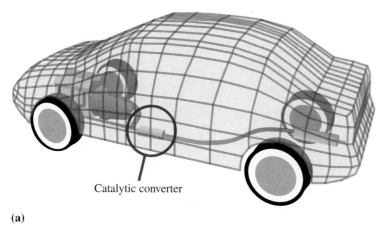

Catalytic converter

(a)

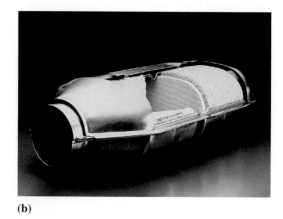

(b)

Figure 1.16

(a) Location of catalytic converter in car. (b) Cutaway view of a catalytic converter. Metals such as platinum and rhodium coat the surface of ceramic beads.

compound if it contains mainly carbon and hydrogen. For example, organic compounds include the hydrocarbons methane and octane mentioned earlier, as well as compounds containing O in addition to C and H, such as sugar. We will discuss organic compounds more fully beginning in Chapter 4.

In the case of tailpipe emissions, **volatile organic compounds (VOCs)** are vapors of incompletely burned gasoline molecules or fragments of these molecules. This incomplete combustion is caused by either insufficient oxygen or insufficient time in the engine cylinders for all the hydrocarbons to be burned to carbon dioxide and water. The exhaust gas still contains oxygen, as not all of it was consumed by burning gasoline in the engine. Catalytic converters also lower the amounts of VOCs emitted by utilizing this oxygen to burn VOCs to form carbon dioxide and water. We will learn more in Section 1.12 about the role of VOCs in forming ozone, a pollutant formed in the atmosphere from auto emissions.

Another success, although much harder won, accompanied the advent of the catalytic converter. For more than 50 years, a lead-containing compound, tetraethyl lead (TEL), was added to gasoline to make it burn more smoothly and eliminate "knocking". About a teaspoon of TEL was added to every gallon of gasoline. TEL worked beautifully in "knocking out the knock," but unfortunately released lead through the tailpipe onto the roadsides and city streets. In many of its chemical forms, lead is highly toxic and acts as a cumulative poison that can cause a wide variety of neurological problems, especially for children. Although its toxicity was well known and documented, the struggles to remove lead from gasoline lasted over 60 years.

The catalytic converter and TEL are connected in that the latter destroys the effectiveness of the former. Therefore, the cars and trucks built with catalytic converters since 1976 were formulated to run on unleaded gasoline (gasoline without TEL). After over 20 year of phasing in unleaded fuel, in 1997 leaded fuel finally was banned by law in the United States. Accordingly, today at the gas pump you see all fuel labeled as unleaded (Figure 1.17). The result has been a dramatic decrease (95%) in lead emissions from vehicles, from 1980 to 1999. Unfortunately, lead has yet to be banned globally, and several dozen countries still allow almost a gram of lead per liter of fuel. Table 1.9 in Section 1.12 shows high levels of lead pollution for Bangkok, Cairo, Jakarta, and Mexico City.

The United States has had less success curbing the emission of nitrogen oxides. Nitrogen and oxygen are always present wherever there is air. And, whenever air is subjected to high temperatures, as in an internal combustion engine or in a coal-fired power plant, N_2 and O_2 combine to form two molecules of NO, nitrogen monoxide, as we saw earlier in equation 1.4. Subsequent reactions with atmospheric oxygen can generate other oxides of nitrogen, including nitrogen dioxide, NO_2.

The air in a pine forest contains VOCs. The wonderful smell comes from volatile compounds emitted by the trees.

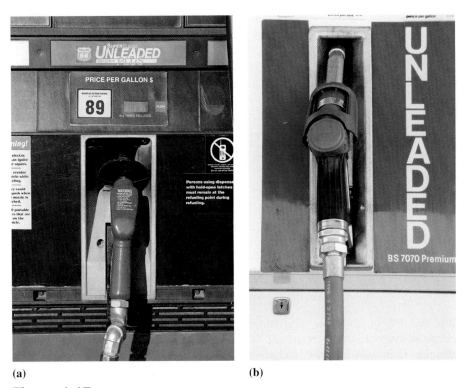

(a)   **(b)**

Figure 1.17

Pumps for unleaded gasoline from **(a)** the United States and **(b)** United Kingdom

> Many countries in the Middle East and Africa still use leaded gasoline.

Unlike N_2, NO is very reactive. It reacts with oxygen to form NO_2.

$$2\,NO + O_2 \longrightarrow 2\,NO_2 \qquad\qquad [1.10]$$

However, this reaction does not occur in a short time (such as while driving your car to work) because it requires high concentrations of NO to proceed quickly. The concentration of NO in polluted air is on the order of 100 ppb, which is not high enough for the NO to quickly react with O_2. How, then, is NO_2 formed from NO? To answer this question, we need to bring in two other players: VOCs (mentioned earlier) and the hydroxyl radical, $\cdot$OH. The latter is a reactive species containing an unpaired electron indicated by the dot. In Chapter 2, you will meet $Cl\cdot$ and other reactive species with unpaired electrons.

> The hydroxyl radical will return in Chapters 6 and 7.

Although both $\cdot$OH and VOCs occur naturally, the concentration of VOCs is considerably higher in polluted air. The following complex chain of events converts NO to NO_2.

$$VOC + \cdot OH \longrightarrow A$$
$$A + O_2 \longrightarrow A'$$
$$A' + NO \longrightarrow A'' + NO_2 \qquad\qquad [1.11]$$

Here, A, A$'$, and A$''$ represent reactive molecules that are synthesized from the VOC molecules. As promised, this reaction is complex! But then again so is the chemistry of air pollution. We emphasize the bottom line: If the air contains sufficient concentrations of NO, O_2, VOCs, and $\cdot$OH, you have the right ingredients to form NO_2. In the process, other reactive molecules (A, A$'$, and A$''$) are formed as well. NO_2 is highly toxic and is a player in the formation of tropospheric ozone, as we will see in the next section. Nitrogen dioxide also contributes to acid rain, the subject of Chapter 6.

The quantity of nitrogen oxides emitted into the atmosphere has increased about 9% since 1980, though they have shown a slight decrease in the last few years. Given the significantly increased number of vehicles and miles driven, any decrease is impressive. Despite early claims from the auto industry that it would be impossible, or too costly, to meet the standards, the industry has, in fact, achieved these goals by using improved catalytic converters, engine designs, and gasoline formulations.

Consider This 1.31   Electric Cars

Some people promote widespread development and use of electric cars as an alternative to the gasoline-powered engine. Such cars are no longer just a hope for the future, but are currently available in some areas. What criteria would you use in deciding whether to buy an electric car?

For more about electric cars and other alternatives to gasoline-powered vehicles, look to Chapter 8.

An obvious way to reduce pollutants is not to have them form in the first place. Over the past decade, an important initiative known as "green chemistry," the use of chemistry to prevent pollution, has taken place. **Green chemistry** is the designing of chemical products and processes that reduce or eliminate the use or generation of hazardous substances. Begun under the EPA's Design for the Environment Program, green chemistry reduces pollution through fundamental chemical breakthroughs in designing and redesigning processes that make chemical products, with an eye toward making them environmentally friendly, that is, "benign by design." In this regard, Dr. Barry Trost, a Stanford University chemist, advocates an "atom economy" approach to the synthesis of commercial chemical products such as pharmaceuticals, plastics, or pesticides. Such syntheses would be designed so that all reactant atoms end up as desired products, not as wasteful by-products. This approach will save money, as well as materials; undesired products would not be produced as waste, which requires disposal.

Dr. Lynn R. Goldman, who joined the EPA in 1993 to head the Office of Prevention, Pesticides, and Toxic Substances, says "Green chemistry is preventative medicine for the environment." Innovative "green" chemical methods already have had an impact on a wide variety of chemical manufacturing processes by decreasing or eliminating the use or creation of toxic substances. For example, the use of green chemical principles has led to cheaper, less wasteful, and less toxic production of ibuprofen, pesticides, new materials for disposable diapers and contact lenses, new dry-cleaning methods, and recyclable silicon wafers for integrated circuits. The research chemists and chemical engineers who developed these and other green chemistry approaches have received the Presidential Green Chemistry Challenge Award. Begun in 1995, it is the only presidential-level award recognizing chemists and the chemical industry for their innovations for a less polluted world; its theme is "Chemistry is not the problem, it's the solution." Since 1995, the products and processes developed by winners of the Presidential Green Chemistry Challenge Award have eliminated more than 102 billion pounds of hazardous chemicals and saved more than 3 billion gallons of water and 26 million barrels of oil. At various places throughout this book, we will discuss applications of green chemistry. They are designated by the Green Chemistry icon.

1.12 Ozone: A Secondary Pollutant

Ozone definitely is a bad actor in the troposphere. As we mentioned earlier, ozone affects the respiratory system, and even at very low concentrations, during periods of exercise it will reduce lung function in normal, healthy people. Low concentrations of ozone also take a toll on vegetation, damaging leaves and the needles of trees. In the previous section, however, we made no mention of ozone coming out of a tailpipe or being produced when coal is burned to generate energy. How then is ozone produced? Before we continue, examine Figure 1.18 and answer the questions in Your Turn 1.32 that accompany it.

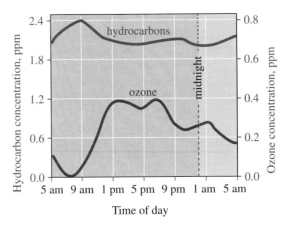

Figure 1.18
Ozone and hydrocarbon levels over time.

Your Turn 1.32   **Ozone Around the Clock**

Figure 1.18 shows how hydrocarbon and ozone concentrations might vary over time in a metropolitan area.

a.   At what time of day are the ozone levels at their highest? Assuming that the sun comes up at 6 AM and goes down at 8 PM, what are the ozone levels like when it is dark?
b.   Why would you expect hydrocarbon levels to rise in the morning rush hour?
c.   Name a compound that could be contributing to the hydrocarbon increase.

Answer

c.   If the hydrocarbons came from automobile engines, an example would be octane or any of the other hydrocarbons in gasoline.

From the previous exercise, you can see that we really have two mysteries to solve: (1) how ozone forms in urban areas, and (2) why daylight and VOCs are linked to its formation.

To see how ozone forms, we begin our story at what might seem like an unlikely point: the oxides of nitrogen. Recall from Section 1.11 that the pollutant nitrogen dioxide does not come directly out of the tailpipe; rather, automobile engines produce nitrogen monoxide. But over time and in the presence of VOCs and $\cdot OH$, NO in the atmosphere is converted to NO_2, as we saw earlier in equation 1.11.

$\cdot OH$ is the hydroxyl radical, a reactive species.

Nitrogen dioxide further reacts following several atmospheric pathways, but the one of concern to us occurs when the Sun gets higher in the sky. The energy provided by sunlight splits one of the bonds in the NO_2 molecule:

$$NO_2 \xrightarrow{\text{sunlight}} NO + O \qquad [1.12]$$

Focus now on the oxygen atom, O, produced in equation 1.12. Like NO_2, this reactive species can suffer many fates in the atmosphere. One is to react with an oxygen molecule to produce ozone.

In Chapter 2, we will refer to O_3 and O_2 as allotropes.

$$O + O_2 \longrightarrow O_3 \qquad [1.13]$$

There you have it! In a sunny metropolitan area, a chemical sequence of events initiated by the pollutants coming out of tailpipes (and other sources of NO) leads to the formation of ozone. Unlike the pollutants described in the previous section, ozone is not directly emitted into the atmosphere. Rather, it is a **secondary pollutant,** that is, it is produced from chemical reactions among two or more other pollutants, in this case, VOCs and NO_2.

Note that equation 1.13 contains three different forms of the element oxygen: O, O_2, and O_3. All three are found in nature, but diatomic oxygen, O_2, is by far the most common as it constitutes about one fifth of the air we breathe. Our atmosphere naturally contains tiny amounts of protective ozone up in the stratosphere, as we shall see in Chapter 2, and even tinier amounts of oxygen atoms that are too reactive to persist very long.

Consider This 1.33   A Summary for O_3

Summarize what you have just learned about ozone formation by developing your own way to arrange these sequentially and in relation to one another: O, O_2, O_3, VOCs, $\cdot OH$, NO, NO_2, and sunlight. Chemicals may appear as many times as you would like.

Because sunlight is involved in ozone formation, you might suspect that the concentration of ozone varies by season and by latitude. Your suspicion would be correct. High levels of tropospheric ozone are much more likely to occur in summer or in regions of the country that are sunny and full of cars. For example, let's look back at the information for Houston presented in Table 1.4. From the data, we see that the air was unhealthy for 26 days in 2003. Closer examination of the data indicates that two pollutants were mainly responsible for the unhealthy air: ozone and $PM_{2.5}$. Depending on the calendar year, one or the other of these was the main pollutant, that is, the one responsible for the highest AQI value. In some years, ozone was the main pollutant practically every day; in others, it predominated for only about a third of the time.

Consider This 1.34   Ozone in a City Near You

Ozone data are available in almost every state. To access the data, go to AIRNOW courtesy of the EPA or use the direct link on the *Online Learning Center*. Select a metropolitan area of your choice. Then see what you can learn about the amount of ozone pollution from the color-coded data provided. Summarize your findings in a paragraph, making a data table to support your points.

Although you may not easily find data on the Web for cities across the globe, ozone pollution nonetheless is an international problem. Couple vehicles with a sunny location anywhere on the planet, and you are likely to find unacceptable levels of ozone. Table 1.9 lists pollutants in cities around the globe. London, with its foggy, cooler days has low ozone levels. In contrast, ozone is a serious pollution problem in Mexico City.

As you might expect, the story of ozone continues after it forms. It affects both animal and plant life. For example, from the National Ambient Air Quality Standards listed in Table 1.5 you saw that if average ozone levels in a 1-hr period exceed 1.20 ppm, the air is harmful to breathe. As mentioned earlier, ozone affects vegetation, crops, and the leaves and needles of trees. Ironically, ozone also attacks rubber, thus damaging the tires of the vehicles that led to its production in the first place. Should you park your car indoors in the garage to minimize possible rubber damage? In fact, should you park yourself indoors if the levels of ozone outside are unhealthy? Check out Section 1.13 to find the answers.

Table 1.9	Air Quality in Megacities Around the World					
City	**SO$_2$**	**PM**	**Lead**	**CO**	**NO$_2$**	**O$_3$**
Bangkok, Thailand	Low	Serious	Moderate to heavy	Low	Low	Low
Beijing	Serious	Serious	Low	...	Low	Moderate to heavy
Cairo	...	Serious	Serious	Moderate to heavy	...	...
Jakarta, Indonesia	Low	Serious	Moderate to heavy	Moderate to heavy	Low	Moderate to heavy
London	Low	Low	Low	Moderate to heavy	Low	Low
Mexico City	Serious	Serious	Moderate to heavy	Serious	Moderate to heavy	Serious
Moscow	...	Moderate to heavy	Low	Moderate to heavy	Moderate to heavy	...
São Paulo, Brazil	Low	Moderate to heavy	Low	Moderate to heavy	Moderate to heavy	Serious
Tokyo	Low	...	Low	Low	Serious	...

Source: Atmospheric Research and Information Centre, Manchester Metropolitan University, Manchester, England (1992 data).

... Data not available

Low pollution: normally meets World Health Organization (WHO) guidelines (on occasion may exceed guidelines short-term).

Moderate to heavy pollution: WHO guidelines exceeded up to twice as much (short-term guidelines regularly exceeded at certain locations).

Serious pollution: WHO guidelines exceeded at greater than twice as much.

1.13 The Inside Story of Air Quality

Most of us sleep, work, study, and play indoors, often spending the overwhelming majority of our time in our dorm rooms, classrooms, offices, shopping malls, restaurants, or health clubs. In spite of where we spend our time, standards have been established for outdoor air, but not for the air inside. Ironically, the levels of air pollutants indoors may far exceed those outdoors. Furthermore, those least able to handle poor air, the very old, newborns, and those who are ill, may seldom get out-of-doors. Consequently, it makes sense to study the chemistry of indoor air pollution with an eye (or nose) to minimizing it.

Indoor air is a complex mixture; nearly a thousand substances typically are detectable at the parts per billion level or higher. If you are in a room where somebody is smoking, add another thousand or so. Although the list of chemicals in the air is lengthy, you will recognize some familiar culprits: VOCs, NO, NO$_2$, SO$_2$, CO, ozone, radon, and PM. Less familiar pollutants include chemicals such as formaldehyde, benzene, and acrolein. Some of these pollutants are present because they are brought in with the air from outside the building; others are generated right inside.

Let's begin by considering the question posed at the end of Section 1.12: Should you move indoors to escape ozone or other pollutants? As you make your decision about a particular pollutant, consider if (1) the pollutant is generated indoors as well, and if not, (2) how reactive the pollutant is, and (3) whether the physical construction of the building will filter it out. For now, let us put aside item (1) when the pollutant in question is also generated indoors (we will discuss this case shortly). In assessing item (2), the chemical reactivity, assume that the more reactive the chemical species, the less likely it is to persist indoors. Thus, for reactive molecules such as ozone, sulfur dioxide, and nitrogen dioxide, we expect much lower levels indoors. The numbers bear this out: The actual ratio of ozone levels outdoors to those indoors is typically between 0.1 to 0.3, meaning that up to 10 times less ozone occurs inside. Thus, all things being equal, one way to escape an ozone alert is to move inside. Similarly, sulfur dioxide and nitrogen dioxide levels are expected to be lower indoors, although the ratio is not quite as favorable as that for ozone.

Carbon monoxide is a different story. Carbon monoxide is a relatively unreactive pollutant. This gas has a long enough atmospheric lifetime to move freely in and out of buildings, either through doors, windows, or through the ventilation system. The same is true for the less reactive VOCs, but not the more reactive ones such as those that give pine forests their scent. Stay outdoors to smell the volatile pine compounds.

Finally, the physical construction of a building determines the level of indoor pollutants. This is especially true for particulate matter. For example, ventilation systems can act as a filter. People who suffer from pollen allergies often escape into air-conditioned buildings, since the pollen levels tend to be lower inside. Similarly, those living near the forest fires on the west coast in the past few years sometimes were able to escape the irritating smoke particles by staying indoors. Only to a certain extent, however, do buildings filter out the fine particles of smoke and ash. Those who lived in New York City in the aftermath of the 2001 attacks on the World Trade Center can attest to the fact that the smell of the smoke pervaded their living and sleeping spaces for a long time.

> Some copy machines and air cleaners generate ozone, in which case the indoor-to-outdoor ratio may not be as favorable.

Your Turn 1.35   Indoor Activities

Name five indoor activities that generate pollutants. To get you started, two examples are pictured in Figure 1.19. Remember that you cannot smell some pollutants, such as carbon monoxide.

Following up on Your Turn 1.35, we now need to examine indoor sources of air pollution. As we said earlier, to decide whether to move indoors to escape a particular pollutant, we need to consider if the pollutant is generated indoors. If so, the rate at which the pollutants are generated indoors will determine how quickly they build up.

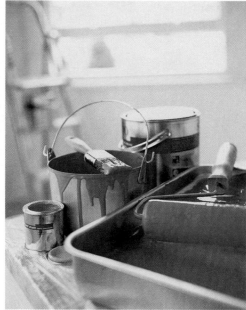

(a)   (b)

Figure 1.19
Activities that pollute indoor air.

As it turns out, several pollutants are present at higher levels indoors, because they are rapidly generated indoors when there is insufficient ventilation.

Let's begin by examining indoor volatile organic compounds. We already have discussed this pollutant in Sections 1.11 and 1.12 in terms of the gasoline compounds and fragments that come out of a tailpipe. Because tailpipes are unlikely to be indoors, we need to look for other sources of VOCs. One source is environmental tobacco smoke (ETS), often referred to as "second-hand smoke." If you know one of the 40 million smokers in the United States (or smoke yourself), you know that the pollutants in cigarette smoke are easily noticeable indoors. As pointed out earlier, cigarette smoke contains over a thousand chemical substances that, taken as a whole, are **carcinogenic,** meaning capable of causing cancer. One VOC that you probably recognize is nicotine; others include benzene and formaldehyde.

But cigarettes are not the only thing we burn indoors and hence not the only source of VOCs. Burning incense and candles also produce VOCs, often with a good deal of soot (particulate matter) as well. Again, the indoor pollutants can be carcinogenic; for example, epidemiological studies have linked the regular burning of incense with some childhood cancers. Wood stoves, fireplaces, and some appliances also emit a variety of VOCs. The risks are high if wood-burning appliances are improperly installed, improperly vented, or simply malfunctioning.

Cigarette smoking in particular and combustion in general are good examples of how indoor sources produce more than one type of pollutant. For example, by burning any carbon-containing fuel you would expect to produce carbon monoxide. Indeed, CO levels from cigarette smoking in bars can reach 50 ppm, well within the range considered unhealthy. Similarly, the high temperatures from burning tobacco or wood would be expected to produce nitrogen oxides. Again for cigarette smoke, NO_2 levels can exceed 50 ppm.

Other indoor activities add pollutants to the air as well. Often your nose alerts you to the source; sometimes your head as well, if you get a headache from the fumes. For example, when you paint, use a brush cleaner, or paint your fingernails, you can smell the VOCs. Paint thinner in particular carries the warning on the label to use with sufficient ventilation. If you use hairspray or something else in a spray can, the odor may linger and be especially noticeable to somebody who enters the room. New carpets and new furniture also emit their own characteristic odors. These and other sources of indoor air pollutants are listed in Table 1.10.

Radon, a noble gas, is a special case of indoor air pollution. It is produced naturally from most soils and subsoils, but only tends to build up in dwellings, particularly

Table 1.10	Selected Indoor Air Pollutants and Their Sources	
Form	**Source**	**Pollutants**
Solid/particulate	Floor tile, insulation	Asbestos
	Pets	Pet dander, dust
	Plants	Molds, mildew, bacteria, viruses
Liquid/gas	Carpet	Styrene
	Cigarette smoke	CO, benzene, nicotine
	Clothes	Dry-cleaning fluid, moth balls
	Electric arcing	Ozone
	Unvented space heaters	CO, NO, NO_2
	Furniture	Formaldehyde
	Glues and solvents	Acetone, toluene
	Paint, paint thinners	Methanol, methylene chloride
	Soils under house	Radon

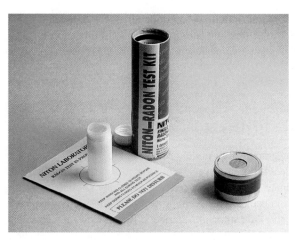

Figure 1.20
A home radon test kit.

basements. Radon also is found in some mines and caves. Like all noble gases, radon is colorless, odorless, tasteless, and chemically unreactive. Unlike the other noble gases, however, radon is radioactive (see Chapter 7). Radon is generated as part of the nuclear decay series of another radioactive element, uranium. Because uranium occurs naturally and is ubiquitous at the surface of our planet, chances are that the rocks and soils beneath your home contain small amounts of uranium. Depending on how your dwelling is constructed and how weather-tight it is, radon may seep into your basement and be trapped indoors. Extended inhalation of radon can result in lung cancer. But the exposure level of radon that creates health problems is controversial. Home radon testing kits are commercially available (Figure 1.20).

A further word on pollutants and building construction is in order. As you would expect, the rates at which outdoor air moves inside and indoor air moves out affects how quickly air pollutants build up indoors. An insufficient exchange of outside air can cause the concentration of indoor air pollutants to build up to troublesome levels. Consider the risk–benefit trade-off in which buildings constructed within the past two decades have been more airtight to increase energy efficiency. Although greater energy efficiency has decreased heating and cooling costs, it has been at the cost of decreasing the circulation of air between the building and the outside. When air exchange is reduced, the levels of indoor air pollutants increase. Therefore, initially what was a benefit (better energy efficiency) can turn into an increased risk (increased pollutant levels). Construction of some large office buildings has been so highly energy efficient that little exchange of outside air occurs within them. In some of these cases, the reduced air exchange has allowed indoor pollutants to reach levels hazardous to the health of some individuals, creating a condition known as "sick building syndrome."

Consider This 1.36   **Radon Testing**

As a public service, local and national agencies provide documents on the Web about radon. Search the Web to bring up information about radon. You might want to add the terms *detection, air quality,* and *EPA* to your search.

a. Find two Web sites about radon provided by government agencies. List the source and the URL for each.

b. How can you measure the radon levels in your home? Search the Web for a company that sells radon test kits. Describe the kit, including its price.

c. Compare the dangers of radon cited on your Web sites from parts **a.** and **b.** Is commercial information about radon any different from that provided as a public service? If so, discuss the differences and suggest reasons why.

Whether we breathe indoor or outdoor air, we inhale (and exhale) a truly prodigious number of molecules and atoms during a lifetime. On a molecular and atomic level, these particles have some fascinating properties, ones that we consider next.

1.14 Back to the Breath—at the Molecular Level

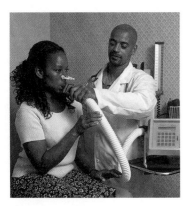

Figure 1.21

A spirometer is used for measuring an individual's breathing capacity.

The maximum concentrations of pollutants specified in Table 1.5 seem very small, and they are. Nine CO molecules out of every 1 million molecules in air is a tiny fraction. But, as we will soon calculate, even this low concentration of CO contains a staggering number of carbon monoxide molecules. This seeming contradiction is a consequence of the minuscule size of molecules and their immense numbers. Recall Consider This 1.1: Take a Breath. If you are an average-sized adult in good physical condition, the total capacity of your lungs is between 5 and 6 liters. You do not exchange this volume of air each time you take a breath. Rather, perhaps now as you are reading this, you are inhaling (and exhaling) only about 500 milliliters (0.500 L) of air, or approximately a half a quart.

Accurately measuring the volume of air you inhale and exhale can be done with the help of a spirometer (Figure 1.21). Determining the number of molecules and atoms in this volume of air is a harder task, but it can be done. From experiments (as well as from theory), we know that a typical breath of 500 mL contains about 2×10^{22} molecules, such as N_2 and O_2, together with atoms such as Ar and He.

Using this number of molecules and atoms in the air (2×10^{22}), we now can calculate the number of CO molecules in the breath you just inhaled. We will assume the breath contained 2×10^{22} molecules and that the CO concentration in the air was the NAAQS of 9 ppm. Thus, out of every million (1×10^6) molecules and atoms in the air, nine will be CO molecules. To compute the number of CO molecules in a breath, multiply the total number of molecules in the air by the fraction that are CO molecules.

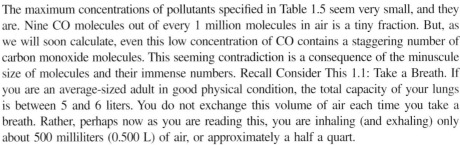

$$\text{\# of CO molecules} = 2 \times 10^{22} \text{ molecules and atoms in air} \times \frac{9 \text{ CO molecules}}{1 \times 10^6 \text{ molecules and atoms in air}}$$

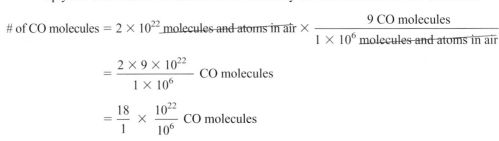

$$= \frac{2 \times 9 \times 10^{22}}{1 \times 10^6} \text{ CO molecules}$$

$$= \frac{18}{1} \times \frac{10^{22}}{10^6} \text{ CO molecules}$$

In writing this out, we carefully retain the labels on the numbers. Not only does this remind us of the physical entities involved, but also it guides us in setting up the problem correctly. The labels "molecules and atoms in the air" cancel each other, and we are left with the label we want: CO molecules.

However, we need to divide 10^{22} by 10^6 to determine a final answer. To *divide* powers of 10, simply *subtract* the exponents. In this case,

$$\frac{10^{22}}{10^6} = 10^{(22-6)} = 10^{16}$$

Thus, a breath contains 18×10^{16} CO molecules.

The preceding answer is mathematically correct, but in scientific notation it is customary to have only one digit to the left of the decimal point. Here we have two: 1 and 8. Therefore, our last step will be to rewrite 18×10^{16} as 1.8×10^{17}. We can make this conversion because $18 = 1.8 \times 10$, which is the same as 1.8×10^1. We *add* exponents to *multiply* powers of 10. Thus, 18×10^{16} CO molecules equals $(1.8 \times 10^1) \times 10^{16}$ CO molecules, which equals 1.8×10^{17} CO molecules in that last breath you inhaled. (If all of this use of exponents is coming at you a little too fast, consult Appendix 2.)

It may sound surprising, but it would be more accurate to round off the answer and report it as 2×10^{17} CO molecules. Certainly 1.8×10^{17} looks more accurate, but the data that went into our calculation were not very exact. The breath contains *about* 2×10^{22} molecules, but it might be 1.6×10^{22}, 2.3×10^{22}, or some other number. The jargon is that 2×10^{22} expresses a physically based property to "one **significant figure,**" that is, a number that correctly represents the accuracy with which an experimental quantity is known. Only one digit, the 2 from the value 2.3 is used, and so 2×10^{22} has only one significant figure. Accordingly, the number of molecules in the breath is closer to 2×10^{22} than to 1×10^{22} or to 3×10^{22}, but anything beyond this level of exactness we cannot say with certainty.

Similarly, the concentration of carbon monoxide is known to only one significant figure, 9 ppm. That 2×9 equals 18 is certainly correct mathematically, but our question about CO is based on physical data. The answer, 1.8×10^{17} CO molecules, includes two significant figures: the 1 and the 8. Two significant figures imply a level of knowledge that is not justified. The accuracy of a calculation is limited by the *least accurate* piece of data that goes into it. In this case, both the concentration of CO and the number of molecules and atoms in the breath were each known only to one significant figure (9 and 2, respectively); two significant figures in the answer are unjustified. The common-sense rule is that you cannot improve the accuracy of experimental measurements by ordinary mathematical manipulations like multiplying and dividing. Therefore, the answer must also contain only one significant figure; hence 2×10^{17}.

Your Turn 1.37   Ozone Molecules

The local news has just reported that today's ground-level ozone readings are right at the acceptable level, 0.12 ppm. How many molecules of ozone, O_3, are in each breath of this air? Assume each breath contains 2×10^{22} molecules and atoms in a breath.

Answer

If ozone is 0.12 ppm, then 0.12 O_3 molecules occur per 10^6 molecules and atoms in air. Multiply the number of molecules and atoms in a breath by this:

$$2 \times 10^{22} \text{ molecules and atoms in a breath of air} \times \frac{0.12 \; O_3 \text{ molecules}}{1 \times 10^6 \; \text{molecules and atoms in air}}$$

$$= 2.4 \times 10^{15} \; O_3 \text{ molecules in a breath}$$

$$= 2 \times 10^{15} \; O_3 \text{ molecules in a breath (to one significant figure)}$$

You may well question the significance of all of this talk about significant figures, but these are important in interpreting numbers associated with physical quantities. It has been observed that "figures don't lie, but liars can figure." Numbers often lend an air of authenticity to newspaper or television stories, so popular press accounts are full of numbers. Some are meaningful and some are not, and the informed citizen must be able to discriminate between the two types. For example, the assertion that the concentration of carbon dioxide in the atmosphere is 375.5537 ppm should be taken with a rather large grain of sodium chloride (salt). The values of 375 ppm and 375.6 ppm (three or four significant figures) better represent what we actually can measure; any assertion with seven significant figures simply is not valid.

Your Turn 1.38   Carbon Monoxide Meters

Carbon monoxide monitors are commercially available to be used in residences, as well as offices and business sites. Figure 1.22 shows a convenient handheld CO detector reading 35 ppm.

a. Would it be more helpful to have one that read 35.0388217 ppm?
b. Would 35.0388217 ppm be more valid? In both cases, explain your reasoning.

(continued on p. 52)

Figure 1.22
Carbon monoxide meter.

Your Turn 1.38   **Carbon Monoxide Meters** (*continued*)

Answer

a. No, it wouldn't be more helpful. The issue with CO is whether it exceeds a certain value, such as 9 ppm over an 8-hr period or 35 ppm over a 1-hr period. At 35 ppm, no matter what number of decimals follow, the value of ppm is too high.

Numbers can introduce ambiguity in other ways as well. You have just encountered some conflicting information. The concentration of CO in air is very small, 9 ppm. Nevertheless, the number of CO molecules in a breath is still large, about 2×10^{17}. Both statements are true. The consequence of these numbers is that it is *impossible* to completely remove pollutant molecules from the air. "Zero pollutants" is an unattainable goal; using the most sophisticated detection methods you still could not even determine whether it had been achieved. At present, our most sensitive methods of chemical analysis are capable of detecting one target molecule out of a trillion. One part per trillion corresponds to: moving 6 inches in the 93 million-mile trip to the Sun; a single second in 320 centuries; or a pinch of salt in 10,000 tons of potato chips. A chemical could be undetectable at this level, and yet a breath might still include 2×10^{10} molecules of the substance.

> Absence of evidence is not the same as evidence of absence. The substance may be present, but in undetectable amounts.

Your Turn 1.39   **Parts Per Million**

To help you comprehend the magnitude of the 2×10^{17} CO molecules in just one of your breaths, assume that they were equally distributed among the 6 billion (6×10^9) human inhabitants of the Earth. Calculate each person's share of the 2×10^{17} CO molecules you just inhaled.

Hint: You are trying to distribute the huge number of molecules in a breath among all the human inhabitants of the Earth. Each person's share can be found by dividing the total number of CO molecules by the total number of humans:

$$\text{Each person's share is } \frac{2 \times 10^{17} \text{ CO molecules}}{6 \times 10^9 \text{ people}}$$

Now see if you can demonstrate that each person's share is 3×10^7, or 30,000,000, molecules of CO per person (to one significant figure).

A breath of air typically contains molecules of hundreds, perhaps thousands, of different compounds, most in minuscule concentrations. For almost all these substances, it is impossible to say whether the origin is natural or artificial. Indeed, many trace components, including the oxides of sulfur and nitrogen, come from both natural sources and those related to human activity. And, as with all chemicals, "natural" is not necessarily good and "human-made" is not necessarily bad. As you read in Section 1.4, what matters is exposure, toxicity, and the assessment of risk.

In addition to being extremely small, the particles in your breath possess other remarkable characteristics. In the first place, they are in constant motion. At room temperature and pressure, a nitrogen molecule travels at about 1000 feet per second and experiences approximately 400 billion collisions with other molecules in that time interval. Nevertheless, relatively speaking, the molecules are quite far apart. The actual volume of the extremely tiny molecules making up the air is only about 1/1000th of the total volume of the gas. If the particles in your 1-liter breath were all squeezed together, their volume would be about 1 mL or about one third of a teaspoon. Sometimes people mistakenly think that air is empty space. It's 99.9% empty space, but the matter that is in it is literally a matter of life and death!

Moreover, it is matter that we continuously exchange with other living things. The carbon dioxide we exhale is used by plants to make the food we eat, and the oxygen that plants release is essential for our existence. Our lives are linked together by the elusive medium of air. With every breath, we exchange millions of molecules with one another. As you read this, your lungs contain 4×10^{19} molecules that have been previously breathed by other human beings, and 6×10^8 molecules that have been breathed by some *particular* person, say Julius Caesar, Mahatma Gandhi, or Joan of Arc. Pick any person, your body almost certainly contains atoms that were once in his or her body. In fact, the odds are very good that right now your lungs contain one molecule that was in Caesar's *last* breath. The consequences are breathtaking!

Sceptical Chymist 1.40   Caesar's Last Breath

We just claimed that your lungs currently contain one molecule that was in Caesar's last breath. That assertion is based on some assumptions and a calculation. Are these assumptions reasonable? We are not asking you to reproduce the calculation, but rather to identify some of the assumptions and arguments we might have used.

Hint: The calculation assumes that all of the molecules in Caesar's last breath have been uniformly distributed throughout the atmosphere.

Consider This 1.41   Growing Interest in Air Pollution

Air pollution did not occur overnight. It has been a growing problem since at least the time of the Industrial Revolution. Why have we as a nation and a world community become so concerned with it lately? Through discussion and library and Web research, identify at least four factors that have combined to make air pollution an important issue at present.

CONCLUSION

The air we breathe has a personal and immediate effect on our health. Our very existence depends on having a large supply of relatively clean, unpolluted air with its essentials for life: the elements, oxygen and nitrogen, and two compounds, water and carbon dioxide. But air can be polluted with potentially toxic substances such as carbon monoxide, ozone, sulfur oxides, and nitrogen oxides. This is true especially in the urban environments of our large cities, the very places where the majority of Americans live. The major pollutants are, for the most part, relatively simple chemical substances. Carbon monoxide and the oxides of sulfur and nitrogen are compounds that exist as molecules made from atoms of their constituent elements. These compounds are formed by chemical reactions, often as unavoidable consequences of our dependence on coal for energy production in power plants and gasoline in internal combustion engines. Over the past 30 years, governmental regulations, industrial participation, modern technology, and green chemistry have resulted in large reductions in many pollutants. But it is impossible to reduce pollutant concentrations to zero because of the minuscule size of atoms and molecules and their immense numbers. Rather we must determine the risk from a given level of pollutant and then decide what level of risk is acceptable for various population groups.

The oxygen-laden air we breathe, whether indoors or out, is, of course, very close to the surface of the Earth. But the Earth's atmosphere extends upward for considerable distance and contains other substances that are also essential for life on this planet. In the next two chapters, we consider two of these substances and how they are changing, perhaps as a result of human activities.

Chapter Summary

The numbers that follow indicate the sections in which the topics are introduced and explained.

Having studied this chapter, you should be able to:

- Describe air in terms of its major components, their relative amounts, and the local and regional variations in the composition of air (1.1–1.3).

- List major air pollutants and describe the effects of each on humans (1.3, 1.11–1.13).

- Compare and contrast indoor and outdoor air, in terms of which pollutants are likely to be present and their relative amounts (1.3, 1.13).

- Interpret values of the color-coded AQI and know how to assess local air quality data from the EPA (1.3).

- Understand the terms NAAQS, exposure, and toxicity, and why the NAAQS are set at different levels for different periods of time (1.3).

- Evaluate conditions significant in risk–benefit analysis (1.4).

- Identify the general regions of the atmosphere with respect to altitude and the relationship of air pressure to altitude (1.5).

- Interpret air quality data in terms of concentration units (ppm, ppb) and pollution levels, including unreasonableness of "pollution-free" levels (1.2–1.3, 1.12, 1.14).

- Relate these terms and differentiate among them: matter, pure substances, mixtures, elements, and compounds (1.6).

- Discuss the features of the periodic table, including the groups it contains, and the locations of metals and non-metals (1.6).

- Understand the difference between atoms and molecules, and between symbols for elements and formulas for chemical compounds (1.7).

- Name selected chemical elements and compounds (1.7).

- Write and interpret chemical formulas (1.8).

- Balance chemical equations, including using sphere equation representations (1.9–1.10).

- Understand oxygen's role in combustion, including how hydrocarbons burn to form carbon dioxide and carbon monoxide (1.9–1.10).

- Discuss the green chemistry initiative (1.11).

- Explain the different pollutants produced by burning coal and gasoline, and how reductions in emissions have occurred (1.11).

- Describe how ozone forms, including how sunlight, NO, NO_2, and VOCs are involved. (1.12).

- Identify the sources and nature of indoor air pollution (1.13).

- Interpret the nature of air at the molecular level (1.14).

- Use scientific notation and significant figures in performing basic calculations (1.4 and 1.14, respectively).

Questions

The questions in this chapter, as well as those in the remaining chapters, are divided into three categories.

- **Emphasizing Essentials** These questions give you the opportunity to practice the fundamental skills to be developed in the chapter. This set of questions relates most closely to the Your Turn exercises in the chapter. Answers are provided in Appendix 5 for questions whose numbers are in blue.

- **Concentrating on Concepts** These questions ask you to focus on the chemical concepts developed in the chapter and their relationships to the topics under discussion. They integrate and apply chemical concepts. This set of questions most closely resembles Consider This activities you have been engaged with throughout the chapter. Answers are provided in Appendix 5 for questions whose numbers are in blue.

- **Exploring Extensions** These questions challenge you to go beyond the information presented in the text. They provide an opportunity for extending and integrating the skills, concepts, and communication abilities practiced in the chapter. Some extension questions are closely related to the type of analysis practiced in the *Sceptical Chymist* activities in the chapter. Questions marked with the Web

icon require using the World Wide Web to obtain further information.

Emphasizing Essentials

1. Calculate the volume of air that an adult person exhales in an 8-hr day. Assume that each breath has a volume of about 1 L and that the person exhales 15 times a minute.

2. Given that air is 78% nitrogen by volume, how many liters of nitrogen are in 500 L of dry air?

3. A 5.0-L mixture of gases is prepared for photosynthesis studies by combining 0.75 L of oxygen, 4.0 L of nitrogen, and 0.25 L of carbon dioxide. Compare the percentage of carbon dioxide gas with that normally found in the atmosphere.

4. Give three examples of particulate matter found in air. What is the difference between $PM_{2.5}$ and PM_{10} in terms of size? In terms of health effects?

5. These gases are found in the troposphere: Rn, CO_2, CO, O_2, Ar, N_2.

   a. Rank them in order of their abundance in the troposphere.

b. For which gases is it convenient to express their concentrations in parts per million?

c. Are any criteria air pollutants, that is, has the EPA set for them permissible levels in the air? If so, which?

d. Are any noble gases (Group 8A)? If so, which?

6. **a.** The concentration of argon in air is approximately 9000 ppm. Express this value as a percent.

b. The smoke inhaled from a cigarette contains about 0.04% CO. Express this concentration in parts per million.

c. The concentration of water vapor in the atmosphere of a tropical rain forest may reach 50,000 ppm. Express this value as a percentage.

7. According to Table 1.2, the percentage of carbon dioxide in inhaled air is *lower* than it is in exhaled air, but the percentage of oxygen in inhaled air is *higher* than in exhaled air. How can you account for these relationships?

8. Cars don't inhale and exhale like humans do. Nonetheless, the air that goes into a car is different from what comes out. In Your Turn 1.28 you listed what comes out of a tailpipe. Now comment on the *differences* between the air that goes into the car engine and that which comes out the tailpipe. For which chemicals have the concentrations noticeably increased or decreased?

9. Express each of these numbers in scientific notation.

a. 1500 m, the distance of a foot race

b. 0.0000000000958 m, the distance between O and H atoms in water

c. 0.0000075 m, the diameter of a red blood cell

d. 150,000 mg of CO, the approximate amount breathed daily

10. Write each of these values in nonscientific notation.

a. 8.5×10^4 g, the mass of air in an average room

b. 1.0×10^7 gallons, the volume of crude oil spilled by the Exxon Valdez

c. $5.0 \times 10^{-3}\%$, the concentration of CO in the air of a city street

d. 1×10^{-5} g, the recommended daily allowance of vitamin D

11. Express each of these numbers in scientific notation.

a. 72000000 cigarettes, an estimate of the number of cigarettes smoked per hour in the United States

b. 15000 °C, the approximate temperature near the spark plug in an automobile engine

c. 0.000000003 g, the number of grams of the insecticide DDT that dissolves in 1 g of water

d. 0.00022 g, the number of grams of NO_2 that can be detected by smell in 1 m^3 of air.

12. Use Figure 1.10 to verify this statement: "Below about 20 km, the air pressure decreases by about 50% for every 5 km increase in altitude." Does this relationship hold throughout the troposphere?

13. Consider this portion of the periodic table and the two groups shaded on it.

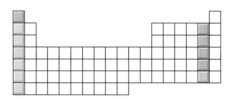

a. What is the group number for each shaded region?

b. Name the elements that make up each group.

c. Give a general characteristic of the elements in each of these groups.

14. Consider this blank periodic table.

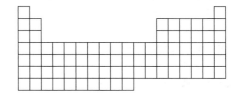

a. Shade the region of the periodic table where metals are found.

b. Six common metals are iron, magnesium, aluminum, sodium, potassium, and silver. Give the chemical symbol for each.

c. Give the name and chemical symbol for five nonmetals (elements that are not in your shaded region).

15. Classify each of these substances as an element, compound, or mixture.

a. a sample of "laughing gas" (dinitrogen monoxide, also called nitrous oxide)

b. steam coming from a pan of boiling water

c. a bar of deodorant soap

d. a sample of copper

e. a cup of mayonnaise

f. the helium filling a balloon

16. Name the compounds formed when these elements combine.

a. potassium and oxygen

b. aluminum and chlorine

c. sodium and iodine

d. magnesium and bromine

17. Write the chemical formula for each of these.

a. "laughing gas," dinitrogen monoxide (also called nitrous oxide)

b. ozone, an air pollutant, also used to purify water

c. sodium fluoride, an ingredient in some toothpastes

d. carbon tetrachloride, formerly used as a dry-cleaning agent

18. These compounds are trace components of the atmosphere. What information does each chemical formula convey in terms of the number and types of atoms present?

    a. CH_2O, formaldehyde

    b. H_2O_2, hydrogen peroxide

    c. CH_3Br, methyl bromide

19. Write balanced chemical equations to represent these reactions. *Hint:* Remember that nitrogen and oxygen are both diatomic molecules.

    a. Nitrogen reacts with oxygen to form nitric oxide (NO).

    b. Ozone decomposes into oxygen and atomic oxygen (O).

    c. Sulfur reacts with oxygen to form sulfur trioxide.

20. Write balanced sphere equations to represent each of the reactions in question 19.

21. Balance these equations in which ethene, C_2H_4, burns in oxygen.

    a. $C_2H_4(g) + O_2(g) \longrightarrow C(s) + H_2O(g)$

    b. $C_2H_4(g) + O_2(g) \longrightarrow CO(g) + H_2O(g)$

    c. $C_2H_4(g) + O_2(g) \longrightarrow CO_2(g) + H_2O(g)$

22. Compare the coefficient for oxygen in the equations from question 21. How does it vary, depending on the products formed?

23. By counting atoms on both sides of the arrow, demonstrate that each of these equations is balanced.

    a. $2\,C_3H_8(g) + 7\,O_2(g) \longrightarrow 6\,CO(g) + 8\,H_2O(l)$

    b. $2\,C_8H_{18}(g) + 25\,O_2(g) \longrightarrow 16\,CO_2(g) + 18\,H_2O(l)$

24. Platinum, palladium, and rhodium are used in the catalytic converters of cars.

    a. What is the chemical symbol for each of these metals?

    b. Where are these metals located on the periodic table?

    c. What can you infer about the properties of these metals, given that they are useful in this application?

25. If a room is 6 m long, 5 m wide, and 3 m high, how many milligrams of formaldehyde are present if the concentration is reported as 40 ppm?

Concentrating on Concepts

26. In Section 1.1, air was referred to as ". . . that invisible stuff. . . ." Is this always true? What factors influence if air appears "invisible" or if you can "see" it?

27. In Consider This 1.1, you calculated the volume of air exhaled in a day. How does this volume compare with the volume of air in your chemistry classroom? Show your calculations. *Hint:* Think ahead about the most convenient unit to use for measuring or estimating the dimensions of your classroom.

28. In Consider This 1.3, you considered how life on Earth would change if the concentration of oxygen were doubled. Now consider the opposite case; discuss how life on Earth would change if the concentration of O_2 were only 10%. Give some specific examples of how burning, rusting, and most metabolic processes in humans and plants would be affected.

29. Carbon monoxide termed the "silent killer." Why? Select two other pollutants for which this name would not apply and explain why not.

30. Consider this table of data from the EPA Office of Air Quality Planning and Standards. The data indicate the number of days metropolitan statistical areas failed to meet acceptable air quality standards (Pollutant Standards Index rating above 100).

Air Quality of Selected U.S. Metropolitan Areas, 1992–1999

Metropolitan Statistical Area	1992	1993	1994	1995	1996	1997	1998	1999
Philadelphia, PA	24	51	26	30	22	32	37	32
Phoenix, AZ	13	16	10	22	17	12	17	12

    a. Prepare a visual representation of these data. Use any type of representation you feel will best convey the information to the general public.

    b. Use your representation to discuss the trends in these two cities from 1992 to 1999.

    c. Can you use your representation to *predict* what the air quality would have been in Philadelphia and Phoenix in 2000? Discuss your reasoning.

31. A certain city has an ozone reading of 0.13 ppm for 1-hr, and the permissible limit is 0.12 for that time. You have the choice of reporting that the city has exceeded the ozone limit by 0.01 ppm or saying that it has exceeded the limit by 8%. What are the advantages of each method?

32. a. Arrange these measurements in order of increasing size: 1 m, 3.0×10^2 m, and 5.0×10^{-3} m.

    b. Draw an analogy between these three length measurements and time. Let 1 year equal 1 meter. How long will the other two measurements be in terms of time expressed in years?

33. Air quality reports are often published in local newspapers, but rarely reported during televised weather reports, unless the level of air pollution is dangerously high. Why do you think this is the case?

34. If risk is related to public perception, what is your feeling about the relative risks associated with each of these items: roller blading, eating raw cookie dough, driving on the expressway, breathing second-hand smoke, not wearing a bike helmet, taking aspirin, and drinking tap water. Rank them in order of your perception of the *most* risky to the *least* risky. Be prepared to explain your choices to your peers.

35. The cabins of commercial airliners flying at 30,000 ft are pressurized. Buildings in Denver, the "mile-high city," do

not need to be pressurized. What is the best explanation for these observations? Figure 1.10 might be helpful in answering this question.

36. In these diagrams, the larger spheres represent one kind of atom and the smaller spheres represent another. Characterize each sample as an element, compound, or mixture and explain your classification.

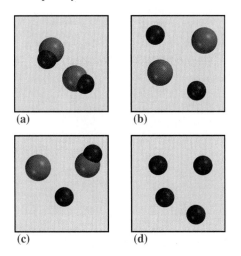

(a) (b)

(c) (d)

37. Consider this representation of the reaction between nitrogen and hydrogen to form ammonia (NH₃).

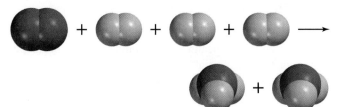

   a. Are the masses of reactants and products the same?

   b. Are the numbers of molecules of reactants and of products the same?

   c. Are the total number of atoms in the reactants and the total number of atoms in the products the same?

38. a. Explain why jogging outdoors (as opposed to sitting outdoors) increases your exposure to pollutants.

   b. Jogging indoors at home can decrease your exposure to some pollutants, but may increase your exposure to others. Explain.

39. Table 1.9 reports moderate to heavy levels of atmospheric lead in some cities of the world. What are the likely sources of this lead pollution? Explain your reasoning.

40. Young adults in Beijing, China, have gone to bars after work, not for glasses of beer or wine, but for fresh air. These "oxygen bars" provide a half-hour of deep breathing for the equivalent of $6.

   a. What does this tell you about air pollution in Beijing?

   b. Consider the information in Table 1.9. If you wanted to establish "oxygen bars" in other cities of the world, which ones would you choose?

41. Air quality in Santiago, Chile, is such a major problem that driving private cars has been severely restricted. Special decals indicate the days on which particular cars can be driven. Some citizens purchased a second car and obtained a decal for that car. However, the increase in the total number of cars will likely make the pollution problem even worse. Write a letter to a friend in Santiago suggesting a possible solution to this problem and defend your suggestion.

42. In urban areas, the concentration of formaldehyde in *outdoor* air is typically about 0.01 ppm, assuming no smog formation. In contrast, the level of formaldehyde *indoors* can average 0.1 ppm, the level at which most people will smell its pungent odor. What factors can lead to formaldehyde accumulation indoors?

Exploring Extensions

43. The percentage of oxygen gas in the atmosphere (21%) is usually expressed as the volume of oxygen gas relative to the total volume of the atmosphere being considered. The percentage can also be reported as the mass of oxygen gas relative to the total mass of the atmosphere being considered; in this case, it is 23%. Offer a possible explanation why these two values are not the same.

44. The EPA oversees the Presidential Green Chemistry Challenge Awards. Use the EPA Web site to find when the program started and to find the list of the most recent winners of the Presidential Green Chemistry Challenge Award. Pick one winner and summarize in your own words the green chemistry advance that merited the award.

45. Recreational scuba divers usually use compressed air that has the same composition as normal air. A mixture being used is called Nitrox. What is its composition, and why is it being used?

46. This table shows information from the EPA Office of Air Quality Planning and Standards. Data indicate the number of days metropolitan statistical areas failed to meet acceptable air quality standards (Pollutant Standards Index rating above 100).

Air Quality of Selected U.S. Metropolitan Areas, 1992–1999

Metropolitan Statistical Area	1992	1993	1994	1995	1996	1997	1998	1999
Boston, MA	9	6	10	8	2	8	7	5
Denver, CO	11	3	1	2	0	0	5	1
Houston, TX	32	28	38	66	26	47	38	50

   a. Do these values show the same types of trends shown in question 30?

   b. What factors influence these values? Offer some reasonable explanations, based on your research or knowledge about these cities.

47. An article in *USA Today* on January 12, 1999, is titled "Taking Technology from Here to the Infinitesimal." By the year 2020, the article predicts, "the age of atomic

engineering, . . . a type of nanotechnology, will dawn." What does this term imply? What kinds of applications will be possible that are not now part of our technology?

48. 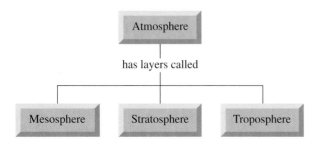 Michael Crichton, in his best-seller *Prey* (2002), entertained his readers with a fearful tale of self-replicating nanorobots. Although his novel fell into the realm of science fiction, nonetheless his point is well taken that a new discovery can have unintended consequences. Along these lines, in 2003 Congress asked for studies to determine the social, economic, and environmental impact(s) of nanotechnology. Find out what has happened since then. Are the reports uniformly positive or have some unfavorable effects of nanotechnology been reported?

49. A concept web or concept map is a convenient way to represent knowledge and connection among ideas. Concept webs are constructed by joining a word or expression to another one by means of linking words. For example, the atmosphere has three layers, the mesosphere, stratosphere, and troposphere.

What advantages or disadvantages does this representation have compared with Figure 1.8? Explain your reasoning.

50. Consider this graph that shows the effects of carbon monoxide inhalation on humans.

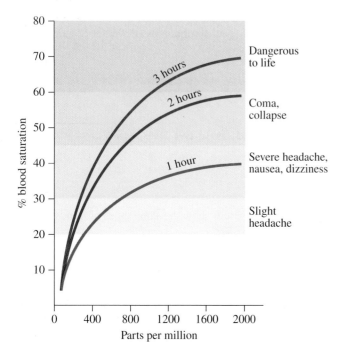

a. Both the amount of exposure and the duration of exposure have an effect on CO toxicity in humans. Use the graph to explain why.

b. Use the information in this graph to write a paragraph to include with a home carbon monoxide detection kit about the health hazards of carbon monoxide gas.

51. You examined this graph in connection with Your Turn 1.32. Return to it and answer the questions that follow.

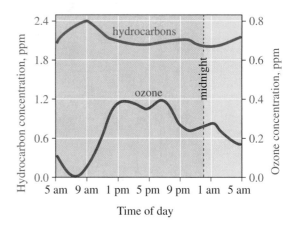

Time of day

a. Interpret the two curves, explaining what they imply about air pollution in an urban area.

b. Where do you think the curve for NO would fit on this graph?

c. What type of health effects would be felt in this major metropolitan area?

52. 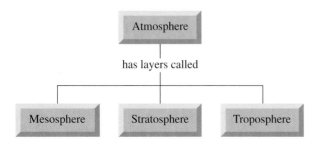 Annually since 1987, the EPA has published the Toxics Release Inventory (TRI), a national directive requiring companies to make available to the public data on the amounts of certain chemicals they have released into the air, water, and land. TRI data are available for each state. Check on the toxic emissions in your state or locale and the progress made in reducing those emissions at the EPA Web site.

a. Compare the current levels of toxic emissions with what they were one or two years ago.

b. Which emissions (if any) have decreased and which have remained the same or increased? Determine, if you can, a reason for the changes.

53. Consider This 1.3 asks you to consider how our world would be different if the oxygen content of the atmosphere were doubled. Develop your answer into an essay. Title your essay "An Hour in the Life of . . ." and describe how things would be different for a person of your choice. If an hour is too short to make your point, substitute "A Morning . . ." or "A Day . . .".

54. Mercury, another serious air pollutant, is not described in this chapter. We probably will add it to the next edition of the book. Meanwhile, if you were a textbook author, what would you include about mercury emissions? Write several paragraphs in a style that would

match that of this textbook. Perhaps even design a Consider This exercise to accompany it. Feel free to send these to one of the authors.

55. The dark color associated with heavy smog can be caused by the presence of particulate matter or a high concentration of brown NO_2 gas (or both). Once in the atmosphere, some NO_2 can form N_2O_4, a colorless gas.

    a. Write a balanced equation for the reaction of 2 molecules of NO_2 to form N_2O_4, a reaction that releases heat energy. Use a double-headed reaction arrow in your equation to indicate that equilibrium is established between the formation of N_2O_4 and its decomposition back to NO_2.

    b. Offer a reason why smog may be darker on a warm summer day than on a cold winter one, even if the levels of nitrogen oxides are the same in both cases.

Protecting the Ozone Layer

Earth Probe TOMS Total Ozone September 26, 2001
Area = 9.8 million miles² Minimum = 99 Dobson Units*

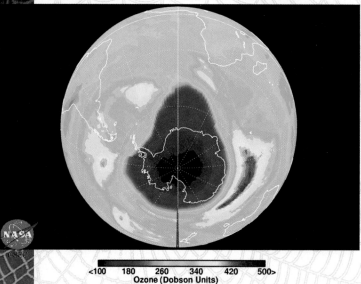

<100 180 260 340 420 500>
Ozone (Dobson Units)

*An image of the **2001** Antarctic stratospheric ozone "hole" (indicated by the dark blue and purple colors) taken from a satellite. The hole extends over about 9.8 million square miles.*

*One Dobson unit (DU), corresponds to about one ozone molecule for every billion molecules and atoms of air.

Earth Probe TOMS Total Ozone September 24, 2002
Area = 8.1 million miles² Minimum = 159 Dobson Units

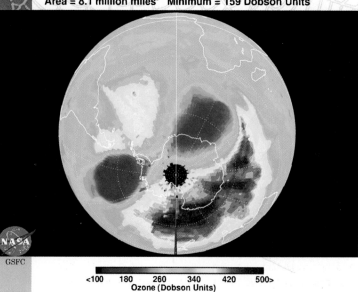

<100 180 260 340 420 500>
Ozone (Dobson Units)

*An image of Antarctic stratospheric ozone from a similar date in **2002** shows that the area in blue has split into two "holes." Together they cover about 8.1 million square miles.*

"What a Difference a Year Makes"—*Headline of the press release from September 30, 2002, issued jointly by the National Aeronautics and Space Administration (NASA) and the National Oceanic and Atmospheric Administration (NOAA).*

Stratospheric ozone plays a vital role in protecting Earth's surface and those who live here from damaging solar radiation. Ever since it was discovered in the 1970s that certain human-produced chemicals could make their way into the upper atmosphere and destroy the protective ozone found there, scientists, policy makers, and indeed concerned citizens worldwide have participated in efforts to control and reverse ozone destruction. Somewhat surprisingly, the most severe depletion has been over Antarctica, and the yearly images of the "ozone hole" have become some of the most widely recognized scientific graphics.

A decrease in the area covered by the Antarctic ozone hole(s) and an increase in the minimum concentration of ozone are both encouraging signs. Both indications revealed that between 2001 and 2002, the amount of protective stratospheric ozone at the time of the observed minima has increased. Scientists responsible for gathering and interpreting these data were quick to attribute the diminished area of ozone depletion over Antarctica in 2002 to unusual stratospheric weather patterns. They cautioned the public against using limited data to predict long-term trends. Craig Long, meteorologist at NOAA's Climate Prediction Center (CPC) summed up the observations this way: "The Southern Hemisphere's stratosphere was unusually disturbed this year. This is the first time we've seen the polar vortex split in September." Their caution was justified when in September 2003 the Antarctic ozone hole reached a minimum of 97 DU and covered the second-largest area ever observed. Later in this chapter, you will have the opportunity to examine past trends and to bring the Antarctic ozone hole story up-to-date.

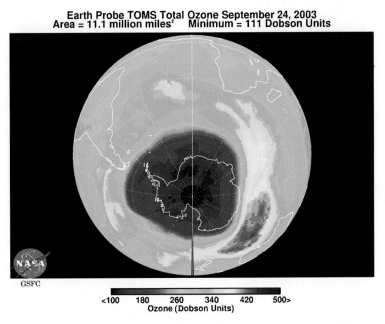

Earth Probe TOMS Total Ozone September 24, 2003
Area = 11.1 million miles²   Minimum = 111 Dobson Units

<100   180   260   340   420   500>
Ozone (Dobson Units)

*An image of Antarctic stratospheric ozone for **2003** shows that the hole covers about 11.1 million square miles. The record hole was 11.5 million square miles in 2000.*

You may be wondering what this story has to do with you, because last time we checked, not many college students were living in Antarctica. Even though the phenomenon of ozone depletion was first documented in the 1970s in that far-away region, it also has been monitored and observed in many other locations on Earth, including over North America. Where you live and the season of the year both influence the amount of stratospheric ozone overhead and how well it provides its protective effects. Take a look at some of the important data for yourself.

Consider This 2.1   Ozone Levels Above Your Spot on Earth

How much protective ozone is above you? How does it compare with the amount of ozone above Antarctica?

a. Use the NASA link at the *Online Learning Center* to access satellite data. Click on your location on the world map or enter your specific latitude and longitude to find the most recent data for total column ozone amount at your location. Also request data for September 26, 2001; September 26, 2002; and September 11, 2003. Values are given in Dobson units (DU).

b. As you will learn later in the chapter, ozone levels vary from 200 to 500 DU over the globe, with 320 DU as the average ozone level over the United States. How do your values compare with the average?

c. How do values at your location compare with those given for Antarctica minima in 2001–2003?

We are now ready to consider many questions involving chemistry and its role in helping understand mechanisms affecting our protective ozone layer. Just what has caused the stratospheric ozone depletion that has already occurred and how serious is this? What has been done to slow down or correct the problem? Are these measures working, and what are the economic and societal costs? Are any new threats to the stratospheric ozone layer emerging that merit careful evaluation?

2.1 Ozone: What and Where Is It?

Ozone is an atmospheric gas found in both the troposphere and the stratosphere. If you have ever been near a sparking electric motor or been in a severe lightning storm, you may have smelled ozone. Ozone's odor is unmistakable, but difficult to describe. Some compare the odor to that of chlorine gas, while others think the odor reminds them of newly mown grass. It is possible for humans to detect concentrations as low as 10 parts per billion (ppb)—10 molecules out of one billion. Appropriately enough, the name *ozone* comes from a Greek word meaning "to smell."

Ozone is oxygen that has changed from the normal diatomic molecule, O_2, to a triatomic form, O_3. A simple chemical equation summarizes the reaction:

$$\text{energy} + 3\,O_2 \longrightarrow 2\,O_3 \qquad\qquad [2.1]$$

Energy must be absorbed for this reaction to occur. This helps explain why ozone forms when oxygen is subjected to electrical discharge, whether from an electric spark or lightning.

Ozone is an allotropic form of oxygen. **Allotropes** are two or more forms of the same element that differ in their molecular or crystal structure, and therefore in their properties. The allotropes diatomic oxygen, O_2, and triatomic ozone, O_3, obviously differ in molecular structure. This variance is responsible for differences in the physical and chemical properties of the two allotropes. For example, ordinary oxygen (O_2) is odorless. It condenses and changes from a colorless gas to a light blue liquid at $-193\ °C$ and a pressure of 1 atmosphere (atm). Ozone liquefies at a higher temperature than does diatomic oxygen, changing its physical state from a gas to a dark blue liquid at $-112\ °C$. Because ozone is chemically more reactive than oxygen, O_3 is often used in the purification of water and the bleaching of paper pulp and fabrics. At one time it was even advocated as a deodorant for air in crowded interiors and continues to be used by some hotels to remove residual smoke from rooms.

In the troposphere, the region of the atmosphere in which we live, somewhere between 20 and 100 ozone molecules typically occur for each billion molecules that make up the air. The highest concentrations are found near Earth's surface, the result of photochemical smog mechanisms. In Chapter 1, we learned that the limits in ambient air are set at very low concentrations, only 0.08 ppm for an 8-hr average. But what is detrimental in one

Diamond and graphite are both familiar allotropic forms of carbon. They have different crystal structures and different properties. Fullerenes (buckyballs), the more recently discovered, but less common carbon allotrope, have a different structure still.

0.080 parts per *million* is equivalent to 80 ozone molecules for every *billion* molecules of air.

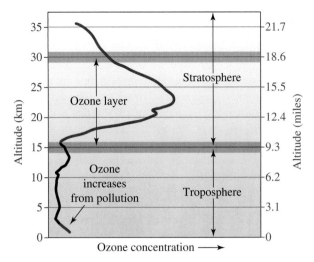

Figure 2.1

Ozone concentrations at different altitudes.

Source: *Scientific Assessment of Ozone Depletion: 2002;* World Meteorological Organization (WMO), United Nations Environmental Program (UNEP).

region of the atmosphere, even at very low concentrations, can be essential in another. The stratosphere, at an altitude of 20 to 30 km, is where ozone performs most of its filtering function on ultraviolet light from the Sun. The concentration of ozone in this region is somewhat greater than in the troposphere, but still very low. At most, there are about 12,000 ozone molecules for every billion molecules and atoms of gases that make up the atmosphere at this level. Most ozone, about 90% of the total, is found in the stratosphere. This region of maximum ozone concentration is often called the **ozone layer.** Figure 2.1 shows the relative location and concentration of ozone in the atmosphere.

12 ppm

Your Turn 2.2   **Finding the Ozone Layer**

Use Figure 2.1 and values from the text to answer these questions.

a. What is the altitude (in kilometers) of maximum ozone concentration?

b. What is the range of altitudes (in kilometers) in which ozone molecules are more concentrated than in the troposphere?

c. What is the maximum number of ozone molecules per billion molecules and atoms of all types found in the stratosphere?

d. What is the maximum number of ozone molecules per billion molecules and atoms of all types found in ambient air just meeting the EPA limit?

e. What is the ratio of maximum number of ozone molecules per billion molecules and atoms of air in the stratosphere to the maximum number in ambient air just meeting the EPA limit?

Answers

a. About 23 km   **b.** Between 16 and 35 km

Because the range of altitudes in which one finds significant ozone is so broad, the concept of the "ozone layer" can be a little misleading. No thick, fluffy blanket of ozone exists in the stratosphere. Many prefer to refer to this region as the ozone "screen" rather than the ozone "layer" to call up a more accurate image. At the altitudes of the maximum ozone concentration, the atmosphere is very thin, so the total amount of ozone is surprisingly small. If all the O_3 in the atmosphere could be isolated and brought to the average pressure and temperature at Earth's surface (1.0 atm and 15 °C), the resulting layer of gas would have a thickness of less than one-half centimeter, or about one-quarter of an inch. On a global scale, this is a minute amount of matter. Yet, this fragile shield protects the surface of the Earth and its inhabitants from the harmful effects of ultraviolet radiation. Because ozone is present in a small and finite quantity, we must protect and preserve it.

Reliable information about atmospheric ozone concentrations can help us understand changes that may occur. The total amount of ozone in a vertical column of air of known volume can be determined with relative ease. The determination can be done from Earth's surface by measuring the amount of UV radiation reaching a detector; the lower the intensity of the radiation, the greater the amount of ozone in the column. G. M. B. Dobson, a scientist at Oxford University, pioneered this measurement method. In 1920, he invented the first instrument to quantitatively measure total atmospheric ozone, and it is fitting that the unit of such measurements is named for him.

Consider This 2.3 | Interpreting Ozone Values

A classmate used the NASA Web site and found the column of ozone above your college to be 417 DU on April 10, and 286 DU on May 10. The student was reassured by these findings, concluding there had been an improvement in protection from damaging UV radiation. Do you agree? Why or why not?

Scientists will continue to measure and evaluate ozone levels using ground observations, weather balloons, and high-flying aircraft. However, since the 1970s, measurements of total ozone have also been made from the top of the atmosphere. Satellite-mounted detectors record the intensity of the ultraviolet radiation scattered by the upper atmosphere. The results are then related to the amount of O_3 present. What has proven more difficult is quantifying ozone concentrations at intermediate altitudes.

The Space Shuttle, *Columbia*, tested a new approach for monitoring ozone. Rather than looking directly downward toward Earth from a satellite, the equipment aboard the Shuttle looks sideways through the thin blue haze that rises above the denser regions of the troposphere and follows the curve of the Earth. This region is known as the Earth's "limb" and is responsible for the name of this new technique, called "limb viewing." Reliable information can be gathered at each level of the atmosphere, particularly allowing scientists to better understand chemistry taking place in the lower regions of the stratosphere. Although the reentry of *Columbia* in February 2003 tragically resulted in the loss of the crew and destroyed the primary data storage devices, data sent back during the mission proved the concept of this new approach and will be used on the next generation of satellites. In January 2004, the National Aeronautics and Space Administration (NASA) planned to launch a new mission called Earth Observing System (EOS) *Aura* to gather additional data about changes in Earth's stratospheric ozone layer.

NASA's EOS *Aura* mission will also collect data about tropospheric air quality (Chapter 1) and global warming (Chapter 3).

The process by which ozone protects us from damaging solar radiation involves the interaction of matter and energy from the Sun. To understand this process will require that we have knowledge about both of these fundamental topics. We turn first to a submicroscopic view of matter, and then examine its interaction with energy from the Sun.

2.2 Atomic Structure and Periodicity

The chemical and physical properties of O_2 and O_3 are closely related to the structure of these allotropes. But, before we can speak in detail about molecular structure, we must consider the atoms from which molecules are formed. Recall from Chapter 1 that each element contains the same type of atoms. During the 20th century, chemists and other scientists made great progress in discovering details about the structure of atoms and the particles that make them up. The physicists have been almost too successful; they have found more than 200 subatomic particles. Fortunately, most chemical behavior can be explained with only three.

We now know that every atom has at its center a minuscule but highly dense **nucleus.** This nucleus is composed of particles called protons and neutrons. **Protons**

Table 2.1	Properties of Subatomic Particles		
Particle	Relative Charge	Relative Mass	Actual Mass, kg
Proton	+1	1	1.67×10^{-27}
Neutron	0	1	1.67×10^{-27}
Electron	−1	0*	9.11×10^{-31}

* The relative mass of the electron is not actually zero, but is so small that it appears as zero when expressed to the nearest whole number.

are positively charged and **neutrons** are electrically neutral, but both have almost exactly the same mass. Indeed, the protons and neutrons in the nucleus account for almost all of an atom's mass. Well beyond the nucleus are the electrons that define the outer boundary of the atom. An **electron** has a much smaller mass than a proton or neutron, approximately 1/2000th the mass. Moreover, an electron has a negative electric charge equal in magnitude to that of a proton, but opposite in sign. The charge and mass properties of these particles are summarized in Table 2.1.

In any electrically neutral atom, the number of electrons equals the number of protons. This number of protons is called the **atomic number.** Each element has its own characteristic atomic number. The atomic number is important because it determines the elemental identity of the atom. For example, the simplest atom is hydrogen, and each hydrogen atom contains one proton, and thus has an atomic number of 1. Helium (He) has an atomic number of 2, hence each atom of He contains two protons. With each successive element, the atomic number increases, right up through element 112, whose atoms contain 112 protons.

Your Turn 2.4   Protons and Electrons

Using the periodic table as a guide, specify the number of protons and electrons in a neutral atom of each of these elements.

a. carbon (C)   **b.** calcium (Ca)   **c.** chlorine (Cl)   **d.** chromium (Cr)

Answers

a. 6 protons, 6 electrons   **b.** 20 protons, 20 electrons

We wish we could show you a picture of a typical atom. However, atoms defy easy representation, and most depictions in textbooks are at best oversimplifications. Electrons are sometimes pictured as moving in orbits about the nucleus, but the modern view of electrons is a good deal more complicated and abstract. For one thing, the relative size of the nucleus and the atom creates serious problems for the illustrator. If the nucleus of a hydrogen atom were the size of a period on this page, the atom's single electron would most likely be found at a distance of about 10 feet from that period. It is true that an atom is mostly empty space. Moreover, electrons do not follow specific circular orbits. In spite of what you may have learned in earlier science courses, an atom is really not very much like a miniature solar system. Rather, the distribution of electrons in an atom is described best using concepts of probability and statistics. A sort of fuzzy cloud in which electrons are more or less likely to occur surrounds the nucleus.

If this sounds rather vague to you, you are not alone. Common sense and our experience of ordinary things are not particularly helpful in our efforts to visualize the interior of an atom. Instead, we are forced to resort to mathematics and metaphors. The mathematics required (a field called quantum mechanics) can be formidable. Chemistry majors do not normally encounter this field until rather late in their undergraduate study. We cannot fully share with you the strange beauties of the peculiar quantum world of the atom, although we can provide some useful generalizations.

In the periodic table, the elements are arranged in order of increasing atomic number. The periodic table also organizes elements so that those with similar chemical and physical properties fall in the same columns (groups). This arrangement shows that the properties of the elements vary in a regular way with increasing atomic number and have a pattern of repeating periodically. Thus, lithium (Li, atomic number 3), sodium (Na, 11), potassium (K, 19), rubidium (Rb, 37), and cesium (Cs, 55) must share something besides their behavior as highly reactive metals. What fundamental feature accounts for these similar properties?

Today we know that the **periodicity of properties** is chiefly the consequence of the number and distribution of electrons in the atoms of the elements. Because the atomic number represents the number of protons in each atom (and electrons in a neutral atom) of each particular element, properties vary with atomic number. And when properties repeat themselves, it signals a repeat in electronic arrangement.

Both experiment and calculation demonstrate that the electrons are arranged in levels about the nucleus. What we are calling "levels" were oftentimes referred to as "shells" when using the earlier solar system model of atomic structure. The electrons in the innermost level are the most strongly attracted by the positively charged nucleus. The greater the distance between an electron and the nucleus, the weaker the attraction is between them. We say that the more distant electron is in a higher energy level, which means that the electron itself possesses more energy.

Each energy level has a maximum number of electrons that can be accommodated and is particularly stable when fully occupied. The innermost level, corresponding to the lowest energy, can hold only two electrons. The second level has a maximum capacity of eight, and the higher levels are also particularly stable when they contain eight electrons.

Table 2.2 shows some important information about electrons in neutral atoms of the first 18 elements. The total number of electrons in each atom is printed in blue and the number of outer electrons is printed in maroon. The number of **outer (valence) electrons** is particularly important because these electrons account for many of the chemical and physical properties of the corresponding elements. Observe that the group designation (1A, 2A, etc.) corresponds to the number of *outer* electrons for the A group elements, one of the great organizing benefits of the periodic table.

Take another look at the first column in Table 2.2. Lithium and sodium atoms both have one *outer* electron per atom, despite having different *total* numbers of electrons. This fact explains much of the chemistry that these two alkali metals have in common. It places them in Group 1A of the periodic table (the 1 indicates one outer electron). Moreover, we would be correct in assuming that potassium, rubidium, and the other elements in column 1A of the periodic table also have a single outer electron in each

> The term **valence electrons** is also used for the number of electrons in the outer energy level.

> The number *above* each atomic symbol is the **atomic number.** It gives the number of protons. It also is the total number of electrons in a neutral atom.

Table 2.2	**Total and Outer Electrons for Atoms of the First 18 Elements**						
Group 1A	**2A**	**3A**	**4A**	**5A**	**6A**	**7A**	**Noble Gases** **8A**
1							2
H							He
1							2
3	4	5	6	7	8	9	10
Li	Be	B	C	N	O	F	Ne
1	2	3	4	5	6	7	8
11	12	13	14	15	16	17	18
Na	Mg	Al	Si	P	S	Cl	Ar
1	2	3	4	5	6	7	8

• Number *above* the atomic symbol is the total number of electrons in a neutral atom.

• Number *below* the atomic symbol is the number of **outer** electrons in a neutral atom.

of their atoms. They are all metals that react readily with oxygen, water, and a wide range of other chemicals. In fact, chemical reactivity and the bonding that holds atoms together to form molecules and crystals are largely consequences of the number of outer electrons in any element. Figure 2.2 shows photographs of Group 1A elements.

The periodic table is a useful guide to electron arrangement in the various elements. In the families or groups of elements marked "A," the number that heads the column indicates the number of outer electrons in each atom. You have already seen that Group 1A elements are characterized by one outer electron. Similarly, the atoms of the Group 2A elements (the "alkaline earths") all have two outer electrons. The same pattern holds true for all of the A groups in the periodic table. Group 3A elements have three outer electrons, Group 4A elements have four, and so on across the table. Seven outer electrons characterize the atoms of the "halogens" that make up Group 7A: fluorine (F), chlorine (Cl), bromine (Br), iodine (I), and astatine (At). The next two exercises provide some practice with elements in the A groups.

> The group number does not *necessarily* indicate the number of outer electrons for elements in B groups, where the situation is a bit more complicated.

Your Turn 2.5   Outer Electrons

Using the periodic table as a guide, specify the group number and number of outer electrons in a neutral atom of each element.

a. sulfur (S)   **b.** silicon (Si)   **c.** nitrogen (N)   **d.** krypton (Kr)

Answers

a. Group 6A; 6 outer electrons   **b.** Group 4A; 4 outer electrons

Your Turn 2.6   Family Features

a. What feature of atomic structure is shared by fluorine (F), chlorine (Cl), bromine (Br), and iodine (I)? To which group do they belong?

b. Give the name and symbol for each element in the A group with two outer electrons. To which A group do they belong?

Answers

a. These elements all have seven outer electrons. They belong to Group 7A, the halogen family.

In addition to electrons and protons, atoms also contain neutrons. The one (and only) exception is an atom of the most common form of hydrogen, which consists of one electron and one proton. But even in pure hydrogen, one atom out of 6700 also has a neutron in its nucleus. Recall our earlier statement that most of the mass of any atom is associated with its nucleus. Because both the proton and the neutron have relative masses of almost exactly 1, the relative mass of an atom of this "heavy hydrogen" very nearly equals 2. This form of hydrogen is called deuterium. It is an example of a naturally occurring isotope of hydrogen. **Isotopes** are two or more forms of the same element (same number of protons) whose atoms differ in number of neutrons, and hence in mass.

Isotopes are identified by their **mass numbers**—the sum of the number of protons and the number of neutrons in an atom. The mass number, indicated by a superscript to the left of an elemental symbol, must be given because it can vary for the same element. The atomic number is often included as a subscript. For example, the full atomic symbol $_1^1H$ represents the most common isotope of hydrogen. Because hydrogen *always* has an atomic number of 1, the subscript is often omitted, making the simplified symbol for the common isotope of hydrogen just ^1H. In text, you may read hydrogen-1, or H-1. Clearly there is redundancy in actually giving the atomic number as well as the elemental symbol, even though it may be convenient for the reader in the case of unfamiliar elements.

Lithium (stored in oil)

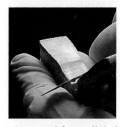

Sodium (removed from oil, being cut)

Potassium (in sealed glass tube)

Rubidium (in sealed glass tube)

Figure 2.2

Selected group 1A elements.

Table 2.3	Isotopes of Hydrogen			
Isotope	Isotopic Symbol	Number of Protons	Number of Neutrons	Sum of Protons and Neutrons
hydrogen, H-1	$_1^1\text{H}$	1	0	1
deuterium, H-2	$_1^2\text{H}$	1	1	2
tritium, H-3	$_1^3\text{H}$	1	2	3

The mass number, indicated by a superscript, must be given because it can vary for the same element. For example, the nucleus of an atom of deuterium contains one proton plus one neutron and is therefore assigned a mass number of 2. Thus, deuterium is designated as hydrogen-2 or H-2; it can also be represented by $_1^2\text{H}$ or simplified to ^2H, where the mass number of 2 is indicated by a superscript. Hydrogen also has a third isotope, called tritium, whose atoms consist of two neutrons in addition to the one proton and one electron characteristic of all neutral hydrogen atoms. Tritium, a radioactive isotope that is rare in nature, has a mass number of 3 (1 proton and 2 neutrons). It can be represented as hydrogen-3, H-3, $_1^3\text{H}$, or ^3H. Table 2.3 summarizes this information about the isotopes of hydrogen.

Your Turn 2.7   **Protons, Electrons, and Neutrons**

Specify the number of protons, electrons, and neutrons in a neutral atom of each isotope.

a.   carbon-14 ($_6^{14}\text{C}$)   **b.**   uranium-235 ($_{92}^{235}\text{U}$)   **c.**   iodine-131 ($_{53}^{131}\text{I}$)

Answers

a.   6 protons, 6 electrons, 8 neutrons   **b.**   92 protons, 92 electrons, 143 neutrons

---

All elements have isotopes, but the number of stable and unstable ones varies considerably. Each element's atomic mass, the number you see on every periodic table, takes the relative abundance of isotopes, as well as their masses, into account. Although the concept of atomic mass is important, we do not require it at this time. Following our general rule of introducing information only as needed, we will defer a discussion of atomic masses to Chapter 3.

*✳ **Mass number** gives the total number of protons and neutrons in a specific isotope. **Atomic mass** refers to a **weighted average** of all naturally occurring isotopes of that element.*

2.3   Molecules and Models

After this excursion into the atom, we come to our primary motivation for studying atoms—understanding molecular structure. The stability of filled electron shells can be invoked to explain why atoms bond to one another to form molecules. The simplest case is H_2, a diatomic molecule we encountered in Chapter 1. A hydrogen atom has only one electron, but if two hydrogen atoms come together, the two electrons become common property. Each atom effectively has a share in both electrons. The resulting H_2 molecule has a lower energy than two individual H atoms, and consequently the molecule with its bonded atoms is more stable than the separate atoms. The two electrons that are shared constitute a **covalent bond.** Appropriately, the name *covalent* implies "shared strength."

If we represent each atom by its symbol and each electron by a dot, the two individual hydrogen atoms might look something like this:

H·   and   ·H

Bringing the two atoms together yields a molecule that can be represented this way.

H:H

A representation showing outer electrons is called a dot structure, or **Lewis structure,** after Gilbert Newton Lewis (1875–1946), an American chemist who pioneered its use. Lewis structures can be predicted for many simple molecules by following a set of straightforward steps. We first illustrate the procedure with hydrogen fluoride, HF, a very reactive compound used to etch glass.

1. Starting with the chemical formula of the compound, note the number of outer electrons contributed by each of the atoms (Remember that the periodic table is a useful guide).

   H·   1 H atom × 1 outer electron per atom = 1 outer electron

   :F̈·   1 F atom × 7 outer electrons per atom = 7 outer electrons

2. Add the outer electrons contributed by the individual atoms to obtain the total number of outer electrons available.

   1 + 7 = 8 outer electrons

3. Arrange the outer electrons in pairs. Then distribute them in such a way as to maximize stability by giving each atom a share in enough electrons to fully fill its outer shell: two electrons in the case of hydrogen, eight electrons for most other atoms.

   H:F̈:

We surrounded the F atom with eight dots, organized into four pairs. The pair of dots between the H and the F represents the electron pair that forms the bond uniting the hydrogen and fluorine atoms. The other three pairs of dots are the three pairs of electrons that are not shared with other atoms and hence not involved in bonding. As such, they are called "nonbonding" electrons or "lone pairs."

When only one pair of shared electrons is involved in a covalent bond, the linkage is called a **single bond.** A line often replaces the electron pair forming a single covalent bond. This line connects the symbols for the two atoms.

   H—F̈:

Sometimes the nonbonding electrons are removed from a Lewis structure, simplifying it still more.

   H—F

Remember that the single line represents one pair of shared electrons. These two electrons plus the six electrons in the three nonbonding pairs mean that the fluorine atom is associated with a total of eight outer electrons, whether or not all the electrons are specifically shown. Remember that the hydrogen atom has no additional electrons other than the single pair shared with fluorine. It is at maximum capacity with 2 electrons, thanks to its small size.

The fact that electrons in many molecules are arranged so that every atom (except hydrogen) shares in eight electrons is called the **octet rule.** This generalization is a useful guide for predicting Lewis structures and the formulas of compounds. Consider the Cl_2 molecule, the diatomic form of elemental chlorine. From the periodic table, we can see that chlorine, like fluorine, is in Group 7A, which means that its atoms each have seven outer electrons. Using the scheme given for HF earlier, we first count and add up the outer electrons for Cl_2.

   2 :C̈l·   2 Cl atoms × 7 outer electrons per atom = 14 outer electrons

For Cl_2 to exist, there must be a bond between the two atoms, which we show by a single line designating a shared electron pair: a single covalent bond. The remaining 12 electrons constitute six nonbonding pairs, distributed in such a way as to give each chlorine atom 8 electrons (2 bonding and 6 nonbonding). This meets the octet rule. Accordingly, this is the Lewis structure for Cl_2.

   :C̈l—C̈l:

Your Turn 2.8   Lewis Structures for Simple
Diatomic Molecules

Use the set of steps just outlined to draw the Lewis structures for these two molecules.

a. HBr   **b.** Br$_2$

Answer

a.   H·   1 H atom × 1 outer electron per atom = 1 outer electron

·B̈r: 1 Br atom × 7 outer electrons per atom = 7 outer electrons

Total = 8 outer electrons

These are the Lewis structures for HBr.

H:B̈r: or   H—B̈r:

So far we have dealt only with molecules having just two atoms. But many compounds contain molecules with more than two atoms; these are called **polyatomic molecules.** The octet rule applies to many of these compounds as well. Here is another generalization just as useful as the octet rule in helping to predict Lewis structures. In most molecules where there is only one atom of one element bonded to two or more atoms of another element (or elements), *the single atom goes in the center of the Lewis structure.* You'll encounter other exceptions to these generalizations, but this is a good place to begin to apply them. We start with a water molecule, H$_2$O, as an example.

Following the same procedures used for two-atom molecules, we first count and add up the outer electrons.

2 H·   2 H atoms × 1 outer electron per atom = 2 outer electrons

·Ö·   1 O atom × 6 outer electrons per atom = 6 outer electrons

Total = 8 outer electrons

We place the O representing the oxygen atom in the center and distribute the eight electrons (dots) around the O, in conformity with the octet rule. Each of the hydrogen atoms is bonded to the oxygen atom with a pair of electrons. The remaining four electrons are also placed on the oxygen, but as two nonbonding pairs. This is the result.

H:Ö:H

A quick count confirms that the O is surrounded by eight dots, representing the eight electrons predicted by the octet rule. Alternatively, we could symbolize the water molecule with lines for the single bonds, with or without oxygen's nonbonding electrons.

H—Ö—H   or   H—O—H

> Each hydrogen atom forms only one bond (two shared electrons). Oxygen, because it can form two bonds, is the central atom.

These Lewis representations provide more information than does the chemical formula, H$_2$O. The formula shows the types and ratio of atoms present, and so does the Lewis structure. The Lewis structure also indicates how the atoms are connected to one another. On the other hand, Lewis structures do not directly reveal the shape of a molecule. From the structures for water given so far, it might appear that the atoms of the water molecule all fall in a straight line. In fact, the molecule is bent. It looks something like this.

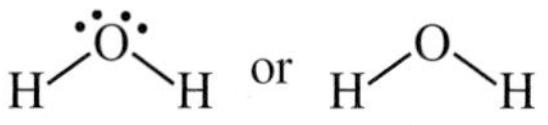

H—Ö—H   or   H—O—H

We will return to this discussion of shape in Chapter 3 and see how the Lewis structure can lead to the prediction of this bent structure. We will examine the experimental evidence for the shape of the water molecule in Chapter 5.

Another example of a molecule with more than two bonded atoms is methane, CH_4. Using the rules and generalizations given earlier, we can write the Lewis structure of methane.

The combustion of methane was discussed in Section 1.10. The geometry of the methane molecule is described in Section 3.3.

4 H·   4 H atoms × 1 outer electron per atom = 4 outer electrons

·Ċ·   1 C atom × 4 outer electrons per atom = 4 outer electrons

Total = 8 outer electrons

The C representing a carbon atom goes in the center and is surrounded by the eight electrons, giving carbon an octet of electrons. Each of the four hydrogen atoms uses two of the electrons to form a shared pair with carbon, for a total of four single covalent bonds. This gives us the Lewis structure of methane.

$$\text{H:}\overset{\cdot\cdot}{\underset{\cdot\cdot}{\text{C}}}\text{:H} \quad \text{or} \quad \text{H}-\overset{\displaystyle \text{H}}{\underset{\displaystyle \text{H}}{\text{C}}}-\text{H}$$

Check the methane structure to be sure that the carbon atom has a share in eight electrons, as would be expected by the octet rule. Remember that H can only accommodate a pair of electrons.

Your Turn 2.9   **Lewis Structures for Polyatomic Molecules**

Use the procedure just outlined to draw the Lewis structures for each of these molecules. Both species obey the octet rule.

a. hydrogen sulfide (H_2S)   **b.** dichlorodifluoromethane (CCl_2F_2)

Answer

a.   2 H·   2 H atoms × 1 outer electron per atom = 2 outer electrons

·S̈·   1 S atom × 6 outer electrons per atom = 6 outer electrons

Total = 8 outer electrons

These are the Lewis structures for H_2S.

$$\text{H:}\overset{\cdot\cdot}{\underset{\cdot\cdot}{\text{S}}}\text{:H} \quad \text{or} \quad \text{H}-\overset{\cdot\cdot}{\underset{\cdot\cdot}{\text{S}}}-\text{H}$$

Given that both S and O are in Group 6A, the Lewis structures for H_2S and H_2O are the same except for the identity of the central element in each structure.

In some structures, single covalent bonds do not allow the atoms to follow the octet rule. Consider, for example, the very important gas oxygen, O_2. Here we have 12 outer electrons to distribute, 6 from each of the Group 6A oxygen atoms. There are not enough electrons to give each of the atoms a share in eight electrons if only one pair is held in common. However, the octet rule can be satisfied if the two atoms share four electrons (two pairs). A covalent bond consisting of two pairs of shared electrons is called a **double bond.** This bond is represented by four dots or by two lines, with or without the nonbonding electrons.

$$\overset{\cdot\cdot}{\underset{\cdot\cdot}{\text{O}}}\text{::}\overset{\cdot\cdot}{\underset{\cdot\cdot}{\text{O}}} \quad \text{or} \quad \overset{\cdot\cdot}{\underset{\cdot\cdot}{\text{O}}}=\overset{\cdot\cdot}{\underset{\cdot\cdot}{\text{O}}} \quad \text{or} \quad \text{O}=\text{O}$$

Double bonds are shorter, stronger, and harder to break than single bonds involving the same atoms. The experimentally measured length and strength of the bond in the O_2 molecule correspond to a double bond. However, oxygen has a peculiar property that is not fully consistent with the Lewis structure just drawn. When liquid oxygen is poured between the poles of a strong magnet, it sticks there like iron filings. Such magnetic behavior implies the presence of unpaired electrons, meaning that the electrons are not as neatly paired as the octet rule would suggest. But this one discrepancy is hardly a reason to discard a useful generalization. After all, simple scientific models seldom if ever explain all phenomena, but they can be helpful approximations. There are other common examples in which the straightforward application of the octet rule leads to discrepancies in interpreting experimental evidence. In the best of all scientific worlds, coming across data that do not conform to existing models may lead to the development of even better ones.

A **triple bond** is a covalent linkage made up of three pairs of shared electrons. Triple bonds are even shorter, stronger, and harder to break than double bonds involving the same atoms. For example, the nitrogen molecule, N_2, contains a triple bond. Each Group 5A nitrogen atom contributes 5 outer electrons for a total of 10. These 10 electrons can be distributed in accordance with the octet rule if 6 of them (three pairs) are shared between the two atoms, leaving 4 of them to form two nonbonding pairs, one on each nitrogen atom.

> The stability of the triple bond linking N atoms in N_2 gas helps explain nitrogen's relative inertness in the troposphere.

$$:\!N\!:\!:\!:\!N\!: \quad \text{or} \quad :\!N\!\equiv\!N\!: \quad \text{or} \quad N\!\equiv\!N$$

The ozone molecule, important in the story of this chapter, introduces another structural feature. We again start with the octet rule. Each of the three oxygen atoms contributes 6 outer electrons for a total of 18. These 18 electrons can be arranged in two ways; each way gives a share in 8 outer electrons to each atom.

$$\ddot{O}\!:\!:\!\ddot{O}\!:\!\ddot{O}\!: \qquad :\!\ddot{O}\!:\!\ddot{O}\!:\!:\!\ddot{O}$$
$$\textbf{a} \qquad\qquad\qquad \textbf{b}$$

Structures **a** and **b** predict that the molecule should contain one single bond and one double bond. In structure **a,** the double bond is to the left of the central atom; in **b** it is to the right. But experiments reveal that the two bonds in the O_3 molecule are identical in length and strength, being somewhere between a single bond and a double bond. Structures **a** and **b** are called **resonance forms,** forms that are hypothetical extremes of electron arrangements that do not exist exactly as represented by any one Lewis structure. The actual structure of the ozone molecule is something like a hybrid of the two resonance forms. A double-headed arrow linking the different forms is used to represent the resonance phenomenon.

$$:\!\ddot{O}\!-\!\ddot{O}\!=\!\ddot{O} \quad \longleftrightarrow \quad \ddot{O}\!=\!\ddot{O}\!-\!\ddot{O}\!:$$

This representation and the word *resonance* seem to imply that the electrons are jumping back and forth between the two arrangements, but this does not happen. Resonance is just another modeling concept invented by chemists to represent the complex microworld of molecules. It is not intended to be the "truth," but rather just a way to describe the structures of molecules that do not exactly fit the octet rule model. Figure 2.3 compares the Lewis structures of several different oxygen-containing species relevant to the chemistry in this and other chapters.

A closer experimental inspection of that microworld reveals that the O_3 molecule is not linear as the simple Lewis structures just drawn would seem to indicate. Remember that Lewis structures tell us only what is connected to what, and do not necessarily show the molecular geometry. The structure of the O_3 molecule is actually bent, as in this representation:

> Observe that both H_2O and O_3 are bent molecules, with O as the central atom.

$$\ddot{O}\!\!\overset{\ddot{O}}{\diagdown}\!\!\ddot{O}\!: \quad \longleftrightarrow \quad :\!\ddot{O}\!\!\overset{\ddot{O}}{\diagup}\!\!\ddot{O}$$

A more complete explanation of why the O_3 molecule is bent will have to wait until Chapter 3. At this point we are more concerned with how bonding in O_2 and O_3 influences their interaction with sunlight.

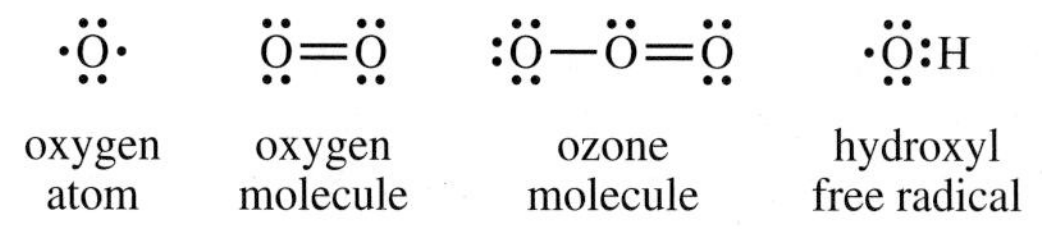

$$\cdot\ddot{O}\cdot \qquad \ddot{O}=\ddot{O} \qquad :\ddot{O}-\ddot{O}=\ddot{O} \qquad \cdot\ddot{O}:H$$

oxygen   oxygen   ozone   hydroxyl
atom   molecule   molecule   free radical

Figure 2.3
Lewis structures for several oxygen species.

Your Turn 2.10   **Lewis Structures with Multiple Bonds**

Draw the Lewis structures for each of these compounds. All follow the octet rule.

a.   carbon monoxide (CO)   **b.**   sulfur dioxide (SO_2)

Answer

a. $\cdot\overset{\cdot}{\underset{\cdot}{C}}\cdot$ 1 C atom $\times$ 4 outer electrons per atom = 4 outer electrons

$\cdot\ddot{\underset{\cdot\cdot}{O}}\cdot$ 1 O atom $\times$ 6 outer electrons per atom = 6 outer electrons

Total = 10 outer electrons

These are the Lewis structures:

$$:C:::O: \quad \text{or} \quad :C\equiv O:$$

Observe that there are 10 outer electrons. The N_2 molecule also has 10 outer electrons, and also forms with a triple bond.

2.4 Waves of Light

The next step in building a better understanding of how stratospheric ozone screens out much of the Sun's harmful radiation is to learn something about the fundamental properties of light. The interaction of sunlight with matter is important in several processes, such as in photosynthesis or in the damage high-energy solar radiation can cause in living organisms. Therefore, we turn now to develop an understanding of light.

Every second, five million tons of the Sun's matter is converted into energy, which is radiated into space. The fact that our eyes are capable of detecting different colors is one indication that the radiation that reaches us is not all identical. Prisms and raindrops break sunlight into a spectrum of colors. Each of these colors can be identified by the numerical value of its wavelength. The word correctly suggests that light behaves something like a wave in the ocean. The **wavelength** is the distance between successive peaks. It is expressed in units of length and symbolized by the Greek letter lambda (λ). Waves are also characterized by a certain **frequency,** the number of waves passing a fixed point in one second. Figure 2.4 shows two waves of different wavelength and frequency.

It is both interesting and humbling to realize that out of the vast array of radiant energies, our eyes are sensitive to only a very tiny portion of the total range: wavelengths between about 700×10^{-9} m (corresponding to red) and 400×10^{-9} m (corresponding to violet). These lengths are very short, so we typically express them in nanometers. One **nanometer (nm)** is defined as one one-billionth of a meter (m).

$$1 \text{ nm} = \frac{1}{1,000,000,000} \text{ m} = \frac{1}{1 \times 10^9} \text{ m} = 1 \times 10^{-9} \text{ m}$$

We can use this equivalence to convert meters to nanometers. For example, how many nanometers are there in 700×10^{-9} m?

$$\text{wavelength } (\lambda) = 700 \times 10^{-9} \text{ m} \times \frac{1 \text{ nm}}{1 \times 10^{-9} \text{ m}} = 700 \text{ nm}$$

The units of meters cancel and we are left with our target unit, nanometers.

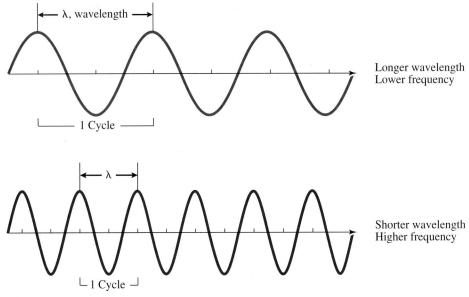

Figure 2.4
Comparison of two different waves.

Your Turn 2.11   Changing Units

Green light in the visible part of the spectrum has a wavelength of 500 nm. Express this wavelength in meters.

Answer

500×10^{-9} m or, expressed in scientific notation, 5.00×10^{-7} m.

Another way of quantifying color is to express the waves associated with each color in terms of frequency. If you were watching waves on the surface of a lake or the ocean, you could measure the distance between successive crests (the wavelength). But you could also determine how often the crests passed your point of observation by counting the number in a particular time interval. That would give you the frequency of the waves. The same idea applies to radiation, as shown later in Figure 2.5. Frequency and wavelength are related; the shorter the wavelength, the higher the frequency, that is, the greater the number of waves that pass the observer in one second. For any wave, as the frequency increases, the wavelength decreases; their values change in opposite directions. In mathematical terms, wavelength and frequency are inversely proportional.

Instead of reporting frequency as "waves per second," the units are shortened to "per second" and written as 1/s or s^{-1}. This unit is also called a hertz (Hz), perhaps familiar to you from radio station frequencies. A companion unit is the megahertz (MHz), one million hertz, or 1×10^{6} s^{-1}. Frequency is represented by the Greek letter nu (ν).

The relationship between frequency and wavelength just described in words can be summarized in a simple equation where ν is the frequency and c represents the constant speed at which light and other forms of radiation travel.

$$\text{Frequency } (\nu) = \frac{c}{\lambda} \qquad [2.2]$$

In metric units, the speed of light is 3.00×10^{8} meters/second or 3.00×10^{8} m $\cdot$ s^{-1}. You may be more familiar with the speed of light being expressed as 186,000 miles/second, but these are the same values expressed in different units. The form of equation 2.2 indicates that wavelength and frequency are *inversely* related: As the value for λ decreases, the value for ν increases, and vice versa. Red light, which has a wavelength

As wavelength ↓, frequency ↑.

This value of constant forward speed is true only for forms of electromagnetic radiation, not for other types of waves.

of 700 nm or 700×10^{-9} m, has a frequency of $4.29 \times 10^{14} \ s^{-1}$. This value can be calculated using equation 2.2.

$$\text{Frequency} = \nu = \frac{c}{\lambda} = \frac{3.00 \times 10^8 \ \text{m·s}^{-1}}{700 \times 10^{-9} \ \text{m}} = 4.29 \times 10^{14} \ s^{-1}$$

Violet light has a shorter wavelength (400 nm) and hence a higher frequency $(7.50 \times 10^{14} \ s^{-1})$ than red light.

Your Turn 2.12   **Green Light, Red Light**

Green light in the visible part of the spectrum has a wavelength of 550 nm.

a. Calculate the frequency of green light, reporting your answer in s^{-1}.
b. Compare the wavelengths and frequencies for green and red light.

Answer
a. $5.45 \times 10^{14} \ s^{-1}$

Consider This 2.13   **Analyzing a Rainbow**

Water droplets in a rainbow act as prisms to separate visible light into its component colors. Which color in the rainbow (Figure 2.5) has the

a. shortest wavelength?   **b.** lowest frequency?

The continuum of waves known as the **electromagnetic spectrum** ranges from very low energy radio waves to very high energy X-rays and gamma rays. Visible light is only a narrow band in this entire range. The term **radiant energy** is used to refer to the entire collection of different wavelengths, each with its own energy. Earth receives its radiant energy from the Sun. Although the wavelengths and frequencies vary considerably, all waves in the electromagnetic spectrum travel at the same forward speed, $3.00 \times 10^8 \ \text{m} \cdot s^{-1}$. Scientists have devised a variety of detectors that are sensitive to the radiation in various parts of this broad band. As a consequence, we can speak with confidence about the regions of the spectrum, even though these regions are invisible to our eyes. Figure 2.6 shows the continuum of types of waves of the electromagnetic spectrum and their relative frequencies and wavelengths.

Figure 2.5
A rainbow of color.

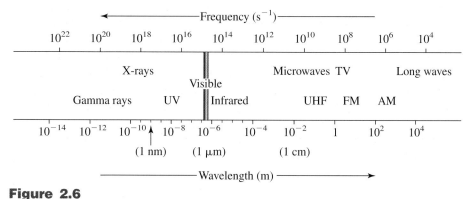

Figure 2.6
The electromagnetic spectrum.

Figures Alive! Visit the *Online Learning Center* to learn more about relationships in the electromagnetic spectrum. Practice, using the interactive exercises. Look for the **Figures Alive!** icon elsewhere in this chapter.

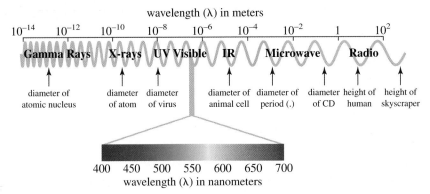

Figure 2.7

The electromagnetic spectrum. The wavelength variation from gamma rays to radio waves is not drawn to scale.

At wavelengths longer than those of red visible light, one first encounters infrared (IR), waves that we cannot see, but can certainly feel. They also are known as heat rays. The microwaves used in radar and to cook food quickly have wavelengths on the order of centimeters. At still longer wavelengths (1–1000 m) are the regions of the spectrum used to transmit your favorite AM and FM radio and television programs. In the next chapter, we will be most concerned with the IR region of the electromagnetic spectrum.

In this chapter we are most concerned with the **ultraviolet (UV) region** that lies at wavelengths shorter than those of the visible color of violet. At still shorter wavelengths are the X-rays used in medical diagnosis and the determination of crystal structure, and gamma rays that are given off in certain radioactive processes. Some gamma rays can have wavelengths as short as 10^{-16} m. Figure 2.7 will help you keep these important regions of the electromagnetic spectrum in perspective.

Our local star, the Sun, emits many types of radiant energy, including infrared, visible, and ultraviolet radiation. However, our Sun does not emit all types with equal intensity. This is evident from Figure 2.8, a plot of the relative intensity of solar radiation as a function of wavelength. The curve represents the spectrum as measured *above* the atmosphere, before there has been opportunity for interaction of radiation with the molecules of the air. The peak indicating the greatest intensity is in the visible region. However, infrared radiation is spread over a much wider wavelength range, with the result that 53% of the total energy emitted by the Sun is radiated to Earth as infrared radiation. This is the major source of heat for the planet. Approximately 39% of the energy comes to us as visible light and only about 8% as ultraviolet. (The areas under the curve give an indication of these percentages.) But in spite of its small percentage, the Sun's UV radiation is potentially the most damaging to living things. To understand why, we need to look at electromagnetic radiation in a different light, this time in terms of its energy.

 Your Turn 2.14   **Wavelength and Frequency**

Consider these four types of radiant energy from the electromagnetic spectrum: infrared, microwave, ultraviolet, visible.

a.   Arrange them in order of *increasing* wavelength.
b.   Arrange them in order of *increasing* frequency.
c.   Are the two arrangements the same or different? Explain your reasoning.

Answer
a.   ultraviolet, visible, infrared, microwave

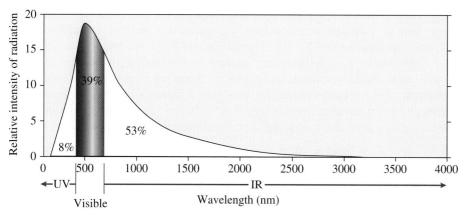

Figure 2.8
Wavelength distribution of solar radiation above Earth's atmosphere.

2.5 Radiation and Matter

The idea that radiation can be described in terms of wave-like character is well established and very useful. However, around the beginning of the 20th century, scientists found a number of phenomena that seemed to contradict this model. In 1909, a German physicist named Max Planck (1858–1947) argued that the shape of the energy distribution curve pictured in Figure 2.8 could only be explained if the energy of the radiating body were the sum of many energy levels of minute but discrete size. In other words, the energy distribution is not really continuous, but consists of many individual steps. Such an energy distribution is called **quantized.** An often-used analogy is that the quantized energy of a radiating body is like steps on a staircase, which are also quantized (no partial steps allowed), not like a ramp, which allows any sized stride. Five years later, in the work that won him his Nobel Prize, Albert Einstein (1879–1955) suggested that radiation itself should be viewed as constituted of individual bundles of energy called **photons.** One can regard these photons as "particles of light," but they are definitely not particles in the usual sense. For example, they have no mass.

Planck and Einstein were both amateur violinists who played duets together.

The wave model is still useful, even with the new development of the quantum theory to explain the particle-like property of energy. Both are valid descriptions of radiation. This dual nature of radiant energy seems to defy common sense. How can light be described in two different ways at the same time, both waves and particles? There is no obvious answer to that very reasonable question—that's just the way nature is. The two views are linked in a simple relationship that is one of the most important equations in modern science. It is also an equation that is relevant to the role of ozone in the atmosphere.

$$E = h\nu = \frac{hc}{\lambda} \qquad [2.3]$$

Here E represents the energy of a single photon. It is *directly* proportional to ν, the frequency of radiation, and *inversely* proportional to the wavelength, λ. Consequently, as the wavelength of radiation gets shorter, its energy increases; as the energy decreases, the wavelength increases. On the other hand, as the frequency of the radiation increases, so does its

Your Turn 2.15   **Color and Energy Relationships**

Arrange these colors of the visible spectrum in order of *increasing* energy per photon: green, red, yellow, violet.

Answer
red < yellow < green < violet

As wavelength ↓, frequency ↑, energy ↑.

energy. These qualitative relationships are summarized in the margin and you may wish to refer to them as you do the next Your Turn. The symbol h in equation 2.3 is Planck's constant, which has a value of 6.63×10^{-34} joule · second (J · s). The **joule** is a unit of energy and is approximately equal to the energy required for one beat of a human heart.

We can also introduce values into equation 2.3 and use them to compare the energy of different photons. For example, the energy of a photon of ultraviolet light with a wavelength of 300 nm and a frequency of $1.00 \times 10^{15} \, s^{-1}$ can be shown to have energy of 6.63×10^{-19} joule (J), a very tiny amount of energy. This is the calculation.

$$E = h\nu = (6.63 \times 10^{-34} \, \text{J} \cdot \text{s})(1.00 \times 10^{15} \, s^{-1}) = 6.63 \times 10^{-19} \, \text{J}$$

By contrast, the photon of a 100-mHz FM radio signal with a wavelength of 300×10^{7} nm has an energy of only 6.63×10^{-26} J. Although these energies are very small, there is a significant difference between the energies of the photon of UV radiation and the photon of the radio signal.

UV radiation: 6.63×10^{-19} J per photon

Radio signal: 6.63×10^{-26} J per photon

The energy of a photon of UV radiation is 10^{7}, or 10 million times larger than the energy of a photon emitted by your favorite radio station. Remember that as wavelength *decreases* (from radio waves to ultraviolet radiation), the energy per photon of radiation *increases*.

A consequence of this large difference in energy is that you cannot damage your skin by listening to the radio—unless you happen to be listening to it outside in the sunlight. Whether or not your radio is turned on, you are continuously bombarded by radio waves. Your body cannot detect them, but your radio can. As we have just seen, the energy associated with each of the radio photons is very low—about 7×10^{-26} J. This energy is not sufficient to produce a local increase in the concentration of the skin pigment, melanin, to cause tanning in people with lighter colored skin. That process involves a quantum jump, an electronic transition that requires approximately 7×10^{-19} J, far more than radio wave photons can supply.

Your body cannot store 10 million of the low-energy photons of radio frequency that would be necessary to equal the energy required for the tanning reaction. It is an either/or situation: either a photon has enough energy to cause a specific chemical change or it does not. Photons of ultraviolet radiation of 300 nm or shorter do have sufficient energy to bring about the changes that result in tanning, burning, or in some cases, skin cancer.

All of this may seem relatively unimportant. But it was essentially this line of reasoning, although applied to a different system, that won Einstein his Nobel Prize in 1905. Moreover, this same logic extends to any interaction of electromagnetic radiation and matter.

Your Turn 2.16   Energy of Green Light

Return once more to the green light of Your Turn 2.12. Calculate the energy of a photon of this radiation, expressing your answer in joules.

Answer
$$E = h\nu = (6.63 \times 10^{-34} \, \text{J} \cdot \text{s}) \, (5.45 \times 10^{14} \, s^{-1}) = 3.61 \times 10^{-19} \, \text{J}$$

The Sun bombards Earth with countless photons—indivisible packages of energy. The atmosphere, the planet's surface, and Earth's living things all absorb these photons. Radiation in the infrared region of the spectrum warms Earth and its oceans, causing molecules to move, rotate, and vibrate. The cells of our retinas are tuned to the wavelengths of visible light. Photons associated with different wavelengths are absorbed and the energy is used to "excite" electrons in biological molecules. The electrons jump to higher energy levels, triggering a series of complex chemical reactions that ultimately

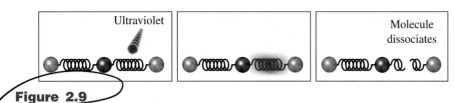

Figure 2.9
Ultraviolet radiation can break chemical bonds. Bonds are represented as springs that hold the atoms together but allow the atoms to move relative to each other.

lead to sight. Compared with animals, green plants capture photons in an even narrower region of the visible spectrum (corresponding to red light) and use the energy to convert carbon dioxide and water into food, fuel, and oxygen in the process of photosynthesis.

Remember that as the wavelength of light *decreases,* the energy carried by each photon *increases.* Consequently, the interaction of shorter-wavelength radiation and matter becomes more energetic. Photons in the UV region of the spectrum are sufficiently energetic to displace electrons within neutral molecules, converting them into positively charged species. The even shorter UV wavelength photons break bonds, causing molecules to come apart. In living things, such changes disrupt cells and create the potential for genetic defects and cancer. This is shown schematically in Figure 2.9.

It is part of the fascinating symmetry of nature that this interaction of radiation with matter explains both the damage ultraviolet radiation can cause and the atmospheric mechanism that protects us from it. We turn next to understanding the ultraviolet shield provided by oxygen and ozone in our stratosphere.

2.6 The Oxygen/Ozone Screen

The presence of oxygen and ozone in Earth's stratosphere guarantees that electromagnetic radiation reaching the planet's surface is different in some important respects from that emitted by the Sun. Solar UV radiation is greatly diminished by passing through oxygen and particularly through ozone in the stratosphere. Different UV wavelengths and energies influence how much UV solar radiation reaches Earth and how much damage it can cause. The UV radiation coming from the Sun can be categorized by its wavelength as UV-A (320–400 nm), UV-B (280–320 nm), or UV-C (<280 nm). Table 2.4 gives information about the three important categories of UV radiation and the role of oxygen and ozone in screening out each type. The ozone screen is shown schematically in Figure 2.10 for UV-A and UV-B.

Consider This 2.17   **The ABCs of Solar UV Radiation**

a. How could UV-C be represented in Figure 2.10?
b. Arrange the three regions of UV in order of increasing wavelength.
c. Arrange the three regions of UV in order of increasing frequency.
d. Will the same order for increasing energy be the same as that for frequency? Explain.
e. Would you buy a sunscreen that claims to protect against UV-C? Why or why not?

As we noted in Chapter 1, about 21% of the atmosphere consists of diatomic oxygen. The forms of life that inhabit our planet are absolutely dependent on the chemical properties of this gas and its interaction with ultraviolet radiation. The strong

Table 2.4		Categories and Characteristics of UV Radiation	
Radiation	**Wavelength Range**	**Relative Energy**	**Comments**
UV-A	320–400 nm	Least energetic of these three UV categories	Least damaging, reaches Earth's surface in greatest amount
UV-B	280–320 nm	More energetic than UV-A, less energetic than UV-C	More damaging than UV-A, less damaging than UV-C, most absorbed by ozone in the stratosphere
UV-C	200–280 nm	Most energetic of these three categories	Most damaging of these three, but not a problem because totally absorbed by oxygen and ozone in stratosphere

covalent bond holding the two oxygen atoms together in the O_2 molecule can be broken by the absorption of a photon of the proper radiant energy. The photon excites a bonding electron to a higher energy level, causing the atoms to come apart. Only photons with energy corresponding to a wavelength of 242 nm or less ($\lambda \leq 242$ nm) have sufficient energy to break the bonds in an O_2 molecule. These wavelengths are found in the UV-C region.

$$\text{photon } (\lambda \leq 242 \text{ nm}) + O_2 \longrightarrow 2\,O \qquad [2.4]$$

Because of this reaction, stratospheric oxygen shields Earth's surface from high-energy radiation. As green plants flourished on the young planet, they released oxygen to the atmosphere. The increasing oxygen concentration led to more effective interception of

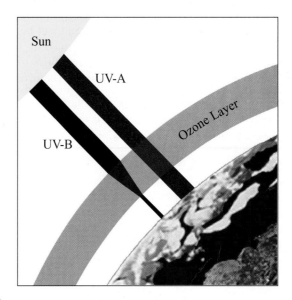

Figure 2.10
The ozone layer's role in protecting Earth from solar UV radiation.

Source: D. W. Fahey, "Twenty Questions and Answers About the Ozone Layer," *Scientific Assessment of Ozone Depletion: 2002,* World Meteorological Organization, United Nations Environmental Program, p. 5.

ultraviolet radiation. Consequently, forms of life evolved that were less resistant to UV radiation than they otherwise might have been.

If O_2 were the only UV absorber in the atmosphere, Earth's surface and the creatures that live on it would still be subjected to damaging radiation in the 242–320-nm range. It is here that ozone plays its protective role. The fact that ozone is more reactive than diatomic oxygen suggests that the O_3 molecule is more easily broken apart than O_2. Recall that the atoms in the O_2 molecule are connected with a strong double bond. Each of the bonds in O_3 is somewhere between a single and double bond in length and in strength. This makes the bonds in O_3 energetically weaker than the double bonds in O_2. Therefore, photons of a lower energy (longer wavelength) should be sufficient to separate the atoms in O_3. This is in fact the case, as radiation of wavelength 320 nm or less induces this reaction.

$$\text{photon } (\lambda \leqslant 320 \text{ nm}) + O_3 \longrightarrow O_2 + O \qquad [2.5]$$

Because of this reaction and that represented by equation 2.4, only a relatively small fraction of the Sun's UV radiation reaches Earth's surface. However, what does arrive can do significant damage.

Sceptical Chymist 2.18   How Are Energies Quantitatively Related to Wavelengths?

It has been stated that it takes higher-energy UV photons, those with wavelengths $\leqslant 242$ nm, to break the double bond in O_2. Given that the bonds in O_3 are somewhat weaker than those in O_2, photons of wavelengths $\leqslant 320$ nm can break those bonds. The Sceptical Chymist does realize that wavelength is inversely proportional to energy, but how does the energy associated with a photon of each of these wavelength limits compare? Just how much greater is the energy of the 242-nm photon than that of a 320-nm photon?

Hint: One approach could be to calculate the ratio of the energies for a 242-nm photon and that of a 320-nm photon and then to compare that to the ratio of their wavelengths.

Every day, 300,000,000 (3×10^8) tons of stratospheric ozone form and an equal mass decomposes. As with any chemical or physical change, new matter is neither created nor destroyed but merely changes its chemical or physical form. In this particular case, the overall concentration of ozone remains constant. The process is an example of a **steady state,** a condition in which a dynamic system is in balance so that there is no net change in concentration of the major species involved. A steady state arises when a number of chemical reactions, typically competing reactions, balance each other.

In the case of stratospheric ozone, the steady state is the net result of four reactions that constitute the **Chapman cycle** (Figure 2.11). This set of reactions is named

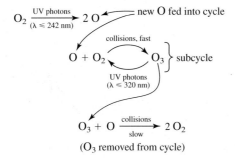

Figure 2.11
The Chapman cycle.

after Sydney Chapman, a physicist who first proposed it in 1929. Equation 2.4 showed the first step, in which diatomic oxygen, O_2, absorbs radiant energy with wavelengths below 242 nm and is split into oxygen atoms, O. A second step is for these oxygen atoms to join with O_2 to form ozone, O_3. Once formed, O_3 can absorb UV radiant energy with wavelengths below 320 nm, causing the molecule to dissociate and regenerate O_2 and O. This step, the third in the series, was shown in equation 2.5. How does nature complete this loop? Occasionally an O_3 molecule collides with an O atom to form two O_2 molecules. This fourth step is a slow reaction capable of removing the "odd oxygen" species O and O_3 from the cycle. This natural process shows both ozone formation and ozone decomposition. The "lifetime" of a given ozone molecule depends strongly on altitude, ranging from days to years. In the center of the ozone layer, an O_3 molecule can persist for several months before it dissociates into O_2 and O.

Your Turn 2.19   The Four Steps of the Chapman Cycle

a. Write an equation for each of the four steps of the Chapman cycle.
  Hint: Equations for steps 1 and 3 are given in the text.
b. Which steps illustrate the formation of O_3 and which the destruction of O_3?

The four reactions of the Chapman cycle constitute a steady state in which the rate of O_3 formation equals the rate of O_3 destruction in the stratosphere. Although reactions are going on, no net change in the concentrations of the reactants or products is observed. The balance point depends on the details of the system. In this particular case, the steady-state concentration of ozone depends on such factors as the intensity of the UV radiation, the concentration of O_2 and other reacting species, temperature, and the rates and efficiencies of the individual steps in the cycle. To further complicate things, all these factors vary with altitude. When these variables are properly evaluated and included, it becomes apparent that the Chapman cycle cannot tell the whole story. It is fundamentally correct in its description of a natural process, but the real world is inevitably more complicated than such idealized constructions. In a later section, we will consider what happens when something disturbs the steady state of the Chapman cycle, leading to destruction of the protective ozone layer.

2.7 Biological Effects of Ultraviolet Radiation

The consequences of UV radiation for plants and animals depend primarily on two factors: the energy associated with the radiation and the sensitivity of the organism to that radiation. The vertical scale of Figure 2.12 indicates the energy intensity of UV solar energy. The unit expresses how many joules of energy fall on a one square meter surface in one second. As should be anticipated from our previous discussion, the graph shows how this energy varies with wavelength. For any wavelength, the total amount of energy is the product of the number of photons striking the surface and the energy per photon. The rather flat upper curve reveals that the energy input above the atmosphere does not depend significantly on wavelength. If Earth's atmosphere did not exist, the planet's surface would be subjected to these extremely high levels of radiant energy. However, the lower curve indicates that the energy reaching Earth's surface varies markedly with wavelength. It starts dropping at 330 nm and falls off sharply as the wavelength decreases, due to the absorption of UV-B radiation by stratospheric ozone.

In fact, the decrease in UV radiation is a good deal more dramatic than the figure at first suggests. The vertical scale in Figure 2.12 is logarithmic, a method of presenting data that permits the inclusion of a wide range of values. On this logarithmic scale, each mark on the axis represents a value (in this case, an energy value) that is 10 times the value corresponding to the mark immediately below it on the vertical axis. Thus, at 320 nm, where ozone starts absorbing, the energy input to

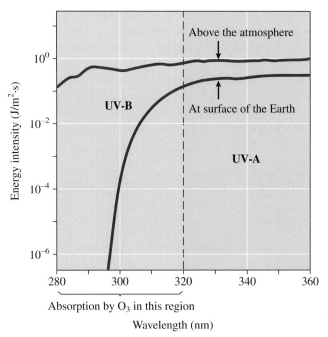

Figure 2.12

Variation of solar energy with wavelength of UV radiation. The UV-C region starts at wavelengths below 280 nm.

Earth's surface is 1×10^{-1}, or 0.1 joules per square meter per second ($J \cdot m^{-2} \cdot s^{-1}$). At 300 nm, the energy has dropped to 1×10^{-4}, or 0.0001 $J \cdot m^{-2} \cdot s^{-1}$. The intensity of UV radiation reaching Earth's surface with 300-nm wavelength has dropped by three powers of ten, 10^3, and now is only one-thousandth (0.001) of the intensity at 320 nm.

We have seen that highly energetic photons can excite electrons and break bonds in biological molecules, rearranging them and altering their properties. Solar radiation at wavelengths below 300 nm is almost completely screened out by O_2 and O_3 in the stratosphere. This is most fortunate, because radiation in this region of the spectrum is particularly damaging to living things. This relationship is evident from Figure 2.13,

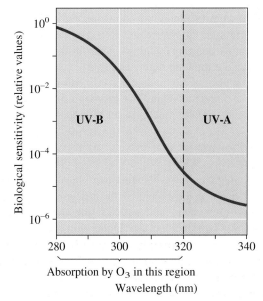

Figure 2.13

Variation of biological sensitivity of DNA with UV wavelength. The UV-C region is below 280 nm.

A discussion of DNA appears in Chapter 12.

As wavelength ↓, frequency ↑, energy ↑, DNA damage ↑.

where biological sensitivity is plotted versus wavelength. As defined here, biological sensitivity is based on experiments in which the damage to deoxyribonucleic acid (DNA), the chemical basis of heredity, is measured at various wavelengths. In the figure, the biological sensitivity is expressed in relative units, once more on a logarithmic scale. Biological sensitivity at 320 nm is about 10^{-5}, or 0.00001 units. But at 280 nm, the sensitivity is 10^0, or 1 unit. This means that radiation at 280 nm is 10^5, or 100,000 times more damaging than radiation at 320 nm. As we have seen, this is because the energy per photon and the potential for biological damage increase as the wavelength decreases.

Consider This 2.20   Relative Biological Sensitivity

Figure 2.13 illustrates that DNA sensitivity falls with increasing wavelength of UV radiation.

a.   What explanation can you propose for this phenomenon?
b.   Does this mean there is no potential for damage at the 340-nm limit of the graph? Explain.

If more high-energy photons reach Earth's surface from the Sun, the potential for significant biological damage increases. All evidence shows that the average stratospheric ozone concentration has dropped significantly in the last 20–30 years. This phenomenon has been documented in many regions of the world, using data gathered from high-flying aircraft, ground-based systems, and satellites. Later in this chapter we will explore the reasons why the observed percent ozone depletion is the greatest in the Antarctic. However, stratospheric ozone depletion means that the ability of the atmosphere to screen out UV radiation with wavelengths below 320 nm has decreased. Although this has happened to varying extents in different regions, living things are now exposed to greater intensities of damaging radiation. Scientists have made calculations predicting that a given percent decrease in stratospheric ozone will increase the effects of biologically damaging UV radiation by twice that percentage. For example, a 6% decrease in stratospheric ozone could mean a 12% rise in skin cancer, especially the more easily treated form, nonmelanoma skin cancers such as basal cell and squamous cell cancers. These conditions are considerably more common among whites than among those with more heavily pigmented skin. People of African and Indian origin are better equipped to withstand the high levels of UV radiation in the intense sunlight that strikes Earth near the equator (Figure 2.14).

Good evidence links the incidence of nonmelanoma skin cancers, the intensity of UV radiation, and the latitude at which you live. For example, the disease generally becomes more prevalent as one moves farther south in the Northern Hemisphere (Figure 2.15). Those who endure the long nights and short days of northern winters are compensated in general by a level of nonmelanoma skin cancer that is only about half that of those who enjoy year-round sunshine. The geographical effect on radiation intensity and skin cancer is, at least to date, much greater than that caused by ozone depletion.

Consider This 2.21   Geography of Skin Cancer

Many generations of immigrants have come to the United States. Some fair-skinned Northern Europeans, for example, have settled in the area around San Antonio, TX; many other equally fair-skinned immigrants have settled in the area around Seattle. Based on Figure 2.15, compare their relative risks of developing skin cancer. Identify several other factors that may affect the risk for any individual in these two populations.

Skin cancer rates continue to rise in all countries, despite increased awareness of the dangers of exposure to UV radiation. This phenomenon is somewhat puzzling, because improvements in early detection and effective treatment have lowered death rates for many

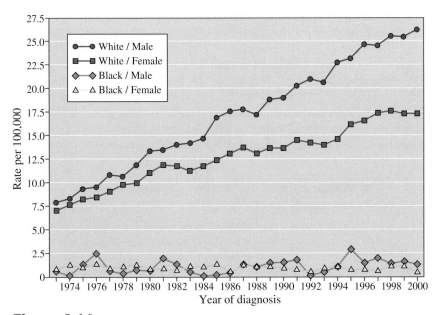

Figure 2.14

Increase in incidence of melanoma skin cancer in the United States, 1973–2000.

Source: Surveillance, Epidemiology, and End Results (SEER) Program of the National Cancer Institute.

cancers, including many skin cancers. Changes in the natural protection afforded by the stratospheric ozone layer are only partially responsible for higher rates of skin cancer. In a study conducted for the National Institutes of Health in 1998, researchers were surprised to find that Vermont was the state with the highest melanoma rate. Clearly, geographic location cannot be the only factor influencing development of skin cancer. Can this be linked to the availability of outside sports such as skiing in winter and hiking in

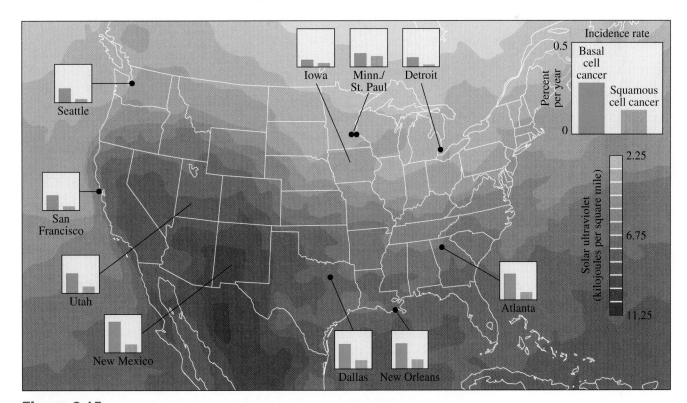

Figure 2.15

Skin cancer risks in selected U.S. locations. Cities are located with dots. Lines without city dots represent the percent per year for the entire state.

the summer? At least one other possible cause needs to be considered. Can the high rate be linked to the fact that Vermont also has a large number of tanning beds per capita?

Tanning, either naturally or in a tanning bed, is clearly a risk/benefit activity. Many light-skinned sun worshippers consider a golden tan a cosmetic benefit worth taking a risk for. The risk undoubtedly exists, as evidenced by the fact that about a million new cases of skin cancers occur each year in the United States, almost as many as the total number of cases of all other cancers. Skin cancers can develop many years after repeated, excessive exposure has stopped. Skin cancers even have been linked to a single episode of extreme sunburn in adolescence with the effects showing up many years later.

Consider This 2.22   Bronze by Choice—Tanning Salons

The indoor tanning industry maintains a constant public relations campaign that highlights positive news about indoor tanning, promoting it as part of a healthy lifestyle. Countering these claims are the studies published in scientific journals that support the view of dermatologists that there is no such thing as a "safe tan." Investigate at least two Web sites that present different points of view and list the specific claims made. Assume you are making a recommendation for a person whose skin has little pigment. Based on your findings, which criteria would you use to decide whether or not to go to an indoor tanning salon?

Fair-haired, fair-skinned individuals are at the highest risk for developing skin cancer from UV-B radiation; fair-skinned whites in Australia have the highest skin cancer rates in the world. Dr. Paul Jelfs, head of the Australian Institute of Health and Welfare, reported that during the 1990s, skin cancer rates in Australia increased by 4.3% per year for men and 1.8% per year for women. One of every two Australians will develop skin cancer sometime during his or her lifetime. The Australian government has acted in several ways to reverse this trend, including banning tanned models from all advertising media.

Wearing protective sunscreen is one way to reduce the risk of skin cancer. Such products contain compounds that absorb UV-B to some extent together with others for absorbing UV-A. The American Academy of Dermatology recommends a sunscreen with a skin protection factor (SPF) of 15 to 30. But wearing a sunscreen does not mean that you are without risk from the Sun's UV rays. Because sunscreens allow you be exposed for a longer time without burning, they may ultimately cause greater skin damage. The Australian product Blue Lizard Suncream (Figure 2.16) uses "smart-bottle" technology for containing and marketing their product. The bottle itself changes color from white to blue in UV light, sending an extra reminder that the dangers of UV light are still present, even if a sunscreen is being used.

Because of the possible damage caused by exposure to UV radiation, the U.S. National Weather Service issues an Ultraviolet Index Forecast that appears nationally in newscasts, in newspapers, and on the Web. UV Index values range from 0 to 15 and are based on how long it takes for skin damage to occur (Table 2.5). As is the case for many

Figure 2.16
Blue Lizard Suncream

Table 2.5	The UV Index		
UV Index	Exposure Level	Minutes for some damage to occur in lighter-skinned people who "never tan"	Minutes for some damage to occur in darker-skinned people who "never tan"
0–2	Minimal	30	>120
3–4	Low	15–20	75–90
5–6	Moderate	10–12	50–60
7–9	High	7–8.5	33–40
10–15	Very high	4–6	20–30

indices designed to communicate with the public, the UV Index may be color-coded to help provide visual information with the numerical value. The UV Index plays an important role in alerting the public to the dangers of increasing exposure to UV radiation.

Consider This 2.23   UV Index Forecasts

The UV index indicates the amount of UV radiation reaching Earth's surface at solar noon (1 P.M. daylight-savings time).

a. The UV Index depends on the latitude, the day of the year, time of day, amount of ozone above the city, elevation, and the predicted cloud cover. How is the UV Index affected by each of these?

b. The UV Index forecast is available on the Web, compliments of a satellite launched by the National Oceanographic and Atmospheric Administration (NOAA). Search either for "Current UV Index Forecast, NOAA" or go to the *Online Learning Center* for a direct link. Account for the range of values that you see on today's map of the United States.

c. Surfaces such as snow, sand, and water intensify your exposure to UV radiation, because they reflect it back at you. What outdoor activities might increase your risk from exposure?

Although the UV Index focuses on skin damage, that is not the only biological effect of UV radiation in humans. Everyone, no matter the pigmentation of their skin, can suffer eye problems caused by UV exposure. Retinal damage can take place, as can photokeratitis, which is sunburn to the eye. Cataracts, a clouding of the lens of the eye, can also be caused by excessive exposure to UV-B radiation. It has been estimated that a 10% decrease in the ozone layer could create up to 2 million new cataract cases globally. However, just as putting on sunscreen often and liberally can cut down on skin damage, wearing optical-quality sunglasses capable of blocking at least 99% of UV-A and UV-B is a sensible action for protecting the eyes. This is particularly important while taking part in water or snow sports, when the danger to unprotected eyes is greatest.

Consider This 2.24   Protecting Your Eyes from UV Rays

Sunglasses make far more than a fashion statement. They can protect your eyes from harmful UV rays. What characteristics would you look for in sunglasses to be used when water skiing or sailing? Check out two or three manufacturers to find out what virtues of their products are stressed in their advertising for this use. What materials are used to make sunglasses for water sports? Did you find that the price of sunglasses was related to the amount of UV protection? Would you purchase and wear the sunglasses you found through your research? Explain your choices.

Human beings are not the only creatures on the globe affected by UV radiation. Increases in UV radiation will bring harm to young marine life, such as floating fish eggs, fish larvae, juvenile fish, and shrimp larvae. There is also experimental evidence of DNA damage in the eggs of Antarctic ice fish. Plant growth is suppressed by UV radiation, and experiments have measured the negative impact that increased UV-B radiation has on phytoplankton. These photosynthetic microorganisms live in the oceans where they occupy a fundamental niche in the food chain. Phytoplankton ultimately supply the food for all the animal life in the oceans, and any significant decrease in their number could have a major effect globally. In a report released in September 1999, an international panel of scientists confirmed that exposure to elevated levels of UV-B radiation affected phytoplankton movement (up and down in water) and their motility (moving through water). Without such movement and motility, phytoplankton cannot achieve proper position in the water and are unable to carry out photosynthesis as effectively. Moreover, these tiny plant-like organisms play an important role in the carbon dioxide balance of the planet by absorbing approximately 80% of the atmospheric CO_2 created by human activities. Thus, it is possible that ozone depletion may influence another atmospheric problem, the greenhouse effect, the topic of Chapter 3.

Decreasing stratospheric ozone and consequences of this reduction are cause for concern and action. But action requires knowledge of the chemistry that occurs 15 miles above Earth's surface; we turn next to consider that.

2.8 Stratospheric Ozone Destruction— A Global Phenomenon

Stratospheric ozone concentrations have been measured over the past 80 years at ground experimental stations spread over the planet and for more than 20 years by satellite-mounted detectors. Figure 2.17 shows measurements made in Antarctica from 1980 to 2003. This graph and others to follow report total ozone levels above Earth's surface in Dobson units (DU). For our purposes, the precise details of obtaining these measurements are less important than the fact that, at this time, a value of 320 DU represents the average ozone level over the northern hemisphere. A value of 250 DU is typical at the equator.

Figure 2.17 shows the dramatic decline in ozone levels observed near the South Pole. The data displayed were collected in the fall of every year from 1980 through 2003 and illustrate the decline in the minimum amount of ozone detected. Indeed, these

> Recall that a Dobson unit corresponds to about one ozone molecule for every billion molecules and atoms of air.

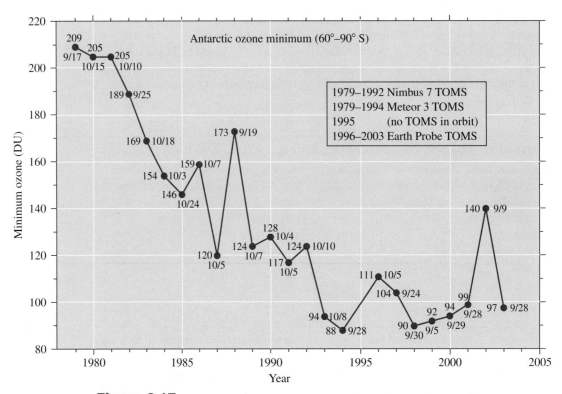

Figure 2.17

Minima in fall stratospheric ozone in Antarctica, 1979–2003. The whole number is the minimum reading in Dobson units. The date is when the minimum occurred that year. TOMS (total ozone-measuring spectrometer) is an analytical instrument.

Source: Climate Prediction Center, NOAA.

changes were so pronounced that, when the British monitoring team at Halley Bay in Antarctica first observed it in 1985, they thought their instruments were malfunctioning. The area covered by ozone levels less than 220 DU, defined as the "ozone hole," was larger in early September 2000, than in any previous year. Total ozone destruction in the hole occurred from an altitude of 15 to 20 km, consistent with measurements taken in recent years. The minimum total ozone value of 98 DU was not the lowest ever recorded. In late September 1994, the total ozone there dropped to 88 DU, the lowest recorded anywhere in the world in 36 years of measurement. Keep in mind that seasonal variation has always occurred in ozone concentration over the South Pole, with a minimum in late September or early October—the Antarctic spring. Unprecedented is the striking decrease in this minimum that has been observed over the 40 years.

More importantly, the stratospheric ozone concentrations are lower than those predicted using the simple Chapman cycle mechanism. That by itself is neither cause for alarm nor proof that the lower stratospheric ozone concentrations are the consequence of human intervention. In fact, many of the factors influencing the stratospheric ozone layer are natural in origin. We know that the processes establishing the steady-state concentration of stratospheric ozone are more complicated than originally believed.

For one thing, the natural concentration of stratospheric ozone is not uniform over all parts of the globe. On average, the total O_3 concentration increases the closer one gets to either pole (with the exception of the seasonal "hole" over the Antarctic). The formation of ozone via steps 1 and 2 of the Chapman cycle is triggered when an O_2 molecule absorbs a photon of UV light. Therefore, ozone production increases with the intensity of the radiation striking the stratosphere, an intensity that is not constant. Intensity varies with the seasons, reaching its maximum (in the Northern Hemisphere) in March and its minimum in October (just the reverse of the Southern Hemisphere). Consequently, stratospheric ozone concentrations also follow this seasonal pattern. In addition, the amount of radiation emitted by the Sun changes over an 11- to 12-year cycle related to sunspot activity. This variation also influences O_3 concentrations, but only by 1–2%. The winds blowing through the stratosphere cause other variations in ozone concentrations, some on a seasonal basis and others over a 28-month cycle. To further complicate matters, seemingly random fluctuations often occur. Finally, it is well established that certain gases from both natural and human-made sources are also responsible for the destruction of stratospheric ozone.

The unique circumstances that produce the Antarctic ozone hole are discussed in Section 2.10.

Extraordinary TOMS images, such as the ones that open this chapter, are color-coded to show stratospheric ozone concentrations in Dobson units. The violet and purple regions indicate where the greatest destruction of O_3 occurred. From the mid-1990s on, the size of the depleted ozone region annually equaled nearly the total area of the North American continent, in some cases exceeding it. Consider This 2.25 gives you the opportunity to examine one series of these images, each taken annually in October.

Consider This 2.25   Purple Octobers

NASA satellites provide stratospheric ozone data over time that can be tabulated in a number of ways, including global images, Antarctica ozone minima, and size of the ozone hole. All three are provided at the *Online Learning Center.*

a. Look first at the global images centered on Antarctica. Describe what is happening with the passage of time.

b. Now look at the graphs that show the minimum ozone levels and the size of the region affected around Antarctica. What information does each plot give you?

c. How have the minimum ozone levels and the size of the region affected around Antarctica changed since 2003, the most recent image shown in the opening pages of this chapter?

d. Use the information from all three views to explain the term *ozone hole.* In your explanation, include references to the region of the globe, area affected, amount of ozone, and time.

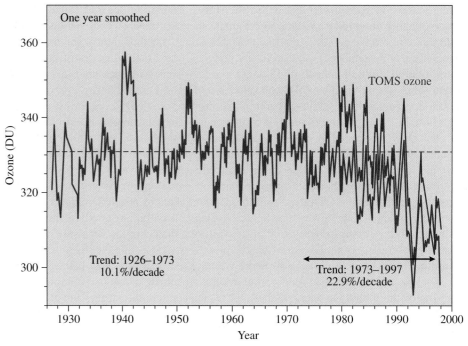

Figure 2.18

Ozone concentrations in Arosa, Switzerland, 1926–1998. The TOMS instrument was carried into the stratosphere by both Earth Probe (EP) satellites and the Japanese Advanced Earth Observations Satellite (ADEOS).

Source: http://www.epa.gov/ozone/science/glob_dep.html

The problem of ozone depletion is not limited to Antarctica. As another example, consider Figure 2.18. Scientists have been gathering data in Arosa, Switzerland, for more than 70 years, including satellite measurements started in the early 1980s. This figure shows that the ozone levels in this Northern Hemisphere location are decreasing. Although levels are lower than those predicted by the Chapman cycle mechanism, the percentage decrease is not nearly as large as that documented in Antarctica.

The major naturally occurring cause of ozone destruction, wherever it takes place around the globe, is a series of reactions involving water vapor and its breakdown products. The great majority of the H_2O molecules that evaporate from the oceans and lakes fall back to Earth's surface as rain or snow. But a few reach the stratosphere, where the H_2O concentration is about 5 ppm. There, photons of UV radiation trigger the dissociation of water molecules into hydrogen atoms (H·) and hydroxyl free radicals (·OH). A **free radical** is an unstable chemical species with an unpaired electron. The unpaired electron is often indicated with a dot, as it is here:

$$\text{photon} + H_2O \longrightarrow H\cdot + \cdot OH \qquad [2.6]$$

Free radicals are also discussed in several other places in this text.
Chpt. 1: smog mechanisms
Chpt. 6: acid rain formation
Chpt. 9: addition polymerization
Chpt. 11: food spoilage

Because of its unpaired electron, a free radical reacts readily. Thus, the H· and ·OH radicals participate in many reactions, including some that ultimately convert O_3 to O_2. This is the most efficient mechanism for destroying ozone at altitudes above 50 km.

Your Turn 2.26   **Free Radicals**

a. How many outer electrons are in the Lewis structure for the ·OH free radical? After deciding, draw the Lewis structure.

b. Can the reaction of H· and ·OH radicals completely account for depletion of ozone in the stratosphere? What is your reasoning?

Water molecules and their breakdown products are not the only agents responsible for natural ozone destruction. Another is NO (nitrogen monoxide, also called nitric oxide). Most of the NO in the stratosphere is of natural origin. It is formed when nitrous oxide, N_2O, reacts with oxygen atoms. The N_2O is produced in the soil and oceans by microorganisms and gradually drifts up to the stratosphere. There is really little that can or should be done to control this process. It is part of a cycle involving compounds of nitrogen and living things.

However, not all the nitric oxide in the atmosphere is of natural origin; human activities can alter steady-state concentrations. That is why, in the 1970s, chemists became concerned about the increase in NO that would result from developing and deploying a fleet of supersonic transport (SST) airplanes. These planes were designed to fly at altitudes of 15–20 km, the region of the ozone layer. The scientists calculated that much additional NO would be generated by the direct combination of nitrogen and oxygen.

$$\text{energy} + N_2 + O_2 \longrightarrow 2\,NO \qquad\qquad [2.7]$$

This reaction requires large amounts of energy, which can be supplied by lightning or the high-temperature jet engines of the SSTs. To evaluate this risk–benefit situation, many experiments and calculations were carried out, leading to predictions about the net effect of a fleet of SSTs. The conclusion was that the risks outweighed the benefits, and the decision was made, partly on scientific grounds, not to build an American fleet. The Anglo–French Concorde was the only commercial plane that operated at this altitude. The Concorde took its last flight on October 24, 2003. Safety concerns and economic factors both played a role in ending the flights of these remarkable jets.

Even when the effects of water, nitrogen oxides, and other naturally occurring compounds are included in stratospheric models, the measured ozone concentration is still lower than predicted. Measurements worldwide indicate that the ozone concentration has been decreasing over the past 20 years. There is a good deal of fluctuation in the data, but the trend is clear. Stratospheric ozone concentration at midlatitudes (60° south to 60° north) has decreased by more than 8% in some cases. These changes cannot be correlated with changes in the intensity of solar radiation, so we must look elsewhere for a more complete explanation.

> NO has an odd number of outer electrons, 11 (5 from N and 6 from O), so it has an unpaired electron. As a free radical, NO reacts with additional oxygen to form NO_2 and N_2O_4, other oxides of nitrogen.

> Nitrogen oxides were discussed in Section 1.11 as air pollutants. They are also prominently featured in Section 6.12 for their role in forming acid rain.

Consider This 2.27 Up and Down the Latitudes

In an earlier exercise, you used the Web to get stratospheric ozone data at a location of your choice. Now go to NASA's archive of satellite data on stratospheric ozone levels to find out how the values have varied between 1979 and 2003 over the lower Northern Hemisphere latitudes. You may wish to coordinate your efforts with other students, so that together you cover a range of years.

a. Obtain values of stratospheric ozone levels at latitudes from +45° north to +0 (the equator) for the year of your choice. Enter −90° west (the middle of the United States) as the longitude and use the satellite Nimbus-7 and a June 15 date. Obtain readings 5° apart. Make a table of the stratospheric ozone values and compute the average.

b. Compare with others in your class data over these 24 years. Note that you may not always be able to use the average as a meaningful comparison, because satellite data may be missing at some latitudes.

We have seen much evidence for the abnormally large decrease in stratospheric ozone. Although natural processes can cause such changes, they are insufficient to cause or explain the magnitude of the depletion. It is time to turn our attention to understanding chlorofluorocarbons, compounds that are an important factor in stratospheric ozone depletion.

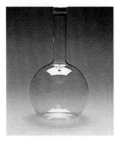

Chlorine

Bromine

Iodine

Figure 2.19

Selected elements from Group 7A, the halogen family.

Freon and Teflon are both trademarks of the DuPont Company.

2.9 Chlorofluorocarbons: Properties, Uses, and Interactions with Ozone

A major cause of stratospheric ozone depletion was uncovered through the masterful scientific sleuthing for which F. Sherwood Rowland, Mario Molina, and Paul Crutzen won the Nobel Prize in 1995. Vast quantities of atmospheric data have been collected and analyzed, hundreds of chemical reactions have been studied, and complicated computer programs have been written to identify the chemical culprit. As with most scientific results, some uncertainties remain, but there is now compelling evidence implicating an unlikely group of compounds: the chlorofluorocarbons (CFCs).

As the name implies, **chlorofluorocarbons** are compounds composed of the elements chlorine, fluorine, and carbon. Fluorine (F) and the more familiar chlorine (Cl) are members of the same chemical family, the halogens (Group 7A). The other halogens are bromine (Br) and iodine (I) (Figure 2.19). At ordinary temperature and pressures, fluorine and chlorine exist as gases made up of diatomic molecules, F_2 and Cl_2. Bromine and iodine also form diatomic molecules, but the former is a liquid and the latter a solid at room temperature. Fluorine is one of the most reactive elements known. It combines with many other elements to form a wide variety of compounds, including those in Teflon and other synthetic materials. Chlorine may be best known to you as a water purifier, but it is also a very important starting material in the chemical industry.

Chlorofluorocarbons do not occur in nature; they are artificially produced. This is an important verification point in the debate over the role of CFCs and stratospheric ozone depletion because there are no known natural sources of CFCs. Other contributors to the destruction of ozone, such as the •OH and •NO free radicals, are formed in the atmosphere from both natural sources and human activities.

Two of the most widely used chlorofluorocarbons have the formulas CCl_2F_2 and CCl_3F, commonly known as CFC-12 and CFC-11, respectively, following a naming scheme developed in the 1930s by chemists at DuPont. Their scientific names and Lewis structures are given in Table 2.6. Note that the scientific names for these two compounds are based on methane, CH_4. The prefixes *di-* and *tri-* specify the number of halogen atoms that substitute for hydrogen atoms.

The introduction of CFC-12 (Freon 12) as a refrigerant in the 1930s was rightly hailed as a great triumph of chemistry and an important advance in consumer safety and environmental protection. This synthetic substance replaced ammonia or sulfur dioxide, two naturally occurring toxic and corrosive refrigerants that made leaks in refrigeration systems extremely hazardous. In many respects, CFC-12 was (and is) an ideal substitute. It has a boiling point in the right range, it is not poisonous, it does not burn, and the CCl_2F_2 molecule is so stable that it does not react with much of anything.

The many desirable properties of CFCs soon led to other uses: propellants in aerosol spray cans, gases blown into polymer mixtures to make expanded plastic

Table 2.6	**Two Important Chlorofluorocarbons**
CFC-11	**CFC-12**
Freon 11	Freon 12
CCl_3F	CCl_2F_2
trichlorofluoromethane	dichlorodifluoromethane

foams, solvents for oil and grease, and sterilizers for surgical instruments. Similar compounds, in which bromine or fluorine atoms replace some or all of the chlorine in CFCs, are known as **halons.** They proved to be very effective fire extinguishers and are used to protect property that would be especially vulnerable to water and other conventional fire-fighting chemicals. Thus, they have found applications in electronic and computer installations, chemical storerooms, aircraft, and rare book rooms.

For better or worse, the synthesis of CFCs has had a major impact on our lives. Because CFCs are nontoxic, nonflammable, cheap, and widely available, they revolutionized air conditioning, making it readily accessible in the United States for homes, office buildings, shops, schools, and automobiles. Oppressive summer heat and humidity became more manageable with the use of low-cost CFCs as coolants. Throughout the American South, beginning in the 1960s and 1970s, air conditioning using CFCs helped to spur the booming growth of cities such as Atlanta, Dallas, and Houston, to be followed by others such as San Antonio, Austin, Charlotte, Phoenix, Memphis, Orlando, and Tampa. Some of these are now among our nation's most populated metropolitan areas. In effect, a major sociological shift occurred because of CFC-based technology that transformed the economy and business potential of an entire region of the country.

By 1985, the combined annual international production of CFC-11 and CFC-12 was approximately 850,000 tons. Venting of refrigerators and air conditioners, evaporation of solvents, and escape during polymer foaming inevitably released some of this material into the atmosphere. In 1985, the ground-level atmospheric concentration of CFCs was about six molecules out of every 10 billion (0.6 ppb), a value that has been increasing by about 4% per year. Of course, the fact that stratospheric ozone levels have been decreasing while CFC levels have been increasing does not *prove* that the two are causally related. However, other evidence suggests a connection. Ironically, the very property that makes CFCs so ideal for so many applications—chemical inertness—ends up posing a threat to the environment.

Chlorofluorocarbons represent a classic case where an apparent virtue becomes a liability. Many of the uses of these compounds capitalize on their low reactivity. The carbon-chlorine and carbon-fluorine bonds in the CFCs are so strong that the molecules can remain intact for long periods. For example, it has been estimated that an average CCl_2F_2 molecule will persist in the atmosphere for 120 years before it is destroyed. In a much shorter time, typically about five years, many CFC molecules penetrate to the stratosphere with their structures intact.

In 1973, Rowland and Molina, motivated largely by intellectual curiosity, set out to study the fate of these stratospheric CFC molecules. They knew that as altitude increases and the concentrations of oxygen and ozone decrease, the intensity of UV radiation increases. Therefore, they reasoned that in the stratosphere high-energy photons, such as UV-C, corresponding to wavelengths of 220 nm or lower, can break carbon-chlorine bonds. Equation 2.8 shows that this reaction releases chlorine atoms from CFC-12, as do similar reactions with other CFCs.

$$\text{photon } (\lambda \leq 220\,\text{nm}) \; + \; \underset{\underset{\textstyle F}{|}}{\overset{\overset{\textstyle Cl}{|}}{F-C-Cl}} \;\longrightarrow\; \underset{\underset{\textstyle F}{|}}{\overset{\overset{\textstyle Cl}{|}}{F-C\cdot}} \; + \; \cdot Cl \qquad [2.8]$$

A chlorine atom has seven outer electrons, six of them paired and one unpaired, making atomic chlorine very reactive. In equation 2.8, we wrote the atom as Cl· to emphasize the unpaired electron. The free radical chlorine atom exhibits a strong tendency to achieve a stable octet by combining and sharing electrons with another atom. Rowland and Molina and subsequent researchers hypothesized that this reactivity would result in a chain of reactions. Although there are several paths known in which CFCs destroy stratospheric ozone, we will illustrate with a typical process known to take place in polar regions.

First, the free radical Cl· pulls an oxygen atom away from the O_3 molecule, forming chlorine monoxide, ClO·, and leaving an O_2 molecule. Equation 2.9 shows this key

Recall that a free radical has an unpaired electron and is very reactive.

first step in the cycle of ozone destruction. The repeated coefficient is not cancelled because we are anticipating the next step, shown in equation 2.10.

$$2\,Cl\cdot + 2\,O_3 \longrightarrow 2\,ClO\cdot + 2\,O_2 \qquad [2.9]$$

The ClO· species is another free radical; it has 13 outer electrons (7 + 6). Recent experimental evidence indicates that 75–80% of stratospheric ozone depletion involves ClO· joining to form ClOOCl, as shown in equation 2.10.

$$2\,ClO\cdot \longrightarrow ClOOCl \qquad [2.10]$$

In turn, ClOOCl decomposes in the two-step sequence shown in equations 2.11 and 2.12.

$$UV\ photon + ClOOCl \longrightarrow ClOO\cdot + \cdot Cl \qquad [2.11]$$

$$ClOO\cdot \longrightarrow Cl\cdot + O_2 \qquad [2.12]$$

We can treat this series of chemical equations as if they were mathematical equations and add them together. Equation 2.13 is the result.

$$2\,Cl\cdot + 2\,O_3 + 2\,ClO\cdot + ClOOCl + ClOO\cdot \longrightarrow$$
$$2\,ClO\cdot + 2\,O_2 + ClOOCl + ClOO\cdot + Cl\cdot + Cl\cdot + O_2 \qquad [2.13]$$

Many of the same species are found on both sides of the combined equation 2.13. Just as is done with mathematical equations, we can eliminate the duplicate Cl·, ClO·, and ClOOCl species from both sides of the chemical equation. What remains, is the net equation showing the conversion of ozone into oxygen gas, equation 2.14.

$$2\,O_3 \longrightarrow 3\,O_2 \qquad [2.14]$$

Thus, the complex interaction of ozone with atomic chlorine provides a pathway for the destruction of ozone.

Consider This 2.28 **A New Proposal: The Deep Space Connection**

Two Canadian scientists proposed that high-energy ionizing radiation from deep space, sometimes called "cosmic rays," may be responsible for releasing chlorine free radicals from CFCs in the stratosphere. A publication in the August 13, 2001, issue of *Physical Review Letters* suggests that these cosmic rays can penetrate ice clouds, knocking electrons loose. These energetic electrons interact with CFCs to liberate the active chlorine atoms. How does this proposal differ from our current understanding of the mechanism for production of chlorine free radicals?

The fact that Cl· appears as a reactant (equation 2.9) and then as a product (equations 2.11 and 2.12) is important. This indicates that Cl· is both consumed and regenerated in the cycle, so there is no net change in its concentration. Such behavior is characteristic of a **catalyst,** a chemical substance that participates in a chemical reaction and influences its speed without undergoing permanent change. Atomic chlorine acts catalytically by being regenerated and recycled to remove more ozone molecules. On average, a single Cl· atom can catalyze the destruction of as many as $1 \times 10^5\,O_3$ molecules before it is carried back to the lower atmosphere by winds.

Interestingly, the mechanism just described for ozone destruction by CFCs in the stratosphere was not the one first proposed by Rowland and Molina. Their initial hypothesis was that Cl· reacted with O_3 to form ClO· and O_2. The second step proposed was that ClO· reacted with oxygen atoms to form O_2 and regenerate Cl· atoms.

$$Cl\cdot + O_3 \longrightarrow ClO\cdot + O_2 \qquad [2.15]$$

$$ClO\cdot + O \longrightarrow Cl\cdot + O_2 \qquad [2.16]$$

The bromine atom, Br·, undergoes a comparable reaction, starting another cycle of ozone destruction.

The term cancellation is often used for removing duplicate terms from two sides of a mathematical equation.

Although this mechanism did not prove to be the correct one in the stratosphere, it did provide a reasonable explanation for why recycling a limited number of chlorine atoms could be responsible for the destruction of a large number of ozone molecules. As is often true in science, hypotheses need to be recast in light of experimental evidence.

Atomic chlorine can also become incorporated into stable compounds that do not react to destroy ozone. Hydrogen chloride, HCl, and chlorine nitrate, $ClONO_2$, are two of these "safe" compounds that are quite readily formed at altitudes below 30 km. Thus, chlorine atoms are fairly effectively removed from the region of highest ozone concentration (about 20–25 km). Maximum ozone destruction by chlorine atoms appears to occur at about 40 km, where the normal ozone concentration is quite low.

Rowland, a professor at the University of California at Irvine, and Molina, then a postdoctoral fellow in Rowland's laboratory, published their first paper on CFCs and ozone depletion in 1974 in the scientific journal *Nature*. At about the same time, other scientists were obtaining the first experimental evidence of stratospheric ozone depletion and CFCs in the stratosphere. Since then, the correctness of the Rowland–Molina hypothesis has been well established. Perhaps the most compelling evidence for the involvement of chlorine and chlorine monoxide in the destruction of stratospheric ozone is presented in Figure 2.20. This figure contains two plots of Antarctic data: one of O_3 concentration and the other of $ClO\cdot$ concentration. Both are plotted versus the latitude at which samples were measured. As stratospheric O_3 concentration decreases, the $ClO\cdot$ concentration increases; the two curves mirror each other almost perfectly. The major effect is a decrease in ozone and an increase in chlorine monoxide as the South Pole is approached. Because $ClO\cdot$, $Cl\cdot$, and O_3 are linked by equation 2.9, the conclusion is compelling. Figure 2.20 is sometimes described as the "smoking gun," the clinching evidence. James G. Anderson of Harvard University, who conducted these measurements, has recently refined his instrumentation so that he can detect pollutants in concentrations as low as one part in 10 trillion, the equivalent of the area of a postage stamp in an area 10 times the size of Texas.

Not all of the chlorine implicated in stratospheric ozone destruction comes from CFCs. Other chlorinated carbon compounds come from natural sources, such as

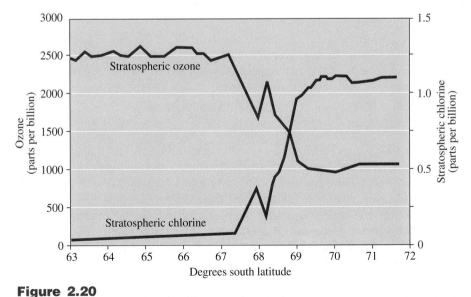

Figure 2.20

Antarctic O_3 and $ClO\cdot$ concentrations. Measurements of ozone and reactive chlorine from a flight into the Antarctic ozone hole, 1987.

Source: United Nations Environment Program. Data from http:www.unep.org/ozone/oz-story

Dr. Susan Solomon, a chemist, headed the team that first gathered stratospheric ClO· and ozone data over Antarctica. The data solidified the causal connection between CFCs and the ozone hole. She was just 30 years old at the time.

seawater and volcanoes. However, the majority of atmospheric scientists agree that most chlorine from natural sources is in water-soluble forms. Therefore, any natural chlorine-containing substances are washed out of the atmosphere by rainfall, long before they would reach the stratosphere. Of particular significance are the data gathered by NASA and by international researchers that establish that high concentrations of HCl and HF (hydrogen fluoride) always occur together. Although some of the HCl might conceivably arise from a variety of natural sources, the only reasonable origin of significant stratospheric HF is CFCs.

Consider This 2.29   Talk Radio Opinion

"And if prehistoric man merely got a sunburn, how is it that we are going to destroy the ozone layer with our air conditioners and underarm deodorants and cause everybody to get cancer? Obviously we're not . . . and we can't . . . and it's a hoax. Evidence is mounting all the time that ozone depletion, if occurring at all, is not doing so at an alarming rate."* Consider the first thing you would ask this talk-show host about these statements. Remember that you need to formulate a short and focused question to get any airtime!

* Limbaugh, R. 1993. *See, I Told You So.* New York: Pocket Books.

2.10   The Antarctic Ozone Hole: A Closer Look

A particularly intriguing question is why the greatest losses of stratospheric ozone have occurred over Antarctica when ozone-depleting gases are present throughout the stratosphere. Why is the effect greatest in polar regions, although less pronounced in the Artic than in the Antarctic? If the ozone-depleting gases are emitted mainly in the more developed Northern Hemisphere, why are their effects felt most strongly in the Southern Hemisphere?

Evidence suggests that CFCs are present in comparable abundance in lower parts of the atmosphere over both hemispheres, driven by global wind circulation patterns. A special mechanism is operative in Antarctica. This mechanism is related to the fact that the lower stratosphere over the South Pole is the coldest spot on Earth. From June to September, during the Antarctic winter, circulator winds blowing around the South Pole prevent warmer air from entering the region. Temperatures get as low as $-90°C$. Under these conditions, the small amount of water vapor present freezes into thin stratospheric clouds, called **polar stratospheric clouds (PSCs),** of ice crystals. The clouds have also been found to contain sulfate particles and droplets or crystals of nitric acid. Atmospheric scientists believe that chemical reactions occurring on the surface of these cloud particles convert otherwise safe molecules that do not deplete ozone, like $ClONO_2$ and HCl, to more reactive species such as HOCl and Cl_2. When the Sun comes out in late September or early October to end the long Antarctic night, the solar radiation breaks down the HOCl and Cl_2, releasing chlorine atoms. The destruction of ozone, which is catalyzed by these atoms, accounts for the missing ozone. Notice the conditions needed for the hole to form: extreme cold and no wind for an extended period to permit ice crystals to provide a surface for the reactions; darkness followed by rapidly increasing levels of sunlight. Figure 2.21 shows the seasonal variation and compares the minimum temperatures above the Artic and the Antarctic.

Changes in ozone above Antarctica closely follow the seasonal temperatures. Typically a rapid ozone decline takes place during spring at the South Pole (September–early November) compared with the summer (January–March). As the sunlight warms the stratosphere, the ice clouds evaporate, halting the chemistry that occurs on the PSCs.

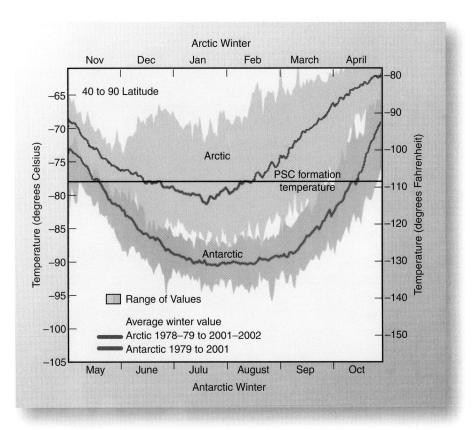

Figure 2.21

Minimum air temperatures in the polar lower stratosphere. Polar stratospheric clouds (PSCs) are thin clouds of frozen water, formed at very low temperatures.

Source: *Scientific Assessment of Ozone Depletion: 2002,* World Meteorological Organization, United Nations Environmental Program.

Then air from lower latitudes flows into the polar region, replenishing the depleted ozone levels. By the end of November, the hole is largely refilled. Although the deepest decrease in the ozone layer over Antarctica occurs during the spring, recent discoveries by British Antarctic Survey researchers indicate that the ozone depletion may begin earlier, as early as midwinter at the edges of the Antarctic, including over populated southern areas of South America.

There is already evidence that the ozone reduction over the Southern Hemisphere is greater than would be predicted solely on the basis of the midlatitude chlorine cycle. Australian scientists believe that wheat, sorghum, and pea production have already been lowered as a result of increased UV radiation. We have already noted that Australian health officials observed significant increases in skin cancers despite a very active public health campaign to alert the population to the danger of exposure to UV radiation. Ultraviolet alerts have even been issued in Australia. Similar effects are also being felt in southern Chile in the area around Punta Arenas, and on the island of Tierra del Fuego at the southernmost tip of South America. Lidia Amarales, Chile's health minister, has warned the 120,000 residents of Punta Arenas not to be out in the sun between 11 A.M. and 3 P.M. during the fall, when ozone depletion reaches its peak. As reported by BBC News in October 2000, Morales said, "If people have to leave their homes, they should wear high-factor suncreams, UV protective sunglasses, wide-brimmed hats and clothing with long sleeves." Despite these warnings, not all of the residents of this area can afford to take these precautions. "I have to go to buy bread and scarcely have money for that, so forget the sunglasses and suncream," said area resident Adriana Cerpa. Those who live in such a stratospheric ozone-depleted area near the South Pole may show the first consequences of ozone destruction in that region.

Decreased stratospheric ozone over the South Pole leads to increased UV-B levels reaching the Earth, causing increased skin cancer rates in Australia.

Figure 2.22
Arctic polar stratospheric clouds.

The general extent of the depletion in the Northern Hemisphere is not as severe as in the Southern Hemisphere. Even the observed downward trend observed through 1994–1995 has reversed in more recent years. Scientists have not classified the ozone depletion over the North Pole as a "hole," but are carefully monitoring the location and intensity of UV-B radiation being received. We have already observed that the main reason for the observed difference between the total ozone changes in the two hemispheres is that the atmosphere above the North Pole usually is not as cold as that over its Southern Hemisphere counterpart. Although polar stratospheric clouds have been observed in the Arctic (Figure 2.22), the air trapped over the Arctic generally begins to diffuse out of the region before the Sun gets bright enough to trigger as much ozone destruction as has been observed in Antarctica. The fact that the stratosphere above the North Pole reached record low temperatures in 1994–1995 may have been a factor in the uncommonly high ozone depletion. Whatever the combination of causes, scientists are giving the situation their close attention.

Low total ozone values were again observed over the Arctic region during the Northern Hemisphere winter of 2002–2003 (Figure 2.23). The area of low ozone readings was larger than for the previous two winters, but not as large as had been observed during the 1990s. NOAA scientists report that in portions of the Arctic region the average values for total ozone were as much as 45% lower than comparable values during the 1980s. Meteorological conditions favorable to ozone destruction and the continued presence of ozone-destroying chemicals in the stratosphere are part of the explanation for the observed values. The situation in the middle latitudes of the Arctic was quite different. There, higher than average values for ozone were recorded, reversing the previously observed downward trend.

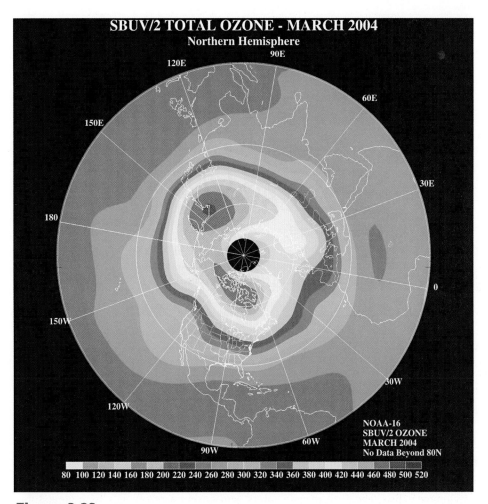

Figure 2.23
Total ozone in the Northern Hemisphere, March 2004.

 Consider This 2.30   Northern Hemisphere Ozone Maps

Figure 2.23 shows total ozone detected in the Northern Hemisphere in March 2004. Another map shows the percent difference in total ozone between March 2004 and the March average from 1979–1986. Visit the *Online Learning Center* for direct access to this other map, courtesy of NOAA's Climate Prediction Center. Compare these two maps and offer some possible advantages for using each.

2.11 Responses to a Global Concern

Once the role of synthetic CFCs in ozone destruction was better understood, the response was surprisingly rapid. Some of the first steps toward reversing ozone depletion were taken by individual countries. For example, the use of CFCs in spray cans was banned in North America in 1978, and their use as foaming agents for plastics was discontinued in 1990. The problem of CFC production and subsequent release, however, is a global one, as it requires international cooperation for actions to be effective. The sequence of events provides a model for solving problems cooperatively before a full-scale global crisis results.

In 1985, in response to the first experimental evidence of an ozone hole, a number of world governments participated in the Vienna Convention on the Protection of the Ozone Layer. Through action taken at the convention, these nations committed

Consider This 2.31 Graffiti with a Message

a. What was the source of humor in this cartoon when it originated in the mid-1970s?

b. Is this cartoon still relevant to the problem of ozone depletion today? Explain your reasoning.

c. If you are so inclined, create your own cartoon dealing with the issue of ozone layer depletion. Be sure that the chemistry is correct!

themselves to protecting the ozone layer and to conducting scientific research to better understand atmospheric processes. A major breakthrough came with the signing, in 1987, of the Montreal Protocol on Substances That Deplete the Ozone Layer. The participating nations agreed to reduce CFC production to one-half the 1986 levels by 1998.

An important provision of the agreement was to plan for future meetings to revise goals as scientific knowledge evolved. Therefore, with growing knowledge of the cause of the ozone hole and the potential for global ozone depletion, atmospheric scientists, environmentalists, chemical manufacturers, and government officials soon agreed that the original Montreal Protocol was not sufficiently stringent. In 1990, representatives of approximately 100 nations met in London and decided to ban the production of CFCs by the year 2000. The phase-out time was further accelerated in the amendments enacted in Copenhagen in 1992 and in Montreal in 1997. The Beijing Amendment in 1999 added bromine-containing halons to the schedule for phase-out, and revised controls on short-term CFC substitutes, HCFCs. **Hydrochlorofluorocarbons (HCFCs)** are compounds of hydrogen, carbon, fluorine, and chlorine. The production of CFCs and other fully halogenated CFCs are to be eliminated by 2010 by all parties to the Montreal Protocol, no matter the basic domestic economic needs.

> HCFCs, allowed during the transition period before a full ban, are discussed in more detail later in this section.

The production of other ozone-depleting compounds was identified for elimination between 2002 and 2010. These include fire-fighting halons (carbon–fluorine–bromine compounds), the solvent carbon tetrachloride (CCl_4), and methyl bromide (CH_3Br), a widely used agricultural fumigant. The United States attempted to gain an exemption on the use of methyl bromide, saying that it was necessary for farm production to remain competitive and to avoid the use of toxic alternatives. This proposal came before the November 2003 meeting held to discuss further amendments to the Montreal Protocol timelines. The representatives of 181 countries met in Nairobi, Kenya, and after due consideration, rejected the U.S. request. The deadline for developed countries to phase out methyl bromide was 2005 when this book went to press. Developing nations have until 2015 to discontinue its use.

> CFCs used as propellants in medical inhalers, such as those used by asthmatics, are exempt from the ban on CFCs for other uses.

A key strategy for reducing chlorine in the stratosphere was to stop production of CFCs. The United States and 140 other countries agreed to a complete halt in CFC manufacture after December 31, 1995. A total of 183 countries have now ratified this agreement. Figure 2.24 indicates that the decline in global CFC production has been dramatic. By 1996, production of CFCs had dropped to 1960 levels. Production and consumption of CFCs fell by 86% overall between 1986 and 1996, and by 95% in

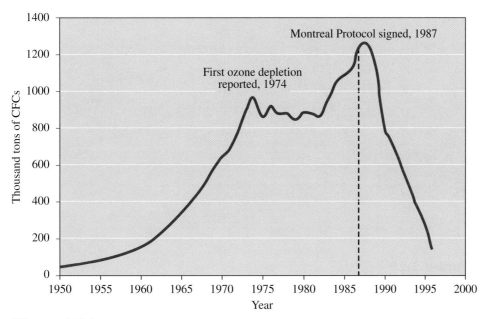

Figure 2.24

Global production of CFCs, 1950–1996.

Source: United Nations Environmental Program, report of May 20, 1999.

industrialized countries by the end of 1998. Halon production and consumption fell by 70% between 1986 and 1996 and by 84% by 1998. Production of the originally controlled halons fell by 99.8% by the end of 1998. It is estimated that without the international action brought to bear by the Montreal Protocol, stratospheric abundances of chlorine found in the stratosphere would have tripled by the middle of the 21st century.

Just because new production of CFCs has been stopped and uses of CFCs have been restricted does not mean that the stratospheric concentration of chlorine instantly will drop. Most Earth systems, including those in the atmosphere, are very complex and respond slowly to change. Slow change, rather than rapid response, is usually beneficial. These observations hold true in the case of CFCs. In fact, atmospheric concentration of ozone-depleting gases continued to rise steadily rising through the 1990s, despite the restrictions of the Montreal Protocol. Many of the CFCs have very long lifetimes in the atmosphere, estimated to be 100 years or more in some cases. However, there are encouraging signs that the Montreal Protocol has already had a significant effect. Decreases are now being observed in the amount of **effective stratospheric chlorine,** a term reflecting both chlorine- and bromine-containing gases in the stratosphere. The values take into account the greater effectiveness but lower concentration of bromine relative to chlorine in depleting stratospheric ozone. Figure 2.25 shows the predicted future abundance of effective chlorine.

Consider This 2.32   Past and Future Effective Chlorine Levels

Use Figure 2.25 to help answer these questions.

a. In approximately what year did effective chlorine concentration peak? What was the reading in that year?

b. How confident are you in assigning the year of the peak from this graph? Explain, using terms such as ±1 year, ±5 year, or whatever is appropriate.

c. In approximately what year will the concentration return to 1980 levels? What will the reading be in that year?

d. How confident are you in assigning the concentration of 1980 levels from this graph? Explain, using terms such as ±100 ppt, ±500 ppt, or whatever is appropriate.

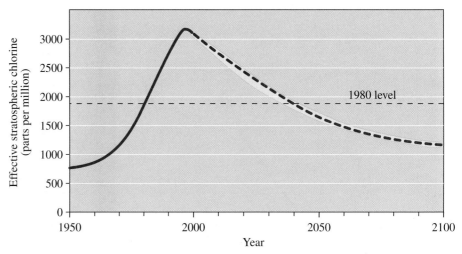

Figure 2.25

Concentrations of effective chlorine, 1950–2100. The height of the yellow band for any year is an estimate of the uncertainty in the prediction.

Source: *Scientific Assessment of Ozone Depletion: 2002,* World Meteorological Organization, United Nations Environmental Program.

Although the Montreal Protocol and its subsequent adjustments set dates for the halt of all CFC production, the sale of existing stockpiles and recycled materials will remain legal until phase-out dates in the future. One reason is, that in the United States alone, 140 million car air conditioners and the majority of home air conditioners are designed to use these compounds. As a result, both the paperwork and the price of legally obtained CFCs have risen sharply. Nevertheless, many U.S. trade groups promote converting to less harmful substitute refrigerants when repairs must be made on older systems. Until 2010, small amounts of CFCs may be produced in developing countries, keeping some CFCS in legal circulation. Unfortunately, this transition period also encourages an increase in the black market. Bootleggers have been tempted to smuggle CFCs into the United States, largely from the Russian Federation, China, India, Eastern Europe, and Mexico. China now holds the dubious honor of being the leading supplier of black-market CFCs. According to U.S. law enforcement officers, CFCs are second only to illicit drugs as the most lucrative illegal import. In 1997 alone, "freon busts" by the U.S. Justice Department led to the confiscation of nearly 12 million pounds of illegal CFCs, bringing in fines of $38 million.

One of the largest CFC-smuggling operations was uncovered in 1999, when federal agents from the EPA enforcement unit intercepted a shipment of six large industrial refrigeration units from Venezuela to the United States. These units only require 3–4 lb of CFC-12 each for operation, but they had been altered to each hold more than 2500 lb of CFC-12. It is still legal in Venezuela to use previously produced CFCs, but the incentive to bring these materials illegally into this country is clear. The street value in the United States of just this shipment was approximately $600,000. This load turned out to be just one part of a very large Freon-smuggling ring, and the federal crackdown on the black market in CFCs was successful in this case.

Sceptical Chymist 2.33   Black Market CFCs

A study reported in the May 2000 issue of *Atmospheric Environment* concluded that illegal trade in CFCs is only a "small threat" to ozone layer recovery. Although the Sceptical Chymist would *like* for this to be true, *is* it? Please provide some current information to either support or refute this statement.

2.12 Looking to the Future

Where do we go from here? Scientists believe that the atmosphere will eventually rid itself of the ozone-destroying agents. However, it will certainly take a long time, given the 100-year lifetime of many CFCs. Analysis of trends in chlorine levels indicates that stratospheric chlorine peaked at 4.1 ppb in the late 1990s and then diminished slowly. This is taken as evidence that the Montreal Protocol and its amendments have slowed the release of CFCs and related ozone-depleting materials. But we are not completely in the clear. Scientists estimate that even under the most stringent international controls on the use of ozone-depleting chemicals, the stratospheric chlorine concentration would not drop to 2 ppb until perhaps 2050 or later. This value is significant because the Antarctic ozone hole first appeared when chlorine levels in that region reached 2 ppb.

To be sure, some have proposed attempting to scrub the atmosphere of chlorine by intentionally introducing other chemicals. For example, a researcher at the University of California, Los Angeles, suggested injecting electrons into the stratosphere. He postulated that these electrons would react with chlorine atoms, converting them into chloride ions (Cl^-). But a recent report argued that this approach would not be effective. Another proposal advocated adding propane (C_3H_8) to the stratosphere, reasoning that the hydrocarbon would react with chlorine atoms and remove them from the ozone-depletion cycle. The author of the scheme estimated that 50,000 tons of C_3H_8 would have to be added each year, at an annual cost of $100 million. This strategy also has been discredited.

Nor can we adopt the solution proposed by the industrialist in Sidney Harris's cartoon (Figure 2.26). The authors of this book calculated that it would take about 17 million planeloads of ozone to replenish 10% of the stratospheric ozone. Sherwood Rowland has estimated that "the energy that would be needed to move the ozone up [to the stratosphere] is about $2\frac{1}{2}$ times all of our current global power use." Even if we could temporarily replace the lost ozone, the steady-state cycle would soon reestablish itself.

"OH, FOR PETE'S SAKE, LET'S JUST GET SOME OZONE AND SEND IT BACK UP THERE!"

Figure 2.26

A solution to ozone depletion?

© Sydney Harris. Reprinted by permission.

Although we cannot undo what already has been done in adding ozone-depleting gases to our atmosphere, we can stop doing it. Key to the process will be the continued success of chemists in finding replacements for CFCs. No one seriously advocates the return to ammonia and sulfur dioxide in home refrigeration units or giving up air-conditioning. In designing replacement molecules, chemists are concentrating on compounds similar to the CFCs. The assumption is that the substitute molecules will include one or two carbon atoms, at least one hydrogen atom, and often fluorine or more tightly bound chlorine atoms. The rules of molecular structure limit the options. For example, each carbon atom forms single bonds to four other atoms in the molecules under consideration.

In synthesizing substitutes for CFCs, chemists must weigh three undesirable properties—toxicity, flammability, and extreme stability—and attempt to achieve the most suitable compromise. Compounds containing only carbon and fluorine (fluorocarbons) are neither toxic nor flammable, and they are not decomposed by UV radiation, even in the stratosphere. Consequently, they would not catalyze the destruction of ozone. This would be ideal, were it not for the fact that the undecomposed fluorocarbons would eventually build up in the atmosphere and contribute to the global warming effect (see Chapter 3) by absorbing infrared radiation.

Introducing hydrogen atoms in place of one or more halogen atoms reduces molecular stability and promotes destruction of the compounds at low altitudes, long before they enter the ozone-rich regions of the atmosphere. However, too many hydrogen atoms increase flammability. Moreover, if a hydrogen atom replaces a halogen atom, the total mass of the molecule is decreased. This results in a decrease in boiling point, making the compounds less ideal for use as refrigerants. A boiling point in the −10 to −30 °C range is an important property for a refrigerant. Too many chlorine atoms seem to increase toxicity and chloroform, $CHCl_3$, would not be a good CFC substitute because of its toxicity. All of these relationships between composition, molecular structure, boiling point, and proposed use must be considered along with toxicity, flammability, and stability for any substitute.

Fortunately, chemists already know a good deal about how these variables are related, and they have used this knowledge to synthesize some promising replacements for CFCs. Table 2.7 shows the formulas, names, and structures for two **hydrofluorocarbons,** compounds of hydrogen, fluorine, chlorine, and carbon. They decompose in the troposphere more readily than CFCs, and hence do not accumulate to the same extent in the stratosphere. HCFC-22 ($CHClF_2$) is the most widely used HCFC, suitable for air conditioners and in the production of foamed fast-food containers. Its ozone-depleting potential is about 5% that of CFC-12 and its estimated atmospheric lifetime is only 20 years, compared with 111 years for CFC-12. HCFC-141b ($C_2H_3Cl_2F$) is also used to form foam insulation.

The Montreal Protocol allows the use of HCFCs as short-term substitutes for CFCs. Because HCFCs themselves have some adverse effects on the ozone layer, they are regarded only as an interim solution. The 1992 Copenhagen amendments to the Montreal Protocol call for a halt in the manufacture of these compounds by 2030. Their

As number of H ↑, flammability ↑.
As number of Cl ↑, toxicity ↑.

Table 2.7	**Two Important Hydrochlorofluorocarbons**
HCFC-22	**HCFC-141b**
$CHClF_2$ chlorodifluoromethane	$C_2H_3Cl_2F$ dichlorofluoroethane

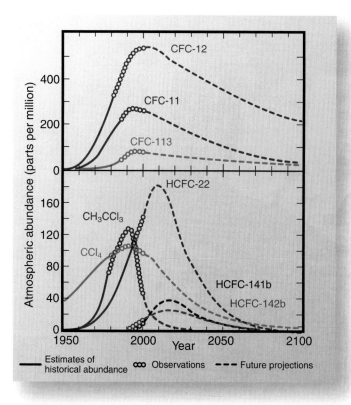

Figure 2.27

Atmospheric CFCs and HCFCs, 1950–2100.

Source: *Scientific Assessment of Ozone Depletion: 2002,* World Meteorological Organization, United Nations Environmental Program.

pattern of accumulation and destruction in the stratosphere is quite different from the pattern observed for the chlorine-containing compounds banned by the Montreal Protocol (Figure 2.27).

Consider This 2.34 **Past and Future, CFCs and HCFCs**

Use Figure 2.27 to help answer these questions.

a. Compare the peaking patterns for CFCs and for HCFCs. How are they the same and how are they different? Offer possible reasons for your observations.

b. Compare the peaking patterns for CFCs and for CCl_4 and CH_3CCl_3. How are they the same and how are they different? Offer possible reasons for your observations.

c. Which halogen-containing compound has shown the largest reduction? Explain your choice.

In the long run, HFCs, compounds of hydrogen, fluorine, and carbon, may be more suitable replacements. HFC-134a ($C_2H_2F_4$), with a boiling point of −26 °C, could prove to be the substitute of choice for CFC-12. HFC-134a has no chlorine atoms to interact with ozone, and its two hydrogen atoms facilitate its decomposition in the lower atmosphere without making it flammable under normal conditions. Other possible replacement refrigerants include hydrocarbons such as propane (C_3H_8) and isobutane (C_4H_{10}). Hydrocarbons do not destroy ozone, but as always, there are tradeoffs in selecting them for use. Hydrocarbons typically are flammable and can contribute to global

The contributions of CFCs, HCFCs, and HFCs to global warming will be discussed in Chapter 3.

Figure 2.28
Pyrocool FEF being applied to subterranean fires at Ground Zero, North Tower at West Street, September 30, 2001.

warming. This interconnection between ozone depletion and global warming mechanisms warrant additional study as more is learned about global atmospheric modeling.

One more class of compounds that must be replaced under the Montreal Protocol is the **halons,** compounds of carbon, fluorine, and bromine. Pyrocool Technologies of Monroe, Virginia, won a 1998 Presidential Green Chemistry Challenge Award for its development of foam that is environmentally benign and yet more effective than the halons it replaces. The product, Pyrocool FEF, can replace halons in fighting even large-scale fires such as those on oil tankers and jet airplanes. A 0.4% solution of Pyrocool FEF was used to extinguish or at least control the spread of fires in the sublevels beneath the collapsed towers of the World Trade Center Towers following the terrorist attack of September 11, 2001 (Figure 2.28). Many of the hot spots buried in the debris were spreading and posing an imminent danger to any rescue operations and to huge tanks used to store Freon for the air conditioning systems. The Pyrocool FEF foam also has a cooling effect that helps firefighters, a useful feature when fighting brush fires as well. Many other companies, nationally and internationally, are working on the challenge of replacing halon compounds.

The Kyoto Accord regulates halons because of their role in global warming. This is discussed in Chapter 3.

Your Turn 2.35   Halon Structures

a.   Draw the Lewis structure of CF_3Br, which is known by the trade name Halon-1301.
b.   Draw the Lewis structure of any halon with two carbon atoms.

Answer
a.

The phase-out of CFCs and the substitution of alternative materials are not without major economic considerations. At its peak, the annual worldwide market for CFCs reached $2 billion, but that was only the tip of a very large financial iceberg. In the United States alone, CFCs were used in or used to produce goods valued at about $28 billion per year. Although the conversion to CFC replacements has had some additional costs associated with it, the overall effect on the U.S. economy actually has been minimal. Companies that produce refrigerators, air conditioners, insulating plastics, and other goods have adapted to using the new compounds. Some substitutes for CFC refrigerants are less energy-efficient, hence increasing energy consumption somewhat. But the conversions provide a market opportunity for innovative syntheses using green chemistry to produce environmentally benign substances.

On a domestic level, the political dimension of CFC regulation raises many issues. What agency establishes the limits? Where is the legislation enacted? Is this a national, state, or local affair? Who will enforce the regulations? What limitations and what time constraints are reasonable, responsible, or even possible to achieve? How much testing is necessary before replacement compounds can be safely introduced? How can the country be confident that those making the political, legal, and economic decisions are getting the best scientific advice and interpreting it correctly? How do the lobbying efforts of big businesses change decision making? There are no easy answers to these questions, maybe not even any right or wrong answers, but Consider This 2.36 gives you an opportunity to struggle with some of them.

Consider This 2.36   Phase-Out of Methyl Bromide

Methyl bromide, CH_3Br, has been widely used since the 1960s for sterilization of soils and fumigation of pests. Under the terms of the 1997 adjustments to the Montreal Protocol, developed countries agreed to a staged reduction starting in 1996 to conclude by 2005. Many American agricultural businesses have asked EPA for "critical-use" exemptions. They fear they will no longer be competitive with agriculture in developing countries, which do not have to eliminate all use by 2015.

a. Was the 2005 deadline for developed countries met or has it been extended?
b. How would granting critical-use exemptions to U.S. agricultural businesses affect efforts to repair the ozone layer?
c. How would exemptions affect development of alternative fumigants?

Developing countries face another set of economic problems and priorities. Chlorofluorocarbons have played an important role in improving the quality of life in the industrialized nations. Few would be willing to give up the convenience and health benefits of refrigeration or the comfort of air conditioning. It is understandable that millions of people over the globe aspire to the lifestyle of the industrialized nations. As an example, over the past decade, the annual production of refrigerators in China has increased from 500,000 to eight million. But, if the developing nations are banned from using the relatively inexpensive CFC-based technology, they may not be able to afford alternatives. "Our development strategies cannot be sacrificed for the destruction of the environment caused by the West," asserts Ashish Kothari, a member of an Indian environmental group. Earlier we noted that one response to these different deadlines was the development of the black market, not a desired outcome but one that exists.

As a result of these international agreements, CFC consumption by industrialized countries dropped between 1986 and 2002. But, during the same interval, use of these compounds by the developing world increased. This suggests that the use of CFCs by developing nations will continue to grow during the next decade. Without further

In 1991, China signed the Montreal Protocol and has begun to take steps to phase out the production and use of CFCs. Starting in 1998, it banned the industrial use of CFCs as aerosol propellants.

restrictions, total CFC emissions by developing countries might easily equal 1 million tons by 2010. Although the result may be industrial and economic progress for one segment of the world's population, it hardly represents progress in the protection of the environment.

Recently, environmentalists have urged that the phase-out schedule for developing countries should be accelerated, but such action is rife with political complications and may well require the infusion of funds from the industrialized nations. There is a precedent. Both India and China originally refused to sign the original Montreal Protocol because they felt that it discriminated against developing countries. To gain the participation of these highly populated nations, the industrially developed nations created a special fund in 1990. Ten donor countries (Austria, Denmark, Finland, Germany, Italy, Japan, Norway, Sweden, United Kingdom, and United States) committed $19 million dollars to the Russian Federation alone to help them close production facilities for CFCs and halons by the year 2000. The United States contributed nearly one fourth of the total to this fund. The World Bank oversees the funds. Developing nations apply to this special fund for grants to underwrite specific projects that lead to discontinuation of CFC use. In 2000, it approved 20 full-project grants, totaling $266 million, and 17 medium-sized grants just over $13 million. The World Bank feels that these investments are having a significant impact in helping countries phase out the use of ozone-depleting substances.

Even this amount of money is insufficient to cover the costs of conversion and phase-out in many nations. To make matters worse, a number of industrialized nations are behind in their payments to the fund, and the United States has not always been supportive of foreign aid. Without financial assistance, the developing nations may not be able or willing to meet a more stringent timetable for discontinuing their use of substances that deplete the ozone layer. Clearly, an understanding of chemistry is necessary to protect the ozone layer, but it is not sufficient. And thus, the second part of the twin challenge—the debate among governments about how best to protect the stratospheric ozone layer—continues in the global political arena.

Consider This 2.37   **Potential Impact of Hydrogen in the Stratosphere**

a. Hydrogen fuel cells are being promoted as a major technological alternative to burning gasoline for transportation and power. Hydrogen is generally considered as an environmentally friendly fuel, only producing water after reacting with oxygen. What impact could the widespread use of hydrogen have on urban air quality?

b. Some scientists are reporting concerns that leakage of hydrogen gas from cars, hydrogen production plants, and fuel transportation could cause problems in the Earth's ozone layer. How significant are these concerns? What is the mechanism by which hydrogen could destroy ozone?

CONCLUSION

Chemistry is intimately entwined with the story of ozone depletion. Chemists created the chlorofluorocarbons whose near-perfect properties only relatively recently revealed their dark side as predators of stratospheric ozone. Chemists worked internationally to discover the mechanism by which CFCs destroy this ozone and warned of the dangers of increasing ultraviolet radiation. And chemists will continue to synthesize the substitutes necessary to ultimately replace CFCs and other related compounds. But the issues involve more than just chemistry. The "Action on Ozone" report for 2000 from the Ozone Secretariat of the United Nations Environmental Program, sums up the global experience with ozone depletion story in this manner: "Perhaps the most important feature of the ozone regime is the way in which it has brought together an array of different participants in pursuit of a common end. Scientists have provided the information, with steadily increasing degrees of precision, on the causes and effects of ozone depletion. Industry, responding to the stimulus provided by the control measures, has developed alternatives far more rapidly and more cheaply than initially thought possible, and has participated fully in the debates over further phase-out. NGOs (nongovernmental organizations) and the media are the essential channels of communication, and education, with the peoples of the world in whose name the measures have been taken. . . . Governments have worked well together in patiently negotiating agreements acceptable to a range of countries with widely varying circumstances, aims, and resources—and showed courage and foresight in putting the precautionary principle into effect before the scientific evidence was entirely clear." These are lessons to remember as we turn to our next topic, the chemistry of global warming.

Chapter Summary

Having studied this chapter, you should be able to:

- Describe the chemical nature of ozone, location of the ozone layer, and factors affecting its existence (2.1, 2.6, 2.8, 2.9–2.12)

- Appreciate the complexities of collecting accurate data for stratospheric ozone depletion and interpreting them correctly (2.1)

- Apply the basics of atomic structure to particular elements (2.2)

- Relate an element's atomic number to its position in the periodic table (2.2)

- Write electron distributions, by levels, for elements in the A groups (2.2)

- Differentiate atomic number from mass number and apply the latter to isotopes (2.2)

- Write Lewis dot structures using the octet rule (2.3)

- Use Lewis dot structures to identify the covalent bonds present in a molecule (2.3)

- Describe the electromagnetic spectrum in terms of frequency, wavelength, and energy, and use appropriate calculations to determine these quantities (2.4, 2.5)

- Interpret graphs related to wavelength and energy, radiation and biological damage, and ozone depletion (2.4, 2.8, 2.11)

- Understand how the ozone layer protects against harmful ultraviolet radiation (2.6, 2.7)

- Differentiate among the energies and biological effects of UV-A, UV-B, and UV-C radiation (2.6, 2.7)

- Discuss the interaction of radiation with matter and changes caused by such interactions, including biological sensitivity (2.6, 2.7)

- Relate the meaning and the use of the UV index (2.7)

- Understand the Chapman cycle and the role of nature in stratospheric ozone depletion (2.8)

- Understand the chemical nature and role of CFCs in stratospheric ozone depletion (2.9, 2.10)

- Appreciate the role of unique circumstances of ozone depletion in the Antarctic (2.10)

- Summarize the political and scientific dimensions the Montreal Protocol and its amendments (2.11, 2.12)

- Evaluate articles on green chemistry alternatives to stratospheric ozone-depleting compounds and recognize the effect that market forces have on the success of these innovations (2.11, 2.12)

- Discuss the factors that will help lead to the recovery of the ozone layer (2.11, 2.12)

Questions

Emphasizing Essentials

1. The text states that the odor of ozone can be detected in concentrations as low as 10 ppb. Will you be able to detect the odor of ozone in any of these air samples?

  a. 0.118-ppm ozone, a concentration reached in the troposphere

  b. 25-ppm ozone, a concentration reached in the stratosphere

2. **a.** What is a Dobson unit?

  b. Does a reading of 320 DU or 275 DU indicate more total ozone overhead?

3. How does ozone differ from oxygen gas in its chemical formula? In its properties?

4. Which of these pairs are allotropes?

   a. diamond and graphite

   b. water, H_2O, and hydrogen peroxide, H_2O_2

   c. white phosphorus, P_4, and red phosphorus, P_8

5. Where is the "ozone layer" found? Answer by giving a range of altitudes.

6. Assume there are 2×10^{20} CO molecules per cubic meter in a sample of tropospheric air. Furthermore, assume there are 1×10^{19} O_3 molecules per cubic meter at the point of maximum concentration of the ozone layer in the stratosphere.

   a. Which cubic meter of air contains the larger number of molecules?

   b. What is the ratio of CO to O_3 molecules in a cubic meter?

7. Using the periodic table as a guide, specify the number of protons and electrons in a neutral atom of each of these elements.

   a. oxygen (O)

   b. nitrogen (N)

   c. magnesium (Mg)

   d. sulfur (S)

8. Consider this periodic table.

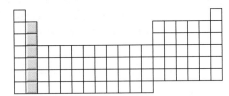

   a. What is the group number of the shaded column?

   b. What elements make up this group?

   c. What is the number of electrons for a neutral atom of each element in this group?

   d. What is the number of outer electrons for a neutral atom of each element of this group?

9. Using the periodic table, give the name and symbol of the element that has each of these number of protons.

   a. 2

   b. 19

   c. 29

10. Give the number of protons, neutrons, and electrons in each of these.

   a. oxygen-18 (O-18, $^{18}_{8}O$)

   b. sulfur-35 (S-35, $^{35}_{16}S$)

   c. uranium-238 (U-238, $^{238}_{92}U$)

   d. bromine-82 (Br-82, $^{82}_{35}Br$)

   e. neon-19 (Ne-19, $^{19}_{10}Ne$)

   f. radium-226 (Ra-226, $^{226}_{88}Ra$)

11. Give the symbol showing the atomic number and the mass number for the element that has

   a. 9 protons and 10 neutrons (an isotope used in nuclear medicine)

   b. 26 protons and 30 neutrons (the most stable isotope of this element)

   c. 86 protons and 136 neutrons (the radioactive gas found in some homes)

12. Write the electron dot structures for an atom of each element.

   a. calcium   b. nitrogen

   c. chlorine   d. helium

13. Assuming that the octet rule applies, write the Lewis structure for each of these molecules. Start by counting the number of available outer electrons. (a) Write the complete electron dot structure; (b) then write the structure representing shared pairs with a dash, showing non-bonding electrons as dots.

   a. CCl_4 (carbon tetrachloride, a substance formerly used as a cleaning agent)

   b. H_2O_2 (hydrogen peroxide, a mild disinfectant; the atoms are bonded in this order: H-to-O-to-O-to-H)

   c. H_2S (hydrogen sulfide, a gas with the unpleasant odor of rotten eggs)

   d. N_2 (nitrogen gas, the major component of the atmosphere)

   e. HCN (hydrogen cyanide, a molecule found in space and a poisonous gas)

   f. N_2O (nitrous oxide, "laughing gas"; the atoms are bonded N-to-N-to-O)

   g. CS_2 (carbon disulfide, used to kill rodents; the atoms are bonded S-to-C-to-S)

14. Several different oxygen species are related to the story of ozone in the stratosphere. These include oxygen atoms, oxygen gas, ozone, and hydroxyl radicals. Compare and contrast the Lewis structure for each of these species.

15. Consider these two waves representing different parts of the electromagnetic spectrum.

   a. How do these two waves compare in wavelength?

   b. How do these two waves compare in frequency?

   c. How do these two waves compare in forward speed?

16. Use Figure 2.6 to specify the region of the electromagnetic spectrum where radiation of each wavelength is

found. *Hint:* Change each wavelength to meters before making the comparison.

a. 2.0 cm   **b.** 400 nm

c. 50 μm   **d.** 150 mm

17. Calculate the frequency that corresponds to each of the wavelengths in question 16.

18. Calculate the energy of a photon for each wavelength in question 16. Which wavelength possesses the most energetic photons?

19. Arrange these types of radiation in order of *increasing* energy per photon: gamma rays, infrared radiation, radio waves, visible light

20. The microwaves in home microwave ovens have a frequency of $2.45 \times 10^9 \, s^{-1}$. What is the wavelength of microwaves in meters?

21. Ultraviolet radiation coming from the Sun is categorized as UV-A, UV-B, and UV-C. Arrange these three regions in order of their increasing:

   a. wavelength

   b. energy

   c. potential for biological damage

22. Consider the Chapman cycle in Figure 2.11 and Your Turn 2.19. Rewrite each step of the cycle using a sphere equation to illustrate the changes that take place.

23. These free radicals all play a role in catalyzing ozone depletion reactions: Cl, NO_2, ClO, and HO.

   a. Count the number of outer electrons available and then draw a Lewis structure for each of these species.

   b. What characteristic is shared by these free radicals that makes them so reactive?

24. In Chapter 1, the role of nitrogen monoxide, NO, in forming photochemical smog was discussed. What role, if any, does NO play in stratospheric ozone depletion? Are NO sources the same in the troposphere and in the stratosphere?

25. **a.** How were the original measurements of increases in chlorine monoxide and the stratospheric ozone depletion over the Antarctic gathered?

   b. How are these measurements made today?

26. Which graph shows how measured increases in UV-B radiation correlate with percent reduction in the concentration of ozone in the stratosphere over the South Pole?

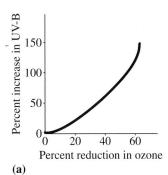

(a)

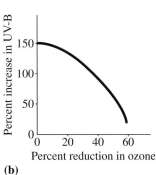

(b)

27. **a.** Most CFCs are based either on methane, CH_4, or ethane, C_2H_6. Use structural formulas to represent these two compounds.

   b. Substituting chlorine or fluorine (or both) for hydrogen atoms, how many different CFCs can be formed from methane?

   c. Which of the compounds that you drew in part **b** has been most successful?

   d. Why weren't all of these compounds equally successful?

Concentrating on Concepts

28. The allotropes oxygen and ozone differ in molecular structure. What differences does this produce in their properties, uses, and significance?

29. Explain why is it possible to detect the pungent odor of ozone after a lightning storm or around electrical transformers.

30. EPA has used the slogan "Ozone: Good Up High, Bad Nearby" in some of its publications aimed at the public. What message is this slogan trying to communicate?

31. How do *allotropes* of oxygen and *isotopes* of oxygen differ? Explain your reasoning.

32. Consider the Lewis structures for SO_2. How are they similar to or different from the Lewis structures for ozone?

33. It is possible to write three resonance structures for ozone, not just the two shown in the text. Verify that all three structures satisfy the octet rule, and offer an explanation as to why the triangular structure is not reasonable.

$$:\ddot{O} = \overset{..}{O} - \ddot{O}: \longleftrightarrow :\ddot{O} - \overset{..}{O} = \ddot{O}: \longleftrightarrow :\ddot{O} - \overset{..}{O} - \ddot{O}:$$

34. The average length of an oxygen-to-oxygen single bond is 132 pm. The average length of an oxygen-to-oxygen double bond is 121 pm. What do you predict the oxygen-to-oxygen bond lengths will be in ozone? Will they all be the same? Explain your predictions.

35. The equation $E = h\nu = \dfrac{hc}{\lambda}$ indicates the relationships among energy, frequency, and wavelength for electromagnetic radiation. Which variables are directly related and which are inversely related?

36. Which of these forms of electromagnetic radiation from the Sun has the lowest energy and therefore the least potential for damage to biological systems? infrared radiation, ultraviolet radiation, visible radiation, radio waves

37. Even if you have skin with little pigment, why can't you get a suntan from standing in front of your radio in your living room or dorm room?

38. The morning newspaper reports a UV index of 6.5. What should that mean to you as you plan your daily activities?

39. UV-C has the shortest wavelengths of all UV radiation and therefore the highest energies. All the reports of the damage caused by UV radiation focus on UV-A and

UV-B radiation. Why is the focus of attention not on the damaging effects that UV-C radiation can have on our skin?

40. If all 3×10^8 tons of stratospheric ozone that are formed every day are also destroyed every day, how is it possible for stratospheric ozone to offer any protection from UV radiation?

41. How is it possible for CFCs gases, which are denser than air, to reach the stratosphere?

42. Which steps in the Chapman cycle (see Figure 2.11) form ozone and which destroy ozone? Explain your choices using equations from Your Turn 2.19.

43. Explain how the small changes in ClO· concentrations in Figure 2.20 (measured in parts per billion) can cause the much larger changes in O_3 concentrations (measured in parts per million).

44. Development of the stratospheric ozone hole has been most dramatic over Antarctica. What set of conditions exist over Antarctica that help to explain why this area is well-suited to studying changes in stratospheric ozone concentration? Are these same conditions not operating in the Artic? Why or why not?

45. Draw the structural formula of HFC-134a. Its condensed chemical formula is $C_2H_2F_4$. *Hint:* You may wish to refer to the structural formula for the two-carbon HCFC-141b given in Table 2.7.

46. The free radical $CF_3O·$ is produced during the decomposition of HFC-134a.
   a. Propose a Lewis structure for this free radical.
   b. Offer a possible reason why this free radical does not cause ozone depletion.

47. One of the mechanisms that helps to break down ozone in the Antarctic region involves the BrO· free radical. Once formed, it reacts with ClO· to form BrCl and O_2. BrCl in turn, reacts with sunlight to break into Cl· and Br·, both of which react with O_3 and form O_2.
   a. Represent this information with a set of equations similar to those shown for the Chapman cycle.
   b. What is the net equation for this cycle?

48. Consider the graphs (top of next column) from the World Meteorological Organization (WMO) and the United Nations Environmental Program (UNEP).
   a. Scientists from the WMO think that if the effective chlorine concentration in the atmosphere were reduced to 1980 level, the ozone hole in the Antarctic would disappear. In what year is this expected to happen according to the original protocol and each of its subsequent amendments?
   b. Discuss how the two graphs in this question are related.
   c. Based on these graphs, do you think there will be future amendments to the Montreal Protocol? Explain your reasoning.

49. Consider the graph in Figure 2.1 showing ozone concentrations at various altitudes.

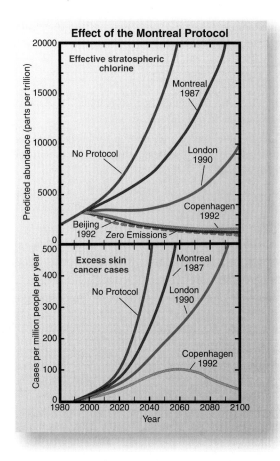

Source: *Scientific Assessment of Ozone Depletion: 2002,* World Meteorological Organization, United Nations Environmental Program.

   a. What does this graph tell you about the concentration of ozone as you travel upward from the surface of the Earth? Write a brief description of the trends shown in the graph, describing the location of the ozone layer.
   b. The *y*-axis in this graph starts at zero. Why doesn't the *x*-axis appear to start at zero?

Exploring Extensions

50. **a.** How does a photograph taken from space differ from either a true or false color image taken from a satellite?
   b. Consider the opening images of this chapter. Are they photographs? How can you tell?

   Hint: You may want to learn something about the field of remote sensing to help in answering this question.

51. Consider this periodic table.

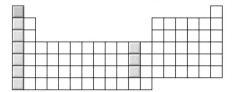

Two groups are highlighted; the first is Group 1A and the second is Group 1B. The text states that although A groups have very regular patterns, the "situation gets a bit more complicated with the B groups." Researching in other resources, find out which of these predictions becomes more complicated.

a. predicted number of electrons for each element in each group

b. predicted number of outer electrons for each element of each group

c. predicted formula when each element combines chemically with chlorine

52. Resonance structures can be used to explain the bonding in charged groups of atoms as well as in neutral molecules, such as ozone. The nitrate ion, NO_3^-, has one additional electron plus the outer electrons contributed by nitrogen and oxygen atoms. That extra electron gives the ion its charge. Draw the resonance structures, verifying that each obeys the octet rule.

53. Although oxygen exists as O_2 and O_3, nitrogen exists only as N_2. Propose an explanation for these facts. *Hint:* Try drawing a Lewis structure for N_3.

54. It has been suggested that the term *ozone screen* would be a better descriptor than *ozone layer* to describe ozone in the stratosphere. What are the advantages and disadvantages to each term?

55. Many different types of ozonators are on the market for sanitizing air, water, and even food. They are often sold with a slogan such as this one from a pool store. "Ozone, world's most powerful sanitizer!"

a. Find out how these devices work.

b. What claims are made for ozonators intended to purify air?

c. What claims are made for ozonators designed to purify water?

d. How do medical ozonators differ from other models?

56. The effect a chemical substance has on the ozone layer is measured by a value called its *ozone-depleting potential, ODP.* This is a numerical scale that estimates the lifetime potential stratospheric ozone that could be destroyed by a given mass of the substance. All values are relative to CFC-11, which has an ODP defined as equal to 1.0. Use those facts to consider these questions.

a. What factors do you think will influence the ODP value for a chemical? Why?

b. Most CFCs have ODP values ranging from 0.6 to 1.0. What range do you expect for HCFCs? Explain your reasoning.

c. What ODP values do you expect for HFCs? Explain your reasoning.

57. Recent experimental evidence indicates that ClO· initially reacts to form Cl_2O_2.

a. Predict a reasonable Lewis structure for this molecule. Assume the order of atom linkage is Cl-to-O-to-O-to-Cl.

b. What impact does this evidence have on understanding the mechanism for the catalytic destruction of ozone by ClO·?

58. Chemical formulas for individual CFCs, such as CFC-11 (CCl_3F), can be figured out from their code numbers. A quick way to interpret the code number for CFCs is to add 90 to the number. In this case, $90 + 11 = 101$. The first number in this sum is the number of carbon atoms, the second is the number of hydrogen atoms, and the third is the number of fluorine atoms. CCl_3F has one carbon, no hydrogen, and one fluorine atom. All remaining bonds are assumed to be chlorine until carbon has the required four single covalent bonds to satisfy the octet rule.

a. What is the chemical formula for CFC-12?

b. What is the code number of CCl_4?

c. Will this "90" method work for HCFCs? Use HCFC-22, which is $CHClF_2$, to explain your answer.

d. Will this method work for halons? Use Halon-1301, which is CF_3Br, to explain your answer.

59. This graph shows the atmospheric abundance of bromine-containing gases from 1950 to 2100.

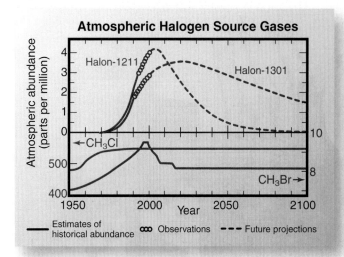

Atmospheric Halogen Source Gases

Source: *Scientific Assessment of Ozone Depletion: 2002,* World Meteorological Organization, United Nations Environmental Program.

a. Compare the patterns for Halon-1211 and Halon-1301. Explain why their concentrations are not peaking at the same time.

b. Are both Halon-1211 and Halon-1301 still legally manufactured and used in the United States? In Mexico?

c. Explain why the level of CH_3Br is predicted to remain high and constant through 2100.

60. **a.** What factors account for the fact that Australia has the highest incidence of skin cancer in the world?

b. Why is their government actively involved in changing this statistic?

c. What facts are being stressed in the public education campaign?

d. Is the rate of skin cancer the same for migrants coming to Australia as it is for white Australians?

e. Is the rate the same for Australian Aborigines as it is for white Australians? Why or why not?

chapter

3

The Chemistry of Global Warming

Construction of the Trans-Amazonian Highway in Brazil has opened millions of square kilometers of previously inaccessible land to development of new communities such as this village. The settlers expand farming, mining, logging, and cattle industries, often with only short-term economic gains. Felled trees cannot carry out photosynthesis, therefore they cannot remove atmospheric CO_2. The practice of clearing by burning adds CO_2 to the atmosphere. These effects can enhance global warming. Changes in the rainforests of the world have many cultural impacts as well.

"The science itself is not in doubt. Of course there are continuing uncertainties about the proportion of natural to human-driven change, but the existence of human-driven change is clear. The conclusions of the Intergovernmental Panel on Climate Change and the main national academies of science (including that of the United States) represent a broad international consensus with little serious dissent."

Editorial: "Communicating Climate Change,"
Crispin Tickell, Chair of the Climate Institute of Washington, DC
Science, 2 August 2002, p. 737

A report released by EPA in May 2002 states that "Greenhouse gases are accumulating in the Earth's atmosphere as a result of human activities, causing global mean surface air temperatures and subsurface ocean temperatures to rise." The interagency "U.S. Climate Action Report" was sent to the United Nations. It included the forecast that total greenhouse emissions by the United States will increase 43 percent between 2000 and 2020. In June 2002, President Bush distanced himself from the report, saying it was "put out by the bureaucracy." New copies of the report have been changed to emphasize scientific uncertainty.

"Bush Pressured to Curb Global Warming," Knight Ridder News Service, 9 June 2002

The introductory sentence "Climate change has global consequences for human health and the environment" was cut and replaced with a paragraph that starts "The complexity of the Earth system and the interconnections among its components make it a scientific challenge to document change, diagnose its causes, and develop useful projections of how natural variability and human actions may affect the global environment in the future."

"Report by the EPA Leaves Out Data on Climate Change," *New York Times,* 19 June 2003

Is global warming a scientific certainty or just a political liability? Why is the continued destruction of the Amazon rainforest relevant to the chemistry of global warming? The answers have to do with how Earth's climate system is regulated and responds to change. There are many parts to the climate puzzle, including incoming solar radiation, outgoing radiation from the Earth, wind and water currents, atmospheric gases, clouds, snow and ice, volcanic gases, and human activities. To more fully understand global warming, we also need to consider the rate at which change is taking place and whether the climate system can respond at a similar rate. For example, we know that the Amazon rainforest is vanishing at three times the rate it was less than 10 years ago. By some estimates, 80% of Brazil's ancient forests are already gone. How will these rapid changes affect the "steady-state" balance of climate regulation?

The role of CO_2 as one of the major players in the debate about global warming is far from obvious. After all, CO_2 is an essential component of the atmosphere, a gas that all animals exhale and green plants absorb. Central to the issue of global warming is the molecular mechanism by which carbon dioxide and other compounds absorb the infrared radiation emitted by the planet, helping to keep it warm. Some knowledge of molecular structure and shape is necessary to understand this mechanism. Global warming has a significant quantitative component; we need numbers to help assess the seriousness of the situation. That need requires introduction and illustration of several quantitative measures. Recognizing that global warming has international implications, we will look

Consider This 3.1   Science Behind the Policy

Policy decisions on global warming are based on scientific research and careful assessment of environmental and economic effects. The United Nations and the World Health Organization sponsor the Intergovernmental Panel on Climate Change (IPCC). In the United States, the Global Change Research Program (USGCRP) carries out climate change research. Use a search engine or the direct links provided at the *Online Learning Center* to learn about these two groups and their work.

a.   What is the membership of each agency?
b.   What are the major responsibilities and activities of each agency?
c.   How do these two groups interact with each other and with political systems around the world?

at parallels between responses to protecting the ozone layer (the Montreal Protocols) and to slowing global warming (the Kyoto Conference Protocols). Current U.S. policy toward restricting emissions of carbon dioxide and other gases implicated in global warming will be examined. Developing understanding of these issues will lead us on a journey into the land of chemistry and its interaction with public policy around the world.

3.1 In the Greenhouse: Earth's Energy Balance

The brightest and most beautiful body in the night sky, after our own moon, is considered by many to be Venus (Figure 3.1). It is ironic that the planet named for the goddess of love is a most unlovely place by earthly standards. Spacecraft launched by the United States and the former Soviet Union have revealed a desolate, eroded surface with an average temperature of about 450 °C (840 °F). The atmosphere surrounding Venus has a pressure 90 times greater than that of Earth, and it is 96% carbon dioxide, with clouds of sulfuric acid. It makes the worst smog-bound day anywhere on Earth seem like a breath of fresh country air. The beautiful blue-green ball we inhabit has an average annual temperature of 15 °C (59 °F).

The point of this little astronomical digression is that both Venus and Earth are warmer than one would expect based solely on their distances from the Sun and the amount of solar radiation they receive. If distance were the *only* determining factor, the temperature of Venus would average approximately 100 °C, the boiling point of water. Earth, on the other hand, would have an average temperature of −18 °C (0 °F), and the oceans would be frozen year around.

The idea that Earth's atmospheric gases might somehow be involved in trapping some of the Sun's heat was first proposed around 1800 by the French mathematician and physicist, Jean-Baptiste Joseph Fourier (1768–1830). Fourier compared the function of the atmosphere to that of the glass in a "hothouse" (his term) or greenhouse. Although he did not understand the mechanism or know the identity of the gases responsible for the effect, his metaphor has persisted. Some 60 years later, John Tyndall (1820–1893) in England experimentally demonstrated that carbon dioxide and water vapor absorb heat radiation. In addition, he calculated the warming effect that would result from the presence of these two compounds in the atmosphere. In the 1890s, Swedish scientist Svante Arrhenius (1859–1927) considered the potential problems that could be caused by CO_2 building up in the atmosphere. Observed warming of surface

Figure 3.1

Computer-generated image of Maat Mons, a volcano on Venus. The image is based from radar data collected by the space probe *Magellan.* The vertical scale has been exaggerated 10-fold.

air temperatures between the 1890s and 1940 led some scientists to suggest that the American Dust Bowl was an early sign of the greenhouse effect at work. Following a 30-year period of slight cooling that started in 1940, U.S. oceanographer Roger Revelle (1909–1991) alerted the world in 1957 to the problems that could be caused by ever-increasing amounts of **greenhouse gases,** those gases capable of absorbing and reemitting infrared radiation, to the atmosphere. Since that time, there has been a steady increase in the amount and reliability of data gathered about the role that CO_2 and other gases play in global warming. We know that molecules of CO_2 absorb heat. We know that the concentration of CO_2 in the atmosphere has increased over the past 150 years, and we know that Earth's average temperature has not remained constant. As we move through the chapter, we will investigate how these observations are interrelated and what other factors come into play.

You may not have personally experienced the warmth of a greenhouse nurturing your prize tropical plants on a cold, winter's day. Almost everyone, however, has had the experience of returning to a car after it has been sitting closed in direct sunlight. The windows of the car allow visible and a relatively small amount of ultraviolet light from the Sun to pass through into the car. Energy is absorbed by the interior of the car, particularly by dark fabrics and surfaces. Some of that energy is reemitted as infrared radiation (IR), but longer wavelength IR cannot escape back through the windows. The heat builds up in the car until, when you return, the meaning of the term *hothouse* is very clear.

 Figures Alive! Visit the *Online Learning Center* to learn more about Earth's energy balance and the greenhouse effect. Practice, using the interactive exercises. Look for the **Figures Alive!** icon elsewhere in this chapter.

> "Dust Bowl" describes a period of severe drought in Texas and Oklahoma during the 1930s.

> Water vapor is the most abundant greenhouse gas in our atmosphere. However, contributions of H_2O from human activity are negligible compared with natural emissions.

> In sunny climates, temperatures in a closed car can quickly exceed 49 °C (120 °F). People or pets should not be left in cars under these conditions.

Your Turn 3.2   **Wavelength and Energy Relationships**

Consider these three types of radiant energy from the electromagnetic spectrum: infrared, ultraviolet, and visible.

a. Arrange them in order of *increasing* wavelength.
b. Arrange them in order of *increasing* energy.
c. How are these arrangements related to the Sun's ability to heat a closed car? Explain your reasoning.

Answer
a. ultraviolet, visible, infrared
b. infrared, visible, ultraviolet

Is heat building up in your car different from heat building up in Earth's greenhouse formed by its atmospheric gases? Every atmospheric process is driven by the enormous power of the Sun, which sends our way an average of 343 watts (W) of energy per every square meter of Earth every day of the year. This is a huge amount, equivalent to the output from 440 million large electrical power plants producing 100 million watts of power every day of the year. Of this incoming solar radiation, about 30% is reflected by the molecules that make up our envelope of air, by the clouds and dust in the atmosphere, or by Earth itself. Another 25% is absorbed in the atmosphere, leaving about 45% to be absorbed by the continents and oceans and warming them. Earth, in turn, radiates some of its absorbed energy back into the atmosphere, where greenhouse gases such as H_2O and CO_2 are very efficient absorbers of this longer-wave IR radiation.

Although a small percentage of the absorbed terrestrial radiation makes it directly back into space from the surface, the greenhouse gas molecules found in our atmosphere absorb and scatter most of the radiation in all directions. Much of this heat is

> A watt (W) is a unit of power. $1\ W = 1\ J \cdot s^{-1}$. You use this unit in selecting light bulbs. A 100-W bulb uses more power to illuminate than a 50-W bulb.

> Absorption of infrared radiation by water vapor, the most common greenhouse gas, explains why humid or cloudy days have a higher "heat index" than dry, clear days of the same air temperature.

redirected and comes back though the lower regions of our atmosphere toward the Earth, rather than being lost to space. Heat is transferred by collisions between neighboring molecules, and these molecules are found in greater abundance in the denser regions of the lower atmosphere. Taking all processes together, about 81% of the energy radiated by Earth is trapped by greenhouse gases and does not directly escape to space. The **greenhouse effect** is the process by which atmospheric gases trap and return a major portion of the heat (infrared radiation) radiated by the Earth. Because of the constant, dynamic exchange between Earth, its atmosphere, and space, a steady state is established, with more or less constant average terrestrial temperature being the result.

This balancing act of energy exchange between our Earth and its atmosphere is a natural and beneficial process that helps maintain the existence of life on our planet. Without the protective layer of our atmosphere, Earth could become very hot from incoming radiation. Without the atmosphere's ability to reflect Earth's radiated heat back toward the surface, our lovely orb could become an ice planet because of the direct loss of heat into space. The current average temperature of our planet, about 15 °C (59 °F), is about 33 °C warmer than what would be expected from its distance from the Sun. It is also much higher than the −270 °C of outer space. Consequently, the Earth acts overall like a global radiator, radiating heat to its frigid surroundings. Figure 3.2 is a schematic representation of Earth's energy balance. Note the role that the atmosphere plays in absorbing and radiating energy both back toward Earth and out into space.

> Disturbance of a steady-state was also important in the Chapter 2 discussion of ozone depletion.

Your Turn 3.3   Earth's Energy Balance

a.   Why doesn't all of the incoming solar radiation reach Earth's surface?
b.   Why doesn't all the heat radiated back from Earth's surface go directly into space?
c.   Why are incoming radiation and outgoing radiation shown in different colors in Figure 3.2? Explain the meaning of the different colors.

Consider This 3.4   Science Fiction Story

Successful writers of science fiction sometimes begin their careers as science majors. Their best work reveals a sound understanding of scientific phenomena and principles. Often a good science fiction story assumes a slightly different scientific reality than the one we know. For example, *Dune,* by Frank Herbert, takes place on a desert planet.

Here is an opportunity to exercise your imagination in a different climate. Make the assumption that the planet has an average temperature of −18 °C (0 °F). What would human life be like? Write a brief description of a day on a frozen planet. (Residents of northern climates should have a great advantage in this exercise.)

Consider This 3.5   Clear or Cloudy?

Why does the temperature drop more on a clear night than on a cloudy night? Explain your reasoning.

Obviously, the greenhouse effect is essential in keeping our planet habitable. However, if having some greenhouse gases in the atmosphere is a good thing, having more is not necessarily better. The term **enhanced greenhouse effect** is often used to refer to an energy return of greater than 81%. An increase in the concentration of infrared absorbers will very likely mean that more than 81% of the radiated energy will be returned to Earth's surface, with an attendant increase in average temperature. Back in 1898, Arrhenius estimated the extent of this effect. He calculated that doubling the concentration of CO_2 would result in an increase of 5–6 °C in the average temperature of the planet's surface. Writing in the *London, Edinburgh, and Dublin Philosophical Magazine* to announce his findings, Arrhenius dramatically described the phenomenon: "We

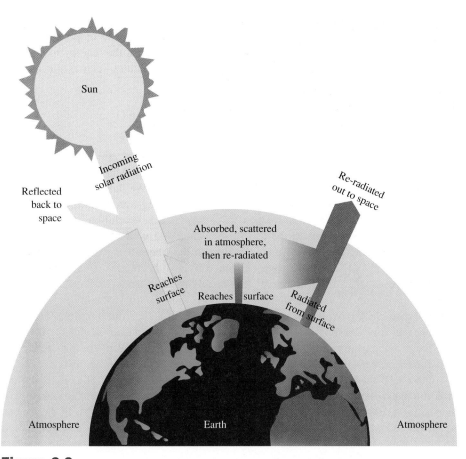

Figure 3.2
The Earth's energy balance. Shorter wavelengths of radiation are shown in yellow, longer in red.

are evaporating our coal mines into the air." At the end of the 19th century, the Industrial Revolution was already well under way in Europe and America, and it was "picking up steam" as well as generating it (and CO_2 also).

Consider This 3.6   **Evaporating Coal Mines**

Although the Arrhenius statement about "evaporating our coal mines into the air" certainly was effective in grabbing attention in 1898, what process do you think he really was referring to in discussing the amount of CO_2 being added to the air? What is your reasoning?

3.2 Gathering Evidence: The Testimony of Time

In the 4.5 billion years that our planet has existed, its atmosphere and climate have varied widely. Evidence from the composition of volcanic gases suggests the concentration of carbon dioxide in Earth's early atmosphere was perhaps 1000 times greater than it is today. Much of the CO_2 that dissolved in the oceans became incorporated in rocks such as limestone, which is calcium carbonate, $CaCO_3$. High concentrations of carbon dioxide all those years ago also made possible a significant event in the history of our planet—the development of life on Earth. Even though the Sun's energy output was 25–30% less than it is today, the ability of CO_2 to trap heat kept Earth sufficiently warm to permit the development of life. As early as 3 billion years ago, the oceans were filled with primitive plants such as cyanobacteria (blue-green bacteria). Like their more sophisticated descendants, these simple plants were capable of photosynthesis. They were able to use chlorophyll to capture sunlight and use this energy to combine carbon dioxide gas and water, forming more complex molecules such as glucose and releasing oxygen (equation 3.1).

$$6\,CO_2 + 6\,H_2O \xrightarrow{\text{chlorophyll}} \underset{\text{glucose}}{C_6H_{12}O_6} + 6\,O_2 \qquad [3.1]$$

Photosynthesis dramatically reduced the concentration of atmospheric CO_2 and increased the amount of O_2 present. The microbiologist Lynn Margulis has called this "the greatest pollution crisis the Earth has ever endured." We, and all past and future generations, are the unknowing beneficiaries of this long-ago pollution crisis. The increase in oxygen concentration helped make possible the evolution of animals. But even 100 million years ago, in the age of dinosaurs and well before humans walked the Earth, the average temperature is estimated to have been 10–15 °C warmer than it is today, and the CO_2 concentration is assumed to have been considerably higher.

How do we know such numbers? Reasonably reliable evidence is available about temperature fluctuations during the past 160,000 years—only yesterday in geological terms. Deeply drilled cores from the ocean floor give us a slice through time. The number and nature of the microorganisms present at any particular level indicate the temperature at which they lived. Supplementing this, the alignment of the magnetic field in particles in the sediment provides an independent measure of time.

Other relevant information comes from the analysis of ice cores. The Soviet drilling project at the Vostok Station in Antarctica has yielded over a mile of ice cores taken from the snows of 160 millennia. The ratio of deuterium, ^2H, to hydrogen, ^1H, in the ice can be measured and used to estimate the temperature at the time the snow fell. Water molecules containing the most abundant form of hydrogen atoms (mass number 1) are lighter than those that contain deuterium (mass number 2). The lighter H_2O molecules evaporate just a bit more readily than the heavier ones. As a result, there is more ^1H and less ^2H in the water vapor of the atmosphere, compared with the amounts in the oceans.

Water molecules that are heavier condense more readily. Therefore, rain or snow that condenses from atmospheric water vapor will be enriched in ^2H. The ratio of ^2H to ^1H in precipitation further varies with average temperature. Higher temperatures tend to increase the deuterium/hydrogen ratio in the rain or snow. This is the key to estimating ancient temperatures by the analysis of the isotopic composition of ice cores.

Isotopes of hydrogen were discussed in Section 2.2.

Figure 3.3
Scientists use data from ice cores to reconstruct greenhouse gas concentration and temperature information going back as far as 160,000 years ago.

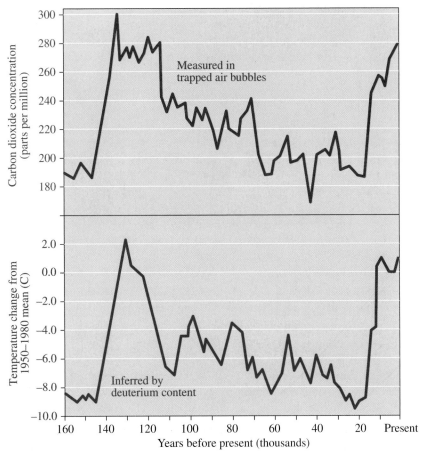

Figure 3.4

Atmospheric CO_2 concentration and average global temperature over 160,000 years (ice core data).

In addition, the bubbles of air trapped in the ice can be analyzed for carbon dioxide and other gases. Figure 3.3 shows drilling for these icy record keepers of the past.

Both carbon dioxide concentration and temperature data are incorporated in Figure 3.4. The upper curve, corresponding to the scale on the left, is a plot of parts per million of carbon dioxide in the atmosphere versus time over a span of 160,000 years. The lower plot and the right-hand scale indicate how the average global temperature has varied over the same period. For example, the figure shows that 20,000 years ago, during the last ice age, the average temperature of Earth was about 9 °C below the 1950–1980 average. At the other extreme, a maximum temperature (just over 16 °C) occurred approximately 130,000 years ago.

Particularly striking about Figure 3.4 is that the temperature and the carbon dioxide concentration follow the same patterns. When the CO_2 concentration was high, the temperature was high. Other measurements show that periods of high temperature have also been characterized by high atmospheric concentrations of methane (CH_4). Such correlations do not necessarily *prove* that elevated atmospheric concentrations of CO_2 and CH_4 caused the temperature increases. Presumably, the converse could have taken place. But both these compounds trap heat, and without doubt, they can and do contribute to global warming.

To be sure, other mechanisms also are involved in the periodic fluctuations of global temperature. Some propose that Earth's climate can sometimes behave "more like a switch than a dial," with abrupt changes taking place over a relatively short period. Temperature maxima seem to come at roughly 100,000-year intervals, with interspersed major and minor ice ages. Over the past million years, Earth has experienced 10 major periods of glacier activity and 40 minor ones. Some of this temperature variation probably is caused by minor changes in Earth's orbit that affect the distance from Earth to the Sun and the angle at which sunlight strikes the planet. However, this hypothesis

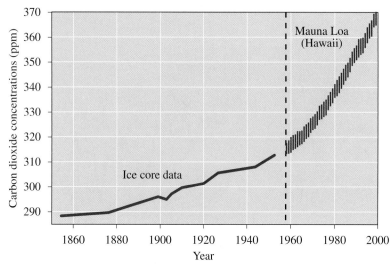

Figure 3.5

The atmospheric level of carbon dioxide rose from 280 to 370 ppm between 1860 and 2000. Increases continue, with 375 ppm reported in 2002 and a record high of 379 ppm in 2003.

 Source: From "The Greenhouse Effect and Historical Emissions," figure 4. Taken from http://clinton2.nara.gov/Initiatives/Climate/greenhouse.html.

cannot fully explain the observed temperature fluctuations. Orbital effects most likely are coupled with terrestrial events such as changes in reflectivity, cloud cover, airborne dust, and carbon dioxide and methane concentration. These factors can diminish or enhance the orbit-induced climatic changes. The feedback mechanism is complicated and not well understood. One thing is clear: Earth is a far different place in the 2000s than it was at the time of our last temperature maximum 130,000 years ago. Our ancestors had discovered fire by then, but they had not learned to exploit it as we have.

The past has provided its testimony, but more recent trends in atmospheric carbon dioxide and average global temperatures are important for assessing the current status of the greenhouse effect. There is compelling evidence that CO_2 concentrations have increased significantly in the past century. The best data are those acquired at Mauna Loa in Hawaii. Figure 3.5 presents values from Antarctic ice cores taken from 1860 to the 1950s, and then adds the Mauna Loa data to show the continuation of the trend. The series of vertical lines from 1960 until 2000 indicates seasonal variation, but the general increase in average annual values from 315 ppm to about 370 ppm in 2000 is clear. Later in this chapter we will examine why scientists believe that much of the added carbon dioxide has come from the burning of fossil fuels.

Projections of changes in CO_2 based on computer modeling will be discussed in Section 3.9.

Consider This 3.7   **The Cycles of Mauna Loa**

The text states the pattern observable in Figure 3.5 after 1960 is due to "seasonal variation."

a.   Estimate the variation in ppm CO_2 within each year.
b.   Explain these seasonal variations in CO_2 concentrations.

Sceptical Chymist 3.8   **Checking the Facts on CO_2 Increases**

a.   A recent government report states that the atmospheric level of CO_2 has increased 30% since 1860. Use the data in Figure 3.5 to either prove or disprove this statement.
b.   A global warming skeptic states that the percent increase in the atmospheric level of CO_2 since 1957 has been only about half as great as the percent increase from 1860 to the present. Comment on the accuracy of that statement and how it could affect policy on global warming.

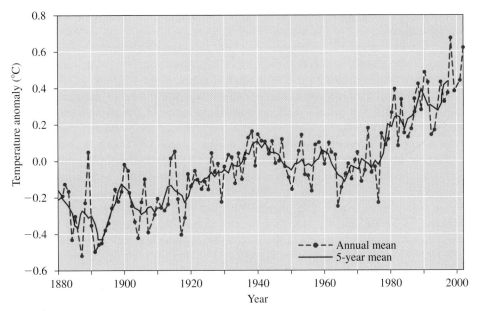

Figure 3.6

Global surface temperature change (1880–2002).

Source: http://www.giss.nasa.gov/research/observe/surftemp

Other measurements indicate that during the past 120 years or so, the average temperature of the planet has increased somewhere between 0.4 and 0.8 °C. Figure 3.6 shows the changes in the air temperature at Earth's surface from 1880 to 2000. Some scientists correctly point out that a century or two is an instant in the 4.5 billion-year history of our planet. They caution restraint in reading too much into short-term temperature fluctuations. In fact, while some areas such as in Alaska and northern Eurasia have warmed by up to 6 °C, cooling has occurred in the North Atlantic and in the central North Pacific. Short-term changes in atmospheric circulation patterns are thought to cause some observed temperature anomalies. Figure 3.6 shows the variability in temperatures from year to year, as well as the longer-term trends. Many scientists conclude that only since about 1970 is

Consider This 3.9   **Winter Woes**

Do you think the comment made in the cartoon is justified? Why or why not?

Pepper . . . and Salt

"This winter has lowered my concerns
about global warming . . ."

Source: From *The Wall Street Journal*. Permission by Cartoon Features Syndicate.

there a demonstrated upward climb in global temperature. From Figure 3.6 we see that the average temperature of the Earth is about 0.6 °C higher now than it was in 1880. Whether this temperature increase is a consequence of the increased CO_2 concentration cannot be concluded with absolute certainty. Nevertheless, experimental evidence implicates carbon dioxide from human-related sources as a cause of recent global warming.

When temperature measurements are extrapolated into the future, predictions made by Arrhenius of a 5–6 °C rise in the average temperature of the planet's surface may need to be revised. Current estimates from the United Nations predict that average temperatures will increase somewhere between 1.4 °C and 5.8 °C (2.5 °F and 10.4 °F) by the year 2100. Other scientists, looking at a possible doubling of CO_2 emissions in the future, estimate a temperature increase between 1.0 °C and 3.5 °C (1.8 °F and 6.3 °F). Future temperature changes can be influenced, at least to a considerable extent, by the human beings who inhabit this planet. We are a long way from the out-of-control hothouse of Venus, but we face difficult decisions. These decisions will be better informed with an understanding of the mechanism by which greenhouse gases interact with radiation to create the greenhouse effect. For that we must again take a submicroscopic view of matter.

Consider This 3.10 CO_2 and You

CO_2 is produced and removed from our atmosphere in several ways. As members of the global community, our activities add to the production of CO_2 and its removal.

a. Review your typical activities and decide which contribute to an increase in the amount of CO_2 in the atmosphere.
b. Which of your activities can hinder the removal of CO_2 from the atmosphere?
c. Which natural activities that are *not* in your control contribute to the amount of atmospheric CO_2?

3.3 Molecules: How They Shape Up

Carbon dioxide, water, and methane are greenhouse gases; nitrogen and oxygen are not. The obvious question is "why?" The not-so-obvious answer has to do with molecular structure and shape. When you encountered Lewis structures in Chapter 2, geometry was not the main consideration. The octet rule that you learned provides a generally reliable method for predicting bonding in molecules, but not shape. In molecules such as O_2 and N_2, the shape is unambiguous, as the two atoms can only be in a straight line.

$$:N:::N: \quad \text{or} \quad :N\equiv N: \quad \text{or} \quad N\equiv N$$

$$\ddot{O}::\ddot{O} \quad \text{or} \quad \ddot{O}=\ddot{O} \quad \text{or} \quad O=O$$

Different shapes become possible with molecules of more than two atoms. Fortunately, knowing where the outer electrons are located within a molecule provides insight into molecular shape. Therefore, the first step in predicting molecular shape is to write the Lewis structure for the molecule. If the octet rule is obeyed throughout the molecule, each atom (except hydrogen) will be associated with four pairs of electrons. Some molecules include nonbonding lone-pair electrons, but all molecules contain some bonding electrons or they would not be molecules!

Bonding electrons can be grouped in pairs to form single bonds. In other molecules, the bonding electrons are involved in double bonds consisting of two pairs of electrons, or in triple bonds made up of three pairs of electrons. A basic rule of electricity is that unlike charges attract and like charges repel. Negatively charged electrons are attracted to a positively charged nucleus in every case. However, the electrons all have the same charge and therefore are found in space as far from each other as possible while still maintaining their attraction to the positively charged nucleus. Groups of negatively charged electrons will repel each other. *The most stable arrangement is one in which the mutually repelling electron groups are as far apart from each other as possible.* In turn, this determines the atomic arrangement and the shape of the molecule.

We illustrate a stepwise procedure for predicting molecular structure with methane, CH_4, a greenhouse gas.

a. **Determine the number of outer electrons associated with each atom in the molecule.** The carbon atom (atomic number 6, Group 4A) has four outer electrons; each of the four hydrogen atoms contributes one electron. There is a total of $4 + (4 \times 1)$, or 8 outer electrons.

b. **Arrange the outer electrons and the atoms in pairs to satisfy the octet rule.** This may require single, double, or triple bonds. Arrange the eight outer electrons in a CH_4 molecule around the central carbon atom in four bonding pairs, each pair connecting the carbon atom to a hydrogen atom. This is the Lewis structure.

$$\begin{array}{ccc} & H & \\ H\!:\!\overset{\displaystyle H}{\underset{\displaystyle H}{\overset{\cdot\cdot}{\underset{\cdot\cdot}{C}}}}\!:\!H & \text{or} & H\!-\!\overset{\displaystyle H}{\underset{\displaystyle H}{C}}\!-\!H \end{array}$$

Although this drawing seems to imply that the CH_4 molecule is flat or planar, this is only because we are restricted to the two dimensions of a sheet of paper. In fact, this is not the case as we will see in the next step.

c. **Assume that the most stable molecular shape has the bonding electron groups attached to any atom as far from one another as possible.** (*Note:* In other molecules we will need to consider nonbonding electrons as well, but there are none in CH_4.)

The four electron pairs around the carbon atom in CH_4 repel one another, and in their most stable arrangement they are as far from one another as possible yet still form C-to-H bonds. Furthermore, because a hydrogen atom is attached to each pair of electrons, the four hydrogen atoms also are as far from one another as possible. One way to describe the shape of a CH_4 molecule is by analogy to the base of a folding music stand. The four C-to-H bonds correspond to the three evenly spaced legs and the vertical shaft of the stand (Figure 3.7). The angle between each pair of bonds is 109.5°. This shape is tetrahedral, because the hydrogen atoms correspond to the corners of a **tetrahedron,** a four-cornered figure with four equal triangular sides. The tetrahedral shape has been experimentally confirmed. Indeed, it is one of the most common atomic arrangements in nature, particularly in carbon-containing molecules.

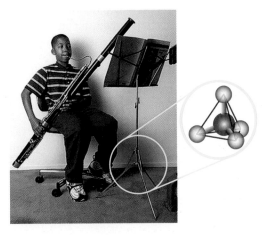

Figure 3.7
The legs and the shaft of this music stand approximate the arrangement of the bonds in a tetrahedral molecule like methane.

Consider This 3.11 Flat or Tetrahedral Methane?

a. If the methane molecule were really two-dimensional as the Lewis structure representation seems to indicate, what would the H-to-C-to-H bond angle be?

b. Offer a reason why the tetrahedral shape, not the two-dimensional flat shape, is more advantageous for this molecule.

c. Consider the music stand shown in Figure 3.7. In the analogy of shape using a music stand, where would the carbon atom be located? Where would each of the hydrogen atoms be?

Answer

a. 90° (and 180° for H atoms across from one another)

Chemists represent molecules in several different ways. The simplest, of course, is the formula itself. In the case of methane, that is simply CH_4. We know that Lewis structures, without further interpretation, provide information on bonding but only two-dimensional information for most molecules. Other representations in Figure 3.8 show some generally accepted methods used by chemists to convey the three-dimensional structure of methane. For example, Figure 3.8a shows a wedge-shaped line. This represents a bond coming out of the paper at an angle generally toward the reader, the dashed wedge represents a bond pointing away from the reader, and the solid lines are assumed to be in the plane of the paper. This is an improvement over the two-dimensional structure, but a better way to visualize molecules is with a molecular modeling program, as in Figures 3.8b,c. You will have a chance to see the results of a modeling program in Consider This 3.14. Seeing and manipulating physical models, either in the classroom or laboratory, can also help you visualize the structure of molecules.

Space-filling models and computer-generated charge-density models both enclose the volume occupied by electrons in a molecule. The charge-density model displays an internal ball-and-stick model to show the location of nuclei. The colors in the charge-density model will help you visualize how the electrons are arrayed within the molecule. Overall, the molecule is neutral. Within the molecule, red hues indicate regions of higher electron density. At the other end of the spectrum, blue hues represent lower electron density. The intensity of the colors reflects how greatly the electrons are pulled from one region of the molecule to another. We will return to this point when discussing the concepts of electronegativity and polarity in Chapter 5.

We can apply the same steps with CCl_3F, a CFC that is a greenhouse gas. Using step 1, we determine that there are a total of 32 outer electrons: 4 for carbon (Group 4A), 7 for fluorine (Group 7A), and 7 for *each* chlorine atom (Group 7A). Applying step 2 reveals that carbon is the central atom and the other four atoms are bonded to it by four single covalent bonds (8 electrons; four shared electron pairs). This satisfies the octet rule for carbon. The remaining 24 electrons serve as nonbonding (unshared or lone) electron pairs on the other bonded atoms, thus achieving an octet around each one (Figure 3.9). Nonbonding electron pairs are shown only in the first two Lewis

Computer-generated charge-density models use a continuum of color. The two extremes are:
• red = higher electron density
• blue = lower electron density

(a)   (b)   (c)

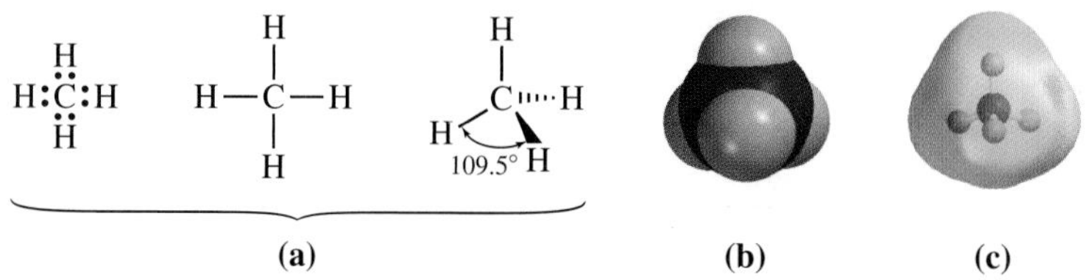

Figure 3.8
Representations of CH_4.
(a) Lewis structure and structural formulas; (b) Space-filling model; (c) Charge-density model.

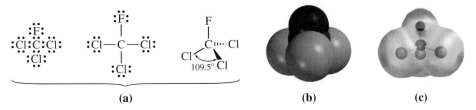

Figure 3.9

Representations of CCl₃F.

(a) Lewis structure and structural formulas; (b) Space-filling model; (c) Charge-density model.

structures. Nonbonding pairs do not usually appear in the 3-D representation showing the bond angle, and they are not explicitly shown in either the space-filling model or the charge-density model. These last two models give better indications of the larger size of the volume occupied by the electrons associated with Cl atoms, relative to either the C or F atoms.

Like the four electron pairs around carbon in CH_4, the four shared electron pairs in CCl_3F also repel one another and are as far apart from one another as possible. The resulting molecular shape is also tetrahedral, with the fluorine and chlorine atoms at the corners of a tetrahedron and carbon in the middle. The F atom is smaller than the Cl atoms, allowing the Cl atoms to spread just a bit beyond the predicted Cl-to-C-to-Cl bond angle of 109.5°. Still, the experimental bond angle is very close to the predicted angle for Cl-to-C-to-Cl, illustrating the usefulness of this method.

In some molecules, the central atom is surrounded by an electron octet, but not all of the electrons are bonding electron pairs. Some are nonbonding (lone) pairs, as in ammonia, NH_3, a refrigerant gas that was largely replaced by CFCs (Figure 3.10).

A nonbonding pair occupies greater space than a bonding pair of electrons. Consequently, the nonbonding pair repels the bonding pairs somewhat more strongly than the bonding pairs repel one another. This stronger repulsion forces the bonding pairs closer to one another, creating an H-to-N-to-H angle slightly less than the predicted 109.5° associated with a regular tetrahedron. The experimental value of 107.3° is close to the tetrahedral angle, again indicating that our model is reasonably reliable.

The shape of a molecule is described in terms of its arrangement of atoms, not electrons. The hydrogen atoms of NH_3 form a triangle with the nitrogen atom above them at the top of the pyramid. Thus, ammonia is said to be a triangular pyramid; it has a triangular pyramidal structure. Going back to the analogy of the folding music stand (Figure 3.7), you could expect to find hydrogen atoms at the tip of each leg of the music stand. This places the nitrogen atom at the intersection of the legs with the shaft, with the nonbonded electron pair forming around the shaft of the stand.

Water, another naturally occurring greenhouse gas, illustrates yet another type of molecular shape. There are eight outer electrons: one from each of the two hydrogen atoms plus six from the oxygen (Group 6A). Its Lewis structure discloses how the eight electrons on the central oxygen atom are distributed: two pairs involved in bonding and two lone pairs (Figure 3.11).

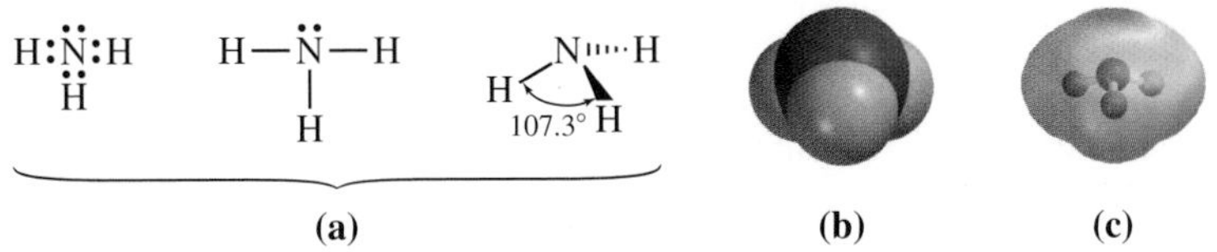

The major use of NH_3 today is for agriculture. The gas is often directly infused into the soil to add nitrogen needed for plant growth.

Figure 3.10

Representations of NH_3.

(a) Lewis structure and structural formulas; (b) Space-filling model; (c) Charge-density model.

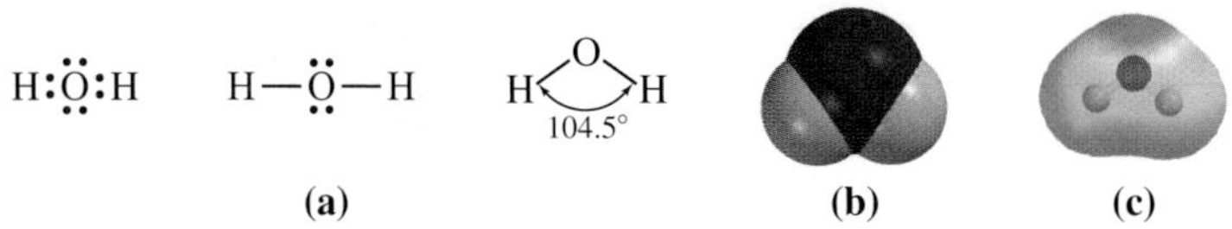

H:Ö:H   H—Ö—H   H⟨O⟩H
                              104.5°

(a)   (b)   (c)

Figure 3.11

Representations of H_2O.

(a) Lewis structure and structural formulas; (b) Space-filling model; (c) Charge-density model.

If these four pairs of electrons are arranged so that they are as far apart as possible, the distribution will be similar to that in methane, and we might predict water to have a tetrahedral shape, with a 109.5° H-to-O-to-H bond angle. Unlike the four bonding pairs in methane, water has two bonding pairs and two nonbonding pairs. The repulsion between the two nonbonding pairs and, in turn, their repulsion of the bonding pairs cause the bond angle to be less than 109.5°. Experiments indicate a value of approximately 104.5°. A water molecule is said to have a *bent* shape.

Your Turn 3.12   Predicting Molecular Shapes, Part 1

Using the strategies just described, predict and sketch the shape of each of these molecules.

a. CCl_4 (carbon tetrachloride)
b. CCl_2F_2 (Freon-12; dichlorodifluoromethane)
c. H_2S (hydrogen sulfide)

Answer

a.   The total number of outer electrons is $4 + 4(7) = 32$. The central atom is C and there will be four single bonds, one to each Cl. The bonding electron pairs and the attached chlorine atoms arrange themselves so that their separation is maximized. There are no nonbonding pairs on the central carbon atom. It follows that the shape of a carbon tetrachloride molecule is tetrahedral, the same as a methane molecule.

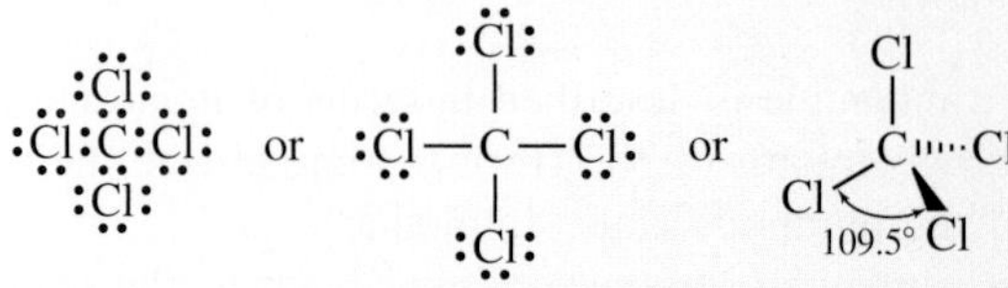

:Cl:
:Cl:C:Cl: or   :Cl—C—Cl: or   Cl⟨C····Cl⟩
:Cl: | 109.5° Cl
                     :Cl:

CH$_4$, H$_2$O, and CFCs are all greenhouse gases.

We have already looked at the structures for several gases important for understanding the chemistry of global warming. What about the structure of that most significant greenhouse gas, carbon dioxide? A count of outer electrons gives a total of 16. The carbon atom contributes 4 electrons and 6 come from each of the two oxygen atoms, for a total of 16 electrons. If only single bonds were involved, there would not be enough electrons to provide 8 electrons for each atom. That would require 20 electrons. However, the octet rule would be obeyed if the central carbon atom shares 4 electrons (a double bond) with each of the oxygen atoms. Thus, two double bonds are formed.

In the CO_2 molecule, two groups of four electrons each are associated with the central atom. These groups of electrons repel one another, and the most stable configuration provides the furthest separation of the negative charges. This occurs when the angle between them is 180° and the molecule is linear. The model predicts that all three atoms in a CO_2 molecule will be in a straight line. This is, in fact, the case as shown in Figure 3.12.

We applied the idea of electron pair repulsion to molecules in which there are four groups of electrons (CH_4, CCl_3F, NH_3, and H_2O) and two groups of electrons (CO_2). Electron pair repulsion also applies reasonably well to molecules that include

(a)   **(b)**   **(c)**

Figure 3.12
Representations of CO_2.
(a) Lewis structure and structural formulas; **(b)** Space-filling model; **(c)** Charge-density model.

three, five, or six groups of electrons. In most molecules, the electrons and atoms are still arranged to keep the separation of the electrons at a maximum. This logic accounts for the bent shape we associated with the ozone molecule in Chapter 2. Remember that according to the octet rule, the O_3 molecule (18 total outer electrons) contains a single bond and a double bond. Remember also that the central oxygen atom carries a nonbonding lone pair of electrons. Thus, there are three groups of electrons on this central atom: the pair that makes up the single bond, the two pairs that constitute the double bond, and the lone pair. These three groups of negatively charged electrons repel one another, and the minimum energy of the molecule corresponds to the furthest separation of these electron groups. This will occur when the electron groups are all in the same plane and at an angle of about 120° from one another. We predict, therefore, that the O_3 molecule should be bent, and the angle made by the three atoms should be approximately 120°. Experiments show the O-to-O-to-O bond angle to be 117°, just slightly smaller than the prediction. This is reasonable if you remember that the nonbonding electron pair on the central oxygen atom occupies greater space than bonding pairs of electrons. The greater repulsion force caused by the nonbonded electron pair causes the slightly smaller bond angle (Figure 3.13).

Your Turn 3.13   **Predicting Molecular Shapes, Part 2**

Using the strategies just described, predict and sketch the shapes of these molecules.

a. SO_2 (sulfur dioxide)   **b.** SO_3 (sulfur trioxide)

Hint (Part a): Note that S and O are in the same group on the periodic table. This means that the structures for SO_2 and O_3 will be closely related.

 Consider This 3.14   **Molecules in Motion**

Three-dimensional representations of molecules can be viewed on the Web with the aid of CHIME, a free plug-in that you can download and install. Following the links on the *Online Learning Center,* use CHIME to view the molecules considered in this section. Has your mental picture of these molecules changed after seeing the 3-D representations? Explain.

(a)   **(b)**   **(c)**

Figure 3.13
Representations of O_3.
(a) Lewis structure and structural formulas; **(b)** Space-filling model; **(c)** Charge-density model.

3.4 Vibrating Molecules and the Greenhouse Effect

Now that we know the molecular shapes of some important greenhouse gases, we can turn to the important phenomenon of how these molecules interact with infrared radiation. You learned in Chapter 2 that if the photon is part of the UV region of the spectrum, it has sufficient energy to disrupt the arrangement of electrons within the molecule. This can cause covalent bonds to break, as in the dissociation of O_2 and O_3 by UV-B and UV-C radiation. Fortunately, this is not the case with photons in the IR range.

Photons in the IR range of the spectrum are not sufficiently energetic to break bonds. However, a photon of IR radiation can enhance the vibrations in a molecule. The covalent bonds holding atoms together can be thought of as springs, and the atoms can move back and forth. Depending on the molecular structure, only certain vibrations are possible, and each of these vibrations has a characteristic set of permissible energy levels. The energy of the photon must correspond exactly to the vibration energy of the molecule for the photon to be absorbed. This means that different molecules absorb IR radiation at different wavelengths and thus vibrate at different energies.

We illustrate these ideas with the carbon dioxide molecule, representing the atoms as balls and the bonds as springs. A molecule of CO_2 can vibrate in the four ways pictured in Figure 3.14. The arrows indicate the direction of motion of each atom when the molecule is vibrating. The atoms can move forward and backward along the arrows. Vibrations **a** and **b** are called stretching vibrations. In vibration **a,** the central carbon atom is stationary and the oxygen atoms move back and forth (stretch) in opposite directions away from the central atom. Alternatively, the oxygen atoms can move in the same direction and the carbon atom in the opposite direction (vibration **b**). Vibrations **c** and **d** look very much alike. In both cases, the molecule bends from its normal linear shape. The bending counts as two vibrations because it can occur in either of two possible planes. Vibration **c** is shown bending in an *xy*-plane, up and down on the plane of the paper on which the diagram is printed. Vibration **d** is moving in an *xz*-plane, in front and in back of the plane of the paper.

In any molecule, the amount of energy required to cause vibration depends on the nature of the motion, the "stiffness" and strength of the bonds, and the masses of the atoms that move. If you have ever examined a spring, you have probably observed that more energy is required to stretch a spring than to bend it. Similarly, more energy is required to stretch a CO_2 molecule than to bend it. This means that more energetic photons, those with shorter wavelengths, are needed to excite stretching vibrations **a** or **b** than to excite bending vibrations **c** or **d**. The two bending motions (**c** and **d**) are both stimulated when the molecule absorbs IR radiation with a wavelength of 15.00 micrometers (μm). Stretching vibration **b** will occur only if radiation of wavelength of 4.26 μm is absorbed. Together, vibrations **b, c,** and **d** account for the greenhouse properties of carbon dioxide.

It turns out that stretching vibration **a** cannot be triggered by the direct absorption of IR radiation. In a CO_2 molecule, the average concentration of electrons is greater on the oxygen atoms than on the carbon atom. This means that the oxygen atoms carry a partial negative charge relative to the carbon atom. As the bonds stretch,

A micrometer is equal to one-millionth of a meter: $1 \ \mu m = 1 \times 10^{-6}$ m.

The property of electronegativity, a measure of an atom's ability to attract bonded electrons, will be discussed in Section 5.5.

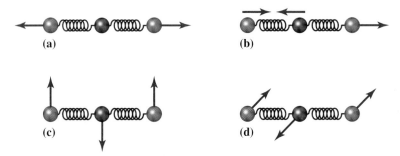

Figure 3.14

Molecular vibrations in CO_2. Each spring represents a C-to-O double bond. Vibrations **a** and **b** are stretching vibrations; **c** and **d** are bending vibrations.

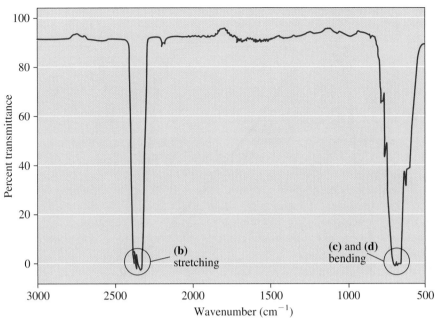

Figure 3.15
Infrared spectrum of carbon dioxide.

the distribution of negatively charged electrons changes. Because of CO_2's linear shape and symmetry, the changes in charge distribution during vibration **a** cancel each other and no infrared absorption occurs.

The infrared (heat) energies absorbed or transmitted by molecules can be measured with an instrument called an infrared spectrometer. Heat radiation from a glowing filament is passed through a sample of the compound to be studied, in this case gaseous carbon dioxide. A detector measures the amount of radiation, at various frequencies, transmitted by the sample. High transmission means low absorbance, and vice versa. This information is displayed graphically, where the relative intensity of the transmitted radiation is plotted versus wavelength. The result is called the *infrared spectrum* of the compound. Figure 3.15 shows the infrared spectrum of CO_2.

Understanding this spectrum and others like it will take just a bit more explanation. The units of the *y*-axis are percent transmittance, as described earlier. The *x*-axis values are expressed in the unit called **wavenumber,** with dimensions of cm^{-1}. Although it would be more comfortable to retain the more familiar units of wavelength, most IR spectra are reported with units of wavenumber. Fortunately, a simple relationship relates wavelength in micrometers to the wavenumber, expressed in cm^{-1}.

$$\text{wavenumber } (cm^{-1}) = \frac{10{,}000}{\text{wavelength } (\mu m)} \qquad [3.2]$$

> Spectroscopy is the field of study that examines matter by passing electromagnetic energy through a sample. Wavelengths absorbed or transmitted are changed to electrical signals in the detector. The visual display of those signals is called a spectrum.

Your Turn 3.15   Relating Wavelength to Wavenumber

Use the relationship given in Equation 3.2 to convert each of these wavelengths to wavenumbers.

a. 4.26 μm **b.** 15.00 μm
c. How do the two values just calculated compare with the locations of lowest transmittance (greatest absorbance) in Figure 3.15?

Answer

a. $\dfrac{10{,}000}{4.26\ \mu m} = 2350\ cm^{-1}$ **b.** $\dfrac{10{,}000}{15.00\ \mu m} = 667\ cm^{-1}$

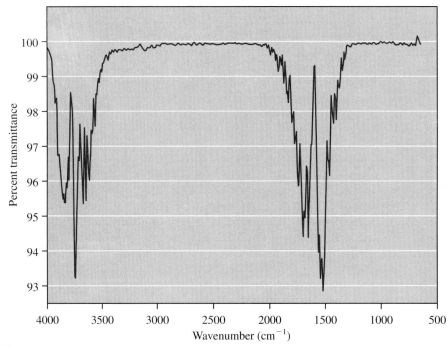

Figure 3.16

Infrared spectrum of water vapor.

The infrared spectrum shown in Figure 3.15 is determined in the laboratory, but the same phenomenon takes place in the atmosphere. CO_2 molecules absorb specific wavelengths of infrared energy, vibrate for a while, and then reemit the energy as heat and return to their normal unexcited, or "ground," state. This is how carbon dioxide captures and returns the infrared radiation coming from Earth's surface, preventing our planet from becoming too cold. This is what makes carbon dioxide a greenhouse gas.

Any molecule that can vibrate in response to the absorption of specific photons of infrared radiation can behave as a greenhouse gas. There are many such substances. Carbon dioxide and water are the most important in maintaining Earth's temperature. Figure 3.16 shows the IR spectrum of H_2O molecules absorbing infrared radiation. However, methane, nitrous oxide, ozone, and chlorofluorocarbons (such as CCl_3F) are among the other substances that help retain planetary heat. Diatomic N_2 and O_2 are not greenhouse gases. Although molecules consisting of two identical atoms do vibrate, the overall electric charge distribution does not change during these vibrations. Hence, these molecules do not absorb infrared radiation.

Nitrous oxide, N_2O, is also called dinitrogen monoxide. You will encounter this gas again in Chapter 6.

Consider This 3.16   Bending and Stretching Water Molecules

a. Use Figure 3.16 to estimate the wavenumber (cm^{-1}) for water's two maximum absorbancies of IR energy.

b. Change each value to wavelength (μm).

c. Which wavelength do you predict represents bending vibrations and which represents stretching? Explain the basis of your predictions.

Hint: Compare the IR spectrum of CO_2 with that for H_2O.

You have encountered two ways that molecules respond to radiation. Highly energetic photons with high frequencies and short wavelengths (such as UV radiation) can break up molecules. The less energetic photons of infrared light cause many molecules to vibrate. Both processes are depicted in Figure 3.17, but the figure also includes another response of molecules to radiant energy that is probably a good deal more

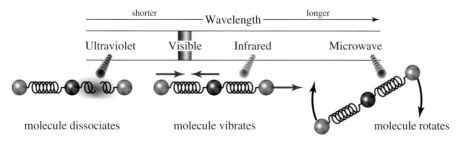

Figure 3.17

Molecular response to types of radiation.

familiar to you. Longer wavelengths than those in the IR range only have enough energy to cause molecules to rotate or spin, not vibrate or dissociate.

Microwave ovens generate radiation that causes water molecules to spin. The radiation generated in such a device is of relatively long wavelength, about a centimeter. Thus the energy per photon is quite low. As the H_2O molecules absorb the photons and spin more rapidly, the resulting friction cooks your food, warms up the leftovers, or heats your coffee. The same region of the spectrum is used for radar. Beams of microwave radiation are sent out from a generator. When the beams strike an object such as an airplane, the microwaves bounce back and are detected by a sensor.

The practical consequences of the interaction of radiation and matter are immense, but there is another application of great significance in our understanding of nature. Spectroscopy provides a means of studying atomic and molecular structure. Electronic, vibrational, and rotational energy are all **quantized,** meaning only certain energy levels are permitted. No matter what region of the spectrum is employed, spectroscopy reveals differences between energy levels. Using the appropriate mathematical model, scientists can translate these energy differences into information about bond lengths, bond strengths, and bond angles. A consequence of looking through a spectroscopic window into atoms and molecules is that chemists can describe the microscopic world with great confidence.

3.5 The Carbon Cycle: Contributions from Nature and Humans

In his book *The Periodic Table,* the late chemist, author, and World War II concentration camp survivor Primo Levi, wrote eloquently about carbon dioxide.

> *"This gas which constitutes the raw material of life, the permanent store upon which all that grows draws, and the ultimate destiny of all flesh, is not one of the principal components of air but rather a ridiculous remnant, an 'impurity' thirty times less abundant than argon, which nobody even notices. . . . [F]rom this ever renewed impurity of the air we come, we animals and we plants, and we the human species, with our four billion discordant opinions, our millenniums of history, our wars and shames, nobility and pride."*

The Periodic Table was written in 1975. About 6 billion people now inhabit Earth.

In the essay from which this quotation is taken, Levi traces a brief portion of the life history of a carbon atom from a piece of limestone (calcium carbonate, $CaCO_3$), where it lies "congealed in an eternal present," to a CO_2 molecule, to a molecule of glucose in a leaf, and ultimately to the brain of the author. And yet that is not the final destination. "The death of atoms, unlike our own," writes Levi, "is never irrevocable." That carbon atom, already billions of years old, will continue to persist into the unimagined future.

This marvelous continuity of matter, a consequence of its conservation, is beautifully illustrated by the carbon cycle. Even without being described with Primo Levi's poetic gifts, the story is fascinating and one that is important to understand to comprehend the danger of the cycle's being changed by human activities. It is certain that without the proper functioning of the carbon cycle, every aspect of life on Earth could undergo dramatic change. Figure 3.18 is one representation of this important cycle.

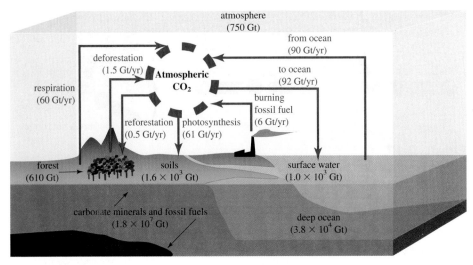

Figure 3.18

The global carbon cycle. The numbers show the quantity of carbon, expressed in gigatons (Gt), that is moving through the system per year.

Source: From Purves, Orians, Heller & Sadava, *Life: The Science of Biology,* 5th Edition, 1998. Reprinted with permission of Sinauer Associates, Inc.

A gigaton (Gt) is a billion metric tons, or about 2200 billion pounds. For comparison, a fully loaded 747 jet airplane weighs about 800,000 lb. It would take nearly 3 million 747s to have a total mass of 1 Gt.

Consider This 3.17   Understanding the Carbon Cycle

a.   What processes add carbon (in the form of CO_2) to the atmosphere?
b.   Compare the processes of deforestation and combustion of fossil fuels. Which adds more carbon to the atmosphere?
c.   How is carbon removed from the atmosphere?
d.   What are the largest reservoirs of carbon?
e.   Which parts of the carbon cycle can be changed by human activities?

The carbon cycle is a dynamic system. All processes illustrated are happening simultaneously, but at different rates. Plants die and decay, releasing CO_2. Other plants enter the food chain where their complex molecules are broken down into CO_2, H_2O, and other simple substances. Animals exhale CO_2, carbonate rocks decompose, and carbon dioxide escapes through the vents of volcanoes. And the cycle goes on and on. Michael B. McElroy of Harvard University estimated, "The average carbon atom has made the cycle from sediments through the more mobile compartments of the Earth back to sediments, some 20 times over the course of Earth's history." Carbon dioxide in the air today may have come from campfires more than a thousand years ago.

The relative amount of carbon found in each location is part of the overall story of the carbon cycle. Table 3.1 highlights some important data to consider when evaluating the possibility of climate change from an enhanced greenhouse effect.

Your Turn 3.18   Comparing Carbon in Reservoirs

a.   What percent of the total fossil fuel carbon reservoir is from coal?
b.   What percent of Earth's total carbon reservoir is in fossil fuels?
c.   What percent of the total carbon reservoir, excluding fossil fuels, is found in the oceans?

As members of the animal kingdom, we *Homo sapiens* participate in the carbon cycle along with our fellow creatures. But we do more than our share. As is true for

Table 3.1	The Earth's Carbon Reservoirs	
		Size (Gt Carbon)
Reservoir		
Atmosphere		750
Forests		610
Soils		1,580
Surface ocean		1,020
Deep ocean		38,100
Total carbon, excluding fossil fuels		42,060
Fossil fuels		
Coal		4,000
Oil		500
Natural gas		500
Total fossil fuel		5,000
Total, all sources		47,060

Source: From James F. Kasting, "The Carbon Cycle, Climate, and the Long-Term Effects of Fossil Fuel Burning," *Consequences, The Nature & Implications of Environmental Change,* Vol. 4, No. 1, 1998. Reprinted with permission.

any animal, we inhale and exhale, ingest and excrete, live and die. But we also have developed processes that permit us to significantly perturb the system. The Industrial Revolution, which began in Europe in the late 18th century, was fueled largely by coal. Coal powered steam engines in mines, factories, locomotives, ships, and later, electrical generators. The subsequent discovery and exploitation of vast deposits of petroleum made possible the development of automobiles and other types of transportation. To a very considerable extent, the Industrial Revolution was a revolution in energy sources and energy transfer.

The result is a large-scale transfer of carbon stored in fossil fuels to carbon dioxide released into the atmosphere. Every year, human activity releases more carbon dioxide to the atmosphere from burning fossil fuels than is removed from the atmosphere by natural cycles. Processes that remove CO_2 are not always as rapid as needed to take up the extra CO_2, so the amounts of atmospheric CO_2 increase. Table 3.2 shows the sources

Table 3.2	Human Perturbations to the Global Carbon Budget	
		Flux (Gt carbon/year)
CO_2 sources		
Fossil fuel combustion and cement production		5.5 ± 0.5
Tropical deforestation		1.6 ± 1.0
Total anthropogenic emissions		7.1 ± 1.1
CO_2 sinks		
Storage in the atmosphere		3.3 ± 0.2
Uptake by the ocean		2.0 ± 0.8
Northern Hemisphere forest regrowth		0.5 ± 0.5
Other terrestrial sinks (CO_2 fertilization, nitrogen fertilization, climatic effects)		1.3 ± 1.5
Total sinks for CO_2		7.1 ± 1.1

Source: From James F. Kasting, "The Carbon Cycle, Climate, and the Long-Term Effects of Fossil Fuel Burning," *Consequences, The Nature & Implications of Environmental Change,* Vol. 4, No. 1, 1998. Reprinted with permission.

and sinks of CO_2 from anthropogenic origin, that is, sources arising from human activities. Table 3.2 presents the optimistic picture that atmospheric, oceanic, and terrestrial sinks will match the sources. Note, however, that nearly half of CO_2 from human-based sources remains in the atmosphere, leading to the observed rise in CO_2.

Consider This 3.19   Uncertainty—A Part of the Problem

Data in Table 3.2 are reported with a plus and minus value that indicates the level of uncertainty in the information. Uptake of CO_2 by the oceans is one of the most difficult sink values to determine accurately.

a.   Offer some possible reasons why the value for uptake of CO_2 by oceans has a high uncertainty.

b.   What would be the effect on global warming if the value for uptake by the oceans were too low? Too high?

c.   Which is the most uncertain anthropogenic source of CO_2? Offer some possible reasons.

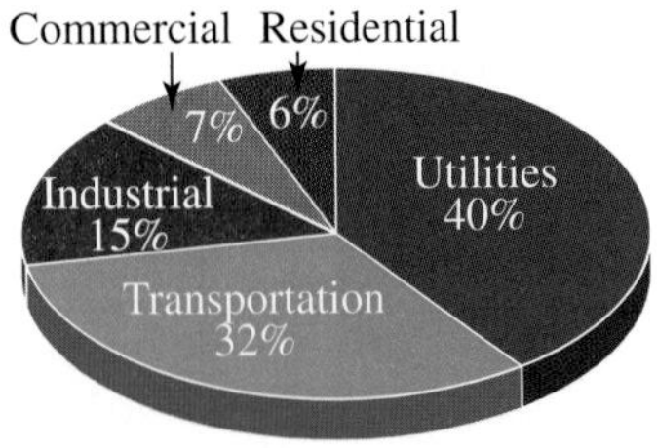

Figure 3.19

Carbon dioxide emission sources from fossil fuel consumption in the United States, 2000.

Source: EIA, *Emission of Greenhouse Gases in the U.S. 2000.*

As the generation of energy and the consumption of fossil fuels has increased, so has the quantity of combustion products such as carbon dioxide released to the atmosphere. Since 1860, the CO_2 concentration has increased from 290 ppm to 375 ppm and the current rate of increase is about 1.5 ppm per year (see Figure 3.5). At the present level, fossil fuels containing more than 5 Gt of carbon are burned annually for a variety of purposes. Most fossil fuel-based CO_2 comes from power utilities (40%) and transportation (32%); much less, for example, comes from home and commercial heating (Figure 3.19).

Consider This 3.20   The CO_2 Emissions—Implications for Policy

Figure 3.19 gives the sources of CO_2 emissions from fossil fuel consumption in the United States for 2000. These values have implications for personal action and for setting control policies.

a.   As an individual, which sources of CO_2 can you control? Explain your reasoning.

b.   Do you think that national priorities for controlling CO_2 emissions are set based on the rank order of percentages in this figure? Why or why not? Explain your reasoning.

Power utilities include coal-burning power plants.

The U.S. Energy Information Administration (EIA) is required by the Energy Policy Act of 1992 to prepare a yearly updated report on greenhouse gas emissions. As part of their continuing review of information and methodology, they began in 2001 to use a new method of reporting emission data. Emissions are no longer reported for power utilities as a separate category. EIA explains the change this way. "Energy-related carbon dioxide emissions now have been revised as part of an agency-wide adjustment to energy consumption data and sectoral allocation." In practical terms, this means that all emissions from utilities have been reassigned based on the end use of the energy. All data since 1990 has been recalculated using this new approach and is shown in Figure 3.20.

The other human perturbation reported in Table 3.2 is deforestation by burning, a practice that releases 0.6–2.6 Gt of carbon to the atmosphere each year. It is estimated that forested land the size of two football fields is lost every second of the day from the rain forests of the world. Although firm numbers are sometimes rather elusive, Brazil continues as the country with the greatest loss of rainforest acreage. In Brazil alone, over 5.4 million acres of Amazon rainforest is vanishing each year. Trees, those very efficient absorbers of carbon dioxide, are removed from the cycle through deforestation.

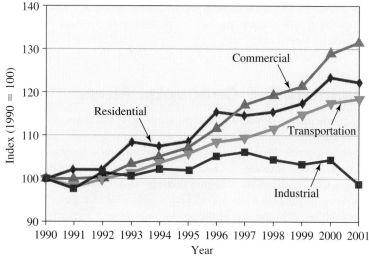

Figure 3.20

U.S. carbon dioxide emissions by sector, 1990–2001.

Source: EIA, *Emission of Greenhouse Gases in the U.S. 2001.*

Consider This 3.21   **Clarity or Confusion?**

Figure 3.20 gives the sources of CO_2 emissions from fossil fuel consumption in the U.S. from 1990 to 2001. The reporting system has changed from that used in Figure 3.19 for 2000.

a. Emissions in Figure 3.20 are all indexed to 1990. There were 1851.8 million metric tons of U.S. carbon dioxide emissions in 1990. Use data in Figure 3.20 to estimate emissions in million metric tons for each reported sector for the year 2000.

b. Use the values estimated in part **a** to calculate the percentage CO_2 emission contributed by each sector in 2000.

c. Compare the percentages calculated in part **b** with those shown in Figure 3.19 for the year 2000. Which sectors changed the most?

d. Do you think the new reporting system will ultimately lead to clarity in reporting emissions data or remain a source of confusion? Explain your reasoning.

> **1 metric ton = 10^3 kg**
> **= 2200 lb**

If the wood is burned, vast quantities of CO_2 are generated; if it is left to decay, that process also releases carbon dioxide, but more slowly. Even if the lumber is harvested for construction purposes and the land is replanted in cultivated crops, the loss in CO_2-absorbing capacity may approach 80%.

Systematic deforestation is not a new phenomenon, nor is it limited to tropical forests. In the course of history, the focus in the 20th century shifted from the heavily deforested regions of Europe and North America to the tropical rainforests of Central and South America, Africa, and Asia. There are actually more trees in the United States now than there were in colonial times and in the later 1800s, although the same cannot be said for European countries.

The total quantity of carbon released by the human activities of deforestation and burning fossil fuels is 6.0–8.2 Gt per year. About half of this is recycled into the oceans and the biosphere. The remainder stays in the atmosphere as CO_2, adding between 3.1 and 3.5 Gt of carbon per year to the existing base of 750 Gt noted in Table 3.1. We are concerned primarily with this *increase* in atmospheric carbon dioxide, because the *excess* carbon dioxide is implicated in global warming. Therefore, it would be useful to know the mass (Gt) of CO_2 added to the atmosphere each year. In other words, what

> Remember that the natural "greenhouse effect" makes life on Earth possible. Problems occur when the amount of greenhouse gases *increases* faster than the sinks can accommodate this increase. The result is the *enhanced* greenhouse effect.

mass of CO_2 contains 3.3 Gt of carbon, the midpoint between 3.1 Gt and 3.5 Gt? To answer this question will require another scenic tour into the land of chemistry, one you will find useful later in this text.

3.6 Quantitative Concepts: Mass

To solve the problem just posed, we need to know how the mass of carbon is related to the mass of carbon dioxide. The approach we will use is to calculate the mass percent of carbon in carbon dioxide, based on the formula of the compound. Regardless of the source of carbon dioxide, its formula is stubbornly the same, CO_2. The mass percent of C in CO_2 is therefore also unwavering. As you work through this and the next section, keep in mind that we are seeking a value for that percentage.

The approach requires the use of the atomic masses of the elements involved. But this raises an important question: How much does an individual atom weigh? Recall from Chapter 2 that most of the mass of an atom is attributable to the neutrons and protons in the nucleus. Thus, elements differ in atomic mass because their atoms differ in composition. Rather than use absolute masses of individual atoms, chemists have found it convenient to employ relative atomic masses—in other words, to relate all atomic masses to some convenient standard. The internationally accepted atomic mass standard is carbon-12, the isotope that makes up 98.9% of all carbon atoms. Carbon-12 has a mass number of 12 because each atom has a nucleus consisting of 6 protons and 6 neutrons plus 6 electrons outside the nucleus. The mass of one of these atoms is arbitrarily assigned a value of exactly 12 atomic mass units (amu). We can thus define the **atomic mass** of an element as the average mass of an atom of that element as compared with an atomic mass of exactly 12 amu for carbon-12. Atoms are so very small that an atomic mass unit is an extremely small unit: 1 amu $= 1.66 \times 10^{-24}$ g.

The periodic table in your text shows that the atomic mass of carbon is 12.01, not 12.00. This is not an error; it reflects the fact that carbon exists naturally as three isotopes. Although C-12 predominates, 1.1% of carbon is C-13, with six protons and *seven* neutrons per atom and an isotopic mass very close to 13. In addition, natural carbon contains a trace of C-14, whose nuclei consist of six protons and *eight* neutrons. The tabulated atomic mass value of 12.01 is a weighted average that takes into consideration the masses of the three isotopes of carbon and their natural abundances. This isotopic distribution and this average atomic mass characterize carbon obtained from any chemical source—a graphite ("lead") pencil, a tank of gasoline, a loaf of bread, a lump of limestone, or your body.

> Isotopes were discussed in Section 2.2.

Carbon-14 is a radioactive isotope that plays a key role in determining the origin of the increasing atmospheric carbon dioxide. In all living things, one out of 10^{12} carbon atoms is a C-14 atom. A plant or animal constantly exchanges CO_2 with the environment, and this maintains the C-14 concentration in the organism at a constant level. However, when the organism dies, the biochemical processes that exchange C stop functioning and the carbon-14 is no longer replenished. This means that after the death of the organism, the concentration of C-14 decreases with time because it undergoes

---

Your Turn 3.22   Isotopes of Nitrogen

Nitrogen (N) is an important element in the atmosphere and in biological systems. It has an atomic mass of 14.01 and an atomic number of 7.

a.   What is the number of protons, neutrons, and electrons in a neutral atom of the most common isotope, N-14?

b.   How does the number of protons, neutrons, and electrons in a neutral atom of N-15 compare with that of N-14?

c.   Given the atomic mass value of 14.01, which isotope do you predict has the greatest natural abundance?

radioactive decay to form N-14. Coal and oil are the fossilized remains of plant life that died millions of years ago. Hence, the level of C-14 is extremely low in fossil fuels, and in the carbon dioxide released when fossil fuels burn. Experiments show that the concentration of C-14 in atmospheric CO_2 has recently decreased. This strongly suggests that the origin of the added carbon dioxide is indeed the burning of fossil fuels, a decidedly human activity.

You will learn to write nuclear equations to represent radioactive decay process in Chapter 7.

Having reviewed the meaning of isotopes and atomic mass, we return to the matter at hand—the masses of atoms and particularly the atoms in CO_2. Not surprisingly, it is impossible to weigh a single atom because of its extremely small mass. A typical laboratory balance can detect a minimum mass of 0.1 mg; that corresponds to 5×10^{18} carbon atoms, or 5,000,000,000,000,000,000 carbon atoms. An atomic mass unit is far too small to measure in a conventional chemistry laboratory. Rather, the gram is the chemist's mass unit of choice. Therefore, scientists use exactly 12 g of carbon-12 as the reference for the atomic masses of all the elements. **Atomic mass** (or atomic weight) can therefore be alternatively defined as the mass (in grams) of the same number of atoms that are found in exactly 12 g of carbon-12. This number of atoms is, of course, *very* large. This important chemical number is named after an Italian scientist with the impressive name of Lorenzo Romano Amadeo Carlo Avogadro di Quaregna e di Ceretto (1776–1856). (His friends called him Amadeo.) **Avogadro's number** is the number of atoms in exactly 12 g of C-12. Avogadro's number, if written out, is 602,000,000,000,000,000,000,000. It is more compactly written in scientific notation as 6.02×10^{23}. Remember, this is the number of atoms in 12 g of carbon, no more than a tablespoonful of soot.

Avogadro's number counts a large collection of atoms, much like the term *dozen* counts a collection of eggs. It does not matter if the eggs are large or small, brown or white, "organic" or not. No matter, for if there are twelve eggs, they are still counted as a dozen. A dozen ostrich eggs will have a greater mass than a dozen quail eggs. Figure 3.21 illustrates this point with a half-dozen tennis and a half-dozen golf balls.

Knowledge of Avogadro's number and the atomic mass of any element permit us to calculate the average mass of an individual atom of that element. Thus, the mass of 6.02×10^{23} oxygen atoms is 16.00 g, the atomic mass from the periodic table. To find the average mass of just one oxygen atom, we must divide the mass of the large collection of atoms by the size of the collection. In chemist's terms, this means dividing

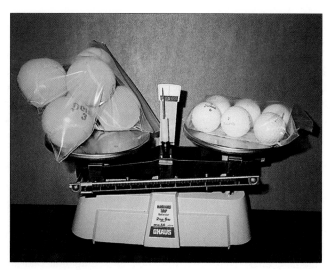

Figure 3.21

Relative masses. Like atoms of different elements, the masses of a tennis ball and a golf ball differ. Six tennis balls have a greater mass than six golf balls. The number of balls is the same in both cases—six in each bag, a half dozen.

Sceptical Chymist 3.23   Marshmallow and Pennies

Avogadro's number is so large that about the only way to hope to comprehend it is through analogies. For example, one Avogadro's number of regular-sized marshmallows, 6.02×10^{23} of them, would cover the surface of the United States to a depth of 650 miles. Or, if you are more impressed by money than marshmallows, assume 6.02×10^{23} pennies were distributed evenly among the more than 6 billion inhabitants of the Earth. Every man, woman, and child could spend $1 million every hour, day and night, and half of the pennies would still be left unspent at death.

Can these fantastic claims be correct? Check one or both of these analogies, showing your reasoning.

the atomic mass by Avogadro's number. Fortunately, calculators help make this job quick and easy.

$$\frac{16.00 \text{ g oxygen}}{6.02 \times 10^{23} \text{oxygen atoms}} = 2.66 \times 10^{-23} \text{ g oxygen / oxygen atom}$$

This very small mass confirms once again why chemists do not generally work with small numbers of atoms. We manipulate trillions at a time. Therefore, practitioners of this art need to measure matter with a sort of chemist's dozen—a very large one, indeed. To learn about it, read on . . . but only after stopping to practice your new skill.

Calculation tip
Predict:
Will the answer be a large number? A small number?
Check:
Does the answer match your prediction? Is it reasonable?

Your Turn 3.24   Calculating Mass of Atoms

a.   Calculate the average mass (in grams) of an individual atom of nitrogen.
b.   Calculate the mass (in grams) of five trillion nitrogen atoms.
c.   Calculate the mass (in grams) of 6×10^{15} nitrogen atoms.

Answer

a. $$\frac{14.01 \text{ g nitrogen}}{6.02 \times 10^{23} \text{ nitrogen atoms}} = 2.33 \times 10^{-23} \text{ g nitrogen / nitrogen atom}$$

3.7   Quantitative Concepts: Molecules and Moles

When used with a number, the unit of measure "mol" is substituted for the term "mole."

Chemists have another way of communicating the number of atoms, molecules, or other small particles present. This is to use the term **mole (mol),** defined as containing an Avogadro's number of objects. The term is derived from the Latin word to "heap" or "pile up." Thus, 1 mol of carbon atoms consists of 6.02×10^{23} C atoms, 1 mol of oxygen gas is made up of 6.02×10^{23} oxygen molecules, and 1 mol of carbon dioxide molecules corresponds to 6.02×10^{23} carbon dioxide molecules.

Moles are fundamental to chemistry because chemistry involves the interaction of individual atoms and molecules. As you already know from Chapter 1, chemical formulas and equations are written in terms of atoms and molecules. For example, reconsider the equation for the reaction of carbon and oxygen.

$$C + O_2 \longrightarrow CO_2 \qquad [3.3]$$

In Chapter 1, we interpreted this expression as stating that one atom of carbon combines with one molecule of diatomic oxygen to yield one molecule of CO_2. The equation reflects the ratio in which the particles interact. Thus, it would be equally correct to say that 10 carbon atoms react with 10 oxygen molecules (20 oxygen atoms) to form 10 carbon dioxide molecules. Or, putting the reaction on a grander scale for that

Table 3.3	Counting Atoms and Molecules	
	$C + O_2 \longrightarrow CO_2$	
# C atoms	**# O_2 molecules**	**# CO_2 molecules**
1	1	1
10	10	10
100	100	100
1000	1000	1000
6.02×10^{23}	6.02×10^{23}	6.02×10^{23}
1 mol	1 mol	1 mol

matter, we could say 6.02×10^{23} C atoms combine with 6.02×10^{23} O_2 molecules (12.0×10^{23} oxygen atoms) to yield 6.02×10^{23} CO_2 molecules. The last statement is equivalent to saying: "one *mole* of carbon plus one *mole* of diatomic oxygen yields one *mole* of carbon dioxide." The point is that *the numbers of atoms and molecules taking part in a reaction are proportional to the numbers of moles of the same substances*. The atomic ratio reflected in a chemical formula or the molecular ratio in a molecular equation is identical to the molar ratio. Accordingly, just as there are two oxygen *atoms* for each carbon *atom* in a CO_2 *molecule*, there are also two *moles* of oxygen atoms for each *mole* of carbon atoms in a *mole* of carbon dioxide. The ratio of two oxygen atoms to one carbon atom remains the same regardless of the number of carbon dioxide molecules, as summarized in Table 3.3.

In the laboratory and the factory, the quantity of matter required for a reaction is usually measured by mass or weight. The mole is a way to simplify matters (and matter) by relating number of particles and mass. Central to this approach is **molar mass,** defined as the mass of one Avogadro's number, or "mole," of whatever particles are specified. In chemistry, molar masses are almost always expressed in grams. Thus, the mass of a mole of carbon atoms, rounded to the nearest tenth of a gram, is 12.0 g. Similarly, a mole of oxygen atoms has a mass of 16.0 g. But we can also speak of a mole of O_2 molecules. Because there are two oxygen atoms in each oxygen molecule, there are two moles of oxygen atoms in each mole of molecular oxygen, O_2. Consequently, the molar mass of O_2 is 32.0 g—twice the molar mass of O. Some books refer to this as the molecular mass or molecular weight of O_2, emphasizing its similarity to atomic mass or atomic weight.

There are 2 mol of oxygen atoms, O, in every mole of oxygen molecules, O_2.

The same logic for the molar mass of O_2 applies to compounds of two or more elements, which brings us, at last, to the composition of carbon dioxide. The formula, CO_2, reveals that each molecule contains one carbon atom and two oxygen atoms. Scaling up by 6.02×10^{23}, we can say that each mole of CO_2 consists of 1 mol of C and 2 mol of O atoms (see Table 3.3). But remember that we are interested in the mass composition of carbon dioxide—the number of grams of carbon per gram of CO_2. This requires the molar mass of carbon dioxide, which we obtain by adding the molar mass of carbon to twice the molar mass of oxygen:

$$\text{Molar mass } CO_2 = (1 \times \text{molar mass C}) + (2 \times \text{molar mass O})$$

Substituting numerical values for the molar masses of the elements gives the desired result.

$$1 \text{ mol } CO_2 = 1 \text{ mol C} + 2 \text{ mol O}$$

$$= \left(1 \text{ mol C} \times \frac{12.0 \text{ g C}}{1 \text{ mol C}} \right) + \left(2 \text{ mol O} \times \frac{16.0 \text{ g O}}{1 \text{ mol O}} \right)$$

$$= 12.0 \text{ g C} + 32.0 \text{ g O}$$

$$1 \text{ mol } CO_2 = 44.0 \text{ g } CO_2$$

This procedure is routinely used in chemical calculations, where molar mass is an important property. Some examples are included in the next activity. In every case, you multiply the number of moles of each element by the corresponding atomic mass in grams, and add the result.

Your Turn 3.25   Molecular Molar Mass

Calculate the molar mass of each of these important atmospheric gases.

a. O_3 (ozone)
b. N_2O (dinitrogen monoxide or nitrous oxide)
c.   Freon-11; trichlorofluoromethane

Answer

a. $\dfrac{3 \text{ mol O}}{1 \text{ mol } O_3} \times \dfrac{16.0 \text{ g O}}{1 \text{ mol O}} = \dfrac{48.0 \text{ g O}}{1 \text{ mol } O_3}$

You may recall that several pages ago we set out to calculate the mass of carbon dioxide that could be produced from burning 3.3 Gt of carbon. We now have all the pieces necessary to solve the problem. Out of every 44.0 g of CO_2, 12.0 g are C. This mass ratio holds for all samples of carbon dioxide, and we can use it to calculate the mass of carbon in any known mass of carbon dioxide. More to the question at hand, we can use it to calculate the mass of carbon dioxide released by any known mass of carbon. It only depends on how we arrange the ratio. The carbon-to-carbon dioxide ratio is $\dfrac{12.0 \text{ g C}}{44.0 \text{ g } CO_2}$, but it is equally true that the carbon dioxide-to-carbon ratio is $\dfrac{44.0 \text{ g } CO_2}{12.0 \text{ g C}}$.

For example, we could compute the number of grams of C in 100.0 g CO_2 by setting up the relationship in this manner.

$$100.0 \text{ g } CO_2 \times \frac{12.0 \text{ g C}}{44.0 \text{ g } CO_2} = 27.3 \text{ g C}$$

Note that carrying along the labels, "g CO_2" and "g C," helps you do the calculation correctly. The label "g CO_2" can be canceled, and you are left with the desired label, "g C." Keeping track of the units and canceling where appropriate are useful strategies in solving many problems. The fact that there are 27.3 g of carbon in 100.0 g of carbon dioxide is equivalent to saying that the mass percent of C in CO_2 is 27.3%.

Calculation tip
Predict:
  Will the answer be larger or smaller than the given value? How will you label the answer?
Check:
  Does the answer match your prediction? Have units cancelled, leaving the label needed for the answer?

Your Turn 3.26   Mass Ratios and Percents

a.   Calculate the mass ratio of sulfur (S) in sulfur dioxide (SO_2).
b.   Find the mass percent of S in SO_2.
c.   Calculate the mass ratio and the mass percent of N in N_2O.

Answers

a.   The mass ratio is found by comparing the molar mass of sulfur to the molar mass of SO_2.

$$\frac{32.1 \text{ g S}}{64.1 \text{ g } SO_2} = \frac{0.501 \text{ g S}}{1.00 \text{ g } SO_2}$$

b.   To find the mass percent of S in SO_2, multiply the mass ratio by 100.

$$\frac{0.501 \text{ g S}}{1.00 \text{ g } SO_2} \times 100 = 50.1\% \text{ S in } SO_2$$

To find the mass of carbon dioxide that contains 3.3 gigatons (Gt) of carbon, we use a similar approach. We could convert 3.3 Gt to grams, but it is not necessary. As long as we use the same units for the mass of C and the mass of CO_2, the same numerical ratio holds. There is one important difference for this problem in how we use the ratio. We are solving for the mass of CO_2, not the mass of carbon. Look carefully at the cancellation of units this time.

The question to be answered is: What mass of CO_2 contains 3.3 Gt of carbon?

$$3.3 \; \cancel{Gt \; C} \times \frac{44.0 \; Gt \; CO_2}{12.0 \; \cancel{Gt \; C}} = 12 \; Gt \; CO_2$$

Once again the labels cancel and the answer comes out with the needed label, Gt CO_2.

Our burning question, "What is the mass of carbon dioxide added to the atmosphere each year from the combustion of fossil fuels?" has finally been answered: 12 gigatons. Of course, our not-so-hidden agenda was to demonstrate the problem-solving power of chemistry and to introduce five of its most important ideas: atomic mass, molecular mass, Avogadro's number, mole, and molar mass. The next few activities provide opportunities to practice your skill with these concepts and manipulations.

Your Turn 3.27   SO$_2$ from Volcanoes

a. It is estimated that volcanoes globally release about 19×10^6 t (19 million metric tons) of SO_2 per year. Calculate the mass of sulfur in this amount of SO_2.
b. If 142×10^6 t of SO_2 is released per year by fossil fuel combustion, calculate the mass of sulfur in this amount of SO_2.

Answer
a. The mass ratio of S to SO_2 is known from Your Turn 3.26.

$$19 \times 10^6 \; \cancel{t \; SO_2} \times \frac{32.1 \times 10^6 \; t \; S}{64.1 \times 10^6 \; \cancel{t \; SO_2}} = 9.5 \times 10^6 \; t \; S$$

Note: There is no need to change the unit of SO_2 from million metric tons to grams or any other unit. The mass ratio expressed in grams to grams is in the same proportion as if expressed in any other comparable unit, as long as the unit is the same.

If you know how to apply these ideas, you have gained the ability to evaluate critically media reports about releases of carbon or carbon dioxide (and other substances as well) and judge their accuracy. One can either take such statements on faith or check their accuracy by applying mathematics to the relevant chemical concepts. Obviously, there is insufficient time to check every assertion, but we hope that readers develop questioning and critical attitudes toward all statements about chemistry and society, even those found in this book.

Sceptical Chymist 3.28   Checking Carbon from Cars

A clean-burning automobile engine will emit about 5 lb of carbon in the form of carbon dioxide for every gallon of gasoline it consumes. The average American car is driven about 12,000 miles per year. Using this information, check the statement that the average American car releases its own weight in carbon into the atmosphere each year. List the assumptions you make in solving this problem. Compare your list, and your answer, with those of your classmates.

3.8 Methane and Other Greenhouse Gases

Concerns about an enhanced greenhouse effect are based primarily, but not solely on increases in atmospheric CO_2. Several other gases are of concern, all of which have been increasing because of human activities. For example, methane, CH_4, is in much lower concentration in the atmosphere than CO_2, but is at least 20 times more effective than CO_2 in its ability to trap infrared energy. Fortunately, CH_4 is quite readily converted to less harmful chemical species by interaction with tropospheric ·OH free radicals. It has a relatively short average atmospheric lifetime of 12 years. This means that most of the CH_4 added to the air in a given year will be gone from the atmosphere 12 years later. Compare that situation with CO_2, a gas with a lifetime ranging from 5 to 200 years. Atmospheric concentration of CH_4 at this time is relatively low, but its current level of 1.8 ppm is estimated to be more than twice that before the Industrial Revolution. Table 3.4 gives a comparison of changes in methane concentration with those of carbon dioxide and nitrous oxide.

Methane comes from a variety of sources. Although the methane cycle is not as well understood as the carbon cycle discussed earlier, about 40% of CH_4 emissions are thought to be from natural sources. Some of the natural sources have been magnified by human activities. For example, because CH_4 is a major component of natural gas, some has always leaked into the atmosphere from rock fissures. But the exploitation of these deposits and the refining of petroleum have led to increased emissions. Similarly, CH_4 has always been released by decaying vegetable matter in wetlands. Its early name, "marsh gas," reflects this origin. Thus, the decaying organic matter in landfills and from the residue of cleared forests generates CH_4. Methane formed in the main New York City landfill is used for residential heating, but at most landfills it simply escapes into the atmosphere.

Another major source of CH_4 is agriculture, particularly cultivated rice paddies. Rice is grown with its roots under water where **anaerobic bacteria,** those that can function without the use of molecular oxygen, produce methane. Most of this methane is released to the atmosphere. Additional agricultural methane comes from an increasing number of cattle and sheep. The digestive systems of these ruminants (animals that chew their cud) contain bacteria that break down cellulose. In the process, methane is formed and released through belching and flatulence—about 500 L per cow per day. The ruminants of the Earth release a staggering 73 million metric tons of methane each year. A similar chemistry is carried on in the guts of termites, making them a major source of methane. And there is more than half a ton of termites for every man, woman, and child on the planet.

There is a possibility that global warming may exacerbate the release of methane from ocean mud, bogs, peat lands, and even the permafrost of northern latitudes. In these areas, a substantial amount of methane appears to be trapped in "cages" made of water molecules. The methane trapped in this way is referred to as methane hydrate. As the temperature increases, the escape of CH_4 becomes more likely. The Commonwealth Scientific and Industrial Research Organization (CSIRO) has been taking a series

Table 3.4	Greenhouse Gases–Concentration Changes and Lifetimes		
	CO_2	CH_4	N_2O
Preindustrial concentration	280 ppm	0.70 ppm	0.28 ppm
2000 concentration	370 ppm	1.8 ppm	0.31 ppm
Rate of concentration change	1.5 ppm/yr	0.010 ppm/yr	0.0008 ppm/yr
Atmospheric lifetime (yr)	5–200*	12	114

*A single value for the atmospheric lifetime of CO_2 is not possible. Different removal mechanisms take place at different rates, leading to variation in atmospheric lifetime.

Ocean Drilling Program

Frozen Methane Hydrate

Figure 3.22

Icy Methane. CSIRO researchers have found evidence to support a theory that an abrupt warming of the Earth 55 million years ago was caused by the sudden release of previously frozen methane from the ocean floor. The Ocean Drilling Program obtained this sample of still frozen methane hydrate from the continental shelf off the coast of Florida.

of ocean core drillings to gather evidence about methane hydrate and its role in global warming. Its findings link periods of historic global warming with the release of methane (Figure 3.22).

The complex details of the generation and fate of atmospheric methane make it difficult to speak with certainty about its future effect on the average temperature of the planet. Current predictions are that methane's effect will be less pronounced than temperature changes caused by CO_2, with methane adding only perhaps a few tenths of a degree to the average temperature of Earth in the next 100 years. This is in sharp contrast with the major effect predicted for CO_2, expected to cause a temperature rise of at least 1.0–3.5 °C by the end of this century.

Nitrous oxide, also known as "laughing gas," has been used as an inhaled anesthetic for dental and medical purposes. Its sources and sinks are not as well established as are those for other greenhouse gases. The majority of N_2O molecules in the atmosphere come from the bacterial removal of nitrates (NO_3^-) from soils, followed by removal of oxygen. Agricultural practices, again linked to population pressures, can speed up the removal of reactive compounds of nitrogen from soils. Other sources include gases from ocean upwelling, and stratospheric interactions of nitrogen compounds with excited oxygen atoms. Major anthropogenic sources of N_2O are automobile catalytic converters, ammonia fertilizers, burning of biomass, and certain industrial processes (nylon and nitric acid production). In the atmosphere, a typical N_2O molecule persists for about 114 years, absorbing and emitting infrared radiation. Over the past decade, atmospheric concentrations of the compound have shown a slow but steady rise.

In addition to its role in the greenhouse effect, nitrous oxide contributes to stratospheric ozone depletion, as discussed in Chapter 2. Near Earth's surface, however, the reactions of nitrogen oxides and hydrocarbons (like methane) lead to the production of ozone, a tropospheric air pollutant. Ozone itself can also act like a greenhouse gas, but its efficiency depends very much on altitude. It appears to have its maximum warming effect in the upper troposphere, around 10 km above the Earth. Depletion of ozone has a slight cooling effect in the stratosphere and it may also promote slight cooling at Earth's surface. Although ozone is a part of the global warming story, depletion of the ozone layer in the stratosphere is clearly not a principal cause of climate change. Ozone depletion and climate change are linked in another important way, and that is through ozone-destroying substances. CFCs, HCFCs, and halons, all implicated in the destruction of stratospheric ozone, also absorb infrared radiation and are greenhouse gases.

Not all greenhouse gases are equally effective in absorbing infrared radiation. This effectiveness is quantified by a **global warming potential (GWP),** a number that

Table 3.5	Global Warming Potential for Three Common Greenhouse Gases		
Substance	Global Warming Potential (GWP)*	Tropospheric Abundance (%)	Tropospheric Abundance (ppm)
CO_2	1 (assigned value)	3.75×10^{-2}	375
CH_4	23	1.8×10^{-4}	1.8
N_2O	296	3.1×10^{-5}	0.31

*GWP values are given for the estimated relative direct and indirect effects over a 100-yr period.

represents the relative contribution of a molecule of the indicated substance to global warming. This number was referred to as the "greenhouse factor" in previous reports, but is now assigned only for greenhouse gases with relatively long lifetimes. Such gases also are fairly evenly distributed throughout the atmosphere. Carbon dioxide is assigned the reference value of 1; all other greenhouse gases are indexed with respect to it. Gases with relatively short lifetimes, such as water vapor, tropospheric ozone, tropospheric aerosols, and other ambient air pollutants, are distributed unevenly around the world. This makes it difficult to quantify their impact and therefore GWP values are not usually assigned. Values of the global warming potential for the three most common long-lived greenhouse gases and their average concentrations in the troposphere are given in Table 3.5.

Your Turn 3.29   Comparing Greenhouse Gas Effectiveness

a. HFC-134a (CF_3CH_2F) has a lifetime of 13.8 years and a GWP value of 1300. Compare its effectiveness as a greenhouse gas to that of CO_2.
b. The average tropospheric abundance of ozone is 4.0×10^{-6}% and its previously assigned greenhouse factor was 2000. Compare the importance of tropospheric ozone as a greenhouse gas to that of methane, CH_4.
c. Freon-12 (CCl_2F_2) was reported (1998 data) to have a tropospheric abundance of 2.6×10^{-8}%. It had a previously assigned greenhouse factor of 25,000. Give some possible reasons why it was not assigned a GWP in the 2001 Report of the Energy Information Administration.

HFCs were discussed in Section 2.12.

Other anthropogenic greenhouse gases that are assigned GWP values include several hydrofluorocarbons (HFCs), two perfluorocarbons (CF_4 and C_2F_6), and sulfur hexafluoride (SF_6). Perfluorocarbons (PFCs) are emitted as a by-product of aluminum smelting and used in the manufacture of semiconductors. Both CF_4 and C_2F_6 have long lifetimes and high GWP values. However, their concentration in the atmosphere is very low at this time, but rising. Sulfur hexafluoride (SF_6), used for electrical insulation in transformers and a cover gas for smelting operations, has a tropospheric lifetime of 3200 years. It is nearly 24,000 times more potent as a greenhouse gas than CO_2, but its atmospheric concentration is very low, 3.5×10^{-10}%. Its GWP is 5700.

3.9 Gathering Evidence: Projecting into the Future

Understanding evidence from the past is important. So is making sense of recent trends. The real challenge, however, lies in understanding the complexities well enough to *predict* climate change. In all computer models, the assumption is that rising concentrations of greenhouse gases will increase the average global temperatures. Rising temperatures, in turn, may produce changes in weather patterns, land use, human health, and alterations in Earth's ecosystems. To accurately model global climate, one must include a number of often incompletely understood astronomical, meteorological, geological, and biological factors. Even the most sophisticated computer program can only succeed if the important factors are identified and weighted in importance. Among these factors are variations in the intensity of the Sun's radiation as a consequence of sunspot activity, winds and air circulation patterns, cloud cover, volcanic activity, dust and soot, aerosols, shifting sea ice and glaciers, the oceans, and the extent and nature of living things, especially human beings. Urban areas create their own "heat islands" further increasing the difficulty of climate modeling on a local rather than global scale. The situation is greatly complicated by the fact that many of these variables are interrelated and cannot be studied independently as in a controlled experiment. Dr. Michael Schlesinger, who directs climate research at the University of Illinois, remarked: "If you were going to pick a planet to model, this is the *last* planet you would choose." Despite all these difficulties, policy decisions must be made on the best possible models. The usefulness and limitations of all models needs to be clearly understood and the level of uncertainty in them quantified. These are tall tasks, to be sure, but essential ones for making informed assessments of climate change.

Important in any model for climate change is the role of the oceans, where over 97% of water on Earth is found. Much of the heat radiated by the greenhouse gases may be going into the oceans, which act as a thermal buffer. Although the oceans are very important in moderating the temperature of the planet, their capacity to do so is limited. We know that increasing the temperature of the oceans will decrease the solubility of CO_2, thus releasing more of it into the atmosphere. You may have witnessed the same effect when a glass of cold sparkling water or soda warms up to room temperature, becoming "flat" as its dissolved gases escape. An increase in the temperature of the oceans may promote the growth of tiny photosynthetic plants called phytoplankton, and hence increase CO_2 absorption. But the result could be just the opposite. Water in a warmer ocean will not circulate as well as it does now, which may inhibit plankton growth and CO_2 fixing. Processes in the ocean vary considerably by depth, with the deep ocean being the largest global carbon reservoir. Turnover time for carbon in the deep ocean is in the range of 2000–5000 years, but 0.1–1 year for marine biomass. These and other rate factors must make their way into the computer model.

Dramatic natural processes can help us better understand factors affecting climate change. Such was the case in 1991 when Mount Pinatubo erupted in the Philippines, throwing vast amounts of debris into the atmosphere (Figure 3.23). The eruption was so massive that aerosols reached even into the stratosphere, partially blocking sunlight from reaching the Earth. The average temperature of the Earth dropped for two years before starting to rise again. The increase in atmospheric CO_2 concentrations

The unique properties of water, including its ability to absorb heat, are discussed in Chapter 5.

Figure 3.23

Secondary eruption on Mt. Pinatubo, May 1994. The cloud is moving toward the camera.

slowed. Although these observations were initially somewhat puzzling, most researchers attributed the slowdown in CO_2 accumulation to reduced plant and soil respiration caused by lower temperatures. Recent data from researchers from the Australian National University in Canberra suggest that increased cloudiness actually results in an increase in photosynthesis, helping to remove CO_2 from the atmosphere. These two effects are not mutually exclusive, however, further complicating the models for climate change.

Consider This 3.30   Climate Questions

Climate-modeling sites on the Web may deluge you with technical terms and numerical analyses. A good place to begin your understanding of climate modeling is to visit the National Climatic Data Center (NCDC), billed as "the world's largest active archive of weather data." A direct link is provided at the *Online Learning Center*. What types of data are provided by NCDC? Propose two or three questions that you might like to investigate, using these data.

Aerosols were defined in Section 1.11 and are discussed further in Section 6.5.

Smoke from all sources may be clouding our view of global warming. One group of researchers, led by Benjamin Santer of Lawrence Livermore National Laboratory, has found that predictions agree more closely with observations if the model includes the cooling effect of atmospheric aerosols. Aerosols are a complex group of materials that include dust, sea salt, smoke, carbon, and compounds containing nitrogen and sulfur. One of the most common aerosols consists of tiny particles of ammonium sulfate, $(NH_4)_2SO_4$. This compound can result if sulfur dioxide (SO_2) reacts with ammonia (NH_3). Both compounds can be released by natural or human-influenced sources. Burning crop wastes, rainforest trees, low-grade fuels such as charcoal, and fossil fuels all produce aerosols capable of blocking sunlight. Most particles in aerosols are smaller than about 4 μm in diameter and are efficient at scattering incoming solar radiation ranging from 0 to 4 μm. Thermal radiation coming from the Earth has wavelengths in the infrared part of the spectrum, ranging from 4 to 20 μm. The smaller aerosol particles are not effective in scattering these wavelengths, allowing the greenhouse gases to continue to absorb terrestrial radiation but at a reduced level because less solar radiation is reaching the surface to be radiated back into space. In addition, aerosol particles serve as nuclei for the condensation of water droplets and hence cloud formation. Thus, aerosols counter the warming effects of greenhouse gases. The temporary drop in average global temperature that followed the eruption of Mount Pinatubo in 1991 may well have been the consequence of the large volume of sulfur dioxide released by the volcano, enhancing aerosol formation.

Participants included Nobel laureate Paul Crutzen and Swedish meteorologist Bert Bolin, former chairman of the UN's Intergovernmental Panel on Climate Change (IPCC).

Top atmospheric scientists meeting in Berlin during the summer of 2003 estimated that aerosols have reduced global warming dramatically. The observed temperature increase of 0.6 °C over the last century might well have been at least 1.8 °C without the rising concentration of aerosols. Have we minimized global warming by emitting increasing amounts of smoke and soot in the atmosphere? Although this might appear fortunate, the bad news is that our climate system may be much more sensitive to greenhouse gas accumulations than previously thought. In any case, the protection provided by anthropogenic aerosols may only be temporary, given that the health effects of many aerosols in the troposphere will mandate declining emissions. Will Steffen of the Swedish Academy of Sciences summed up the implications for policy makers this way: "We need to get on top of the greenhouse gas emissions problem sooner rather than later."

Consider This 3.31   Nobel Laureate Paul Crutzen

a.   Why was Paul Crutzen honored with a share of a 1995 Nobel Prize in chemistry? *Hint:* See Section 2.9.

b.   What is the connection between his earlier scientific work and his current interest in climate change caused by global warming?

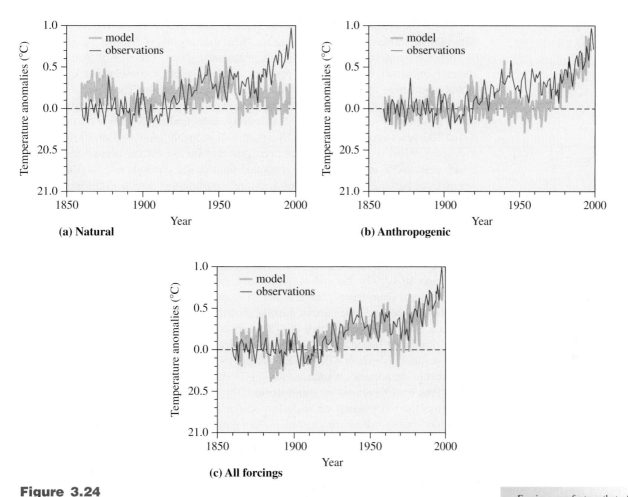

Figure 3.24

Simulated annual global mean surface temperatures. Simulating Earth's temperature variations and comparing the results with measured changes can provide insight into the underlying causes of the major changes.

Source: *Intergovernmental Panel on Climate Change (IPCC) Report, 2001.*

> *Forcings* are factors that affect the annual global mean surface temperature. It is a term commonly used by scientists studying climate change.

Scientists have developed increasingly detailed computer programs to model Earth's climate in spite of all the uncertainties. As supercomputers have become more powerful, models have become more sophisticated. For example, the oceans are usually represented as a multilayer circulating system and the model atmosphere is assumed to contain 10 or more interacting layers. Typically, the surface of the planet is divided into about 10,000 cells, not enough to provide detailed predictions, but sufficient to include general patterns of weather development. One test of these simulations of global climate is how well they predict the 0.6 ± 0.2 °C temperature increase observed over the last century when CO_2 concentrations increased by 25–30%. Figure 3.24 shows how the current data match the most complete computer models.

Consider This 3.32   **Comprehending Computer Models**

a. How successful was the IPCC model shown in Figure 3.24 in correlating natural forcings with observed temperature change?

b. How successful was the IPCC model shown in Figure 3.24 in correlating anthropogenic forcings with observed temperature change?

c. What are some of the forcings included in the part (c) graph?

Given the complexity of the global system, considerable uncertainty is associated with climate models. There is no wonder, then, that experts sometimes disagree. First of all is the matter of projected levels of greenhouse gases. The *rate* of their emission currently increases by about 1.5% per year. This is largely a consequence of growing global population, agricultural production, and industrialization. The population of the planet tripled during the 20th century and it is expected to double or triple again before reaching a plateau sometime in the 21st century. Industrial production is 50 times what it was 100 years ago. In the next 50 years, it will probably grow to 5 or 10 times what it is today. Most of this growth has been powered by the combustion of fossil fuels. Every year, 2–3% more energy is generated than in the previous one, and most of it comes from burning coal. If these rates of fuel consumption continue, the atmospheric concentration of CO_2 will be twice its 1860 level sometime between the years 2030 and 2050.

All models predict that this doubling will increase the average global temperature, but the magnitude of that increase is estimated in different ways. Many predictions, including the most recent one by the 2001 IPCC, fall in the range of 1.4 °C and 5.8 °C (2.5 °F and 10.4 °F) by the end of the century. Such an increase in average global temperature might seem trivial, but it is not. A temperature drop of just a few degrees is the difference between the current average global temperature and that of a much chillier epoch, the last ice age, about 20,000 years ago.

A significant uncertainty in any such projections is trying to predict whether the rate of population growth will stabilize during the next 100 years. During the 20th century, worldwide population increased from 2 billion people to approximately 6 billion today. Projected increases of atmospheric CO_2 until 2100 are shown in Figure 3.25. The projections are based on differing scenarios, each assuming that no dedicated efforts are made to diminish greenhouse gas emissions. Because increased numbers of people translate into increased energy use and greater greenhouse gas emissions through burning fossil fuels, the scenarios incorporate different assumptions about population growth rates and economic growth. The low-end scenario assumes a world population in the year 2100 of 6.4 billion and an annual economic growth rate of 1.2%. The midrange projection is based on 11.3 billion people with a 2.3% annual economic growth rate, nearly twice that of the low-end scenario. In the high-end projection, the year 2100 population is 11.3 billion as in the midrange calculation, but annual economic growth would occur at 3.0%. The models represent a truly global research effort,

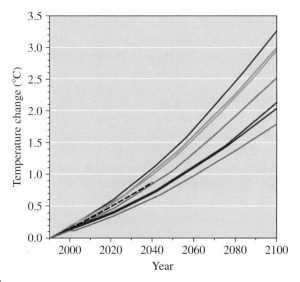

Figure 3.25

Projecting temperature change resulting from greenhouse gas emissions. The dashed line represents the median of projected anthropogenic warming, 1991–2041. The colored lines represent predictions from different computer-based models.

Your Turn 3.33 — Reading the Curves

a. The top line (red) in Figure 3.25 represents a model from NOAA. What temperature increase is predicted for the period 2005–2040?

b. How does the 2005–2040 increase estimated by the NOAA model compare with that predicted by DOE, shown by the bottom (pink) line?

c. If the average temperature of the Earth is 15 °C at this time, what will the predicted temperature be in 2100 according to the NOAA model? The DOE model?

with projections made by U.S., Canada, Australian, UK, and German scientists. All models seek to reconcile complex climate simulations with recent observed climate change information.

The many hundreds of scientists from all over the world who participated in preparing the 2001 IPCC report used words that quantified the probability that their estimates of change were valid. This was to help policy makers and the public at large better understand the inherent uncertainty and reliability of the data. These terms continue to be used in all subsequent updates to this report. Table 3.6 gives the scientists' definitions.

Using these terms, the IPCC came to conclusions and assigned a range of probabilities that each was true. It was judged very unlikely that all of the observed global warming was due to natural climate variability. Rather, the scientific evidence strongly supports the position that human activity is a significant factor causing the increase in average global temperature observed over the last century. Some of IPCC's conclusions, together with their assigned probabilities, are shown in Table 3.7.

Since the time of the IPCC report, NASA has reported that 2002 became the second warmest year on record, just behind 1998 and just ahead of 2001. The trend toward increasing temperatures appears to be extending into this century. One clear effect of observed climate change is shown in Figure 3.26. Decreased snow and ice cover, which would attend global warming, would lower the amount of sunlight reflected from Earth's surface. The resultant increase in absorbed radiation would promote a further increase in temperature.

Changes in sea ice, snow, and glaciers all can contribute to changes in sea level. Global mean sea level is projected to rise by 9–88 cm (3.5–34.6 in.) between 1990 and 2100. Predictions of rising sea levels, which have been scaled back from earlier IPCC predictions of 13–92 cm (5.1–36.2 in.), are caused by thermal expansion of warmer water as well as by melting of frozen precipitation. Sea level rises of 9–88 cm would endanger New York, New Orleans, Miami, Venice, Bangkok, Taipei, and other coastal cities. Millions of people might have to relocate. Dr. Richard Williams of the U.S. Geological Survey states: "If the ice sheets on Greenland melted, that alone could raise sea

Table 3.6 — Judgmental Estimates of Confidence

Term Used	Probability That a Result is True
Virtually certain	> 99%
Very likely	90–99%
Likely	66–90%
Medium likelihood	33–66%
Unlikely	10–33%
Very unlikely	1–10%

Source: *Summary for Policymakers, A Report of Working Group 1 of the Intergovernmental Panel on Climate Change,* Shanghai: IPCC, January 1, 2001.

Table 3.7   IPCC Conclusions

Very Likely	Likely	Very Unlikely
The 1990s were the warmest decade and 1998 the warmest year since 1861.	Temperatures in the Northern Hemisphere during the 20th century are *likely* to have been the highest of any century during the past 1000 years.	The observed warming over the past 100 years is due to climate variability alone, providing new and even stronger evidence that changes must be made to stem the influence of human activities.
Higher maximum temperatures are observed over nearly all land areas.	Arctic sea-ice thickness declined about 40% during late summer to early autumn in recent decades.	
Snow cover decreased about 10% since the 1960s (satellite data); in the 20th century there was a reduction of about two weeks in lake and river ice cover in the mid- and high-latitudes of the Northern Hemisphere (independent ground-based observations).	An increase in rainfall, similar to that in the Northern Hemisphere, has been observed in tropical land areas falling between 10°N and 10°S.	
Increased precipitation has been observed in most of the Northern Hemisphere continents.	Increased summer droughts are *likely* in a few areas.	

level by 20 feet." It is far from certain that major increases in sea level will occur. Even if they do, they will take place over many years, providing considerable time for preparation and protection.

Some climatologists also feel that an increase in the average temperature of the oceans could cause more weather extremes including storms, floods, and droughts. In the Northern Hemisphere, summers are predicted to be drier and winters wetter. The regions of greatest agricultural productivity could change. Drought and high temperatures could reduce crop yields in the American Midwest, but the growing range might extend farther into Canada. It is also possible that some of what is now desert could get sufficient rain to become arable. One region's loss may well become another locale's gain, but it is too early to tell.

Global warming is already having observable effects on plants, insects, and animal species around the world. Species as diverse as the California starfish, Alpine herbs, and checkerspot butterflies have all exhibited changes in either their ranges or their habits. Dr. Richard P. Alley, a Pennsylvania State University expert on past climate shifts, sees particular significance in the fact that animals and plants that rely on each other will not necessarily change ranges or habits at the same rate. Referring to affected species, he said, "You'll have to change what you eat, or rely on fewer things to eat, or travel farther to eat, all of which have costs." The result in decades to come could be substantial ecological disruption, local losses of wildlife, and possible extinctions.

In other respects, we may all be losers in a warmer world. Recently, physicians and epidemiologists have attempted to assess the costs of global warming in terms of public health. An increase in average temperatures is expected to increase the geographical range of mosquitoes, tsetse flies, and other insects. The result could be a significant increase in diseases such as malaria, yellow and dengue fevers, and sleeping sickness in new areas, including Asia, Europe, and the United States. Indeed, it has been suggested that the deadly 1991 outbreak of cholera in South America is attributable to a warmer Pacific Ocean. The bacteria that cause cholera thrive in plankton. The growth of plankton and bacteria are both stimulated by higher temperatures.

There are many different species of checkerspot butterflies. This one is found in parts of Wisconsin.

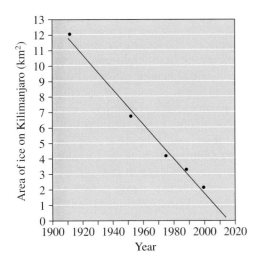

Figure 3.26

The snows of Kilimanjaro. In 1912, Kilimanjaro in Tanzania had significant snow cover on its peak (top photo). By 2001, about 82% of the massive ice field atop Kilimanjaro had been lost (bottom photo). If the measured rate of retreat of the snow cover continues unchanged, the snow cover will have disappeared by 2015.

Source of Graph: *Science* Vol. 293, 6 July 2001, p. 47.

Even more uncertainty is associated with the regional weather patterns predicted by the various models. Most climate models lack enough detail to predict changes in small-scale phenomena. Thunderstorms, tornadoes, hail, and lightning are not included in most climate models. One of the more controversial forecasters is James E. Hansen, Head of NASA's Goddard Institute for Space Studies. Hansen has estimated that doubling the concentration of greenhouse gases would mean that New York City could expect 48 days a year with temperatures above 90 °F instead of the current 15. In Dallas, the number of days per year with temperatures above 100 °F would increase from 19 to 87. Many scientists have questioned Hansen's estimates, and they have not been endorsed by IPCC.

Consider This 3.34 **An Alternative Scenario**

Dr. James E. Hansen supports the scientific findings of IPCC. However, he does not agree that "economically wrenching" steps are necessary to slow the rate of global warming. Instead, he proposes taking actions that make economic sense, independent of global warming. Use the resources of the *Online Learning Center* or directly research his "alternate scenario." Report what actions he proposes, and why his recommendations are controversial.

3.10 Responding to Science with Policy Changes

The debate over climate change has subtly shifted in the last 10 years. The focus is no longer on whether there is an observable increase in global temperatures. Climate scientists agree that global warming is occurring and that very likely there is an anthropogenic link. The focus now is on understanding the causes of such increases and means of preventing projected changes. There are two related questions. What *can* we do and what *should* we do about the possibility of significant climate change caused by global warming? One thing is clear: Given the recent results from improved climate modeling, we will start seeing even more definitive climatic changes within a decade or so. But can we prudently wait that long, or is prompt action essential? Whether to act and how to act are not just scientific issues. What determines our response is a complicated mix of science, perception of risk, societal values, politics, and economics.

Several lines of thought are possible for dealing with what we can and should do, although any classification system is somewhat arbitrary when applied to any complicated issue. Many believe that the case has been made for action now and are already implementing changes to reduce greenhouse gases. Others prefer to study the situation in more detail before deciding on a course of action. Perhaps there are even those who would not act at all, believing that climate change is inevitable and just part of a long-term natural cycle.

> Plans to change from fossil fuels to alternative energy sources in the United States will be discussed in Chapter 8.

The most obvious strategy for dealing with global warming would be to reduce our reliance on fossil fuels. Such action would be very difficult, not only because this energy source is so important to our modern economy, but also because of its implications for international energy policy. Although the developing countries may well become the major producers of carbon dioxide and other greenhouse gases in the not-too-distant future, the developed countries have a prodigious lead. The annual carbon output (in the form of CO_2) of the United States is about 5.5 metric tons for each of its inhabitants per year (Figure 3.27). On a per capita basis, the United States leads the world for developed countries. This creates a mandate for the country to help lead the way in reducing emissions, but to do so without sacrificing the economy.

> CO_2 emissions from fossil fuel combustion in China have decreased by 9% from 1995 to 2000. By contrast, CO_2 emissions from fossil fuel combustion in Western Europe increased by 4.5%, in the United States by 6.3%, and in Japan by 3.0%.

Comparable per capita values for China and India are about 0.6 and 0.3 tons, respectively. Even so, the Peoples' Republic of China ranks second behind the United States in *total* carbon dioxide emissions from fossil fuels. If China were to succeed in raising its per capita gross national product to only 15% of the U.S. figure, the increase in CO_2 production would approximately equal the current American annual emissions from burning coal. The IPCC has estimated that by 2010, the developing countries plus the nations of the former Soviet Union will produce about half the world's CO_2. It is

---

Consider This 3.35   The Top Emitters

The Carbon Dioxide Information Analysis Center (CDIAC) provides data, analysis, and graphics about CO_2 emissions. Use Figure 3.27 and the direct link to the CDIAC Web site provided on our *Online Learning Center* to answer these questions.

a. This text cites the United States and Peoples' Republic of China as the leaders in total CO_2 emissions from fossil fuels in 2000. What countries rank third, fourth, and fifth in 2000?

b. Which countries were in the top five in per capita CO_2 emissions in 1950? Explain any differences between the 1950 and 2000 lists.

c. How has the contribution of the United States to the percentage of world emissions changed in the time interval 1950–2000? What are the reasons for the change?

d. From the CDIAC Web site, pick any country *not* in the 2000 top 20 list. Where in the world is that country located? How have this country's CO_2 emissions changed in the time interval reported? What are the major sources for their emissions?

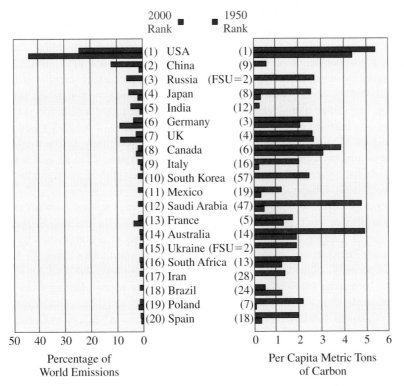

Figure 3.27

Top 20 per capita CO_2 emissions for 2000 for selected nations. Values are compared with 1950 and given as a percentage of world emissions. FSU stands for the former Soviet Union, now called the Russian Federation.

Source: Carbon Dioxide Information Analysis Center, http://cdiac.ornl.gov/trends/emis/graphics/top20_2000.gif

unrealistic for the developed countries to expect the nations of the Third World to abandon their hopes for economic growth and become good, nonpolluting global citizens.

To just keep using fossil fuels at the same rate until we run out is not a reasonable approach. In fact, it is already too late for this to be a viable option. Significant time lags occur from the production of CO_2, until its accumulation in the atmosphere, and to the eventual return of the CO_2 into an environmental sink. Thus, waiting until we run out of fossil fuels to take action will not solve the problem. From a realistic perspective, we will continue to depend heavily on fossil fuels into the near future. Currently, more than 85% of the world's energy needs are supplied by fossil fuels, making it difficult to develop sufficient alternative technological capacity soon enough to mitigate climate change.

Some advocate delaying action regarding global warming, feeling that further study is needed. During the 2000 presidential campaign, candidates George W. Bush and Al Gore agreed that carbon dioxide is a pollutant and that legislation should be passed to help reduce CO_2 emissions from power plants. Once elected President, George W. Bush's proposed energy policy no longer included mandatory reductions. The Bush administration argues that any shift from coal to cleaner-burning natural gas would mean higher prices for electricity, a change that could damage our economy. Others argue that uncertainties in our predictive powers and climate models are so great that money and effort would be wasted now to undertake preventive or ameliorative action. Without more knowledge, we run the risk of making more mistakes.

In 2001, the U.S. National Academy of Sciences was asked by the U.S. Domestic Policy Council to review the report and identify areas of uncertainty in the science of climate change. The Academy, in turn, entrusted this responsibility to the U.S. National Research Council. Their report generally agreed with the assessment of human-caused climate change presented in the IPCC scientific report, but issued this caution. "Because

Time lags were also important to the story of ozone depletion discussed in Chapter 2. Although CFCs are no longer being added to the atmosphere, it will be many years before their concentration in the stratosphere is reduced to earlier desirable levels.

there is considerable uncertainty in current understanding of how the climate system varies naturally and reacts to emissions of greenhouse gases and aerosols, current estimates of the magnitude of future warming should be regarded as tentative and subject to future adjustments (either upward or downward)." The "not yet" attitude is found unacceptable by many, including Dr. Michael Oppenheimer, chief scientist of the Environmental Defense Fund, who stated: "Scientific uncertainties, which are substantial, of course, [regarding global warming] are not a reason to put off action. In fact, we only have one Earth to experiment on." James Hansen, Head of NASA's Goddard Institute for Space Studies, also addressed this point in his testimony before the U.S. Senate's Committee on Commerce, Science, and Transportation. "Doubt and uncertainty are the essential ingredient[s] in science. They drive investigation and hypotheses, leading to predictions. Observations are the judge."

Other approaches to dealing with global warming involve various ways of putting the carbon dioxide back into the Earth, thus sequestering the carbon. **Sequestration** literally means keeping something apart. Chemically this is accomplished by forming stable bonds between the sequestering agent and the substance "trapped". If carbon dioxide is properly sequestered, it cannot reach the troposphere and contribute to global warming. Increasing rates of carbon sequestration may mean planting trees as environmental sinks that absorb CO_2. For example, British and Malaysian scientists and volunteers planted 120,000 trees in logged-over rainforest in Malaysian Borneo during 2003. This is the largest-ever experiment to explore how tree diversity can influence both timber production and the storage of carbon. Although reforestation, saving old-growth forests, and improved land management are good for humankind, they may turn out to have little long-term impact on the CO_2 in the atmosphere. This is because much of the carbon tied up in forests and soils moves back into circulation in 30–60 years, according to Dale Simbeck, Vice President of Technology for the consulting firm SFA Pacific.

Sequestering agents form stable bonds with minerals that could cause discoloration or cloudiness in soft drinks and other prepared food products.

Consider This 3.36 Trees as Carbon Sinks

Some researchers have concluded that new forest plantations are not very efficient at sequestering carbon. What evidence is there for this conclusion? Does it make a difference if the new plantings replace other trees or cropland? Present your findings in a written report.

Oil deposits are usually accompanied by natural gas. The principal components in natural gas are methane and ethane, but CO_2 is also in the mixture. Oil companies propose separating CO_2 and pumping it directly back underground into nearby depleted natural gas fields. The recovered CO_2 could also be used to flush residual crude oil from shrinking reservoirs. These practices are in use already in the United States and could be expanded. Gardiner Hill, the CO_2 program manager for British Petroleum, estimates that in the United States alone, 40–50 billion metric tons of CO_2 could be stored in depleted oil reservoirs, 80–100 billion metric tons of CO_2 could be stored in gas reservoirs, and 15–20 billion metric tons could be sequestered in coal beds no longer suitable for mining. Although many technical questions remain, there is a body of experience for successfully managing storage of fluids and gases in geologic formations.

Consider This 3.37 Drop in the CO_2 Bucket?

How do these billions of metric tons of sequestered CO_2 compare with the total CO_2 emissions per year in the United States? Show your reasoning. *Hint:* Figure 3.27, combined with the population figure for the United States, would provide a basis for comparison.

Another proposed form of carbon sequestration is based on capturing CO_2 from stationary sources, such as an electric power plant. The CO_2 can then be liquefied and pumped deep into the ocean. Proponents of such carbon sequestration technology point out that much of the carbon dioxide now emitted eventually finds its way into the oceans anyway, although this happens over a much longer period. Carbon sequestration is currently being implemented off the coast of Norway where the Sleipner natural gas rig pumps CO_2 1000 m below the ocean's surface, a project driven by concern for global warming and financial incentive. Norway's Statoil oil company is projected to save millions of dollars by using carbon sequestration. The company built an $80 million dollar at-sea facility to separate CO_2 from natural gas for two reasons: first because they could not sell their natural gas to European customers without first removing carbon dioxide from it, and second because sequestering the CO_2 avoids a stiff "carbon tax" imposed by Norway. This tax would have cost Statoil about $50 for every ton of CO_2 emitted, thus saving about $50 million a year in taxes. Critics cite experimental evidence that increased oceanic carbon dioxide could damage coral reefs.

Consider This 3.38 Disappearing Coral Reef Color

The brilliant beauty of coral reefs has begun to disappear in several parts of the world. What evidence is there for this statement? What other factors are placing stress on the world's coral reefs. Present your findings in a written report.

Several other techniques are being investigated for sequestering CO_2. One goal is to bring the cost of carbon sequestration down from the current $100 per ton to $10 per ton. David Keith, a professor of engineering and public policy at Carnegie Mellon University, argues that the total cost of using fossil fuels for electricity production and sequestering the CO_2 in geologic formations is now in "more or less" the same ballpark as electricity produced by renewable means. At a 2002 meeting of the National Academy of Engineering, Dr. Keith made his point this way. "We often hear that the cost of solving the climate problem will be very high," he said, "but I am skeptical."

The magnitude of the problems associated with global warming makes some conclude that nothing can be humanly done to halt or reverse the process. Their outlook is that the generation of energy, and with it carbon dioxide, is an essential feature of modern industrial life, and we must therefore learn to live with its consequences and begin adapting to our warm new world. We encourage you to debate this and other options. Clearly, your opinions and the evidence you marshal to support them will help shape future policy. However, as authors of this text, we hope that at least some of you will develop a compromise strategy from the best characteristics of the responses to global warming just described. There are, after all, elements of truth in each possible response.

Consider This 3.39 Three Reactions to Global Warming

The text identifies three reactions to the problem of global warming: act now, continue to gather data before making any decisions, and take no action at all. Review the three positions and their consequences. Discuss the situation with others and prepare to state and defend your position to the rest of the class.

3.11 The Kyoto Protocol on Climate Change

More than a century ago Arrhenius first proposed that carbon dioxide emissions could accumulate in the atmosphere and lead to global warming. The world may have been slow in responding, but the last few decades have seen considerable progress. Rising concentrations of CO_2 were detected in the early 1960s, and data gathered for other greenhouse gas concentrations in the 1970s. Mainly only atmospheric scientists knew of these trends before the mid-1980s, when international workshops and conferences brought the situation to United Nations agencies, particularly the UN Environmental Program and the World Meteorological Organization (WMO). The stage was set in 1988 when the IPCC was established to gather available scientific research on climate change and provide advice to policy makers. A series of international conferences led to the "Earth Summit" in Rio de Janeiro, Brazil, in June of 1992. More than 160 countries, including the United States, adopted the Framework Convention on Climate Change (FCCC) by the close of that meeting. This document presented scientific evidence that increasing temperatures were a global concern and suggested effective ways of responding.

In 1997, nearly 10,000 participants from 161 countries gathered in Kyoto, Japan. They established goals to stabilize and reduce atmospheric greenhouse gas concentration levels that are more environmentally responsible. The result is what has come to be known as the Kyoto Protocol to the Framework Convention, or simply the Kyoto Protocol. Binding emission targets based on five-year averages were set for 38 Annex I nations to reduce their emissions of six greenhouse gases from 1990 levels. Accomplishing these goals between the years 2008 to 2012 could decrease emissions from industrialized nations overall by about 5%. Under the Kyoto Protocol, the United States was expected to reduce emissions to 7% below its 1990 levels, the European Union nations 8%, and Canada and Japan 6%. No new binding emission targets were established for the Annex II developing countries, a contentious issue then and now. Developed nations are permitted to trade emission credits to meet their targets. That is, countries that have emissions lower than their targets can sell the residual amounts to countries exceeding their targets. Developed nations also can receive further credits for investments and projects to help developing countries reduce their emission of greenhouse gases through better technologies. The gases regulated include carbon dioxide, methane, nitrous oxide, hydrofluorocarbons (HFCs), perfluorocarbons (PFCs), and sulfur hexafluoride.

> The Framework Convention divided all the signing countries into two groups. **Annex I** countries are those considered economically developed and **Annex II** countries are those in the developing world.

Consider This 3.40   **Kyoto Conference Humor**

What is the humor in this cartoon? Would everyone find it amusing? Explain your reaction to this cartoon, including whether you feel it is trying to communicate a certain point of view.

Cartoon by Jeff MacNelly. Copyright © Newsday. Reprinted with permission of Tribune Media Services.

Consider This 3.41   Why Not Water?

Water is the largest contributor to the greenhouse effect. Why were no emission targets set for water under the Kyoto Protocol?

The Kyoto Protocol will go into effect when 55 countries *and* when Annex I signatories with CO_2 emissions totaling 55% of the total 1990 CO_2 emissions have ratified it. As of June 2003, 110 countries had ratified or otherwise accepted the protocol. The Annex I countries that have ratified the protocol account for 44.2% of total Annex I CO_2 emissions in 1990, short of the 55% required for the protocol to become fully effective. Notable among the holdouts is the United States.

Former President Clinton made it clear while president that he would not submit the treaty to the U.S. Senate for ratification until key developing countries such as China and India agreed to limit their greenhouse gas emissions. This step is important because the rates of greenhouse gas emissions of developing nations are increasing faster than those of industrialized countries and are expected to grow even faster (Figure 3.28).

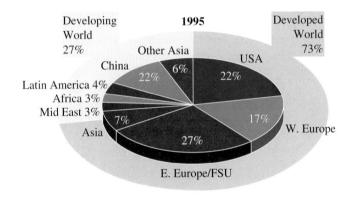

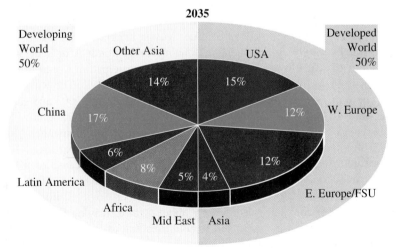

Figure 3.28

Total world CO_2 emissions for 1995 (6.46 billion tons) and 2035 (projected to be 11.71 billion tons). Contributions from the developed world are shown in light blue, the developing world in light yellow. Notice that parts of Asia fall into each category. The abbreviation FSU stands for the former Soviet Union, now called the Russian Federation.

President George W. Bush stated he does not support the agreements reached in Kyoto, calling it "fatally flawed," and again citing the problem with lack of restrictions for developing nations. President Bush also believes that complying with the Kyoto Protocol would cause serious harm to the U.S. economy. Instead, the United States is pursuing several alternatives to participation in the Kyoto Protocol. In February 2002, President Bush announced a Global Climate Change Initiative. The proposal sets greenhouse gas emission goals based on units of gross domestic product. From 2002 to 2012, a reduction of 18% based on this measure is predicted, attained by totally voluntary efforts.

Specifics on the proposals to reduce U.S. dependence on burning fossil fuels will be discussed in Chapter 8.

Your Turn 3.42   Changing Contributions

a. How will the total tons of world CO_2 emissions change from 1995 to what is projected for 2035?

b. How will the percentage of total world CO_2 emissions from the developed world change from 1995 to what is projected for 2035?

c. How will the total tons of world CO_2 emissions from the developed world change from 1995 to what is projected for 2035?

Consider This 3.43   Developing Countries; Developed Countries

Consider how your position on controlling emissions of carbon dioxide would change if you were a student in a developing country rather than in the United States. To help your consideration, pick a country in the developing world. Then use a search engine or the direct access provided on the *Online Learning Center* to find out if that country has signed and ratified the Kyoto Protocol. Also find the total tons of carbon dioxide emitted and the per capita emission for that country. Compare these data with those of the United States and comment on the differences and their possible effect on policy.

Given that the United States has essentially abandoned the Kyoto Protocol, the Russian Federation has ended up playing a pivotal role. It is now the only country that can bring the Kyoto Protocol into force. Russia announced, at the World Summit on Sustainable Development held in Johannesburg, South Africa, in September 2002, that they did plan to ratify the protocol. Their participation would bring Annex I ratification from 44.2% to 61.2%, well over the required 55%. The Russian Federation has yet to make good on their promise and seemed, during 2003 and early 2004, to be backing away from making a commitment. Mukhamed Tsikanov, Russia's deputy minister of economics, has explained their position in this manner. "Russia is not against the Kyoto Protocol . . . [but] as of today there are no particular economic incentives for Russia to ratify it." Their reluctance is partly because of a historical twist of fate concerning the year 1990. That was the last year that the factories were operating at top speed before the collapse of the former Soviet Union. Since then, greenhouse gas emissions have dropped by about a third, leaving Russia with many emissions credits they had hoped to sell to countries that exceeded their limits. Japan and the European Union are still potential purchasers of these credits. However, without the participation of the United States, the prospects for financial gain are somewhat more limited.

The Russian Federation's position changed again in mid-2004 when the European Union announced they would back the Federation's entry into the World Trade Organization (WTO), an important step for Russia's economic growth. The negotiation hinged on Russia's promise to support environmental policies of the Union, particularly agreeing to work toward ratification of the Kyoto Protocol. "Russia clearly traded its support for Kyoto in exchange for some concessions on WTO entry terms," said Christopher Weafer, chief equity strategist with Alfa Bank in Moscow. Both houses of the Russian Parliament have now ratified the Kyoto Protocol, setting the stage for it to finally come into effect in early 2005.

3.12 Global Warming and Ozone Depletion

Global warming and ozone depletion both involve the atmosphere, and both are much in the news. The casual reader of newspaper accounts may easily mix them up. Sometimes, the authors of the articles themselves get confused! We want to avoid such mix-ups and for that reason, will conclude this chapter by returning to one of the common misconceptions. The question is often posed this way: Is the depletion of the ozone layer the principal cause of climate change? Perhaps this seems logical because if stratospheric ozone is destroyed, more solar radiation might be able to reach Earth and warm it. This is not the case, however. In fact, stratospheric ozone causes a small negative forcing, standing in contrast to tropospheric ozone, which shows a larger positive forcing. This negative forcing arises from the absorption of UV radiation by O_3 during its destruction in the stratosphere. This effect is small compared with the observed effects of the major greenhouse gases. Figure 3.29 shows the relative contributions, or forcings, from changes in atmospheric gases. Even though the ozone in the stratosphere contributes little to global warming, nonetheless the issues of ozone depletion and global warming are linked. Human activities have led to the accumulation in the atmosphere of several long-lived greenhouse gases. This list includes ozone and many of the CFCs that have been held responsible for stratospheric ozone depletion. It even includes some of the compounds used to replace CFCs, such as HFCs. We end this chapter by summarizing in Table 3.8 some of the important differences between global warming and stratospheric ozone destruction. Such a tabulation invites oversimplification, but it can be a useful summary of some of the important aspects of these two environmental problems and the responses to them by the United States and the international community. The next Consider This activity provides an opportunity to express your informed opinion about their relative significance.

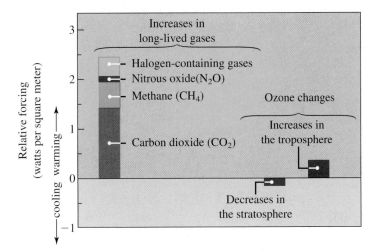

Figure 3.29

Radiative forcing of climate change from atmospheric gas changes (1750–2000).

Source: *Scientific Assessment of Ozone Depletion: 2002;* World Meteorological Organization, United Nations Environmental Program.

Table 3.8	Global Warming and Ozone Depletion: Some Characteristics

	Global Warming	Stratospheric Ozone Depletion
Region of atmosphere involved	Stratosphere	Mostly troposphere
Major substances involved	H_2O, CO_2, CH_4, N_2O	O_3, O_2, CFCs
Interaction with radiation	When molecules absorb IR radiation, they vibrate and return heat energy to Earth.	When molecules absorb UV radiation, they break apart into smaller molecules or atoms.
Nature of problem	Increasing concentrations of greenhouse gases are apparently increasing average global temperature.	Decreasing concentration of O_3 is increasing exposure to UV radiation.
Major sources	Release of CO_2 from burning fossil fuels, deforestation; CH_4 from agriculture. Natural sources of H_2O	Release of long-lived CFCs from past uses as solvents, foaming agents, air conditioners. CFCs release Cl· that destroys O_3.
Credible consequences	Altered climate and agricultural productivity, increased sea level, effects on health	Increased incidence of skin cancer, damage to phytoplankton
Possible remedies	Decrease use of fossil fuels, slow deforestation; change agricultural practices	Eliminate use of CFCs, find suitable replacements
International response	Kyoto Protocol, 1997 and later amendments	Montreal Protocol, 1987 and later amendments
U.S. response	Signed Kyoto Protocol in 1998; not submitted to Senate for ratification, therefore not bound by its provisions; alternative proposals	Signed 1987; full participation in Protocol and its amendments

Consider This 3.44   **Air Quality, Ozone Depletion, or Global Warming**

Now that you have studied air quality (Chapter 1), stratospheric ozone depletion (Chapter 2), and global warming (Chapter 3), which do you believe poses the most serious problem for you in the short run? In the long run? Discuss your reasons with others and draft a 1-page report on this question.

CONCLUSION

Is our world warming to unintended climatic effects? To assess and reverse such effects, much will depend on the quality of information gathered and how it is used to make sound economic and environmental decisions. The development of technology to exploit fossil fuel energy resources has been a double-edged sword, leading both to dramatic improvements in our lives and to a whole new generation of problems. During the 20th century, the atmosphere on which our very existence depends was subjected to repeated assaults. The fact that most of these environmental insults were unintentional and, in some cases, the unexpected consequences of social progress does not alter the problems we face. We have only relatively recently recognized the harm that air pollution, stratospheric ozone depletion, and global warming can bring to our personal, regional, national, and global communities. To reverse the damage already done and to prevent more, all these communities must respond with intelligence, compassion, commitment, and wisdom. It is instructive that even in the absence of threats such as global warming, many suggestions for change would contribute to sound, prudent, and responsible stewardship of our planet.

A central theme to many of the issues explored in *Chemistry in Context* is the production of electricity. In an editorial comment in *Science* (4 April 2003), Editor emeritus Philip Abelson reminds us of the importance of looking at the larger picture and longer timeline of our dependence on fossil fuels.

> *"The United States has large resources of coal that now supply the energy for more than half of its electricity. Nuclear energy furnishes another 20%. In principle, future U.S. needs for liquid fuels and electricity could be met. However, . . . the greenhouse effect will cause added future problems for the use of coal. Construction and testing of new electrical generating plants cannot be achieved quickly."*

We will return to the theme of energy in the next chapter, titled "Energy, Chemistry, and Society". Nuclear energy is considered in Chapter 7, and alternative energy sources such as fuel cells and solar power are explored in Chapter 8. The many energy-related challenges and uncertainties faced by our modern society are daunting. They also are the wages of our success.

Chapter Summary

Having completed this chapter you should be able to:

- Understand the different processes that take part in Earth's energy balance (3.1)
- Realize the difference between Earth's natural greenhouse effect and the enhanced greenhouse effect (3.1)
- Understand the major role that certain atmospheric gases play in the greenhouse effect (3.1, 3.2)
- Explain the methods used to gather past evidence for global warming (3.2)
- Relate Lewis structures to molecular geometry, including bond angles (3.3)
- Understand how molecular geometry is related to absorption of infrared radiation (3.4)
- Know which molecular species are greenhouse gases because of their molecular structures and shapes (3.4)
- Explain the roles that natural processes play in the carbon cycle and through it, to global warming (3.5)
- Summarize the contributions human activities make to the carbon cycle and through it, to global warming (3.5)
- Understand how molar mass and atomic weight are defined and used (3.6)
- Use Avogadro's number to calculate the average mass of an atom (3.6)

- Develop an understanding of the mole and its relationship with molecules (3.7)
- Assess the sources, relative emission quantities, and effectiveness of greenhouse gases other than CO_2 (3.8)
- Recognize the successes and limitations of computer-based models in predicting climatic change (3.9)
- Summarize the different levels of confidence in drawing conclusions about climate change (3.9)
- Consider the global and national implications of a rise in Earth's average temperature (3.10)
- Identify the single most effective strategy for reducing CO_2 emissions (3.10)
- Describe ways in which CO_2 emissions can be sequestered (3.10)
- Explain world and U.S. policy concerning the Kyoto Protocol (3.11)
- Clarify reasons for projected changes in the relative amounts of CO_2 emissions for developed and developing countries (3.11)
- Compare how the issue of global warming is both similar to and different from the issue of ozone depletion (3.12)

- Read and hear news stories on global warming with some measure of confidence in your ability to interpret the accuracy and conclusions of such reports

- Take an informed position with respect to issues surrounding global warming

Questions

Emphasizing Essentials

1. **a.** Is the greenhouse effect taking place now? Explain your reasoning.

   b. Is the enhanced greenhouse effect taking place now? Explain your reasoning.

2. Concentrations of CO_2 in Earth's early atmosphere were much higher than today. What happened to this CO_2?

3. Using the analogy of a greenhouse to understand the energy radiated by Earth, what are the "windows" of Earth's greenhouse made of?

4. Consider equation 3.1 for the photosynthetic conversion of CO_2 and H_2O to form glucose, $C_6H_{12}O_6$, and O_2.

   a. Demonstrate that the equation is balanced by counting atoms of each element on either side of the arrow.

   b. Is the number of molecules on either side of the equation the same? Why or why not?

5. What is the difference between climate and weather?

6. **a.** Incoming solar radiation is 343 watts/m². Our atmosphere reflects 30% of it. How much energy enters our atmosphere?

   b. Under steady-state conditions, how much energy leaves our atmosphere?

7. Consider Figure 3.4.

   a. How does the concentration of CO_2 in the atmosphere at present compare with its concentration 20,000 years ago? How does the present concentration of CO_2 compare with the concentration 120,000 years ago?

   b. How does the temperature at present compare with the 1950–1980 mean temperature of the atmosphere? How does the temperature 20,000 years ago compare? How does each of these values compare with the average temperature 120,000 years ago?

   c. Do your answers to parts **a** and **b** indicate causation, correlation, or no relation? Explain.

8. Understanding Earth's energy balance is essential to understanding the issue of global warming. For example, the solar energy striking the Earth's surface averages 168 watts/m², but the energy leaving Earth's surface averages 390 watts/m². Why isn't the Earth cooling rapidly?

9. Explain each of the observations.

   a. If a car is left in the Sun, it may become hot enough to endanger the lives of pets or small children.

   b. Clear winter nights tend to be colder than cloudy ones.

   c. A desert shows much wider daily temperature variation than a moist environment.

10. Using the Lewis structures for H_2 and H_2O as examples, show how the Lewis structure for a molecule can allow an unambiguous prediction of molecular geometry for some molecules, but not for others.

11. Use a molecular model kit to build a methane molecule, CH_4. (If a kit is not available, this model can be made using Styrofoam balls or gumdrops to represent the atoms and toothpicks to represent the bonds.) Demonstrate that the hydrogen atoms are farther from one another in a tetrahedron than they would be if all the atoms of methane were in the same plane (square planar).

12. Use the stepwise procedure given in the text to predict the shape of each of these molecules.

   a. CH_2Cl_2

   b. CO

   c. PH_3

13. **a.** Write the Lewis structure for H_3COH, which is methanol (wood alcohol).

   b. Based on this structure, predict the hydrogen-to-carbon-to-hydrogen bond angle. Explain the reason for your prediction.

   c. Based on this structure, predict the hydrogen-to-oxygen-to-carbon bond angle. Explain the reason for your prediction.

14. **a.** Write the Lewis structure for H_2CCH_2, ethene, a simple hydrocarbon with a C-to-C double bond.

   b. Based on this structure, predict the hydrogen-to-carbon-to-hydrogen bond angle. Explain the reason for your prediction.

   c. Sketch the molecule showing the predicted bond angles.

15. The text states that a UV photon can break chemical bonds, but that an IR photon can cause only vibration in the bonds.

   a. Calculate the energy associated with each of these processes by assuming a wavelength of 320 nm for the UV photon and a wavelength of 5000 nm for the IR photon.

   b. What is the ratio of the energy that breaks bonds to the energy that causes vibration?

16. Three different modes of vibration of a water molecule are shown at the top of the next page. Imagine the atoms being connected by bonds that act as springs. Each vibration can be visualized by moving the atoms in the directions indicated and then back again.

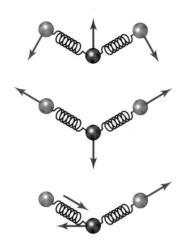

Which of these modes of vibration contributes to the greenhouse effect? Explain your reasoning.

17. If a carbon dioxide molecule interacts with certain photons in the IR region, the molecule vibrates. For CO_2, the major wavelengths and their corresponding wavenumbers of absorption occur at 4.26 μm (2350 cm^{-1}) and 15.00 μm (667 cm^{-1}).

   a. What is the energy corresponding to each of these IR photons?

   b. What happens to the energy in the vibrating CO_2 species?

18. Water vapor makes up about 1% of our atmosphere, but it is not regulated as a greenhouse gas. Why not?

19. What effect would each of these changes have on global warming?

   a. volcanic eruptions

   b. CFCs in the troposphere

   c. CFCs in the stratosphere

20. One of the biochemical processes that releases carbon dioxide to the atmosphere is the fermentation of sugar to produce alcohol. For example, when glucose, $C_6H_{12}O_6$, undergoes fermentation, ethanol (C_2H_5OH) and CO_2 are produced. Yeast is used to catalyze this conversion. Write a balanced chemical equation for this reaction.

21. Consider Figure 3.19.

   a. What is the major source of CO_2 emission from fossil fuel combustion?

   b. What alternatives exist for each of the major contributors to CO_2 emissions?

22. Silver has an atomic mass of 107.9 and an atomic number of 47.

   a. What is the number of protons, neutrons, and electrons in a neutral atom of the most common isotope, Ag-107?

   b. How do the numbers of protons, neutrons, and electrons in a neutral atom of Ag-109 compare with those of Ag-107?

23. There are just two naturally occurring isotopes of silver. If silver-107 accounts for 52% of natural silver, what is the mass number of the other isotope of silver?

24. a. Calculate the average mass (in grams) of an individual atom of silver.

   b. Calculate the mass (in grams) of 10 trillion silver atoms.

   c. Calculate the mass (in grams) of 5.00 $\times$ 10^{45} silver atoms.

25. Calculate the molar mass of each of these important gases in atmospheric chemistry.

   a. H_2O

   b. CCl_2F_2 (Freon-12)

   c. NO

26. a. Calculate the mass percent of chlorine in CCl_3F (Freon-11).

   b. Calculate the mass percent of chlorine in CCl_2F_2 (Freon-12).

   c. What is the maximum mass of chlorine that could be released in the stratosphere by 100 g of each compound?

   d. How many molecules of chlorine correspond to the masses calculated in part c?

27. The total mass of carbon in living systems is estimated to be 7.5 $\times$ 10^{17} g. Given that the total mass of carbon on Earth is estimated to be 7.5 $\times$ 10^{22} g, what is the ratio of carbon atoms in living systems to the total carbon atoms on Earth? Report your answer in percent and in ppm.

28. Consider the information presented in Table 3.4. Calculate these changes.

   a. What is the percent increase in CO_2 when comparing 2000 concentrations with preindustrial concentrations?

   b. Considering CO_2, CH_4, and N_2O, which has shown the greatest percentage increase when comparing 2000 concentrations with preindustrial concentrations?

29. Consider the information presented in this graph.

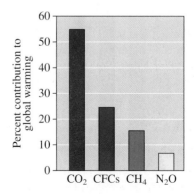

   a. Which substance makes the largest percentage contribution to global warming?

b. Use these percentages together with the global warming potentials for CO_2, CH_4, and N_2O given in Table 3.5 to decide which of these gases has the largest effect on global warming. Explain your reasoning. *Hint:* One approach is to calculate "net effectiveness" by finding the product of % contribution and GWP for each gas.

30. Is the tropospheric abundance of methane given in Table 3.5 the same as that given in Table 3.4? Explain your reasoning.

Concentrating on Concepts

31. The text makes a distinction between the *correlation* of two events and the *causation* of one by the other. Identify each of these pairs as an example of correlation, causation, or no relationship. Explain your reasoning.

   a. metric tons of coal burned | metric tons of CO_2 emitted

   b. national per capita income | per capita emission of CO_2

   c. number of cigarettes smoked per day | increase in the incidence of lung cancer

   d. number of bonds between two oxygen atoms | length of oxygen-to-oxygen bond

   e. building a greenhouse | raising beautiful tropical plants

   f. buying a pair of roller blades | breaking your leg

32. Why do people sometimes bring living plants, rather than cut flowers, to help a friend along the road to recovery from an illness?

33. Given that direct measurements of Earth's atmospheric temperature over the last several thousands of years are not available, how can scientists know the fluctuations in the temperature in the past?

34. Consider Figure 3.5 showing atmospheric concentrations of CO_2 at Mauna Loa, and Figure 3.6, showing temperature changes on Earth's surface. Why is the pattern with the trend so regular in Figure 3.5, but not as regular in Figure 3.6?

35. Consider this figure showing sources of CO_2 emissions in the United States for 1998.

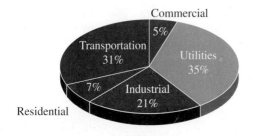

a. Compare this figure to Figure 3.19 in the text, which gives CO_2 emissions for 2000. Which values have increased? Which have decreased?

b. If you wanted to reduce your personal contribution to CO_2 emissions, what changes could you make that would be the most effective? Explain your reasoning.

c. Would the entire quantity of CO_2 emitted end up in the atmosphere? Why or why not?

36. The text makes the statement that Lewis structures show linkages (what is connected to what) but they do not show shape. Explain this statement, using the molecule H_2O as an example.

37. Carbon dioxide gas and water vapor both absorb IR radiation. Do they also absorb visible radiation? Offer some evidence based on your everyday experiences to help explain your answer.

38. How do you think that the energy required to cause IR-absorbing vibrations in CO_2 would change if the carbon and oxygen atoms were connected with single rather than with double bonds?

39. Explain why water contained in a glass cup is quickly warmed in a microwave oven, but the glass cup warms much more slowly, if at all.

40. Ethanol, C_2H_5OH, can be isolated from sugars and starches in crops such as corn or sugar cane. The ethanol is used as a gasoline additive and when burned, it combines with O_2 to form H_2O and CO_2.

   a. Write a balanced equation for the complete combustion of C_2H_5OH.

   b. How many moles of CO_2 are produced from each mole of C_2H_5OH completely burned?

   c. How many moles of O_2 are required to burn 10 mol of C_2H_5OH?

41. Hexane, C_6H_{14}, and octane, C_8H_{18}, are both hydrocarbons that will burn with oxygen gas, O_2. The products are H_2O and CO_2.

   a. Write a balanced equation showing the reaction when hexane burns.

   b. Write a balanced equation showing the reaction when octane burns.

   c. Compare the number of moles of CO_2 produced when 1 mol of each hydrocarbon burns.

42. Why is the atmospheric lifetime of a potential greenhouse gas important?

43. CO_2 has a greater density than N_2. Why don't these gases settle out into layers in the atmosphere?

44. One of the first radar devices developed during World War II used microwave radiation of a specific wavelength that triggers the rotation of water molecules. Why do you think this design was not successful?

45. It is estimated that Earth's ruminants, such as cattle and sheep, produce 73 million metric tons of CH_4 each year. How many metric tons of carbon are present in this mass of CH_4?

46. The three warmest years in the United States in the last 100 years have been 1998, 2001, and 2002. Does this prove that the enhanced greenhouse effect (global warming) is taking place? Explain your reasoning.

47. A possible replacement for CFCs is HFC-152a, with a lifetime of 14 years and a GWP of 120. Another is HFC-23, with a lifetime of 260 years and a GWP of 12,000. Both of these possible replacements have a significant impact as greenhouse gases and are regulated under the Kyoto Protocol.

    a. Based on the given information, which appears to be the better replacement? Consider only the potential for global warming.

    b. What other considerations are there in choosing a replacement?

48. The quino checkerspot butterfly is an endangered species with a small range in northern Mexico and Southern California. Evidence reported in 2003 indicates that the range of this species is even smaller than previously thought.

    a. Propose an explanation why this species is being pushed north, out of Mexico.

    b. Propose an explanation why this species is being pushed south, out of southern California.

    c. Propose a plan to prevent further harm to this endangered species.

49. This figure shows global emissions of CO_2 in metric tons (tonnes) per person per year.

Who pumps out the most CO₂?

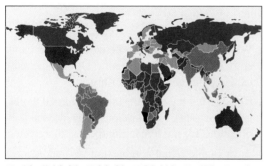

■ <1.0 ■ 1.0–2.9 ■ 3.0–6.9 ■ 7.0–14.9 ■ >15.0 ■ no data

Metric tons per person

Source: From *New Scientist Supplement*, April 28, 2001. Reprinted with permission.

    a. Which countries lead the world in CO_2 emissions?

    b. The values in this figure are reported as metric tons of CO_2, not as metric tons of C in CO_2. How are these values related? *Hint:* Think about the mass relationships developed in Section 3.6.

c. Does the value for U.S. CO_2 emissions in this figure agree with the value given in Figure 3.27? Explain your reasoning.

50. This figure shows who uses the most energy and who gains the greatest value for money spent on energy.

Who uses most energy? And who gets most value for money?

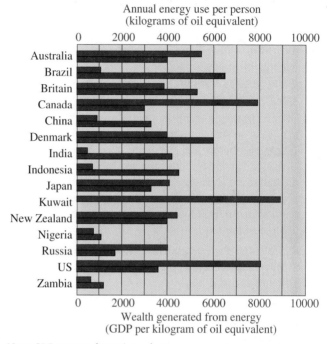

Note: GDP = gross domestic product

Source: From *New Scientist Supplement*, April 28, 2001. Reprinted with permission.

    a. What units are used to report annual energy use per person?

    b. Which countries lead the world in energy use?

    c. What units are used to report wealth generated from energy?

    d. Which countries lead the world in wealth generated from energy?

    e. What implications for U.S. policy are suggested by this figure? Explain your reasoning.

Exploring Extensions

51. The molecule BF_3 is planar with boron in the center and fluorine atoms forming a triangle. It has triangular planar 120° fluorine-to-boron-to-fluorine bond angles, whereas the NF_3 molecule is a triangular pyramid with fluorine-to-nitrogen-to-fluorine bond angles of about 103°. Account for the different geometries and bond angles. *Hint:* The boron atom is too small to obey the octet rule. It is stable with only six outer electrons.

52. a. Assuming only single bonds are present, how many outer electrons are needed to form a three-atom

molecule from atoms of X, Y, and Z? Assume that X, Y, and Z follow the octet rule and that Y is in the center.

    b. What shape is the molecule formed from X, Y, and Z in part **a,** and what is the X-to-Y-to-Z bond angle? Explain your reasoning.

    c. How would the answers to parts **a** and **b** change if you removed the restriction on using only single bonds? Explain.

53. Figure 3.6 shows the temperature increase on Earth's surface from 1880 to 2000. Imagine you are in charge of extending this graph to include the present year. What kind of information would you need and how might you gather such data? *Hint:* Consider the source of the data in Figure 3.6.

54. If a water molecule interacts with certain photons in the IR region, the molecule vibrates (see Figure 3.16). In Consider This 3.16, you estimated the wavenumber, expressed in cm^{-1}, for water's two maximum absorbancies. Then, you converted those values to wavelengths, expressed in μm. (Complete this if you have not already done so.)

    a. Calculate the energy associated with each maximum absorbance. *Hint*: See equation 2.3 for the appropriate relationship.

    b. Which absorption peak represents higher energy?

    c. Does the higher energy absorption correspond to bending or stretching in the water molecule?

55. Data taken over time reveal an increase in CO_2 in the atmosphere. The large increase in the combustion of hydrocarbons since the Industrial Revolution is often cited as a reason for the increasing levels of CO_2. However, an increase in water vapor has *not* been observed during the same period. Remembering the general equation for the combustion of a hydrocarbon, does the difference in these two trends *disprove* any connection between human activities and global warming? Explain your reasoning.

56. In the energy industry, 1 standard cubic foot (SCF) of natural gas contains 1196 mol of methane (CH_4) at 15.6 °C (60 °F).

    a. How many moles of CO_2 could be produced by the complete combustion of 1 SCF of natural gas?

    b. How many kilograms (kg) of CO_2 could be produced?

    c. How many metric tons of CO_2 could be produced? *Hint:* See Appendix 1 for conversion factors.

    d. How many pounds (lb) of CO_2 could be produced? *Hint:* See Appendix 1 for conversion factors.

57. Consider this information about three greenhouse gases. These data represent the change from preindustrial time until 2000. Then compare these data with those given in Table 3.4, which updates the information with changes

suggested by the 2001 report of the United Nations Intergovernmental Panel on Climate Change.

Greenhouse Gases–Concentration Changes and Lifetimes

	CO_2	CH_4	N_2O
Preindustrial conc.	280 ppm	0.70 ppm	0.28 ppm
2000 conc.	370 ppm	1.8 ppm	0.31 ppm
Rate of conc. change	1.5 ppm/yr	0.010 ppm/yr	0.0008 ppm/yr
Atmospheric lifetime (yr)	50–200	12	120

    a. What is different between these two sets of data? Offer a possible explanation for any differences observed.

    b. Use these data and those in Table 3.4 in the text to write a commentary for your local newspaper explaining which gases are experiencing the greatest percent increases. Explain some of the reasons for the observed increases and explain how knowing the atmospheric lifetime of a greenhouse gas is an important piece of information for setting control strategies for greenhouse gases.

58. The per capita CO_2 emissions for several countries are shown in Figure 3.27. This list gives the estimated 2002 gross domestic product (GDP) per capita for some of those countries.

U.S.	$36,300
Canada	$29,400
Japan	$28,000
Australia	$27,000
Brazil	$ 7,400
China	$ 4,600
India	$ 2,540

    a. What is the relationship between the per capita CO_2 emissions and GDP per capita? Offer some reasonable explanation for the relationship.

    b. Considering the relationship between GDP per capita and per capita CO_2 emissions, what are the policy implications for implementing the Kyoto agreement?

59. Incomplete fuel combustion also adds greenhouse gases to the atmosphere. Researchers are studying emissions from charcoal stoves, ceramic stoves, and simple three-stone stoves with a metal grate to hold the cooking pot in an agricultural community in central Kenya. Results have been published in 2003 by Bailis and colleagues in *Environmental Science and Technology* and reported briefly in *Science* on 18 April 2003. Find out more about this research, including which greenhouse gases and air pollutants are emitted and which stoves are best. Also discuss some of the policy implications of this research.

60. The world community responded differently to the atmospheric problems described in Chapters 2 and 3. The evidence of ozone depletion was met with the Montreal Protocol, a schedule for decreasing the production of ozone-depleting chemicals. The evidence of global warming was met with the Kyoto Protocol, a plan calling for targeted reduction of greenhouse gases.

a. Suggest reasons why the world community dealt with the issue of ozone depletion before that of global warming.

b. Compare the current status of the two responses. When was the latest amendment to the Montreal Protocol? How many nations have ratified it? Has the level of chlorine in the stratosphere dropped as a result of the Montreal Protocol? How many nations have ratified the Kyoto Protocol? Has it gone into effect? Have any other initiatives been proposed? Have levels of greenhouse gases dropped as a result of the Kyoto Protocol?

Catherine's favorite

chapter

4

Energy, Chemistry, and Society

The U.S. uses prodigious amounts of energy daily from burning gasoline and natural gas. Under what circumstances can we expect unlimited availability and guaranteed low prices?

"As a country, we have demanded more and more energy. But we have not brought on line supplies needed to meet that demand. . . . We can explore for energy, we can produce energy and use it, and we can do so with a decent regard for the natural environment."

Vice-President Richard Cheney, "Reliable, Affordable, Environmentally Sound Energy for America's Future," *May 2001*

"From a national perspective, there is overwhelming evidence that renewable energy can provide reliable, safe, affordable energy. . . . Even with conservative assumptions, a 10% renewable base energy supply would lower the nation's energy bills by $15 billion per year by 2020 compared to a heavily fossil based supply mix."

George Sterzinger, Executive Director, Renewable Energy Policy Project, "Energy: Maximizing Resources, Meeting Our Needs and Retaining Jobs," Testimony to the House Government Reform Committee Subcommittee on Energy Policy, Natural Resources and Regulatory Affairs, *June 2002*

"The 'Climate Stewardship Act of 2003' . . . would address the supposed problem of 'global warming' by suppressing America's use of energy. . . . The government sets a mandatory cap on total greenhouse gas emissions. . . . By suppressing the amount of energy that can be produced, the Lieberman/McCain plan will inevitably lead to energy scarcity and, thus, to higher energy prices."

Bonner R. Cohen, Senior Fellow at The National Center for Public Policy Research, in National Policy Analysis, *September 2003*

The citizens of this country generate a tremendous need for energy and do so with an almost casual attitude about its availability and cost. The opening images remind us of various facets of energy production and use. Fossil fuels (coal, oil, and natural gas) account for about 70% of U.S. electricity and almost 85% of all of the nation's energy needs. After the country experienced shortages of petroleum products in the middle and late 1970s that were precipitated by political turmoil in the Middle East, our national policy focused on decreasing dependence on oil from that part of the world. The changes that resulted from new policies did produce a lower demand overall and a redistribution of our suppliers. However, in the last two decades our imports have risen from about 35% of our needs to over 50%.

Every year the use of and demand for energy increases in the United States and at an even higher rate in other parts of the world. To some extent the oil suppliers respond to those needs, but estimates suggest that at current rates of consumption, oil reserves will be depleted in less than 50 years and coal in over a century and a half. Renewable energy sources continue to attract the attention of policy makers and the energy industry alike. For example, traditional suppliers of fossil fuels are the largest investors in wind power, electricity generated by farms of windmills. An increasing amount of grain produced in this country is finding its way into alternative fuels: ethanol from the fermentation of corn that also acts as an oxygenate in gasoline to reduce emissions and biodiesel produced from the oil isolated from soybeans.

The increasing use of ethanol as an alternative fuel points to another aspect of this country's voracious appetite for energy. We burn fossil fuels and the alternative ones to release their energy. The combustion of these carbon-containing materials to carbon dioxide is what releases the molecules' huge energy content. But carbon dioxide, and the accompanying oxides of sulfur and nitrogen, are the basis of monumental problems: global warming, acid rain, and other environmental issues. How do we approach the seemingly intractable relationship among all of these factors? Conservation measures seem rational, particularly if they do not have negative economic implications. Increasing our renewable energy sources would certainly have positive outcomes. But major changes to the way we use fuels and the ways in which they are supplied raise questions about who is willing and able to bear the costs. Before we can understand these issues, let's look at some fundamentals in chemistry.

Besides factors like availability and costs, materials to be used as fuels are selected by virtue of their chemical structures and the resulting properties. Therefore, we must

Air quality was discussed in Chapter 1 and the contribution of carbon dioxide to global warming in Chapter 3.

understand some of the molecular aspects of the fuels. We start by defining some forms of energy, namely heat and work, and a law that tells us that energy is never created or destroyed, just changes forms. Inefficiency becomes an issue in our ability to harness the energy or produce it completely in a form we want. We will learn how chemical structure is responsible for the storage of energy, reactions that describe its release, and a method to predict such a process. Once we understand these principles, we turn our attention to energy consumption and to the major forms of fossil fuels—coal and petroleum—and to manipulating them into useful forms. The story ends with a discussion about policies and the possibilities of conservation.

Consider This 4.1   **Energy in the News**

a. Examine the Web site of your local newspaper for recent articles about local, regional, and national aspects of energy production, use, and policies.

b. Search national or international news sites for citations about the same issue. Do the items in both sets of sources overlap? Do the "big picture" items find their way into the local news?

4.1 Energy, Work, and Heat

Energy is a word we hear every day, but whose precise chemical meaning is not well understood by the public. Unfortunately the scientific definition, the capacity to do work, contains another word whose colloquial use does not match the chemical one. To a scientist, **work** is done when movement occurs against a restraining force. Mathematically, work is equal to the force multiplied by the distance over which the motion occurs. Thus, when you lift a book against the force of gravity, you are doing work. When you hold that same book motionless above your head, you are not, strictly speaking, working. If you hold it there all day, you are again doing something that will not happen by itself. Your arms will suggest to you that it does require energy.

Much of the work done on our planet comes from the most common form of chemical work, the expansion of gases like those produced in an internal combustion engine, or from another familiar form of energy—heat. The formal definition of the latter sounds a little strange: **heat** is energy that flows from a hotter to a colder object. When we grab a hot pan on the stove, we immediately experience heat. Temperature is a property of heat; it defines the degree of hotness (or coldness) on a specified scale. Another definition of temperature sounds awkward and circular: **temperature** is a property that determines the direction of heat flow. However, we know that temperature and heat are not the same thing. When two bodies are in contact, heat always flows from the object at the higher temperature to that at the lower temperature. Your bottle of water and the Pacific Ocean may be at the same temperature but the ocean contains and can transfer far more heat than the bottle of water. Indeed, bodies of water can affect the climate of an entire region as a consequence of the transfer of heat.

Heat is a consequence of motion at the molecular level. For example, when liquid water in a pan absorbs heat, its molecules move more rapidly. Temperature is a statistical measure of the average speed of that motion. Hence, temperature rises as the amount of heat energy in a body increases. Consider two containers of water, each at room temperature (25 °C). One contains 100 mL of water; the other holds 200 mL of water. Because the two water samples are at the same temperature, the average speed of their water molecules is the same, but their heat content is not the same. Starting at the same lower temperature, it takes twice as much heat energy to raise the temperature of 200 mL of water to 25 °C than it does to reach that temperature for 100 mL. Therefore, the 200 mL of water has twice the heat energy of the smaller volume of water.

In order to continue our discussion of energy, we need a unit in which to express it. Historically, there have been many, but recently there has been an international agreement to make the common unit the joule. To put this unit into context, one joule (1 J) is approximately equal to the energy required to raise a 1-kg book 10 cm against the force of gravity. On a more personal basis, each beat of the human heart requires about 1 J of energy. In the next few sections, we will use energy units that range from kilojoules (1 kJ = 1000 J) in discussions about chemical bonds to exajoules (1 EJ = 10^{18} J) when we consider the energy usage of the entire world in one year.

The calorie was introduced as a measure of heat with the metric system in the late 18th century. Originally, the **calorie** was defined as the amount of heat necessary to raise the temperature of exactly one gram of water by one degree Celsius. It has been redefined as exactly 4.184 J. Calories are perhaps most familiar when used to express the energy released when foods are metabolized. The values tabulated on package labels and in cookbooks are, in fact, kilocalories (1 kcal = 1000 cal = 1 Calorie); when Calorie is written with a capital "C," it generally means kilocalorie (Figure 4.1). Thus, the energetic equivalent of a donut is 425 Cal (425 kcal, 425,000 calories). For most purposes we will use joules and kilojoules in this chapter, but when it seems more appropriate or more easily understandable, we express energy in calories or kilocalories. We will not worry about British Thermal Units (Btu), ergs, or foot-pounds, but you are cautioned that the world also expresses energy in these units.

> 1 kg = 2.2 lb
> 10 cm = 3.9 in.

> More exactly, the calorie is the heat required to raise the temperature of one gram of water from 14.5 °C to 15.5 °C.

Figure 4.1
The energy in food is listed in Calories or kilojoules.

Your Turn 4.2   Simple Energy Calculations

a.   When a donut is metabolized, 425 kcal (425 food Calories) are released. Convert this amount to kilojoules.

b.   Calculate the number of 2-lb books you could lift to a shelf 6 ft off the floor with the amount of energy from metabolizing one donut.

c.   A 12-oz can of a soft drink has an energy equivalent of 92 kcal. Convert the energy released when metabolizing the soft drink, expressing the answer in kilojoules.

d.   Assume that you use this energy to lift concrete blocks that weigh 22 lb (10 kg) each. How many blocks could you lift to a height of 4 ft with this quantity of energy?

Answers

a.   Recall that 1 kcal is equivalent to 4.184 kJ.

$$425 \text{ kcal} \times \frac{4.184 \text{ kJ}}{1 \text{ kcal}} = 1.78 \times 10^3 \text{ kJ}$$

b.   Earlier the text stated that 1 J is approximately equal to the energy required to raise a 2-lb book a distance of 4 in. against Earth's gravity. We can use this information to calculate the number of 2-lb books that could be lifted 6 ft. First note that 6 ft is 72 in. Next, calculate the energy (joules) required to lift a 2-lb book:

$$72 \text{ in.} \times \frac{1 \text{ J}}{4 \text{ in.}} = 18 \text{ J}$$

Then, express this value in kilojoules.

$$18 \text{ J} \times \frac{1 \text{ kJ}}{10^3 \text{ J}} = 0.018 \text{ kJ}$$

Use this value to make the final calculation.

$$1.78 \times 10^3 \text{ kJ} \times \frac{1 \text{ book}}{0.018 \text{ kJ}} = 9.9 \times 10^4 \text{ books}$$

In round numbers, about 100,000 books! Lots of exercise is required to work off one donut.

Sceptical Chymist 4.3   Checking Assumptions

A simplifying (and erroneous) assumption was made in doing the calculations in parts **b** and **d** of the previous Your Turn. What was the assumption, and is it reasonable? Is the answer based on this assumption too large or too small? Give a reason for your answer.

4.2 Energy Transformation

Essentially all the fuels we will consider in this chapter—coal, oil, alcohol, and garbage—give up their energy through combustion. Each of the fuels burns to generate heat, liberating simpler molecules like carbon dioxide and water. Energy stored in the chemical bonds of the fuel molecules is released during combustion. The **first law of thermodynamics,** also called the law of conservation of energy and mass, states that energy is neither created nor destroyed. Energy often changes forms as it did in the combustion reaction, but the energy of the universe is constant.

> The law of conservation of energy and mass is first defined in Chapter 1.

For the most part, heat is not the form in which the energy is ultimately used. Although heat is nice to have around on a cold winter day, it is a cumbersome form of energy. It is dangerous if uncontrolled, difficult to transport, and hard to harness. The industrialization of the world's economy began with the invention of devices to convert heat to work. Chief among these was the steam engine, developed in the latter half of the 18th century. The heat from burning wood or coal was used to vaporize water, which in turn was used to drive pistons and turbines. The resulting mechanical energy was used to power pumps, mills, looms, boats, and trains. The smoke-belching mechanical monsters of the English midlands soon replaced humans and horses as the primary source of motive power in the Western world.

A second energy revolution occurred early in the 1900s with the commercialization of electrical power. Today, most of the electrical energy produced in the United States is generated by the descendants of those early steam engines. Figure 4.2 illustrates a modern power plant. Heat from the burning fuel is used to boil water, usually under high pressure. The elevated pressure serves two purposes: it raises the boiling point of the water

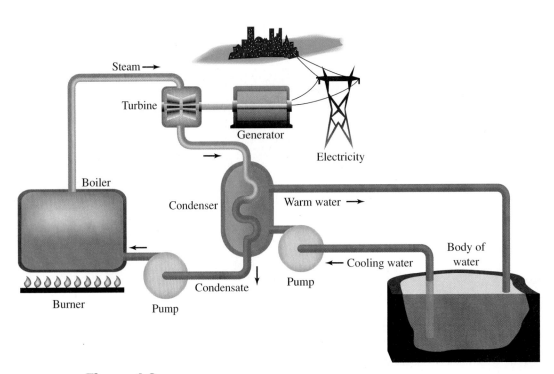

Figure 4.2
Diagram of a power plant for the conversion of heat to work to electricity.

Figure 4.3

Energy transformation in an electrical power plant.

and it compresses the water vapor. The hot, high-pressure vapor is directed at the fins of a turbine. As the gas expands and cools, it gives up some of its energy to the turbine, causing it to spin like a pinwheel in the wind. The shaft of the turbine is connected to a large coil of wire that rotates within a magnetic field. The turning of this dynamo generates an electric current—a stream of electrons that represents energy in a new and particularly convenient form. Meanwhile, the water vapor leaves the turbine and continues in its closed cycle. It passes through a heat exchanger where a stream of cooling water carries away the remainder of the heat energy originally acquired from the fuel. The water condenses into its liquid state and reenters the boiler, ready to resume the energy transfer cycle.

This process of energy transformation is summarized in the three steps diagrammed in Figure 4.3. **Potential energy,** the form of energy related to the positions of atoms and molecular structure and stored in the chemical bonds of fossil fuels, is first converted to *heat energy.* The heat released in combustion is absorbed by the water, vaporizing it to steam. This heat is then transformed into *mechanical energy* in the spinning turbine that turns the generator that changes the mechanical energy into *electrical energy.* In compliance with the first law of thermodynamics, energy is conserved throughout these transformations. To be sure, no new energy is created, but none is lost, either. We may not be able to win, but we can at least break even . . . or can we?

The last question is not as facetious as it might sound. In fact, we cannot break even. No power plant, no matter how well designed, can completely convert heat into work. Inefficiency is inevitable, in spite of the best engineers and the most sincere environmentalists. There are, of course, energy losses due to friction and heat leakage that can be corrected, but these are not the major problems. The chief difficulty is nature; more specifically, the nature of heat and work.

Table 4.1 lists the efficiencies of a number of steps in energy production. The overall efficiency is the product of the efficiencies of the individual steps: the individual efficiencies are multiplied. The net result is that today's most advanced power plants operate at an overall efficiency of only about 42%. Consider, for example, the case of electrical home heating, sometimes touted as being clean and efficient. We will assume that the electricity is produced by a methane-burning power plant with a maximum theoretical efficiency of 60%. The efficiencies of the boiler, turbine, electrical generator, and power transmission lines are given in Table 4.1; converting the electrical energy back into heat in the home is 98% efficient.

Efficiencies are multiplicative. To find the overall efficiency of the electricity-generation to home-heating sequence, we multiply the efficiencies of the individual steps, expressed as their decimal equivalents.

Overall efficiency = efficiency of (power plant) × (boiler) × (turbine)
× (electrical generator) × (power transmission) × (home electric heater)

= 0.60 × 0.90 × 0.75 × 0.95 × 0.90 × 0.98 = 0.34

> In this calculation we have used the maximum theoretical efficiency of a powerplant, not its actual operating efficiency.

Table 4.1	**Some Typical Efficiencies in Power Production**
Maximum theoretical efficiency	55–65%
Efficiency of boiler	90%
Mechanical efficiency of turbine	75%
Efficiency of electrical generator	95%
Efficiency of power transmission	90%

The overall efficiency of 0.34 indicates that only 34% of the total heat energy derived from the burning of methane at the power plant is available to heat the house. If the electrically heated house requires 3.5×10^7 kJ (a typical value for a northern city in January), how much methane (in grams) has to be burned at the power plant to furnish the heat needed? The combustion of 1 gram of methane releases 50.1 kJ. Remember that only 34% of the energy from the burned methane is available to heat the house. So, because of inefficiencies, far more methane has to be burned than the amount to release 3.5×10^7 kJ. The total quantity of heat that must be used can be calculated.

$$\text{Heat needed} = \text{Heat used} \times \text{efficiency}$$

$$3.5 \times 10^7 \text{ kJ} = \text{Heat used} \times 0.34$$

$$\text{Heat used} = \frac{3.5 \times 10^7 \text{ kJ}}{0.34} = 1.0 \times 10^8 \text{ kJ}$$

Because each gram of burning methane yields 50.1 kJ, 2.0×10^6 g of methane must be burned to furnish 1.0×10^8 kJ.

$$1.0 \times 10^8 \text{ kJ} \times \frac{1 \text{ g } CH_4}{50.1 \text{ kJ}} = 2.0 \times 10^6 \text{ g } CH_4$$

You can compare the efficiencies of heating a home with electricity or a natural gas furnace by completing the following Sceptical Chymist activity.

Sceptical Chymist 4.4   **Clean Electric Heat**

Is electric heat clean and efficient? The electricity must first be generated, usually by a fossil-fuel power plant.

a.   The house could also have been heated directly with a gas furnace burning methane at 85% efficiency. Calculate the number of grams of methane required in January using this method of heating.

b.   On the basis of your answer to part **a** and the discussion preceding this exercise, comment on the claim that electric heat is "clean" and efficient.

Answer

a.   Because the only inefficiency is that of the furnace, we can do a calculation similar to that just done but using 0.85 as the efficiency. The energy required is 4.1×10^7 kJ, which corresponds to 8.2×10^5 g of methane.

Your Turn 4.5   **Energy and Efficiency**

A coal-burning power plant generates electrical power at a rate of 500 megawatts (5.00×10^8 J/s). The plant has an overall efficiency of 0.375 for the conversion of heat to electricity.

a.   Calculate the total quantity of electrical energy (in joules) generated in one year of operation and the total quantity of heat energy used for that purpose.

b.   Assuming the power plant burns coal that releases 30 kJ/g, calculate the mass of coal (in grams and metric tons) that will be burned in 1 year of operation (1 metric ton = 1×10^3 kg = 1×10^6 g).

Answers

a. 1.58×10^{16} J generated; 4.20×10^{16} J used

b. 1.40×10^{12} g; 1.40×10^6 metric tons

Although we asserted that heat cannot be completely converted into work, we offered little evidence and no explanation for this fact of nature. To illustrate the problem, push this book off the desk, and wait for it to come back up by itself. Be prepared to wait quite a while! The idea of a book picking itself off the ground and rising against the force of gravity is so bizarre, it is unbelievable. In fact, it is just unlikely—extremely unlikely. To understand why, let us examine the process in a little more detail.

You probably recall that a book resting on a table has potential energy by virtue of its position above the floor. When the book is dropped, the potential energy is converted to **kinetic energy,** the energy of motion. When the book strikes the floor, the kinetic energy is released with a bang. Some of it goes into the shock wave of moving air molecules that transmit the sound. Most of the energy goes to increase the motion of the atoms and molecules in the book and in the floor beneath it. This microscopic motion is the origin of what we call *heat*. A careful measurement would show that the book and the floor and the air immediately around them are all very slightly warmer than they were before the impact. Energy has been conserved, but it has also been dissipated. Heat, or **thermal energy,** is characterized by the random motion of molecules. They move chaotically in all directions.

> The potential energy of a molecule that reacts or burns is stored in its chemical bonds. The potential energy of an object (or molecule) that can move or fall is stored in its position.

Now consider what would be necessary for the book to rise by itself, and thus to work against the force of gravity. All the molecules in the book would have to move upward at the same time. At that instant, all the molecules in the floor under the book would also have to move in an upward direction, giving the book a little shove. Needless to say, such agreement among the 10^{25} or so molecules involved is very unlikely. Yet, such a change is necessary to convert heat into work—to transform random thermal motion into uniform motion. The first law of thermodynamics may confidently assert that all forms of energy are equal, but the fact remains that some forms of energy are more equal than others! The chaotic, random motion that is heat is definitely low-grade energy.

Both the inability of a power plant to convert heat into work with 100% efficiency or for dropped books to spontaneously rise are both manifestations of the same law of nature, the **second law of thermodynamics.** The second law has many versions, but all describe the directionality of the universe. One version states that it is impossible to completely convert heat into work without making some other changes in the universe. Another observes that heat will not of itself flow from a colder to a hotter body. The falling book provides another version of the second law. That, too, is a transformation of ordered kinetic energy into random heat energy. Like all naturally occurring changes, it involves an increase in the disorder or randomness of the universe. This randomness in position or energy level is called **entropy.** The most general statement of the second law of thermodynamics is that the entropy of the universe is increasing. This means that organized energy, the most useful kind for doing work, is always being transformed into chaotic motion or heat energy.

A helpful way to look at the increase in universal entropy that characterizes all changes is in terms of probability. Disordered states are more probable than ordered ones, and natural change always proceeds from the less probable to the more probable. Let's suppose you define perfect order as a beautifully organized sock drawer, all the socks matched, folded, and placed in rows. This would represent a condition of low entropy (little randomness). If you are like most people, this is probably a rather unlikely arrangement. It certainly did not occur by itself; it took work to organize the socks. Without the continuing work of organization, it is quite possible that, over the course of a week or a month or a semester, the entropy and disorder of that sock drawer would increase. The point is that the socks can be mixed up in lots of ways, and therefore disorder is more probable than order. Conversely, it is not very likely that you will open your drawer some morning and find that the previously jumbled socks are in perfect order and the entropy in that particular part of the universe has suddenly and spontaneously decreased without any external intervention. That sort of change from disorder to order is essentially what is involved in the conversion of heat to work. Henry Bent, a renowned chemist, has estimated that the probability of the complete conversion of one calorie of heat to work is about the same as the likelihood of a bunch of monkeys typing Shakespeare's complete works 15 quadrillion times (15×10^{15}) in succession without a mistake.

Perhaps by now some Sceptical Chymist in the class has objected that there are many earthly instances in which order increases. Sock drawers do get organized, power plants convert heat to work, dropped objects get picked up, water can be decomposed into hydrogen and oxygen, refrigerators transfer heat from a colder to a hotter body, and students learn chemistry. All of these are "unnatural" events—**nonspontaneous** in the vocabulary of thermodynamics. They will not occur by themselves; they require that work be done by someone or something. An input of energy is necessary to reduce the entropy and increase the order. And in every case, the work that is done generates more entropy somewhere in the universe than it reduces in one small part of it. Even when entropy appears to decrease in a spontaneous change, for example, the freezing of water at temperatures below 0 °C, there are balancing increases in entropy. In this particular case, the heat given off by the freezing water adds to the disorder of its surroundings. In short, when the entire universe is considered, entropy always increases. As the saying goes, there is no free lunch!

One word of caution: you need to be a little careful about the scientific meaning of *spontaneous* and *nonspontaneous*. Spontaneous is often taken to mean a process that occurs all by itself, with no apparent initiation, as in the tabloid headline "SLEEPING MAN BURSTS INTO FLAME." In scientific usage, a spontaneous change is one that *could* occur; in other words, it is thermodynamically possible. But it might not take place all by itself because a large energy barrier must be overcome to start the reaction. Let us return to our sleeping man. Human beings are thermodynamically unstable with respect to combustion products such as carbon dioxide and water. So the burning of a human is a spontaneous change in the scientific sense of that term. But, fortunately for us, there is an energy barrier to the process that is so high that we do not have to worry about bursting into flames without any provocation. In the case of the reported sleeping man, it would be a good idea to look for an outside agent, perhaps a disgruntled friend, who might have helped him over that barrier with a can of gasoline and a match.

Consider This 4.6   Humpty Dumpty

All around us are examples of the natural tendency for things to get messed up. Some have even been enshrined in literature: "All the king's horses and all the king's men, couldn't put Humpty Dumpty together again." Cite some examples of your own.

Consider This 4.7   Entropy Decrease–Entropy Increase

During midterm time, many students become very serious about their studying, and, for hours on end, will concentrate on the plays of Shakespeare or the causes of World War II. This decrease in "intellectual entropy" is often associated with an increase in the entropy of the student's room. Identify another process in which entropy appears to decrease but is actually coupled with an increase in entropy elsewhere in the universe.

4.3 From Fuel Sources to Chemical Bonds

What makes some substances such as coal, gas, oil, or wood usable as fuels, while many others are not? To find an answer, we must consider the molecular properties of fuels and how energy is released from them. The most common energy-generating chemical reaction is burning, or combustion. **Combustion** is the combination of the fuel with oxygen to form products. In such a chemical transformation, the potential energy of the reactants is greater than that of the products. Because energy is conserved, the difference in energy is given off, primarily as heat.

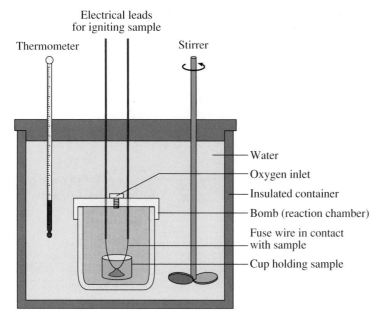

Electrical leads
for igniting sample

Thermometer Stirrer

Water
Oxygen inlet
Insulated container
Bomb (reaction chamber)
Fuse wire in contact
with sample
Cup holding sample

Figure 4.4
Schematic drawing of a bomb calorimeter.

We illustrate the process with the combustion of methane, CH_4, the principal component of natural gas, a major home heating fuel. The products are carbon dioxide and water. In Chapter 1 you encountered this combustion equation.

$$CH_4(g) + 2\,O_2(g) \longrightarrow CO_2(g) + 2\,H_2O(g) + \text{energy} \qquad [4.1]$$

The reaction is said to be **exothermic,** a term applied to any chemical or physical change accompanied by the release of heat. The quantity of heat energy released in a combustion reaction such as this can be experimentally determined with a device called a **calorimeter** (Figure 4.4). Not surprisingly, the amount of heat generated depends on the amount of fuel burned. Therefore, a known mass of fuel and an excess of oxygen are introduced into a heavy-walled stainless steel "bomb." The bomb is then sealed and submerged in a bucket of water. The reaction is initiated with an electric current that burns through a fuse wire. The heat evolved by the exothermic reaction flows from the bomb to the water and the rest of the apparatus. As a consequence, the temperature of the entire calorimeter system increases. The quantity of heat given off by the reaction can be calculated from this temperature rise and the known heat-absorbing properties of the calorimeter and the water it contains. The greater the temperature increase, the greater the quantity of energy evolved.

Experimental measurements of this sort are the source of most tabulated values of heat of combustion. As the name suggests, the **heat of combustion** is the quantity of heat energy given off when a specified amount of a substance burns in oxygen. Heats of combustion are typically reported as positive values in kilojoules per mole (kJ/mol), kilojoules per gram (kJ/g), kilocalories per mole (kcal/mol), or per gram (kcal/g). The energy equivalents of various foods are also usually determined by calorimetry. In the case of methane, experiment shows its heat of combustion to be 802.3 kJ. This means that 802.3 kJ of heat is given off when one mole of $CH_4(g)$ reacts with two moles of $O_2(g)$ to form one mole of $CO_2(g)$ and two moles of $H_2O(g)$ (see equation 4.1). We can also calculate the number of kilojoules released when one gram of methane is burned. The molar mass of CH_4, calculated from the atomic masses of carbon and hydrogen, is 16.0 g/mol. The heat of combustion per gram of methane gas (kJ/g) is obtained as follows:

Heats of combustion, by convention, are tabulated as positive values even though all combustion reactions *release* heat.

$$\frac{802.3\ \text{kJ}}{1\ \text{mol}\ CH_4} \times \frac{1\ \text{mol}\ CH_4}{16.0\ \text{g}\ CH_4} = 50.1\ \text{kJ/g}\ CH_4$$

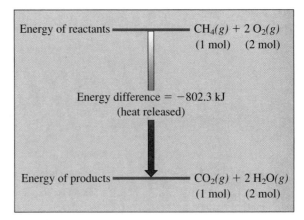

Figure 4.5
Energy difference in an exothermic reaction.

The heat evolved signals a decrease in the energy of the chemical system during the reaction. In other words, the reactants (methane and oxygen) are at higher energy than the products (carbon dioxide and water). The burning of methane is thus somewhat like a waterfall or a falling object. In all these changes, potential energy decreases and is manifested in some other form of energy (heat, sound, etc.). This decrease is signified by the negative sign that is traditionally attached to the energy change for *all* exothermic reactions. For the combustion of methane, the energy change is listed as -802.3 kJ/mol. Figure 4.5 is a schematic representation of this process. The downward arrow indicates that the energy associated with 1 mol of $CO_2(g)$ and 2 mol of $H_2O(g)$ is less than the energy associated with 1 mol of $CH_4(g)$ and 2 mol of $O_2(g)$. The energy difference between the products and the reactants is thus a negative quantity, as is the case for all exothermic reactions. In the combustion of methane, the energy difference is -802.3 kJ.

$Energy_{products} - Energy_{reactants} < 0$ for an exothermic reaction.

Your Turn 4.8   Methane by the Cubic Foot

According to information in this section, the heat of combustion of methane is 802.3 kJ/mol. Methane is usually sold by the standard cubic foot (SCF). One SCF contains 1.250 mol of methane. Calculate the energy (in kilojoules) that is released by burning 1.000 SCF of methane.

Answer
1003 kJ released

We still need to adequately explain the origin of the energy released in an exothermic reaction. To do that, we investigate the structure of the molecules involved. We can write structures for all of the molecular species in the combustion of methane.

$$\underset{H}{\overset{H}{\underset{|}{C}}}\text{H} + 2\,\ddot{O}{=}\ddot{O} \longrightarrow \ddot{O}{=}C{=}\ddot{O} + 2\,\underset{H}{\overset{..}{O}}\diagdown_H \qquad [4.2]$$

The reaction represented by this and any other chemical equation is a rearrangement of atoms. Chemical bonds are formed and broken. Energy is required to break bonds, just as energy is required to break wood or tear paper. Bond breaking is thus an **endothermic process,** a term applied to any chemical or physical change that absorbs energy. On the other hand, the formation of chemical bonds is an exothermic process in which energy is released. The overall energy change associated with a chemical reaction depends on the net effect of the bond breaking and bond making. If the energy required to break the bonds in the reactants (endothermic) is greater than the energy released (exothermic) when the products form, the overall reaction is *endothermic;* energy is absorbed. If, on the other hand, the situation is reversed, the exothermic bond-making energy of the products is greater than the endothermic bond breaking in the reactants, then the net energy change is *exothermic;* energy is released by the reaction.

Endothermic Reaction	Exothermic Reaction
$Energy_{products} > Energy_{reactants}$	$Energy_{products} < Energy_{reactants}$
Energy change is positive.	Energy change is negative.
Energy is absorbed.	Energy is released.

The potential energy associated with any specific chemical species, for example, a CH_4 molecule, is in part a consequence of the interaction of the atoms via chemical bonds. When methane or any other fuel burns in oxygen, the energy released in bond formation exceeds the energy absorbed in bond breaking. The net result is the evolution of energy, mostly in the form of heat. Another way to look at such exothermic reactions is as a conversion of reactants involving weaker bonds (for example, CH_4 and O_2) to products involving stronger ones (CO_2 and H_2O). In general, the products are more stable and less reactive than the starting substances.

Although the chemical reactions used to generate energy are all exothermic, many naturally occurring endothermic reactions such as photosynthesis absorb energy as they occur. You already encountered two that are very important in atmospheric chemistry. One is the decomposition of O_3 to yield O_2 and O, and the other is the combination of N_2 and O_2 to yield two molecules of NO. Both reactions require energy, which can be in the form of electrical discharge, high-energy photons, or high temperatures. It is possible to experimentally determine the energy changes associated with many reactions, either exothermic or endothermic. But sometimes it is easier to calculate values. We illustrate the process in the next section.

4.4   Energy Changes at the Molecular Level

We just observed that the energy release from the combustion of a fuel like methane could be quantified and is a direct result of the structure of the reactants and products. Which of the chemical bonds in those structures give rise to most of the energy content or release? Can we even answer this question? The combustion of the simple molecule and potential fuel, hydrogen, provides the answers. There is much interest in hydrogen as a fuel because of the large amount of energy per gram produced when it burns.

We can calculate the total energy change associated with the combustion of hydrogen to form water, as represented by equation 4.3.

$$2\,H_2(g) + O_2(g) \longrightarrow 2\,H_2O(g) + energy \qquad [4.3]$$

A simple approach for such calculations is to assume that all the bonds in the reactant molecules are broken and then the individual atoms are reassembled into the product molecules. In fact, the reaction does not occur that way. But we are interested in only the overall (net) change, not the details. Therefore, we proceed with our convenient plan and see how well our calculated result agrees with the experimental value.

The numbers we need in the computation are given in Table 4.2, a listing of the bond energies associated with a large variety of covalent bonds. **Bond energy** is the amount of energy that must be absorbed to break a specific chemical bond. Thus, because energy must be absorbed, breaking bonds is an endothermic process, and all the bond energies in Table 4.2 are positive. Obviously, the amount of energy required depends on the number of bonds broken: more bonds take more energy. Typically, bond energies are expressed in kilojoules per mole of bonds. Note that chemical symbols for

Table 4.2		Bond Energies (in kJ/mol)						
	H	**C**	**N**	**O**	**S**	**F**	**Cl**	**Br**	**I**

Single bonds

	H	C	N	O	S	F	Cl	Br	I
H	436								
C	416	356							
N	391	285	160						
O	467	336	201	146					
S	347	272	—	—	226				
F	566	485	272	190	326	158			
Cl	431	327	193	205	255	255	242		
Br	366	285	—	234	213	—	217	193	
I	299	213	—	201	—	—	209	180	151

Multiple bonds

C=C	598	C=N	616	C=O	803 in CO_2
C≡C	813	C≡N	866	C≡O	1073
N=N	418	O=O	498		
N≡N	946				

Source: Data from Darrell D. Ebbing, *General Chemistry,* Fourth Edition, 1993 Houghton Mifflin Co. Data originally from *Inorganic Chemistry: Principles of Structure and Reactivity,* Third Edition by James E. Huheey, 1983, Addison Wesley Longman.

elements appear both across the top of Table 4.2 and down the left side. The number at the intersection of any row and column is the energy (in kilojoules) needed to *break a mole of bonds* linking the atoms of the two elements thus identified. For example, the bond energy of a H-to-H bond, as in the H_2 molecule, is 436 kJ/mol. Similarly, the energy required to break 1 mol of O-to-O double bonds is 498 kJ, as noted from the bottom part of the table. Bond energies for other double bonds, as well as for triple bonds, are also given in the table.

Because we are doing energetic bookkeeping, we need to keep track of the energy change involved in each step and whether the energy is taken up or given off. To do this, we assume that energy that is absorbed carries a positive sign, like a deposit to your checkbook. On the other hand, energy given off is like money spent; it bears a negative sign. Bond energies are positive because they represent energy absorbed when bonds are broken. But the formation of bonds releases energy, and hence the associated energy change is negative. For example, the bond energy for the O-to-O double bond is 498 kJ/mol. This means that when 1 mol of O-to-O double bonds is broken, the energy change is +498 kJ; correspondingly, when 1 mol of O-to-O double bonds is formed, the energy change is −498 kJ.

Now we are finally ready to apply these concepts and conventions to the burning of hydrogen gas, H_2. First we need to determine how many moles of bonds are broken and how many moles of bonds are formed. We can do so using structural formulas relating them to equation 4.4.

$$2\,H\!-\!H \;+\; \ddot{\underset{\cdot\cdot}{O}}\!=\!\ddot{\underset{\cdot\cdot}{O}} \;\longrightarrow\; 2\;{}_{H}\!\nearrow^{\ddot{O}\cdot}\!\searrow_{H} \qquad\qquad [4.4]$$

Remember that chemical equations can be read in terms of moles. Equation 4.4 indicates "2 mol of $H_2(g)$ plus 1 mol of $O_2(g)$ yields 2 mol of gaseous water (water vapor)." But, to use bond energies, we need to count the number of moles of *bonds* involved. Because each H_2 molecule contains one H-to-H bond, 1 mol of H_2 must contain 1 mol of H-to-H bonds. Similarly, equation 4.4 indicates that 1 mol of O_2 contains 1 mol of O-to-O double bonds. Each mole of water contains 2 mol of H-to-O bonds; thus, 2 mol of water contain 4 mol of H-to-O bonds. Therefore, we now have the total number of moles of bonds to be broken (2 mol of H-to-H and 1 mol of O-to-O double) and those to be formed (4 mol of H-to-O). These number of bonds are then *multiplied* by the representative bond energy, using the appropriate sign convention (+ for bonds broken; − for bonds made).

Molecule	Bonds per molecule	Moles	Total Number of Bonds	Bond Process	Energy per bond	Total Energy
H—H	1	2	1 × 2 = 2	breaking	+436 kJ	2 × (+436) = +872 kJ
O=O	1	1	1 × 1 = 1	breaking	+498 kJ	1 × (+498) = +498 kJ
H—O—H	2	2	2 × 2 = 4	making	−467 kJ	4 × (−467) = −1868 kJ

Consequently, the *overall* energy change in breaking bonds (872 kJ + 498 kJ = 1370 kJ) and forming new ones is (−1868 kJ) is −498 kJ.

A schematic representation of this calculation is presented in Figure 4.6. The energy of the reactants, 2 H_2 and O_2, is set at zero, an arbitrary but convenient value. The green arrows pointing upward signify energy absorbed to break bonds and convert the reactant molecules into individual atoms: 4 H and 2 O. The red arrow on the right pointing downward represents energy released as these atoms are reconnected with new bonds to form the product molecules: 2 H_2O. The shorter red arrow corresponds to the net energy change of −498 kJ signifying that the overall combustion reaction is strongly exothermic. The *release* of heat corresponds to a *decrease* in the energy of the chemical system, which explains why the energy change is *negative*.

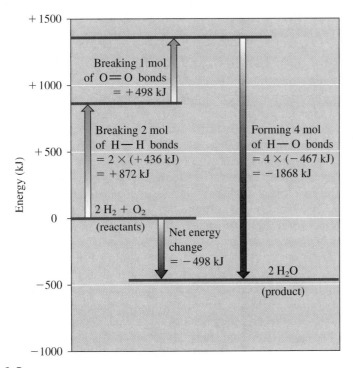

Figure 4.6
The energetics of the combustion of hydrogen to form water.

Figures Alive! Visit the *Online Learning Center* to learn more about the energetics of this reaction. Look for the Figures Alive! icon in this chapter as a guide to related activities.

We can also use bond energies from Table 4.2 to calculate the energy change for the combustion of methane.

$$CH_4(g) + 2\,O_2(g) \longrightarrow CO_2(g) + 2\,H_2O(g) + \text{energy}$$

One mole of methane contains 4 mol of C-to-H bonds, each with a bond energy value of 416 kJ. Breaking 2 mol of O-to-O double bonds requires 996 kJ ($2 \times +498$ kJ). Bonds formed in the products are 2 mol of C-to-O double bonds in 1 mol of CO_2 (2×-803 kJ), and 4 mol of H-to-O bonds in 2 mol of water (4×-467 kJ). Note again that bond formation is exothermic and the associated bond energies have minus signs.

Molecule	Bonds per molecule	Moles	Total Number of Bonds	Bond Process	Energy per bond	Total Energy
CH_4	4	1	$4 \times 1 = 4$	breaking	$+416$ kJ	$4 \times (+416) = +1664$ kJ
O_2	1	2	$1 \times 2 = 2$	breaking	$+498$ kJ	$2 \times (+498) = +996$ kJ
CO_2	2	1	$2 \times 1 = 2$	making	-803 kJ	$2 \times (-803) = -1606$ kJ
H_2O	2	2	$2 \times 2 = 4$	making	-467 kJ	$4 \times (-467) = -1868$ kJ

Total energy change in breaking bonds = $(+1664\ \text{kJ}) + (+996\ \text{kJ}) = +2660\ \text{kJ}$

Total energy change in making bonds = $(-1606\ \text{kJ}) + (-1868\ \text{kJ}) = -3474\ \text{kJ}$

Net energy change = $(+2660\ \text{kJ}) + (-3474\ \text{kJ}) = -814\ \text{kJ}$

Heats of combustion, by convention, are listed as positive values. Thus, the heat of combustion of methane calculated using bond energies is +814 kJ.

The energy changes we just calculated from bond energies, -506 kJ for the burning of 1 mol of hydrogen and -814 kJ for 1 mol of methane combustion, compare favorably with the experimentally determined values. This agreement justifies our rather unrealistic assumption that all the bonds in the reactant molecules are first broken, then all the bonds in the product molecules are formed. This is not at all what actually happens. But the energy change that accompanies a chemical reaction depends on the energy *difference* between the products and the reactants, not on the particular process, mechanism, or individual steps that connect the two. This is an extremely powerful idea for understanding chemical energetics and doing related calculations.

Not all calculations come out as well as this one did. For one thing, the bond energies of Table 4.2 apply only to gases, so calculations using these values agree with experiment only if all the reactants and products are in the gaseous state. Moreover, tabulated bond energies are really average values. The strength of a bond depends on the overall structure of the molecule in which it is found; in other words, on what else the atoms are bonded to. Thus, the strength of an O-to-H bond is slightly different in HOH (H_2O), HOOH (H_2O_2), and CH_3OH. Nevertheless, the procedure illustrated here is a useful way of estimating energy changes in a wide range of reactions. The approach also helps illustrate the relationship between bond strength and chemical energy.

This analysis also helps clarify why the H_2O or CO_2 formed in combustion reactions cannot be used as fuels. There are no substances into which these compounds can be converted that have stronger bonds and are lower in energy; we cannot run a car on its exhaust.

Generally, experimental values differ somewhat from those calculated using bond energies. The experimental heat of combustion of methane is 802.3 kJ; the calculated value, 814 kJ, differs by 1.5%.

Your Turn 4.9　　Heat of Combustion of Propane

Use the bond energies in Table 4.2 to calculate the heat of combustion of propane, C_3H_8. Report your answer both in kilojoules per mole (kJ/mol) of C_3H_8 and kilojoules per gram (kJ/g) C_3H_8. This is the equation for the reaction, written with Lewis structures.

$$H-\overset{\overset{\displaystyle H}{|}}{\underset{\underset{\displaystyle H}{|}}{C}}-\overset{\overset{\displaystyle H}{|}}{\underset{\underset{\displaystyle H}{|}}{C}}-\overset{\overset{\displaystyle H}{|}}{\underset{\underset{\displaystyle H}{|}}{C}}-H + 5\,\ddot{O}{=}\ddot{O} \longrightarrow 3\,\ddot{O}{=}C{=}\ddot{O} + 4\,H{-}\overset{..}{\underset{..}{O}}{\diagdown}H$$

Hint: 8 mol of C-to-H bonds, 2 mol of C-to-C bonds, 5 mol of O-to-O double bonds, 6 mol of C-to-O double bonds and 8 mol of O-to-H bonds are involved in the reaction.

Answer

Energy change = −2024 kJ/mole C_3H_8 or −46.0 kJ/g C_3H_8
Heat of combustion = 2024 kJ/mol C_3H_8 or 46.0 kJ/g C_3H_8.

Your Turn 4.10　　Heat of Combustion of Ethanol

Use the bond energies in Table 4.2 to calculate the heat of combustion of ethanol, C_2H_5OH, one of the components of "gasohol" fuel. This is the structure of ethanol:

$$H-\overset{\overset{\displaystyle H}{|}}{\underset{\underset{\displaystyle H}{|}}{C}}-\overset{\overset{\displaystyle H}{|}}{\underset{\underset{\displaystyle H}{|}}{C}}-\overset{..}{\underset{..}{O}}-H$$

Answer

Energy change = −1281 kJ/mol C_2H_5OH or −27.8 kJ/g C_2H_5OH
Heat of combustion = 1281 kJ/mol C_2H_5OH or 27.8 kJ/g C_2H_5OH

Your Turn 4.11　　UV Radiation Absorption

Ultraviolet radiation absorbed by O_2 has a shorter wavelength than that absorbed by O_3. Why is that? Use the bond energies in Table 4.2 plus information from Chapter 2 to explain. *Hint:* Look at Lewis structures to decide which bond is broken. Consider the relationship between energy and wavelength.

4.5　Energy as a Barrier to Reaction

Just because two substances can react in an exothermic process does not mean that they will do so, even if they are in close contact. For example, if you turn on the burner of a gas stove, methane and oxygen will be present in a combustible mixture. But they will not react unless a spark, a flame, or some other source of energy is supplied (Figure 4.7). (Stoves often have an electric sparker or pilot light.) This turns out to be a fortunate feature of matter. Wood, paper, and many other common materials including our own bodies are energetically unstable and capable of exothermic conversion to water, carbon dioxide, and other simple molecules, but they do not suddenly burst into flame.

The energy necessary to initiate a reaction is called its **activation energy.** Figure 4.8 is a schematic of the energy changes that might occur in a typical exothermic reaction. It looks a little like the cross section of a hill. The activation energy corresponds to a peak over which a boulder must be pushed before it will roll downhill. Although energy must be expended to get the reaction (or the boulder) started, a good deal more energy is given off as the process proceeds to a lower potential energy state. Generally, reactions that occur rapidly have low activation energies; slower reactions have

Figure 4.7

Natural gas from a stove burner ignites to produce a hot flame.

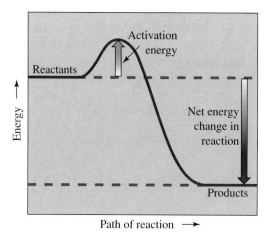

Figure 4.8
Energy–reaction pathway diagram.

higher activation energies. However, there is no direct relationship between the height of the activation barrier and the net change in energy in the reaction. In other words a very exothermic reaction can have a very large or very small activation energy.

Activation energy is also involved in another aspect of chemical reactions that determines whether a given substance can be used as a fuel. Useful fuels react at rates that are neither too fast nor too slow. Slow reactions are of little use in producing energy because the energy is released over too long a time. For example, it would not be very practical to try to warm your hands over a piece of rotting wood, even though the overall reaction is similar to burning and forms CO_2 and H_2O. On the other hand, fast reactions can release energy too rapidly to be put to convenient use. In fact, such reactions often lead to explosions because the reaction is out of control.

One way to speed up the rate of a reaction is to divide the fuel into small particles. This principle is used in fluidized-bed power plants in which pulverized coal is burned in a blast of air. The fine coal dust is quickly heated to the kindling point and the large surface area means that oxygen reacts rapidly and completely with the fuel. The combustion actually occurs at a lower temperature than that required to ignite larger pieces of coal. As a result, the generation of nitrogen oxides is minimized. If finely divided limestone (calcium carbonate) is mixed with the powdered coal, sulfur dioxide is also removed from the effluent gas. Thus, the amount of pollution is reduced while the efficiency of coal combustion is enhanced.

Increasing temperature also increases the rates at which reactions occur. The added heat helps the reactants assemble the energy needed to go over the activation energy barrier. Catalysts, including those used in automobile catalytic converters (Section 1.11) and in petroleum refining (Section 4.8), increase reaction rates by providing alternative reaction pathways with lower activation energies.

> Explosions occur in grain elevators (storage facilities) when an inadvertent spark or flame causes finely powdered grain to burn explosively.

Consider This 4.12   **The Striking Case of a Single Match**

The electrical power may go out in a residential area during a summer thunderstorm. Imagine that this happened in your house just as you were about to cook dinner. You scrambled around in the camping gear looking for an alternative to your electric stove. You found a portable gas stove, one match, a small grill, and a bag of charcoal briquettes. Because you have only one match and no other sources of fire, you must choose between using the portable gas stove or the charcoal grill. Which will you choose and why? Justify your decision based on what you have learned about combustion.

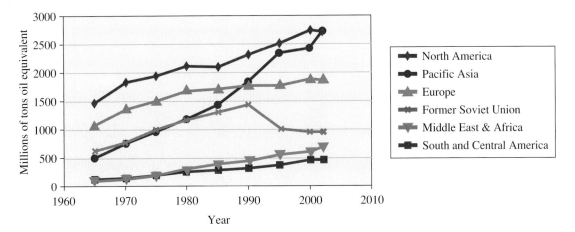

Figure 4.9

Total Energy Consumption. Oil, gas, coal, nuclear, and hydroelectricity converted to millions of tons oil equivalent.

Source: Data from BP *Statistical Review of World Energy 2003.* June 2003, British Petroleum Company.

4.6　Energy Consumption

In the United States, we burn a prodigious quantity of fossil fuels to generate energy. Figure 4.9 compares our annual total energy use with that of other parts of the world. The energy comes from many different sources, but the graph expresses it as if all of it were generated from oil. Thus, in 2002 the energy share of North America was roughly 30% of the world supply. In India, the amount of energy derived from traded fuel sources was about 3.5% of the world supply or about a tenth the values for the United States and Canada. Perhaps an equal quantity of India's energy was derived from traditional fuels such as wood, grass, or animal dung, but the international energy imbalance remains staggering. It is no coincidence that the nations with the highest consumption in Figure 4.9 are industrialized and wealthy, and that those with the lowest are struggling with poverty. Energy appears to drive industrial and economic progress; gross national product correlates well with energy production and use. So do life expectancy, infant mortality, and literacy.

The great burst of energy consumption is of relatively recent origin. Two million years ago, before our ancestors learned to use fire, the sources of energy available to an individual were that of his or her own body or that from the Sun. Earliest hominids probably consumed the equivalent of 2000 kcal per day and expended most of it finding food. This daily energy use corresponds to that used by a 100-watt light bulb burning for 24 hr. The discovery of fire and the domestication of beasts of burden increased the energy available to an individual by about six times. Hence, we estimate that about 2000 years ago, a farmer with an ox or donkey had roughly 12,000 kcal at his or her disposal each day. The Industrial Revolution brought another five- or sixfold increase in the energy supply, most of it from coal via steam engines. Yet another energy jump occurred during the 20th century. By the end of the 20th century, the total energy used in the United States (from all sources and for all purposes) corresponded to about 650,000 kcal per person per day. This translates to an annual equivalence of 65 barrels of oil or 16 tons of coal for each American. Or, to put it in human terms, the energy available to each resident of the United States would require the physical labor of 130 workers. Yet, there are still people on the planet whose energy use and lifestyle closely approximate those of 2000 years ago.

The history of increasing energy consumption is closely related to changing energy sources and the development of devices for extracting and transforming that energy. Figure 4.10 displays the average American energy consumption from a variety of sources over a 200-year period. The data start in 1800 and extend to 2000. The graph indicates that wood was originally the major energy source in the United States, and it

Fossil fuels were also discussed in Chapter 3.

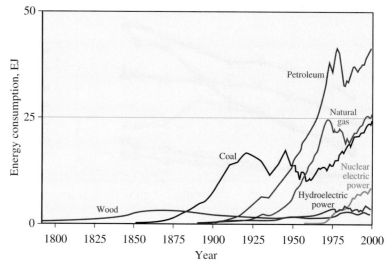

Figure 4.10

History of United States energy consumption by source, 1800–2000 (1 EJ $= 10^{18}$ J).

Source: *Annual Energy Review 2001*, Department of Energy/EIA.

continued to be so until the late 1880s, when it was surpassed by coal. Coal provided more than 50% of the nation's energy from then until about 1940. By 1950, oil and gas were the source of more than half of the energy used in this country. Falling water has long been used to power mills and, more recently, to generate electricity, but it provides only a small percentage of our total energy output. Nuclear fission, once hailed as an almost limitless source of energy, has not achieved its full potential for a variety of reasons.

Nuclear fission is discussed in Chapter 7.

Consider This 4.13 U.S. Sources of Energy Over Time

Many changes in our sources of energy are represented in Figure 4.10. Concentrating on the period from 1950 to 2000 shown in the figure, which sources of energy have shown steady growth and which have not? Propose reasons for the observed trends.

Wood, waste, alcohol, geothermal, wind, and solar sources are combined into a sliver marked "Other" in Figure 4.11. This pie chart indicates the percentage of the U.S. 2002 energy consumption derived from various sources. Some of the currently underutilized alternative energy sources will be considered in Chapters 7 and 8.

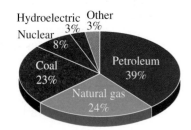

Figure 4.11

Annual U.S. energy consumption by source, 2002.

Source: Department of Energy/EIA.

The role of carbon dioxide in global warming was discussed in Chapter 3.

Consider This 4.14 Now and in the Future

Dr. Ronald Breslow, a Columbia University chemistry professor and a recent president of the American Chemical Society, speaking at a February 2001 symposium on "Sustainability Through Science," remarked: "Succeeding generations are going to curse us for burning their future raw materials, and they are right. Not only are we using up valuable resources—petroleum and coal—but we are adding pollution and carbon dioxide which may be contributing to global warming." Comment on Dr. Breslow's remarks.

So how do we select fuels? Materials such as coal, oil, and natural gas possess many of the properties needed in a fuel. They contain substantial energy content, and typical values appear in Table 4.3. Vast quantities appear to be available as "natural" resources, to be harvested almost at will. From where do these reserves come?

Table 4.3	Energy Content of Fuels	
	Source	**kJ/g**
	Wood (oak)	14
	Gasoline	48
	Coal (hard)	31
	Methane	56
	Propane	51
	Hydrogen	140
	Ethanol	30

Most of the energy that drives the engines of our economy comes from these remnants of the past. In a very real sense, these fossil fuels are sunshine in the solid, liquid, and gaseous state. The sunlight was captured millions of years ago by green plants that flourished on the prehistoric planet. The reaction is the same one carried out by plants today.

$$2800 \text{ kJ} + 6 CO_2(g) + 6 H_2O(l) \xrightarrow{\text{chlorophyll}} \underset{\text{glucose}}{C_6H_{12}O_6(s)} + 6 O_2(g) \qquad [4.5]$$

This conversion of carbon dioxide and water to glucose and oxygen is endothermic. It requires the absorption of 2800 kJ of sunlight per mole of $C_6H_{12}O_6$ or 15.5 kJ/g of glucose formed. The reaction could not occur without the absorption of energy and the participation of a green pigment molecule called chlorophyll. The chlorophyll interacts with photons of visible sunlight and uses their energy to drive the photosynthetic process.

You are already aware of the essential role of photosynthesis in the initial generation of the oxygen in Earth's atmosphere, in maintaining the planetary carbon dioxide balance, and in providing food and fuel for creatures like us. In our bodies, we run the preceding reaction backward, like living internal combustion engines.

$$C_6H_{12}O_6(s) + 6 O_2(g) \longrightarrow 6 CO_2(g) + 6 H_2O(l) + 2800 \text{ kJ} \qquad [4.6]$$

We extract the 2800 kJ released per mole of glucose "burned" and use that energy to power our muscles and nerves, though we do not do it with perfect efficiency (see Sceptical Chymist 4.3). The same overall reaction occurs when we burn wood, which is primarily cellulose, a polymer composed of repeating glucose units. In equation 4.6, an *exo*thermic reaction, energy is shown as a product (2800 kJ) because energy is *given off* by the reaction. In an *endo*thermic reaction, energy is noted as a reactant because the reaction *absorbs* energy. Chemical equations are often written without including energy as a reactant or product. Including the energy is a way of emphasizing the energy change associated with the reaction.

When plants die and decay, they also are largely transformed into CO_2 and H_2O. However, under certain conditions, the glucose and other organic compounds that make up the plant only partially decompose and the residue still contains substantial amounts of carbon and hydrogen. Such conditions arose at various times in the prehistoric past of our planet, when vast quantities of plant life were buried beneath layers of sediment in swamps or on the ocean bottom. There, these remnants of vegetable matter were protected from atmospheric oxygen, and the decomposition process was halted. However, other chemical transformations occurred in Earth's high-temperature and high-pressure reactor. Over millions of years, the plants that captured the rays of a young Sun were transmuted into the fossils we call coal and petroleum. Jacob Bronowski, in his book *Biography of an Atom—And the Universe,* aptly describes the cycle by saying, "You will die but the carbon will not; its career does not end with you . . . it will return to the soil, and there a plant may take it up again in time, sending it once more on a cycle of plant and animal life."

Respiration is the process by which humans exchange the oxygen necessary for metabolism with the carbon dioxide produced by it.

Polymers are discussed in Chapter 9.

4.7 Coal

The great exploitation of fossil fuels began with the Industrial Revolution, about two centuries ago. The newly built steam engines consumed large quantities of fuel, but in England, where the revolution began, wood was no longer readily available. Most of the forests had already been cut down. Coal turned out to be an even better energy source than wood because it yields more heat per gram. Burning 1 g of coal releases approximately 30 kJ, compared with 10–14 kJ per gram of wood. This difference in heat of combustion is a consequence of differences in chemical composition. When wood or coal burn, a major energy source is the conversion of carbon to carbon dioxide. Coal is a better fuel than wood because it contains a higher percentage of carbon and a lower percentage of oxygen and water.

Coal is a complex mixture that naturally occurs in varying grades. Although coal is not a single compound, it can be approximated by the chemical formula $C_{135}H_{96}O_9NS$. This formula corresponds to a carbon content of 85% by mass. The carbon, hydrogen, oxygen, and nitrogen atoms come from the original plant material. In addition, samples of coal typically contain small amounts of silicon, sodium, calcium, aluminum, nickel, copper, zinc, arsenic, lead, and mercury.

Soft lignite, or brown coal, is the lowest grade. The vegetable matter that makes it up has undergone the least amount of change, and its chemical composition is similar to that of wood or peat. Consequently, the heat of combustion of lignite is only slightly greater than that of wood (Table 4.4). The higher grades of coal, bituminous and anthracite, have been exposed to higher pressures in the earth. In the process, they lost more oxygen and moisture and have become a good deal harder—more mineral than vegetable. The percentage of carbon has increased, and with it, the heat of combustion. Anthracite has a particularly high carbon content and a low concentration of sulfur, both of which make it the most desirable grade of coal. Unfortunately, the deposits of anthracite are relatively small, and the United States supply is almost exhausted. We now rely more heavily on bituminous and subbituminous coal.

Generally speaking, the less oxygen a compound contains, the more energy per gram it will release on combustion because it is higher up on the potential energy scale. This explains why burning one mole of carbon to form carbon dioxide yields about 40% more energy than obtained from burning one mole of carbon monoxide. To be sure, coal is a mixture, not a compound, but the same principles apply. Anthracite and bituminous coals consist primarily of carbon. Their heat of combustion is, gram for gram, about twice that of lignite, which contains a much lower percentage of carbon.

Although the global supply of coal is large and it remains a widely used fuel, it has some serious drawbacks. Coal is difficult to obtain, and underground mining is dangerous and expensive. For *each* of the first 45 years of the 20th century, more than 1000 coal miners were killed in mining accidents. In many years there were more than 2000 deaths, and in just the month of December, 1907, there were more than 3000 coal-mining deaths. In 1961 there were 293 deaths; in 1981 there were 153, and in 2002 there were 29. In *Invention and Technology,* Summer 1992, Mary Blye Howe reported that since 1900 more than 100,000 workers have been killed in American coal mines by accidents, cave-ins, fires, explosions, and poisonous gases. Many thousands more

The energy content of other fuels appeared in Table 4.3.

Table 4.4	Fuel Value of Various U.S. Coals	
Type of Coal	**State of Origin**	**Heat Content (kJ/g)**
Anthracite	Pennsylvania	30.5
Bituminous	Maryland	30.7
Subbituminous	Washington	24.0
Lignite (brown coal)	North Dakota	16.2
Peat	Mississippi	13.0
Wood	various	10.4–14.1

have been injured or incapacitated by respiratory diseases. Although mining sounds safer now, far fewer miners are currently involved than a century ago. Furthermore, if the coal deposits lie sufficiently close to the surface, safer open-pit mining can be used. In this method, the overlying soil and rock are stripped away to reveal the coal seam, which is then removed by heavy machinery. But underground mining still is used extensively where labor is cheap, and large numbers of fatalities and injuries are being realized in China, Poland, and Ukraine.

Mining must be done carefully to prevent serious environmental deterioration. Current regulations in the United States require the replacement of earth and topsoil and the planting of trees and vegetation at mine sites. But, in the past, these regulations were not in place to prevent the great holes in the earth and heaps of eroding soil that still dot regions of abandoned strip mines. Once the coal is out of the ground, its transportation is complicated because it is a solid. Unlike gas and oil, coal cannot be pumped unless it is finely divided and suspended in a water slurry.

Coal is a dirty fuel. It is, of course, physically dirty, but its dirty combustion products are more serious. The unburned soot from countless coal fires in the 19th and early 20th centuries blackened buildings and lungs in many cities. Less visible but equally damaging are the oxides of sulfur and nitrogen formed when certain coals burn. If these compounds are not trapped, they contribute to the acid precipitation that forms the subject for Chapter 6. Although mercury is only present in coal in minor amounts (50–200 ppb), it becomes concentrated in the fly ash that escapes as particulate matter into the atmosphere or the "bottom" ash that remains and must be disposed of after burning. In addition, coal suffers from the same drawback of all fossil fuels; the greenhouse gas carbon dioxide is an inescapable product of its combustion.

In spite of these less-than-desirable properties, the world's energy dependence on coal is likely to increase rather than decrease. The recoverable world supply of coal is estimated as 20–40 times greater than world petroleum reserves. As the latter become

---

Your Turn 4.15   Calculations Concerning Coal

a. Assuming the composition of coal can be approximated by the formula $C_{135}H_{96}O_9NS$, calculate the mass of carbon (in tons) contained in 1.5 million tons of coal. This is the quantity of coal that might be burned by a power plant in one year.

b. What mass of CO_2 will be produced by the complete combustion of 1.5 million tons of this coal?

c. Compute the amount of energy (in kilojoules) released by burning this mass of coal. Assume the process releases 30 kJ per gram of coal. Recall that 1 ton = 2000 lb and that 1 lb = 454 g.

Answers

a. Use the formula to calculate the approximate molar mass of coal. The subscripts for each element give the number of moles:

$$135 \text{ mol C} \times \frac{12.0 \text{ g C}}{1 \text{ mol C}} = 1620 \text{ g C}$$

$$96 \text{ mol H} \times \frac{1.0 \text{ g H}}{1 \text{ mol H}} = 96 \text{ g H}$$

$$9 \text{ mol O} \times \frac{16.0 \text{ g O}}{1 \text{ mol O}} = 144 \text{ g O}$$

$$1 \text{ mol N} \times \frac{14.0 \text{ g N}}{1 \text{ mol N}} = 14.0 \text{ g N}$$

$$1 \text{ mol S} \times \frac{32.1 \text{ g S}}{1 \text{ mol S}} = 32.1 \text{ g S}$$

(continued on p. 192)

Your Turn 4.15   Calculations Concerning Coal (*continued*)

The sum of these elemental contributions for $C_{135}H_{96}O_9NS$ is 1906 g/mol. Every 1906 g of coal has 1620 g C. The mass-to-mass relationship stays the same as long as the same mass unit is used for both; the ratio is just as useful expressed in tons.

$$\text{Mass of carbon} = 1.5 \times 10^6 \text{ tons } C_{135}H_{96}O_9NS \times \frac{1620 \text{ tons C}}{1906 \text{ tons } C_{135}H_{96}O_9NS}$$

$$= 1.3 \times 10^6 \text{ tons C} = 1.3 \text{ million tons C}$$

b.   4.8 million tons   **c.** 4.1×10^{13} kJ

depleted, reliance on coal will increase unless alternative energy sources are developed. It is, however, possible that coal will not be burned in its familiar form, but rather converted to cleaner and more convenient liquid and gaseous fuels. After we discuss petroleum, we will consider alternative fuels.

4.8   Petroleum

Most people in the average American city or town would be hard-pressed to find lumps of coal. Indeed, many of you may never have seen coal, but you have undoubtedly seen gasoline. Around 1950, petroleum surpassed coal as the major energy source in the United States. The reasons are relatively easy to understand. Petroleum, like coal, is partially decomposed organic matter, but it has the distinct advantage of being liquid. It is easily pumped to the surface from its natural, underground reservoirs, transported via pipelines, and fed automatically to its point of use. Moreover, petroleum is a more concentrated energy source than coal, yielding approximately 40–60% more energy per gram. Typical figures are 48 kJ/g for petroleum and 30 kJ/g for coal.

The major component extracted from petroleum (crude oil) is gasoline. Although it has been extracted since the mid-1800s, gasoline became valuable and important only with the advent of the automobile and the internal combustion engine early in the 20th century. That fuel and engine partnership has led to our seemingly insatiable appetite for gasoline. In 1998, 125 billion gallons of gasoline was burned in more than 203 million

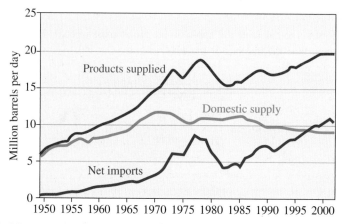

Figure 4.12

U.S. petroleum product use, domestic production, and imports. At present, more than 50% of the total oil used in the United States is imported, and projections show oil imports will continue to increase.

Source: Department of Energy, Energy Information Administration, *Annual Energy Review 2002.*

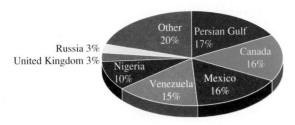

Figure 4.13

Sources of crude oil and petroleum products imported by the U.S. (August 2003).

Source: *Department of Energy/EIA.*

American automobiles, SUVs, and light trucks traveling an astounding 2.6 trillion miles, the equivalent of 1400 round trips to the Sun. For 2003, gasoline consumption in the United States is expected to be 180 billion gallons. However, our capacity to consume gasoline has far outstripped our ability to produce it from crude oil extracted in this country. With 5% of the world's population, we consume 25% of the oil produced worldwide. By the mid-1970s, the United States was producing only about two thirds of the crude oil it required to power its automobiles and factories, heat its homes, and lubricate its machines (Figure 4.12).

Our fragile dependency on oil from abroad continues today, increasing from 4.3 million barrels per day in 1985 to 10.4 million barrels per day in 2002. Our dependency on foreign oil rose from 29% in 1985 to 57% in 2002. But, over the past decade, we have shifted our sources of imported oil from the politically volatile Middle East; the Persian Gulf supplies less than 20% of our imports (Figure 4.13). Saudi Arabia provides the bulk (> 70%) of oil imported from the Middle East.

Consider This 4.16 **Shipping Oil**

Figure 4.13 identifies some regional and specific sources of U.S. oil imports in 2003. Use the Web to find the 12 countries that supply the most oil to us.

a. List these 12 countries in order of the amount of oil provided.
b. Which of those 12 countries are not in North or South America?
c. Which countries, if any, surprised you as sources of our imported oil? Why?

As a nation, our voracious appetite for oil was met by consuming an average of nearly 20 million barrels of oil *daily* in 2002. Two thirds of this was for transportation. Unlike coal, however, crude oil is not ready for immediate use when it is extracted from the ground. Crude oil must first be refined, a process that has given gainful employment to many chemists and chemical engineers (and quite a few others). It has also provided an amazing array of products. Petroleum is a complex mixture of thousands of different compounds. The great majority are **hydrocarbons,** molecules consisting of only hydrogen and carbon atoms. Hydrocarbons in petroleum can contain from one to as many as 60 carbon atoms per molecule. A set of **alkanes,** hydrocarbons with only single bonds between carbons, is shown in Table 4.5. Although petroleum contains compounds with elements other than carbon and hydrogen, concentrations of sulfur and other contaminating elements are generally quite low. This minimizes polluting combustion products.

The oil refinery has become an icon of the petroleum industry (Figure 4.14). During one step in the refining process, the crude oil is separated into fractions that consist of compounds with similar properties. This fractionation is accomplished by a physical process called distillation. **Distillation** is a purification, or separation, process in which a solution is heated to its boiling point and the vapors are condensed and collected. The crude oil is pumped into an industrial-sized container or still, and

Table 4.5	**Alkanes with One to Eight Carbons**

H—C—H
methane

H—C—C—H
ethane

H—C—C—C—H
propane

H—C—C—C—C—H
butane

H—C—C—C—C—C—H
pentane

H—C—C—C—C—C—C—H
hexane

H—C—C—C—C—C—C—C—H
heptane

H—C—C—C—C—C—C—C—C—H
octane

the mixture is heated. As the temperature increases, the components with the lowest boiling points are the first to vaporize. The gaseous molecules escape from the liquid in the container and move up a tall distillation column or tower. There the cooled vapors recondense into the liquid state, only this time in a much purer condition. All fractions are not produced in equal proportions. Market demand dictates that some components should be maximized. To accomplish this, a second kind of refinery

Figure 4.14

An oil refinery, symbol of the petroleum industry.

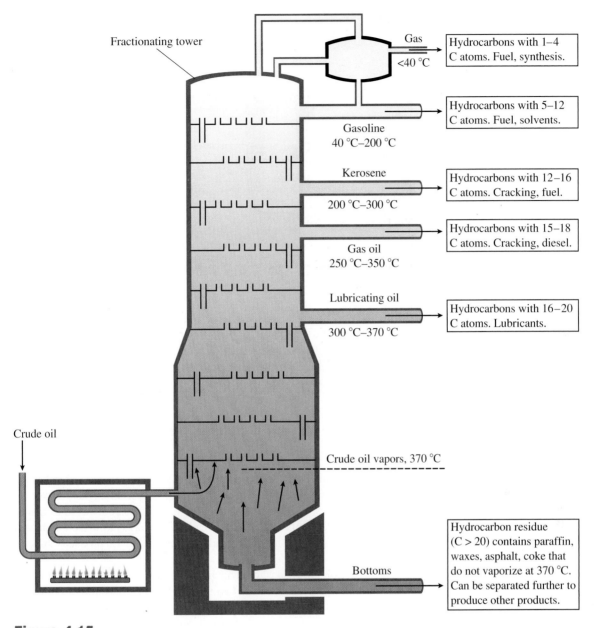

Figure 4.15

Diagram of a fractionating tower showing various fractions and some typical uses.

process (see Section 4.9) causes oil molecules to either break larger ones or combine smaller ones.

Figure 4.15 illustrates a distillation tower and lists some of the fractions obtained. They include gases such as methane, liquids such as gasoline and kerosene, waxy solids, and a tarry asphalt residue. Note that the boiling point goes up with increasing number of carbon atoms in the molecule and hence with increasing molecular mass and size. Heavier, larger molecules are attracted to each other more than are lighter, smaller molecules. In turn, higher temperatures are required to vaporize the compounds of higher molecular mass and size.

Because of differences in properties, the various fractions distilled from crude oil have different uses. Indeed, the great diversity of products obtained has made petroleum a particularly valuable source of matter and energy. The most volatile components of the fractionating tower boil far below room temperature and are called **refinery gases.** In the past, flames at the tops of refinery towers signaled that the refinery

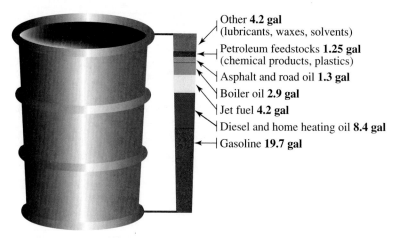

Other **4.2 gal**
(lubricants, waxes, solvents)

Petroleum feedstocks **1.25 gal**
(chemical products, plastics)

Asphalt and road oil **1.3 gal**

Boiler oil **2.9 gal**

Jet fuel **4.2 gal**

Diesel and home heating oil **8.4 gal**

Gasoline **19.7 gal**

Figure 4.16

Products from the refining of 1 barrel (42 gallons) of crude oil.

gases were being burned off, a waste of energy and another source of greenhouse gases. Currently, refinery gases are used as fuels to operate the distillation towers or made into other molecules. The gasoline fraction, containing hydrocarbons with 5 to 12 carbon atoms per molecule, is particularly important to our automotive civilization. Efforts at designing and mass producing self-propelled vehicles were largely unsuccessful until petroleum provided a convenient and relatively safe liquid fuel. The kerosene fraction is somewhat higher boiling, and it finds use as a fuel in diesel engines and jet planes. Still higher boiling fractions are used to fire furnaces and as lubricating oils.

The refining of a barrel of crude oil (42 gal) provides an impressive array of products, the vast majority of which is gasoline (Figure 4.16). A staggering 35 gal of the 42 in a barrel of crude oil is simply burned for heating and transportation. The remaining 7 gal is used nonfuel purposes, including only 5 qt (1.25 gal) set aside to serve as nonrenewable starting materials (reactants, commercially called *feedstocks*) to make the myriad of plastics, pharmaceuticals, fabrics, and other carbon-based industrial products so common in our society.

A discussion of petroleum also should include natural gas. This fuel, which is mostly methane, currently provides heat for two thirds of the single-family homes and apartment buildings in the United States. Recently, interest has increased in using natural gas as an energy source for generating electricity and for powering cars and trucks. A distinct advantage of natural gas is that it burns much more completely and cleanly than other fossil fuels. Because of its purity, it releases essentially no sulfur dioxide. Natural gas emits only very low levels of unburned volatile hydrocarbons, carbon monoxide, and nitrogen oxides; and it leaves no residue of ash or toxic metals like mercury. Moreover, natural gas produces 30% less carbon dioxide than oil and 43% less carbon dioxide than coal per joule of energy produced.

Typically, natural gas is composed of 87–96% methane, 2–6% ethane, and smaller quantities of larger hydrocarbons, nitrogen, carbon dioxide, and oxygen.

Consider This 4.17   **A Mix of Products from a Barrel of Oil**

Chemical research has contributed to increasing the amount of gasoline derived from a barrel of crude oil. For example, in 1904 a barrel of crude oil produced 4.3 gal gasoline, 20 gal kerosene, 5.5 gal fuel oil, 4.9 gal lubricants, and 7.1 gal miscellaneous products. By 1954, the products were 18.4 gal gasoline, 2.0 gal kerosene, 16.6 gal fuel oil, 0.9 gal lubricants, and 4.1 gal miscellaneous products. Compare these values with those shown in Figure 4.16 and offer some reasons why the products have changed over time.

4.9 Manipulating Molecules to Make Gasoline

Many of the compounds distilled from crude oil are not ideally suited for the desired applications; nor does the normal distribution of molecular masses of the distilled fractions correspond to the prevailing use pattern. For example, the demand for gasoline is considerably greater than that for higher boiling fractions. Gasoline that comes directly from the fractionating tower represents less than 50% of the original crude oil. Heavier and lighter crude oil fractions can undergo chemical reactions to form still more gasoline. **Cracking** is a chemical process by which large molecules are broken into smaller ones suitable to be used in gasoline. For example, a hydrocarbon with 16 carbons can be cracked into two almost equal fragments,

$$C_{16}H_{34} \longrightarrow C_8H_{18} + C_8H_{16} \qquad [4.7]$$

or into different sized ones.

$$C_{16}H_{34} \longrightarrow C_5H_{12} + C_{11}H_{22} \qquad [4.8]$$

Note that the numbers of carbon and hydrogen atoms are unchanged from reactants to products; the larger reactant molecules simply have been fragmented into smaller, more economically important molecules. Previously, **thermal cracking** was achieved by heating the starting materials to a high temperature. Currently, valuable energy is saved when catalysts are used to promote molecular breakdown at lower temperatures in an operation called **catalytic cracking.** Important cracking catalysts have been developed by chemists at all major oil companies, and the research continues to find more selective and inexpensive processes.

> The catalysts employed in catalytic cracking are chemically similar to the ion exchangers used in water softening, a topic that appears in Chapter 5.

If refining produces more small molecules than needed but not enough intermediate-sized ones essential for gasoline, catalytic combination can be used. In this process, smaller molecules are joined to form useful intermediate-sized molecules.

$$4\,C_2H_4 \longrightarrow C_8H_{16} \qquad [4.9]$$

The refining process can also rearrange the atoms within a molecule. It turns out that not all the molecules with a single chemical formula are necessarily identical. For example, octane, an important component of gasoline, has the formula C_8H_{18}. Careful analysis discloses 18 different compounds with this formula. Although the chemical and physical properties of these forms are similar, they are not identical. For example, the substance called *octane* has a boiling point of 125 °C, whereas that of the compound commonly known as *isooctane* is 99 °C. Different compounds with the same chemical formula are called **isomers.** Isomers differ in molecular structure—the way in which the constituent atoms are arranged. The structures of octane and isooctane are illustrated here:

> Chapter 10 includes more information about isomers and how to interpret these structures.

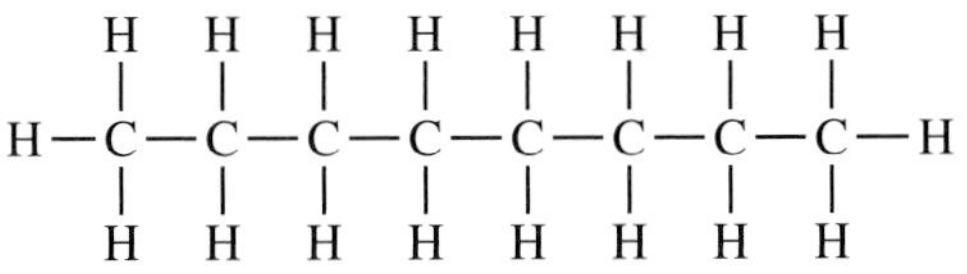

octane 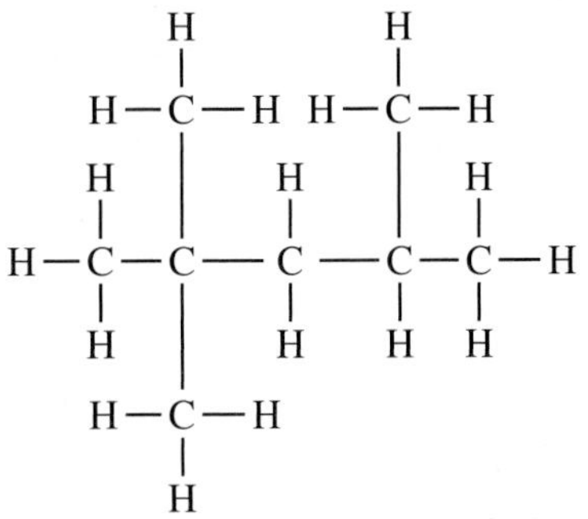   isooctane

The molecules of both isomers consist of eight carbon atoms and 18 hydrogen atoms, but these atoms are arranged differently in the two compounds. In octane all the carbon atoms are in an unbranched ("straight") line; the two end carbons are only attached to one carbon, whereas the six interior ones are attached to two other carbons. In isooctane the carbon chain is branched and the carbons of the backbone are attached to one, two, three or four other carbons. The structures shown here illustrate the same C_8H_{18} isomers with hydrogens omitted for clarity.

$$-\overset{|}{\underset{|}{C}}-\overset{|}{\underset{|}{C}}-\overset{|}{\underset{|}{C}}-\overset{|}{\underset{|}{C}}-\overset{|}{\underset{|}{C}}-\overset{|}{\underset{|}{C}}-\overset{|}{\underset{|}{C}}-\overset{|}{\underset{|}{C}}- \qquad -\overset{|}{\underset{|}{C}}-\overset{\overset{|}{C}-}{\underset{|}{C}}-\overset{|}{\underset{\underset{|}{C}-}{C}}-\overset{\overset{|}{C}-}{\underset{|}{C}}-\overset{|}{\underset{|}{C}}-$$

Both octane and isooctane have essentially the same heat of combustion, but the latter compound has a different shape and ignites much more easily. In a well-tuned car engine, gasoline vapor and air are drawn into a cylinder, compressed by a piston, and ignited by a spark. Normal combustion occurs when the spark plug ignites the fuel–air mixture, and the flame front travels across the combustion chamber rapidly and smoothly until the fuel is consumed. But compression alone may be often enough to ignite the fuel before the spark occurs. This premature firing is called preignition. It results in lower engine efficiency and higher fuel consumption because the piston was not in its optimal location when the burned gases expanded. Knocking, a violent and uncontrolled pressure that may be several times the usual value for the engine, occurs after the spark ignites the fuel, causing the unburned mixture to burn at supersonic speed with an abnormal rise in pressure. Knocking produces an objectionable metallic sound, loss of power, overheating, and engine damage when severe.

In the 1920s, knocking was shown to be due the burning characteristics of gasoline components. The "octane rating" was developed to designate a particular gasoline's resistance to knocking. The performance of isooctane in an automobile engine has been measured and arbitrarily assigned an octane rating of 100. Heptane is a straight-chain hydrocarbon with one CH_2 fewer than octane. Heptane has a similar boiling point to that of isooctane but has a much higher tendency to cause knocking than does isooctane (Table 4.6). When you go to the gasoline pump and fill up with 87 octane, the gasoline has the same knocking characteristics as a mixture of 87% isooctane (octane number 100) and 13% heptane (octane number zero). Higher-grade gasolines are also available: 89 octane (regular plus) and 92 octane (premium); these contain a greater percentage of compounds with higher octane ratings (Figure 4.17).

Again remember that gasolines are very complex mixtures, and only a few components appear in Table 4.6. Although octane has a poor rating, it is possible to rearrange or "reform" octane to isooctane, thus greatly improving its performance. This is accomplished by a chemical reaction that occurs when octane passes over a catalyst consisting of rare and expensive elements such as platinum (Pt), palladium (Pd), rhodium (Rh), or iridium (Ir). Reforming isomers to improve octane rating became important starting in the late 1970s because of the nationwide efforts to ban tetraethyl-lead (TEL) in gasoline. Oxygenated fuels like ethanol are octane boosters and will be discussed in the next section.

Figure 4.17
Gasoline is commonly available in 87, 89, and 92 octane.

Table 4.6	Octane Ratings of Several Substances
Compound	**Octane Rating**
Octane	−20
Heptane	0
Isooctane	100
Methanol	107
Ethanol	108
MTBE	116

Consider This 4.18   Gasoline Additives

Modern gasoline pumps are often still marked "Unleaded Fuel" although tetraethyllead (TEL) is no longer used in the United States. Use the resources of the Web to find out what alternatives are currently used to replace TEL and in what amounts. Do Canada and Mexico use the same replacements? Do any countries in North or South America still use TEL?

4.10   Newer Fuels and Other Sources

The ubiquitous role of the automobile in U.S. culture and the consequential need for gasoline have given rise to some additional issues. Elimination of TEL as an octane enhancer necessitated finding substitutes that were inexpensive, easy to produce, and environmentally friendly. Several are used, including ethanol and MTBE (*methyl tertiary-butyl ether*), each with an octane rating greater than 100 (see Table 4.6).

$$
\begin{array}{cc}
\underset{\text{ethanol}}{
\begin{array}{c}
\text{H} \quad \text{H} \\
| \quad\quad | \\
\text{H}-\text{C}-\text{C}-\overset{..}{\underset{..}{\text{O}}}-\text{H} \\
| \quad\quad | \\
\text{H} \quad \text{H}
\end{array}}
&
\underset{\text{MTBE}}{
\begin{array}{c}
\text{H} \quad\quad \text{CH}_3 \\
| \quad\quad\quad | \\
\text{H}-\text{C}-\overset{..}{\underset{..}{\text{O}}}-\text{C}-\text{CH}_3 \\
| \quad\quad\quad | \\
\text{H} \quad\quad \text{CH}_3
\end{array}}
\end{array}
$$

Fuels with these additives also are referred to as **oxygenated gasolines,** blends of petroleum-derived hydrocarbons with oxygen-containing compounds such as MTBE, ethanol, or methanol (CH_3OH). Because they already contain oxygen, oxygenated gasolines burn more cleanly by producing less carbon monoxide than their nonoxygenated counterparts, thereby reducing CO emissions. The Winter Oxyfuel Program, originally implemented in 1992 as part of the Clean Air Act Amendments, targeted cities with excessive wintertime carbon monoxide emissions to use oxygenated gasolines that contain 2.7% oxygen by weight. Ethanol is the primary oxygenate used in this program. About 40 cities were mandated to participate, but by 2001 about half were no longer implementing the winter oxy program.

> Oxygenated gasoline is required to be sold in Los Angeles, New York, Baltimore, Chicago, Hartford, Houston, Milwaukee, Philadelphia, Sacramento, and San Diego, cities with air quality problems.

Your Turn 4.19   Atomic Composition

The molecular formula of MTBE is $C_5H_{12}O$; that of ethanol is C_2H_6O. Calculate the percent (by mass) of oxygen in each of these oxygenated fuels. *Hint:* Section 3.6 contains examples of such calculations.

a. MTBE   **b.** ethanol

Consider This 4.20   Ethanol and Politics

A section of the Clean Air Act Amendments of 1990 requires gasoline to contain 2% of oxygenates such as ethanol. This section was championed by Senators Tom Daschle of South Dakota and Bob Dole of Kansas. Why did Senators Daschle and Dole have so much interest in this legislation?

Since 1995, the Year-round Reformulated Gasoline Program has been mandated by the Clean Air Act Amendments of 1990 or used voluntarily by about 90 cities or areas with the worst ground-level ozone. **Reformulated gasolines (RFGs)** are oxygenated gasolines that also contain a lower percentage of certain more volatile hydrocarbons such as benzene found in nonoxygenated conventional gasoline. RFGs cannot have greater than 1% benzene (C_6H_6) and must be at least 2% oxygenates. Because of their composition, reformulated

> The molecular structure of benzene is discussed in Section 10.3.

gasolines evaporate less easily than conventional gasolines, and produce less carbon monoxide emissions. The more volatile hydrocarbons (benzene, etc.) in conventional gasoline also are involved in tropospheric ozone formation, especially in high-traffic metropolitan areas. Currently, about 30% of U.S. gasoline is reformulated (that is, RFG) of which nearly 90% contains MTBE. The Clean Air Act Amendments do not require the use of MTBE, but refiners have chosen it as the main oxygenate in RFGs for economic reasons and its blending characteristics. MTBE can be shipped through existing pipelines and it is less volatile than ethanol, making it easier to meet the emission standards.

The use of RFGs and oxygenated gasolines exemplifies a risk-benefit situation. The potential benefits are considerable. The use of RFGs compared with conventional gasolines has resulted in substantial reductions of smog-forming pollutants (volatile organic compounds and nitrogen oxides) and toxic molecules like benzene. Starting with the second phase of the RFG program in January 2000, the EPA estimates an annual reduction of at least 100 thousand tons of smog-forming pollutants and over 20 thousand tons of toxics. Yet, RFGs and oxygenated gasolines may not be risk free; they have not been used long enough for possible long-term adverse effects, if any, to arise. Currently, risk levels are presumed to be low, with benefits outweighing risks. In January 2004, the National Institute of Environmental Health Sciences reported the human health effects of short-term exposure to large or small amounts of MTBE are not known. Animal studies at high doses (much higher than human exposure) have shown adverse effects on the nervous system, ranging from hyperactivity and uncoordination to convulsions and unconsciousness.

Although MTBE is now a major component of oxygenated and reformulated gasolines with annual production of about 4 billion gallons in 1999, its continued use for this purpose is in doubt. MTBE has leaked from underground gasoline storage tanks at gas stations into ground water. MTBE has considerable solubility in water and thus is finding its way into drinking water. Fortunately, an EPA drinking water advisory states that there is little likelihood that MTBE will cause adverse health effects at concentration of ca. 40 ppb or below; above these levels most people can detect its presence by taste or odor. In March 1999, California's governor released an executive order that would phase out the use of MTBE in gasoline in that state by the end of 2002 (Figure 4.18). Although the deadline was extended, many fuel producers were already in compliance by the end of 2004. Several other states, many in the Northeast, also are phasing out or restricting the use of MTBE.

> Volatile organic compounds and other air pollutants were discussed in Section 1.11.

Figure 4.18
MTBE notice on a California gasoline pump.

Consider This 4.21   What's Up with MTBE?

Use the Web to find this information.

a. The status of the MTBE phase out in California and any suits by oil companies that seek to postpone or eliminate it.

b. Which other states are phasing out or plan to phase out MTBE in gasoline.

c. How the phaseout plans for these states resemble or differ from that in California.

Consider This 4.22   Your Contribution to Air Quality

According to the EPA, driving a car is "a typical citizen's most 'polluting' daily activity."

a. Do you agree? Why or why not?

b. What pollutants do cars emit?

Information on automobile emissions provided by the EPA (together with the information in this text) can help you fully answer this question. Check the *On-Line Learning Center* for a direct link.

c. RFGs play a role in reducing emissions. Where in the country are RFGs required? Check the current list published on the Web by the EPA.

d. Explain which emissions RFGs are supposed to lower.

In addition to environmental issues like the ones caused by tetraethyllead and now by MTBE, the United States talks about wanting to reduce its reliance on imports of crude oil. For example, the Energy Policy Act of 1992 (EPAct) requires certain fleets of vehicles to include alternative-fuel vehicles (AFVs) that are capable of operating on nonpetroleum fuels. Synthetic fuels and ones derived from (current and not ancient) biological sources are being developed.

Because the world's coal supply far exceeds the available oil reserves, research and pilot studies are being conducted on converting coal into gaseous and liquid fuels that are identical with or similar to petroleum products. As a matter of fact, some of the appropriate technology is quite old. Before large supplies of natural gas were discovered and exploited, cities were lighted with water gas. This is a mixture of carbon monoxide and hydrogen, formed by blowing steam over hot coke (the impure carbon that remains after volatile components have been distilled from coal):

$$\underset{\text{coke}}{C(s)} + H_2O(g) \longrightarrow \underset{\text{water gas}}{CO(g) + H_2(g)} \qquad [4.10]$$

This same reaction is the starting point for the Fischer–Tropsch process for producing synthetic gasoline. The carbon monoxide and hydrogen are passed over an iron or cobalt catalyst, which promotes the formation of hydrocarbons. These can range from the small gas molecules like methane, CH_4, to the medium-sized molecules (containing five to eight carbon atoms) typically found in gasoline. This process, developed during the 1930s in Germany, is economically feasible only where coal is plentiful and cheap, and oil is scarce and expensive. This is the case in South Africa, where 40% of gasoline is obtained from coal. In the future, such technology may also become competitive in other parts of the world.

Concerns about the dwindling supply of petroleum have also led to the use of renewable energy sources. This generally means **biomass**—materials produced by biological processes. One such source, wood, was much touted during the 1970s energy crisis. But the energy demands of our modern society cannot possibly be met by burning wood. Burning trees would also destroy effective absorbers of carbon dioxide while adding that greenhouse gas and other pollutants to the atmosphere.

Ethanol (C_2H_5OH, or CH_3CH_2OH) is another alternative fuel produced from renewable biomass (Figure 4.19). It is formed by a method known since ancient times,

A structure of ethanol appears at the beginning of Section 4.10. Octane appears in Table 4.5 and Section 4.9.

Figure 4.19
Billboard advertisement for gasohol (Iowa Corn Grower's Association).

the fermentation of starch and sugars in grains such as corn. Enzymes released by yeast cells catalyze the reaction typified by this equation.

$$C_6H_{12}O_6 \longrightarrow 2\,C_2H_5OH + 2\,CO_2 \qquad [4.11]$$

glucose   ethanol

Ethanol can also be prepared commercially in large quantities by the reaction of water (steam) with ethylene, C_2H_4 (CH_2CH_2).

$$CH_2CH_2(g) + H_2O(g) \longrightarrow CH_3CH_2OH(l) \qquad [4.12]$$

ethylene

When the second method is used to produce ethanol for oxygenated fuels, any residual water must scrupulously be removed so that it will not create problems in an automobile engine. Making ethanol by fermenting grains creates a fuel that, unlike gasoline, is renewable because crops can continue to be planted. Thus, the source of the fuel can be replaced.

The burning of ethanol releases 1367 kJ per mole of C_2H_5OH.

$$C_2H_5OH(l) + 3\,O_2(g) \longrightarrow 2\,CO_2(g) + 3\,H_2O(l) + 1367\,kJ \qquad [4.13]$$

The energy output of 1376 kJ/mol of ethanol burned corresponds to 29.7 kJ/g. This value is less than the 47.8 kJ/g produced by burning C_8H_{18}, because the ethanol is already partially oxidized. Nevertheless, ethanol is being mixed with gasoline to form "gasohol." At the usual concentration of 10% ethanol, gasohol can be used without modifying standard automobile engines. More than 3 million flexible fuel vehicles (FFVs) already sold in the United States can use E85 (85% ethanol and 15% gasoline), gasoline or any mixture of the two. It is likely that the buyers of many of those 3 million FFVs, which include sedans, minivans, SUVs and pick-up trucks, remain unaware that they can fuel with E85. Of the 13 million vehicles in Brazil, more than 4 million use pure ethanol (made from fermented sugar cane), and the remainder of the cars operate on a mixture of ethanol and gasoline.

Racing engines have been modified to run on pure methanol (methyl alcohol), CH_3OH.

However, ethanol as a fuel is not without its drawbacks and critics. Some point to the fact that a gram of ethanol does not produce as much energy as a gram of gasoline. In addition, those opposed to ethanol as a fuel question whether valuable farmland, normally used to grow crops such as corn that feed people and animals, should be used to produce grain for ethanol. Therefore, the use of agricultural products for the production of fuel must depend on supply and demand, surpluses and shortages. Currently, the United States produces a significant surplus of corn and other grains that could be converted to ethanol. But, in a 2002 paper in the journal *Food Policy*, the Danish economist and journalist Bernard Gilland estimates that meeting a small percentage of the current primary energy demand with alcohol would require that a substantial portion of the world's cropland be removed from food and feed production. Clearly, there are limits to the amount of energy we can obtain from the fermentation of crops.

The use of ethanol as a petroleum substitute or as an oxygenating agent for gasoline has become a political hot potato in the United States. The issues involve not only ethanol, but also other oxygen-containing compounds such as MTBE. In March 2004 California lawmakers urged the EPA to approve the state's request for a waiver to use the fuel additives ethanol and MTBE to meet federal rules. The letter was signed by 52 of the 53 members of California's congressional delegation and follows the waiver request submitted by Gov. Arnold Schwarzenegger in January 2004. California has sought a waiver for five years although it has already banned MTBE. They argue that oil refiners have better and cheaper ways to blend cleaner-burning gasoline without using oxygenates. The state believes that ethanol is too expensive to ship there. Under the Clinton administration the EPA decided to allow the waiver, convinced that doing so would reduce emissions; however, President Clinton did not complete action on the waiver. In June 2001, the EPA, under the George W. Bush administration, denied the waiver, saying: "After an extensive analysis, the Agency concluded that there is significant uncertainty over the change in emissions that would result from a waiver. California has not clearly demonstrated what the impact on smog would be from a waiver of the oxygen mandate." The EPA ruled that, if California banned the use of MTBE, the state would have to replace it specifically with ethanol. If ethanol replaces MTBE in California, estimates are that the state would use about 580 million gallons of ethanol for fuel per year, roughly 20% of the 2.81 billion gallons of ethanol produced in the United States in 2003. In anticipation of a shift to a greater use of ethanol as a fuel, the nation's ethanol producers are building their capacity to produce 3.7 billion gallons by 2005.

Consider This 4.23   To MTBE or Not to MTBE?

California has attempted to waive the Clean Air Act requirement for oxygenates for the last several years. In the Consider This 4.21, you were asked to find what other states are in the process of or planning to phase out MTBE.

a.   What impact have these waiver attempts had on the production and consumption of MTBE since 1999?

b.   Select the state in which you live or another one of interest to you. Find references to gasoline spills or MTBE environmental issues in this state. Does this appear to be a serious concern to the citizens of this state?

c.   Does the state you have selected have a vested interest in the arguments about MTBE or ethanol as oxygenates? You may wish to consider factors like facilities or sites in the state that produce these molecules, distance from the sources to your state, and from where gasoline consumed in your state comes.

Sceptical Chymist 4.24   Keeping Track of Ethanol

Senator Tom Harkin (D-Iowa), chair of the Senate Agriculture Committee, predicts a need for 400 million additional gallons of ethanol per year to replace MTBE for auto use in California. It is estimated that the resulting gasohol mixture would be sufficient to power California vehicles for about 8 billion miles of driving per year, or over 4600 miles per vehicle. California has approximately 17 million passenger vehicles. Assume that the needed ethanol is used exclusively for gasohol and that 10 gal of gasohol contains 1 gal of ethanol and 9 gal of gasoline. State any additional assumptions you make and then show calculations to support or refute this estimate.

The battle lines regarding the use of ethanol as a fuel are clearly drawn, largely on the basis of self-interest. Supporting the greater use of ethanol are the EPA, Archer Daniels Midland (ADM, a huge agribusiness), and over 20 farm groups, including the

Figure 4.20
An advertisement for gasohol made using ethanol.

National Corn Growers Association (Figure 4.20). In opposition are the American Petroleum Institute, the petroleum refiners and gasoline companies, and the Sierra Club. U.S. senators stake out positions on the issue depending on whether they are from agricultural states or ones tied closely to oil. There is a good deal at stake—among other things, 100 million to 200 million bushels of corn per year. Those favoring petroleum use respond that, even with extensive farm support programs, ethanol is significantly more expensive per gallon and per kilojoule than conventional gasoline. The Sun is not the only source involved. Energy is required to plant, cultivate, and harvest the corn; to produce and apply the fertilizers; to distill the alcohol from the fermented mash; and to manufacture tractors. Hard, accurate data are difficult to get, but some sources claim that more energy goes into producing a gallon of ethanol than can be obtained by burning it.

Consider This 4.25   Gasohol

Unlike fuels obtained from petroleum, a nonrenewable source, ethanol can be produced from renewable resources. One method of producing ethanol is through the fermentation of crops such as sugar cane, potatoes, and corn.

a.   Draw up a list of reasons that argue for using crops for food and also a list of reasons for converting crops to ethanol for use in fuels.
b.   What factors determine how you decide which is the optimal use of these crops?

Biodiesel pricing will require Federal subsidies. For example, feedstock costs alone are at least $1.50 per gallon for biodiesel made from soy beans.

The alternative fuel, biodiesel (Figure 4.21), has grown dramatically during the last few years. Biodiesel is made from natural, renewable resources such as new and used vegetable oils and animal fats. It can be used as a pure fuel or blended with petroleum products and used in diesel engines that have no major modifications. The use of biodiesel results in a substantial reduction of unburned hydrocarbons compared with traditional diesel fuel with a minimum increase in cost. The Energy Policy Act (EPAct) was amended by the Energy Conservation Reauthorization Act of 1998 to include biodiesel as a way for federal, state, and public utility vehicles to meet requirements for alternative fuels. According to the American Biofuels Association, biodiesel sales could reach about 2 billion gallons per year or replace about 8% of conventional diesel fuel consumption. The National Biodiesel Board reports that this fuel is available in all 50 states but the distributors are concentrated in the agricultural production areas in the Midwest.

Figure 4.21

Biodiesel at the pump (left) and being formulated by a regional vendor from recycled restaurant vegetable oil (right).

The newest alternative fuel to be formulated for flexible fuel vehicles (FFVs) is designated by the registered trademark, "P-series." P-series are liquid blends of hydrocarbons ("pentanes plus," a mixture of 20 light hydrocarbon liquids extracted from natural gas), ethanol, and methyltetrahydrofuran (MeTHF) produced from biomass with high cellulose content). P-series are designed to be used alone or in any proportion with gasoline in conventional vehicles.

Yet another potential energy source is a commodity that is cheap, always present in abundant supply, and always being renewed—garbage. Other than in a movie, no one is likely to design a car that will run on orange peels and coffee grounds, but approximately 140 power plants in the United States do just that. One of these, pictured in Figure 4.22, is the Hennepin Energy Resource Company (HERC) in Minneapolis, Minnesota. Hennepin County produces about 1 million tons of solid waste each year. One truckload of garbage (about 27,000 lb) generates the same quantity of energy as 21 barrels of oil. HERC converts 365,000 tons of garbage per year into enough

Figure 4.22

Hennepin County Resource Recovery Facility, a garbage-burning power plant.

to provide power to the equivalent of 25,000 homes. In addition, over 11,000 tons of iron-containing metals are recovered from the garbage and recycled. Elk River Resource Recovery Facility, the second in Hennepin County, converts another 235,000 tons of garbage to electricity. The emissions at both sites are significantly below state and federal standards.

This resource recovery approach, as it is sometimes called, simultaneously addresses two major problems: the growing need for energy and the growing mountain of waste. The great majority of the trash is converted to carbon dioxide and water and no supplementary fuel is needed. The unburned residue is disposed of in landfills, but it represents only about 10% of the volume of the original refuse. Although some citizens have expressed concern about gaseous emissions from garbage incinerators, the incinerator's stack effluent is carefully monitored and must be maintained within established limits. Both Japan and Germany are making considerably greater use of waste-to-energy technology than the United States.

Perhaps the ultimate example of using waste as an energy source is provided by methane generators. Rural China and India have over one million reactors in which animal and vegetable wastes are fermented to form biogas. This gas, which is about 60% CH_4, can be used for cooking, heating, lighting, refrigeration, and generating electricity. The technology lends itself very well to small-scale applications. The daily manure from one or two cows can generate enough methane to meet most of the cooking and lighting needs of a farm family. Two thirds of China's rural families use biogas as their primary fuel.

Consider This 4.26   **Building a Waste-Burning Plant in Your City**

Imagine that you were the administrator of a city of a million residents charged with producing a proposal to your city council outlining the pros and cons of a waste-burning plant. Use the Hennepin County facilities in Minnesota as a model and their Web site to collect information and examples. Prepare slides that outline issues like a master plan, resource recovery, potential concerns of residents, and any other questions that should be addressed.

4.11   The Case for Conservation

A fundamental feature of the universe is that energy and matter are conserved. However, the process of combustion converts both energy and matter to less useful forms. For example, the energy stored in hydrocarbon molecules is eventually dissipated as heat when those molecules are converted to carbon dioxide and water—essential compounds, to be sure, but unusable as fuels. As residents of the universe, we have no choice; we must obey its inexorable laws. Nevertheless, we have many options within those constraints. One of the most important is to make a human contribution to the conservation of energy and matter.

The planet's store of fossil fuels is finite (Figure 4.23), although our appetites for them seem infinite. Worldwide demand for oil is increasing at more than 2% per year. Over the past 10 years, energy use is up in Latin America (32%), Africa (27%), and Asia (30%). In the mid-1990s, China became an oil importer rather than an exporter as it had been. Projections are for China to increase oil imports from 1 million barrels per day currently to 5 to 8 million barrels per day in the next two decades, predominantly from Middle East sources. Oil forecasts by the U.S. Energy Information Administration project a 60% increase in global demand for oil by 2020 to a whopping 40×10^9 barrels per year. To meet this growth, the market share of oil from Middle Eastern countries will likely rise beyond 30%, approaching the levels that produced the oil crises of 1974 and again in 1979. The conventional oil reserves $(2.183 \times 10^{12}$ barrels) are calculated to last 43 years at the current rate of consumption. In addition, enough petroleum is locked up in heavy crude oil, bitumen,

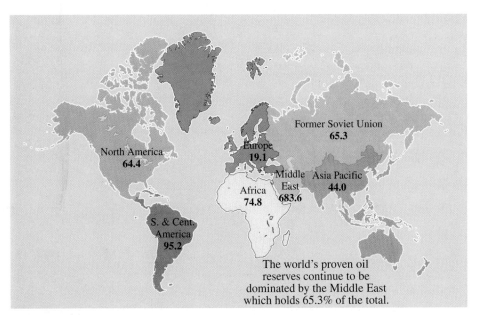

Figure 4.23

World proven reserves of oil at the end of 2000 (billions of barrels).

Source: *BP Statistical Review of World Energy 2001*, June 2000. Taken from www.bp.com/centres/energy/.

and oil shale to provide for another 170 years, though this reserve will be more difficult and more expensive to extract. Global coal reserves appear to be considerably greater, but they too are limited. And in any event, we can be certain that the world's energy consumption will not remain at its current level.

Oil is the prime energy source for all regions of the world except for the former Soviet Union, where the prime fuel is natural gas (Figure 4.24). The Organization for Economic Co-operation and Development, an organization of western European nations, has estimated that between 1990 and 2010 world energy consumption will increase by 50%, oil consumption will increase by 40%, coal consumption by 45%, and natural gas consumption by 66%.

One consequence of this increased use of energy will be a 50% rise in global CO_2 emissions. Not surprisingly, the greatest increases are expected to occur in the developing

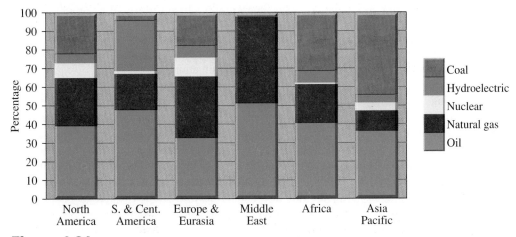

Figure 4.24

World energy consumption by region, 2002.

Source: *BP Statistical Review of World Energy 2002*, August 2000. Taken from www.bp.com/centres/energy/.

Figure 4.25
The Three Gorges Dam will supply China with significant hydroelectric power. The first generators went online in 2003.

To continue to meet such increases, it is estimated that China will need to open one medium-sized power plant every week for a year.

The use of nuclear power, as discussed in Chapter 7, is another strategy for reducing dependency on fossil fuels for energy.

countries that, by 2010, will account for over half of the energy consumed. Growing populations, migration to the cities, and industrialization will drive up the energy demand of the Third World to unprecedented highs. The pattern has already been established. In China, energy utilization in 1993 was 22 times what it was in 1952. From 1990 to 1999, electricity consumption in China nearly doubled (97%). The construction of the immense and controversial Three Gorges dam and hydroelectric power station in China is one attempt by that country to meet its exploding electrical energy demands (Figure 4.25).

The demands of conventional power plants for coal, oil, and gas are tremendous. But fossil fuels are so important as feedstocks for chemical synthesis that it is a great waste to burn them. Late in the 19th century, Dmitri Mendeleev, the great Russian chemist who proposed the periodic table of the elements, visited the oil fields of Pennsylvania and Azerbaijan. He is said to have remarked that burning petroleum as a fuel "would be akin to firing up a kitchen stove with bank notes." Mendeleev recognized that oil could be a valuable starting material for a wide variety of chemicals and the products made from them. But he would, no doubt, be amazed at the fibers, plastics, rubber, dyes, medicines, and pharmaceuticals currently produced from petroleum. Yet, we continue to ignore Mendeleev's warning and burn nearly 85% of the oil pumped from the ground. This is clearly a risk–benefit situation.

In short, the arguments for conserving energy and the fuels that supply it are compelling. Fortunately, some promising strategies are available, and considerable savings have already been realized. Although energy production and fuel consumption have increased since the 1974 oil crisis, the efficiency with which the fuels are used has also significantly improved. The production of electricity by power plants is the major use of energy in the United States, making up 38% of the total. The conversion of heat to work is, of course, limited by the second law of thermodynamics, but power plants currently operate well below the thermodynamic maximum efficiency. Better design will bring power plants closer to that upper limit, perhaps to overall efficiencies of 50 or 60%. A particularly appealing approach is an integrated system that uses "waste" heat from a power plant to warm buildings.

Once the electricity is generated, great savings can be realized in its use. Estimates of the technically feasible savings in electricity range from 10 to 75%. The wide range in these predictions is worrisome, but specific data are encouraging. For example, in an article in *Scientific American* in September 1990, Arnold P. Fickett, Clark W. Gellings, and Amory B. Lovins made the following statement: "If a consumer replaces a single 75-watt bulb with an 18-watt compact fluorescent lamp that lasts 10,000 hours, the consumer can save the electricity that a typical United States power plant would make from 770 pounds of coal. As a result, about 1600 pounds of carbon dioxide and 18 pounds of sulfur dioxide would not be released into the atmosphere." In the process, about $100 would be saved in the cost of generating electricity. Improvements in the design of electric motors and refrigeration units also hold considerable potential for increased efficiency.

Some utility companies now promote consumer education and provide financial inducements for conserving electricity. Sophisticated economic planning, new financing arrangements, and pricing policy are all part of efforts to save energy. One particularly important concept is "payback time," the period necessary before a private consumer, an industry, or a power company recaptures in savings the initial cost of a more efficient refrigerator, manufacturing process, or power plant. A business or an individual consumer are not likely to invest money unless they can recoup the savings within a reasonable timeframe.

Sceptical Chymist 4.27   Lightbulbs Revisited

The quotation from Fickett, Gellings, and Lovins provides a marvelous opportunity for Sceptical Chymist to apply her or his knowledge of chemistry. For example, let us check their assertion that replacement of a 75-watt (W) bulb with an 18-W fluorescent lamp will save the electricity made from 770 lb of coal. That is, assume that 770 lb of coal, containing 65% carbon and 2% sulfur, will not be burned.

First, we note that the difference in the rate of energy consumption of the two bulbs is 75 W − 18 W = 57 W or 57 J/s. The total projected energy savings over the life of the bulb (10,000 hr) is obtained in this way:

$$\text{Energy savings} = 10{,}000 \text{ hr} \times \frac{60 \text{ min}}{1 \text{ hr}} \times \frac{60 \text{ s}}{1 \text{ min}} \times \frac{57 \text{ J}}{\text{s}} = 2.05 \times 10^9 \text{ J}$$

The energy comes from coal, and coal typically yields 30 kJ/g or 30×10^3 J/g. To determine the mass of coal that must be burned to obtain 2.05×10^9 J, the following operation is performed.

$$\text{Mass coal} = 2.05 \times 10^9 \text{ J} \times \frac{1 \text{ g coal}}{30 \times 10^3 \text{ J}} \times \frac{1 \text{ lb coal}}{454 \text{ g coal}} = 150 \text{ lb coal}$$

This is a significant discrepancy from the quoted value of 770 lb. Possibly Fickett, Gellings, and Lovins made an error, or perhaps they made an assumption that we neglected. Explore the latter possibility, and suggest what the assumption might have been.

Your Turn 4.28   Gases Released from Burning Coal

Now it is your turn to exercise your computational skills by checking the other two claims. Calculate the mass of CO_2 and SO_2 that would *not* be released into the atmosphere if a 75-W bulb were replaced by an 18-W compact fluorescent lamp. Assume that 770 lb of coal, containing 65% carbon and 2% sulfur, would not be burned.

Sceptical Chymist 4.29   Shedding Light on Assumptions

Fickett, Gellings, and Lovins also claim that the bulb replacement we have been discussing would save about $100 in the cost of generating electricity. What value are they assuming for the cost of electricity?

(continued on p. 210)

Sceptical Chymist 4.29   **Shedding Light on Assumptions** (*continued*)

Electricity is generally priced per kilowatt-hour (kWh), so we need to know the number of kilowatt-hours saved over the 10,000-hr lifetime of the bulb. First multiply the 57 W saved by 10,000 hr:

$$57 \text{ W} \times 10,000 \text{ hr} = 570,000 \text{ watt hr (Wh)}$$

Then convert the answer to kilowatt-hours, recognizing that 1 kWh = 1000 Wh:

$$570,000 \text{ Wh} = \frac{1 \text{ kWh}}{1000 \text{ Wh}} = 570 \text{ kWh}$$

If 570 kWh of electricity costs $100, as the writers imply, what is the cost per kilowatt-hour? Once you have calculated the answer, find the cost of electricity to consumers in your city.

In 1999 the U.S. Department of Energy announced an aggressive, new research and development program, Vision 21, whose goals according to a 2003 National Research Council (NRC) report include ". . . facilities [that] will be able to convert fossil fuels (e.g., coal, natural gas, and petroleum coke) into electricity, process heat, fuels, and/or chemicals more effectively, with very high efficiency and very low emissions, including of the greenhouse gas carbon dioxide." For example, the targets for emissions include a 40–50% reduction in CO_2 by efficiency improvement and essentially a 100% reduction if the carbon dioxide is separated and sequestered. Vision 21 activities are oriented toward achieving revolutionary rather than evolutionary improvements that would be ready for deployment in 2015. The NRC report goes on to state that the highest priority technology-related findings and recommendations pertain to gasification, gas purification, turbines, and fuel cells. The Clean Coal Power Initiative, recently authorized by Congress, can provide support for the construction of high-risk, early commercial plants to implement the new technologies needed to achieve near-zero emissions or carbon dioxide management.

Fuel cells are discussed in Section 8.4.

Consider This 4.30 Vision 21 Progress

The Vision 21 program made an initial investment of about $50 million in the research and development of new technologies to begin to accomplish its goals. Much of that work and subsequently funded projects are in progress.

a. Use the Department of Energy Web site to check on the status of research in the Vision 21 program. How has the initial $50 million been spent? Is there additional funding and have the priorities changed?
b. The Clean Coal Power Initiative and the National Energy Technology Laboratory (NETL) have demonstration projects on coal utilization and fuel systems. Use the Web to look at their efforts and reports of success. Make a list of the five most significant contributions from each program.

Recent advances in information technology and data processing also make sizeable energy savings possible. "Smart" office buildings or homes feature a complicated system of sensors, computers, and controls that maintain temperature, airflow, and illumination at optimum levels for comfort and energy conservation. Similarly, the computerized optimization of energy flow and the automation of manufacturing processes have brought about major transformations in industry. Over the past 20 years, industrial production in the United States has increased substantially, but the associated energy consumption has actually gone down. A case in point is the low-pressure, gas-phase process developed by Union Carbide chemical engineers for making poly-ethylene, which is the world's most common plastic. This new process uses only one quarter the energy required by previous high-pressure methods. Although the capital investment associated with such conversions is often substantial, consumers and man-ufacturers may ultimately enjoy financial savings and increased profits. Energy conservation also results from recycling materials, especially aluminum. Because of high energy costs required to extract aluminum metal from its ore, recycling the metal yields an energy saving of about 70%. To put things in perspective, you could watch television for 3 hr on the energy saved by recycling just one aluminum can.

One final area where energy conservation has a direct impact on lifestyle is transportation. About 20% of the total energy used and one half of the world's oil production goes to power motor vehicles. But even here, we are making some progress in conservation. From the mid-1970s to the early 1990s, gasoline consumption in the United States dropped by one half. Much of this saving was attributable to lighter-weight vehicles, thanks to the use of new materials, and to new engine designs. Yet this environmentally friendly trend is under assault as the 21st century begins. Since 1988 there has been an overall decline in vehicle fuel economy. It currently stands at 24.0 miles per gallon (mpg), the lowest since 1980 and 1.9 mpg less than the 25.9 mpg highest value obtained in 1988. A major factor for the decline in fuel economy has been the significant increase in the proportion of vehicles classified as light-duty trucks—sport utility vehicles (SUVs), vans, and pickup trucks—which now command a substantial part (46%) of the traditional car market.

Two decades ago, the average light-duty truck and car weighed about the same and had similar engines. That is no longer the case; light-duty trucks are now more than 15% heavier and have nearly 80% more horsepower than those of 20 years ago. Although car manufacturers must meet a federally mandated 27.5 mpg average fuel consumption for cars, the requirement is only 20.7 mpg for light-duty trucks. SUVs are classified as light trucks, not as cars. Therefore, SUVs need not meet the higher mpg requirements for cars. The lower mpg standards for trucks and SUVs have allowed their weight and horsepower to increase. A full-size SUV or a pickup truck emits much more CO_2 than even a large automobile (SUV, 80% more; pickup truck, 67% more). Passenger cars have an average fuel economy of 28.1 mpg; that of light-duty (pickup) trucks is only 20.1 mpg (Figure 4.26). SUVs average 20.0 mpg; vans and minivans, 22.5 mpg.

Senator Diane Feinstein (D–California) introduced a bill that would require SUVs and pickup trucks to meet the same fuel economy standards as automobiles by 2007. Feinstein asserts: "Simply put, this is the single most effective action we can take to limit our dependency on foreign oil, to save consumers at the pump, and to reduce global warming." Ford Motor Company has pledged to increase the fuel economy of

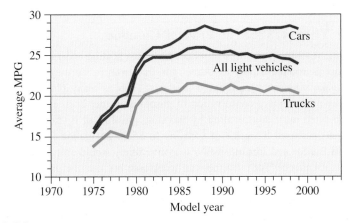

Figure 4.26

Fuel economy by model year and vehicle type.

Source: U.S. EPA. Taken from http://www.epa.gov/otag/fetrends.htm

its SUVs by 25% by the 2005 model year, which would result in its vehicles able to get 23 mpg, up from the current 18 mpg.

On the plus side, research to develop cars with higher gas mileage continues. The 2004 Honda Insight is the most fuel efficient vehicle sold in the United States since 1975, with a rating of 60 mpg in the city and 66 mpg on the highway. The two-seater Insight is a hybrid vehicle that uses a combination of gasoline and battery power, like the Toyota Prius, which gets 60 mpg (city) and 51 mpg (highway) (Figure 4.27). Methane and propane are being used to power a growing number of cars and trucks, and Chapter 8 explores such alternative energy sources as hydrogen, electric batteries, fuel cells, and photovoltaic cells.

Such potential improvements in fuel economy are impressive, but the fact remains that the automobile is an energy-intensive means of transportation. A mass transit system is far more economical, provided it is heavily used. In Japan, 47% of travel is by public transportation, compared with only 6% in the United States. Of course, Japan is a compact country with a high population density. The great expanse of North America is not ideally suited to mass transit, although some regions, such as the population-dense Northeast are. One must also reckon with the long love affair between Americans and their automobiles.

Hybrid vehicles are discussed in Section 8.7.

Figure 4.27

The Toyota Prius is a gasoline/battery hybrid car.

Consider This 4.31   Mass Transit

Advocates propose mass transit as a way to reduce fossil fuel consumption in the United States. Suggestions include trains to interconnect a set of small cities or a metropolitan center like Chicago to a new huge airport far beyond the suburbs.

a.   List reasons for and against the two suggestions.
b.   What arguments would you expect from citizens concerning more extensive use of mass transit?
c.   What other creative measures could be taken to reduce the use of personal vehicles?

Consider This 4.32   Gasoline Rationing

Imagine that you were transported to the future in the United States or were living in Europe where you must pay $5.00 per gallon for gasoline. Your budget allows $250 per month for gasoline. Assume that your vehicle averages 20 mpg. Describe the number of people in your family, how many of them go to school or work each day, and how far each must travel. Use this information to prepare a detailed gasoline budget of how you and your family would use the $250 for gasoline in a month.

Consider This 4.33   The Price of Gasoline

Oil is a valuable resource, even beyond its use for home heating and gasoline. If we continue to use our petroleum supply to extract gasoline from it, we may lose our starting materials to make other petroleum-based products, such as many pharmaceuticals and plastics. Until now, voluntary conservation of gasoline has not been effective. The government could force more conservation by rationing gasoline or by heavily taxing it. If our government increased the price of gasoline to $5.00 per gallon (a price typical of that in Western Europe and Japan), sales of gasoline likely would drop. This price would have serious consequences for the American workforce because the price of a gallon of gasoline would be raised relative to the current minimum hourly wage.

Suppose a bill had been introduced in Congress to raise the price of gasoline to $5.00 per gallon. Draft a letter to a friend at another college, either supporting or protesting the bill. Include the reasons for your position and your opinion on what should be done with this new revenue if the bill were to be passed.

CONCLUSION

To a considerable extent, choice ultimately influences what technology can do to conserve energy. As individuals and as a society, we must decide what sacrifices we are willing to make in speed, comfort, and convenience for the sake of our dwindling fuel supplies and the good of the planet. The costs might include higher taxes, more expensive gasoline and electricity, fewer and slower cars, warmer buildings in summer and cooler ones in winter, perhaps even drastically redesigned homes and cities. During the 1970s a series of energy crises occurred because of a dramatic rise in the cost of imported crude oil, principally from the Middle East. Although we have broadened our sources of imported crude oil to other regions of the world, our reliance on imported fossil fuels remains high. This ongoing dependence keeps alive the specter of another energy crisis, perhaps on a global basis. One thing seems to be clear. The best time to examine our options, our priorities, and our will is before we face another full-blown energy crisis. Quite obviously energy, chemistry, and society are closely intertwined. This chapter is an attempt to untangle them.

Chapter Summary

Having studied this chapter, you should be able to:

- Distinguish between energy and heat, and be able to convert among energy units: joules, kilocalories, Calories (4.1)
- Apply the terms exothermic, endothermic, and activation energy to chemical systems (4.2–4.5)
- Relate the energy potentially available from a process with the efficiency of that process (4.2)
- Use entropy as a concept to explain the second law of thermodynamics (4.2)
- Understand potential energy stored in a chemical compound and how reactions like combustion release that energy (4.3)
- Interpret chemical equations and basic thermodynamic relations to calculate heats of reaction, particularly heats of combustion (4.4–4.5)
- Use bond energies to describe the energy content of materials (4.5)
- Describe the factors related to the United States' dependency on fossil fuels for energy (4.6)

- Evaluate the risks and benefits associated with petroleum, coal, and natural gas as fossil fuel energy sources (4.6–4.8)
- Relate energy use to atmospheric pollution and global warming (4.7, 4.8)
- Understand the physical and chemical principles associated with petroleum refining (4.8–4.9)
- Describe *octane rating* and how refining, leaded gasoline, ethanol, and MTBE relate to it (4.9)
- Discuss approaches to alternative (supplemental) automobile fuels (4.9–4.10)
- Describe why reformulated and oxygenated gasolines are used (4.9–4.10)
- Take an informed stand on what energy conservation measures are likely to produce the greatest energy savings (4.11)
- With confidence, examine news articles on energy crises and energy conservation measures to interpret the accuracy of such reports (4.11)

Questions

Emphasizing Essentials

1. **a.** List three fossil fuels.
    b. What is the origin of fossil fuels?
    c. Are fossil fuels a renewable resource?
2. Consider the water in each container.

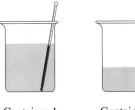

Container 1
80 g H_2O
70 °C

Container 2
40 g H_2O
70 °C

The temperature of the water is the same in each. Is the heat content of the water the same in each of these containers? How do you know?

3. The Calorie, used to express food heat values, is the same as a kilocalorie of heat energy. If you eat a chocolate bar from the United States with 600 Calories of food energy, how does the energy compare with eating a Swiss chocolate bar that has 3000 kJ of food energy? (*Note:* 1 kcal = 4.184 kJ)

4. A single serving bag of Granny Goose Hawaiian Style Potato Chips has 70 Calories. Assuming that all of the energy from eating these chips goes toward keeping your heart beating, how long can these chips sustain a heartbeat of 80 beats per minute? *Note:* 1 kcal = 4.184 kJ and each human heart beat requires 1 J of energy.

5. Three power plants have been proposed, operating at these power plant efficiencies.

Plant	Power plant efficiency
I	81%
II	66%
III	41%

   a. Calculate the maximum (overall) efficiency of each plant using the efficiencies of boiler, turbine, electrical generator, and power transmission in Table 4.1.

   b. Identify the factors that affect the efficiency.

   c. Discuss the practical limits that govern such efficiencies. Which plant would be most likely to be built? If plant III only costs half of plant I or II to operate, which would be most likely to be built?

6. Which is the better analogy for a state of high entropy—an unopened deck of playing cards or a plate of cooked spaghetti? Explain your reasoning.

7. Equation 4.1 shows the complete combustion of methane.

   a. Write a similar chemical equation for the complete combustion of ethane, C_2H_6.

   b. Represent this equation with Lewis structures.

   c. Represent this reaction with a sphere equation.

8. The heat of combustion for ethane, C_2H_6, is 52.0 kJ/g. How much heat would be released if 1 mol of ethane undergoes combustion?

9. a. Write the chemical equation for the complete combustion of heptane, C_7H_{16}.

   b. The heat of combustion for heptane is 4817 kJ/mol. How much heat would be released if 250 kg of heptane undergoes complete combustion?

10. Figure 4.5 shows energy differences for the combustion of methane, an exothermic chemical reaction. The combination of nitrogen gas and oxygen gas to form nitrogen monoxide is an example of an endothermic reaction:

$$180 \text{ kJ} + N_2(g) + O_2(g) \longrightarrow 2 \text{ NO}(g)$$

Analogous to Figure 4.5, sketch an energy difference diagram for this reaction.

11. One way to produce ethanol for use as a gasoline additive is the reaction of water vapor with ethylene:

$$H_2C=CH_2 + H_2O \longrightarrow CH_3CH_2OH$$

   a. Rewrite this equation using Lewis structures.

   b. Was it necessary to break *all* the chemical bonds in the reactants to form the product ethanol? Explain your answer.

12. From personal experience, state whether these processes are endothermic or exothermic. Give a reason for each.

   a. A charcoal briquette burns.

   b. Water evaporates from your skin.

   c. Ice melts.

   d. Wood burns.

13. Use the bond energies in Table 4.2 to estimate the energy change associated with this reaction.

$$2 \text{ C}{\equiv}\text{O} + \text{O}{=}\text{O} \longrightarrow 2 \text{ O}{=}\text{C}{=}\text{O}$$

14. Analogous to Figure 4.6, sketch a diagram for the reaction in question 13.

15. Use the bond energies in Table 4.2 to explain why

   a. chlorofluorocarbons, CFCs, are so stable.

   b. it takes less energy to release Cl atoms than F atoms from CFCs.

16. Use the bond energies in Table 4.2 to calculate the energy changes associated with each of these reactions. Lewis structures of the reactants and products may be useful for determining the number and kinds of bonds. Label each reaction as endothermic or exothermic.

   a. $2 \text{ C}_5H_{12}(g) + 11 \text{ O}_2(g) \longrightarrow 10 \text{ CO}(g) + 12 \text{ H}_2O(l)$

   b. $H_2(g) + Cl_2(g) \longrightarrow 2 \text{ HCl}(g)$

   c. $N_2(g) + 3 \text{ H}_2(g) \longrightarrow 2 \text{ NH}_3(g)$

17. Use the bond energies in Table 4.2 to calculate the energy changes associated with each of these reactions. Label each reaction as endothermic or exothermic.

   a. $H_2(g) + O_2(g) \longrightarrow H_2O_2(g)$

   b. $2 \text{ H}_2(g) + O_2(g) \longrightarrow 2 \text{ H}_2O(g)$

   c. $2 \text{ H}_2(g) + CO(g) \longrightarrow CH_3OH(g)$

18. Use Figure 4.11 to compare the sources of U.S. energy consumption. Arrange the sources in order of decreasing percentage and comment on the relative rankings.

19. Table 4.3 lists the energy content of some fuels in kilojoules per gram (kJ/g). Calculate the fuel energy in kilojoules per mole (kJ/mol) for methane CH_4, propane C_3H_8, hydrogen H_2, and ethanol C_2H_6O. Visit the *Online Learning Center* to explore other comparisons of fuel energy.

20. Mercury is present in minor amounts (50–200 ppb) in coal. Use the amount of coal burned by a power plant in Your Turn 4.15 to determine how much Hg is released by that plant. Calculate the amount based on the lower (50 ppb) and higher (200 ppb) limits.

21. Figure 4.12 shows U.S. oil production and oil imports.

   a. Calculate the percentage of total oil products supplied by domestic oil production in 1970, 1980, 1990, and 2000. Also calculate the predicted value for 2005.

   b. How are these values changing with time?

   c. How have the sources of oil changed from 1970 to 2000? What is predicted for 2005?

22. An energy consumption of 650,000 kcal per person per day is equivalent to an annual personal consumption of 65 barrels of oil or 16 tons of coal. Use this information to calculate the amount of energy available in each of these quantities.

   a. one barrel of oil

   b. one gallon of oil (42 gallons per barrel)

c. one ton of coal

d. one pound of coal (2000 pounds per ton)

23. Use the information in question 22 to find the ratio of the quantity of energy available in one pound of coal to that in 1 lb of oil. *Hint:* One pound of oil has a volume of 0.56 qt.

24. Pentane, C_5H_{12}, has a boiling point of 36.1 °C. Octane, C_8H_{18}, has a boiling point of 125.6 °C. Will pentane and octane be gases at room temperature (25 °C)?

25. Table 4.5 shows the structures of alkanes containing one to eight carbons.

    a. Draw the structure for decane, $C_{10}H_{22}$.

    b. Use the formulas and structures for octane and decane to predict the formulas for nonane, the alkane with nine carbons, and dodecane, the alkane with 12 carbons.

    c. The structures in Table 4.5 are two-dimensional. Use the bond angle information in Chapter 3 to predict the C-to-C-to-C and H-to-C-to-H bond angles in decane.

26. Consider this reaction representing the process of cracking.

    $$C_{16}H_{34} \longrightarrow C_5H_{12} + C_{11}H_{22}$$

    a. Which bonds are broken and which bonds are formed in this reaction? Use Lewis structures to help answer this question.

    b. Use the information from part **a** and Table 4.2 to calculate the energy change during this cracking reaction.

27. How many isomers does butane, C_4H_{10}, have? Draw the Lewis structure for each isomer. *Hint:* Be careful not to repeat isomers. Lewis structures show how atoms are linked, but not their spatial arrangement.

28. A premium gasoline available at most stations has an octane rating of 92.

    a. What does the octane rating tell you about the knocking characteristics of this gasoline?

    b. What does this tell you about whether the fuel contains oxygenates?

Concentrating on Concepts

29. How might you explain the difference between temperature and heat to a friend? Use some practical, everyday examples and assume your friend has not taken a chemistry course.

30. How is the statement that "energy is neither created nor destroyed in a chemical reaction" related to the law of conservation of energy and the first law of thermodynamics?

31. In each of these pairs, select the substance that has the greater entropy. Explain the reasons behind your choice.

    a. $H_2O(g)$ at 100 °C and $H_2O(l)$ at 100 °C

    b. a solid piece of iron and an equal mass of iron powder

    c. peanuts or an equal mass of peanut butter

32. State whether the entropy increases or decreases in each of these.

    a. Liquid water is converted to ice.

    b. Solid sodium chloride is dissolved in water.

    c. A hydrocarbon with 16 carbons is cracked.

33. One mole of diamond has an entropy of 2.4 J/°C at 25 °C; 1 mol of methanol, $CH_3OH(l)$, has an entropy of 127 J/K at 25 °C. What generalization can be drawn from these two values?

34. A friend tells you that hydrocarbons containing larger molecules are better fuels than those containing smaller molecules.

    a. Use these data, together with appropriate calculations, to discuss the merits of this statement.

Hydrocarbon	Heat of Combustion
Octane, C_8H_{18}	5450 kJ/mol
Butane, C_4H_{10}	2859 kJ/mol

    b. Considering your answer to part **a**, do you expect the heat of combustion per gram of candle wax, $C_{25}H_{52}$, to be more or less than the heat of combustion per gram of octane? Do you expect the molar heat of combustion of candle wax to be more or less than the molar heat of combustion of octane? Justify your predictions.

35. Halons are synthetic chemicals similar to CFCs, but they also include bromine. Although halons are excellent materials for fire fighting, they are more effective at ozone depletion than CFCs. This is the structural formula for halon-1211.

    $$\overset{\displaystyle :\!\overset{\displaystyle \cdot\cdot}{Br}\!:}{\underset{\displaystyle :\!\overset{\displaystyle \cdot\cdot}{Cl}\!:}{\overset{\displaystyle |}{\underset{\displaystyle |}{:\!\ddot{F}\!-\!C\!-\!\ddot{F}\!:}}}}$$

    a. Which bond in this compound is broken most easily? How is that related to the ability of this compound to interact with ozone?

    b. C_2HClF_4 is a compound being considered as a replacement for halons as a fire extinguisher. Draw the Lewis structure for this compound and identify the bond broken most easily. How is that related to the ability of this compound to interact with ozone?

36. During the distillation of petroleum, kerosene and hydrocarbons with 12–18 carbons used for diesel fuel will condense at position C marked on this diagram.

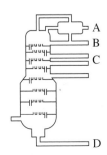

a. Separating hydrocarbons by distillation depends on the hydrocarbons having differences in a specific physical property. Which property is that?

b. How will the number of carbon atoms in the hydrocarbon molecules separated at A, B, and D compare with those separated at position C? Explain your prediction.

c. How will the uses of the hydrocarbons separated at A, B, and D differ from those separated at position C? Explain your reasoning.

37. Imagine you are at the molecular level, looking at what happens when liquid ethylene, C_2H_4, boils. Consider a collection of four ethylene molecules, representing each molecule with this sphere formula.

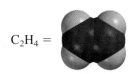

$C_2H_4 =$

a. Draw a representation of ethylene in the liquid state and then in the gaseous state. How will the collection of molecules change?

b. Estimate the temperature at which this transition from liquid to gas is taking place. What is the basis for your estimation?

38. Crude oil can undergo distillation or further change through cracking.

a. Explain why cracking is necessary.

b. Hydrocarbons undergo physical changes during distillation in a fractionating tower. Does cracking also involve physical changes? Explain your reasoning. *Hint:* A physical change is one in which a property of the molecule is varied but the structure stays the same. In a chemical change, the molecular structure is different.

39. Catalysts are used to speed up cracking reactions in oil refining and allow them to be carried out at lower temperatures.

a. Draw a sketch similar to Figure 4.8, in which you illustrate the energy changes for such a reaction in the absence and in the presence of the catalyst. Explain how your sketch illustrates the effect of the catalyst.

b. What examples of catalysts were given in the first two chapters of this text?

40. The octane rating is used to classify gasoline. Octane ratings of several substances are listed in Table 4.6

a. What evidence can you give that the octane rating is or is not a measure of the energy content of a gasoline?

b. Octane ratings are measures of a fuels ability to minimize or prevent engine knocking. Why is the prevention of knocking important?

c. Why are higher octane rating gasolines more expensive than lower ones?

41. One risk of depending on oil imports is the shortage of gasoline in the event of unfavorable international events. Does a gasoline shortage affect only individual motorists? What are some of the ways that a gasoline shortage could affect your life?

42. Here are three possible isomers of octane, C_8H_{18}; the hydrogen atoms and C—H bonds have been omitted for simplicity.

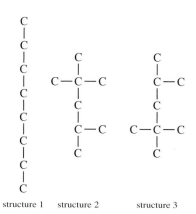

structure 1   structure 2   structure 3

a. Fill in the missing hydrogen atoms and confirm that these structures all represent C_8H_{18}.

b. Do any of the representations refer to identical isomers? If so, which ones?

c. Obtain a model kit and construct one of the structures. What are the C-to-C-to-C bond angles in the structure?

d. If you were to build a different one, would the C-to-C-to-C bond angles change? Why or why not?

e. Draw the structural formula of two more isomers of octane.

43. How is the growth in oxygenated gasolines related to

a. restrictions on the use of lead in gasoline?

b. federal and state air quality regulations?

44. a. Do oxygenated fuels have a higher energy content than nonoxygenated fuels? Explain your reasoning.

b. Do oxygenated fuels have a higher entropy than comparable nonoxygenated fuels? Explain your reasoning.

45. Reformulated gasolines (RFGs) were mandated or suggested for voluntary use in about 100 cities by federal legislation starting in 1995.

a. Has the definition of RFGs or suggested contents for them changed since 1995?

b. How many states are currently involved in the Year-round Reformulated Gasoline Program?

46. Your neighbor is shopping for a new family vehicle. The salesperson identified a van of interest as a flexible fuel vehicle (FFV).

a. Explain what is meant by FFV to your neighbor.

b. What does it mean for the van be able to use E85 fuel?

c. Would your neighbor and his family be particularly interested in using E85 fuel depending on what region of the country they live?

47. Find information or a map that shows the availability of biodiesel fuel distributors in the United States.

a. Why are a majority of the distributors located where they are?

b. According the National Biodiesel Board, their distributors will ship the fuel anywhere in the country, particularly to operators of fleets of trucks or cars. Would trucking companies in Florida and Oklahoma both be equally interested? Make a list of factors that would be important in such a decision.

48. The newest alternative fuels are the "P-series."

a. Use the Internet to find information on the increase in production of these fuels. Do P-series fuels appear to be making a serious impact as an alternative fuel for vehicles?

b. Can you find places in the country to buy it? What about distributors where you live?

49. China's large population has increased energy consumption as the standard of living increases.

a. Report information about China's increasing number of automobiles over the last 10 years.

b. What evidence suggests that the increase in the number of vehicles has affected air quality? What interventions, if any, does the Chinese government have underway?

Exploring the Extensions

50. Section 4.10 states that RFGs burn more cleanly by producing less carbon monoxide than nonoxygenated fuels. What evidence supports this statement?

51. Consider this diagram.

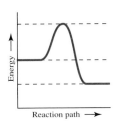

a. Does this representation show an exothermic or endothermic reaction? Explain your reasoning by commenting on the shape of the curve.

b. Sketch this type of energy diagram for the type of reaction, exothermic or endothermic, *not* shown in the diagram.

52. Bond energies such as those in Table 4.2 are sometimes found by "working backward" from heats of reaction. A reaction is carried out and the heat absorbed or evolved is measured. From this value and known bond energies, other bond energies can be calculated. For example, the energy change associated with this reaction is -81 kJ.

$$NBr_3(g) + 3\ H_2O(g) \longrightarrow 3\ HOBr(g) + NH_3(g)$$

Use this information and the values found in Table 4.2 to calculate the energy of the N-to-Br bond. Assume all atoms other than hydrogen obey the octet rule.

53. Explain why a fractionation tower can separate a mixture of hydrocarbons into different fractions, but it is not possible to separate seawater, also a complex mixture, into all of its different fractions.

54. Section 4.9 states that both octane and isooctane have essentially the same heat of combustion. How is that possible if they have different structures? Explain your thinking.

55. Why do you think that countries are willing to go to war over energy issues, but not over other environmental issues? Write a brief op-ed piece for your school newspaper discussing this issue.

56. What relative advantages and disadvantages are associated with using coal and with using oil as energy sources? Which do you see as the better fuel for the 21st century? Give reasons for your choice.

57. What are the advantages and disadvantages of replacing gasoline with renewable fuels such as ethanol? Indicate your personal position on the issue and state your reasoning.

58. C. P. Snow, a noted scientist and author, wrote an influential book called *The Two Cultures,* in which he stated: "The question, 'Do you know the second law of thermodynamics?' is the cultural equivalent of 'Have you read a work of Shakespeare's?'" How do you react to this comparison? Discuss these questions in light of your own educational experiences.

The Water We Drink

Many types of bottled waters are available to consumers in the United States and around the world.

"Americans have long felt safe drinking their water out of the tap, and with good reason. For the most part, public water supplies and even well water throughout the country are highly unlikely to make anyone acutely ill. The occasional failures of public systems to operate properly are few and attract national attention . . . But in recent years, the concern has been mounting about less obvious potential hazards in drinking water, especially residues of pesticides and industrial wastes, lead from old pipes, and compounds formed by chlorine and organic matter that are believed to be cancer-causing. The concern about water safety has prompted millions of Americans to reject the water that comes straight from the tap, resulting in two new growth industries: bottled water and filtration systems."

Jane E. Brody,
"On Tap or Bottled, Pursuing Purer Water" Personal Health Column,
New York Times, July 18, 2000

"Clean water is one of the areas of sharpest contrast between industrialized nations and the developing world. When New Jersey faced drought restrictions last summer, suburbanites complained of dying lawns, but they still had plenty of clean water to drink. In developing countries, on the other hand, diseases caused by unsafe drinking water, insufficient sanitation, and poor hygiene related to inadequate water supply kill 5 million people each year, according to the United Nations."

"Turning on the Tap," Report from World Summit on Sustainable Development, Johannesburg,
South Africa, 26 August–4 September 2002 Reported in *Chemical and Engineering News,*
18 November 2002

"On June 11, 2003, the Campaign for Safe and Affordable Drinking Water and its member groups participated in the release of a Natural Resources Defense Council (NRDC) study of drinking water quality which found deteriorating water infrastructure and pollution threaten the municipal drinking water supplies of 19 U.S. cities. The NRDC report, 'What's on Tap? Grading Drinking Water in U.S. Cities,' also concluded that recent . . . proposals to weaken the Clean Water Act and other laws would aggravate these risks."

"Drinking Water Right to Know Reports"
Campaign for Safe and Affordable Drinking Water,
11 June 2003

Arguably, water is the most important chemical compound on the face of the Earth. In fact, it covers about 70% of that face, giving the planet the lovely blue color in the famous "blue marble" photos taken from outer space. Water is essential to all living species; without it humans would die within a week. A child's body is about 75% water; adult bodies are approximately 50–65% water. A human brain is 75% water, blood is 83% water, and lungs are approximately 90% water. Even our bones, seemingly so solid, are 22% water. Water is so important to life that speculation about life elsewhere in the universe hinges first and foremost on the availability of water. Water refreshes and sustains us, dominates weather systems, and gives us aesthetic and relaxing pleasures.

Although we generally take water for granted, it is a remarkable chemical compound with unique properties that account for its essential life-supporting role. In this chapter, we will consider water from the perspective of those who drink it. There is more to know about water, however, than can be seen or tasted. Unseen impurities in water, depending on their identities and amounts, can impart a crisp, fresh taste or produce an unpleasant illness. Water is incredibly versatile, dissolving many substances and suspending others. To better appreciate how water works its magic, we will use chemical concepts such as electronegativity, polarity, and hydrogen bonding to understand the properties of water molecules.

Some of the most critical questions about drinking water deal with its safety and how that safety is ensured. The Safe Drinking Water Act, as amended in 1996, mandates that each water supplier deliver a right-to-know report once a year to consumers from any community water system, public or private. These reports, also called Consumer Confidence Reports, are available online for many larger water districts. They may be distributed in pamphlet form with water bills. As a person studying chemistry, you are in a good position to understand the meaning of the measurements being reported and the standards of quality that must be met. Knowing the quality of tap water can then help consumers to make wise choices about purchasing bottled or filtered water as alternatives. You will have the opportunity to explore the quality of water

at your college or in your hometown, once we have considered many of the questions of chemistry and public policy involved in safe drinking water. But first, we invite you raise your glass of tap water, bottled water, or filtered tap water and prepare to "take a drink."

Consider This 5.1   Take a Drink of Water

Obtain a glass of tap water, bottled water, and filtered tap water if available. Answer these questions.

a. Carefully describe each water sample—its taste, odor, appearance, and other observable characteristics. A table will help to organize your observations.

b. What do you like or dislike about each water?

c. Make a list of five main qualities you expect in your drinking water.

d. What concerns, if any, do you have about the safety of the tap water, bottled water, or filtered water?

5.1 Water from the Tap or Bottle

Chapter 1 began with an invitation to "Take a breath of air." Without thinking, we do it automatically about 12–16 times a minute. We would die in less than 15 minutes without air and its life-supporting oxygen. We generally don't have any choice about *what* air we breathe; we must rely on that which surrounds us. On the other hand, we generally do have choices with water. We can make decisions about *how frequently* we drink, *how much* we drink, and about the *source* of the water we drink. Municipal tap water may be consumed, with or without further home filtering. Some may have access to well water. We may choose bottled water or water in beverages. If out on the trail, we may look for water from nearby streams or collect rainwater. If **potable water,** water that is fit for human consumption, is not available, we are in danger. Our bodies can go weeks without food but only 5–7 days without water. If the water in our bodies is reduced by just 1%, thirst will develop. When the loss reaches 5%, muscle strength declines. At a 10% loss, delirium and blurred vision occur, and a 20% reduction results in death.

Nothing could be more familiar than this clear, colorless, and (usually) tasteless liquid. And yet, we generally take water and water quality for granted in most regions of this country. Unless a water emergency occurs, brought on by drought or contamination of our municipal water supply, we seldom think about where the water comes from, what it contains, how pure it is, or how long the supply will last. We turn on a faucet for a drink or a shower and simply expect a sufficient quantity of water to come flowing out of the tap. Most Americans obtain their drinking water from a water faucet or a drinking fountain (Figure 5.1a). This marvelous liquid is remarkably inexpensive, costing only about 1/10 of a penny per quart.

But not everyone drinks tap water. An increasing number of Americans are drinking bottled water instead (Figure 5.1b). Indeed, bottled water is big business and is now the fastest growing segment of the beverage industry in the world. Current estimates, based on information provided by the International Bottled Water Association, are that annual revenues worldwide from the sale of bottled water are over 40 billion $U.S. per year. Sales in the United States alone have now topped $7.7 billion annually. Growth rates in selling bottled water exceed 12% yearly, far outstripping growth rates seen for soft drinks, fruit beverages, or beer. In Europe, bottled water is now the biggest selling "soft drink." In 2002, the bottled water consumption level of 21.2 gal per capita was an 11% increase from the previous year. It is a common sight on campuses worldwide to see students carrying bottled water. Consumers 18–24 years old are the major users of bottled water, particularly in pint-sized plastic bottles.

Restoration of safe municipal drinking water was a major concern after the massive power blackout of August 2003 in the Northeast, Hurricane Isabel in September 2003, and several hurricanes of the 2004 season. Many cities were under a "boil before using" order for several days. Local supermarkets experienced brisk sales of bottled water.

(a)   (b)

Figure 5.1

(**a**) We usually take the safety of drinking water for granted in the United States. (**b**) Some prefer the taste and convenience of bottled water.

You have only to walk down a grocery aisle in the United States, Canada, Europe, or Asia to take a tour of bottled water brands. Advertisements for bottled water tout its purity, often using words or images that conjure up nature, purity, and pristine beauty. For example, the label for Dasani bottled water, the brand owned by Coca-Cola, makes these statements.

> *"Purified Water. Enhanced with Minerals for a Pure, Fresh Taste."*

The Web sites for bottled water also try to convey images of purity and natural processes. This is Evian's statement.

> *"Every drop of Evian Natural Spring Water begins as rain and snow falling high in the pristine and majestic French Alps."*

Some Web sites, such as that for LeBleu UltraPure Drinking Water, even try to teach a little chemistry!

> *"Water, the fluid of life and the shaper of the earth, is made from the simplest and most abundant element in the universe, hydrogen, joined to the vital gas oxygen. Two atoms pair with a single oxygen atom to establish the triple structure water, H_2O. Water, the universal solvent, given sufficient time, will dissolve or suspend almost any material on earth."*

Although highly popular, bottled water is also very expensive, relative to tap water. Typically, bottled water in large-volume containers costs from $1.00–3.00 or more per gallon. The price can be much higher for individually sized bottles. These costs are approximately 1000 times more expensive than the same volume of tap water. Bottled water is far more expensive, drop for drop, than West Texas crude oil, milk, or lemonade (Figure 5.2). Are there important reasons why consumers are willing to ante up so

"HEY, JOEY. FORGET THE LEMONADE... THIS IS WHERE THE BIG MONEY IS."

Figure 5.2

Bottled water and Dennis the Menace.

much more for bottled water? One of the goals of this chapter is to help you learn enough about drinking water quality to make intelligent choices.

Consider This 5.2   Bottled Water and You

Why do people buy bottled water, despite its high cost relative to tap water?

a. List what you perceive to be the advantages and the disadvantages of drinking bottled water.

b. Rank your lists in decreasing order of importance (most to least) for your personal decision about whether to drink bottled water.

Consider This 5.3   Finding Out About Bottled Water

If you search for "bottled water" on the Web, you will get over a million "hits." Select two sites to explore, one provided by a supplier and the other provided as a source of consumer information. The former may flood you with statistics about the benefits of bottled water; the latter may raise questions, such as "Is bottled water safer?" or "Is it worth the cost?" For each site, list the title, source, URL, and two things that you learned about water from the site.

> Growth of the bottled water industry has had a direct impact on CO_2 emissions (Chapter 3), energy costs for production and transportation (Chapter 4), and plastics production (Chapter 9).

5.2   Where Does Drinking Water Come From?

What journey does water take to get from its natural source to your tap or bottle? Water is widely distributed on planet Earth (Figure 5.3). On the surface, it is found in oceans, lakes, rivers, snow, and glaciers (Figure 5.4). In the atmosphere it exists as water vapor and as tiny droplets in clouds that replenish surface water by means of rain and snow.

> Aquifers are often major sources of potable water in other regions of the world, as well as in certain parts of the United States.

Water is found underground in **aquifers,** great pools of water trapped in sand and gravel 50–500 ft below the surface. Some aquifers are enormous, such as the Ogallala Aquifer in the center of the United States that underlies parts of eight states from South Dakota to Texas (Figure 5.5). Keeping these underground resources free from

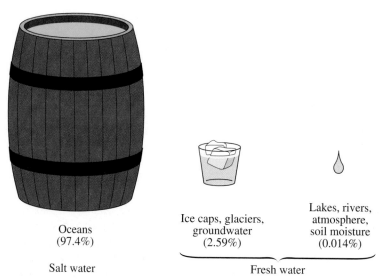

Figure 5.3

Distribution of water on Earth.

contamination is an important consideration in conserving sources of drinking water for the future. If an aquifer becomes contaminated, it may take decades to become clean again.

Water that can be made suitable for drinking comes from either surface water or groundwater. This is true whether considering municipal supplies or the source for bottled water. **Surface water,** water from lakes, rivers, and reservoirs, frequently contains substances that must be removed before it can be used as drinking water. By contrast, **groundwater,** water pumped from wells that have been drilled into underground aquifers, is usually free of harmful contaminants. Large-scale water supply systems for cities tend to rely on surface water resources. Smaller cities, towns, and private wells tend to rely on groundwater, the source of drinking water for a little over half the U.S. population.

 Consider This 5.4 **Your Home's Drinking Water Source**

When you turn on the tap at home, do you know where your drinking water comes from? You can find out by using the Safe Drinking Water Information System (SDWIS) of the EPA Web site either directly or through the direct link on the *Online Learning Center.* Search for the source of your home's drinking water by entering the geographic area in which you live. Then answer these questions.

a.   What is the name of the water system that services your home?
b.   What is the primary water source for that system?
c.   What information is given about the quality of that system's drinking water?

 Consider This 5.5 **Water Quality in Your State**

Each state has different concerns about its surface and groundwater.

a.   Draw up a list of the issues in your state. For example, you might consider if agricultural runoff, leaking storage tanks, or pollution from other human activities has affected the quality of water in your state.
b.   Read the state fact sheet provided by the Office of Water at the EPA through the direct link at the *Online Learning Center.* What does the sheet say about the surface and groundwater quality in your state?
c.   The EPA notes that some states have lakes, rivers, and streams that support no aquatic life. What are the reasons behind this lack of life and what percent of the surface water in your state falls in this category?

Figure 5.4
Lakes and reservoirs provide much of our drinking water.

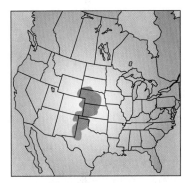

Figure 5.5
The Ogallala Aquifer is shown in dark blue on this map.

The Earth was not always as wet as it is now. Scientists believe that much of the water now on this planet originally was spewed as vapor from thousands of volcanoes that pocked Earth's surface. The vapor condensed as rain and the process repeated over the ages. Water molecules cycled from sea to sky and back again. About three billion years ago, primitive plants, and then animals, extracted water from and contributed water to the cycle, a cycle that continues today. During an average year, enough precipitation falls on the continents to cover all the land area to a depth of more than 2.5 ft. As we are well aware, this precipitation does not fall all at once, nor is it distributed uniformly. The wettest place on record is Mount Waialeale on Kauai, Hawaii, which receives an annual average rainfall of 460 inches. In contrast is the yearly average of 0.03 inches of rain in Arica, Chile. Data have been gathered in this Chilean desert for the past 61 years. For 14 of those years, it did not rain at all! Closer to home, an average of 1.5×10^{13} L of water falls daily on the continental United States—enough to fill about 400 million swimming pools.

This sounds like a great deal of water, but on a global scale, the amount of fresh water is relatively quite small. The great majority of Earth's water, 97.4% of the total, is in the salty oceans, water that is undrinkable without expensive purification. The remaining 2.6% of Earth's total water is all the fresh water we have. The majority of even this relatively small supply of fresh water is permanently frozen in glaciers and polar ice caps. Only about 0.01% of Earth's total water is conveniently located in lakes, rivers, and streams as fresh water. Consequently, the world's drinking water supplies are quite limited, varying widely depending on locale. In the United States, 80% of the fresh water is used to irrigate crops and to cool electrical power plants.

5.3   Water as a Solvent

A major reason we must consume water is that it is an excellent solvent for many of the chemicals that make up our bodies, as well as for a wide variety of other substances. In this capacity, water acts as a **solvent,** a substance capable of dissolving other substances. **Solutes** are those substances that dissolve in a solvent. The resulting mixture is called a **solution,** a homogeneous mixture of uniform composition. Furthermore, **aqueous solutions** are solutions in which water is the solvent. Later in this chapter we will examine why certain kinds of substances dissolve in water and others do not. For now, we simply note that a remarkable variety of substances can dissolve in water and that this has important consequences for living organisms as well as for the environment. Table 5.1 summarizes some examples of water acting as a solvent.

Your Turn 5.6	Common Sense Solubility

Based on your experience, which of these dissolves in water? Use relative terms such as very soluble, partially soluble, or insoluble in describing the solubility.

a. salt   **b.** sugar   **c.** sidewalk chalk
d. grape Kool Aid   **e.** cooking oil   **f.** aspirin

Because water is such a good solvent, drinking water is rarely, if ever, just "pure" water. You can be assured that it almost certainly contains other substances. Municipal water companies provide information about the dissolved mineral content, the solutes, for tap water. An analysis of tap water in a Midwest home revealed the information in Table 5.2.

Table 5.1	Importance of Water as a Solvent

In our bodies:
- Blood plasma is an aqueous solution containing a variety of life-supporting substances.
- Inhaled oxygen dissolves in blood plasma in the lungs, allowing O_2 to combine with hemoglobin.
- Blood plasma carries dissolved CO_2 to the lungs to be exhaled.
- Blood plasma transports nutrients into all the cells and organs.
- Water helps to maintain a chemical balance by carrying wastes away.

In the environment:
- Water can transport toxic substances into, within, and out of living organisms.
- Water-soluble toxic substances, such as some pesticides, lead ions, and mercury ions, can be widely distributed.
- Water may reduce the concentrations of pollutants to safe levels by dilution or by carrying them away (or both).
- Rainwater carries substances, including those responsible for acid rain, from the atmosphere down to Earth.

Ions, unlike atoms, are electrically charged. They are discussed in detail in Section 5.8.

Table 5.2		**Mineral Composition of Tap Water, mg/L**		
Calcium	66	Sulfates	42
Magnesium	24	Chlorides	48
Sodium	18	Nitrates	6
		Fluorides	1

The minerals shown in Tables 5.2 and 5.3 are in ionic form, as we will see in Section 5.7.

Similar information can often be found on the labels or at the Web sites for commercial bottled water. For example, a label on Evian bottled water includes the information in Table 5.3.

Most of the solutes in Tables 5.2 and 5.3 will be discussed in this chapter. The number given with each dissolved ion indicates how much of that substance (in milligrams) is present in 1 L of water. This raises a reasonable question: Should we be concerned about the amounts of any of these substances? Calcium ions, for example, have a definite health benefit in producing stronger bones. Milk and milk products, not Evian water, is the preferred source for calcium ion; you would have to drink 4 L of Evian water to get the same amount of calcium ion as that in one 8-oz glass of milk. In contrast, the nitrate ion, depending on its concentration, can be dangerous, especially for infants. The other substances listed for Evian bottled water are not likely to cause a health problem. Elsewhere on the label it is noted that sodium (sodium ions), a health concern for some people, is present at less than 5 mg per 500-mL bottle.

1.00 liter (L) = 1.06 quart

Your Turn 5.7   Bottled Water for Dietary Calcium?

One 500-mL bottle of Evian water provides 4% of the recommended daily requirement of calcium (in the form of calcium ion, Ca^{2+}).

a. Use the label information to calculate the approximate number of milligrams of calcium recommended per day.
b. Will this recommended value apply to everyone? Why or why not?
c. How many 500-mL bottles of Evian water would you have to drink to obtain your total daily supply of calcium?

Sceptical Chymist 5.8   Bottled Water and Claims of Purity

The Web site for Penta Ultra Premium Purified Drinking Water has addressed the question of what makes their water unique.

"Penta is simply the purest drinking water you can buy. Penta is created through a seven-step process that removes all impurities and chemicals. . . . Penta water is 100% free of chemicals, solids and other contaminants."

As a Sceptical Chymist, what is your opinion of these claims? Be sure to explain the reasoning behind your opinions.

Table 5.3		**Mineral Composition of Evian, mg/L**		
Calcium	78	Bicarbonates	357
Magnesium	24	Sulfates	10
Silica	14	Chlorides	4
		Nitrates (N)	1

Perhaps you have never considered drinking a glass of water as a risk–benefit act, yet it is. We usually consider water that has been chemically analyzed and treated to have important benefits with very low risk. Overwhelmingly, this is a valid assumption. But, however useful each tap water analysis or bottled water label may be, it necessarily is incomplete. As already noted, no information appears about health risks for the given concentrations of solutes. The information indicates nothing about whether other substances, if any, are present in the water. It does not indicate how much of each of the other substances is present or whether the substance might be harmful. For example, even though a tiny amount of lead is found in almost all water samples, it is usually in such low amount as not to be a health problem. If the water has been chlorinated to purify it, the water almost certainly has trace amounts of some chlorination by-products. Indeed, we rarely stop to think about what trace amounts of substances may be in the water, because we tend to assume that the water is safe to drink. In part, this is because extensive federal and state regulations and standards govern municipal water quality to protect the public. Most bottled water is regulated as well, often by self-imposed industry standards.

In assessing the health-giving or risk-taking aspects of drinking water, it is not sufficient to only know what substances are present in the water and how toxic they are. We also need to know how much of each substance is present in a particular amount of the water. In other words, we need to be able to understand what is meant by the concentration of a solute and the usual ways of expressing it. We now turn to these topics.

> Chlorination and its by-products will be examined later in this chapter.

> Absence of evidence is not the same as evidence of absence. A substance may be present, but in undetectable amounts. It just means that it can't be detected at current levels of analytical chemical technology. You may recall that this same point was made for air pollutants in Chapter 1.

> Concentrations for major components of air or of an aqueous solution are often expressed using percentages.

> Concentrations for minor components, such as for air pollutants, are often expressed in parts per million (ppm) or even parts per billion (ppb).

5.4 Solute Concentration in Aqueous Solutions

The concept of concentration was first introduced in Chapter 1 in relation to the composition of air. We used concentration units again in Chapters 2 and 3, looking at concentrations of chlorine compounds in the stratosphere or greenhouse gases accumulating in the troposphere. Now we will revisit this concept in terms of substances dissolved in water.

Although concentrations of components found in air might be a bit hard to visualize, solute concentrations in aqueous solutions are more familiar and therefore more easily imagined. For example, if you were asked to dissolve 1 teaspoon of an ingredient in 1 cup of water, a solution of a specific concentration would result: 1 tsp per cup (1 tsp/cup). Note that you would have the same 1 tsp/cup concentration if you also dissolved 2 tsp of the ingredient in 2 cups of water, 4 tsp in 4 cups, or 1/2 tsp in 1/2 cup. Even though you used larger or smaller quantities of the ingredient, the number of cups of water increased or decreased proportionally. Therefore, the **concentration,** the ratio of amount of ingredient to amount of water solution, would be the same in each case: 1 tsp per 1 cup (1 tsp/cup). Expressing solute concentrations in aqueous solutions follow the same pattern, but are often expressed in different units. We will use four ways of expressing concentration: percent; parts per million; parts per billion; and molarity. Three of these units are familiar to you from their use in earlier chapters, and molarity uses the mole concept introduced in Chapter 3. Each unit has particular application in various circumstances.

> The mass of the solution is determined by the mass of the solvent for low-solute concentrations.

Percent The most familiar way of expressing concentration is percent, defined in Chapter 1 as parts per hundred. For example, a solution containing 5 g of sodium chloride (NaCl) in 100 g of solution would be a five percent (5%) solution by weight. Hydrogen peroxide (H_2O_2) solutions, often found as an antiseptic in medicine cabinets, are usually 3% H_2O_2, indicating that they contain 3 g of H_2O_2 in 100 g of solution (or 6 g in 200 g of solution, etc.).

Ppm and ppb Concentrations of dissolved substances in drinking water are normally far lower than 1% (1 part per hundred, pph). Correspondingly, different units are used to express such low concentrations. Parts per million (ppm) is the most common way of expressing the concentration of a solute in drinking water. A 1-ppm solution of calcium ion in drinking water contains 1 g of calcium ion in 1 million (1,000,000, or 10^6) g of that sample of drinking water, actually a dilute solution. The same concentration, 1 ppm, could be applied to a solution with 2 g of calcium ion in 2×10^6 g of water, 5 g in 5×10^6 g of water, or 5 mg (5×10^{-3} g) in 5000 (5×10^3) g of water. Although parts per million is a very useful concentration unit, measuring one million grams of water is not very convenient. Therefore, we look to find an easier but equivalent way to establish ppm. We find it using the unit **liter (L),** the volume occupied by 1000 g of water at 4 °C. It is far easier to measure 1 L, a volume unit, rather than 1×10^{-6} g of water. Now we can say that 1 ppm of any substance in water equals 1 mg of that substance per 1 L of water.

$$1 \text{ ppm} = \frac{1 \text{ g solute}}{1,000,000 \text{ g water}} = \frac{1 \text{ mg solute}}{1,000 \text{ g water}} = \frac{1 \text{ mg solute}}{1 \text{ L water}}$$

Drinking water contains substances naturally present at concentrations in the parts per million range, as illustrated on the Evian bottled water label. Toxic water pollutants also may be present in the parts per million concentration range. For example, the acceptable limit for nitrate ion, often found in well water in some agricultural areas, is 10 ppm; and the limit for the fluoride ion is 4 ppm.

Some pollutants are of concern at concentrations much lower than even parts per million so are reported as parts per billion (ppb). One part per billion of mercury (Hg) in water means 1 g Hg in 1 billion (1×10^9) g of water. In more convenient terms, this means 1 microgram (1×10^6 g, abbreviated as 1 μg) Hg in 1 L (1×10^3 g) of water. For example, the acceptable limit for mercury in drinking water is 2 ppb.

$$2 \text{ ppb Hg} = \frac{2 \text{ g Hg}}{1 \times 10^9 \text{ g H}_2\text{O}} = \frac{2 \times 10^{-6} \text{ g Hg}}{1 \times 10^3 \text{ g H}_2\text{O}} = \frac{2 \text{ μg Hg}}{1 \text{ L H}_2\text{O}}$$

One part per million is a tiny concentration. Several analogies to a concentration of 1 ppm were given in Section 1.2, including that 1 ppm corresponds to 1 second in nearly 12 days. A similar analogy can be offered for parts per billion: 1 ppb corresponds to 1 second in 33 years, or approximately 1 inch on the circumference of the Earth.

> Although strictly true only at 4 °C, 1 L is close to being the volume of 1000 g (1×10^3 g) of H_2O throughout its liquid range, including at room temperature.

> 1 ppm = 1 mg/L

> Soluble forms of mercury ions, not elemental Hg, are implied by these concentration values.

> 1 ppb = 1 μg/L

Your Turn 5.9   **Lead Ion Concentrations**

a. If 80 μg of lead were detected in 5 L of water, what would be the concentration of lead? Express your answer first in parts per billion and then in parts per million.
b. If the maximum lead concentration in drinking water allowed by the federal government were 15 ppb, would the sample in part **a** be in compliance with federal limits? Explain.

Molarity Another useful concentration unit in chemistry is molarity, which is based on the unit of the chemical mole. **Molarity (M)** is defined as the number of moles of solute present in one liter of solution.

$$\text{Molarity (M)} = \frac{\text{moles of solute}}{\text{liter of solution}}$$

The great advantage of molarity is that a 1 molar (1 M) solution of any solute contains exactly the same number of chemical units (atoms or molecules) as any other 1 molar solution. The mass of solute may vary depending on the molar mass, but the number of chemical units will be the same for all 1 M solutions. Methods of chemical analysis of water (Section 5.15) frequently use molarity to express concentration. For now, we simply want to develop some familiarity with molarity itself.

The molar mass of NaCl, 58.5 g, is the sum of the mass of 1 mol of sodium, 23.0 g, plus 1 mol of chlorine, 35.5 g. See Section 3.7 to practice molar mass calculations.

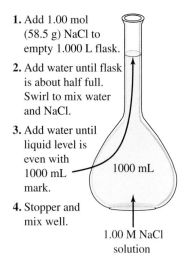

1. Add 1.00 mol (58.5 g) NaCl to empty 1.000 L flask.

2. Add water until flask is about half full. Swirl to mix water and NaCl.

3. Add water until liquid level is even with 1000 mL mark.

4. Stopper and mix well.

1000 mL

1.00 M NaCl solution

Figure 5.6

Preparing a 1.00 M NaCl solution.

As an example, consider a solution of NaCl in water. The molar mass of NaCl is 58.5 g; therefore, 1 mol of NaCl weighs 58.5 g. If we were to dissolve 58.5 g of NaCl in some water and then added enough water to make exactly 1.000 L of solution, we would have a 1.00 M NaCl solution (Figure 5.6). Note the use of a **volumetric flask,** a type of glassware that contains a precise amount of solution when filled to the mark on its neck. But, there are many ways to make a 1 M NaCl solution. Another possibility, among many others, would be to use 0.500 mol NaCl (29.2 g) in 0.500 L of solution. This would require the use of a 500.0-mL volumetric flask, rather than the 1.000-L flask shown in Figure 5.6.

$$1 \text{ M NaCl} = \frac{1 \text{ mol NaCl}}{1 \text{ L solution}} \text{ or } \frac{0.5 \text{ mol NaCl}}{0.5 \text{ L solution}}, \text{etc.}$$

Your Turn 5.10   **Moles and Molarity**

a. Consider a 1.5 M and a 0.15 M NaCl solution. How many moles of solute are present in 500 mL of each?

b. A solution is prepared by adding enough water to 0.50 mol NaCl to form 250 mL of solution. A second solution is prepared by adding enough water to 0.60 mol NaCl to form 200 mL of solution. Which solution is more concentrated? Explain your reasoning.

So far in this chapter we have developed some ideas about drinking water, some of the substances that may be present in it, and how to express the concentrations of those substances. We shift now to a more detailed examination of water at the molecular level. Our aim is to understand water's unique properties, including its excellence as a solvent.

5.5   Water's Molecular Structure and Physical Properties

"Water has never lost its mystery. After at least two and a half millennia of philosophical and scientific inquiry, the most vital of the world's substances remains surrounded by deep uncertainties. Without too much poetic license, we can reduce these questions to a single bare essential: What exactly is water?"

Philip Ball, in *Life's Matrix: A Biography of Water,*
University of California Press,
Berkeley, CA, 2001, p. 115

This section brings us to try and answer this important question—what *is* water? It is clear that water is essential to our lives and that water is an excellent solvent. What may not be as clear is that our dependence on water is possible only because water has a number of unusual properties. In fact, the physical properties of water are quite peculiar, and we are very fortunate that they are. If water were a more conventional compound, we would be very different creatures.

This most common of liquids is full of surprises, not the least of which is its physical state. Water is a liquid and not a gas at room temperature (about 25 °C) and normal atmospheric pressure. This is surprising because almost all other compounds with similar molar masses to water's 18.0 g/mol are gases under similar conditions of temperature and pressure. Consider three common atmospheric gases (N_2, O_2, and CO_2) whose molar masses are 28, 32, and 44 g/mol, respectively. All have molar masses greater than that of water, yet they are gases to breathe rather than liquids to drink.

In general, as molar mass increases in covalently bonded molecules, the boiling point also increases.

Figure 5.7

Representations of H_2O.

(a) Lewis structure and structural formulas;

(b) Space-filling model;

(c) Charge-density model.

Not only is water a liquid under these conditions, it also has an anomalously high boiling point of 100 °C. This temperature is one of the reference points for the Celsius temperature scale. The other is the freezing point of water, 0 °C. And when water freezes, it exhibits another somewhat bizarre property—it expands. Most liquids contract when they solidify. These and other unusual properties derive from water's chemical composition and molecular structure. We will continue to explore the reasons for water's unusual behaviors as this section continues.

To better understand the chemical and physical properties of water, we need information about its chemical composition and molecular structure. The chemical composition is known to practically everyone. Indeed, the formula for water, H_2O, is very likely the world's most widely known bit of chemical information. Recall from Chapter 2 that water is a covalently bonded molecule (Section 2.3). Then, in Chapter 3 (Section 3.3), water's molecular shape was illustrated by means of a ball-and-stick model and a space-filling model. These representations are shown again in Figure 5.7.

The electrons being shared between oxygen and hydrogen atoms, forming the covalent bond, are not shared equally. Experimental evidence indicates that the oxygen atom attracts the shared electron pair more strongly than does the hydrogen atom. To use the appropriate technical term, oxygen is said to have a higher electronegativity than hydrogen. **Electronegativity (EN)** is a measure of an atom's attraction for the electrons it shares in a covalent bond. The greater the electronegativity, the more an atom attracts bonding electrons to itself. Table 5.4 shows a periodic table of electronegativity values for the first 18 elements, all in "A" subgroups.

An examination of Table 5.4 reveals some useful generalizations about electronegativity. The highest electronegativity values are associated with nonmetallic elements such as fluorine and chlorine. These halogens, members of Group 7A, have atoms with seven outer electrons. Recall (Section 2.3) that each of these atoms has a strong tendency to bond with another atom in such a way as to acquire a share in an additional electron, thus completing a stable octet of electrons. A similar argument explains the high electronegativity values of other nonmetals. For example, oxygen, with six outer electrons per atom, also exhibits a relatively strong attraction for shared electrons. Conversely, the lowest electronegativity values are associated with the metals found in

> The periodic table was introduced in Section 1.6 and is found inside the front cover.

Table 5.4		Electronegativity Values, Arranged by Group Number					
1A	**2A**	**3A**	**4A**	**5A**	**6A**	**7A**	**8A**
H 2.1							He —
Li 1.0	Be 1.5	B 2.0	C 2.5	N 3.0	O 3.5	F 4.0	Ne —
Na 0.9	Mg 1.2	Al 1.5	Si 1.8	P 2.1	S 2.5	Cl 3.0	Ar —

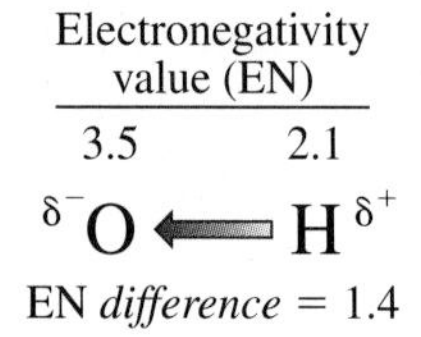

Electronegativity value (EN)

3.5   2.1

$\delta^- \text{O} \Longleftarrow \text{H} \delta^+$

EN *difference* = 1.4

Figure 5.8

Polar covalent bond between hydrogen and oxygen atoms. The electrons are displaced toward the more electronegative oxygen atom.

Compare:
• *Intramolecular* forces are *within* molecules.
• *Intramural* sports are played *within* a college with different units competing.

Groups 1A and 2A. Atoms of these metallic elements have much weaker attractions for electrons than do nonmetals. In general, electronegativity values increase as you move across a row of the periodic table from left to right (from metals to nonmetals) and decrease as you move down a group of the table.

According to Table 5.4, the electronegativity of oxygen is 3.5; that of hydrogen is 2.1. Because of these electronegativity differences, the shared electrons are actually pulled by the more electronegative oxygen to itself and away from the less electronegative hydrogen. This unequal sharing gives the oxygen end of the O-H bond a partial negative charge and the hydrogen end a partial positive charge. The result is a **polar covalent bond,** a covalent bond in which the electrons are not equally shared, but rather displaced toward the more electronegative atom. The greater the electronegativity difference of the elements involved, the more polar the bond. A polar covalent bond is an example of an **intramolecular force,** a force that exists within a molecule. In Figure 5.8, an arrow is used to indicate the direction in which the electron pair is displaced. The δ^+ and δ^- symbols indicate partial positive and partial negative charges, respectively.

Your Turn 5.11   Polar Bonds

For each pair, which is the more polar bond? To which of the atoms in the bond will the electron pair be more strongly attracted?
Hint: Use the values for electronegativity given in Table 5.5.

a. H-to-F or H-to-Cl
b. N-to-H or O-to-H
c. N-to-O or S-to-O

The EN difference for C-to-H bonds is quite small:
2.5 − 2.1 = 0.4
These bonds are considered nonpolar. All hydrocarbons are therefore nonpolar compounds.

If covalent bonds are nonpolar, a molecule containing such bonds must also be nonpolar. This is why diatomic molecules such as Cl_2 or H_2 are nonpolar molecules. If covalent bonds are polar, a molecule may or may not be polar. This depends on the geometry of the molecule. The specific case of the water molecule is shown in Figure 5.9. Note that the hydrogen atoms are partially positive and that a partial negative charge appears to be concentrated on the oxygen atom. Many of the unique properties of water are a consequence of both the polarity of bonds within each molecule and the overall shape of its molecules.

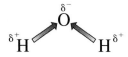

Figure 5.9

H_2O, polar covalent bonds forming a polar covalent molecule.

Consider This 5.12   Polar or Nonpolar Molecules

We know that H_2O contains polar bonds and is a polar molecule. What about CO_2?

a. Are the covalent bonds in CO_2 polar or nonpolar? *Hint:* Use the electronegativity values given in Table 5.4.
b. Draw a figure similar to Figure 5.9 for CO_2. Be sure to consider the bond angle. *Hint:* Consider Figure 3.12.
c. Offer a possible explanation why the H_2O molecule is polar, but the CO_2 molecule is not.

5.6   The Role of Hydrogen Bonding

Polar covalent bonds can help us understand some of the unusual properties of water. Consider what happens at the molecular level when two water molecules approach each other. Because opposite charges attract, one of the partially positively charged hydrogen atoms of one water molecule is attracted to one of the regions of partial negative charge associated with the nonbonding electron pairs of the other water molecule. This is an **intermolecular force,** a force that occurs between molecules. The fact that each H_2O molecule has two hydrogen atoms and two nonbonding pairs of electrons increases the opportunities for intermolecular attraction (Figure 5.10). The bond that forms is known as a **hydrogen bond,** an electrostatic attraction between an atom bearing a partial positive charge in one molecule and an atom bearing a partial negative charge in a neighboring molecule. Hydrogen bonds typically are only about one tenth as strong as the covalent bonds that connect atoms together *within* molecules; they are also longer than the covalent bonds. The effect of hydrogen bond formation is central to understanding water's unusual properties.

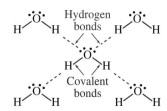

Figure 5.10
Hydrogen bonding in water (distances not to scale).

    Figures Alive! Visit the *Online Learning Center* to learn more about hydrogen bonds. Look for the Figures Alive! icon as a guide to related activities.

Your Turn 5.13   Water's Hydrogen Bonds

a. How many hydrogen bonds are shown in Figure 5.10 forming around the central water molecule?
b. Are hydrogen bonds intermolecular or intramolecular forces? Explain.

Very large molecules, such as DNA (Chapter 12) can form hydrogen bonds within different regions of the *same* molecule.

Although hydrogen bonds are not as strong as covalent bonds, hydrogen bonds are quite strong compared with other types of intermolecular forces. For example, to boil water, the H_2O molecules must be separated from their relatively close contact in the liquid state and moved into the gaseous state, where they are much farther apart. In other words, their intermolecular hydrogen bonds must be broken. If the hydrogen bonds in water were weaker, water would have a much lower boiling temperature and require less energy to boil. If water had no hydrogen bonding at all, it would boil at about −75 °C, a prediction based on its molar mass. This would make life very uncomfortable, if not impossible. Because of hydrogen bonding, almost all of our body's water, whether in cells, blood, or other body fluids, is in the liquid state, well below the boiling point. Our very existence depends on hydrogen bonding; without it, we would be a gas!

Compare:
• *Intermolecular* forces are *between* molecules.
• *Intercollegiate* sports are played *between* colleges.

Consider This 5.14   Bonds Within and Between Water Molecules

Use Figure 5.10 to help explain what bonds are broken when water boils. Draw a diagram to help illustrate your understanding. *Hint:* Diagrams and sphere equations were used extensively in Chapter 1 to help illustrate chemical reactions. Start with molecules of water in the liquid state. Then show what happens to those molecules when water boils.

The phenomenon of hydrogen bonds is not restricted to water. There is evidence for similar intermolecular attraction in many molecules that contain hydrogen atoms covalently bonded to oxygen, nitrogen, or fluorine atoms. In each of these cases, polar covalent bonds form within each molecule, the necessary requirement for the formation of intermolecular hydrogen bonds. Hydrogen bonding is also important in stabilizing the shape of large biological molecules, such as proteins and nucleic acids. In proteins, which are major components of skin, hair, and muscle, hydrogen bonding occurs between hydrogen atoms and oxygen or nitrogen atoms. The coiled, double-helical structure of DNA (deoxyribonucleic acid) is stabilized by thousands of hydrogen bonds formed between particular segments of the linked DNA strands. So in this respect, too, hydrogen bonding plays an essential role in the life process.

The molecular structures of proteins and nucleic acids such as DNA are discussed in Chapters 11 and 12, respectively.

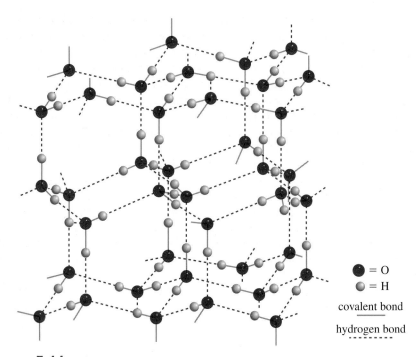

= O

= H

covalent bond ——————

hydrogen bond ----------

Figure 5.11

The hydrogen-bonded lattice structure of the common form of ice. Note the open channels between "layers" of water molecules that cause ice to be less dense than liquid water.

Hydrogen bonding also explains why ice cubes and icebergs float in water. Ice is a regular array of water molecules in which every H_2O molecule is hydrogen-bonded to four others. The pattern is shown in Figure 5.11. Note that the pattern includes a good deal of empty space in the form of hexagonal channels. When ice melts, this regular array begins to break down and individual H_2O molecules can enter the open channels. As a result, the molecules in the liquid state are, on the average, more closely packed than in the solid state. Thus, a volume of one cubic centimeter (1 cm^3) of liquid H_2O contains more molecules than 1 cm^3 of ice. Consequently, liquid water has a greater mass per cubic centimeter than ice. This is simply another way of saying that the **density,** the ratio of mass per unit volume, of liquid water is greater than that of ice.

For water, mass is often expressed in grams and the "unit of volume" is cubic centimeters (cm^3), which is identical to a milliliter (mL). Furthermore, 1.00 cm^3 of liquid water weighs 1.00 g. In other words, its density is 1.00 g/cm^3, or 1.00 g/mL. On the other hand, 1 cm^3 of ice weighs 0.92 g, so its density is 0.92 g/cm^3, or 0.92 g/mL.

People often confuse density with mass. For example, you may hear someone say that iron is "heavy" or that lead is "very heavy." Large pieces of iron and lead are indeed often quite heavy, but it is more accurate to say that iron has a high density (7.9 g/cm^3) and that lead has an even higher density (11.3 g/cm^3). On the other hand, popcorn has a low density; we are likely to say that even a large bag of popcorn feels "light."

For the great majority of substances, the solid state is denser than the liquid. The fact that water shows the reverse behavior means that lakes freeze from the top down, not the bottom up. This topsy-turvy behavior is convenient for aquatic plants, fish, and ice skaters. On the other hand, it is not so convenient for people whose water pipes and car radiators burst when the water inside them expands as it freezes.

> For any liquid at any temperature, 1 cm^3 = 1 mL.
> The statement that 1.00 cm^3 of liquid water has a mass of 1.00 g is only true for water. Strictly speaking, it is valid only at 4 °C, but is a useful approximation for water at room temperature.

Consider This 5.15   Oil and Water

Relative densities have practical consequences for water when it mixes (but does not dissolve) with other substances in the environment. Crude oil has a density of approximately 0.8 g/mL; salt water has a density more than 1.0 g/mL. What implications do these relative densities have for cleaning oil spills in the ocean?

Finally, we want to examine another of water's unusual properties, namely, its uncommonly high capacity to absorb and release heat. This property is expressed by **specific heat,** the quantity of heat energy that must be absorbed to increase the temperature of 1 g of a substance by 1 °C. The specific heat of liquid water is 1.00 cal/g·°C, which means that 1 cal of energy will raise the temperature of one gram of liquid water by 1 °C. In fact, the calorie was originally defined in this manner. Conversely, when the temperature of one gram of liquid water falls 1 °C, one calorie of heat is given off. The specific heat of water can also be expressed as 4.18 J/g·°C. Liquid water has one of the highest specific heats of any known liquid. Because of this, it is an exceptional coolant used to carry away excess heat in chemical industry, power plants, and the human body. Most other compounds have significantly lower specific heats.

On a global scale, water's high specific heat helps determine worldwide climates. By absorbing vast quantities of heat, the oceans and the droplets of water in clouds help mediate global warming. The specific details of these processes are among the uncertainties that complicate efforts to model global warming accurately. We do know that heat is absorbed when water evaporates from seas, rivers, and lakes. Heat is also released when water condenses as rain or snow. These changes between the solid, liquid, and gaseous forms of water create the great thermal engine that helps drive weather patterns in the short term and regulates climates over longer periods of time. But, when not changing its physical state, liquid water absorbs more energy than the ground if equal masses are used. This happens because water has a higher capacity to store heat than do rocks and dirt. As a consequence, when the weather turns colder, the ground has less stored heat to lose than the water and therefore, cools more quickly. The water retains more heat and is able to provide more warmth for a longer time to the areas bordering it. Such properties should be familiar to anyone who has ever jumped into a warm lake or pool on a cool day.

The unusually high heat capacity of water is a consequence of strong hydrogen bonding and the resultant degree of order that exists in the liquid. When molecules are strongly attracted to one another, a good deal of energy is required to overcome these intermolecular forces and enable the molecules to move more freely. Such is the case with water. On the other hand, intermolecular forces are much weaker in nonhydrogen-bonded liquids such as the hydrocarbon benzene (C_6H_6) and the forces are much easier to overcome. Consequently, the specific heat for benzene is only 0.406 cal/g·°C, less than half that of water.

> Joules and calories, units of heat energy, were defined in Section 4.1.

> Global warming was discussed in Chapter 3.

Consider This 5.16   Showering Yourself with Heat

The high specific heat of water has important consequences for energy consumption or conservation in residences, where large amounts of energy are required to heat water for bathing and washing clothes and dishes. Suppose that the water enters the water heater in your residence at 20 °C (68 °F), and the heater is set to heat the water to 50 °C (122 °F).

a. How many calories of heat energy are needed to heat the 100 L of water that are used in a typical 5-min shower?

b. To conserve energy, the heater is reset to 40 °C (104 °F). How many calories of heat energy would be saved during that 5-min shower?

Answer

a. $100 \text{ L} \times \dfrac{10^3 \text{ g}}{1 \text{ L}} \times \dfrac{1 \text{ cal}}{\text{g} \cdot {}^\circ\text{C}} \times (50\,{}^\circ\text{C} - 20\,{}^\circ\text{C}) = 3.0 \times 10^6 \text{ cal}$

5.7   Water as a Solvent: A Closer Look

One of the most important properties of water was discussed in Section 5.3, namely, that water is an excellent solvent for a wide variety of substances. A great deal of chemistry occurs in **aqueous solution,** one in which water is the solvent. Because aqueous solutions are so important, we need some understanding of how substances dissolve in water.

(a)

(b)

(c)

Figure 5.12

Conductivity in water and aqueous solutions.

(a) Distilled water (nonconducting)

(b) Sugar dissolved in water (nonconducting)

(c) Salt dissolved in water (conducting)

Ions in aqueous solution are indicated with *(aq).*

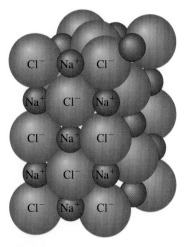

Figure 5.13

The arrangement of Na^+ and Cl^- ions in a crystal of sodium chloride.

A sugar solution and salt water illustrate two main classes of aqueous solutions. A significant difference between the two can be demonstrated experimentally with a **conductivity meter,** an apparatus that produces a signal to indicate that electricity is being conducted (Figure 5.12). Two wires attach a battery to a lightbulb. As long as the two separate wires do not touch, the electrical circuit is not completed. If the separated wires are placed into "distilled" water or a solution of sugar in distilled water, the bulb will not be illuminated. However, if the separate wires are placed into an aqueous solution of salt, the bulb illuminates. Perhaps the light has also gone on in the mind of the experimenter! Pure water or a solution of sugar in water do not conduct electricity and therefore do not complete the electrical circuit; the light does not glow. Sugar and other nonconducting solutes in aqueous solution are called **nonelectrolytes.** On the other hand, an aqueous solution of common table salt, NaCl, is an electrical conductor and the lightbulb lights. Sodium chloride and other conducting solutes are classified as **electrolytes,** defined as conducting solutes in aqueous solution.

What makes salt in solution behave any differently from sugar in solution or pure water? The observed flow of electric current through a solution involves the transport of electric charge. Therefore, the fact that aqueous NaCl solutions conduct electricity suggests they contain some charged species capable of serving to move electrons through the solution. When solid NaCl dissolves in water, it separates into electrically charged $Na^+(aq)$ and $Cl^-(aq)$. **Ions** are electrically charged species that carry current in aqueous solution. The term is derived from the Greek for "wanderer." Na^+ is an example of a **cation,** a positively charged ion. Cl^- is an example of an **anion,** a negatively charged ion. No such separation occurs with covalently bonded sugar or water molecules, making these liquids unable to carry electric charge. Although many hydrogen bonds are present in both the water and sugar solutions, even polar covalent bonds do not have enough charge separation to allow the transport of electric charge.

It may be a little surprising to learn that Na^+ and Cl^- ions exist in the solid crystals of salt such as those in a saltshaker, as well as in a water solution of salt. Solid sodium chloride is a three-dimensional cubic arrangement of sodium and chloride ions occupying alternating positions. The attractions between cations and anions in the crystal are called **ionic bonds** and hold the crystal together. In an ionic compound, such as NaCl, there are no true covalently bonded molecules, only positively charged cations and negatively charged anions held together by electrical attractions. In the case of NaCl, each Na^+ ion is surrounded by six oppositely charged Cl^- ions. Likewise, each Cl^- ion is surrounded by six positively charged Na^+ ions. A single, tiny crystal of sodium chloride consists of many billions of Na^+ and Cl^- ions held together in the arrangement shown in Figure 5.13.

Table 5.5	Comparison of a Sodium Atom with a Sodium Ion
Sodium Atom	**Sodium Ion**
Na	Na$^+$
11 protons	11 protons
11 electrons	10 electrons
Net charge: zero	*Net* charge: 1+

We have described the structure and some of the properties of ionic compounds, but not explained *why* certain atoms lose or gain electrons to form ions. Not surprisingly, the answer involves the distribution of electrons within atoms. Recall that a sodium atom, with an atomic number of 11, has 11 electrons and 11 protons. Sodium, like other metals in Group 1A, has only one electron in its outer energy level. This electron is rather loosely attracted to the nucleus and can be easily removed from the atom by absorbing a small amount of energy. When this happens, the Na atom becomes a Na$^+$ ion by losing an electron (e$^-$), a process represented by the equation 5.1.

$$Na \longrightarrow Na^+ + e^- \qquad [5.1]$$

A Na$^+$ ion has a 1+ charge because it contains the 11 protons of the Na atom, but only 10 electrons. These 10 electrons are in a configuration that is essentially the same as the 10 electrons in an atom of the inert element neon (Ne). Table 5.5 compares the sodium atom and the sodium ion.

A Na$^+$ ion, like a Ne atom, has two inner electrons and eight outer electrons. This is a particularly stable arrangement. We may generalize by saying that **metals** tend to form cations by losing electrons. Metals are the largest category of elements and are found in the left and middle blocks of the periodic table.

By contrast, a chlorine atom has a tendency to gain an electron. The electrically neutral Cl atom includes 17 electrons and 17 protons. It has seven outer electrons. Because of the stability associated with eight outer electrons, it is energetically favorable for a Cl atom to acquire an extra electron, such as one from a sodium atom, to become a Cl$^-$ ion. Equation 5.2 shows this change.

$$Cl + e^- \longrightarrow Cl^- \qquad [5.2]$$

This ion has 18 electrons and 17 protons; thus the net charge is 1$-$ (Table 5.6). Because elemental chlorine consists of diatomic Cl$_2$ molecules, we also can write this gain of electrons in the following fashion.

$$Cl_2 + 2\,e^- \longrightarrow 2\,Cl^- \qquad [5.3]$$

In general, **nonmetals** gain electrons to form anions. They are found on the right-hand side of the periodic table. The elements in Group 8A, the noble gases, are exceptions to this generalization. Some of the Group 8A nonmetals, such as helium, neon, and argon do not combine chemically with any elements.

Metals and nonmetals were introduced in Section 1.6.

Table 5.6	Comparison of a Chlorine Atom with a Chloride Ion
Chlorine Atom	**Chloride Ion**
Cl	Cl$^-$
17 protons	17 protons
17 electrons	18 electrons
Net charge: zero	*Net* charge: 1$-$

When sodium metal and chlorine gas react, electrons are transferred from sodium atoms to chlorine atoms with the release of a considerable amount of energy. The result is the aggregate of Na^+ ions and Cl^- ions known as sodium chloride. In the formation of an ionic compound such as sodium chloride, the electrons are actually transferred from one atom to another, not simply shared, as they would be in a covalent compound.

Is there evidence for electrically charged ions in pure sodium chloride? Experimental tests show that crystals of sodium chloride do not conduct electricity, but when these crystals are melted, the resulting liquid conducts electricity. This provides evidence that Na^+ and Cl^- ions from the solid NaCl also exist in the liquid state, without the presence of water. Crystals of NaCl and other ionic compounds are hard and brittle. When hit sharply, they shatter rather than being flattened, as would be true for a substance consisting of molecules with weak forces between them. This suggests the existence of strong forces that extend throughout the ionic crystal. Literally speaking, there is no such thing as a specific, localized "ionic bond" analogous to covalent bonds in molecules. Rather, generalized ionic bonding holds together a large assembly of ions.

Other elements form ions and ionic compounds, not just sodium and chlorine. Electron transfer to form cations and anions, respectively, is likely to occur between metallic elements (those elements in the left and middle blocks of the periodic table) and nonmetallic elements (those elements on the right side of the periodic table, except the noble gases, Group 8A). Sodium, lithium, magnesium, and other metallic elements have a strong tendency to give up electrons and form positive ions. On the other hand, chlorine, fluorine, oxygen, and other nonmetals have a strong attraction for electrons and readily gain electrons to form negative ions. Therefore, **ionic compounds** are formed when elements from opposite sides of the periodic table exchange electrons. Potassium chloride (KCl) and sodium iodide (NaI) are two of many such compounds. Because ordinary table salt (NaCl) is such an important example of an ionic compound, chemists frequently refer to other ionic compounds simply as "salts," which are ionic crystalline solids.

The idea of ion formation, where atoms gain or lose electrons, can help explain and even determine the formulas of many compounds. Back in Chapter 1 you were told that the formula of the compound formed from calcium and chlorine is $CaCl_2$, but no explanation was given. The reason should now be apparent. An atom of calcium, a member of Group 2A, readily loses its two outer electrons to form a Ca^{2+} ion.

$$Ca \longrightarrow Ca^{2+} + 2\,e^- \qquad [5.4]$$

Chlorine, as we have already seen, forms Cl^- ions. For the electric charges to be balanced, two Cl^- ions are required for each Ca^{2+} ion. Hence, the formula of calcium chloride is $CaCl_2$. In an ionic compound, the sum of the positive charges equals the sum of the negative charges. Your Turns 5.17 and 5.18 give you opportunities to apply similar logic to other elements and compounds.

Your Turn 5.17   **Predicting Ionic Charge**

Predict the charge on the ion that will form from each of these atoms. Draw the Lewis structure for each atom and for its ion, clearly labeling the charge on the ion.

a. Br   **b.** Mg   **c.** O   **d.** Al

Hint: Use the periodic table to find the number of outer electrons. Then determine how many electrons must be lost or gained to achieve stability with an octet of electrons.

Answer

a. Bromine is in Group 7A of the periodic table and gains one electron to form a stable ion with a charge of $1-$, just as was the case for chlorine. These are the Lewis structures.

$$:\!\overset{..}{\underset{.}{Br}}\!\cdot \quad \text{and} \quad \left[:\!\overset{..}{\underset{..}{Br}}\!:\right]^-$$

Table 5.7		**Common Polyatomic Ions**	
Name	**Formula**	**Name**	**Formula**
acetate	$C_2H_3O_2^-$	nitrite	NO_2^-
bicarbonate*	HCO_3^-	phosphate	PO_4^{3-}
carbonate	CO_3^{2-}	sulfate	SO_4^{2-}
hydroxide	OH^-	sulfite	SO_3^{2-}
hypochlorite	OCl^-	ammonium	NH_4^+
nitrate	NO_3^-		

* Also called the hydrogen carbonate ion.

Your Turn 5.18   Writing Formulas and Names for Ionic Compounds

Write the formulas of the ionic compounds that will form from each pair of elements. Also name each compound. *Hint:* When naming ionic compounds, it is not necessary to use the prefixes such as di- or tri- as we did when naming covalently bonded compounds such as carbon dioxide or dinitrogen pentoxide.

a. Ca and Br   **b.** K and F   **c.** Li and O   **d.** Sr and Br

Answer

a. Ca forms Ca^{2+} ions and Br forms Br^- ions. To be electrically neutral, the compound must have two Br^- ions for each Ca^{2+} ion. The formula is $CaBr_2$ and the name is calcium bromide.

Some ionic compounds include **polyatomic ions,** ions that are themselves made up of more than one atom or element. A case in point is sodium sulfate, Na_2SO_4. This compound consists of Na^+ and SO_4^{2-} ions. In the sulfate ion, the four oxygen atoms are covalently bonded to a central sulfur atom in a symmetric tetrahedral arrangement. Counting the electrons in the Lewis structure (Figure 5.14) reveals that there are 32 electrons, 2 more than the 30 valence electrons provided by one neutral sulfur atom (6) and four neutral oxygen atoms ($4 \times 6 = 24$). The "extra" two electrons give the sulfate ion a charge of $2-$.

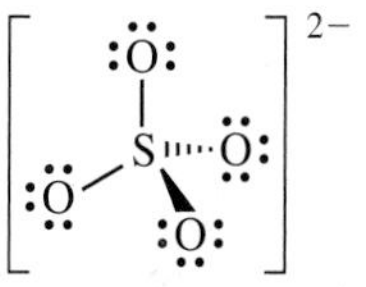

Table 5.7 is an alphabetical list of some common polyatomic ions. Most are anions, but polyatomic cations also are possible, as in the case of the ammonium ion, NH_4^+. Note that some elements (carbon, sulfur, and nitrogen) form more than one polyatomic anion with oxygen.

The rules for writing formulas and names of compounds containing polyatomic ions are similar to those that apply to simple compounds of two elements. Consider, for example, aluminum sulfate, an ionic compound used in water purification. The compound is formed from Al^{3+} and SO_4^{2-} ions. As is true for all ionic compounds, the positive ion is named first and the sum of the positive charges must equal the sum of the negative charges. Table 5.8 illustrates the ratio in which these two ions can join. Notice that the subscript 3 applies to the *entire* SO_4^{2-} ion that is enclosed in parentheses. The formula of the compound thus represents two aluminum ions along with three sulfate ions containing a total of 3 S atoms and 12 O atoms.

Figure 5.14
Structure of the sulfate ion, SO_4^{2-}.

Table 5.8	**Writing the Formula for an Ionic Compound**	
Aluminum Cation	**Sulfate Anion**	**Aluminum Sulfate**
Al^{3+}	SO_4^{2-}	
Al^{3+}	SO_4^{2-}	$Al_2(SO_4)_3$
	SO_4^{2-}	
$2(+3) = +6$	$3(-2) = -6$	Electrically neutral

Your Turn 5.19   Writing Formulas for Ion Pairs

Write the formula for the ionic compound formed from each pair of ions.

a. Na^+ and SO_4^{2-}
b. Mg^{2+} and OH^-
c. Al^{3+} and $C_2H_3O_2^-$

Answers
a. Na_2SO_4
b. $Mg(OH)_2$

Your Turn 5.20   Naming Ionic Compounds

Give the correct name for each of these compounds:

a. KNO_3 **b.** $(NH_4)_2SO_4$ **c.** $NaHCO_3$ **d.** $CaCO_3$ **e.** $Mg_3(PO_4)_2$

Answers
a. potassium nitrate
b. ammonium sulfate

Your Turn 5.21   Writing Formulas for Ionic Compounds

Write the formula for each of these compounds.

a. calcium hypochlorite (used in bleaches)
b. lithium carbonate (treatment of bipolar disorders)
c. potassium nitrate (matches and fireworks)
d. barium sulfate (medical X-rays)

Answer
a. $Ca(OCl)_2$ The sums of the negative and positive charges must be equal. Therefore, two hypochlorite ions are needed to equal the charge on a calcium ion, balancing the charges at 2+ and 2−.

5.8 Water Solutions of Ionic Compounds

We are now in a position to understand one of the most important properties of ionic compounds, namely, why many are quite soluble in water. Recall from Section 5.6 that water molecules are polar: They have partially positive hydrogen atoms and a partially negative oxygen atom. When a solid sample of an ionic compound is placed in water, the polar H_2O molecules are attracted to the individual ions. The partial negative charge on the oxygen atom of a water molecule is attracted to the positively charged cations of the solute. At the same time, hydrogen atoms in H_2O, with their partially positive charges, are attracted to the negatively charged anions of the solute. Thus, the ions are surrounded by water molecules, diminishing the anion–cation attraction in the solid. The substantial attraction between the ions and H_2O molecules results in the surrounding water molecules literally plucking the ions out of the solid and into solution. In dissolving, the ionic compound separates into its component cations and anions. Equation 5.5 and Figure 5.15 represent this process for sodium chloride and water.

$$NaCl(s) \xrightarrow{H_2O} Na^+(aq) + Cl^-(aq) \qquad [5.5]$$

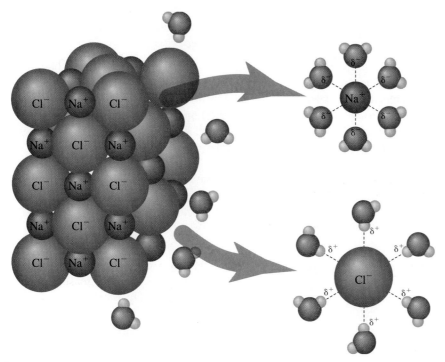

Figure 5.15
Dissolving sodium chloride in water.

When compounds containing polyatomic ions dissolve in water, the polyatomic ions remain intact. For example, when sodium sulfate dissolves in water, the sodium ions and sulfate ions separate, but the SO_4^{2-} ions remain intact.

$$Na_2SO_4(s) \xrightarrow{H_2O} 2\,Na^+(aq) + SO_4^{2-}(aq) \qquad [5.6]$$

What has just been described for sodium chloride and sodium sulfate dissolving in water is true for many other ionic compounds. Indeed, this behavior is so common that the chemistry of ionic compounds is largely that of their behavior in aqueous solutions. Conversely, almost all naturally occurring water samples contain various amounts of ions. Even our body fluids contain significant concentrations of ions.

Consider This 5.22 **Electricity and Water Don't Mix**

Small electric appliances, such as a hair dryer or curling iron, carry a warning label prominently advising the consumer of the hazard associated with using the appliance near water. Why is this a problem if water does not conduct electricity? What is the best course of action if a plugged-in hair dryer does accidentally fall into a sink full of water?

In principle, the dissolving process in water, as just described, ought to be true for any ionic compound. Indeed, many ionic compounds are highly soluble in water. But some are at best only slightly soluble, and some have extremely low solubility in water. The reasons for this range of behavior involve the sizes and charges of the ions, how strongly the ions attract one another, and how strongly the ions are attracted to water molecules. Despite the uncertainties, a few generalizations are quite useful for predicting the solubility of common ionic compounds (Table 5.9).

It is possible to use Table 5.9 and a periodic table of the elements to determine the solubility (or insolubility) of many compounds. For example, calcium nitrate, $Ca(NO_3)_2$, is soluble in water because all nitrates are soluble. On the other hand,

Table 5.9	**Generalizations About the Solubility of Ionic Compounds in Water**

All **sodium, potassium,** and **ammonium** (NH_4^+) compounds are soluble.
All **nitrates** are soluble.
Most **chlorides** are soluble (except silver, some mercury, and lead chlorides).
Most **sulfates** are soluble (except strontium, barium, and lead sulfate).
Most **carbonates** are insoluble* (except those with Group 1A or NH_4^+ cations).
Most **hydroxides** and **oxides** are insoluble (except those with Group 1A or NH_4^+ cations).
Most **sulfides** are insoluble (except those with Group 1A or NH_4^+ cations).

* "Insoluble" means that the compounds have extremely low solubility in water (less than 0.01 M). All ionic compounds have at least a very small solubility in water.

calcium carbonate, $CaCO_3$, is insoluble because most carbonates are insoluble, and calcium is not one of the exceptions for carbonates. By similar reasoning, copper hydroxide, $Cu(OH)_2$, is insoluble, but copper sulfate, $CuSO_4$, is soluble.

Your Turn 5.23   Solubility of Ionic Compounds

From the solubility generalizations in Table 5.8, which of these compounds are soluble?

a.   ammonium nitrate, NH_4NO_3 (used in fertilizers)
b.   sodium sulfate, Na_2SO_4 (used as an additive in detergents)
c.   mercury sulfide, HgS (the mineral cinnabar)
d.   aluminum hydroxide, $Al(OH)_3$ (used in some antacid tablets)

Answer
a.   Soluble. All nitrates and all ammonium compounds are soluble.

Table 5.10		**Environmental Consequences of Solubility**
Source	**Ions**	**Solubility and Consequences**
Salt deposits	sodium and potassium halides*	These salts are soluble. Over time, they dissolve from the land and wash into the sea. Thus, oceans are salty and sea water cannot be used for drinking without expensive purification.
Agricultural fertilizers	nitrates	All nitrates are soluble. The runoff from fertilized fields carries nitrates into surface and groundwater. Nitrates are toxic, especially for infants.
Metal ores	sulfides and oxides	Most sulfides and oxides are insoluble. Minerals containing iron, copper, and zinc are often sulfides and oxides. If these minerals had been soluble in water, they would have been washed out to sea long ago.
Mining waste	mercury, lead	Most mercury and lead compounds are insoluble. They are leached slowly from waste piles into rivers and lakes where they contaminate water supplies.

* Halides are the anions in Group 7, such as Cl^- and I^-.

The landmasses on Earth are made up largely of minerals consisting of ionic compounds that have extremely low solubility in water. If that were not the case, most would have dissolved long ago. Table 5.10 summarizes some environmental consequences of the differing solubility of minerals and other substances in water.

5.9 Covalent Compounds and Their Solutions

From the previous discussion, you might get the impression that only ionic compounds dissolve in water. But, other kinds of compounds dissolve as well. Common experience tells us that ordinary table sugar dissolves readily in water. But table sugar, chemically known as sucrose, contains no ions; it is a covalent or molecular compound. Like water, carbon dioxide, chlorofluorocarbons, and many of the other compounds you have been reading about, table sugar molecules consist of covalently bonded atoms. The formula for sucrose is $C_{12}H_{22}O_{11}$, and it exists as individual covalently bonded molecules consisting of 45 atoms, with its structure as shown in Figure 5.16.

When sugar dissolves in water, its molecules become uniformly dispersed among the H_2O molecules. As in all true solutions, the mixing is at the most fundamental level of the solute and solvent—the molecular or ionic level. The $C_{12}H_{22}O_{11}$ molecules remain intact and do not separate into ions. Evidence for this is the fact that aqueous sucrose solutions do not conduct electricity, as was shown in Figure 5.12. However, the sugar molecules do interact with the water molecules. In fact, solubility is always promoted when a net attraction exists between the solvent molecules and the solute molecules or ions. This suggests a general solubility rule: *Like dissolves like.* Compounds with similar chemical composition and molecular structure tend to form solutions with each other. The intermolecular attractive forces between similar molecules are high, promoting solubility. Dissimilar compounds do not dissolve in each other.

Consider, for example, three familiar covalently bonded compounds, all of which are highly soluble in water: sucrose; ethylene glycol (the main ingredient in antifreeze); and ethanol (ethyl alcohol, the "grain alcohol" found in alcoholic beverages). Like all alcohols, they contain an —OH group (Figures 5.17 and 5.18.)

We start with the simplest, ethanol, C_2H_5OH. Its oxygen atom is covalently bonded to a hydrogen atom and to a carbon atom. The —OH group of a C_2H_5OH molecule can form hydrogen bonds with H_2O molecules (see Figure 5.18). This hydrogen bonding is the reason that water and ethanol have a great affinity for each other, a conclusion consistent with the fact that they form solutions in all proportions. Ethylene glycol is also an

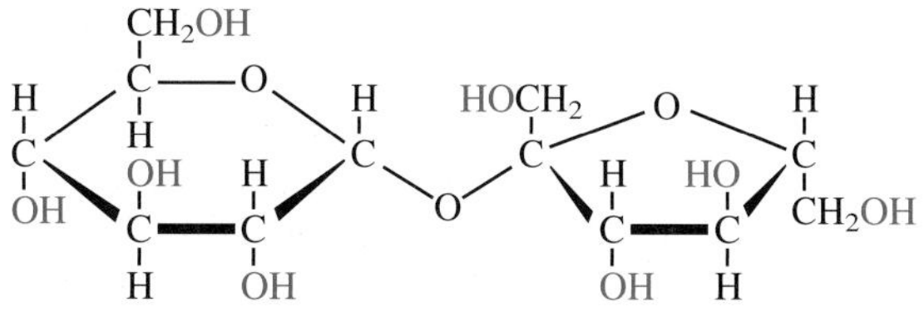

"Like dissolves like" is a useful generalization.

Covalently bonded compounds are also called "molecular" compounds.

Figure 5.16
Molecular structure of sucrose. The —OH groups are shown in red.

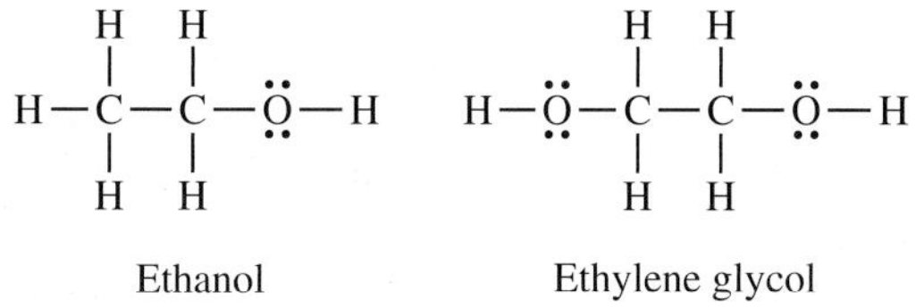

Ethanol   Ethylene glycol

Figure 5.17
Structures of ethanol and ethylene glycol.

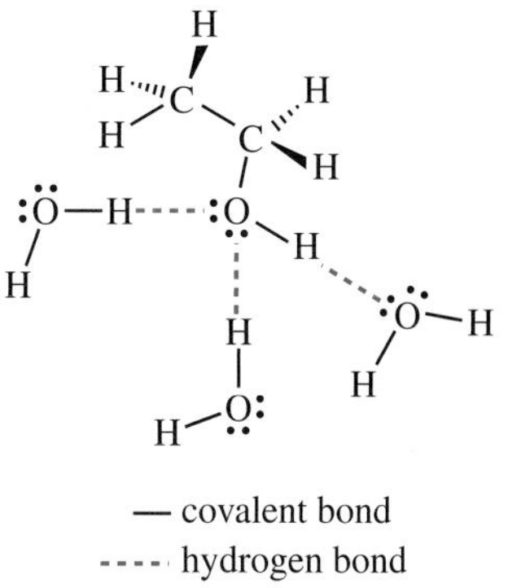

— covalent bond
---- hydrogen bond

Figure 5.18
Hydrogen bonding of ethanol with water.

Figure 5.19
Oil and water do not
dissolve in each other.

alcohol with two —OH groups available for hydrogen bonding with H_2O. Therefore, ethylene glycol is highly water-soluble, a necessary property for an antifreeze ingredient.

Your Turn 5.24   **Hydrogen Bonding—Ethylene Glycol and Water**

Sketch a diagram to show hydrogen bonding between ethylene glycol and water.

Finally, we consider sucrose, the compound that introduced this section. Examination of its structure (see Figure 5.16) discloses that the sucrose molecule contains eight —OH groups and three additional oxygen atoms that can also participate in hydrogen bonding. This helps explain the high solubility of sugar in water.

Consider This 5.25   **Three-Dimensional Representations of Molecules**

Three-dimensional representations of molecules can be viewed on the Web using CHIME, a free plug-in that you can download and install. Three-dimensional representations of ethanol, ethylene glycol, and sucrose are available on the site. Use these molecular representations to identify the places in each compound where hydrogen bonding occurs. Has your mental picture of these molecules changed after seeing these 3-D representations? Explain.

"Like dissolves like" is a useful generalization. Implied is the fact that covalently bonded compounds that differ in composition and molecular structure do not attract each other strongly. It has often been observed that "oil and water don't mix." They don't mix because they are structurally very different. Water is a highly polar molecule, whereas oil consists of nonpolar hydrocarbon molecules. When placed in contact, these molecules remain apart in separate layers (Figure 5.19). Even if shaken vigorously, the oil and water return to their own layers. But oily, nonpolar compounds generally dissolve readily in hydrocarbons or chlorinated hydrocarbons. For this reason, the latter have often been used in dry cleaning solvents.

The tendency of nonpolar compounds to mix with other nonpolar substances affects how fish and animals store certain highly toxic substances such as PCBs (polychlorinated biphenyls) or the pesticide DDT. PCB and DDT molecules are nonpolar, and so when fish absorb them from water, the molecules are stored in body fat (which is also nonpolar) rather than in the blood (which is a highly polar aqueous solution).

PCBs are organochlorine chemicals that were widely used as cooling agents in electrical transformers. Careless disposal has caused serious environmental problems.

Solvents used to dry-clean clothes are usually chlorinated compounds such as tetrachloroethylene, C_2Cl_4, also known as "perc" (perchlorinated ethylene). Perc is a human **carcinogen,** a compound capable of causing cancer. These materials also have serious environmental consequences. Dr. Joe DeSimone of the University of North Carolina–Chapel Hill has discovered a substitute for chlorinated compounds by synthesizing cleaning detergents that work in liquid carbon dioxide. The key to the process are the detergents, whose molecules are designed so that one end of the molecule is soluble in nonpolar substances like grease and oil stains, while the other end dissolves in the liquid CO_2. The new process recycles carbon dioxide produced as a waste product from industrial processes. Replacing large volumes of perc by using recycled CO_2 reduces perc's negative impact on the workplace and the environment. The breakthrough process is paving the way for designing replacements for conventional halogenated solvents currently used in manufacturing and industries making coatings. For his work, Professor DeSimone received the 1997 Presidential Green Chemistry Challenge Award.

5.10 Protecting Our Drinking Water: Federal Legislation

We can now apply to drinking water what we know about the structure and properties of pure water and aqueous ionic solutions. What dissolves in drinking water determines its quality and the potential for adverse health effects. Keeping public water supplies safe has long been recognized as an important public health issue. In 1974, the U.S. Congress passed the Safe Drinking Water Act (SDWA) in response to public concern about findings of harmful substances in drinking water supplies. The aim of the SDWA, as amended in 1996, is to provide public health protection to all Americans who get their water from community water supplies (over 200 million people). Contaminants that may be health risks are regulated by EPA as required by the SDWA. The EPA sets legal limits for such contaminants that reflect knowledge about health effects and risk calculations (Table 5.11). These limits also take into account the practical realities of the concentration of contaminants likely to be present in drinking water sources, and the ability of water supply utilities to remove the offending contaminants by using available technology.

For each contaminant, the EPA has established a health goal, or maximum contaminant level goal (MCLG). The **MCLG** is the level, expressed in parts per million or parts per billion, at which a person weighing 70 kg (154 lb) could drink 2 L (about 2 qt) of water containing the contaminant every day for 70 years without suffering any ill effects. Each MCLG includes built-in safety factors for uncertainties in the standardizing data and for individual differences in sensitivity to the contaminant. An MCLG is not a legal limit with which water systems must comply; it is based solely on considerations of human health. For known carcinogens, the EPA has set the health goal at zero, under the assumption that *any* exposure to the substance could present a cancer risk.

The mere presence of a contaminant does not necessarily mean a serious health problem exists. Before any regulatory action is initiated, the concentration of the impurity must exceed the maximum contaminant level (MCL). The **MCL,** expressed in parts per million or parts per billion, sets the legal limit for the concentration of a contaminant. The EPA sets legal limits for each impurity as close to the MCLG as possible, keeping in mind the practical realities of technical and financial barriers that may make it difficult to achieve the goals. Except for contaminants regulated as carcinogens, for which the MCLG is zero, most legal limits and health goals are the same. Even when they are less strict than the MCLGs, the MCLs provide substantial public health protection.

Consider This 5.26   Understanding MCLGs and MCLs

Most people are unfamiliar with these terms from the Safe Drinking Water Act. Assume you are making a presentation in another class to explain what these acronyms mean and how the information helps to safeguard our drinking water. Prepare a short outline of what you will say. Be prepared to answer questions from the audience, particularly dealing with why MCLs are not set to zero for all carcinogens.

Table 5.11	**MCLGs and MCLs (in ppm) for Selected Pollutants in Drinking Water**	
Pollutant	**MCLG**	**MCL**
Cadmium (Cd^{2+})	0.005	0.005
Chromium (Cr^{3+}, CrO_4^{2-})	0.1	0.1
Lead (Pb^{2+})	0	0.015
Mercury (Hg^{2+})	0.002	0.002
Nitrate (NO_3^-)	10	10
Benzene (C_6H_6)	0	0.005
Trihalomethanes ($CHCl_3$, etc.)	0	0.080

Because of improved detection and quantitative analytical methods, the number of regulated contaminants in drinking water increases each time Congress updates the legislation. Lower limits for MCL values have been established as more accurate risk information has become available. Currently, more than 80 contaminants are regulated; they fall into several major categories: metals (e.g., cadmium, chromium, copper, mercury, and lead), a few nonmetallic elements (e.g., fluoride and arsenic), pesticides, industrial solvents, compounds associated with plastics manufacturing, and radioactive materials. Depending on the particular contaminant, MCLs vary from around 10 ppm to less than 1 ppb. Some contaminants interfere with liver or kidney function. Others can affect the nervous system if ingested over a long period at levels consistently above the legal limit (MCL). Pregnant women and infants are at particular risk for some contaminants because of their effects on a developing fetus or the digestive system of an infant. In Section 5.15 we look at two contaminants: lead and trihalomethanes (THMs).

In addition to contaminants that can pose chronic health problems, other substances in drinking water present acute health risks. For example, nitrate (NO_3^-) and nitrite (NO_2^-) ions limit the blood's ability to carry oxygen. Even when consumed in tiny doses, these ions cause immediate health effects for infants. Therefore, the EPA limit for nitrate and nitrite ions in drinking water specifically protects infants. Another acute health risk is biological, not chemical—from bacteria, viruses, and other microorganisms, including *Cryptosporidium* and *Giardia*. News media warnings announcing a "boil-water emergency" are typically the result of a "total coliform" violation. Coliforms are a broad class of bacteria, most of which are harmless, that live in the digestive tracts of humans and other animals. The presence of high coliform concentration in water usually indicates that the water-treatment or distribution system is not working properly. Diarrhea, cramps, nausea, and vomiting, the symptoms of coliform-related illness, are not serious for a healthy adult, but can be life-threatening for the very young, the elderly, or those with weakened immune systems.

Consider This 5.27   **A Drink of Water—What Is in It?**

Table 5.11 is merely a starting point for the wealth of information available about possible pollutants in drinking water. At the EPA Office of Ground Water and Drinking Water, you can access a consumer fact sheet on each of these pollutants, as well as for dozens more. A consumer version and a technical version are available, and the latter is recommended. Look up a pollutant listed in Table 5.11 to find out how the pollutant gets into the water supply and how you would know if it were in your drinking water. Is your state listed as one of the top states that releases the contaminant?

Hint: Arsenic, cadmium, lead, chromium, mercury, and nitrate/nitrite ions are found under the section on Inorganic Chemicals. Benzene is listed under Volatile Organic Chemicals. No THM such as $CHCl_3$ is currently listed, but you can find a variety of other chlorinated compounds found in water, such as CCl_4 and CH_2Cl_2.

Consider This 5.28   **MTBE in Groundwater**

The gasoline additive MTBE was discussed in Chapter 4 as one of the mandated oxygenates in certain metropolitan areas to reduce air pollution. Concern about the accumulation of MTBE in surface and groundwater, with the potential to compromise the quality of drinking water supplies, has led to some changes in policy in some areas of the United States. Consult the *Online Learning Center* or other resources to find the current status of the use of MTBE.

In addition to the Safe Drinking Water Act, other federal legislation also controls pollution of surface waters, including lakes, rivers, and coastal areas. The Clean Water Act (CWA), passed by Congress in 1972 and amended several times, provided the foundation for dramatic progress in reducing surface water pollution over the past three decades. The CWA establishes limits on the amounts of pollutants that industries can discharge into surface waters, resulting in actions that have removed over a billion pounds of toxic pollution from U.S. waters every year. Improvements in surface water quality have at least two major beneficial effects: They reduce the amount of clean-up needed for public drinking water supplies, and they result in a more healthful natural environment for aquatic organisms. In turn, a more healthful aquatic ecosystem has many indirect benefits for humans. In keeping with the new trend toward green chemistry, industries are finding ways to convert these waste materials into useful products, as well as to initially design processes so that they neither use nor produce substances that degrade water quality.

Consider This 5.29 *Cryptosporidium*

As of January 1, 2002, EPA's surface water-treatment rules require large systems using surface water, or groundwater under the direct influence of surface water, to remove or deactivate 99% of *Cryptosporidium*.

a. What is *Cryptosporidium?*
b. What are the sources of this contaminant in drinking water?
c. What are the potential health effects from drinking water contaminated with *Cryptosporidium?*
d. Why was this rule issued in 2000, but not required until 2002 for large systems (those serving more than 10,000 customers)?
e. For systems serving fewer than 10,000 customers, when does this rule take effect?

5.11 Treatment of Municipal Drinking Water

Just because a supply of water is large is no guarantee that it is fit to drink. Coleridge's shipwrecked ancient mariner knew this all too well, surrounded as he was by "Water, water every where, nor any drop to drink." So how is water treated to make it potable, that is, fit for human consumption?

The first step in a typical municipal drinking water-treatment plant is to pass the water through a screen that excludes larger objects both natural (fish and sticks) and artificial (tires and beverage cans). The usual next step is to add two chemicals, aluminum sulfate, $Al_2(SO_4)_3$, and calcium hydroxide, $Ca(OH)_2$. These compounds react to form a sticky gel of aluminum hydroxide, $Al(OH)_3$, that collects suspended clay and dirt particles on its surface (Equation 5.7). The $Al(OH)_3$ gel settles, slowly carrying with it the suspended particles down into a settling tank. Any remaining particles are removed as the water is filtered through gravel and then sand.

Calcium hydroxide is also called by its common name, slaked lime.

$$Al_2(SO_4)_3(aq) + 3\,Ca(OH)_2(aq) \longrightarrow 2\,Al(OH)_3(s) + 3\,CaSO_4(aq) \qquad [5.7]$$

The next step, disinfection to kill disease-causing organisms, is the most crucial one for making drinking water safe. In the United States, this is most commonly done by chlorination. Chlorine is usually added in one of three forms: chlorine gas, Cl_2; sodium hypochlorite, $NaOCl$; or calcium hypochlorite, $Ca(OCl)_2$. The antibacterial agent generated in solution by all three substances is hypochlorous acid, $HOCl$. The degree of chlorination is adjusted so that a very low concentration of $HOCl$, between 0.075 and 0.600 ppm, remains in solution to protect the water against further bacterial contamination as it passes through the pipes to the user.

$NaOCl$ is used in Clorox and other brands of laundry bleach. $Ca(OCl)_2$ is commonly used to disinfect swimming pools.

Before chlorination was used, thousands died in epidemics spread via polluted water. In a classic study, John Snow was able to trace a mid-1800s cholera epidemic in London to water contaminated with the excretions of victims of the disease. A more contemporary example occurred in Peru in 1991. This cholera epidemic was traced to bacteria in shellfish growing in estuaries polluted with untreated fecal matter. The bacteria found their way into the water supply, where they continued to multiply because of the absence of chlorination.

Chlorination, however, is not without some drawbacks. The taste and odor of residual chlorine may be objectionable to some and is a reason commonly cited as why people drink bottled water or use home water filters to remove residual chlorine at the tap. A possibly more serious drawback is the reaction of residual chlorine with other substances in the water to form by-products at potentially toxic levels. The most widely publicized of these are the trihalomethanes (THMs) such as chloroform, $CHCl_3$, which we will consider in more detail later in this chapter.

Many European and a few U.S. cities use gaseous ozone (O_3) to disinfect their water supplies. Chapter 1 discussed tropospheric ozone as a serious air pollutant. Chapter 2 described the beneficial effects of the stratospheric ozone layer. In water treatment, the toxic property of ozone is used for a beneficial purpose. The degree of antibacterial action necessary can be achieved with a smaller concentration of ozone than chlorine, and ozone is more effective than chlorine against water-borne viruses. But ozonation is more expensive than chlorination and becomes economical only for large water-treatment plants. An additional major drawback of ozone is that it decomposes quickly and hence does not protect the water from contamination after the water leaves the treatment plant. Consequently, a low dose of chlorine is added to ozonated water as it leaves the treatment plant.

Another disinfection method gaining in popularity is the use of ultraviolet (UV) radiation. In Chapter 2 it was pointed out that UV radiation is dangerous for living species, including bacteria. UV disinfection is very fast; leaves no residual by-products, and is economical for small installations (including rural homes with unsafe well water). Like ozonation, UV disinfection does not protect the water from contamination after it leaves the treatment site unless a low dose of chlorine is added.

Depending on local conditions, one or more additional purification steps may be carried out at the water-treatment facility after disinfection. Sometimes the water is sprayed into the air to remove volatile chemicals that create objectionable odors and taste. If the water is sufficiently acidic to cause problems such as corrosion of pipes or leaching of heavy metals from pipes, calcium oxide (lime) is added to partially neutralize the acid. If little natural fluoride is present in the water supply, many municipalities add about 1 ppm of fluoride (as NaF) to protect against tooth decay. In water, sodium fluoride, NaF, dissociates into $Na^+(aq)$ and $F^-(aq)$ ions. In teeth, fluoride ions are incorporated into a calcium compound called fluorapatite, which is more resistant to dental decay than apatite, the usual tooth material.

 Consider This 5.30   **Purifying Water Away from Home**

How can you purify water when you are hiking? Use the resources of the Web to explore some of the possibilities. What are the relative costs and effectiveness of these alternatives? Are any of the methods similar to those used to purify municipal water supplies? Why or why not?

5.12 Effects of Dissolved Materials on Water Quality

It should be clear by now that water is an excellent solvent for many different substances. Some solutes in water, such as oxygen, can be beneficial. Some solutes can change the properties of water. Some are highly toxic and are cause for concern. In this section, we consider some water-quality issues that you may face as a consumer. Our aims are to understand the nature of the impurity, how serious a health threat it poses, and how it gets into water. Different methods for measuring solute concentration in water will be described. Finally, as a consumer, you will need to know what can or should be done to reduce the impurity.

5.12.1 How Hard is Your Water?

Although not usually a health concern, having water with a high concentration of dissolved minerals, unless removed from solution, can be a nuisance and a reason for increased costs. Such mineral-rich water can affect the way water tastes, build up deposits on fixtures and pipes, make it difficult to produce soap lather, and leave unattractive soap scum behind on your laundry. Most water is tested for total dissolved solids (TDS), but two ions are of particular interest in reports of water hardness. **"Hard water"** contains high concentrations of dissolved calcium and magnesium ions. These ions are present in water in the form of their soluble chloride, bicarbonate, and sulfate compounds. In contrast, **"soft water"** contains few dissolved calcium or magnesium ions.

Because calcium ions, Ca^{2+}, are generally the largest contributors to hard water, hardness is usually expressed in parts per million of calcium carbonate ($CaCO_3$) by mass. This method of reporting does not mean that the water sample actually contains $CaCO_3$ at the indicated concentration. Rather, it specifies the mass of solid $CaCO_3$ that could be formed from the Ca^{2+} in solution, provided sufficient CO_3^{2-} ions were also present:

$$Ca^{2+}(aq) + CO_3^{2-}(aq) \longrightarrow CaCO_3(s) \qquad [5.8]$$

Thus, a hardness of 10 ppm indicates that 10 mg of $CaCO_3$ could be formed from the Ca^{2+} ions present in 1 L of water. Table 5.12 provides more quantitative information about water hardness classifications.

If you live in a hard water region, you have probably noticed a white deposit inside a teakettle or other utensil used to heat water. In hot water pipes, the build-up can cause serious interference with water flow (Figure 5.20). In water heaters, the build-up interferes with heat transfer, resulting in wasted energy and greater cost for water heating. Mg^{2+} can behave in similar ways to Ca^{2+} in forming part of the deposit. If Fe^{3+} is also present, either from the water source itself or possibly rusty pipes, it may also cause discoloration to the scale inside of pipes.

The source of most hard water is limestone rock, which is composed of calcium carbonate or a mixture of calcium carbonate and magnesium carbonate. Limestone

Fe^{3+} may also be present in hard water and can change the color of colorless tap water to a distasteful yellow.

Table 5.9 indicates that carbonate salts are generally insoluble in water.

Table 5.12	**Classification of Water Hardness**	
Classification	**mg/L (ppm)**	**grains/gal***
Soft	0–17.1	0–1
Slightly hard	17.1–60	1–3.5
Moderately hard	60–120	3.5–7.0
Hard	120–180	7.0–10.5
Very Hard	180 & over	10.5 & over

* One grain of hardness per gallon equals 17.1 mg/L (ppm). Many water-softening companies will test your water and report its hardness in grains/gallon.

Figure 5.20
A pipe with hard-water scale build-up.

was formed from ancient inland seas in which calcium carbonate slowly deposited over millions of years. When surface or groundwater flows over or through limestone rock, a small amount of calcium and magnesium carbonate dissolves in the water. Hard water results when sufficient magnesium and calcium ions dissolve. Water hardness in the United States varies from nearly 0 ppm in mountainous regions with mostly granite rock to over 400 ppm in parts of the Midwest, where limestone is prevalent.

Consider This 5.31   Is Your Water Hard?

Use the resources of the *Online Learning Center* to locate the water-quality report for your hometown. You may also wish to consult with a local water-softening company to find what level of hardness they typically find in your area. Report both the TDS and the water hardness. How does the hardness of your water compare with the classifications in Table 5.12?

The usual laboratory method for measuring water hardness depends on the reaction of Ca^{2+} with a solution of EDTA. The acronym EDTA is a convenient shorthand for the chemical name, *ethylenediaminetetraacetic* acid. To carry out the reaction with Ca^{2+} in the aqueous phase, EDTA is actually in a solution as a salt with the sodium ion, Na_2EDTA. The reaction is shown in Equation 5.9. Note that the product is not a precipitate, but rather stays in aqueous solution. Special chemical indicators are used to tell when the one-to-one mole ratio of Ca^{2+} and $EDTA^{2-}$ has been reached. By measuring the volume of water tested and the concentration and volume of the reacting Na_2EDTA solution needed to reach the end point of the reaction, the concentration of the Ca^{2+} can be determined.

Na_2EDTA is also used as a food additive in soft drinks such as Sprite. Na_2EDTA reacts with any metal ions to prevent cloudiness, discoloration, or changes in taste.

$$Ca^{2+}(aq) + EDTA^{2-}(aq) \longrightarrow CaEDTA(aq) \qquad [5.9]$$

Probably the most common manifestation of water hardness is the way in which calcium ions and magnesium ions interfere with soaps. The Mg^{2+} and Ca^{2+} ions react with soap to form an insoluble compound that separates from solution. This insoluble compound is the stuff of bathtub rings and the scum deposited on clothing washed in hard water. Because much of the soap is tied up in the precipitate, more soap is required to form suds and cleanse things in hard water than in soft water.

The action of soap illustrates the generalization that "like dissolves like" (Section 5.9).

As shown in Figure 5.21, the soap "molecule" contains two parts: sodium ion (Na^+) and a long hydrocarbon chain with a negatively charged ionic end (the "soap ion"). In aqueous solution, soap releases Na^+ ions and the negative soap ions. Soaps are used for

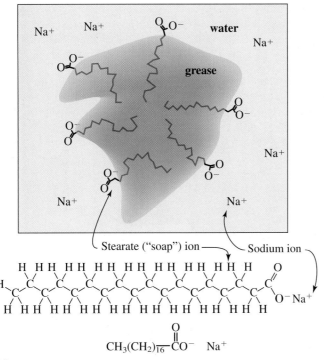

Figure 5.21

Soap and its interaction with grease.

cleaning because their long, nonpolar hydrocarbon tails dissolve readily in materials that are predominantly nonpolar, such as grease, chocolate, or gravy. The negatively charged ionic ends stick out of the surface of a grease globule because ionic substances are not soluble in nonpolar media. Thus, the grease becomes covered with negative charges. These negatively charged grease particles interact favorably with the partially positive regions of water molecules, are temporarily made soluble, and get carried away with the rinse water.

One obvious way to avoid the formation of the precipitate is to remove the ions, in other words, to "soften" the water. Softening can be accomplished by adding sodium carbonate (Na_2CO_3), commonly called washing soda, together with the soap. The carbonate ions (CO_3^{2-}) react with the Ca^{2+} to form insoluble calcium carbonate that is rinsed away (see equation 5.8). Other water-softening compounds, such as sodium tetraborate ($Na_2B_4O_7$, known by the common name of borax) and sodium phosphate (Na_3PO_4), work in a similar fashion. Calgon water softener contains hexametaphosphate ions, $P_6O_{18}^{6-}$. These ions are very effective in tying up calcium and magnesium ions as large, soluble ions.

Another way to soften hard water is to remove Ca^{2+} and Mg^{2+} ions before they get to the washing machine or shower. **Ion exchange** is a process in which ions are interchanged, usually between a solution and a solid. For softening water, the ions responsible for hard water are swapped for Na^+ ions, accomplishing the conversion of hard water to soft water. A water softener typically contains a zeolite, a clay-like mineral made up of aluminum, silicon, and oxygen. These atoms are bonded into a rigid, three-dimensional structure bearing many negative charges. All these charges must be balanced by positive charges, usually supplied by Na^+ ions associated with the zeolite. However, when water containing Ca^{2+}, Mg^{2+}, or Fe^{3+} is passed through the zeolite, these ions displace the Na^+ ions because the multiply charged ions are more strongly attracted to the negatively charged zeolite than are the singly charged Na^+ ions. In other words, ions associated with "hard water" are exchanged for sodium ions that are associated with "soft water." If we represent the zeolite as simply "Z," we can write an equation representative of the process.

Na_3PO_4 is often called by the name of trisodium phosphate, or TSP.

$$Na_2Z(s) + Ca^{2+}(aq) \longrightarrow CaZ(s) + 2\,Na^+(aq) \qquad [5.10]$$

zeolite (sodium form) — (in hard water) — zeolite (calcium form) — (in soft water)

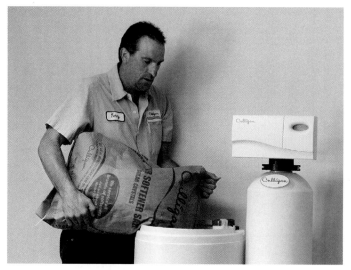

Figure 5.22
Adding salt to recharge an ion exchange water softener.

The Na^+ ions rather than Ca^{2+} ions flow through the tank and into the pipes of the residence. Sodium ions do not interfere with the function of soap or result in the build-up of scale, and so the problem of hardness has been left behind on the zeolite ion exchanger. When the exchanger becomes saturated with Mg^{2+}, Ca^{2+}, and other undesirable ions, it is back-flushed with a concentrated solution of NaCl. This is easily accomplished by simply adding salt to the ion exchange unit, as shown in Figure 5.22. The high Na^+ ion concentration of this solution displaces the Mg^{2+} and Ca^{2+} from the zeolite, reversing equation 5.10, and they are flushed down the drain as $MgCl_2$ and $CaCl_2$. The ion exchanger is left in its fully charged sodium form, ready to soften more hard water. Note that zeolite softening adds sodium ions to the water, something that is not beneficial to those on restricted sodium diets.

Consider This 5.32   **Softening Water at Home**

a.   Assume you decide that water softening is desirable at your home. What are the options for it in your area? Which option is the most cost-effective?

b.   Are any health risks associated with using hard water? With using soft water?

c.   Why do you think that municipal water supplies in hard water areas are not routinely softened at the water-treatment facility?

d.   Has EPA set standards for any of the ions responsible for hard water? Explain.

5.12.2   Is There Lead in Your Drinking Water?

Lead is one of the most serious pollutants that can make its way into drinking water. Concentrations may be low, but still can cause harm. The story of looking for lead in drinking water illustrates that standards are of little value unless accurate methods are available to measure the concentrations of such substances. Newer analytical methods, in turn, have made possible more careful studies of pollutant concentrations in water and their toxicological effects. Chemical analysis of water involves a variety of analytical methods, depending on the impurity in question.

All the metals near lead on the periodic table are toxic. Several of them, including lead, mercury, and cadmium may be encountered in drinking water. Their positive ions (Pb^{2+}, Hg^{2+}, and Cd^{2+}) form ionic compounds that are soluble and of concern. Because Pb^{2+} is the most common of these three and poses the most serious health risk due to its widespread occurrence, we examine its story in some detail. The source of the problem frequently arises within individual homes. Unless proper precautions are taken, lead from drinking water can have serious long-term health effects, especially tragic for young children.

Consider This 5.33 Lead, Mercury, or Cadmium in Your Drinking Water

Find out whether lead, mercury, or cadmium ions are a significant problem in drinking water where you live or on your campus. You might begin with the map of local drinking water systems provided by EPA's Office of Ground Water and Drinking Water. Your local water utility company or state drinking water program should be able to provide information as well.

a. If these ions are present, what are some likely sources?
b. Are the concentrations of these ions in your water above the MCLG or MCL values? Compare the values reported for your water with the values in Table 5.11.

MCLs and MCLGs are discussed in Section 5.10.

In its metallic form, lead is very dense (50% more dense than iron or steel). Because lead is an abundant, soft, and easily worked metal that does not rust, it has been used since ancient times for water pipes and roofs. Romans were likely the first to use lead for water pipes and as a lining for wine casks. Some historians attribute lead poisoning from such extensive use as a major factor contributing to the fall of the Roman Empire.

The symbol Pb comes from the Latin name for lead, *plumbum*, the origin of our word *plumbing*.

In more modern times, most U.S. homes built before 1900 had lead water pipes, now replaced over time by copper or plastic ones. Until 1930, lead pipes were commonly used to connect homes to public water mains. There is no accurate way of knowing how many people suffered permanent health damage from living in residences with lead pipes. But, there are a few recorded cases of fatalities caused by lead poisoning in which the victim over many years habitually prepared a morning beverage using the "first draw" of water that had been standing in lead pipes overnight.

An interesting chemical connection exists between water hardness and how much lead dissolves from a lead pipe. Hard water forms a protective coating of calcium carbonate inside lead pipes that prevents water from coming into direct contact with the lead. On the other hand, with naturally soft water no such protective coating forms, and a small amount of Pb^{2+} can be released into the water. Soft water also tends to be more acidic, which causes additional lead to dissolve. The effect is most severe in hot water pipes.

Some Pb^{2+} can get into drinking water even where there are no lead pipes. Solder used to join copper pipes contains 50–75% lead. Some drinking fountains were designed with a holding tank to store chilled water, and the seams in the tank and connections from it to the fountain may have been made with lead-based solder. Water for drinking fountains may stand in the tank for many hours, thus providing more contact time for lead from the solder to dissolve into the water.

When ingested, lead causes severe and permanent neurological problems in humans. This is particularly tragic for children, who may suffer mental retardation and hyperactivity as a result of lead exposure, even at relatively low concentrations. Severe exposure in adults causes irritability, sleeplessness, and irrational behavior, including loss of appetite and eventual starvation. Unlike many other toxic substances, lead is a cumulative poison and is not transformed into a nontoxic substance. Once it enters the body, it accumulates in bones and the brain.

Lead toxicity is a particular problem for children because Pb^{2+} can be incorporated rapidly into bone along with Ca^{2+}. In children, who have less bone mass than adults, the Pb^{2+} remains in the blood longer, where it can damage cells, especially in the brain. Besides lead in drinking water, young children are exposed to large amounts of lead from chewing on paint that contains lead. This is especially the case in older houses where the paint is chipped and flaking. A national program monitoring blood lead levels in children is aimed at identifying children at risk. Health officials are required to investigate cases in which children are known to exceed the currently acceptable blood level of 15 μg/dL (micrograms per deciliter). The EPA estimates that as many as one of six U.S. children under six years of age has a blood lead level above this limit.

> 1 deciliter (dL) = 0.1 L

Your Turn 5.34   Comparing Lead Content

Two samples of drinking water were compared for their lead content. One had a concentration of 20 ppb and the other had a concentration of 0.003 mg/L.

a. Explain which one contains the higher concentration of lead.
b. Compare each sample to the current acceptable limit.

Since the 1970s, the federal government has had regulations for acceptable levels of lead in water and foods. These limits have gradually become more restrictive with the development of better analytical methods for measuring extremely low concentrations and as more has been learned about the health effects of lead. Lead is so widespread in the environment that older measurements suffered from unintentional contamination of both the equipment and the reference standards. Until recently, the maximum contaminant level (MCL) for Pb^{2+} in drinking water was 15 ppb. In 1992, the U.S. EPA converted this to an "action level," meaning that the EPA will take legal action if 10% of tap water samples exceed 15 ppb. The hazard from lead is so great that the EPA has established an MCLG of 0, even though lead is not a carcinogen.

The good news is that very little lead is present in most public water supplies. Amounts exceeding allowable limits are estimated to be present in less than 1% of public water supply systems and they serve less than 3% of the U.S. population. Most lead in drinking water comes from corrosion of plumbing systems, not from the source water itself. When lead is reported, consumers are advised to take simple steps to minimize exposure, such as letting water run before using and using only cold water for cooking. Both actions minimize the chances of ingesting dissolved Pb^{2+}.

> Pb^{2+} is slightly more soluble in hot water than in cold.

Timely notification of consumers of high lead levels in drinking water was an issue in 2004 in our nation's Capitol. Lawmakers faulted the Army Corps of Engineers, operator of the reservoir and water treatment plants; EPA, monitor of water quality; and the District of Columbia Water and Sewage Authority, the water distributor, for being negligent in informing consumers that tests found over two thirds of more than six thousand homes had unacceptable lead levels, some as high as 20 times the 15-ppb limit. The major cause of contamination was aging lead pipes, although the ensuing scandal was caused by the failures of the three agencies to notify consumers and promptly work toward remediation of the problems.

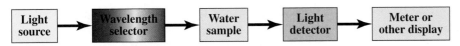

Figure 5.23

Features common to spectrophotometers used for water analysis.

 Consider This 5.35   Regulating Arsenic in Drinking Water

Another toxic metal that can find its way into public water supplies is arsenic. Early in January 2001, the Clinton administration issued a 10-ppb standard for arsenic in drinking water, replacing the standard of 50 ppb set in 1962. The Bush administration soon after recalled the rule before it could take effect, thus reverting to the 50 ppb standard, a controversial decision.

a. What was the reasoning behind each administration's decision?
b. What has been the response to each administration's decision?
c. Determine whether 50 ppb is still the standard for arsenic.

The almost universal method for Pb^{2+} analysis in water utilizes a spectrophotometric technique. The general features of a spectrophotometer are shown in Figure 5.23. Light of a specific wavelength passes through the sample and strikes a special detector where the light intensity is converted into an electrical voltage. The voltage is displayed on a meter or sent to a computer or other recording device. The amount of that wavelength of light absorbed by the solution, and which therefore does not reach the detector, is proportional to the concentration of the species being tested. The higher the concentration of the species, the more light that is absorbed by the sample.

Low concentrations of Pb^{2+} can be analyzed using furnace atomic absorption (AA) spectrophotometry. A small water sample is vaporized at very high temperature into a beam of UV light coming from a lead-containing lamp. Radiation unique to lead atoms is emitted from the hot lead atoms in the lamp and absorbed by lead atoms in the vaporized water sample. Conventional versions of AA spectrophotometers, in which the Pb^{2+} is heated in a flame, can measure Pb^{2+} concentrations in the parts per million range but cannot gather data in the range of 15 ppb, the current action level for Pb^{2+} in drinking water. An even more sophisticated AA spectrophotometer can measure lead at well below 1 ppb. However, many smaller communities cannot afford the appropriate equipment to make such sensitive measurements.

Spectrophotometric measurements from the sample being tested must be compared with absorbance data taken for known concentrations of the same species. This is done by use of a **calibration graph,** a graph that is made by carefully measuring the absorbancies of several solutions of known concentration for the species being analyzed. An example of a calibration graph for Pb^{2+} analysis at low ranges of concentration is shown in Figure 5.24. Pb^{2+} concentration is shown on the horizontal axis and absorbance at a wavelength of 283.3 nm is shown on the vertical axis. For example, if a water sample gives an absorbance measurement of about 0.24, an analyst can use that value to read directly from the graph that the concentration of Pb^{2+} is just at the 15 ppb regulatory limit (see dashed lines, Figure 5.24).

> A spectrophotometer is often called simply a spectrometer. Its use was discussed in Section 2.8 for total ozone measurements and in Section 3.4 to measure infrared radiation.

Your Turn 5.36   Using the Pb^{2+} Calibration Graph

Use Figure 5.24 to estimate the concentration of Pb^{2+} in each water sample being analyzed.

a. Absorbance = 0.50   **b.** Absorbance = 0.05   **c.** Absorbance = 0.30

Answer
a. Approximately 37–38 ppb Pb^{2+}.

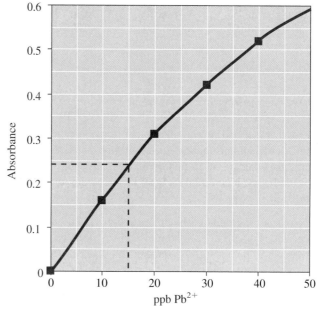

Figure 5.24

Calibration graph for furnace AA spectrophotometric analysis of Pb^{2+} at a wavelength of 283.3 nm.

Figure 5.24 illustrates a caution about water analysis: The accuracy of the analysis is only as good as the accuracy of the calibration graph. Some uncertainty is present in each of the measurements. That leads to a small uncertainty in the analysis of any water sample compared with the calibration graph.

Consider This 5.37   **Shifting Limits for Lead**

Before 1962, the recommended limit for lead in drinking water was 100 ppb. In 1962, the limit was lowered to 50 ppb. In 1988, the MCL was lowered to 15 ppb, where it remains today.

In addition to the current action level of 15 ppb of lead in tap water in residences, the EPA recommends that source water from water utilities should contain no more than 5 ppb lead, and water in school drinking fountains should contain no more than 20 ppb. Suggest possible reasons for these differences.

a.   Suggest possible reasons for these differences.
b.   If stricter limits are set, who do you think should pay the costs? Give the reasons behind your opinions.

5.12.3   Are There Trihalomethanes in Your Drinking Water?

Another type of drinking water impurity is a class of compounds known as **trihalomethanes (THMs),** in which any three halogen atoms replacing three of the four H atoms in methane, CH_4. The THM most likely to cause problems in your drinking water is chloroform, $CHCl_3$. Figure 5.25 gives several representations of the chloroform molecule.

Recall from Section 5.11 that chlorination leaves some residual HOCl in the water. This HOCl can lead to formation of THMs by its reaction with humic acids, which are organic breakdown products of plant or animal materials. Humic acids are almost always present in surface waters, and therefore formation of THMs in chlorinated surface water is unavoidable. However, the concentration of THMs in drinking water is normally far less than 1 ppm. However, they contribute an unpleasant taste to chlorinated water even in very low concentrations. The presence of THMs can sometimes be detected by their chlorine-like odor, especially in a hot shower. Recent

Halogens are Group 7A elements in the periodic table.

Humic acids are complex and variable materials resulting from partial decomposition of plant or animal matter; they form the organic portion of soils.

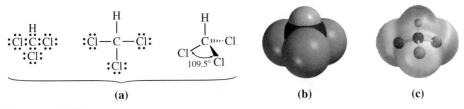

Figure 5.25

Representations of chloroform, $CHCl_3$.

(a) Lewis structure and structural formulas; (b) Space-filling model; (c) Charge-density model.

EPA standards require municipal water-treatment facilities to reduce the concentration of humic acids in water prior to its chlorination, which will decrease the likelihood of THM formation.

In addition to $CHCl_3$, other THMs that may find their way into your drinking water supply are $CHCl_2Br$, $CHClBr_2$, and $CHBr_3$. The bromine-containing THMs are also by-products of drinking water treatment, particularly if sodium hypochlorite, NaOCl or calcium hypochlorite, $Ca(OCl)_2$ are used as the disinfectants. These salts contain traces of hypobromite ion and also react with humic acids.

The primary health concern is that chloroform, a THM, is suspected of causing liver cancer. Some evidence points to slightly greater cancer rates (including bladder and rectal cancer) in people living in communities with chlorinated drinking water compared with those that do not. The current MCL for total THMs established in 1998 by the EPA is 80 ppb (0.080 ppm), down from the previous level of 100 ppb. Most drinking water samples meet that standard. The national average for THMs is 51 ppb for municipal drinking water that comes from surface water. Well water has a lower THM concentration because it usually contains little or no plant material and is not chlorinated.

Consider This 5.38 **Health Risk From Chloroform**

Chloroform can cause cancer, but chloroform is also an effective cough suppressant that has been used in many over-the-counter medications. Children's cough syrups used to contain several percent of chloroform. To what extent does this provide convincing evidence that chloroform in drinking water is not a health hazard and therefore that chlorination of drinking water is safe?

A reliable analytical method is needed to measure THMs in the low-concentration parts per billion range found in drinking water. The method of choice for THMs is called gas chromatography (GC), a powerful technique for measuring trace amounts of various molecular substances in water. In GC, molecular solutes in water are first extracted from a large sample of water into a smaller volume of a nonpolar liquid such as octane. This extraction concentrates the solutes to be analyzed. A very small portion of the extract is injected into a flowing gas stream that passes down a long, heated tube coated with an absorbing material; a detector is at the far end (Figure 5.26a). Components in the sample move down the tube at various rates, thus reaching the detector at different times. The signal from the detector is displayed on a recorder, as shown in Figure 5.26b; each peak corresponds to a different substance. For each plotted peak, the time required for the substance to reach the detector identifies the substance, while peak area measures its concentration. Gas chromatography is the normal method for analyzing trace amounts of a wide variety of toxic substances in drinking water, including pesticides, PCBs, dioxins, industrial solvents, and gasoline or other petroleum products.

THMs in chlorinated drinking water create a classic risk–benefit situation. On the one hand, chlorination of drinking water is an efficient, inexpensive, effective method that greatly reduces the health risk from bacteria and other disease-causing organisms in the water. On the other hand, chlorination results in the formation of relatively low concentrations of THMs that may cause cancer.

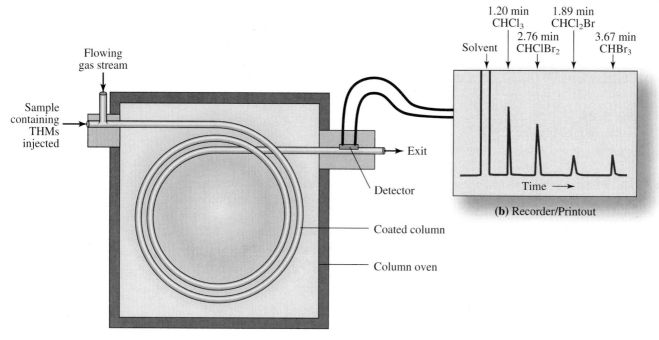

(a) A simplified gas chromatography apparatus.

Figure 5.26

(a) A simplified gas chromatography apparatus. (b) Gas chromatography analysis of a mixture of THMs.

5.13 Consumer Choices: Tap Water, Bottled Water, and Filtered Water

You now have a better chemical background to make informed choices about the water you drink. As always, decisions depend on risk–benefit analysis and your personal preferences, but must be grounded in factual information.

Tap Water

1. Is safe drinking water generally available in the United States? The answer, a resounding "yes," is due to high standards mandated by federal regulations for drinking water provided by public water supply utilities. As important, the treatment technology is available to achieve these high standards; without such technology, standards would be merely hollow gestures. Very few people in our country suffer acute illness from drinking contaminated water unless they are using water from a private well that has not been properly tested. It should be noted that the Safe Drinking Water Act Amendments of 1996 enhanced protection, including increased requirements for notifying consumers promptly of any problems with water safety.

2. Is tap water "pure" water? Certainly not; it almost surely contains small amounts of sodium, calcium, magnesium, chloride, sulfate, and bicarbonate ions, as well as trace amounts of other ions. Tap water also contains dissolved air, which is a mixture that includes N_2, O_2, CO_2, and air-borne particles.

3. What problems are likely to exist? In some parts of the country, extremely hard water can cause problems, at least economically. Water softening options are readily available, if desired. Some tap water may contain dangerous Pb^{2+} concentrations, although lead is normally a problem only in buildings with lead pipes, and then only if the water is naturally soft. Other heavy-metal elements,

such as mercury and cadmium, may be present at dangerous concentrations, although this is extremely unlikely. Chlorinated tap water from surface water sources will contain a small amount of residual chlorine. It may also contain small amounts of trihalomethanes, by-products of chlorination. Depending on its source, the water may contain low concentrations of mercury, nitrate, pesticide residues, PCBs, and industrial solvents. By now you should understand that the presence of such substances in drinking water is not likely a cause for alarm. Rather, the crucial question is "How much?" If pollutant concentrations are below the maximum contaminant levels (MCLs), the EPA regards the water as safe, with an adequate margin of safety.

Bottled Water

This chapter began with a look at bottled water, drunk by people for a variety of reasons: taste, convenience, or the belief that it is more healthful than tap water. Whatever its source, bottled water is expensive. We can raise the same questions about bottled water as those asked about tap water.

1. Is it safe? The same laws that apply to public water supplies do not regulate bottled water. However, it is highly regulated by other regulations, both government- and industry-imposed. Considered a food, bottled water is regulated by the Food and Drug Administration (FDA). Bottled water must meet standards of quality, comply with labeling regulations, and meet good manufacturing practices. A provision of the SDWA amendments of 1996 requires the FDA to develop bottled water standards that are equal to EPA drinking water standards. In years past, critics have questioned the safety of bottled water. However, member companies of the International Bottled Water Association (IBWA) produce more than 85% of the bottled water currently sold in the United States. The member companies must meet higher water-quality standards than those imposed by the FDA (Figure 5.27).

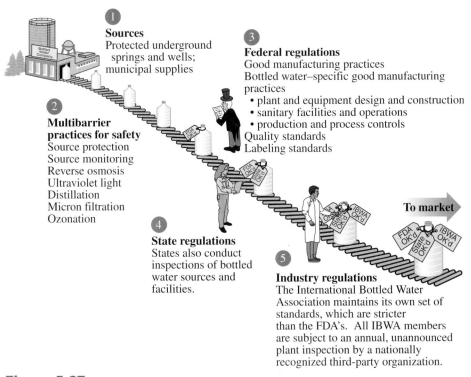

Figure 5.27

Bottled water's path to market. The International Bottled Water Association illustrates the process its members' products follow from the source to the consumer's satisfaction. Federal, state, and industry regulations guarantee safety and quality.

Springs and underground aquifers that do not require disinfection are the principal sources of bottled water. If disinfection is required, it is done with ozone or UV radiation, rather than with chlorine, thus leaving no objectionable taste and no THMs. In addition, most bottled water is subjected to filtration, reverse osmosis, or distillation (see Section 5.11). The absence of chlorine, THMs, and the probable absence of various trace pollutants found in surface water provide much of the argument for bottled water as a more healthful alternative to tap water. In the majority of all bottled water sold in the United States, the source is municipal tap water that has been subjected to further purification. Interestingly enough, if the municipal water meets processing standards allowing it to be labeled "distilled" or "purified," the water does not need to divulge its municipal tap water source.

2. Is bottled water pure? Because bottled water often comes from springs or wells, we can be sure that it contains ions, dissolved as the water percolates through the surrounding rocks. In fact, bottled water from some well-known spas, such as Bath in England, Baden-Baden in Germany, and White Sulfur Springs in West Virginia, contains relatively large amounts of calcium and other ions, as well as dissolved carbon dioxide. In a few cases, dissolved hydrogen sulfide gas provides a characteristic "sulfur" odor, thought to be a positive virtue by some connoisseurs of bottled water.

Filtered Water

1. Is filtered water safe? Yes, certainly it is as safe as the tap water supply being filtered. Water from such a unit is free of objectionable taste and odor and should be free of most hazardous substances. Most filters reduce the concentrations of ions responsible for hard water (Ca^{2+}, Mg^{2+}) and toxic metal ions (Pb^{2+}, Cu^{2+}). Although these ions are not necessarily totally removed, their concentrations will be well below those of concern for human health.

2. How do the filters work? These units generally attach to a kitchen faucet, purifying water for drinking or cooking using two methods. The first is "activated carbon," a special form of charcoal with a very high surface area that absorbs most of the molecular solutes, including residual chlorine, THMs, pesticide residues, solvents, and other similar substances. The second component is an ion exchange resin (see Section 5.12.1) that removes ions responsible for hard water or those that can cause toxicity.

3. Are filters cost-effective? Tap water remains the least expensive choice for drinking water. Filtered water systems generally treat only the water for drinking and cooking, bringing costs to less than 20% of the costs for purchasing bottled water used for the same purposes.

Consider This 5.39 **Evaluating Your Drinking Water Choices**

In Consider This 5.1 and 5.2 activities, you were asked to think about which characteristics of drinking tap, bottled, and filtered water were important to you. Having now studied this chapter, check your lists. Would the order of importance be the same? Explain how your reasoning may have changed based on the information and understanding gained in the study of this chapter.

5.14 International Needs for Safe Drinking Water

Those who live in the United States are privileged to have drinking water choices available. We can select from tap, bottled, or filtered water, all generally of high quality. Such is not the case for people in most of the rest of the world. The reality is that more than a billion people (one in six), principally in developing nations, lack access to safe

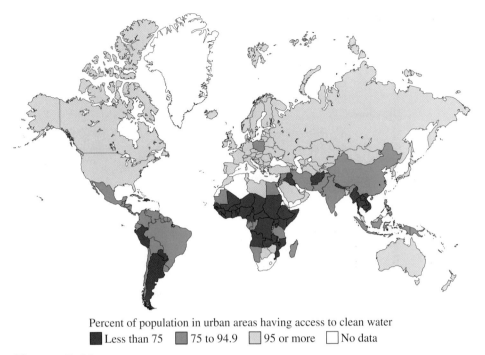

Percent of population in urban areas having access to clean water

■ Less than 75   ■ 75 to 94.9   ■ 95 or more   □ No data

Figure 5.28

Access to safe drinking water varies widely across the world.

Source: The World Resources Institute, data based on surveys of national governments. http://www.sciam.com/ 1197issue/1197scicit5.html

drinking water. About 1.8 billion people do not have adequate sanitary facilities. One estimate, made by *Scientific American,* is that it would cost $68 billion dollars over the next 10 years to provide safe water and decent sanitation facilities to everyone. Lack of access to safe water poses a particular risk to infants and young children. Whereas bottled water is a discretionary option for many in the United States, the majority of the world's population does not have that option. Figure 5.28 shows how access to clean water in urban areas varies worldwide.

For those living in arid regions, such as the Middle East, fresh water is scarce. Sea water is readily available in many such areas, but its high salt concentration makes it unfit for human consumption. Coleridge's ancient mariner is more than just a poetic fantasy; it is a physiological reality. Ocean water contains 3.5% salt compared with only about 0.9% salt in body cells. Consequently, sea water can be drunk only after most of the salt is removed. Fortunately, there are ways to do this, but they require large amounts of energy. Collectively, the methods are known as **desalination,** a broad general term describing any process that removes ions from salty water.

One desalination method is distillation, an old and remarkably simple way of purifying water for laboratory and other uses. **Distillation** is a separation process in which a solution is heated to the boiling point and the vapors are condensed and collected. Distilled water is used in steam irons, some car batteries, and other devices whose operation can be impaired by dissolved ions. An apparatus such as that shown in Figure 5.29 is used. Impure water is put into a flask, pot, or other container and heated to its boiling point, 100°C. As the water vaporizes, it leaves behind most of its dissolved impurities. The water vapor passes through a condenser where it cools and reverts back into a liquid, now free of contaminants. Not surprising, the product is usually called "distilled water." If distillation is done very carefully, extremely pure water, with no detectable amounts of contaminants, is produced.

Energy is required for the distillation of any liquid, but recall from Section 5.5 that water has an unusually high specific heat and an unusually large amount of heat required for evaporation. Both result from the uniquely extensive hydrogen bonding in water. High energy costs for water purification by distillation suggests that it is

A process similar to distillation takes place as part of the natural hydrologic cycle. Water evaporates and then condenses and falls as rain or snow.

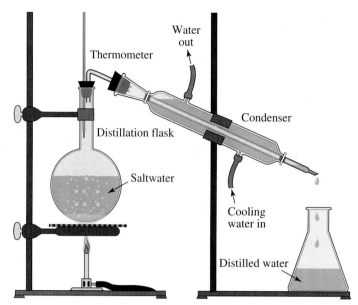

Figure 5.29
Water purification by distillation.

economically practical only for countries or regions with abundant and cheap energy. Ion exchange, described in Section 5.12.1, is another method for desalination, although it is not very practical for large-scale desalination.

Another desalination technique gaining in popularity is reverse osmosis. To understand this method, we need to know that **osmosis** is the natural tendency for a solvent (water in this case) to move through a membrane from a region of higher solvent concentration to a region of lower solvent concentration. This tendency to equalize concentrations is involved in many biological processes. In this particular instance, the net effect is that cells lose water rather than gain it. However, osmosis can be reversed. If sufficient pressure is applied to the saltwater side, water molecules can be forced through the membrane, leaving ions behind. This is **reverse osmosis.** Figure 5.30 is a schematic representation of this process.

The world's largest desalination plant, located at Jubail, Saudi Arabia, provides 50% of that country's drinking water using reverse osmosis to desalinate Persian Gulf water. Although most such installations are in the Middle East, the number of

Solar-powered reverse osmosis desalination units providing 400 L/day were developed at Murdoch University in Perth, Australia. Twenty-five units are now in operation in Australia and Asia. Larger units providing 15,000 L/day are being tested.

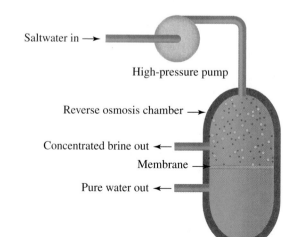

Figure 5.30
Water purification by reverse osmosis.

reverse osmosis plants is increasing in the United States. Florida has over 100 reverse osmosis desalination facilities, including the one that furnishes the city of Cape Coral with 15 million gallons of fresh water every day from brackish underground supplies. Small reverse osmosis installations are used in spot-free car washes and individual units are available for boaters. Figure 5.31 shows a small unit, suitable for use on a sailboat. Generally reverse osmosis desalination is too expensive for use in most developing nations. It is an often-used method of purification for bottled water, particularly high-end "designer" waters. Using reverse osmosis is meant to impress the customer and help them justify the cost because the quality of the water is so high.

Figure 5.31

A small reverse osmosis apparatus for converting sea water to potable water.

CONCLUSION

This chemical compound H_2O is a very unusual substance, with many unique properties that contribute to its life-supporting role. Like the air we breathe, water is central to life, and we humans require large quantities of it. We take for granted that our drinking water, whether it is straight from the tap, filtered tap water, or bottled water, typically is free of harmful contaminants. This chapter has focused almost exclusively on the quality of drinking water—its sources, substances dissolved in it, and potential contaminants and determination of their concentrations. Federal and state regulations help make our drinking water safe. In three cases, we considered how particular substances in water can be analyzed and treated. In the next chapter, we examine rainwater and the ways in which substances dissolved in rain can adversely affect the environment.

Chapter Summary

Having studied this chapter you should be able to:

- Describe the desirable properties of drinking water (5.1, 5.13)

- Explain some of the reasons why bottled water is so popular (5.2, 5.13)

- Recognize the sources and distribution of water (5.2)

- Discuss why water is such an excellent solvent for ionic and some covalent compounds (5.3, 5.7, 5.8, 5.9)

- Describe the factors involved in providing pure drinking water (5.3, 5.11, 5.14)

- Use concentration units: percent, ppm, ppb, and molarity (5.4)

- Discuss the relationship between the properties of water and its molecular structure (5.5, 5.6)

- Describe the specific heat of water and compare it with that of other substances (5.5, 5.6)

- Understand how electronegativity and bond polarity are related to the structure of water (5.5)

- Describe hydrogen bonding and its importance to the properties of water (5.6)

- Describe how the densities of ice and water are related to the molecular structure of water (5.6)

- Determine the formulas for ionic compounds, including those with common polyatomic ions (5.7)

- Explain how ionic substances dissolve in water (5.8)

- Explain how covalent substances dissolve in water (5.9)

- Understand the role of federal legislation in protecting safe drinking water (5.10)

- Discuss the maximum contaminant level goal (MCLG) and the maximum contaminant level (MCL) established by the EPA to ensure water quality (5.10)

- Discuss how drinking water is made safe to drink (5.11, 5.12, 5.14)

- Relate chlorination with water purification (5.11, 5.12.3)

- Know the causes and effects of water hardness (5.12.1)

- Cite water-softening methods (5.12.1)

- Describe atomic absorption spectrophotometry and gas chromatography as methods for analyzing contaminants in water (5.12.2, 5.12.3)

- Compare and contrast tap water, bottled water, and filtered water in terms of water quality (5.13)

- Understand distillation and reverse osmosis (5.14)

- Appreciate the relative availability of pure drinking water in the United States and compare with international needs (5.14)

Questions

Emphasizing Essentials

1. a. The text states that 50–65% of an adult body weight is water. How many pounds of water will this be for a 150-lb adult? Report your answer as a range of values.

    b. Given that a gallon of water weighs about 8 lb, how many gallons of water will this be for a 150-lb adult? Report your answer as a range of values.

2. a. What is an aquifer?

    b. Why is it important to prevent unwanted substances from reaching a clean aquifer?

3. If the water in a full 500-L drum was representative of the world's total supply, how many liters could be suitable for drinking? *Hint:* See Figure 5.3.

4. Based on your experience, what is the solubility of each of these substances in water? Use terms such as very soluble, partially soluble, or not soluble. Cite some supporting evidence for your decisions.

    a. orange juice concentrate

    b. liquid clothes-washing detergent

    c. household ammonia

    d. chicken broth

    e. chicken fat

5. a. Bottled water consumption was reported to be 21 gal per person in the United States in 2002. The last census reported 2.9×10^8 people in the United States, what is the total bottled water consumption estimated to be?

    b. If the per capita consumption of bottled water has increased 20% in the last 10 years, what was the per capita consumption 10 years ago.

6. a. A certain bottled water lists a calcium concentration of 55 mg/L. What is its calcium concentration expressed in parts per million?

    b. How does this concentration compare with that for Evian listed in Table 5.3?

7. A certain vitamin tablet contains 162 mg of calcium and supplies 16% of the recommended daily amount of calcium required by a person on a typical 2000-Calorie diet. How many 500-mL bottles of Evian bottled water would you have to drink each day to obtain the same mass of calcium? *Hint:* See Your Turn 5.7 and Table 5.3.

8. The acceptable limit for nitrate, often found in well water in agricultural areas, is 10 ppm. If a water sample is found to contain 350 μg/L, does it meet the acceptable limit? Show a calculation to support your answer.

9. One reagent bottle on the shelf in a laboratory is labeled 12 M H_2SO_4 and another is labeled 12 M HCl.

    a. How does the number of moles of H_2SO_4 in 100 mL of 12 M H_2SO_4 solution compare with the number of moles of HCl in 100 mL of 12 M HCl solution?

    b. How does the number of grams of H_2SO_4 in 100 mL of 12 M H_2SO_4 solution compare with the number of grams of HCl in 100 mL of 12 M HCl solution?

10. A student weighs out 5.85 g of NaCl to make a 0.10 M solution. What size volumetric flask will she need? *Hint:* See Figure 5.6.

11. Both methane, CH_4, and water are compounds in which hydrogen atoms are bonded with a nonmetallic element. Yet, methane is a gas at room temperature and pressure and water is a liquid. Offer a molecular explanation for the difference in properties.

12. Explain why the term *universal solvent* is applied to water.

13. Consult Table 5.4 to help answer this question.

    a. Calculate the electronegativity difference between each pair of atoms.

        N and C

        S and O

        N and H

        S and F

    b. A single covalent bond forms between the atoms in each pair. Identify the atom that attracts the electron pair in the bond more strongly.

    c. Arrange the bonds in order of increasing polarity.

14. NaCl is an ionic compound, but chlorine and silicon are joined by covalent bonds in $SiCl_4$.

    a. Use Table 5.4 to determine the electronegativity difference between chlorine and sodium, and between chlorine and silicon.

    b. What correlations can be drawn about the difference in electronegativity between bonded atoms and their tendency to form ionic or covalent bonds?

    c. How can you explain on the molecular level the conclusion reached in part **b**?

15. Consider a molecule of ammonia, NH_3.

    a. Write the Lewis structure for NH_3.

    b. Are there polar bonds in NH_3?

    c. Is the NH_3 molecule polar? *Hint:* Consider the geometry of the ammonia molecule.

    d. Is NH_3 soluble in water? Explain.

16. This diagram represents two water molecules in a liquid state. What kind of bonding force does the arrow indicate? Is this an *inter*molecular or *intra*molecular force?

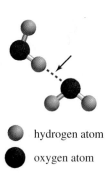

○ hydrogen atom

● oxygen atom

17. The density of liquid water at 0 °C is 0.9987 g/cm^3; the density of ice at this same temperature is 0.917 g/cm^3.

    a. Calculate the volume occupied at 0 °C by 100. g of liquid water and by 100. g of ice.

    b. Calculate the percentage increase in volume when 100. g of water freezes at 0 °C.

18. Consider these liquids.

Liquid	Density, g/mL
dishwashing detergent	1.03
maple syrup	1.37
vegetable oil	0.91

    a. If you pour equal volumes of these three liquids into a 250-mL graduated cylinder, in what order will you add the liquids to create three separate layers? Explain your reasoning.

    b. If an unknown liquid were poured into the cylinder and it formed a layer that was on the bottom of the other three layers, what can you tell about one of the properties of the unknown liquid?

    c. What would happen if a volume of water equal to the other liquids were poured into the cylinder in part **a** and then the contents are mixed vigorously? Explain.

19. Why is there the possibility of a water pipe breaking if the pipe is left full of water during extended frigid weather?

20. Calculate the quantity of heat absorbed (+) or released (−) during each of these changes.

    a. 250 g of water (about 1 cup) is heated from 15 °C to 100 °C

    b. 500 g of water is cooled from 95 °C to 55 °C

    c. 5 mL of water at 4 °C is warmed to 44 °C

21. Which ion most likely forms from each of these atoms? Use the Lewis structure of each atom and its corresponding ion to show how the ion obeys the octet rule. *Hint:* Consider Tables 5.5 and 5.6.

    a. Cl

    b. Ba

    c. S

    d. Li

    e. Ne

22. Write the formula and give the name of the ionic compound formed by the reaction of each pair of elements.

    a. Na and S

    b. Al and O

    c. Ga and F

    d. Rb and I

    e. Ba and Se

23. Write the formula for each compound.

    a. calcium bicarbonate   c. magnesium chloride

    b. calcium carbonate   d. magnesium sulfate

24. Name each compound.

    a. $KC_2H_3O_2$

    b. $Ca(OCl)_2$

    c. LiOH

    d. Na_2SO_4

25. Solutions can be tested for conductivity using this type of apparatus.

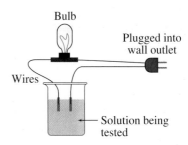

    Predict what will happen when each of these dilute solutions is tested for conductivity. Explain your predictions briefly.

    a. $CaCl_2(aq)$

    b. $C_2H_5OH(aq)$

    c. $H_2SO_4(aq)$

26. What ions are present in each of these solutions?

    a. $Ca(OCl)_2(aq)$

    b. $C_2H_5OH(aq)$

27. Based on the generalizations in Table 5.9, which compounds are likely to be water-soluble?

    a. $KC_2H_3O_2$

    b. $Ca(NO_3)_2$

    c. LiOH

    d. Na_2SO_4

28. How would each water sample's hardness or softness be classified based on the classifications given in Table 5.12?

    a. 55 ppm $CaCO_3$

    b. 225 mg $CaCO_3$/L

    c. 0.05 grains $CaCO_3$/gal

    d. 9.0 grains $CaCO_3$/gal

29. This represents the structural formula of a soap:

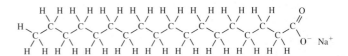

    How is the ability of this soap to remove grease from clothes related to the structure of this soap?

30. Explain why desalination techniques, despite proven technological effectiveness, are not used more widely to produce potable drinking water.

Concentrating on Concepts

31. We take water for granted. How would you explain to a friend that we should value water as a unique substance, one essential for life?

32. Why is the concentration of calcium often given on the label for bottled water?

33. The label on Evian bottled water lists a magnesium concentration of 24 mg/L. The label of a popular brand of multivitamins lists the magnesium content as 100 mg per tablet. Which do you think is a better source of magnesium? Explain your reasoning.

34. A new sign is posted at the edge of a favorite fishing hole that says "Caution: Fish from this lake may contain over 1.5 ppm Hg." Explain to a fishing buddy what this unit of concentration means, and why the caution sign should be heeded.

35. This periodic table contains four elements identified by numbers.

   a. Based on trends within the periodic table, which of the four elements would you expect to have the highest electronegativity value? Explain.

   b. Based on trends within the periodic table, rank the other three elements in order of decreasing electronegativity values. Explain your ranking.

36. A diatomic molecule XY that contains a polar bond *must* be a polar molecule. However, a triatomic molecule XY_2 that contains a polar bond *does not necessarily* form a polar molecule. Use some examples of real molecules to help explain this difference.

37. Imagine you are at the molecular level, looking at what happens when gaseous water condenses.

   a. Consider a collection of four water molecules. Sketch each water molecule using a space-filling representation similar to this one.

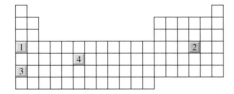

   Draw a representation of water in the gaseous state and then in the liquid state. How does the collection of molecules change?

   b. Discuss what happens to the bonding at the molecular level when water condenses to a liquid.

   c. What happens at the molecular level when water changes from a liquid to a solid?

38. Propose an explanation for the fact that NH_3, like H_2O, has an unexpectedly high specific heat. *Hint*: See question 15 for the Lewis structure and H-to-N-to-H bond angle in NH_3.

39. **a.** What type of bond holds together two hydrogen atoms in the hydrogen molecule, H_2.

   b. Explain why the term "hydrogen bonding" does *not* apply to the bond within H_2.

40. Hydrogen bonding has been offered as a reason why ice cubes and icebergs float in water. Consider ethanol, C_2H_5OH.

   a. Draw its Lewis structure and use it to decide if pure ethanol will exhibit hydrogen bonding.

   b. A cube of solid ethanol sinks rather than floats in liquid ethanol. Explain this behavior in view of your answer in part a.

41. The unusually high heat capacity of water is very important in regulating our body temperature and keeping it within a normal range despite time, age, activity, and environmental factors. Consider some of the ways that the body produces heat, and some of the ways that it loses heat. How would these functions differ if water had a much lower heat capacity?

42. How do generalizations about solubility influence water quality? Suppose that you are in charge of regulating an industry in your area that manufactures agricultural pesticides. How will you decide if this plant is obeying necessary environmental controls? What criteria affect the success of this plant?

43. Health goals for contaminants in drinking water are expressed as MCLG, or maximum contaminant level goals. Legal limits are given as MCL, or maximum contaminant levels. How are MCLG and MCL related for a given contaminant?

44. Why is the hardness of water reported as parts per million $CaCO_3$ and not parts per million $CaHCO_3$?

45. A 100.0 mL water sample was analyzed for hardness using a 0.0100 M solution of $EDTA^{2-}$. If 27.3 mL of $EDTA^{2-}$ were required for the indicator to change color, what was the hardness of the water sample? Report your answer in ppm hardness (mg $CaCO_3$/L). *Hint:* See equation 5.9.

46. Borax, $Na_2B_4O_7$, can be used to soften hard water. Write an equation to represent borax reacting with calcium ions in hard water.

47. An ion exchange resin totally charged with Na^+ ions holds 0.35 g of Na^+ per gram of resin. Use this information to determine the mass of water with a hardness of 40 ppm that could be deionized by 1 g of this resin.

48. Water quality in the chemistry building on a campus was continuously monitored because testing indicated water from drinking fountains in the building had dissolved lead levels above those established by the Safe Drinking Water Act.

   a. What is the likely major source of the lead in the drinking water?

b. Does the chemical research carried out in this chemistry building account for the elevated lead levels found in the drinking water? Why or why not?

Exploring Extensions

49. Most people turn on the water tap with little thought about where the water comes from. In Consider This 5.4 and 5.5 activities, you investigated the source of your drinking water. Now take a more global view. Where does drinking water come from in other areas of the world? Investigate the source of drinking water in a desert country, in a developed European country, and in an Asian country. How do these sources differ?

50. One of the large aquifers in the United States is under the pine barrens of New Jersey.

a. Where are the pine barrens in New Jersey?

b. Why are there increasing political pressures to use the water in this aquifer?

51. Aquifers are also important in providing clean drinking water in other parts of the world. Recently four countries in South America have reached an historic agreement to share the immense Guarani aquifer. This is particularly significant because although surface waters often provoke discord and result in agreements, underground sources are routinely not considered at least by international law.

a. Where is the Guarani aquifer and what four countries are part of this international "underground concordat"?

b. What concerns did each country have about this aquifer that led to the agreement?

52. Is there any such thing as "pure" drinking water? Discuss what is implied by this term, and how the term's meaning might change in different parts of the world.

53. In the mid-1990s, researchers in Canada and Australia reported that consumption of drinking water with more than 100 ppb aluminum can lead to neurological damage, such as memory loss and perhaps to a small increase in the incidence of Alzheimer's disease. Has further research substantiated these findings? Find out more about this topic, and write a brief summary of your findings. Be sure to cite the sources of your information.

54. The text states that hydrogen bonds are only about one tenth as strong as the covalent bonds that connect atoms within molecules. Check out that statement with this information. Hydrogen bonds vary in strength from about 4 to 40 kJ/mol. Given that the hydrogen bonds between water molecules are at the high end of this range, how does the strength of a hydrogen bond between water molecules compare with the strength of an hydrogen-to-oxygen covalent bond within a water molecule? *Hint:* Consult Table 4.2 for covalent bond energies.

55. The text states that mass and density are often confused. Here is an example of that potential misunderstanding of terms.

a. What do you think the term *heavy metal* implies when talking about elements on the periodic table?

b. Compare the scientific definitions of this term that you may find in different sources, and discuss whether each definition is related to relative density or to relative mass.

56. We all have the amino acid glycine in our bodies. This is its structural formula.

$$\begin{array}{ccccc} & H & & :O: & \\ & | & & || & \\ H-&\overset{..}{N}-&\overset{|}{C}-&C-&\overset{..}{\underset{..}{O}}-H \\ & | & | & & \\ & H & H & & \end{array}$$

a. Is glycine a polar or nonpolar molecule? Use electronegativity differences to help answer this question.

b. Can glycine exhibit hydrogen bonding? Explain your answer.

c. Is glycine soluble in water? Explain.

57. How hard is the water in your local area? One way to answer this question is to determine the number of water-softening companies in your area. Use the resources of the Web, as well as ads in your local newspapers and yellow pages, to find out if your area is targeted for marketing water softening devices.

58. The calibration curve shown in Figure 5.24 is useful for Pb^{2+} concentrations between 0 and 40 ppm. What are your options if a water sample is expected to contain a much higher concentration of Pb^{2+}?

59. Some areas have a higher than normal amount of trihalomethanes, THMs, in the drinking water. Suppose that you are considering moving to such an area. Write a letter to the local water district asking relevant questions to be answered before deciding to move.

60. PCBs (polychlorinated biphenyls) are very useful chemicals that may end up in the wrong place, causing long-term damage to birds and mammals. What are the uses of PCBs that made them desirable, and what are some of the negative effects of these materials?

Neutralizing the Threat of Acid Rain

"In the beginning, there was acid rain, millions of year ago. Then it got better. Then it got worse as people came along. Then it got better, especially in 1990 for the U.S. with the Clean Air Act Amendments. Now we see it is getting somewhat worse, as we increased our understanding of the effects of acid rain and, most of all, the linkages of acid rain with other issues."

Professor James Galloway
University of Virginia, Department of Environmental Sciences
Conference in 2001, Acid Rain: Are the Problems Solved?

Acid rain is nothing new to our planet. Millions of years ago, when volcanoes were dumping tons of sulfur dioxide into the early atmosphere of our planet, no humans were around to gasp breaths of the acidic air. But since the dawn of the age when humans walked the planet, the acidity has been worse at certain times and in certain places. No matter which particular time or place you pick, you will find only a few major sources of acidic precipitation. In fact, you can count them while drawing a single breath: volcanoes, bolts of lightning, wild fires, releases from microbes, and, of course, human activities. It is the last of these categories—humans and our activities—that may cause you to exhale with a sigh, or perhaps with a cough, before the chapter is over.

To understand the recent history of acid rain in the United States, think in terms of decades. The 1970s and 1980s stood as a period of passionate concern over our forests and lakes, with outcries over the damage to fragile ecosystems. In 1980, Congress created the National Acid Precipitation Assessment Program (NAPAP), which in 1990 released a comprehensive study on the causes and effects of acid rain. That year also marked the beginning of more stringent air quality legislation. The 1990 Clean Air Act Amendments, passed by the joint efforts of Congress and then-President George H. W. Bush, included the creation of the Acid Rain Program. Phase I of the Clean Air Act Amendments commenced in 1995, when 110 electric utility plants in 21 states began phased reductions of emissions that produced acid rain. The 2000s began expectantly, when these and even more utilities began Phase II cuts in emissions. However, in the past few years, rather than decreases there have been small *increases* in the amounts of nitrogen oxides, a major contributor to acid rain. Although the United States has clearly achieved reductions in the big scheme of things, reducing emissions remains a challenge. Furthermore, we still await the signs of ecological recovery.

As you might surmise, our knowledge of acid rain predates these recent legislative acts by over a century. The acidity of rain apparently was first studied in detail in 1852 by a British chemist named Angus Smith. Twenty years later, he wrote a book entitled *Air and Rain,* but the book and Smith's ideas soon fell into obscurity. Then, in the 1950s, the effects of acid rain were rediscovered by scientists working in the northeastern United States, in Scandinavia, and in the English Lake District.

Reports of damage attributed to acidic precipitation grew dramatically over the next three decades. Dozens of books, scientific papers, and popular articles described the damage already observed and made dire predictions of what was yet to come (Figure 6.1).

Comments about air being "acidic" were made as early as the beginning of the 1700s.

Figure 6.1
Acid rain has been in the news for some time (*The Cleveland Press,* June 23, 1980).

Similar reports came from every part of the world. Lakes in Norway and Sweden were reported as being effectively "dead," without fish or any other living things. Trees in northern Germany had been stripped of their leaves. The sculptures adorning the exteriors of cathedrals in Europe and prehistoric sites in Mexico and Central America were eroding away. In all these instances, acid rain was blamed as one of the major causes of the damage.

Acid rain differs in scope from the issues discussed in earlier chapters. Ambient air quality, as we saw in Chapter 1, is largely a *localized* problem and most serious in cities and urban areas. The destruction of stratospheric ozone and the greenhouse effect, as described in Chapters 2 and 3, are both *global* atmospheric phenomena. Acid rain, however, tends to be *regional* in character; that is, it falls on areas neighboring its source. Acidic gases that originate in the Midwest, especially in the Ohio River valley, are carried to the northeast by prevailing winds and fall as acidic rain or snow on New York, New England, and eastern Canada. They are also carried to the southeast, producing acidic precipitation in Tennessee, North Carolina, and Virginia. In Europe, acids generated in Germany, Poland, and the United Kingdom are carried northward into Norway and Sweden.

The problem of acid rain is compounded by what may be the environmental paradigm of our times: By trying to use a technological fix for one problem, we inadvertently create another. Taller smokestacks were built to eject pollutants high into the atmosphere, where they could be carried away and thus improve local air quality; in effect, an "out of sight out of mind" policy. But what goes up must come down, and it does somewhere, sometimes hundreds of miles away. Thus, potential local pollution problems are converted into regional ones. Because acid rain does not respect state or national boundaries, it becomes entangled with intense political controversies and accusations. Once again, we are all caught in the same web.

Consider This 6.1   The Clean Air Act

If you want to read over 400 pages of text, access the 1990 Clean Air Act online. A far friendlier (and shorter) version, courtesy the EPA, is "The Plain English Guide to the Clean Air Act." Search for this document online or use the link provided at the *Online Learning Center.* The document explains the role of the federal government and that of the states. How do these roles differ? Summarize the Clean Air Act's program to reduce acid air pollutants.

6.1 What Is an Acid?

The topic of acid rain nicely brings together water (Chapter 5) and atmospheric pollutants (Chapter 1). Quite obviously, we need to define acids to understand this linkage. Like many other chemical concepts, acids can be defined either by their observable properties or conceptually, by theories of chemical structure. Chemists usually use both types of definitions, which is what we will do here.

Historically, chemists identified acids by their common properties—sour taste, color changes with indicators, and reactions with carbonate-containing materials. Although tasting is not a safe way to identify chemicals, you undoubtedly know the sour taste of vinegar, which contains acetic acid. The sour taste of lemons and limes comes from acids as well (Figure 6.2). You may even be a fan of the incredibly sour candies that pucker your mouth. These candies usually contain citric acid and malic acid (Figure 6.3).

Other tests for acids rely on their chemical reactivity. A familiar example is the litmus test. Litmus, a vegetable dye, changes from blue to pink in the presence of acids.

Figure 6.2

Citrus fruit contains both citric acid and ascorbic acid.

Figure 6.3
Sour candy that contains malic acid and citric acid.

Indeed, the litmus test is so well known that it is used as a figure of speech, even for things unrelated to chemistry. For example, a politician might say "The litmus test for this judicial candidate is . . ." Another simple chemical test that works for more concentrated acids is to add them to a carbonate-containing material such as marble or eggshell. Doing so releases bubbles of carbon dioxide that may appear as "fizz." We will return to this chemical reaction later.

One way to define an **acid** is as a substance that releases hydrogen ions, H^+, in aqueous solution. You will recall from Chapter 5 that an ion is any atom or group of atoms bonded together that has a net electric charge. But if an atom or molecule gains one or more electrons, it becomes negatively charged. Likewise, it becomes positively charged if it loses one or more electrons. For example, a hydrogen atom consists of one electron and one proton and is electrically neutral. If this electron is lost, the hydrogen atom acquires a positive charge and is designated as H^+. Since only a proton remains, the hydrogen ion sometimes is referred to as a proton.

As an example, consider hydrogen chloride gas, which, at room temperature, is made up of covalent HCl molecules. This gas is quite soluble in water and dissolves to release two ions, H^+ and Cl^-.

$$HCl(g) \xrightarrow{H_2O} H^+(aq) + Cl^-(aq) \qquad [6.1]$$

This reaction could be described in other ways: HCl *dissociates* into H^+ and Cl^-, or HCl *ionizes* to form H^+ and Cl^-. Either way, essentially no undissociated HCl molecules remain in solution. Thus, we say that HCl is an acid that ionizes (or dissociates) completely.

There is a slight complication with the definition of acids as substances that release H^+ ions (protons) in aqueous solutions. By themselves, H^+ ions are much too reactive to exist, so they attach to something else, such as to water molecules. When dissolved in water, each HCl donates a proton (H^+) to an H_2O molecule, forming H_3O^+, the hydronium ion. The Cl^- (chloride) ion remains unchanged. The overall reaction is

$$HCl(g) + H_2O(l) \longrightarrow H_3O^+(aq) + Cl^-(aq) \qquad [6.2]$$

The resulting solution represented in both equations 6.1 and 6.2 is called hydrochloric acid and has the characteristic properties of an acid because of the presence of H_3O^+ ions. Chemists often simply write H^+ when referring to acids (such as in equation 6.1), but understand this to mean H_3O^+ in aqueous solutions.

The carbonate ion is CO_3^{2-}. (Section 5.12) Marble and eggshells both contain calcium carbonate, $CaCO_3$.

The notation *(aq)*, short for *aqueous*, represents a species dissolved in water.

The Lewis dot structure for the hydronium ion is

$$\left[\begin{array}{c} H \\ H:\overset{\cdot\cdot}{\underset{\cdot\cdot}{O}}:H \end{array} \right]^+$$

It obeys the octet rule.

Your Turn 6.2   Acidic Solutions

Here are four acids. For a molecule of each one, write a chemical equation to show the ionization (dissociation) of a hydrogen ion in aqueous solution. Use equation 6.1 as your model. *Hint:* Be sure to include the charges on the ions.

a.   HI (hydroiodic acid)
b. HNO_3 (nitric acid)
c. H_2SO_4 (sulfuric acid)
d. H_3PO_4 (phosphoric acid)

Answer

d. $H_3PO_4(aq) \longrightarrow H^+(aq) + H_2PO_4^-(aq)$

Consider This 6.3   Are All Acids Harmful?

Although the word *acid* may conjure up all sorts of pictures in your mind, every day you eat or drink certain types of acids. Check the labels of foods or beverages and make a list of the acids you find. For each acid, what do you think the function is?

6.2 What Is a Base?

No discussion of acids would be complete without mentioning their chemical opposites—bases. For our purposes, a **base** is any compound that produces hydroxide ions, OH^-, in aqueous solution. Bases have their own characteristic properties attributable to the presence of OH^-. Unlike acids, bases lack any appeal as food items as they generally taste bitter. When dissolved in water, bases have a slippery, soapy feel. Common examples of bases are aqueous solutions of the gas ammonia, NH_3, and of sodium hydroxide, NaOH, sometimes known as lye. The cautions on household drain cleaners that contain lye (Figure 6.4) give dramatic warning that some bases, like some acids, can cause severe damage to eyes, skin, and clothing.

We can visualize how hydroxide ions are produced in solution from sodium hydroxide, an ionic compound. When the solid NaOH dissolves in water, the sodium and hydroxide ions are released from the crystalline lattice.

$$NaOH(s) \xrightarrow{H_2O} Na^+(aq) + OH^-(aq)$$ [6.3]

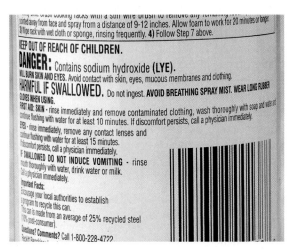

Figure 6.4
Oven cleaner contains NaOH.

Your Turn 6.4 | **Basic Solutions**

For these solids, write a chemical equation showing the release of hydroxide ions when each one dissolves in water.

a. KOH   **b.** LiOH   **c.** $Ca(OH)_2$

Answer

c. $Ca(OH)_2(s) \longrightarrow Ca^{2+}(aq) + 2\,OH^-(aq)$

Some bases are a bit more complicated. For example, the cleaning agent "household ammonia" is a basic solution. To make household ammonia, the gas ammonia, NH_3, is dissolved in water in an amount that makes about a 5% solution by mass (not very concentrated). We call the resulting solution "aqueous ammonia" and represent it as $NH_3(aq)$.

$$NH_3(g) \xrightarrow{H_2O} NH_3(aq) \tag{6.4a}$$

But there is more to the story. A hydrogen ion, H^+, is transferred to the aqueous ammonia molecule to form the ammonium ion, NH_4^+.

$$NH_3(aq) + H^+(aq) \longrightarrow NH_4^+(aq) \tag{6.4b}$$

Assuming that the H^+ in this equation came from a water molecule, the reaction of ammonia and water can be represented as the formation of NH_4OH, or ammonium hydroxide.

$$NH_3(aq) + H_2O(l) \longrightarrow NH_4OH(aq) \tag{6.4c}$$

The source of the hydroxide ion in household ammonia now should be obvious. Ammonium hydroxide dissociates to form the hydroxide ion (together with the ammonium ion). This reaction occurs only to a limited extent; that is, only tiny amounts of the two ions are formed in an aqueous solution of ammonia. Nonetheless, this is enough to produce a basic solution.

> The ammonium ion, NH_4^+ is formed in a manner analogous to the hydronium ion, H_3O^+.

6.3 Neutralization: Bases Are Antacids

Acids and bases react with each other. We illustrate this familiar process with hydrochloric acid and sodium hydroxide. If all the HCl and NaOH react, the products are sodium chloride and water.

$$HCl(aq) + NaOH(aq) \longrightarrow NaCl(aq) + H_2O(l) \tag{6.5}$$

This is called **neutralization,** a chemical reaction in which the hydrogen ions from an acid combine with the hydroxide ions from a base to form molecules of water. The formation of water can be represented by this equation.

$$H^+(aq) + OH^-(aq) \longrightarrow H_2O(l) \tag{6.6}$$

What about the sodium and chloride ions? Recall from equations 6.1 and 6.3 that HCl and NaOH completely dissociate into ions when dissolved in water. We can rewrite equation 6.5 to show this:

$$H^+(aq) + Cl^-(aq) + Na^+(aq) + OH^-(aq) \longrightarrow Na^+(aq) + Cl^-(aq) + H_2O(l) \tag{6.7}$$

Notice that the Na^+ ions and the Cl^- ions don't take part in the neutralization reaction. They remain unchanged on both sides of the equation. Canceling them produces equation 6.6.

> We use an acid–base neutralization reaction when we put lemon juice, which contains an acid, on fish. The acid neutralizes the ammonia-like compounds that produce a "fishy smell."

> Recall from Section 5.8 that NaCl is an ionic compound. It dissolves in water to produce two ions, Na^+ and Cl^-.

Your Turn 6.5   Neutralization Reactions

Write chemical equations showing the complete reaction of each acid and base pair. First write the balanced equation, and then rewrite it in ionic form.

a. $HBr(aq)$ and $Ba(OH)_2(aq)$

b. $H_2SO_4(aq)$ and $NaOH(aq)$

c. $H_3PO_4(aq)$ and $Mg(OH)_2(aq)$

Answer

a. $2\,HBr(aq) + Ba(OH)_2(aq) \longrightarrow BaBr_2(aq) + 2\,H_2O(l)$

$2\,H^+(aq) + 2\,Br^-(aq) + Ba^{2+}(aq) + 2\,OH^-(aq) \longrightarrow$
$$Ba^{2+}(aq) + 2\,Br^-(aq) + 2\,H_2O(l)$$

$$2\,H^+(aq) + 2\,OH^-(aq) \longrightarrow 2\,H_2O(l)$$

or by dividing both sides of the equation by 2 to simplify it:

$$H^+(aq) + OH^-(aq) \longrightarrow H_2O(l)$$

Neutral solutions are neither acidic nor basic, that is, they have equal concentrations of H^+ and OH^- ions. This condition exists in pure water. It also can exist in some salt solutions, such as when NaCl is dissolved in pure water. In contrast, in acidic solutions, the concentration of H^+ ions is greater than that of OH^- ions. As you might guess, the concentration of OH^- ions is greater than that of H^+ ions in basic solutions.

It may seem strange that acidic solutions contain OH^- ions and likewise that basic solutions contain H^+ ions. But when water is involved, it is not possible to have H^+ without OH^-, or vice versa. A simple, useful, and very important relationship exists between the concentration of hydrogen ions and the concentration of hydroxide ions in any aqueous solution.

$$[H^+][OH^-] = 1 \times 10^{-14} \qquad [6.8]$$

> The product $[H^+][OH^-]$ is temperature-dependent. The value of 1×10^{-14} holds true only at 25°C.

Take note of the square brackets. These indicate that the ions are expressed in the unit of molarity. For example, $[H^+]$ is read as "the concentration of hydrogen ion" and has a value such as 1×10^{-3} M. When $[H^+]$ and $[OH^-]$ are multiplied together, the product is a constant, with a value of 1×10^{-14} at 25°C, as shown in mathematical expression 6.8. This expression also tells us that the concentrations of H^+ and OH^- are linked to each other. When $[H^+]$ increases, $[OH^-]$ decreases. And when $[H^+]$ decreases, $[OH^-]$ increases. Both ions are always present in aqueous solutions.

Knowing the numerical value of either $[H^+]$ or $[OH^-]$, we can use expression 6.8 to calculate one from the other. For example, if a rain sample has a H^+ concentration of 1×10^{-5} M, then the OH^- concentration can be calculated. Substituting in 1×10^{-5} M for $[H^+]$ gives us

$$1 \times 10^{-5} \times [OH^-] = 1 \times 10^{-14}$$

$$[OH^-] = \frac{1 \times 10^{-14}}{1 \times 10^{-5}}$$

$$[OH^-] = 1 \times 10^{-9}\,M$$

Since the hydroxide ion concentration (1×10^{-9} M) is smaller than the hydrogen ion concentration (1×10^{-5} M), the solution is acidic.

> Acidic solution $[H^+] > [OH^-]$
> Neutral solution $[H^+] = [OH^-]$
> Basic solution $[H^+] < [OH^-]$

In pure water or a neutral solution, the molarities of hydrogen and hydroxide ions are equal, each with a value of 1×10^{-7} M. Applying mathematical expression 6.8, we can see that $[H^+][OH^-] = (1 \times 10^{-7})(1 \times 10^{-7}) = 1 \times 10^{-14}$.

Your Turn 6.6   Acidic and Basic Solutions

Classify these solutions as acidic, neutral, or basic at 25 °C. Then, for parts **a** and **c**, calculate $[OH^-]$. For **b**, calculate $[H^+]$.

a. $[H^+] = 1 \times 10^{-4}$ M
b. $[OH^-] = 1 \times 10^{-6}$ M
c. $[H^+] = 1 \times 10^{-10}$ M

Answer

a. The solution is acidic because $[H^+] > [OH^-]$.

$[H^+][OH^-] = 1 \times 10^{-14}$. Solving, $[OH^-] = 1 \times 10^{-10}$ M.

Your Turn 6.7   Ions in Acidic and Basic Solutions

Classify each solution as acidic, basic, or neutral. Then list all of the ions present in order of decreasing concentration, starting with the most abundant for each of these solutions.

a. $KOH(aq)$ **b.** $HNO_2(aq)$ **c.** $H_2SO_3(aq)$ **d.** $Ca(OH)_2(aq)$

Answer

d. The solution is basic, with far more OH^- than H^+ in solution.

$$OH^-(aq) > Ca^{2+}(aq) \gg H^+(aq)$$

When calcium hydroxide dissociates, two hydroxide ions are released for every calcium ion.

We now need a convenient way of reporting how acidic or basic a solution is. As you will see, the pH scale is just such a tool. It relates the acidity of a solution to its H^+ concentration.

6.4 Introducing pH

The term pH may already be familiar to you. Test kits for soils and for the water in swimming pools report the acidity in terms of pH. Shampoos claim to be pH-balanced (Figure 6.5). And, of course, articles about acid rain make reference to pH. The notation pH is always written with a small p and a capital H and stands for "power of hydrogen." In the simplest terms, **pH** is a number, usually between 0 and 14, that indicates the acidity of a solution.

A pH of 7.0, midpoint on the pH scale, separates acidic from basic solutions. Solutions with a pH of less than 7.0 are acidic; those with a pH greater than 7.0 are described as alkaline or basic; and those with a pH of 7.0 are neutral. Stomach acid is highly acidic and can have a pH less than 2. How does the acidity of milk compare with this? Check Figure 6.6 to see pH values for a number of common substances.

You may be surprised to learn how many acids we eat, drink, and produce through metabolism. The naturally occurring acids in foods contribute distinctive tastes. For example, the sharp taste of vinegar (pH $\cong$ 2.5) comes from acetic acid. The tangy taste of McIntosh apples, pH $\cong$ 3.3, is from malic acid. And the sour taste of lemons (pH $\cong$ 2.3) is from citric acid. Tomatoes are well known for their acidity, but in fact are usually less acidic (pH $\cong$ 4.5) than most fruits. Yogurt (pH $\cong$ 4.0) gets its sour taste from lactic acid, and cola soft drinks (pH $\cong$ 2.6) contain several acids, including phosphoric acid.

The pH of a sample may be lower than 0 (or higher than 14) for highly concentrated solutions of acids (or bases).

Figure 6.5

Shampoo that is "pH-balanced." Soaps tend to be basic, which can be irritating to the skin. The pH of some shampoos is adjusted to be closer to neutral.

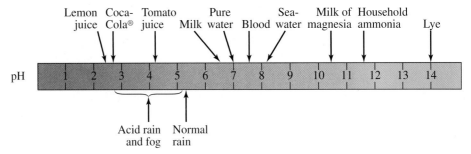

Figure 6.6
Common substances and their pH values.

Figures Alive! Visit the *Online Learning Center* to learn more about acids, bases and the pH scale.

Consider This 6.8   **Acidity of Foods**

a. Put vinegar, tomatoes, lemons, apples, cola, pure water, and yogurt in order of increasing acidity. See Figure 6.6.

b. List any four foods of your own choosing in order of increasing acidity. Then look up the actual pH values on the Web, using the link provided at the *Online Learning Center* or at a site of your own choosing.

c. Why do you suppose there are so few foods with pH greater than 7?

The mathematical relationship is $pH = -\log[H^+]$. More information can be found in Appendix 3.

For solutions in which the hydrogen ion concentration is some power of 10, the pH is the exponent with its sign changed. For example, if $[H^+] = 1 \times 10^{-3}$ M, then the pH is 3. Similarly, if $[H^+] = 1 \times 10^{-9}$ M, then the pH is 9. A more detailed description of the relation between pH and $[H^+]$ is provided in Appendix 3.

Two aspects of the pH scale may confuse you. First, as the pH value *decreases,* the acidity *increases.* For example, a sample of water with a pH of 5.0 is more acidic than one with a pH of 6.0. Second, as the pH *decreases* by one unit, the H^+ concentration *increases* by a factor of 10. For example, a solution in which $[H^+] = 0.0001$ M has a pH of 4. By contrast, a solution in which $[H^+] = 0.00001$ M has a pH of 5. This second solution is *less* acidic with only 1/10 the concentration of hydrogen ion as a solution of pH 4. This same pattern involving powers of 10 holds throughout the entire pH scale shown in Figure 6.7.

Note that "normal" rain is naturally slightly acidic, with a pH value between 5 and 6. **Acid rain** is more acidic than "normal" rain and has a lower pH value. In contrast, pure water is neutral and has a pH of 7.0 (Figure 6.6). So the obvious inference is that rain is not pure H_2O. You will soon see what "impurities" make all rain acidic.

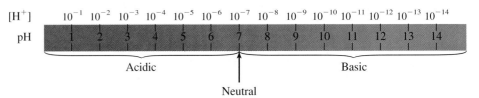

Figure 6.7
The relationship between pH and the concentration of the H^+ ion. As the pH increases, the $[H^+]$ decreases.

For each pair of pH values, state which value is more acidic and how much more acidic. Assume all sample sizes are the same.

a. A sample of rain with a pH of 5 and a sample of lake water with a pH of 4.

b. A tomato juice sample with a pH of 4.5 and a sample of milk with a pH of 6.5.

Answer

b. The juice with pH 4.5 is 100 times more acidic and has 100 times more H^+ than the milk with a pH of 6.5.

Consider This 6.10   On the Record

A legislator from a state in the Midwest is on record for making an impassioned speech in which he argued that the environmental policy of the state should be to bring the pH of rain all the way down to zero. Assume that you are a legislative aide to this legislator. Draft a tactful memo to your boss to save him from public embarrassment.

Having established the pH scale as a measure of acidity, we now apply the concept of pH to acid rain and its causes.

6.5   Measuring the pH of Rain

Rain is only one of several ways that acids can be delivered to Earth's surface and waters. Snow and fog are obviously others. A more inclusive term is acid deposition. **Acid deposition** includes wet forms such as rain, snow, fog, and cloud-like suspensions of microscopic water droplets often more acidic and damaging than acid rain. It also includes the "dry" forms of acids. For example, during dry weather, tiny solid particles (aerosols) of the acidic compounds ammonium nitrate and ammonium sulfate can settle on surfaces. This "dry deposition" has been shown to be almost as important as the wet deposition of the acids in rain, snow, and fog. These aerosols also contribute to haze, as we will see in Section 6.11.

Ammonium nitrate NH_4NO_3
Ammonium sulfate $(NH_4)_2SO_4$

The pH of a rain sample or of other aqueous solutions is usually determined with a pH meter. This device includes a special probe capped with an H^+-sensitive membrane that is immersed in the sample. The H^+ ions in a sample create a voltage across the membrane. The meter measures this voltage and converts it to pH, which is indicated on a dial or digital display (Figure 6.8).

In Section 8.4, you will see that hydrogen fuel cells also work by creating a voltage across a membrane.

Figure 6.8
A pH meter with a digital display.

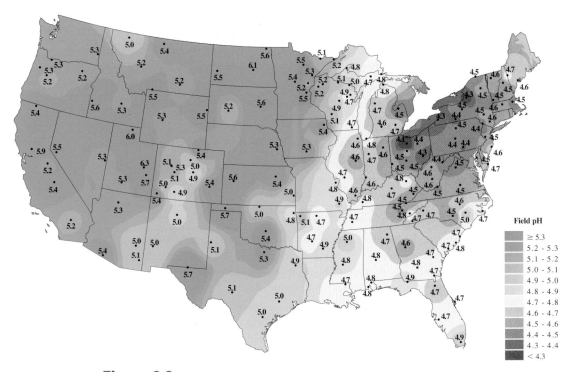

Figure 6.9

The pH of rain samples. Measurements made at the Central Analytical Laboratory, 2002.

Source: National Atmospheric Deposition Program National Trends Network. http://nadp.sws.uiuc.edu

It is fairly easy to measure the pH of rain samples, although certain precautions are necessary to obtain good results. The use of scrupulously clean collection containers is crucial, and the containers must be placed high enough to prevent "soil splash" that would contaminate rain samples collected at ground level. The pH meter also must be calibrated carefully with solutions of known pH to ensure that it is operating correctly.

Rain samples have been collected at selected sites in the United States and Canada since about 1970. A more systematic study has been under way since 1978, with over 225 sites at which weekly samples are collected. The pH is measured immediately, and then all samples are sent to a central laboratory in Illinois for further analysis. Figure 6.9 was prepared from such data. It is a map showing the pH of precipitation during 2002.

The different colored regions represent areas of the country with average pH values within a given range. This map contains a great deal of useful information, and we will return to it several times in this chapter.

From the data in Figure 6.9, it appears that all rain is at least slightly acidic. At first thought, this might seem surprising. If rain is pure water (as we tend to assume), we would expect it to be neutral and have a pH of 7.0. But pure, unpolluted rain always contains a small amount of dissolved carbon dioxide from air. Recall that CO_2 is a natural component of Earth's atmosphere present in low concentration, about 375 ppm or 0.0375%. Carbon dioxide dissolves to a slight extent in water and reacts with it to produce a slightly acidic solution containing H^+ and HCO_3^- ions:

$$CO_2(g) + H_2O(l) \longrightarrow H^+(aq) + HCO_3^-(aq) \qquad [6.9]$$

> Under high pressure, large amounts of carbon dioxide can be forced to dissolve in water. The result is carbonated water or "soda water" with a pH of about 4.7.

> $HCO_3^-(aq)$ is the hydrogen carbonate ion (bicarbonate ion). See Table 5.7.

The reaction shown in equation 6.9 occurs only to a limited extent; that is, only tiny amounts of H^+ and HCO_3^- are formed. But the hydrogen ions formed are enough to account for most of the acidity in "pure" rainwater. At 25 °C, a sample of water exposed to the normal atmospheric concentration of carbon dioxide has a pH of 5.6.

Figure 6.6 indicated that normal rain has a pH of about 5.3. It follows that CO_2 cannot be the sole source of H^+ in rainwater. Small amounts of other natural acids, including formic acid and acetic acid, are almost always present in rain and contribute to its acidity. However, even these additional acids cannot account for the fact that rain frequently may have a pH significantly below 5.3 (see Figure 6.9). We are now ready to search for the source of this extra acidity.

Consider This 6.11 **The Rain in Maine . . . or Ohio or Vermont**

All 50 states plus Puerto Rico and the Virgin Islands have at least one precipitation monitoring site, as part of the National Atmospheric Deposition Program/National Trends Network. Figure 6.10 depicts the sites in the Northeast. Some sites have been collecting data since the 1970s, and the data are posted on the Web. Search or use the link at the *Online Learning Center* to answer these questions.

a. How many monitoring sites are in your state? Select one and find out who operates it. What is a typical pH value at the site? What is the trend in acidity over the past few years?

b. How does the acidity of the precipitation in your state compare with that in another state? Make a prediction and then look up the data for this state. Again look at the trend over time.

Figure 6.10

The precipitation monitoring sites in the Northeast. Precipitation data for each site is available.

Source: National Atmospheric Deposition Program National Trends Network. http://nadp.sws.uiuc.edu

6.6 In Search of the Extra Acidity

According to Figure 6.9, the most acidic rain falls in the eastern third of the United States, with the region of lowest pH being roughly the states along the Ohio River valley. The extra acidity must be originating somewhere in this heavily industrialized part of the country. Analysis of rain for specific compounds confirms that the chief culprits are the oxides of sulfur and nitrogen: sulfur dioxide (SO_2), sulfur trioxide (SO_3), nitrogen monoxide (NO), and nitrogen dioxide (NO_2). These compounds are collectively designated SO_x and NO_x and often referred to as "sox and nox."

If this interpretation of the origins of acid precipitation is correct, the geographical regions with the most acidic rain should have the largest depositions of the oxides of SO_x and NO_x. Earlier we mentioned that acid deposition could be either wet or dry. Figure 6.11 shows wet deposition, usually called acid rain, but also includes the other forms of precipitation for which you might need an umbrella, such as snow, sleet, or even hail. On this graph, SO_x is reported in the form of the sulfate ion, for reasons you will see shortly in equations 6.10, 6.11, and 6.12. Similarly, NO_x is reported the form of the nitrate ion.

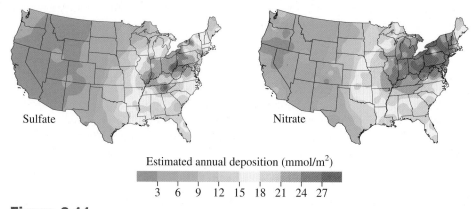

Figure 6.11

Annual Wet Deposition (estimated) 1995–1999.

Source: *EPA, Nitrogen: Multiple and Regional Impacts*, 2002, p. 10.

As you might suspect, sulfur dioxide emissions are highest in regions with many coal-fired electric power plants, steel mills, and other heavy industries that rely on coal. Allegheny County, in western Pennsylvania, is just such an area, and in 1990 it had the dubious distinction of leading the United States in atmospheric SO_2 concentration. Although power plants also generate nitrogen oxides, the highest NO_x emissions are generally found in states with large urban areas, high population density, and heavy automobile traffic. Therefore, it is not surprising that in 1990 (and still today) the highest levels of atmospheric NO_2 were measured over Los Angeles County, the car capital of the country. Figure 6.11 does not show these high levels of nitrate, because the deposition in the arid west is often dry, rather than wet. Nonetheless, the emissions are significant. For example, the vegetation in Joshua Tree National Park, east of L.A. County, is damaged by the dry deposition.

The circumstantial evidence linking acid precipitation with the oxides of sulfur and nitrogen appears compelling, but at this stage the Sceptical Chymist should be raising an important question. Given the definition of an acid as a substance that contains and releases H^+ ions in water, how can SO_2, SO_3, NO, and NO_2 qualify? These compounds don't contain hydrogen! The explanation is that SO_x and NO_x react with water to release H^+ ions. Although they are not acids themselves, the oxides of sulfur and nitrogen are **acid anhydrides,** literally "acids without water." When an acid anhydride is added to water, an acid is generated. For example, sulfur dioxide dissolves in water to form sulfurous acid.

Chemical equations 6.10 and 6.11 are analogous to the reaction of CO_2 with water.	

$$SO_2(g) + H_2O(l) \longrightarrow \underset{\text{sulfurous acid}}{H_2SO_3(aq)} \qquad [6.10]$$

Similarly, sulfur trioxide reacts with water to form sulfuric acid.

$$SO_3(g) + H_2O(l) \longrightarrow \underset{\text{sulfuric acid}}{H_2SO_4(aq)} \qquad [6.11]$$

In water, sulfuric acid is a source of H^+ ions.

$$H_2SO_4(aq) \longrightarrow H^+(aq) + HSO_4^-(aq) \qquad [6.12a]$$

The hydrogen sulfate ion (HSO_4^-) formed also dissociates to yield another H^+ ion.

Sulfate ion SO_4^{2-}
Hydrogen sulfate ion HSO_4^-

$$HSO_4^-(aq) \longrightarrow H^+(aq) + SO_4^{2-}(aq) \qquad [6.12b]$$

Adding equations 6.12a and 6.12b shows that sulfuric acid dissociates to yield two hydrogen ions and a sulfate ion.

Not all of the HSO_4^- ions are converted to SO_4^{2-} ions.

$$H_2SO_4(aq) \longrightarrow 2\,H^+(aq) + SO_4^{2-}(aq) \qquad [6.12c]$$

Your Turn 6.12   Sulfurous Acid

Write equations for the formation of two H^+ ions from sulfurous acid, analogous to chemical equations 6.12a, 6.12b, and 6.12c for sulfuric acid.

In a similar, but more complicated way, NO_2 can form nitric acid, HNO_3, when dissolved in water. Here, a molecule of oxygen is required.

$$4\,NO_2(g) + 2\,H_2O(l) + O_2(g) \longrightarrow \underset{\text{nitric acid}}{4\,HNO_3(aq)} \qquad [6.13]$$

Like sulfuric acid, nitric acid also dissociates to release the H^+ ion.

$$HNO_3(aq) \longrightarrow \underset{\text{nitrate ion}}{H^+(aq) + NO_3^-(aq)} \qquad [6.14]$$

Now that we see how oxides of sulfur and nitrogen contribute to acid rain formation, we need to get a closer look at how these oxides are formed and released into the atmosphere.

6.7 Sulfur Dioxide and the Combustion of Coal

Thus far, this chapter has established a relationship involving coal burning, atmospheric sulfur dioxide, and acid rain formation. Moreover, the fact that SO_2 and SO_3 react with water to yield acidic solutions is indisputable. What is not yet clear is why the combustion of coal should yield SO_2, the choking gas formed from burning sulfur (Figure 6.12). To answer this, we need to know something about the chemical nature of coal. At first glance, coal appears to be just a black solid, not very different from charcoal or black soot, both of which are essentially pure carbon. When carbon is burned, it forms carbon dioxide and liberates large amounts of heat (which of course is the reason for burning it).

$$C\text{(}in\ coal\text{)} + O_2(g) \longrightarrow CO_2(g) \qquad [6.15]$$

As you learned in Chapter 4, coal is a complex substance. No two coal samples have exactly the same composition. Although coal is not a pure chemical compound, we can approximate its composition with the formula $C_{135}H_{96}O_9NS$. In addition to these five elements, coal also contains small amounts of silicon and various metal ions such as sodium, calcium, aluminum, nickel, copper, zinc, arsenic, lead, and mercury. When coal is burned, oxygen reacts with *all* the elements present to form oxides of those elements. Because carbon and hydrogen are the most plentiful elements in coal, large quantities of gaseous CO_2 and H_2O are produced when coal is burned. But the element sulfur is of primary interest right now. The combustion of sulfur produces sulfur dioxide, a poisonous gas with an unmistakable choking odor.

$$S(s) + O_2(g) \longrightarrow SO_2(g) \qquad [6.16]$$

You may be wondering why coal contains sulfur and why the percentage of sulfur in coal varies. Coal formed 100–400 million years ago from decaying vegetation in swamps, peat bogs, or other areas. Because sulfur is present in all living things, when the ancient plants decayed, some sulfur was left behind in the material that eventually became coal. However, most of the sulfur in coal originated from the sulfate ion (SO_4^{2-}) naturally present in seawater. Millions of years ago, bacteria on sea floors utilized the sulfate ion as an oxygen source, removing the oxygen and releasing the sulfide ion (S^{2-}). In turn, the sulfide ion became incorporated into the ancient rocks (including coal) that were in contact with seawater. In contrast, the coal formed in freshwater peats has a lower sulfur content.

While coals from different regions can vary considerably in their sulfur content, the combustion of almost all coals produces sulfur oxides. This fact is central to the acid rain story. In large coal-burning electrical power-generating stations and industrial plants, the sulfur dioxide goes up the smokestack (unless control measures are used) along with the carbon dioxide, water vapor, and various metal oxides. Once in the atmosphere, SO_2 can react with more oxygen to form sulfur trioxide, SO_3.

$$2\,SO_2(g) + O_2(g) \longrightarrow 2\,SO_3(g) \qquad [6.17]$$

This reaction is fairly slow, but it is catalyzed (speeded up) by the presence of finely divided solid particles, such as the ash that goes up the stack along with the SO_2. Once SO_3 is formed, it reacts rapidly with any water vapor or water droplets in the atmosphere to form sulfuric acid (equation 6.11). A variety of other agents and pathways are available for the conversion of sulfur dioxide into sulfuric acid. Of particular importance are tropospheric ozone, hydrogen peroxide, H_2O_2, and the hydroxyl radical $\cdot OH$, which is formed from ozone and water in the presence of sunlight. The reaction of SO_2 with $\cdot OH$ accounts for 20–25% of the sulfuric acid in the atmosphere. The reaction goes faster in intense sunlight, and thus is more important in summer and at midday.

We can use a chemical calculation to better appreciate the vast quantities of SO_2 produced by coal-burning power plants. Such plants typically burn 1 million metric tons of coal a year, where 1 metric ton is equivalent to 1000 kg or 1×10^3 kg.

$$1 \times 10^6 \text{ metric tons coal/yr} = 1 \times 10^9 \text{ kg coal/yr} = 1 \times 10^{12} \text{ g coal/yr}$$

Figure 6.12
Sulfur burns to produce SO_2. If dissolved in water, this gas produces an acidic solution.

Burning coal releases compounds of mercury, arsenic and lead into the environment. These compounds are toxic.

In ancient times sulfur was known as brimstone, thus the biblical admonition about "fire and brimstone."

Sulfur's movement through the biosphere should remind you of the carbon cycle from Chapter 3.

Sulfur trioxide plays a role in aerosol formation, as we will see in Section 6.11.

Metric tons and short tons differ. Emissions data usually are either in metric tons (1000 kg, 2200 lb) or in short tons (2000 lb). To add to the confusion, short tons are sometimes simply called tons.

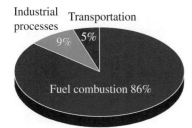

Figure 6.13

U.S. sulfur dioxide emission sources, 2002.

Source: EPA, Air Trends, http://www.epa. gov/air/airtrends/sulfur2.html

> See Section 3.7 for a review of molar mass and moles.

We will assume a low-sulfur coal that contains 2.0% sulfur; that is, 2.0 g sulfur per 100 g coal. First we can calculate the grams of sulfur released each year from one million metric tons (1×10^{12} g) of coal.

$$\frac{1 \times 10^{12} \text{ g coal}}{\text{yr}} \times \frac{2.0 \text{ g S}}{100 \text{ g coal}} = \frac{2.0 \times 10^{10} \text{ g S}}{\text{yr}}$$

Next, we use the fact that one mole of sulfur reacts with oxygen to form one mole of SO_2 (equation 6.16). The molar mass of sulfur is 32.0 g, and the molar mass of SO_2 is 64.0 g, that is, 32.0 g + 2(16.0 g). Therefore, 32.0 g of sulfur burn to produce 64.0 g of SO_2.

$$\frac{2.0 \times 10^{10} \text{ g S}}{\text{yr}} \times \frac{64.0 \text{ g } SO_2}{32.0 \text{ g S}} = \frac{4.0 \times 10^{10} \text{ g } SO_2}{\text{yr}}$$

This mass of SO_2 is equivalent to 40,000 metric tons or 88 million pounds of SO_2 per year. Power plants burning a higher sulfur coal emit more than twice this amount!

The connection between coal and sulfur dioxide emissions in the United States is evident in Figure 6.13. Most of the emissions arise from burning fossil fuels, chiefly coal, for the generation of electrical power. A small amount is produced from vehicles, because sulfur is present in small amount in some fuels. The coal burned in industrial processes contributes much of the remainder.

One other source of SO_2 emissions included in industrial processes is the production of metals. Both copper and nickel ores are sulfides, that is, compounds of the metal and sulfur. When nickel sulfide is heated to high temperatures in a smelter, the ore decomposes and sulfur dioxide is released. Although the large-scale production of nickel and copper contributes only a few percent to the total emissions, huge quantities of SO_2 are generated in particular regions. For example, the world's largest smelter in Sudbury, Ontario, converts nickel sulfide to nickel. The bleak, lifeless landscape in the immediate vicinity of the plant stands in mute testimony to earlier uncontrolled releases of SO_2. Today, after a major renovation in 1993, the two major smelters in the area have reduced their sulfur dioxide emissions substantially. Nonetheless, in 1997 over 250,000 metric tons of SO_2 still was released, some of it up a 1250-ft smokestack. The fact that this is the world's tallest smokestack (equal in height to the Empire State Building) simply means that the emissions were carried farther away from Sudbury by the prevailing winds (Figure 6.14). Lest we point any fingers, Canadians report that more than half of acid depositions in the eastern portion of their country originate in the United States. The quantity of sulfur dioxide that drifts northward over the border into Canada is estimated to be 4 million tons per year.

Your Turn 6.13   Coal Calculations

a. Assume the composition of coal can be represented by $C_{135}H_{96}O_9NS$. Calculate the fraction and percent (by mass) of sulfur in the coal.

b. A power plant burns one million (1×10^6) tons of coal per year. Assuming the sulfur content calculated in part **a,** calculate the number of tons of sulfur released per year.

c. Calculate the number of tons of SO_2 formed from this mass of sulfur.

d. Once it is released into the atmosphere, the SO_2 is likely to react with oxygen to form SO_3. What will happen next if this gas encounters water droplets?

Answers

a. 0.0168, or 1.68%   **b.** 1.68×10^4 tons S (16,800 tons)

Figure 6.14

The 1250-ft smokestack in Sudbury, Ontario, is the world's tallest.

6.8 Nitrogen Oxides and the Acidification of Los Angeles

The combustion of coal has been indicted as a major environmental offender, contributing sulfur dioxide to the atmosphere and to acid deposition. But SO_2 is not the only cause of acid precipitation, and another guilty party has been identified in California and other areas. The concentration of SO_2 in the smoggy air above the Los Angeles metropolitan area is relatively low, yet the pH of rain is still quite acidic. For example, in January 1982, fog near the Rose Bowl in Pasadena was found to have a pH of 2.5. Breathing it must have been like breathing a fine mist of vinegar. This level is at least 500 times more acidic than normal, unpolluted precipitation and 10 times more acidic than required to kill all fish in lakes. And in December 1982, fog at Corona del Mar, on the coast south of Los Angeles, was 10 times more acidic, with a registered pH of 1.5. In both cases, something other than sulfur dioxide was involved.

To solve the mystery, we turn to the cars and trucks that jam the Los Angeles freeways day and night. At first glance, it may not be obvious how these thousands of vehicles contribute to acid precipitation. Gasoline burns to form CO_2 and H_2O. But gasoline contains almost no sulfur, and therefore its combustion yields practically no SO_2. Consequently, we must look for another source of acidity.

Nitrogen oxides have already been identified as contributors to acid rain, but gasoline doesn't contain nitrogen. Therefore, logic (and chemistry) asserts that nitrogen oxides cannot be formed from burning gasoline. Literally, that is correct. Remember, however, that 78% of air consists of N_2 molecules. These molecules are remarkably stable and for the most part are unreactive. Thankfully, that is why nitrogen remains unchanged as we breathe it in and out of our lungs. Nevertheless, if the temperature is high enough, nitrogen can and does react directly with a few elements. One of these elements is oxygen. As we saw in Chapter 1, all that is needed is sufficient energy in the form of a high temperature or an electric spark. Under these conditions, the two elements combine to form nitrogen monoxide (nitric oxide), NO.

$$N_2(g) + O_2(g) \xrightarrow{\text{high temperature}} 2\,NO(g) \qquad [6.18]$$

Because air is a mixture of nitrogen and oxygen, it is always a potential source for the production of nitrogen monoxide. The energy necessary for the reaction can come from lightning bolts or from the "lightning" inside an internal combustion engine. In such an engine, gasoline and air are drawn into the cylinders and compressed. With the higher pressure, the N_2 and O_2 molecules are closer together and even more likely to react. Once ignited in the engine, the gasoline burns rapidly. The energy released powers the vehicle. But the unfortunate truth is that the energy also triggers reaction 6.18.

The reaction of N_2 with O_2 to form NO is not limited to lightning and automobile engines. The same reaction occurs when air is heated to a high temperature in the furnace of a coal-burning electrical power plant. Hence, such plants contribute vast amounts of *both* sulfur oxides and nitrogen oxides that acidify precipitation. On a national basis, the combustion of fuel (such as coal) in electrical utility plants and by industry releases just over a third of the nitrogen oxides (Figure 6.15). Transportation sources (including motor vehicles, aircraft, trains) account for over half. In urban environments, however, vehicles account for a far greater proportion of the atmospheric NO.

In the early 1990s, a green chemistry solution to reducing NO emissions and energy consumption was introduced into U.S. glass manufacturing by Praxair Inc. of Tarrytown, NY. Their award-winning technology substitutes oxygen for air in the large furnaces used to melt and reheat glass. Switching from air (78% nitrogen) to pure oxygen reduces NO production by 90% and cuts energy consumption by up to 50%. Glass manufacturers using the Praxair Oxy-Fuel technology save enough energy annually to meet the daily needs of one million Americans.

Once formed, nitrogen monoxide is very reactive. As we noted in Chapter 1, through a series of steps it reacts with oxygen, the hydroxyl radical, and volatile organic compounds (VOCs) to form NO_2.

$$VOC + \cdot OH \longrightarrow A + O_2 \longrightarrow A' + NO \longrightarrow A'' + NO_2 \qquad [6.19]$$

Again from Chapter 1, A, A′ and A″ represent reactive molecules with unpaired electrons (free radicals) that are synthesized from the VOC molecules. As you can see, the production of acid rain is connected to the same trace compounds in the atmosphere that you met in Chapter 1.

Recall from Chapter 1 that combustion is not always complete. Especially in older cars, some CO and unburned hydrocarbons escape in the exhaust.

Gasoline is a mixture of hydrocarbons. See Section 4.8.

In Section 6.12, you will learn how nitrogen-fixing bacteria are able to convert N_2 from the air to nitrogen-containing compounds such as NH_3. This is part of the nitrogen cycle.

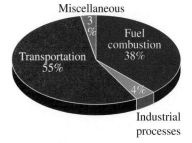

Figure 6.15

U.S. nitrogen oxide emission sources, 2001.

Source: EPA Report 454/R-00-002, *National Air Pollutant Emission Trends: 1900-1998*, Figure 2.2.

The hydroxyl radical, $\cdot OH$, was introduced in Section 1.11.

Nitrogen dioxide, NO_2, is a highly reactive, poisonous, red-brown gas with a nasty odor. For our purposes, its most significant reaction is the one that converts it to nitric acid, HNO_3. You saw a simplification of this conversion in equation 6.13. Actually, a series of steps is involved. These steps take place in the air above Los Angeles, Phoenix, and other sunny metropolitan areas. Sunlight is required. Again, VOCs (some released in the incomplete combustion of gasoline) and the hydroxyl radical, $\cdot OH$ (formed from ozone, another pollutant) are involved. The hydroxyl radicals rapidly react with nitrogen dioxide to yield nitric acid.

$$NO_2(g) + \cdot OH(g) \longrightarrow HNO_3(l) \qquad [6.20]$$

As you have already seen (equation 6.14), HNO_3 dissociates completely in water to release H^+ and NO_3^- ions. The result is the alarmingly low pH values occasionally found in Los Angeles' rain and fog.

6.9 SO_2 and NO_x—How Do They Stack Up?

Now that we have identified SO_2 and NO_x as the two major contributors to acid precipitation, we should examine their sources, both in this country and across the globe. Currently, their annual U.S. anthropogenic emissions are of roughly equal magnitude, about 20 million tons for SO_2 and about 24 million tons for NO_x. Most of the sulfur dioxide emissions can be traced to coal-burning electrical utility plants. But these same utilities account for only about a third of the nitrogen oxides released (see Figure 6.15). The combustion engines that power various forms of transportation emit more than half of the NO_x.

However, these pollutants were not always equal contributors. In years past, far less NO_x was present in rain, fog, and snow. Current NO_x values are the result of a relentless increase in emissions until about 1975, when the amounts leveled off. And unlike SO_2 emissions, today the nationwide NO_x emissions do not show the dramatic decreases resulting from caps and controls. Although SO_2 emissions have *decreased* about 40% since 1970, a tribute to the Clean Air Act and its amendments, NO_x emissions have not fared as well. Between 1970 and 1992, there was about a 15% *increase*. Only over the period 1992–2001 has a decrease been evident—a mere 3%. In the final two sections of this chapter, we will look more closely at the costs, the control strategies, and the politics that accompany these changes.

> 1 ton ("short ton") = 2000 lb = 0.9072 metric tons.

> The chemistry of NO in the atmosphere is complicated. NO can destroy ozone, as seen in Chapter 2. But remember from Chapter 1 that NO can react with O_2 to form NO_2. In turn, NO_2 can react in sunlight to produce ozone.

> **Your Turn 6.14** SO_2 and NO_x Emissions
>
> Figure 6.16 shows four sources of NO_x emissions in the United States. Which two did not change much between 1940 and 1995? Which two did? Similarly, for SO_2 in this same period, which sources led to the large increase in emissions in the 1970s?

How do U.S. emissions stack up against those of the rest of the world? We cannot easily answer this question for NO_x, as these emissions originate in large part from millions of small, unregulated, and mobile sources. In contrast, sulfur dioxide emissions can be estimated with the help of national data on fossil fuel consumption and on the refining of metal ores that contain sulfur, such as copper ore. To get an estimate, we start with the amount of fossil fuels (together with their sulfur content) produced in a country, then add in imports of fossil fuels, and finally subtract out exports. Metal refining is a bit trickier to estimate, as the amount of sulfur released depends on the technologies used (which are not always known). Nonetheless, using these types of data, a 1999 analysis estimated that in 1990, approximately half of the world's sulfur dioxide emissions came from the combined smokestacks of the United States, USSR, and China. European countries and Japan followed in turn.

But again, it is important to view the data over time. Not surprisingly, just as SO_2 emissions have dropped over the past decade in the United States, the same is true in

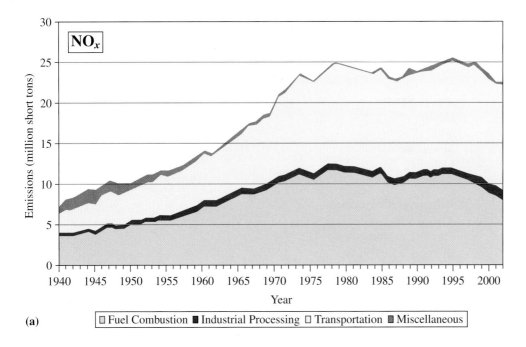

(a) ☐ Fuel Combustion ■ Industrial Processing ☐ Transportation ■ Miscellaneous

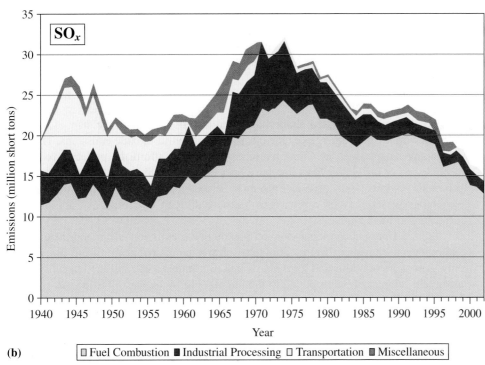

(b) ☐ Fuel Combustion ■ Industrial Processing ☐ Transportation ■ Miscellaneous

Figure 6.16

(a) U.S. nitrogen oxide emissions 1940–2002. **(b)** U.S. sulfur dioxide emissions 1940–2002. *Note:* Fuel combustion is fossil fuel combustion, such as coal. On-road vehicles include cars, vans, trucks, motorcycles and buses. Non-road sources include engines such as those in lawn mowers, tractors and boats.

Source: EPA/OAR, *National Air Pollutant Emission Trends, 1900–1998*, with recent data added.

both western Europe and the former Soviet Union. These decreases occurred for different reasons: environmental regulations in Europe in contrast to economic depression in Russia. But such decreases have been more than offset by a massive increase in SO_2 emissions by the rapidly developing countries. For example, in 1970, the United States emitted about 30 million tons of sulfur dioxide and China emitted 10 million tons. In

Table 6.1	**Estimated Global Emissions of Sulfur and Nitrogen Oxides**	
	SO$_2$*	**NO$_x$†**
Natural Sources		
Oceans‡	25	
Soil		5.6
Volcanoes	10	
Lightning		5.0
Subtotal	35	10.6
Anthropogenic Sources		
All sources	69	
Fossil fuel combustion		33
Biomass combustion		7.1
Aircraft		0.7
Subtotal	69	40.8
Total	104	51.4

Source: *Climate Change 2001: The Scientific Basis, Contribution of Working Group I to the Third Assessment Report of the Intergovernmental Panel on Climate Change,* Cambridge University Press, 2001, p. 315 and p. 260. Reprinted with permission.

* In units of 10^{12} g sulfur/year.

† In units of 10^{12} g nitrogen/year.

‡ Sulfur is emitted from oceans in the form of dimethyl sulfide rather than SO$_2$.

> Oceans contribute S to the atmosphere in the form of dimethyl sulfide. This compound is naturally converted to sulfur dioxide by the hydroxyl radical, •OH.

1990, both countries released about 22 million tons of SO$_2$. With the start of the year 2000, however, China emerged as the clear leader in sulfur dioxide emissions. Thus far, developing nations such as China have been unable to afford the pollution-reduction technologies or low-sulfur fuels that have been adopted by the wealthier nations. Nitrogen oxide emissions may pose an even more serious long-range problem. Although the technology exists, they have proven to be more difficult to control and appear to be increasing in most countries.

Table 6.1 presents a global view of SO$_2$ and NO$_x$ emissions from both natural and anthropogenic (human) sources. This table clearly points out that humans are not the only generators of sulfur and nitrogen oxides. But humans are still the largest contributors. The amount of sulfur added to the atmosphere by humans is twice that of natural sources such as volcanoes and oceans. The amount of nitrogen added as NO$_x$ by humans is roughly four times that of natural sources such as lightning and the bacteria found in soils. Interpret these data with care. Natural emissions are inherently variable and difficult to estimate. Research studies have shown that the tons of NO$_x$ formed by lightning vary widely by region (tending to be higher near the equator) and by month (higher during July in the north, January in the south). In addition, the local concentrations of NO$_x$ are subject to the updrafts and downdrafts of storms.

Occasionally, major geological events alter the pattern. The June 1991 eruption of Mount Pinatubo in the Philippines is a case in point. This eruption, the largest in a century, injected between 15 and 30 million tons of sulfur dioxide into the stratosphere. There the SO$_2$ reacted to form small droplets of sulfuric acid. For more than two years, much of this H$_2$SO$_4$ aerosol remained suspended in the atmosphere, reflecting and absorbing sunlight. The temporary drop in average global temperature observed in late 1991 and that continued through 1992 has been attributed to the effects of the Mount Pinatubo eruption. Indeed, when the cooling effects of the Mount Pinatubo eruption are included in the computer programs used to model global temperature changes, the predictions agree well with the observations, validating the models. Evidence also indicates that droplets and frozen crystals of H$_2$SO$_4$ formed as a

> NO$_x$ is produced by bacteria in soils. As mentioned earlier, this is part of the nitrogen cycle. Look for more about the nitrogen cycle in Section 6.12.

> The effect of Mount Pinatubo's eruption was also discussed in Section 3.9.

result of the eruption provided many new microsites for chemical reactions leading to the destruction of stratospheric ozone. Quite obviously, the topics of this text are tightly interwoven.

6.10 Acid Deposition and Its Effects on Materials

As we have already seen, much of the rain, mist, and snow in the United States is more acidic than unpolluted precipitation. On a regional basis, the acidity of precipitation has increased significantly since the Industrial Revolution, and few regions have escaped. In the worst cases, fog and dew can have a pH of 3.0 or lower. But does this all really matter? To answer this question, we need to know something about the effects of acid deposition and how serious they really are.

National studies can help us. During the 1980s the U.S. Congress funded a national research effort called the National Acid Precipitation Assessment Program (NAPAP). Over 2000 scientists were involved, with a total expenditure of $500 million. The project was completed in 1990 and the participating scientists prepared a 28-volume set of technical reports (NAPAP, *State of the Science and Technology*, 1991). Some of the material in the remainder of this chapter is drawn from the NAPAP report and from a report from a conference in 2001. This conference was entitled "Acid Rain: Are the Problems Solved?" and was sponsored by the Center for Environmental Information. Its purpose was to "put the acid rain problem squarely back on the forefront of the public agenda."

And we agree—acid rain should remain on the public agenda. One reason is the seriousness of the damage done by acid rain, as listed in Table 6.2. In this section, we focus on damage to materials such as metals, statues, and monuments. The effects of acid deposition on human health will be explored in the section that follows.

To begin our discussion about the damage done to metals, remember that metals make up most of the periodic table. Of the over 100 elements that we now know, about 80% are metals. Typically metals are shiny and silvery in appearance; at least, they are shiny before they have been tarnished or rusted by acid rain. Although acid rain (pH = 3–5) does not affect all metals, unfortunately iron is one that is affected.

The jewelry metals, gold, silver and platinum, are all too inert to react with acids in acid rain's pH range.

Your Turn 6.15 **Metals and Nonmetals**

Using the periodic table as your guide, classify these elements as metals or nonmetals and give the elemental symbol for each.

a. iron   **b.** aluminum   **c.** fluorine
d. calcium   **e.** zinc   **f.** oxygen

As you can tell by looking around you, iron plays an important role as a structural metal. Iron is the major component in steel. Bridges, railroads, and vehicles of all kinds depend on iron and steel. Rods of steel are used to strengthen concrete. In many parts of the country, decorative iron fences and iron latticework both ornament and protect city homes.

The problem is that iron rusts to form the familiar reddish brown material "iron oxide," or Fe_2O_3. The chemical reaction can be represented this way.

$$4\,Fe(s) + 3\,O_2(g) \longrightarrow 2\,Fe_2O_3(s) \qquad [6.21]$$

But iron directly combines with oxygen only if you heat or ignite it, such as with a sparkler on the Fourth of July. At room temperature, iron requires the presence of hydrogen ions to rust. The role of H^+ is evident in equation 6.22, which represents the first of a two-step process. Here, metallic iron, Fe, reacts with oxygen and H^+ to yield the Fe^{2+} ion.

$$4\,Fe(s) + 2\,O_2(g) + 8\,H^+(aq) \longrightarrow 4\,Fe^{2+}(aq) + 4\,H_2O(l) \qquad [6.22]$$

Table 6.2	Effects of Acid Rain and Recovery Benefits
Effects	**Recovery Benefits**
Human Health	
Sulfur dioxide and nitrogen oxides in the air increase deaths from lung disorders (asthma and bronchitis) and impair the cardiovascular system.	Fewer visits to the emergency room, fewer hospital admissions, and fewer deaths.
Surface waters	
Acidic surface waters decrease the survivability of animal life in lakes and streams. In more severe instances, acidity eliminates some or all types of fish and other organisms.	Lower levels of acidity in the surface waters, and a restoration of animal life in the more severely damaged lakes and streams.
Forests	
Acid deposition contributes to forest degradation by impairing trees' growth and increasing their susceptibility to winter injury, insect infestation, and drought. It also causes leaching and depletion of natural nutrients in forest soil.	Less stress on trees, thereby reducing the effects of winter injury, insect infestation, and drought. Less leaching of nutrients from soil, thereby improving the overall forest health.
Materials	
Acid deposition contributes to the corrosion and deterioration of buildings, cultural objects, and cars. This decreases their value and increases the costs of correcting and repairing damage.	Less damage to buildings, cultural objects, and cars, therefore lowering the cost in the future of correcting and repairing such damage.
Visibility	
In the atmosphere, sulfur dioxide and nitrogen oxides form sulfate and nitrate particles that impair visibility and affect enjoyment of national parks and other scenic views.	Reduced haze, therefore the ability to view scenery at a greater distance and with greater clarity.

Source: Adapted from Emission Trends and Effects in the Eastern U.S., United States General Accounting Office, Report to Congressional Requesters, March 2000.

Even pure water (pH = 7) has sufficient H^+ concentration to promote slow rusting. In the presence of acid, the rusting is greatly accelerated.

The Fe^{2+} produced by the reaction of iron with oxygen and hydrogen ions reacts with more oxygen to produce iron oxide:

$$4 \, Fe^{2+}(aq) + O_2(g) + 4 \, H_2O(l) \longrightarrow 2 \, Fe_2O_3(s) + 8 \, H^+(aq) \qquad [6.23]$$

Iron oxide, Fe_2O_3, is the familiar reddish brown material that we call rust.

Your Turn 6.16   **Rust Adds up**

Show that rust formation, as represented in equation 6.21, is the sum of equations 6.22 and 6.23.

Your Turn 6.17   Careful with the Charges

Fe, Fe^{2+}, and Fe^{3+} are three different chemical forms of iron, element number 26 on the periodic table (26 protons). These species are metallic iron, the $2+$ iron ion, and the $3+$ iron ion, respectively, and differ only in the number of electrons. Explain the differences in numbers of electrons.

Because iron is inherently unstable when exposed to the natural environment, enormous sums of money are spent annually to protect exposed iron and steel in bridges, cars, and ships. Paint is the most common means of protection, but even paint degrades, especially when exposed to acidic rain and gases. Coating the iron with a thin layer of a second metal such as chromium (Cr) or zinc (Zn) is another means of protection. Iron coated with zinc is called **galvanized iron.** Galvanized iron is still susceptible to the presence of acid rain. Because of acid rain, galvanized structures must be replaced more frequently than in the past.

Automobile paint also can be spotted or pitted by acid deposition. To prevent this, automobile manufacturers are using acid-resistant paints on new vehicles. It is an irony that automobiles emit NO, an air pollutant that forms acidic compounds that can attack the gleaming finish that many vehicle owners work so hard to maintain.

Acidic rain also damages statues and monuments. For example, those in the Gettysburg National Battlefield and in New York City parks are victims of irreparable damage (Figure 6.17). These statues are made of marble, a form of limestone composed mainly of calcium carbonate, $CaCO_3$. Limestone and marble slowly dissolve in the presence of H^+ ions.

$$CaCO_3(s) + 2\ H^+(aq) \longrightarrow Ca^{2+}(aq) + CO_2(g) + H_2O(l) \qquad [6.24]$$

Your Turn 6.18   Damage to Marble

Marble may contain magnesium carbonate as well as calcium carbonate. Write a chemical equation analogous to equation 6.24 for the reaction of acid rain with magnesium carbonate.

In 1944   At present

Figure 6.17

Acid rain damaged this limestone statue of George Washington that was first put outside in New York City in 1944.

Figure 6.18
Acid rain knows no geographic or political boundaries. Acid rain has eroded Mayan ruins
at Chichén Itzá, Mexico.

Your Turn 6.19   **Damage from SO_2**

Suppose that the acid represented in equation 6.24 by H^+ is sulfuric acid. Write the balanced chemical equation for the reaction of sulfuric acid on marble.

Visitors to the Lincoln Memorial in Washington learn that the huge stalactites
growing in chambers beneath the Memorial are the result of acid rain eroding the marble (calcium and/or magnesium carbonates). Other monuments and structures in the
eastern United States are suffering similar fates. Some limestone tombstones are no
longer legible. Worldwide, many priceless and irreplaceable marble and limestone statues and buildings are also being attacked by airborne acids (Figure 6.18). The
Parthenon in Greece, the Taj Mahal in India, and the Mayan ruins at Chichén Itzá
show signs of acid erosion. Much of the acid deposition at these sites is due to the
NO_x produced by traffic, including tour buses and numerous vehicles without emission control.

 Consider This 6.20   **Deterioration and Damage**

Reexamine Figure 6.17. Although it may be tempting to blame the damage on acid
rain, other agents act as well. View the different types of deterioration for yourself by
taking a photo tour of our nation's capitol, courtesy of the United States Geological
Survey. Locate the Web site either by searching for "USGS," "acid rain" and "capitol,"
or use the direct link at the *Online Learning Center*. What kinds of damage do the
photos show? What factors promote damage by acid rain? What else has caused the
deterioration?

 Consider This 6.21   **Acid Rain Across the Globe**

The issues and concerns of acid rain vary around the globe. Many countries in North
America and Europe have Web sites dealing with acid rain. Either locate one by searching or use the links provided at the *Online Learning Center*. What are the concerns in
the country you selected? Does part of the acid deposition originate outside the country
you picked?

6.11 Acid Deposition, Haze, and Human Health

The more obvious effects of acid deposition often can be observed simply by looking out the window. Anyone living in the eastern half of the United States is familiar with the summer haze that may settle over the landscape. (Ironically, you become more aware of it on the occasional clear day when it really does seem that you can see forever.) Jet passengers, as they peer down from 30,000 feet, notice that the features and colors of the landscape become a blur. And if you visit the Great Smoky Mountains National Park, you can view a prominent display of photographs of the haze (Figure 6.19).

The causes of haze are well understood. In the east, coal-burning power plants, such as those in the Ohio Valley, produce the smoke and particulate matter that produce the haze. In the west, other particulates (such as soil dust and the soot of wood-burning stoves) add to the haze produced by burning coal.

What you see is what you breathe. The power plants are emitting both NO_x and SO_x. Although both contribute to haze, for the purposes of illustrating acid deposition we will focus on the latter. As we mentioned earlier, coal contains a few percent sulfur, and when the coal is burned, a steady stream of sulfur dioxide is released into the atmosphere. Since sulfur dioxide is colorless, this gas is not what we are peering through as "haze." But, as you might guess, SO_2 is the precursor to this haze.

Let's focus on a molecule of SO_2 as it exits the tall smokestack of a power plant. As it moves downwind, it eventually forms an aerosol of sulfuric acid. The first step is the reaction of SO_2 with oxygen to form SO_3, as we saw earlier in equation 6.17. Sulfur trioxide is another colorless gas, but it has the property of being **hygroscopic,** that is, it seeks water and readily absorbs it. As we saw in equation 6.11, a molecule of SO_3 can react rapidly with a water molecule to form sulfuric acid.

> Aerosols consist of tiny particles that remain suspended in our atmosphere (Section 1.11).

Your Turn 6.22   **Droplets of Acid**

As a review of the sulfur chemistry just described, write chemical equations that start with the elemental sulfur in coal and show how (through several steps) sulfuric acid can be produced.

Visual range 20 miles   Visual range 100 miles

Figure 6.19

A hazy day and a clear day from Look Rock Tower in the Great Smoky Mountains National Park.

The tiny, tiny droplets of sulfuric acid then coagulate to produce larger droplets. These droplets form an aerosol with particles about a micrometer (1×10^{-6} m) in size. These particles of sulfuric acid absorb no sunlight. Rather, they scatter (reflect) sunlight, reducing visibility. The aerosols of sulfuric acid can travel hundreds of miles downwind, which is why the haze is so widespread. In fact, these fine particles of acid are stable enough that they enter our buildings and become part of the air that we breathe indoors as well as outdoors.

> Sulfuric acid aerosols are thought to persist for up to five days.

You may also hear of sulfate aerosols. Remember that sulfuric acid, H_2SO_4, contains both the hydrogen ion, H^+, and the sulfate ion, SO_4^{2-}. Both of these can be measured in an aerosol, and you will find both reported. But the sulfuric acid may react with bases to produce salts that contain the sulfate ion. Typically this base is ammonia, or in aqueous form, ammonium hydroxide. Thus, the aerosol particles may be a mixture of sulfuric acid, ammonium sulfate, $(NH_4)_2SO_4$, and ammonium hydrogen sulfate, NH_4HSO_4. By reporting the concentration of sulfate ion, rather than just the acidic concentration (pH), you get a better indication of how much sulfuric acid was initially present.

> Similarly, acidic aerosols of ammonium nitrate form from NO_x.

Your Turn 6.23   Sulfate Aerosols

As a review of the acid–base chemistry just described, write balanced chemical equations for these processes.

a.   The reaction of 1 mol of sulfuric acid with 1 mol of ammonium hydroxide to form ammonium hydrogen sulfate and water.

b.   The reaction of 1 mol of sulfuric acid with 2 mol of ammonium hydroxide to form ammonium sulfate and water.

> Each summer, wild fires also contribute to the haze seen over parts of the western part of the United States.

The haze is most pronounced in summer when there is more sunlight to accelerate the photochemical reactions leading to sulfuric acid. As a consequence of this haze, average visibility in the east is now about 20 miles, and occasionally as low as 1 mile. By contrast, visibility in the western states is now lessened from the natural visual range of about 200 miles to 100 miles or less. Where you formerly might have been able to see the mountains 100 miles away, these mountains may now have disappeared into the haze. National parks such as the Grand Canyon, Yellowstone, Glacier, Rocky Mountain, and Zion have all been affected. In the last few days of his presidency, Bill Clinton signed a bill authorizing the EPA to issue regulations to help clear the skies in national parks and wilderness areas. The bill required hundreds of older power plants that emit vast quantities of SO_2, NO_x, and particulates to retrofit their operations with pollution controls. As of 2004, President George W. Bush was not moving forward with this plan.

Consider This 6.24   Hazy in Yellowstone?

What is the latest news in the battle for clear skies? If you search for keywords such as "haze," "national parks," and possibly "new source review," you will be rewarded by a variety of documents. Find out:

a.   The basis of any current controversies
b.   Any new legislation to reduce emissions
c.   The policies of our current administration

In extreme cases, inhaling sulfate and sulfuric acid aerosols can cause illness and even death. The acidic droplets, when inhaled, attack sensitive lung tissue. The elderly, the ill, and those with preexisting respiratory problems such as asthma, emphysema, and cardiovascular disease are especially susceptible. People with preexisting bronchitis and pneumonia may exhibit increased mortality rates. Even those in good health experience irritation from the acidic aerosols.

(a)   (b)

Figure 6.20

(a) A 1948 news headline from Donora, PA. (b) Donora at noon during the deadly smog of 1948.

One of the worst recorded instances of pollution-related respiratory illness occurred in London in 1952. Periods of foggy bad air were nothing unusual to the British Isles, as factory chimneys had belched smoke into the air for several hundred years. But in December, 1952, the weather was colder than usual and people were burning large quantities of sulfur-rich coal in their home fireplaces. Due to unusual weather conditions, a deep layer of fog developed that trapped all the smoke and pollutants for five days, dropping visibility to practically zero. The deadly aerosol caused more than 4000 deaths, during its peak claiming 900 lives daily.

In 1948, a similar incident occurred in Donora, PA, a steel mill town south of Pittsburgh. Again a layer of fog trapped industrial pollutants close to the ground. By noon, the skies had darkened with a choking aerosol of fog and smoke (Figure 6.20). An 81-year-old fireman who took oxygen door-to-door to the victims reported, "It may sound dramatic or exaggerated, but you could barely see." High concentrations of sulfuric acid and other pollutants soon caused widespread illness. During the fog, 17 people died, to be followed by 4 more later. To be sure, Donora and London were both extreme and unusual situations from the past. But the U.S. EPA and the World Health Organization estimate that 625 million people are still being exposed to unhealthy levels of SO_2 released by burning fossil fuels.

Although acidic fogs such as these are immediately hazardous to health, concern is growing over the indirect effects of acid deposition. For example, the solubilities of some toxic heavy-metal ions, including lead, cadmium, and mercury, are significantly increased in the presence of acids. These elements are naturally present in the environment, but normally they are tightly bound in the minerals that make up soil and rock. Dissolved in acidified water and conveyed to the public water supply, these metals can pose serious health threats. Elevated concentrations of heavy metals have already been discovered in some of the major reservoirs in western Europe.

Clearly there is a connection between burning fossil fuels, acidic precipitation, and human health. An article written in the journal *Science* in 2001 by an international team of authors bluntly assessed the situation, "For every day that policies to reduce fossil-fuel combustion emissions are postponed, deaths and illness related to air pollution will increase." Studies by the EPA have estimated that the reductions in SO_2 and associated acid aerosols pollution called for by the Clean Air Act Amendments of 1990 could result in saving billions of dollars in health care costs over time. The savings would come principally from reduced costs to treat pulmonary diseases such as asthma and bronchitis and from a decrease in premature deaths. But there is another connection between acidic precipitation and humans that may be less obvious. To find it, we need to return to NO_x.

6.12 NO$_x$—The Double Whammy

A slice of pizza? A cool glass of lemonade? A green salad with oil and vinegar? Rarely a day goes by that you don't ingest calories in one form or another. Clearly, you must eat in order to stay alive. And clearly, men and women across the globe produce food by planting fields of grain, by harvesting fruits and vegetables by the truckload, and perhaps even by growing a bit of oregano or chives on a windowsill. To their credit, humans have become quite expert in raising both plants and animals. A complication, however, is that producing a pizza, just like driving a car (perhaps the one you used to pick up the pizza), adds to the acidity of the environment.

In earlier sections, we discussed the connection between energy production and the acidic emissions of SO$_x$ and NO$_x$. In this section, we will see that food production and NO$_x$ also are connected. This connection stems from a key difference between compounds of nitrogen and compounds of sulfur in the environment; namely, that nitrates act as fertilizers and promote plant growth. Actually plants depend on sulfur as well, as they do on many other elements such as carbon, hydrogen, phosphorus, and potassium. Except for nitrogen, however, these other elements tend to be readily available in the biosphere for uptake by plants. More than any other nutrient, nitrogen is the one in short supply and hence we often must add it in the form of fertilizers.

> All living things (not just plants) require nitrogen.

The Sceptical Chymist might wonder how N$_2$ can be in short supply when it makes up about four fifths of our atmosphere. The explanation lies in the chemical behavior of nitrogen. The nitrogen molecule, N$_2$, is not in a chemical form that can be used by plants. As pointed out in Chapter 1, nitrogen gas is very unreactive compared with oxygen gas.

Your Turn 6.25   Unreactive Nitrogen

To help your understanding of this section, please review the following information.

a. Nitrogen is a major constituent of our atmosphere. Approximately what percent is it?
b. The Lewis structure of N$_2$ has a triple bond. Draw it.
c. Compared with other bonds, is the triple bond in N$_2$ easy or hard to break?

> Some scientists designate reactive nitrogen as Nr, where the r stands for reactive. We do not use this representation in this book, as Nr resembles the chemical symbol for an element (and no element has this symbol).

In order to grow, plants need a more reactive form of nitrogen, such as the ammonium ion, ammonia, or the nitrate ion. These and other reactive forms of nitrogen are listed in Table 6.3. We refer to them collectively as **reactive nitrogen.** These compounds of nitrogen are biologically active, chemically active, or active with light in our atmosphere. As you might suspect, NO and NO$_2$ are among them. Some nitrogen compounds react so vigorously as to be explosive. All of these forms of nitrogen occur naturally and, until recently, all were present on our planet *in relatively small amounts.*

Table 6.3	Some Reactive Forms of Nitrogen	
Name	**Chemical Formula**	**Naturally Occurring?**
Nitrogen monoxide	NO	yes
Nitrogen dioxide	NO$_2$	yes
Nitrous oxide	N$_2$O	yes
Nitrate ion	NO$_3{}^-$	yes
Nitrite ion	NO$_2{}^-$	yes
Nitric acid	HNO$_3$	yes
Ammonia	NH$_3$	yes
Ammonium ion	NH$_4{}^+$	yes

Other forms of reactive nitrogen exist as well, but we will introduce these later when we need them for the study of polymers, proteins and DNA in later chapters.

Figure 6.21
Nodules on the root of a soya plant that contain nitrogen-fixing bacteria.

Your Turn 6.26   Reactive Nitrogen

Select three forms of reactive nitrogen from Table 6.3.

a. For each, write one or more chemical reactions that illustrate the reactive nature of the chemical. *Note:* In the case of ions, select a compound that contains the ion.
b. For each, draw the Lewis structure. *Hint:* Remember to add an extra valence electron for negatively charged ions and to subtract one for positively charged ions.
c. Do any of the structures from part **b** have unpaired electrons? What does this mean in terms of their reactivity?

Although we categorized N_2 as unreactive, one reaction involving the nitrogen molecule is of utmost importance: biological nitrogen fixation. Here, certain plants such as alfalfa, beans, and peas remove (or "fix") N_2 from the atmosphere (Figure 6.21). To be more accurate, it is not the plants themselves, but rather the bacteria living on or near the roots of these plants. As part of their metabolism, **nitrogen-fixing bacteria** remove nitrogen from the air and convert it to ammonia. If the ammonia dissolves in water, it releases the ammonium ion (equation 6.4b). This is one of two forms of reactive nitrogen that most plants can absorb. Here is the pathway:

$$N_2 \xrightarrow[\text{nitrogen fixation}]{} NH_3 \xrightarrow{H_2O} NH_4{}^+ \qquad [6.25]$$

Represents bacteria responsible for chemical change. Bacteria of the genus *Nitrosomonas* convert NH_3 to $NO_2{}^-$. Bacteria of the genus *Nitrobacter* convert $NO_2{}^-$ to $NO_3{}^-$.

The other form of reactive nitrogen that plants can absorb is the nitrate ion. **Nitrification** is the process of converting ammonia in the soil to the nitrate ion. Two different types of bacteria are involved and the pathway continues:

$$NH_4{}^+ \xrightarrow[\substack{\text{bacteria in} \\ \text{the soil}}]{} NO_2{}^- \xrightarrow[\substack{\text{bacteria in} \\ \text{the soil}}]{} NO_3{}^- \qquad [6.26]$$

Finally, to come full circle, **denitrification** occurs, that is, the process of converting nitrates back to nitrogen gas. Again, this is accomplished by bacteria. By so doing, these bacteria harness the energy available in forming the stable N_2 molecule. Depending on the soil conditions, the pathway back to N_2 may be stepwise, including NO and N_2O. Thus, these reactive forms of nitrogen may be released from the soil as well.

$$NO_3{}^- \xrightarrow[\substack{\text{bacteria in} \\ \text{the soil}}]{} NO \xrightarrow[\substack{\text{bacteria in} \\ \text{the soil}}]{} N_2O \xrightarrow[\substack{\text{bacteria in} \\ \text{the soil}}]{} N_2 \qquad [6.27]$$

N_2O, nitrous oxide, is the oxide of nitrogen that is emitted naturally in the greatest abundance. It is a potent greenhouse gas.

All of these pathways are part of the **nitrogen cycle,** a set of chemical pathways whereby nitrogen moves through the biosphere. Figure 6.22 assembles pathways 6.25, 6.26, and 6.27 into a simplified version of the nitrogen cycle. In this cycle, all species are forms of reactive nitrogen except for N_2.

But we need to get back to the story of acidification. Again, reactive forms of nitrogen are needed for plant growth. Since the bacteria in the soil cannot supply ammonia, the ammonium ion and/or the nitrates as fast as crops need them for growth, farmers use fertilizers. A few centuries ago, fertilizers were obtained by mining deposits of saltpeter (ammonium nitrate from the deserts of Chile) or by collecting guano, a nitrogen-rich deposit from bird and bat droppings in Peru. Neither source, however, was sufficient to meet the demand. An additional drain on the supply of nitrates was that they were used to make gunpowder and other explosives such as TNT. Thus, in the early 1900s, the search was on for a synthetic source of reactive nitrogen compounds.

The word *guano*, from the Inca civilization, means "the droppings of sea birds." Guano from Peru served as a fertilizer.

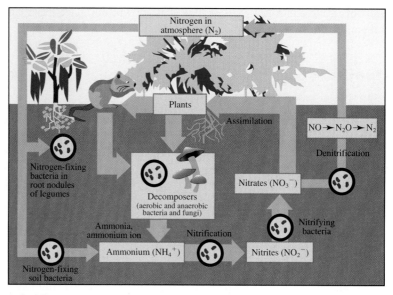

Figure 6.22

The nitrogen cycle (simplified).

How are fertilizers now obtained in the large quantities needed for present-day agriculture? The answer lies in a second important reaction of N$_2$, one that literally captures it out of the air to synthesize ammonia:

$$N_2(g) + 3\,H_2(g) \longrightarrow 2\,NH_3(g) \qquad [6.28]$$

This famous chemical reaction is known as the Haber–Bosch process. It allows the economical production of ammonia on a large scale, and hence the large-scale production of fertilizers (and nitrogen-based explosives). The large increase in reactive nitrogen from the Haber–Bosch process is represented by the green line in Figure 6.23.

But also notice the purple line on the graph. Clearly, the burning of fossil fuels is another very large source of reactive nitrogen in our environment. The high temperatures in engines and in electrical utility plants convert N$_2$ to NO. The top blue line for population, of course, is no surprise. The large increase in reactive nitrogen from burning fossil fuels (energy production) and the large increase from fertilization (food production) closely parallel the growth in world population (people production).

Now we can understand the double whammy of NO$_x$ emissions. The first problem, as we saw earlier, is that NO$_x$ emissions are increasing, or at very best leveling off. They contribute to acid deposition that in turn forms haze and diminishes air quality. The extra acidity also damages ecosystems and compromises human health. The oxides of nitrogen also form ground-level ozone in the presence of sunlight, contributing to photochemical smog, as we saw in Chapter 1.

The second problem is that NO$_x$ emissions are a form of reactive nitrogen, just like the fertilizers used for food production. Both NO$_x$ and fertilizers are disturbing the balances within the nitrogen cycle on our planet. In addition, the nitrogen players in this cycle continuously move from one position to another. Thus, the ammonia that starts out as a fertilizer may end up as NO, in turn increasing the acidity of the atmosphere and soil. Or the NO may end up as N$_2$O, a greenhouse gas that is currently rising in atmospheric concentration. Or the ammonium ion, instead of being tightly bound to the soil, may end up being leached out as the nitrite or nitrate ion, in turn contaminating a water supply. Thus, the drops of acid rain that fall can unleash a raging torrent of effects in the biosphere.

With too much reactive nitrogen, an ecosystem may become overloaded. The origin of the reactive nitrogen doesn't matter—it could be from acidic deposition or it could be from excess fertilization. Regardless of the source, the build-up of reactive nitrogen has devastating consequences for ecosystems. In the next section, we consider the effects of this excess in the context of acidic deposition in lakes.

Fritz Haber received the Nobel Prize in chemistry in 1918, and Carl Bosch in 1931.

Ammonia is either directly applied to the soil as a fertilizer or applied in the form of ammonium nitrate or ammonium phosphate.

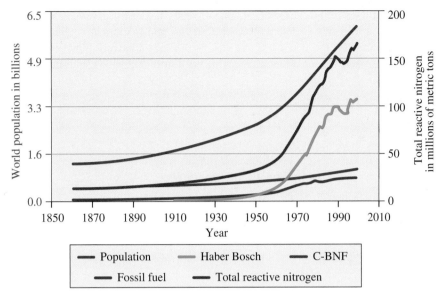

Figure 6.23

Global changes in reactive nitrogen (million metric tons, scale on the right). The top line is the world's population (billions, scale on the left).

Note: C-BNF is the reactive nitrogen created from the cultivation of legumes, rice, and sugarcane.

Source: *BioScience,* April 2003 vol. 53. No. 4 page 342.

6.13 Damage to Lakes and Streams

As mentioned earlier in Table 6.2, acidification of surface waters is another of the effects of acidic deposition. Healthy lakes have a pH of 6.5 or slightly above. As the pH is lowered below 6.0, fish and aquatic life are affected (Figure 6.24). Only a few hardy species survive below pH 5.0; and at pH 4.0, a lake is essentially dead.

Numerous studies have reported the progressive acidification of lakes and rivers in certain geographic regions, along with reductions in fish populations. In southern Norway and Sweden, where the problem was first observed, one fifth of the lakes no longer contain any fish, and half of the rivers have no brown trout. In southeastern Ontario, the average pH of lakes is now 5.0, well below the pH of 6.5 required for a healthy lake. In Virginia, more than a third of the trout streams are episodically acidic or at risk of becoming so.

Many areas of the Midwest have no problem with acidification of lakes or streams, even though the Midwest is a major source of acidic precipitation. This apparent paradox can be explained quite simply. When acidic precipitation falls on or runs off into a lake, the pH of the lake will drop (become more acidic) unless the acid is neutralized or somehow utilized by the surrounding vegetation. In some regions, the surrounding soils contain bases that can neutralize the acid. The capacity of a lake or other body of water to resist a decrease in pH is called its **acid-neutralizing capacity (ANC).** The surface geology of much of the Midwest is limestone, $CaCO_3$. As a result, lakes

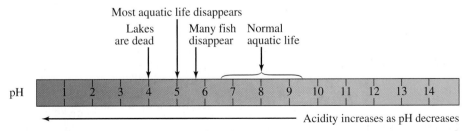

Figure 6.24
Aquatic life and pH.

in the Midwest have a high acid-neutralizing capacity because limestone slowly reacts with acid rain, as we saw earlier with marble statues and monuments (equation 6.24).

More importantly, the lakes and streams also have a relatively high concentration of calcium hydrogen carbonate as a result of reaction of limestone with carbon dioxide and water.

$$CaCO_3(s) + CO_2(g) + H_2O(l) \longrightarrow Ca^{2+}(aq) + 2\,HCO_3^-(aq) \qquad [6.29]$$
$$\text{calcium ion}\quad\text{hydrogen carbonate ion}$$

Your Turn 6.27   **The Bicarbonate Ion**

The product of equation 6.29, the hydrogen carbonate ion, can also accept a H^+ ion, thus acting to neutralize acids. Write the chemical equation.

Answer

$$HCO_3^-(aq) + H^+(aq) \longrightarrow H_2CO_3(aq) \longrightarrow CO_2(g) + H_2O(l)$$

Because acid is consumed by the carbonate and hydrogen carbonate ions, the pH of the lake will remain more or less constant.

In contrast to the Midwest, many lakes in New England and northern New York (as well as in Norway and Sweden) are surrounded by granite, a hard, impervious, and much less reactive rock. Unless other local processes are at work, these lakes have very little acid-neutralizing capacity. Consequently, many show a gradual acidification.

Experimental evidence indicates that fish populations most likely are affected through a chain of events that starts with acid rain and ends with the biological uptake of aluminum ions (Al^{3+}). Aluminum is the third most abundant element in Earth's crust (after oxygen and silicon). Granite contains aluminum ions, and soil includes complex aluminum silicate structures. Natural aluminum compounds have a very low solubility in water, but in the presence of acids, their solubilities increase dramatically. Thus, when the pH of a lake drops from, say, 6.0 to 5.0, the aluminum ion concentration in the lake may increase 1000-fold. When fish are exposed to a high concentration of aluminum ions, a thick mucus forms on their gills and the fish suffocate. Additionally, aluminum ions (Al^{3+}) react with water molecules to generate H^+ ions, increasing the acidity, which in turn dissolves more aluminum ions to further aggravate the problem.

$$Al^{3+}(aq) + H_2O(l) \longrightarrow H^+(aq) + Al(OH)^{2+}(aq) \qquad [6.30]$$

As it turns out, understanding the acidification of lakes is a good deal more complicated than simply measuring pH and acid-neutralizing capacities. One level of complexity is added by variations over the years. For example, in some years the heavy winter snowfalls persist into the spring and melt suddenly. As a result, the runoff may be more acidic than usual, as it contains all the acidic deposits locked away in the winter snows. A surge of acidity may enter the waterways at just the time when fish are spawning or hatching from eggs and are more vulnerable. In the Adirondacks, about 70% of the sensitive lakes are at risk for episodic acidification, in comparison to a far smaller percent that are chronically affected (19%). In the Appalachians, the number of episodically affected lakes (30%) is seven times those chronically affected.

Another level of complexity is added by build-up of reactive nitrogen species, such as the nitrate ion or the ammonium ion. **Nitrogen saturation** occurs when an area is overloaded with "nitrogen," that is, when the reactive forms of nitrogen entering an ecosystem exceed the system's capacity to absorb the nitrogen. The patterns of nitrogen absorption depend on both the age of the vegetation (in general, younger, growing forests absorb nutrients more than older ones) and the time of year (plant growth stops in the winter). But nitrogen absorption seems to have its limits. Once nitrogen saturation develops, the nitrate ion accumulates with an accompanying rise in acidity. As a result, the soils have little ability to neutralize acidic precipitation before it runs off into the lakes and streams.

When, if ever, will the lakes recover? The good news is that the SO_2 emissions have been declining in recent years, and we have seen a corresponding decrease in the sulfate ion concentrations in the lakes of the Adirondacks. However, even though NO_x emissions have remained fairly constant, the amount of nitrates in the Adirondacks is increasing in more lakes than not. Thus, it appears that nitrogen saturation has occurred in the surrounding vegetation, with more of the acidity ending up in the lakes. The soil in the region of these lakes most likely has lost some of its acid-neutralizing capacity.

Recent predictions are fairly gloomy. A March 2000 report to Congress puts it bluntly, "The lakes in the Adirondack Mountains are taking longer to recover than lakes located elsewhere and are likely to recover less or not recover, without further reductions of acid deposition." Even with the implementation of the 1990 Clean Air Act Amendments, more lakes are likely to become acidic, both in the Adirondacks and elsewhere.

6.14 Control Strategies

With the Clean Air Act Amendments of 1990, many hoped that the problems of acid rain would be solved. The Acid Rain Program, established as part of the Clean Air Act Amendments of 1990, made reducing NO_x and SO_2 emissions a national priority. Although as a nation we have made significant reductions, we are still challenged to clean up regions of polluted and acidic air.

For NO_x, the Acid Rain Program set a target of reducing the annual emissions by 2 million tons by 2000. Phase I of the NO_x program applied to about 170 coal-fired boilers that produce electricity, specifying an emission rate of either 0.50 or 0.45 pounds of NO_x per million Btu of heat input, depending on the type of boiler. Flexibility was built in, so that emission rates could be averaged over several units. Phase II began in 2000, tightening these emission standards and applying standards to still other types of boilers.

> Btu is a British thermal unit, the amount of heat needed to raise one pound of water one degree Fahrenheit.

In spite of these efforts, the goal for NO_x emissions has not yet been achieved. Although emissions by electrical utilities (generating about a quarter of the NO_x) declined, NO_x emissions increased elsewhere, such as by the increasing number of trucks and automobiles on our highways. Reduction of nitrogen oxides from these vehicles is particularly challenging, because as sources they are small, individually owned, and by design, mobile. And there are more than 200 million motor vehicles in the United States (about 1 billion worldwide). Of these, the biggest contributors to NO_x pollution continue to be diesel engines (Figure 6.25).

Consider This 6.28 Less Dirty Diesels?

In April 2003, the EPA announced a new plan to reduce emissions from 850,000 bulldozers, tractors, portable generators, and other "off road" diesel engines. According to EPA estimates, collectively, these sources account for about 12% of the NO_x emissions and over a third of the soot emissions. How successful are the current EPA efforts to reduce diesel emissions? Check the *Online Learning Center* for a link to the EPA.

Sceptical Chymist 6.29 Tractors and Cars

According to former EPA administrator, Christine Todd Whitman, a large bulldozer produces 800 pounds of pollution per year, the equivalent of 26 cars. From this, how many pounds of pollution is she crediting to each car per year? Which pollutants is a car producing and does her number seem to be within bounds? You may want to assume 10,000 miles driven per year at 20 miles per gallon as a basis for your calculations. *Hint:* Carbon dioxide is not considered a pollutant by the EPA.

To reduce NO_x, a variety of techniques bearing a range of price tags are in use. It is chemically possible to reduce the NO emitted by cars and trucks by fitting them with catalytic converters and other emissions control devices. We already mentioned one of the

Figure 6.25

Sooty exhaust from a diesel truck. *Note:* NO_x emissions are also present, but not visible.

Figure 6.26
Clean School Bus Program of
the EPA.

The Clean Coal Power Initiative
was discussed in Section 4.11.

functions of these catalysts: converting CO and unburned hydrocarbon fragments to CO_2. Other catalysts, typically in other parts of the catalytic converter, promote the reversal of the combination of nitrogen and oxygen that occurs in the engine at high temperatures. As the exhaust gases cool, the NO tends to decompose into its constituent elements.

$$2\,NO(g) \longrightarrow N_2(g) + O_2(g) \qquad [6.31]$$

Normally, this reaction proceeds slowly, but the appropriate catalyst can significantly increase its rate and thus decrease the amount of NO emitted. Currently, a program is underway at the EPA to fund cleaner school buses (Figure 6.26). Using catalysts is one of several strategies employed to reduce school bus emissions.

Coal-fired utility plants, another major source of NO_x emissions, demonstrate other new technologies. For example, the Clean Coal Technology (CCT) Demonstration Program has developed and installed low-NO_x burners on numerous coal-fired plants. These burners decrease the amount of air during the combustion process, so that with less oxygen present, less NO_x is produced. In 2003, the U.S. Department of Energy reported that low-NO_x burners are now on 75% of the coal-burning power stations. Another CCT option is "reburning," in which additional fuel is injected into the combustion products to strip away the oxygens from the NO_x. Both these new technologies are sufficiently complex that an artificial intelligence system may be used to optimize their performance. The successes in reducing emissions reported by one power station using new CCT technologies are shown in Figure 6.27.

 Consider This 6.30　CCT Demonstration Program

The CCT Program is funded both by government and industry and seeks technologies that meet the needs of our environment. What is new in coal-cleaning technology? Use the map at the Clean Coal Technology Compendium Web site to access a demonstration site of your choice. A link is provided the *Online Learning Center*.

The Acid Rain Program also called for a 10-million-ton reduction of SO_2 emissions by the year 2000. Phase I, begun in 1995, required 263 mostly coal-burning boiler units at 110 electrical utility power plants (located in 21 different states) to reduce their

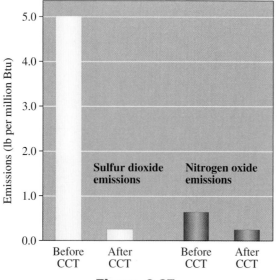

Figure 6.27
Changes in emissions at the Milliken Station power plant in Lansing, NY.
Note: CCT stands for clean coal technology.

Source: U.S. Office of Fossil Energy, http://www.fossil.energy.gov/programs/powersystems/index.html

emissions. Phase II, begun in 2000, further tightened the emissions on these plants. This phase also set further restrictions on power plants fired by natural gas and oil to encompass over 2000 boiler units. To date, the SO_2 emissions program has met with success. The fact that most anthropogenic SO_2 comes from a limited number of point sources (coal-burning power plants and factories) made the SO_2 problem easier to attack. As we already saw from Figure 6.16, great strides have occurred in reducing U.S. SO_2 emissions.

Three major strategies have been employed to decrease SO_2 emissions: (1) switch to "clean coal" with lower sulfur content, (2) clean up the coal to remove the sulfur before use, and (3) use chemical means to neutralize the acidic sulfur dioxide in the power plant. We briefly consider the effectiveness and the cost of each of these.

Coal switching is an option because coals vary widely in their sulfur content and their heat content. Anthracite, or "hard," coal, found mainly in Pennsylvania, yields the greatest amount of energy and the smallest amount of sulfur. But the anthracite supply is practically exhausted and more expensive. Bituminous, or "soft," coal, abundant in the Midwest, has nearly the same heat content as anthracite but usually contains 3–5% sulfur. Western states have enormous deposits of low-sulfur sub-bituminous coal and lignite (brown coal); however, this coal has a low heat content and may contain up to 40% water.

Coal cleaning is relatively easy to do, and the technology is available. The coal is crushed to a fine powder and washed with water so that the heavier sulfur-containing minerals sink to the bottom. But the process removes only about half of the sulfur, and it is expensive—from $500 to $1000 per ton of SO_2 eliminated.

An alternative to using coal switching or coal cleaning is to chemically remove the SO_2 during or after combustion in the power plant. The chief method for doing this is called *scrubbing*. The stack gases are passed through a wet slurry of powdered limestone, $CaCO_3$. Calcium carbonate neutralizes the acidic SO_2 to form calcium sulfate, $CaSO_4$.

$$2\,SO_2(g) + O_2(g) + 2\,CaCO_3(s) \longrightarrow 2\,CaSO_4(s) + 2\,CO_2(g) \qquad [6.32]$$

Limestone is cheap and readily available. Although the process is highly efficient, installing scrubbers is expensive, so that the cost of this method has been estimated at $400–600 per ton of SO_2 removed. Part of the expense is associated with the disposal of the $CaSO_4$ formed. We simply cannot avoid the law of conservation of matter. The sulfur must end up somewhere; either it goes up the stack as SO_2 or gets trapped as $CaSO_4$.

Consider This 6.31 Emissions Close to Home

Thanks to the EPA, you now can find the acid rain emissions data for the power plants in your state. Visit the EPA's Web site for Clean Air Market Programs or use the link provided at the *Online Learning Center.* Select a plant of your choice and report:

a. the name of the plant and the type(s) of fuel it burns
b. whether emissions controls are installed
c. the tons of SO_2 and NO_x emitted
d. the trend in emissions, by looking at previous years

The principal reason compliance with the 1990 Clean Air Act Amendments regulations was achieved and even bettered was coal switching, in which high-sulfur coal was replaced by low-sulfur coal. By the early 1990s, the use of a new rail carrier and favorable railway tariffs made vast deposits of cheaper low-sulfur coal (even less than 1% S) in Montana and Wyoming available at costs lower than that for midwestern or eastern low-sulfur coal. In 1991, western low-sulfur coal averaged just \$1.30 per million Btus; eastern low-sulfur coal was \$1.60–1.70 per million Btus. High-sulfur eastern coal cost \$1.35–1.55 per million Btus. Given this price advantage, it is not surprising that nearly 60% of SO_2 reduction came from switching to low-sulfur western coal rather than using more expensive alternatives, such as scrubbing.

But this conversion to low-sulfur coal has hidden costs. It ignores the social and economic impact on the states that produce high-sulfur coal. Since 1990, it has been estimated that coal switching has caused a 30% decline in employment in areas where high-sulfur coal is mined. This includes regions of Pennsylvania, Kentucky, Illinois, Indiana, and Ohio, although half of the drop can be attributed to automation and other market factors. Western states now produce nearly 33% of the coal mined in the United States, up from only 6% in 1970.

The shift to low-sulfur western coal has another side to it. Because the coal produces less heat per gram than eastern coals, power plants must burn more of it to generate the same amount of electricity. Burning more coal may release more pollutants. For example, mercury and other trace metals are more prevalent in coal from western states. If more coal is burned, more metals are released unless steps are taken to remove them before they go up the smokestacks (a costly proposition).

6.15 The Politics of Acid Rain

The neutralization of acid rain will require more than chemistry. As we have seen throughout this textbook, industrial leaders, state officials, politicians, and citizens across the nation are all important players in finding workable solutions. It may come as no surprise, then, that the electrical power industry and the producers of high-sulfur coal found tactics to resist acid rain legislation and controls. For example, a "grandfather" clause in the Clean Air Act of 1970 exempted certain older dirty plants from making improvements. This made sense, because these plants would have been expensive to retrofit with pollution controls, and presumably they would soon be replaced with newer plants. However, the power companies kept patching up the older plants rather than bringing newer, cleaner ones on line.

In 1977, Congress updated the Clean Air Act with an eye toward making industry bring these older plants into compliance. The New Source Review (NSR) clause was introduced that allowed the older, dirtier plants to continue to operate, but now required the emissions controls at the time when these plants were "substantially modified." A detailed and complex set of rules defined what fell into the category of substantial modifications.

Some newer and cleaner electrical utility plants were built, ones that burned natural gas instead of coal. But the emissions from the dirty, older coal-fired plants remained. To avoid the stricter standards of NSR, the power industry billed any work they did on the plants as "routine maintenance." To those on the receiving end of the pollutants, this appeared as an abuse of the New Source Review.

A clause in the Clean Air Act allows citizens to sue for violations. The attorney general of New York State (with help from the EPA) did just that, suing power plants in other states with the contention that much of the acidic pollution in New York was blowing in from places where controls were lax. In November 2000, a large, polluting Virginia power company lost a suit to New York and agreed to cut emissions from eight coal-burning plants. In 2003, the price tag was set at over \$1 billion to install the needed new technology. Christine Todd Whitman, former EPA administrator reported, "These settlement efforts show this administration's firm commitment to fully enforcing our environmental laws."

This commitment continues to be questioned. The Bush administration quietly moved the New Source Review to the sidelines in 2002 and 2003. President Bush offered new rules that allow "routine repairs and upgrades without enormous costs and endless disputes." For example, the Monroe Plant in Michigan operated by Detroit Edison is one of the nation's dirtiest. Speaking at this plant, President Bush pointed out that "We simplified the rules. We made them easy to understand. We trust the people in this plant to make the right decisions." Thus, at this point, the law suits have been put aside. In 2003, Christine Todd Whitman (whom we quoted in the previous paragraph) resigned as the head of the EPA.

A unique feature of the Clean Air Act Amendments of 1990 was to set up a national "cap and trade" system. The SO_2 emissions are capped to meet ever-decreasing environmental goals. For example, in 2001 the release of SO_2 was set at 10.6 million tons from electrical utilities; in 2010 the releases will be lowered to 8.95 million tons. In order to reach the environmental goal, each utility company operates with a *permit* that caps the pollution it can legally release per year. Exceeding this maximum carries fines of up to $25,000 per day.

The "trade" part of the cap and trade system works through a system of allowances. Companies are assigned emission *allowances* that authorize a source to emit one ton of SO_2, either during the current year or any year thereafter. At the end of a year, each company must have sufficient allowances to cover its actual emissions. If it has extra allowances, it can sell them or save them for a future year. If it has insufficient allowances, it must purchase them. Most of the allowance trading has taken place in the Ohio Valley.

An example of the cap and trade system is shown in Figure 6.28. With no controls, 20,000 tons is emitted from each of two units. Each one is capped at 10,000 tons, but one, through greater efficiency, performs better. The unit with emissions below its cap is assigned a credit for each ton of SO_2 saved. These credits can be sold to power plants that cannot efficiently meet their emission allowances. There is thus a financial incentive for power producers to achieve significant reductions of acidic oxide emissions. On the other hand, the purchase of credits by those who cannot yet meet the more stringent standards allows them to continue operation, at or below the permit level, while the plant works to reduce emissions.

The first official trade of emissions allowances under the provisions of the new law occurred in 1993. Since then, allowances have been bought and sold in private

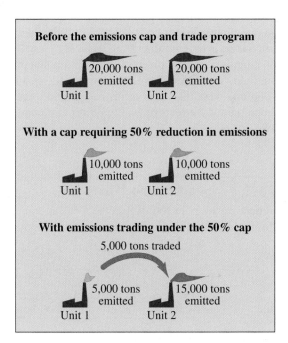

Figure 6.28

The emissions cap and trade concept.

Source: EPA, *Clearing the Air, The Facts about Capping and Trading Emissions,* 2002, page 3.

http://www.epa.gov/airmarkets/articles/clearingtheair.pdf

transactions and at public auctions. The Chicago Board of Trade even has a commodity trading market in emission allowances. Prices have ranged between $400 (1993) to $68 (1996) per allowance, nowhere near the $1000 predicted by utility officials. At the 2003 acid rain allowance auction, the lowest successful bid for allowances usable this year was $171, with the highest at $250. In his State of the Union message in 2002, President George W. Bush introduced a new Clear Skies Initiative that continues a mandatory cap and trade system. Bush commented that "a market-based cap and trade approach . . . rewards innovation, reduces costs, and guarantees results." However, as this book went to press, the Clear Skies Initiative had not been passed. In part, the resistance has come because the new caps proposed by President Bush were higher and thus allowed *more* emissions than the currently existing law. Eric Schaeffer, who headed civil enforcement for the EPA from 1997 to 2002, commented "We can do better under current law than what they're putting on the table."

Consider This 6.32   Up for Auction

The year 2005 marks the 13th annual auction for sulfur dioxide allowances conducted for the EPA by the Chicago Board of Trade (CBOT). How have allowances sales been going? You can learn more about emissions credits at the Web sites of both the EPA and the CBOT. For example, you can find recent information about allowance auctions and price trends at the EPA's site, "Acid Rain Program." Do some detective work on the Web and see if you can find out:

a.   Are the allowances more costly or less costly this year than last?
b.   How many allowances were auctioned last year?
c.   Are most companies still achieving compliance without having to buy credits?

A national emission trading program has not been finalized for NO_x. Since 1999, a NO_x Budget Program has been in operation in the northeast. Similar to the SO_2 emission trading program, an allowance can be bought, sold, or banked to emit one ton of NO_x. This program, however, is quite distinct from the acid rain reduction requirements and is aimed at alleviating tropospheric ozone, a secondary pollutant formed from NO_x in the presence of sunlight. The Clear Skies Initiative would set up two trading zones for NO_x—one east and one west. The caps in the east would represent greater reductions, reflecting the more harmful effects of acidity on the east coast. No trading would occur between the two zones.

Your Turn 6.33   Summer and Winter NO_x

An emissions cap and trade program for NO_x is likely to focus on summer months when NO_x emissions lead to the formation of ozone.

a.   As a review, write the chemical equations involving NO_x that show the formation of ozone. *Hint:* This reaction takes place in the presence of sunlight. See Chapter 1.
b.   Why do winter emissions of NO_x still need to be addressed? *Hint:* Review the episodic acidity of lakes caused by NO_x described in the previous section.

As these programs demonstrate, industry has been offered both flexibility and economic incentives to reduce its emissions. In turn, the public has been rewarded both with cleaner air and home electrical utility bills that have not risen significantly because of the Clean Air Act Amendments, even in eastern and midwestern states. Through a marriage of technology with economic forces, the average price charged for electricity by many large utilities has remained reasonably constant over the past decade.

CONCLUSION

If you have learned anything from this chapter, we hope it has been skepticism, prudence, and the recognition that complex problems cannot be solved by simple or simplistic strategies. "Acid rain" is not the dire plague once described by environmentalists and journalists. Nor is it a matter to be ignored. It is sufficiently serious that federal legislation, the Clean Air Act Amendments of 1990, has been enacted to reduce SO_2 and NO_x emissions, precursors to acid deposition. The Act already has helped to clean the air.

Current research indicates that acid deposition is a complex and tenacious problem, especially for certain watersheds and ecological niches. Any failure to acknowledge the intertwined relationships involving the combustion of coal and gasoline, the production of sulfur and nitrogen oxides, and the reduced pH of fog and precipitation is to deny some fundamental facts of chemistry. Knowledge of ecology and biological systems is needed as well, so that acid deposition can be understood in the context of entire ecosystems, a task that requires that experts from several disciplines work together.

One response that we as individuals and as a society might make to the problems of acid precipitation has hardly been mentioned in this chapter, yet it is potentially one of the most powerful. It is to conserve energy. Sulfur dioxide and nitrogen oxides are by-products of our voracious demand for energy—energy for electricity and energy for transportation. And, of course, carbon dioxide is an even more plentiful product. If our personal, national, and global appetite for fossil fuels continues to grow unchecked, our environment may well become a good deal warmer and a good deal more acidic, especially if developing countries fail to put into place environmental policies that restrict emissions from fossil fuel combustion. Moreover, the problem may be intensified as petroleum and low-sulfur coals are consumed and we become even more reliant on high-sulfur coal.

There are other sources of energy—nuclear fission, water and wind, renewable biomass, and the Sun itself. All of them are already being used, and their use will no doubt increase. We explore nuclear fission in the next chapter. But we conclude this chapter with the modest suggestion that, for a multitude of reasons, the conservation of energy by industry and collectively by individuals could have profoundly beneficial effects on our environment.

Chapter Summary

Having studied this chapter, you should be able to:

- Define and apply the definitions of acid and base (6.1–6.3)
- Use chemical equations to represent the dissociation of acids and bases (6.1–6.2)
- Write neutralization reactions for acids and bases (6.3)
- Describe solutions as acidic, basic, or neutral based on their pH or concentrations of H^+ and OH^- (6.3–6.4)
- Calculate pH values given hydrogen or hydroxide ion in whole-number concentrations (6.4)
- Describe the differences between the pH of water, the pH of ordinary rain, and the pH of acid rain, and locate on a map of the United States where the most acidic rain falls.
- Explain the role of sulfur oxides and nitrogen oxides in causing acid rain (6.7–6.8)
- List the different sources of NO_x and of SO_2 and explain the variations in the levels of these pollutants over the past 30 years (6.9)
- Explain how acidic aerosols are produced and their effects on building materials and on human health (6.10–6.11)
- Explain why N_2 is so inert, and describe how nitrogen converts between different reactive forms on our planet, including the role of microbes in nitrogen fixation (6.12)

- Describe how the industrial production of ammonia and the acidic deposition of nitrates both contribute to the build-up of reactive nitrogen on our planet (6.12)

- Describe nitrogen saturation and its consequences for lakes (6.13)

- Discuss the 1990 Clean Air Act Amendments and the cap and trade program. Describe the impact these continue to have on SO_2 emissions (6.13–6.14)

- Describe how NO_x emissions have been controlled differently from SO_2 emissions (6.13)

- Outline different ways to control acid rain, noting the cost–benefit considerations involved (6.13–6.14)

- Explain why acid rain control continues to be a politically sensitive issue (6.15)

Questions

Emphasizing Essentials

1. **a.** Give the names and chemical formulas for any four acids. Do not include any of those listed in question 3.

   b. Name three properties associated with acids.

2. **a.** Rewrite equation 6.1 using Lewis structures.

   b. Rewrite equation 6.1 using sphere representations.

3. Write a chemical equation showing the release of one hydrogen ion from a molecule of each of these acids.

   a. HBr (hydrobromic acid)

   b. H_2SO_3 (sulfurous acid)

   c. $HC_2H_3O_2$ (acetic acid)

4. **a.** Give names and chemical formulas for four bases. Do not include any of those listed in question 6.

   b. Name three properties associated with bases.

5. **a.** Rewrite equation 6.4 using Lewis structures.

   b. Rewrite equation 6.4 using sphere representations.

6. Write a chemical equation showing the release of ions as each base dissolves in water.

   a. RbOH(s)

   b. $Ba(OH)_2(s)$

7. Give chemical formulas for the nitrate ion, the sulfate ion, the carbonate ion, and the ammonium ion.

8. For each of the ions in the previous question, write a chemical equation where the ion (in aqueous form) appears as a product.

9. Write a balanced chemical equation for each acid–base reaction.

   a. Potassium hydroxide is neutralized by nitric acid.

   b. Hydrochloric acid is neutralized by barium hydroxide.

   c. Sulfuric acid is neutralized by ammonium hydroxide.

10. Classify each of these aqueous solutions as acidic, neutral, or basic.

    a. HI(aq)   **b.** NaCl(aq)

    c. NH_4OH(aq)   **d.** $[H^+] = 1 \times 10^{-8}$ M

    e. $[OH^-] = 1 \times 10^{-2}$ M   **f.** $[H^+] = 5 \times 10^{-7}$ M

    g. $[OH^-] = 1 \times 10^{-12}$ M

11. Refer back to the previous question. For parts **d** and **f,** calculate the $[OH^-]$ that corresponds to the given $[H^+]$. Similarly, for parts **e** and **g,** calculate the $[H^+]$.

12. Again referring back to question 10, calculate the pH for the concentrations given in parts **d–g.**

13. What is the difference in the $[H^+]$ between:

    a. a solution of pH = 6 and a solution of pH = 8?

    b. a solution of pH = 5.5 and a solution of pH = 6.5?

    c. a solution in which $[H^+] = 1 \times 10^{-8}$ M and one in which $[H^+] = 1 \times 10^{-6}$ M?

    d. a solution where $[OH^-] = 1 \times 10^{-2}$ M and one in which $[OH^-] = 1 \times 10^{-3}$ M?

14. Mammoth Cave National Park in Kentucky is in close proximity to the coal-fired electric utility plants in the Ohio Valley. Noting this, the National Parks Conservation Association (NPCA) reports that this national park has the poorest visibility of any in the country.

    a. What is the connection between coal-fired plants and poor visibility?

    b. The NPCA reports "the average rainfall in Mammoth Cave National Park is 10 times more acidic than natural." From this information and that in your text, estimate the pH of rainfall in the park.

15. You just purchased a new car and are worried about whether its paint will be damaged by acid rain. Consult Figure 6.9 to find the data necessary to answer these questions.

    a. Which of these cities—Chicago, Atlanta, Seattle, or San Francisco—would have the least damaging atmosphere for your car? Why?

    b. How does the average pH of rain for each of those cities compare with the average pH of rain where you live?

16. Forget the car from the previous question. Instead, suppose you have a new mountain bike and somebody just accidentally spilled a can of carbonated soft drink all over the metallic handle bars and paint.

    a. Soft drinks are more acidic than acid rain. About how many times more acidic? *Hint:* Consult Figure 6.6.

    b. In spite of the higher acidity, this spill is unlikely to damage your handle bars and paint (although the sugar probably isn't great on your gears). Why don't you have to worry about damage?

17. Write a balanced chemical equation for the chemical reaction involving sulfur shown in Figure 6.12.

18. The text states that the reaction of SO_2 with an $\cdot OH$ free radical accounts for 20–25% of the sulfuric acid in the atmosphere.

    a. Write a balanced chemical equation for this reaction.

    b. What is the source of free radical in the atmosphere?

19. Give the formulas for the acid anhydride of each of these acids.

    a. carbonic acid, H_2CO_3

    b. sulfurous acid, H_2SO_3

20. Assume that coal can be represented by the formula $C_{135}H_{96}O_9NS$.

    a. What is the percent of nitrogen by mass in coal?

    b. If 3 tons of coal were burned completely, what mass of nitrogen in NO would be produced? Assume that all of the nitrogen in the coal is converted to NO.

    c. Would the mass of nitrogen in NO calculated in part **b** be the *maximum* amount of nitrogen in NO produced in this combustion reaction? Why or why not?

21. The United States burned 980 million tons of coal in 2002. Assuming that this coal was 2% sulfur by weight, how many tons of sulfur dioxide was emitted?

22. Acid rain can damage marble statues and limestone building materials. Write a balanced chemical equation to represent this destruction.

23. **a.** What does the phrase "pH-balanced" imply on the label of a shampoo bottle?

    b. Will the presence of the phrase "pH-balanced" influence your decision to buy a particular shampoo? Why or why not?

24. Several gases are associated with exhaust from jet engines including CO, CO_2, O_3, NO, NO_2, SO_2, and SO_3.

    a. Of these gases, which would you expect jet engines to emit *directly*?

    b. Of these gases, which would you expect to form secondarily, that is, as a result of the emissions of part **a?**

25. Figure 6.13 offers information about the percentage of SO_2 emissions from fuel combustion, mainly for electrical power production, and the percentage of SO_2 produced by transportation. Figure 6.15 offers information about the percentage of NO_x from fuel combustion and the percentage of NO_x produced by transportation. Is the relative importance of fuel combustion and transportation the same for emissions of both SO_2 and NO_x? Why might they differ?

26. Almost equal *masses* of SO_2 and NO_x are produced by human activities in the United States.

    a. How does their production compare based on a *mole* basis? Assume that all the NO_x is produced as NO_2.

    b. Suggest reasons why the U.S. percentage of global emissions is greater for NO_x than for SO_2.

27. Reactive nitrogen compounds affect the biosphere both directly and indirectly (through other compounds that they form).

    a. Name a direct effect of reactive nitrogen compounds that is a benefit.

    b. Name two direct effects of reactive nitrogen compounds that are harmful to human health.

    c. Ozone formation is a harmful *indirect* effect. Explain the connection between reactive nitrogen compounds and the formation of ozone.

28. Calculate the mass of $CaCO_3$ (in tons) necessary to react completely with 1.00 ton of SO_2 according to the reaction shown in equation 6.32.

29. A garden product called dolomite lime is composed of tiny chips of limestone that contain both calcium and magnesium carbonates. This product is "intended to help the gardener correct the pH of acid soils," as it is "a valuable source of calcium and magnesium."

    a. Write the chemical formulas for magnesium carbonate and calcium carbonate. In which form is the calcium in this compound, calcium ion or calcium metal?

    b. Write a chemical equation that shows why limestone "corrects" the pH of acid soils.

    c. Will the addition of dolomite lime to soils cause the pH to rise or fall?

    d. Plants such as rhododendrons, azaleas, and camellias should not be given dolomite lime. Why?

30. **a.** The Clean Air Act has been discussed in this chapter, the Montreal Protocol in Chapter 2, and the Kyoto Accord in Chapter 3. What principal issue is each of these important pieces of legislation or international agreements trying to address?

    b. Place each of these important legislative or international agreements, together with any significant amendments, on an appropriate time line. You may choose any format for the time line, but it should communicate the maximum amount of information.

Concentrating on Concepts

31. James Galloway, the professor quoted at the start of the chapter, writes "Human activity is not making the world acidic, rather it is making the world *more* acidic."

   a. Explain why the world is naturally acidic.

   b. Explain why humans are making the world more acidic.

   c. One large part of the natural world is basic. Which is it? *Hint:* Consult Figure 6.6.

32. Judging by the taste, do you think there are more hydrogen ions in a glass of orange juice or in a glass of milk? Explain your reasoning.

33. The formula for acetic acid, the acid present in vinegar, is commonly written as $HC_2H_3O_2$. Many chemists write the formula as CH_3COOH.

   a. Draw the Lewis structure for acetic acid.

   b. Show that the two formulas both represent acetic acid.

   c. What are the advantages and disadvantages of each formula?

   d. How many hydrogen atoms can be released as hydrogen ions per acetic acid molecule? Explain.

34. Television and magazine advertisements remind us about the need for antacid tablets. A friend suggests that a good way to get rich quickly will be to market "antibase" tablets. Explain to your friend the purpose of an antacid tablet and offer some advice about the potential success of the "antibase" tablets.

35. In Your Turn 6.7, you listed the ions present in aqueous solutions of acids, bases, and common salts. Now add water, a molecular species, to this list.

   a. List all molecular and ionic species in order of decreasing concentration in a 1.0 M aqueous solution of NaOH.

   b. List all molecular and ionic species in order of decreasing concentration in a 1.0 M aqueous solution of HCl.

36. Which of these has the *lowest* concentration of hydrogen ions: 0.1 M HCl, 0.1 M NaOH, 0.1 M H_2SO_4, pure water? Explain your answer.

37. Explain why rain is naturally acidic, but not all rain is classified as "acid rain."

38. Taken together, do Figures 6.9 and 6.11 prove that emissions of SO_2 and NO_x cause the pH of rain to drop (a *causal relationship*)? Or is there merely a *correlation* between the pH drop and the emissions? Explain your answer. *Hint:* For a discussion of cause and correlation, see question 3.31. You will find examples from everyday life.

39. Ozone in the troposphere is an undesirable pollutant, but stratospheric ozone is beneficial. Does nitric oxide, NO, have a similar dual personality in these two atmospheric regions? Explain your thinking. *Hint:* Consult Chapter 2.

40. The mass of CO_2 emitted during combustion reactions is much greater than the mass of NO_x or SO_x, but there is less concern about the contributions of CO_2 to acid rain than from the other two oxides. Suggest two reasons for this apparent inconsistency.

41. The average pH of precipitation in New Hampshire or Vermont is relatively low, even though these states have low levels of vehicular traffic and virtually no industry that emits large quantities of air pollutants. How do you account for this low pH?

42. Equation 6.19 shows NO reacting to form NO_2 in the atmosphere, involving intermediate species A′ and A″. Here are examples of A′ and A″.

   a. What does the dot (·) represent in each structure? Draw in the other electrons around the atom with the dot (both bonding and nonbonding) to show that this atom does not have an octet of electrons.

   b. Name a chemical property that A′ and A″ have in common.

43. a. Efforts to control air pollution by limiting the emission of particulates and dust can sometimes contribute to an increase in the acidity of rain. Offer a possible explanation for this observation. *Hint:* These particulates may contain basic compounds of calcium, magnesium, sodium, and potassium.

   b. In Chapter 2, stratospheric ice crystals in the Antarctic were involved in the cycle leading to the destruction of ozone. Is this effect related to the observations in part **a**? Why or why not?

44. a. Several strategies to reduce SO_2 emissions are described in the text. The most effective ones in the last 10 years have been coal switching and stack gas scrubbing. Prepare a list of the advantages and disadvantages associated with each of these methods.

   b. Explain why coal cleaning has not been an effective strategy.

45. Discuss the validity of the statement, "Photochemical smog is a local problem, acid rain is a regional one, and the enhanced greenhouse effect is a global one." Describe the chemistry behind each of these air-quality problems and explain why the problems affect different geographical areas.

Exploring Extensions

46. The text makes this statement. "By trying to use a technological fix for one problem, we inadvertently create another."

   a. Explain how the problems associated with the build-up of reactive nitrogen in the environment fit this statement.

b. Pick another example from any of the issues explored in Chapters 1–5. Briefly explain how your choice fits the statement as well.

47. Here are two substances that both contain OH in their chemical formulas. Explain why you cannot write a reaction similar to equation 6.3 for either one.

a. $Al(OH)_3$ *Hint:* Consult a solubility table.

b. C_2H_5OH *Hint:* Consider bond energy and the nature of the bonds.

48. In Your Turn 6.7, you listed ions present in aqueous solutions of acids, bases, and common salts. In question 35, you added molecular substances to the list. To quantify this list,

a. Calculate the molar concentration of all molecular and ionic species in a 1.0 M solution of NaOH.

b. Calculate the molar concentration of all molecular and ionic species in a 1.0 M solution of HCl.

49. A representative of the Electric Power Research Institute, making a presentation in a workshop to establish research priorities and criteria on factors that govern precipitation chemistry, made this statement: "If whatever control strategy is hit upon is successful in cutting the acidity in half, an evil conspiracy of chemists will only allow the pH of precipitation to increase by 0.3."

As a Sceptical Chymist in attendance at this workshop, how would you respond to this statement? Explain the reasons for your response. *Hint:* See Appendix 3 on logarithms.

50. Equation 6.18 shows that energy (in the form of a hot engine or other source of heat) must be added to get N_2 and O_2 to react to form NO. A Sceptical Chymist wants to check this assertion and determine how much energy is required. Show the Sceptical Chymist how this can be done. *Hint:* Draw the Lewis structures for the reactants and products and then check Table 4.2 for bond energies.

51. The text describes a green chemistry solution to reducing NO emissions for glass manufacturers.

a. Identify the strategy.

b. Use the Web to research what other industries might use this green chemistry strategy. Write a report to summarize your findings.

52. Many things have been suggested to help reduce acid rain, and some examples of what an individual can do are given here. For each item, explain the connection between what you would or wouldn't do and the generation of acid rain.

a. Hang your laundry to dry it.

b. Walk, ride a bicycle, or take public transportation to work.

c. Run the dishwasher and washing machine only with full loads.

d. Add additional insulation on hot water heaters and hot water pipes.

e. Buy locally produced food and other items.

f. Fertilize your lawn frequently (but not enough to kill it).

53. How do researchers determine whether the negative effects of acid deposition on aquatic life are a direct consequence of low pH or the result of Al^{3+} released from rocks and soil? Find at least one article that gives the details of such a study. In your own words write a summary of the experimental plan and its results.

54. One way to compare the acid-neutralizing capacity of different substances is to calculate the mass of the substance required to neutralize 1 mol of hydrogen ion, H^+.

a. Write a balanced equation for the reaction of $NaHCO_3$ with H^+, and use it to calculate the acid-neutralizing capacity for $NaHCO_3$.

b. Determine the cost to neutralize 1 mol of H^+ with $NaHCO_3$ if this compound costs $9.50/kg.

55. Why are developing countries likely to emit an increasingly higher percentage of the global amount of SO_2? Pick a nation, research its current emissions of SO_2 and calculate its percentage of global emissions. Speculate on whether increasing emissions are likely to continue in the future and offer an explanation for your prediction.

56. Like diesel trucks, sport utility vehicles (SUVs) emit more than their share of pollutants. Do SUVs emit NO_x, SO_2, or both? What are the current proposals to clean up their emissions? Use the resources of the Web or an owner's manual to research this question.

57. Some local weather maps give forecasts for pollen, UV index, and air quality. Why do you suppose that no forecast for acid rain is provided?

58. Blue-baby syndrome (methemoglobinemia) can occur as a result of nitrate ion in the drinking water. What are the current guidelines for the nitrate ion in drinking water? What happens to infants and young children who ingest too much nitrate? What is the likely source of nitrate ion in the drinking water? The nitrite ion is involved as well. How? Use the resources of the Web to answer these questions. *Hint:* Include "nitrate ion" in your search or you will bring up congenital causes of methemoglobinemia.

59. Was the Clear Skies Initiative put into law? If so, was the cap that it set higher or lower than existing caps? In 2004, a lengthy article in the *New York Times* charged that "Clear Skies allowed 50% more sulfur dioxide, nearly 40 percent more nitrogen oxides and three times as much mercury as the Clean Air Act—rigorously enforced—called for." (Bruce Barcott, *New York Times Magazine,* April 4, 2004.) Alternatively, is other air quality legislation pending in its place?

The Fires of Nuclear Fission

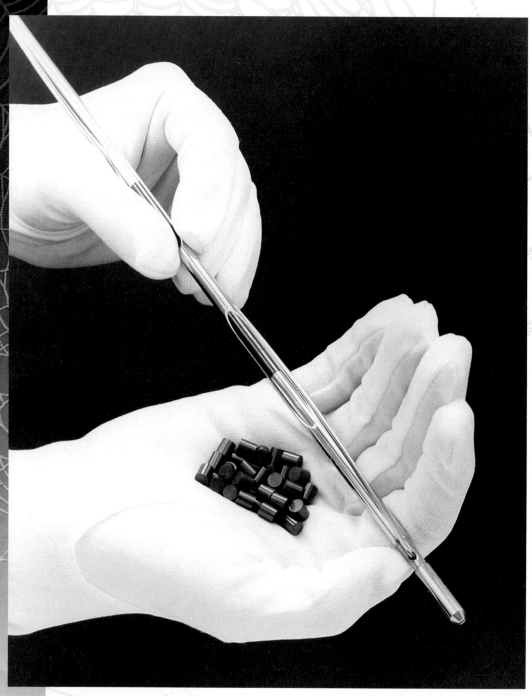

A worker is holding simulated fuel pellets to be used in a nuclear power reactor. The long stainless steel fuel pin contains pellets of the actual size used in one type of reactor, the fast breeder reactor.

Nuclear fuel pellets are small and quite ordinary looking. They are formed from a black, ceramic material that is cool to the touch. If you were to pick up a few and examine them, you would have no clue as to the extravagant amount of energy packed within them. But when 7–10 million fuel pellets are put together under the proper conditions in a nuclear reactor, these pellets become a highly concentrated energy source. Over 100 nuclear power reactors across our nation are fueled by pellets such as the ones pictured.

Although each pellet contains a mere 4 grams of uranium oxide and is only slightly shorter than a dime (Figure 7.1), one fuel pellet is roughly equivalent to the energy released when a ton of coal, a cord of wood, or 150 gallons of gasoline is burned. The current price of a pellet—under $3.00 each—is economically competitive with other fuel prices. The cost of the nuclear fuel is low compared with the cost of 150 gallons of gasoline, and would be an even better bargain if gas prices were to triple. However, once fuel pellets and the fuel rods that house them have been used to generate power in a nuclear reactor, they become highly radioactive. The cost for a fuel pellet does *not* include the price of handling it after it is "spent" and becomes nuclear waste. Similarly, the price of a gallon of gasoline does not include any costs to clean up the air pollutants released by burning it in an automobile engine.

Figure 7.1
Nuclear fuel pellets and a U.S. dime.

The word *nuclear* carries a tremendous baggage of disturbing associations, including the bombing of Hiroshima and Nagasaki, radioactive fallout from bomb tests, radiation-induced cancer and birth defects, the risks of accidents and meltdowns (Three Mile Island and Chernobyl), the difficulties of disposing of high-level radioactive wastes, and the ultimate threat of nuclear annihilation. Probably no subject in all of physical science is more likely to provoke an emotional response. And yet, people recognize that many benefits also spring from the nucleus: radiation therapy for the treatment of cancer, medical X-rays and nuclear diagnostic scans that can be done without anesthesia and surgery, and of course, the production of electricity by nuclear power plants. The applications of nuclear phenomena, harmful at one extreme and beneficial at the other, present us with a dilemma of risks and benefits. The atom holds a dual challenge, as Dr. Hans Blix, former Director General of the International Atomic Energy Agency (and more recently UN chief weapons inspector in Iraq) notes: "to exploit in peace, not explode in war."

We hope this talk about nuclear phenomena has piqued your curiosity. How does nuclear fission produce energy? What are the safeguards against a meltdown? Why is uranium enriched? Can a nuclear power plant explode like an atom bomb? Can fuel pellets be diverted to make nuclear weapons? As you study nuclear fission and radioactive decay, you will come to understand how fuel pellets are used in a nuclear power plant to generate electricity. It should become clear why the radiation emitted by the fuel pellets is low *before* they are used to produce energy, but very high afterward. And finally, by knowing both the chemical and the nuclear processes involved, we hope you will understand why it is so difficult to find a solution for handling the nuclear waste produced by nuclear reactors. As has been the case in earlier chapters, your understanding will be based on a complex interplay of issues involving chemistry, energy, and societal issues.

We begin by examining the prospects for nuclear power in the years to come. But, before we start, we ask you to consider your own position regarding nuclear power by completing this exercise.

Consider This 7.1   **Your Opinion of Nuclear Power**

a.   Given a choice between purchasing electricity being generated by a nuclear power plant and a traditional coal-burning plant, which would you choose, and why?

b.   Under what circumstances, if any, would you be willing to change your position on the use of nuclear power for generating electricity?

Save your answers to these questions, because you will be asked to revisit them at the end of the chapter.

Figure 7.2

We have a high demand for electricity (and caffeine).

7.1 A Comeback for Nuclear Energy?

Most people mindlessly switch on lights, giving no thought to the source of the energy that makes the bulbs glow. But for anybody who has lost electrical power because of a storm or repeated power blackouts, flipping a light switch may trigger a set of memories. Is the power back on? Will it be another evening without electricity? Can I brew my coffee in the morning (Figure 7.2)? The answers depend on the choices we make to produce our electrical energy now and in the years to come.

Let's assume that you have power, and that you start your electric coffee maker. The odds are 1 in 5 that the electricity heating the water for your coffee came from a nuclear power plant. In 2004, about 20% of the electrical power in the United States was being produced by the 103 nuclear reactors licensed by the Nuclear Regulatory Commission (NRC). These reactors, such as the one shown in Figure 7.3, are at 65 sites in 31 states. However, *no* new nuclear plants have been licensed since 1978; a moratorium was placed on their construction after the Three Mile Island accident in 1979. Furthermore, nine nuclear plants ceased their operations, some of them before their licenses even expired (Table 7.1). They include what was once the nation's largest nuclear plant, the Zion nuclear power station on the shores of Lake Michigan. Reasons cited for plant closings included the competition of natural gas and the competitive pressures of energy deregulation. Therefore, when you brew your coffee a decade from now, from where will the electricity come?

 Consider This 7.2   Nuclear Power State by State

Choose a state. Is a nuclear power plant operating there? If so, what percent of the state's electrical energy needs is furnished by nuclear power? Check the interactive map at the Nuclear Energy Institute. Either search to locate it or use the direct link at the *Online Learning Center*. In addition, identify a state where more than half of the energy is produced by nuclear power.

The largest nonmilitary application of nuclear energy is the generation of electricity by nuclear power plants. When it was first demonstrated that electricity could be obtained from the splitting of atoms, a new age appeared to be dawning. The first commercial nuclear power generating station in this country was completed in 1957 at Shippingport, Pennsylvania, along the Ohio River near Pittsburgh. This new source held the promise of unlimited, cheap electricity. During the early 1960s, proponents of nuclear power were plentiful, including the former head of the U.S. Atomic Energy Commission who suggested naively that electricity produced by this method would be so inexpensive that it would be inconsequential to even meter consumers' use of it.

Figure 7.3

Watts Bar nuclear plant, Watts Bar, TN.

Table 7.1	Nuclear Plant Closings Since 1990		
Nuclear Plant	**State**	**License Issued**	**Date Shut Down**
Millstone 1	Connecticut	1966	1998
Zion 1, Zion 2	Illinois	1973	1998
Big Rock Point	Missouri	1962	1997
Maine Yankee	Maine	1972	1997
Haddam Neck	Connecticut	1967	1996
San Onofre 1	California	1967	1992
Trojan	Oregon	1975	1992
Yankee–Rowe	Maine	1960	1991

Source: From Environmental Law & Policy Center, http://www.elpc.org/energy/nuclear_closings.html. Reprinted with permission.

There would be plenty of electricity for everyone! In spite of the optimism, the prediction of costless electricity has not come true.

It is hard to predict how long the nuclear plants near you (or in neighboring states) are likely to continue to operate. In the late 1990s, the decommissioning of a dozen or so plants seemed a foregone conclusion, and the list in Table 7.1 was expected to grow. Some worried that so few nuclear plants would renew their licenses that the electrical power generated by them would drop from the current 20% to less than 10%. But, given the recent increased demands for electricity in the United States, a renaissance in nuclear power is now being predicted. For example, the Oyster Creek plant in New Jersey was scheduled for shutdown in 2000, but the plant is still in operation, purchased by an international power company. Calvert Cliffs Nuclear Power Station in Maryland was the first to successfully complete the lengthy and costly process necessary to renew their operating licenses for another 20 years, followed shortly thereafter by Oconee Nuclear Power Station in South Carolina. A total of 10 reactor units have received extensions until 2033–2036. As of 2004, the possibility of constructing a new plant in the United States by 2010 was on the horizon. Although it is too soon to tell, nuclear plants may be becoming a hot item on the energy market.

The construction and continued operation of nuclear plants is not only a matter of energy supply and demand, but also a matter of public acceptance. Depending on your age, you may have little recollection of the controversy that surrounded some nuclear power plants when they were proposed or being constructed. People have been lining up on one side or the other of the nuclear fence for quite some time, such as those protesting the construction of the Seabrook nuclear power plant in New Hampshire. Figure 7.4 shows one of the many protests that took place at Seabrook.

The decommissioning (or shutdown) of a nuclear plant is a complex operation. All parts of the plant must be analyzed and removed according to strict hazardous waste criteria. We say more about nuclear waste in Sections 7.10 and 7.11.

Figure 7.4
Signs held at the construction of the Seabrook nuclear power plant.

Consider This 7.3   Reporting a Controversy

The Seabrook demonstrations took place in 1977. Now, about 30 years later, an older nuclear plant near you may be up for relicensing or a new one may be on the drawing board. A public meeting is being held. If you were to attend and hold a sign, what would it say? If a reporter were to interview you about your position, list three of the points that you would wish to convey.

Before we explore nuclear fission and the energy it produces, consider these two statements.

"We are going to see a revival of nuclear energy; we don't really have an option."

Dr. Gregory Choppin
Distinguished Professor of Chemistry
Florida State University
221st American Chemical Society Meeting, San Diego, CA, 2001

"There has never been much middle ground for nuclear power—people either like it or they don't."

Jeffrey W. Johnson
Senior Editor, Washington News Bureau
Chemical & Engineering News, October 2, 2000

These two public figures speak to distinct but related issues. Nuclear energy *is* an option that stands poised for a revival. But regarding its use, people are not of one mind. In the section that follows, we will examine the process of nuclear fission, thus taking the first step in explaining both the controversies and the hopes for nuclear energy as a power source.

7.2 How Does Fission Produce Energy?

The key to answering this question is probably the most famous equation in all of the natural sciences, $E = mc^2$. This equation dates from the early years of the 20th century and is one of the many contributions of Albert Einstein (1879–1955). It summarizes the equivalence of energy, E, and matter or mass, m. The symbol c represents the speed of light, 3.0×10^8 m/s, so c^2 is equal to 9.0×10^{16} m^2/s^2. The large value of c^2 means that it should be possible to obtain a tremendous amount of energy from a small amount of matter, whether in a power plant or in a weapon.

For over 30 years, Einstein's equation was a curiosity. Scientists believed that it described the source of the Sun's energy, but as far as anyone knew, no one had ever observed on Earth a transformation of a substantial fraction of matter into energy. But in 1938, two German scientists, Otto Hahn (1879–1968) and Fritz Strassmann (1902–1980), discovered otherwise. When they bombarded uranium with neutrons, they found what appeared to be the element barium (Ba) among the products. The observation was unexpected because barium has an atomic number of 56 and an atomic mass of about 137. Comparable values for uranium are 92 and 238, respectively. At first, the scientists were tempted to conclude that the element was radium (Ra, atomic number 88), a member of the same group in the periodic table as barium. But Hahn and Strassmann were good chemical researchers, and the chemical evidence for barium was too compelling.

The German scientists were unsure of how barium could have been formed from uranium, so they sent a copy of their results to their colleague, Lise Meitner (1878–1968), for her opinion (Figure 7.5). Dr. Meitner had collaborated with Hahn and Strassmann on related research, but was forced to flee Germany in March 1938 because

Figure 7.5
Lise Meitner (left) at Bryn Mawr College (1959).

of the anti-Semitic policies of the Nazi government. When she received their letter, she was living in Sweden. She discussed the strange results with her physicist nephew, Otto Frisch (1904–1979), as the two of them went walking in the snow. In a flash of insight, she understood. Under the influence of the bombarding neutrons, the uranium atoms were splitting into smaller ones such as barium. The nuclei of the heavy atoms were dividing, like biological cells undergoing fission.

That word from biology is applied to a physical phenomenon in the letter that Meitner and Frisch published on February 11, 1939, in the British journal *Nature*. In the letter, entitled "Disintegration of Uranium by Neutrons: A New Type of Nuclear Reaction," the authors state the following:

> *Hahn and Strassmann were forced to conclude that isotopes of barium are formed as a consequence of the bombardment of uranium with neutrons. At first sight, this result seems very hard to understand. . . . On the basis, however, of present ideas about the behavior of heavy nuclei, an entirely different . . . picture of these new disintegration processes suggests itself. . . . It seems therefore possible that the uranium nucleus . . . may, after neutron capture, divide itself into two nuclei of roughly equal size. . . . The whole "fission" process can thus be described in an essentially classical way.*

Although just over a page long, this letter was immediately recognized for its significance. In fact, it would be difficult to think of a more important scientific communication. Niels Bohr (1885–1962), an eminent Danish physicist, learned of the news directly from Frisch and brought it to the United States on an ocean liner several days before its publication. Within a few weeks of Meitner and Frisch's letter in *Nature*, scientists in a dozen laboratories in various countries confirmed that the energy released by the fission of uranium atoms was that predicted by Einstein's equation. Lise Meitner's contributions to the discovery of nuclear fission were honored by naming element 109 meitnerium. Earlier, element number 96, curium, had been named to honor Marie Curie, another woman who was a nuclear pioneer (Section 7.7).

Nuclear fission is the splitting of a large nucleus into smaller ones with the release of energy. Energy is released because the total mass of the products is slightly less than the total mass of the reactants. In spite of what you may have been taught, neither matter nor energy is *individually* conserved. Matter disappears and an equivalent quantity of energy appears. Alternatively, one can view matter as a very concentrated form of energy; nowhere is it more concentrated than in the atomic nucleus. Remember that an atom is mostly empty space. If a hydrogen nucleus were the size of a baseball, then its electron would travel in a sphere half a mile in diameter. Because almost all the mass of an atom is associated with its nucleus, the nucleus is incredibly dense. Indeed, a pocket-sized matchbox full of atomic nuclei would weigh over 2.5 billion tons! Given the energy–mass equivalence of Einstein's equation, the energy content of all nuclei is, relatively speaking, immense.

Only the nuclei of certain elements undergo fission and these only under certain conditions. The relative factors to consider are the size of the nucleus, the numbers of protons and neutrons it contains, and the energy of the neutrons used to initiate the fission. For example, relatively light and stable atoms such as oxygen, chlorine and iron do not split. Extremely heavy nuclei may fission spontaneously. And the familiar heavy atoms, such as uranium and plutonium, will split if hit hard enough with a neutron. Some (but not all) isotopes of uranium will even fission with a more gentle nudge, such as in the conditions of a nuclear power plant.

Let's examine uranium more closely. *All* uranium atoms contain 92 protons that are accompanied by 92 electrons if the atom is electrically neutral. But beyond this, uranium atoms can differ. In nature, uranium is found as three different isotopes. Approximately 99.3% of the uranium atoms contain 146 neutrons. The mass number of this isotope of uranium is 238, that is, 92 protons plus 146 neutrons. We represent this isotope as uranium-238, or more simply as U-238. The remaining uranium atoms in nature (about 0.7%) are predominantly U-235.

> The mass number is the number of protons plus the number of neutrons. See Section 2.2.

> Isotopes of an element differ in the number of neutrons. See Section 2.2.

Your Turn 7.4   Other Uranium Isotopes

a. A trace amount of uranium-234 is also found in nature. How many protons are in an atom of uranium-234? How many neutrons?

b. Fifteen (or so) other uranium isotopes have been synthesized by humans. How many protons does each of these isotopes have?

As we will see momentarily, it can be useful to include both the mass number and the atomic number with an isotope. By convention, we usually write the mass number (as a superscript) and the atomic number (as a subscript) to the left of the chemical symbol. Using this convention, uranium-238 becomes:

$$\text{Mass number = number of protons + number of neutrons} \longrightarrow {}^{238}_{92}\text{U}$$
$$\text{Atomic number = number of protons} \longrightarrow {}^{238}_{92}\text{U}$$

Similarly, uranium-235 is written ${}^{235}_{92}\text{U}$. The difference between these two isotopes is a mere three neutrons, but in nuclear terms this difference can be significant. For example, under the conditions present in a nuclear reactor where the neutrons are of relatively low energy, U-238 does *not* undergo fission, yet U-235 does. Small differences in the nucleus can mean large differences in *nuclear* behavior.

The process of fission usually requires neutrons to initiate it and always releases neutrons, as can be seen by this nuclear equation with uranium-235.

$$ {}^{1}_{0}\text{n} + {}^{235}_{92}\text{U} \longrightarrow [{}^{236}_{92}\text{U}] \longrightarrow {}^{141}_{56}\text{Ba} + {}^{92}_{36}\text{Kr} + 3\,{}^{1}_{0}\text{n} \qquad [7.1]$$

> Uranium-235 typically fissions into two smaller nuclei, but a three-way split also is possible.

Nuclear equations are similar to, but not the same as the chemical equations that you have seen in earlier chapters. Let's look at the components, from left to right. Initially, a neutron hits the nucleus of U-235. The subscript for a neutron (designated n) is 0, indicating zero charge; the superscript is 1 because the mass number of a neutron is one. The nucleus of ${}^{235}_{92}\text{U}$ captures the neutron, forming a heavier isotope of uranium, ${}^{236}_{92}\text{U}$. This isotope is written in square brackets indicating that it exists only momentarily. Uranium-236 immediately splits into two smaller atoms (Ba and Kr) with the release of three more neutrons.

In a nuclear equation, the sum of the subscripts on the left side must equal that of the subscripts on the right side. Likewise, the sum of superscripts on each side of the equation is equal. Coefficients in nuclear equations, such as the 3 preceding the $_0^1$n in equation 7.1, are treated the same way as in chemical equations, multiplying the term that follows it. When no coefficient is given, a 1 is understood. We demonstrate these features with nuclear equation 7.1.

Left	**Right**
Superscripts: $1 + 235 = 236$	$141 + 92 + (3 \times 1) = 236$
Subscripts: $0 + 92 \ = 92$	$56 + 36 + (3 \times 0) = 92$

A wide array of fission products can be formed when the nucleus of an atom of U-235 is struck with a neutron. Your Turn 7.5 and Your Turn 7.6 give other possibilities.

Your Turn 7.5   Other Examples of Fission

With the help of a periodic table, write these two nuclear equations. For both, U-235 is first hit by a neutron of appropriate energy to initiate the fission process.

a. U-235 fissions to form Ba-138, Kr-95, and releases neutrons.

b. U-235 fissions to form an element (atomic number 52, mass number 137), another element (atomic number 40, mass number 97), and neutrons.

Answer

a. $_0^1$n $+$ $_{92}^{235}$U $\longrightarrow$ $_{56}^{138}$Ba $+$ $_{36}^{95}$Kr $+$ 3 $_0^1$n

---

Your Turn 7.6   Fission and Sr-90

Strontium-90 is a radioactive fission product that entered the biosphere after the first atmospheric nuclear bomb tests in the United States. This isotope is formed from the fission of U-235 in a reaction that produces three neutrons and another element. Write the nuclear equation.

> You will learn more about strontium-90 and its health effects in Section 7.9.

Look again at nuclear equation 7.1. Both sides contain neutrons. Why don't we cancel them out? Although you would do this in a mathematical equation, nuclear equations are not handled in the same manner. The neutrons on both sides of the equation are important: One initiates fission and others are produced from the fission process. With this net production of neutrons, a **chain reaction** can occur in which the fission reaction becomes self-sustaining. Each neutron produced can in turn strike another U-235 nucleus, cause it to split, and release a few more neutrons. The result is a rapidly branching chain reaction that spreads in a fraction of a second (Figure 7.6). A **critical mass** is the amount of fissionable fuel required to sustain a chain reaction, providing that the fuel is held together long enough for the reaction to proceed. For example, the critical mass of uranium-235 is about 15 kg, or 33 lb. Were this mass of pure U-235 to be brought together in one place, fission would spontaneously occur. Nuclear weapons work on this principle. But as you will soon see, the uranium fuel in a nuclear power plant is far from pure U-235 and is unable to explode like a nuclear bomb.

We mentioned earlier that energy is given off during fission because the mass of the products is slightly less than that of the reactants. However, from the nuclear equations we have just written no mass loss is apparent, because the sum of the mass numbers is the same on both sides. In fact, the actual mass does decrease slightly. To understand this, remember that the actual masses of the nuclei are not the mass numbers (the sum of the number of protons and neutrons); rather, they have measured values with many decimal places. For example, an atom of uranium-235 weighs 235.043924 atomic mass units. Were you to keep all seven decimal places and compare the masses on both

> Atomic mass units are convenient for weighing atoms. Each is exactly 1/12 of the mass of a C-12 atom or 1.66×10^{-27} kg.

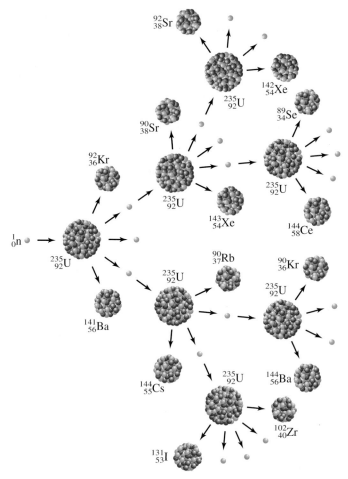

Figure 7.6
A chain reaction of U-235.

Figures Alive! Visit the *Online Learning Center* to learn more about nuclear fission and chain reactions.

sides of the nuclear equation for the fission of U-235, you would find that the products' mass is less by about 0.1%, or 1/1000th. As a consequence, the energy of the products is less than that of the reactants. This difference corresponds to the energy released.

How much energy would be released if all the nuclei in 1.0 kg (2.2 lb) of U-235 were to undergo fission? We can calculate an answer by using an equation closely related to $E = mc^2$, namely, $\Delta E = \Delta mc^2$. Here the Greek letter delta (Δ) means "the change in", so now with a change in mass we can calculate a change in energy. Since 1/1000 of this mass is lost, the value for Δm, the change in mass, is 1/1000 of 1.0 kg, which is 1.0 g or 1.0×10^{-3} kg. Now substitute this value and $c = 3.0 \times 10^8$ m/s into Einstein's equation.

$$\Delta E = \Delta mc^2 = (1.0 \times 10^{-3}\,\text{kg}) \times (3.0 \times 10^8\,\text{m/s})^2$$

$$\Delta E = (1.0 \times 10^{-3}\,\text{kg}) \times (9.0 \times 10^{16}\,\text{m}^2/\text{s}^2)$$

Completing the calculation gives an energy change in what may appear to be unusual units.

$$\Delta E = 9.0 \times 10^{13}\,\text{kg} \cdot \text{m}^2/\text{s}^2$$

The unit $kg \cdot m^2/s^2$ is identical to a joule (J). Therefore, the energy released from the fission of an entire kilogram of uranium-235 is a whopping 9.0×10^{13} J or 9.0×10^{10} kJ.

To put things into perspective, 9.0×10^{13} J is the amount of energy released by the explosion of 33,000 tons (33 kilotons) of TNT. Alternatively, you could get this much energy by burning 3300 tons of coal. This is enough energy to raise about 700,000 cars six miles into the sky or to turn 8.7 million gallons of water into steam. Yet, this massive amount of energy comes from the fission of a single kilogram of U-235, in which only one gram (0.1% mass change) is actually transformed into energy.

As it turns out, one cannot fission a kilogram or two of pure U-235 in one fell swoop. In an atomic weapon, for example, the energy that is released in a fraction of a second blasts the fissionable fuel apart and stops the chain reaction before all the nuclei can undergo fission. Nonetheless, this energy released is considerable, and on the order of 10 kilotons of TNT for a small atomic bomb such as the one dropped on Hiroshima. However, the energy of nuclear fission can be *continually* released under *controlled conditions* in a nuclear power plant. In the next section, we will see how.

> As described in Section 4.1, the joule (J) is a unit of energy.
> $1 \text{ J} = 1 \text{ kg} \cdot m^2/s^2$

Your Turn 7.7   Nuclear Power

At full capacity, the nuclear plant at Seabrook, New Hampshire, generates 1160 million joules of electrical energy per second. Calculate the total amount of electrical energy produced per day and the loss of mass of U-235 each day.

Hint: Start by calculating the quantity of energy generated not per second, but per day. Then use the equation $\Delta E = \Delta m c^2$ and solve for the change in mass, Δm. Report the mass lost in grams.

7.3 How Does a Nuclear Reactor Produce Electricity?

Chapter 4 described how a conventional power plant burns a fuel such as coal or oil to produce heat. The heat is then used to boil water, converting it into high-pressure steam that turns the blades of a turbine. The shaft of the spinning turbine is connected to large wire coils that rotate within a magnetic field, thus generating electrical energy. A nuclear power plant operates in much the same way, except that the water is heated not by fossil fuel, but by the energy released from the fission of nuclear "fuel" such as U-235. Like any power plant, a nuclear power plant is subject to the efficiency constraints imposed by the second law of thermodynamics. The theoretical efficiency for converting heat to work depends on the maximum and minimum temperatures between which the plant operates. This thermodynamic efficiency, typically 55–65%, is significantly reduced by other mechanical, thermal, and electrical inefficiencies.

> Section 4.2 first introduced the concept of operating efficiency.

A nuclear power station consists of two parts: a nuclear reactor and a nonnuclear portion (Figure 7.7). The nuclear reactor is the hot heart of the power plant. The reactor, together with one or more steam generators and the primary cooling system, is housed in a special steel vessel within a separate reinforced concrete dome-shaped containment building. The nonnuclear portion contains the turbines that run the electrical generator. It also contains the secondary cooling system. In addition, the nonnuclear portion must be connected to some means of removing heat from the coolants. Accordingly, a nuclear power station will have one or more cooling towers or be located near a sizeable body of water (or both). Look back at Figure 4.2, a diagram of a fossil-fuel power plant. This plant needs a means of removing heat as well, as shown by the stream of cooling water.

The uranium fuel in the reactor core is in the form of uranium dioxide (UO_2) pellets, each about the size of a large pencil eraser, as shown earlier in Figure 7.1. These pellets are placed end to end in tubes made of a special metal alloy, which in turn are grouped into stainless-steel clad bundles (Figure 7.8). There are at least 200 pellets per tube.

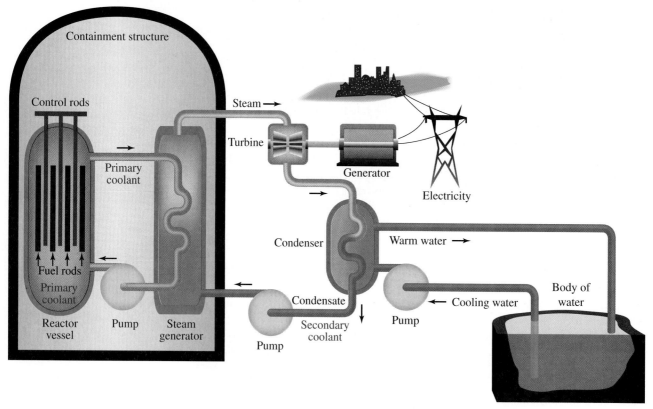

Figure 7.7

Diagram of a nuclear power plant.

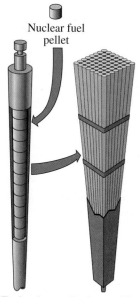

Nuclear fuel
pellet

Fuel rod   Fuel assembly

Figure 7.8

Fuel pellet, fuel rod, and fuel
assembly making up the core of
a nuclear reactor.

Cadmium absorbs neutrons,
but itself does not undergo
nuclear fission.

Although a fission reaction, once started, can sustain itself by a chain reaction, neutrons are needed to induce the process (equation 7.1, Figure 7.6). One means of generating the required neutrons is to use a source that produces neutrons. Such a source may contain beryllium-9 and a heavier element such as plutonium or americium. The heavier element releases alpha particles (helium-4 nuclei, or $^4_2$He, as we will discuss in Section 7.7). When an alpha particle ($^4_2$He) strikes a beryllium atom, the two nuclei combine to form $^{12}_6$C, a neutron, and a gamma ray, $^0_0\gamma$.

$$^{238}_{94}\text{Pu} \longrightarrow ^{234}_{92}\text{U} + ^4_2\text{He} \qquad [7.2]$$

$$^4_2\text{He} + ^9_4\text{Be} \longrightarrow ^{12}_6\text{C} + ^1_0\text{n} + ^0_0\gamma \qquad [7.3]$$

The neutrons produced in this way can initiate the nuclear fission of uranium-235 in the reactor core.

Your Turn 7.8   Poo-bee and Am-bee

A neutron source constructed with Pu and Be is called a PuBe, or "poo-bee," source. Similarly, the "am-bee" or AmBe source is constructed from americium and beryllium. Analogous to the PuBe source, write the set of reactions that produce neutrons from an AmBe source. Start with Am-241.

Once fission is initiated, the rate of fission and the amount of heat generated by the fission are controlled using a principle employed in the first controlled nuclear fission reaction, which took place at the University of Chicago in 1942. Rods composed primarily of the element cadmium, an excellent neutron absorber, are interspersed among the fuel elements. Modern control rods also contain silver and indium. As long as these rods are in place, the reaction cannot become self-sustaining because neutrons are absorbed by the control rods. When the rods are withdrawn, the reactor "goes critical"; that is, the

Figure 7.9
Cooling tower (with its cloud of condensed water vapor) at the Union Electric Callaway (MO) nuclear power plant. The dome is the containment building for the nuclear reactor.

fission chain reaction becomes self-sustaining. But the rods can be rapidly adjusted to halt the self-sustaining chain reaction in the event of an emergency (see Figure 7.7).

The fuel bundles and control rods are bathed in the **primary coolant,** a liquid that comes in direct contact with the nuclear reactor to carry away heat. In the Callaway nuclear reactor shown in Figure 7.9, and in many others, the primary coolant is an aqueous solution of boric acid, H_3BO_3. The boron atoms absorb neutrons and thus control the rate of fission and the temperature. The solution also serves as a **moderator** for the reactor, slowing the speed of the neutrons and making them more effective in causing fission. Another major function of the primary coolant is to absorb the heat generated by the nuclear reaction. Because the primary coolant solution is at a pressure more than 150 times normal atmospheric pressure, it does not boil. It is heated far above its normal boiling point and circulates in a closed loop from the reaction vessel to the steam generators, and back again. This closed primary coolant loop thus forms the link between the nuclear reactor and the rest of the power plant (see Figure 7.7).

The heat from the primary coolant is transferred to what is sometimes referred to as the **secondary coolant,** the water in the steam generators that does not come in contact with the reactor. At the Callaway nuclear plant, more than 30,000 gallons of water is converted to vapor each minute. The energy of this hot vapor turns the blades of turbines that are attached to an electrical generator. To continue the heat transfer cycle, the water vapor is then cooled and condensed back to a liquid and returned to the steam generator. In many nuclear facilities the cooling is done using large cooling towers that commonly are mistaken for the reactors. The reactor is actually housed in a relatively small dome-shaped building (see Figure 7.9).

Cooling towers are also used in coal-fired plants.

Your Turn 7.9   Clouds (not mushroom-shaped)

Some days you can see a cloud coming out of the cooling tower of a nuclear power plant, as shown in Figure 7.9. What causes the cloud? Does it contain any radioisotopes produced from the fission of U-235?

Nuclear power plants also use water from lakes, rivers, or the ocean to cool the condenser. For example, at the Seabrook nuclear power plant in New Hampshire, every minute 398,000 gallons of ocean water flows through a huge tunnel (19 feet in diameter and 3 miles long) bored through rock 100 feet beneath the floor of the ocean. A similar tunnel from the plant carries the water, now 22 °C warmer, back to the ocean. Special nozzles distribute the hot water so that the observed temperature increase in the immediate area of the discharge is only about 2 °C. The ocean water is two loops away from the fission reaction and its products. The primary coolant (water with boric acid) circulates through the reactor core inside the containment building. However, this boric acid solution is kept isolated in a closed circulating system, which makes the transfer of radioactivity to the secondary coolant water in the steam generator highly unlikely. Similarly, the ocean water does not come in direct contact with the secondary system, so the ocean water is well protected from radioactive contamination. It should be obvious that the electricity generated by a nuclear power plant is identical to the electricity generated by a fossil-fuel plant; the electricity is not radioactive, nor can it be.

Sceptical Chymist 7.10   Watching the Watts

Consider the statistics quoted for the Seabrook plant. The energy generated in one day is said to be the equivalent of 10,000 tons of coal. Perform a calculation to determine if this value is consistent with the quoted power rating of 1160 megawatts (1160×10^6 J/s). Assume that burning coal releases 30 kJ/g or 30,000 J/g.

Hint: Start by rereading Your Turn 7.7, or completing it if you did not do so earlier. Show that the quantity of energy per day would be equivalent to burning less than the 10,000 tons of coal quoted here. How big is this discrepancy and what important factor can account for such a discrepancy?

7.4 Could There Be Another Chernobyl? Safeguards Against a Meltdown

A 1979 film called *The China Syndrome* told the story of a near-disaster in a fictitious nuclear power plant. The heat-generating fission reaction almost got out of control. If such a thing were to happen, the intense heat might cause a meltdown of the uranium fuel and the reactor housing. Fancifully, the underlying rock might even melt "all the way to China." But in spite of various human and instrumental errors, the safety features of the system worked in the film and fictional disaster was averted. Seven years later on April 26, 1986, the engineers of the very real Chernobyl power plant in the Ukraine, then part of the Soviet Union, were far less fortunate (Figure 7.10). This plant consisted of four reactors, two built in the 1970s and two more in the 1980s, all near the town of Chernobyl (pop. 12,500). Water from the nearby Pripyat River was used to cool the reactors. Although the surrounding region was not heavily populated, nonetheless approximately 120,000 people lived within a 30 km radius.

Chernobyl stands as the world's worst nuclear power plant accident to date. What went wrong there? During an electrical power safety test at Chernobyl reactor 4,

Chernobyl is the transliteration of the Russian pronunciation; Chornobyl is the Ukrainian word.

Figure 7.10

Chernobyl, in Ukraine of the former Soviet Union.

operators deliberately interrupted the flow of cooling water to the core as part of the test. The temperature of the reactor rose rapidly. In addition, the operators had left an insufficient number of control rods in the reactor (that couldn't be reinserted quickly enough), and the steam pressure was too low to provide coolant (due to both operator error and faulty design). A chain of events quickly produced a disaster. An overwhelming power surge produced heat, rupturing the fuel elements, and releasing hot reactor fuel particles. These, in turn, exploded on contact with the coolant water and the reactor core was destroyed in seconds. The graphite used to slow neutrons in the reactor caught fire in the heat. When water was sprayed on the burning graphite, the water and graphite reacted *chemically* to produce hydrogen gas, which exploded when it chemically reacted with oxygen in the air.

> The explosion at Chernobyl was produced by combustion—a chemical reaction, not a nuclear process.

$$2\ H_2O(l) + C(graphite) \longrightarrow 2\ H_2(g) + CO_2(g) \qquad [7.4]$$

$$2\ H_2(g) + O_2(g) \longrightarrow 2\ H_2O(g) \qquad [7.5]$$

The explosion blasted off the 4000-ton steel plate covering the reactor (Figure 7.11). Although a "nuclear" explosion never occurred, the fire and explosions of hydrogen blew vast quantities of radioactive material out of the reactor core into the atmosphere.

Fires started in what remained of the building. In a short time, the plant lay in ruins. The head of the crew on duty at the time of the accident wrote: "It seemed as if the world was coming to an end . . . I could not believe my eyes; I saw the reactor ruined by the explosion. I was the first man in the world to see this. As a nuclear engineer I realized the consequences of what had happened. It was a nuclear hell. I was gripped with fear." (*Scientific American,* April 1996, p. 44.)

The disaster continued in the countryside for over a week. As the reactor burned, it continued to spew large quantities of radioactive fission products into the atmosphere for 10 days (Figure 7.12). The release of radioactivity was estimated on the order of

Figure 7.11

The Chernobyl 4 reactor after the chemical explosion.

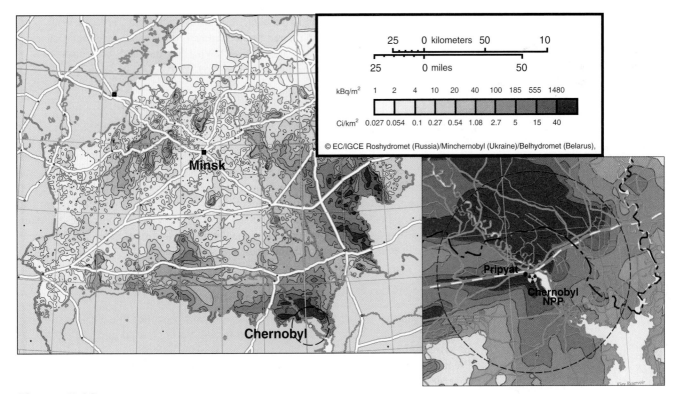

Figure 7.12

Radioactive fallout from the Chernobyl accident, in Belarus and in the vicinity of Chernobyl. Red areas range from 5 to over 40 curies per square kilometer. The curie, Ci, is a unit of radiation explained in Section 7.8.

Source: R. Stone, 20 April 2001 *Science* "Living in the Shadow of Chornobyl"

about 100 of the atomic bombs dropped on Hiroshima and Nagasaki. People in nearby regions reported an odd, bitter, and metallic taste as they inhaled the invisible particles. The radioactive dust cut a swath across Ukraine, Belarus, and up into Scandinavia. Nearly 150,000 people living within 60 km of the power plant were permanently evacuated after the meltdown.

The human toll was immediate and continues to grow. Several people working at the plant were killed outright, and another 31 firefighters died in the cleanup process from acute radiation sickness, a topic we take up in a later section. An estimated 250 million people were exposed to levels of radiation that may ultimately shorten their lives. Included in this figure are 200,000 "liquidators," people who buried the most hazardous wastes and constructed a 10-story concrete structure ("the sarcophagus") to surround the failed reactor. As of 2001, more than 700 children in Belarus, a neighboring country, have been treated for thyroid cancer; most have survived. Presumably this illness is caused by radioactive iodine-131, one of the fission products blown from the reactor. As Dr. Akira Sugenoya, a Japanese physician who volunteered his expertise in Belarus to treat the children suffering from thyroid cancer, remarked "The last chapter of the terrible accident is far from written."

> The thyroid gland incorporates iodide ion to manufacture thyroxine, a hormone essential for growth and metabolism.

Given the demonstrable problems with their design, the four reactors at Chernobyl have been shut down. On Friday, December 15, 2000, the control rods slid into the core at unit 3, the last remaining reactor operating at Chernobyl, permanently shutting it down. Ukranian President Leonid Kuchma reported "This decision came from our experience of suffering. We understand that Chernobyl is a danger for all of humanity and we forsake a part of our national interests for the sake of global safety."

Consider This 7.11   Nuclear Neighbors

The Chernobyl reactor site and the land near it continue to be among the most highly radioactive places on Earth. Ukranian President Kuchma is reported to have said, "We shall continue to bear this. This is our fate." Use the Web to find out what kinds of humanitarian aid continues to be offered to the victims of Chernobyl. Even now, about 20 years later, what cleanup tasks are required? What are the medical needs? What avenues exist for you or your class to give assistance?

The Ukrainian government estimates that a replacement gas-burning power station and associated expenses will cost $4 billion. Thus far, Western countries have pledged $2.3 billion toward the project, and the United States is providing financial and technical assistance to establish an international nuclear safety and environmental research center near Chernobyl. But there are many hidden costs. A study by a Russian economist estimates the total cost of the Chernobyl meltdown at $358 billion, a figure that includes the expense of the cleanup and the loss of farm production. Present plans include encasing the ruined reactor in additional concrete and steel. No plans have been made to deal with the radioactive materials and dust inside.

Consider This 7.12   After Chernobyl

As of 2004, thirty reactors were operating in Russia, accounting for roughly 10% of the electricity produced. Since the late 1990s, Russia also has been exporting reactor technology to China, Iran, and India. What is the current status of nuclear power in Russia? Use the resources of the Web to research this question. If possible, comment on any controversies and on how the former Soviet public views nuclear power today.

This recounting of the solemn facts concerning the Chernobyl tragedy leads to an inevitable question: "Could it happen here?" America's closest brush with nuclear disaster occurred in March 1979, when the Three Mile Island power plant near Harrisburg,

Figure 7.13
Seabrook nuclear power plant. The dome is part of the containment building that houses
the reactor.

Pennsylvania, lost coolant and only a partial meltdown occurred. No fatalities and no
serious release of radiation occurred. In spite of the initial failure, the system held and
the damage was contained. Since then, refinements in design and safety have been made
to existing reactors and those under construction. Nuclear engineers agree that no commercial nuclear reactors in the United States have the design defects that led to the
Chernobyl catastrophe.

Consider, for example, the Seabrook nuclear power plant that has been hailed as
an example of state-of-the-art engineering. The energetic heart of the station is the
400-ton reaction vessel with 44-foot high walls made of 8-inch carbon steel. It is surrounded by a reinforced concrete, dome-shaped containment building, a feature the
Chernobyl plant did not have, but all reactors operating in the United States must
have. As the name suggests, this structure is built to withstand accidents of natural
or human origin and prevent the release of radioactive material. It is clearly visible
in the photograph (Figure 7.13). The inner walls of the building are 4.5 feet thick
and made of steel-reinforced concrete; the outer wall is 15 inches thick. Information
supplied by North Atlantic Energy Service Corporation (NAESCO), the company that
manages the Seabrook station, states that the containment building is constructed to
withstand hurricanes, earthquakes, 360-mph winds, and the direct crash of a U.S. Air
Force FB-111 bomber. At the time, however, he had no reason to worry about a direct
crash with a larger Boeing 757 commercial jet. Look ahead to Consider This 7.13
for more about this.

Although it seems unlikely that a nuclear accident of the proportions of Chernobyl
could occur in the United States, the question still remains: Could a nuclear disaster
happen in some other region of the world? That possibility does exist, because such
disasters result from the complex interplay of faulty plant design, human error, and
political instability. All three of these must be minimized to keep a nuclear power plant
operating safely. Although the nuclear units in many parts of the world get high rankings on all three factors, this is not the case everywhere. For example, in July 2001 the
German government urged the closing of a Czech nuclear power plant near the German
border because of safety concerns. Several reactors in Russia have long histories of
safety violations and raise similar concerns, both for operation and the handling of

waste (the topic of Section 7.10). A plume of radioactive dust easily crosses international boundaries, and so the concerns of neighboring nations are well placed.

A related and far scarier question relates to acts by terrorists. Are nuclear reactors being considered as targets? This question must be taken seriously. Fortunately, as this book went to press, no incidents have occurred. Even before the terrorism, however, nuclear reactors were built to withstand earthquakes. About a fifth of the reactors worldwide are in regions of seismic activity (such as on the Pacific rim). For safety, reactors are fitted with seismic detectors that can quickly shut the reactor down if a tremor occurs. The impact of a fully fueled commercial jet on a containment dome, however, is another matter entirely. In such a case, the results could truly be catastrophic, exceeding the radioactive releases of Chernobyl.

Consider This 7.13   After September 11th

In regards to terrorist attacks on reactors, Steven P. Kraft, the Director for Waste Management at the Nuclear Energy Institute (NEI) gave a report to the National Academy of Sciences on December 3, 2003. He stated "Nuclear power plants are robust structures, and while not designed specifically for aircraft crash, are highly resistant to the potential effects." Either refute or support his statement. A starting point might be to check out the general policy statements of the NEI, to see what point of view this institute takes.

7.5 Can a Nuclear Power Plant Undergo a Nuclear Explosion?

Many people ask this question. The devastation and destruction that atomic bombs brought to Hiroshima and Nagasaki at end of World War II are painfully etched in the memory of anyone who has even seen the pictures of those cities and their survivors. Therefore, it is reassuring that the answer to the question is an emphatic "No."

Obviously, nuclear power plants and nuclear bombs are constructed for different purposes. Although both derive energy from nuclear fission, each requires a different rate of reaction. A nuclear power plant requires a slow, controlled energy release; in a nuclear weapon, the release is rapid and uncontrolled. In both cases, the fuel is U-235 and the fission reaction is essentially the same. The difference lies in the fuel. Commercial nuclear power plants typically operate with fuel that is 3–5% U-235, and atomic weapons use fuel that may be as high as 90% U-235. Both fuels are examples of **enriched uranium,** that is, uranium that has a higher percent of U-235 than its natural abundance of about 0.7%. The fuel in weapons is sometimes referred to as highly enriched, or military grade, uranium.

Most commercial reactors worldwide use enriched uranium as fuel. However, some reactors by the British and the Canadians are designed to run on natural (unenriched) uranium.

Your Turn 7.14   Enriched Uranium

The fuel pellets in a nuclear power plant contain 3–5% uranium-235.

a. What other isotope of uranium is present in far greater percentage in the pellet? Is this isotope fissionable under the conditions of a nuclear reactor?

b. After being used in a reactor, these pellets may contain elements such as strontium, barium, krypton, and iodine. Explain why.

In a nuclear reactor, the concentration of fissionable U-235 is low. Most of the neutrons given off during fission of U-235 are absorbed by U-238 nuclei and elements such as cadmium and boron. As a consequence, the neutron stream cannot build up sufficiently to establish an explosive chain reaction, such as that in an atomic weapon. In contrast to the nuclear power plants, atomic weapons use highly enriched uranium in which neutrons are likely to encounter another U-235 nucleus.

As we noted earlier, a spontaneously explosive nuclear chain reaction (the explosion of an atomic bomb) will occur only if a critical mass of U-235 is quickly assembled in one place. Fortunately for our troubled world, pure U-235 is not easily plucked from the shelf of any military arsenal. Because U-235 and U-238 behave essentially the same in all chemical reactions, the separation of these two isotopes is *extremely* difficult.

However, U-235 and U-238 differ by the mass of three neutrons, and this mass difference can be exploited to achieve a separation. How? On average, lighter gas molecules move faster than heavier ones. So gas molecules containing U-235 should travel slightly more rapidly than their analogs containing U-238. One way to separate molecules of different masses is by **gaseous diffusion,** a process in which a gas is forced through a series of permeable membranes. Lighter gas molecules diffuse more rapidly through the membrane than heavier ones.

But uranium ore clearly is not a gas; rather, it is a mineral that contains UO_3 and UO_2. Most other uranium compounds are solids as well. However, the compound uranium hexafluoride (UF_6) has a notable property. Known as "hex," this compound is a solid at room temperature, but readily vaporizes when heated to 56 °C (about 135 °F). To produce hex, the uranium ore is converted to UF_4, which in turn is reacted with more fluorine gas.

$$UF_4(g) + F_2(g) \longrightarrow UF_6(g) \qquad [7.6]$$

On average, a $^{235}UF_4$ molecule travels about 0.4% faster than a $^{238}UF_6$ molecule. If the gaseous diffusion is allowed to occur over and over through a long series of permeable membranes, significant separation of the isotopes can be achieved. For more than four decades, uranium isotopes were separated by gaseous diffusion at the Oak Ridge National Laboratory in Tennessee.

In the United States, one commercial plant in Kentucky currently carries out uranium enrichment (Figure 7.14); another recently closed at Portsmouth, Ohio. As of 2004, large commercial enrichment plants are operating in the United Kingdom, the Netherlands, France, Germany, and the former Soviet Union with smaller ones elsewhere. Today, one other commercial separation method is in use, the centrifugation of UF_6 molecules. A third involving laser excitation is being developed in Australia.

By any enrichment method, once fissionable U-235 has been separated from U-238, depleted uranium remains. Nicknamed DU, **depleted uranium** contains almost entirely U-238 and has been depleted of the small amount of U-235 that it once naturally contained. Over 700,000 metric tons of depleted UF_6 is estimated to be in storage at the Paducah site in Kentucky. This compound is hazardous and requires safe storage both because of its *chemical* properties (fluorides are toxic) and because of its *nuclear* products. All isotopes of uranium are radioactive, a topic we discuss in Section 7.7. The military has also employed DU in the form of uranium metal.

The critical mass of U-235 is about 33 lb. See Section 7.2.

Equation 7.6 is a *chemical* equation, not a *nuclear* equation.

Figure 7.14

The transportation of "hex" at Paducah, Kentucky. A cylinder of enriched uranium product from the Paducah plant is being loaded into an autoclave of the transfer and shipping facility. Cylinders are weighed and selectively sampled for purity.

Consider This 7.15 Depleted Uranium

Depleted uranium, in the form of uranium metal, was used to construct antitank shells that were first used in the Gulf War in 1991 and later in several armed conflicts around the world, including Kuwait and Bosnia. In the Gulf War alone, the military estimates that over 300 tons of DU was dispersed. Which properties of depleted uranium make it useful to tip bullets and shells? Create a list of issues and points of controversy surrounding the use of DU.

7.6   Could Nuclear Fuel Be Diverted to Make Weapons?

Given the amount of processing required to produce highly enriched U-235 from lower reactor-grade fuel, diverting fuel from peaceful to military use is both difficult and costly. A more likely fissionable material for clandestine weapons' manufacturing is plutonium-239 (Pu-239), formed in a conventional reactor when the plentiful nuclei of U-238 absorb a neutron. The nuclear reaction proceeds in several steps. The first is similar to equation 7.1 in which U-235 absorbs a neutron and forms an unstable species. Rather than undergoing nuclear fission, the U-238 becomes more stable by a simple beta decay, which it does in a matter of an hour or so:

See the next section for more about beta decay.

$$\ _{0}^{1}\text{n} + \ _{92}^{238}\text{U} \longrightarrow [\ _{92}^{239}\text{U}] \longrightarrow \ _{93}^{239}\text{Np} + \ _{-1}^{0}\text{e} \qquad [7.7]$$

The new element formed, neptunium-239, is also a beta emitter. It undergoes radioactive decay to form plutonium-239, a fissionable isotope.

$$\ _{93}^{239}\text{Np} \longrightarrow \ _{94}^{239}\text{Pu} + \ _{-1}^{0}\text{e} \qquad [7.8]$$

See Section 7.12 for more about SNF and how it is handled as nuclear waste.

This transformation was discovered early in 1940. The chemical and physical properties of plutonium were determined with an almost invisible sample of the element on the stage of a microscope. The chemical processes devised on such minute samples were scaled up a billion-fold and used to extract plutonium from the spent fuel pellets from a reactor built on the Columbia River at Hanford, Washington. **Spent nuclear fuel (SNF)** is the radioactive material remaining in fuel rods after the fuel has been used to generate power in a nuclear reactor. The reactor was called a breeder reactor because it was designed primarily to convert U-238 to fissionable Pu-239 by means of neutron capture (equation 7.7). The plutonium was chemically separated from the uranium and used in the first test explosion of a nuclear device on July 16, 1945, in Alamogordo, New Mexico. The bomb dropped on Nagasaki a little less than a month later also was fueled by plutonium.

You will learn more about the radiation emitted by Pu-239 in Your Turn 7.16.

Plutonium-239 can also fuel nuclear reactors. Thus, a **breeder reactor** is one that creates both energy and new fuel (Pu-239) as it fissions the old fuel (U-235). This seems like a dream come true to an energy-hungry planet. France, the United Kingdom, Russia, Japan, and the United States have conducted research on breeder reactors that permit recovery of plutonium from spent fuel. However, such reactors represent another example of the very mixed blessings of modern technology. The problems are largely associated with the product, Pu-239. The radiation emitted by Pu-239 cannot penetrate the skin. Moreover, solid metallic plutonium is not easily absorbed into the body. But when plutonium is exposed to air, it reacts with oxygen to form plutonium oxide, PuO_2, a powdery compound. The PuO_2 dust can be easily dispersed and inhaled. Once it enters the body, plutonium is one of the most toxic elements known. A few micrograms (10^{-6} g) of PuO_2, lodged in the lungs, can induce lung cancer. Plutonium oxide can also slowly dissolve in the blood and be transported to other parts of the body, especially bone and liver, where its long-lived radioactivity can do serious damage.

Plutonium-239 poses an international problem because the plutonium produced in nuclear power reactors could possibly wind up in bombs (Figure 7.15). It has been widely speculated that the 1981 bombing of a nuclear facility in Iraq by war planes from Israel was done to prevent Iraq from being able to produce plutonium-containing nuclear weapons. If Iraq had succeeded in developing a nuclear arsenal, the outcome of Operation Desert Storm in 1991 might have been quite different. A more recent international crisis involved efforts to dissuade North Korea from building a reactor to produce plutonium. And, in 1998, United Nations weapons inspectors were denied access to sites in Iraq suspected of being nuclear weapons production facilities. Given the risks associated with Pu-239 and U-235, it is essential that the supplies and distribution of these isotopes be carefully monitored nationally and around the world. The United States for many years banned the reprocessing of commercial fuel elements. That ban was lifted in 1981, but no plutonium is currently being recovered from commercial reactors in this country. The price of uranium is currently so low that plutonium recovery is not competitive.

Figure 7.15
A smuggled canister of military grade Pu-239 captured in Germany.

Safeguarding nuclear materials has taken on a new meaning since the end of the Cold War and the demise of the former Soviet Union. One part of the problem is the plutonium and highly enriched uranium in Russia's nuclear arsenal (about 20,000 warheads). At present, though, these warheads are stored with relatively good security. Furthermore, any thief would find it difficult to remove the plutonium and uranium from the warheads. In contrast, Russia's legacy from the Cold War, a stockpile of highly enriched uranium and plutonium (about 600 metric tons), is far more accessible and hence far more threatening to world security. The fissionable materials stored in labs, research centers, and shipyards across the former Soviet Union are far more vulnerable to theft. These 600 tons of fissionable material translate into the capacity to construct approximately 40,000 new nuclear weapons.

The dangers of nuclear trafficking and the need for effective safeguards are now well recognized by the world community. Anita Nilsson, head of the Vienna-based International Atomic Energy Agency's Office of Physical Protection and Material Security, warned in 2001 that nuclear trafficking "has emerged as a real and dangerous threat." Since September 11, 2001, the danger has increased substantially, as it appears that terrorists will use whatever weapons they can get their hands on. As the main hurdle in constructing a nuclear device, even a primitive one, is acquiring a fissionable fuel, it is essential to safeguard all highly enriched uranium against theft. The future safety of nations, if not of our planet, may depend on our ability to safeguard (and ultimately recycle) highly enriched fissionable fuel.

The topics of enriched uranium, spent reactor fuel, fissionable stockpiles, and nuclear waste all rest on an understanding of the topic of radioactivity. We now turn to this topic.

Consider This 7.16 **The Reality of Reprocessing**

The supply of uranium in the United States (and the rest of the world) is large, but not limitless. By failing to reprocess spent nuclear fuel, we are discarding a potential source of energy that some European countries are presently tapping. Is the current American practice justified? Why is it different from the European practice? List arguments on both sides of this issue and then take a stand.

7.7 What Is Radioactivity?

Radioactivity was discovered accidentally in 1896 by Antoine Henri Becquerel (1852–1908). The French physicist found that when a uranium-containing mineral sample was placed on a photographic plate that had been wrapped in black paper, the plate's light-sensitive emulsion darkened. It was as though the plate had been exposed to light. Becquerel immediately recognized that the mineral itself was emitting a powerful form of radiation that penetrated the light-proof paper. Further investigation by the Polish scientist Marie Curie (1867–1934) (Figure 7.16) revealed that the rays were coming from the element uranium, a constituent of the mineral. In 1899, Marie Curie applied the term **radioactivity** to this spontaneous emission of radiation by certain elements. Subsequent research by Ernest Rutherford (1871–1937) in Canada and England led to the identification of two major types of radiation. Rutherford named them after the first two letters of the Greek alphabet, alpha (α) and beta (β).

Alpha and beta radiation have strikingly different properties. A **beta particle (β)** is a high-speed electron emitted from a nucleus. A beta particle has a negative electrical charge ($1-$) and only a tiny bit of mass, about 1/2000 that of a proton or a neutron. In contrast, an **alpha particle (α)** is positively charged ($2+$) and consists of the nucleus of a helium atom—two protons and two neutrons. It is far heavier than an electron, and since it has no electrons to accompany the helium nucleus, it has a $2+$ charge.

It was subsequently discovered that a third type of radiation, gamma radiation, frequently accompanies the emission of alpha or beta radiation. Unlike alpha and beta

Figure 7.16

Marie Skłodowska Curie won two Nobel Prizes—one in chemistry, the other in physics—for her basic research on radioactive elements.

Table 7.2		**Types of Nuclear Radiation**		
Type	**Symbol**	**Consists of**	**Charge**	**Change to nucleus that emits it**
Alpha	$_2^4He$	2 protons 2 neutrons	2+	The mass number decreases by 4, and the atomic number decreases by 2.
Beta	$_{-1}^0e$	an electron	1−	The mass number does not change, and the atomic number increases by 1.
Gamma	$_0^0\gamma$	photon of energy	0	No change in either the mass number or in the atomic number.

Gamma rays were introduced as part of the electromagnetic spectrum in Chapter 2.

radiation, gamma rays do not consist of particles. Rather, **gamma rays (γ)** are made up of high-energy, short-wavelength photons of energy. Just like infrared (IR), visible, and ultraviolet (UV) light rays, gamma rays are part of the electromagnetic spectrum. Gamma rays have no charge and no mass. In terms of their energy, they are similar in energy to X-rays (and can be even higher), and you will learn more about them in connection with food preservation in Chapter 11.

Table 7.2 summarizes the properties of these three types of radiation. Watch out for the word *radiation*. In a scientific conversation, people often just say "radiation" and assume that from the context that the listener will understand whether they mean electromagnetic radiation or nuclear radiation. *Electromagnetic radiation* refers to all the different types of light, including visible light and ultraviolet light. In fact, it is perfectly correct to say visible radiation instead of visible light (and the same for ultraviolet radiation). Gamma rays also are a type of electromagnetic radiation. However, *nuclear radiation* refers to the radiation emitted by nuclei, such as alpha, beta, or gamma radiation. To add to the confusion, gamma rays are both a type of electromagnetic radiation and a type of nuclear radiation.

Your Turn 7.17   Radiation

Nuclear radiation or electromagnetic radiation? For each sentence, decipher the meaning of the word *radiation* from the context.

a. Name a type of radiation that has a shorter wavelength than visible light.
b. Gamma radiation can penetrate right through the walls of your home.
c. Watch out for UV rays! If you have lightly pigmented skin, this type of radiation can give you a sunburn.
d. Rutherford detected the radiation emitted by uranium.

Answer
a. Electromagnetic radiation (UV, gamma rays, and X-rays all have a shorter wavelength than visible light.)

Whenever an alpha or beta particle is emitted during radioactive decay, a remarkable transformation occurs: the emitting atom changes its identity. For example, when an atom of uranium-238 emits an alpha particle, it becomes an atom of thorium-234. Such a change was long sought, but never achieved by the ancient alchemists who wanted to turn lead and other common metals into gold. You saw a nuclear equation for an alpha emission earlier with the PuBe neutron source (Section 7.3). Similarly, the alpha emission by uranium-238 to form thorium-234 can be represented by a nuclear equation.

$$\ce{^{238}_{92}U} \longrightarrow \ce{^{234}_{90}Th} + \ce{^{4}_{2}He} \qquad [7.9]$$

For any alpha emission, the mass number of the product will be 4 lower. The loss of an alpha particle—2 protons and 2 neutrons—accounts for this. The nuclear equation for alpha emission shows that the sum of the mass numbers on both sides is equal: $234 + 4 = 238$. The same is true for the sum of the atomic numbers: $90 + 2 = 92$.

Sometimes the product formed in a nuclear equation still is radioactive. For example, the thorium-234 produced in equation 7.10 is a beta emitter.

$$\ce{^{234}_{90}Th} \longrightarrow \ce{^{234}_{91}Pa} + \ce{^{0}_{-1}e} \qquad [7.10]$$

Just as with alpha emission, beta emission transforms one element into another. Here, Th-234 (atomic number 90) decays to produce Pa-234 (atomic number 91). The number of protons increases by one when a beta particle is ejected from a nucleus. You saw this same increase earlier (equation 7.8) when neptunium (atomic number 93) decayed to plutonium (atomic number 94) by beta decay.

You may be wondering how an electron (a beta particle) can be emitted from the nucleus when there *are* no electrons in the nucleus. One model that can help you make sense of this is to regard a neutron as consisting of a proton and an electron. The loss of an electron (a beta particle) occurs when a neutron is "broken down" into a proton and an electron, as shown in equation 7.11.

$$\ce{^{1}_{0}n} \longrightarrow \ce{^{1}_{1}p} + \ce{^{0}_{-1}e} \qquad [7.11]$$

The mass number (neutrons plus protons) in the nucleus remains constant during beta emission because the loss of the neutron is balanced by the formation of a proton. In nuclear equation 7.10, both the thorium and the protactinium have the same mass number of 234. When the beta particle was emitted, a neutron in thorium "became" a proton in protactinium. Again, regard this as a useful way to think of the process of beta emission rather than the actual process that occurs.

Your Turn 7.18   Alpha and Beta Decay

a. Write a nuclear equation for the beta decay of rubidium-86 (Rb-86), a radioisotope that can be produced by the fission of U-235.

b. Plutonium-239, a toxic isotope that causes lung cancer, is an alpha emitter. Write the nuclear equation.

c. Iodine-131 is a radioisotope used medically to measure thyroid gland activity. I-131 is a beta emitter. Write the nuclear equation.

Answer

a. $\ce{^{86}_{37}Rb} \longrightarrow \ce{^{86}_{38}Sr} + \ce{^{0}_{-1}e}$

How do you know if a particular isotope is stable or radioactive? *All* isotopes of *all* elements with atomic number 84 (polonium) and higher are radioactive. These radioactive elements include uranium, plutonium, radium, and radon, as well as others that we will discuss later in this chapter. And for a particular isotope such as uranium-238, thorium-234 or plutonium-239, how do you know if it is an alpha, beta, or gamma emitter? To answer this question, you have to look up the information up in a table of isotopes.

What about all the rest of the elements with atomic numbers less than 84? Most of the atoms that make up our planet are *not* radioactive, that is, they are stable. They are here today, and you can count on their being here tomorrow, although possibly not located in the same spot you last saw them (such as the atoms that make up your car keys). Nonetheless, some naturally occurring lighter atoms on our planet are radioactive,

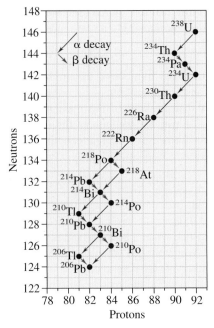

Figure 7.17
The U-238 radioactive decay series.

such as carbon-14, hydrogen-3 (tritium), and potassium-40. Whether a particular isotope is radioactive (a **radioisotope**) or stable (a nonradioactive isotope) depends on the ratio of neutrons to protons in its nucleus. By emitting alpha, beta, or other radiation, radioactive nuclei change this neutron-to-proton ratio until a stable ratio is achieved, and the nucleus is no longer radioactive.

In some cases, radioisotopes may require several decay steps to produce a stable isotope. For example, the radioactive decay of U-238 and Th-234 (equations 7.9 and 7.10) are the first two steps in a much longer series. Naturally occurring U-238 decays through 14 steps until it reaches Pb-206, a stable isotope (Figure 7.17). This is called a **radioactive decay series,** that is, a characteristic pathway of radioactive decay for a set of radioisotopes. Radon, a radioactive gas, is produced midway in this series. Thus, wherever uranium is present, radon will be present as well. Radon was discussed earlier as an indoor air pollutant in Chapter 1.

Similarly, the radioactive decay series of U-235 goes through 11 steps to produce a stable isotope of Pb-207.

7.8 What Hazards Are Associated with Radioactivity?

It may come as a surprise that the radiation exposure an American citizen receives from living near a nuclear power plant is about one tenth the radiation he or she would get during a coast-to-coast trip on a commercial flight. Even under adverse circumstances, radiation exposure from nuclear power plants still can be low. One nuclear accident stands as a notable exception—the explosion at Chernobyl discussed in Section 7.4.

The evidence of the past makes clear that it would be a serious mistake to dismiss radioactivity as harmless. Unfortunately, some of the first scientists to study radioactivity, including Marie Curie, were not fully aware of the dangers inherent in the phenomenon. Madame Curie died of a type of leukemia that most likely was induced by her exposure to radiation. Often alpha and beta particles or gamma rays have sufficient energy to ionize the atoms and molecules they strike. The resulting changes in molecular structure can have profound effects on living things. Rapidly growing cells are particularly susceptible to damage, a fact that makes radiation ideal to treat certain kinds of cancer. But other rapidly dividing cells are damaged as well, such as those in the bone marrow, the skin, hair follicles, stomach, and intestines. **Radiation sickness** is the

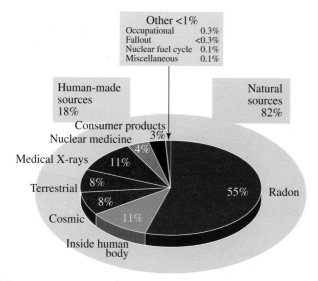

Figure 7.18

U.S. sources of background radiation. Natural sources of radiation account for about 82% of all natural exposure, and human-made sources account for the remaining 18%.

Source: National Council on Radiation Protection and Measurements (NCRP) Report No. 93, *Ionizing Radiation Exposure of the Population of the United States,* 1987.

illness produced by exposure to large amounts of radiation, characterized by early symptoms of anemia, malaise, and susceptibility to infection. Radiation sickness affected those near Chernobyl and the survivors of the atomic blasts at Hiroshima and Nagasaki. Radiation-induced transformations of DNA also can produce harmful genetic mutations, some leading to cancer and birth defects.

Today, considerable care is taken to shield medical, laboratory, and other workers from nuclear radiation. Protective shielding made of lead and other dense metals is used to absorb much of the radiation in the workplace. However, it is impossible to be fully protected from exposure to radiation. In fact, one might not even want to be completely shielded, as tiny amounts of radiation may play a therapeutic role. In either case, radioactive substances are a natural part of our world. **Background radiation** is the average amount of radiation we each are exposed to daily (Figure 7.18). About 80% of all background radiation is natural in origin. The largest natural source is radon, a radioactive gas released in the radioactive decay series of uranium in the soils and rocks (see Figure 7.17).

> Radon was described earlier in Section 1.14 as an indoor air pollutant.

Your Turn 7.19   Radon

Radon-222 is an alpha emitter. Write the nuclear equation for the decay process. Would you expect the product to be radioactive?

Answer

$$^{222}_{86}\text{Rn} \longrightarrow {}^{218}_{84}\text{Po} + {}^{4}_{2}\text{He}$$

The product, polonium-218, is radioactive, as are all elements above atomic number 84.

The amount of background radiation you receive depends on where you live, what type of residence you live in, the number of people you live with, and how close you get to them. Earth itself, the building materials quarried or manufactured from it, and our food contain some naturally occurring radioactive atoms. The late Isaac Asimov, a prolific science writer, pointed out in one of his many books that a human body contains approximately 3.0×10^{26} carbon atoms, of which 3.5×10^{14} are radioactive carbon-14 atoms. With each breath you inhale about three and a half million (3.5×10^{6}) carbon-14 atoms.

Sceptical Chymist 7.20 | **Radioactive Carbon in Your Body**

Assume that Isaac Asimov's figures are correct, and that 3.5×10^{14} of the 3.0×10^{26} carbon atoms in your body are radioactive. Calculate the fraction of carbon atoms that are radioactive carbon-14.

To better evaluate the hazards of radioactive substances, we need units to measure radiation and the damage it can cause. One simple way to measure radioactivity of a substance is in terms of the number of nuclear decays (alpha, beta, or gamma) in a given period. The **curie (Ci)** is a measure of radioactivity and is equivalent to the number of decays per second from one gram of radium.

$$\text{One curie} = 1 \text{ Ci} = 3.7 \times 10^{10} \text{ disintegrations/second}$$

This unit was named in honor of Marie Curie, and the disintegrations refer to alpha, beta, and gamma emissions.

Radium is a highly radioactive substance and one curie is a large amount of radiation. Accordingly, people typically measure radioactivity in *milli*curies (mCi), *micro*curies (μCi), *nano*curies (nCi) or even *pico*curies (pCi). For example, household radon measurements are quoted in picocuries, as you will see in Your Turn 7.24. Chemists or biochemists using radioactive samples in the lab typically work with millicuries or microcuries of radioactive substances. If a laboratory worker had a spill as large as 100 millicuries, serious cleanup procedures would be needed. In contrast, a spill of 100 microcuries would not.

To put these values in perspective, the explosion at Chernobyl spewed 100–200 million curies into the atmosphere. In terms of radioactivity, this is the equivalent of dispersing 100–200 million grams of radium. At the time of the accident, radiation near Chernobyl was measured from 5 to over 40 Ci per square kilometer, as shown by the map in Figure 7.12. The amount of radiation released by the atomic bombs that exploded on Nagasaki and Hiroshima was 1/100 as much—lower by two orders of magnitude.

Other units can help evaluate the hazards as well. For example, rather than focusing on the radioactive sample (as the curie does), other units focus on the biological damage. In part, this depends on the total amount of energy absorbed by the tissue. A unit called the **rad,** short for radiation absorbed dose, is defined as the absorption of 0.01 joule of radiant energy per kilogram of tissue. Thus, if a 70 kg person were to absorb 0.70 J of energy, he or she receives a dose of 1 rad. Although this is not very much energy, the energy is localized right where the radiation hits and can damage molecules.

But the biological damage to an organism is more than simply a matter of the energy deposited. Some types of radiation are more destructive because of how energetically they hit the molecules in cells and tissues. Therefore, to estimate the potential physiological damage, another unit is created by multiplying the number of rads by a factor Q that is set by the type of radiation, the rate the radiation is delivered, and the type of tissue. Less damaging types, including beta, gamma, and X-rays, are arbitrarily assigned a Q of 1. Highly damaging types of radiation, such as alpha particles and high-energy neutrons, have a Q as high as 20. The **rem,** short for "roentgen equivalent man," is Q multiplied by the number of rads.

$$\text{Number of rems} = Q \times (\text{number of rads})$$

Thus, a 10-rad dose of alpha radiation may be as high as 200 rem, because of the damage done by alpha particles. For beta and gamma radiation, a dose of 10 rads is 10 rems, the same. The number of rems in a dose of radiation exposure is thus a measure of the power of the radiation to cause damage to human tissue. It would take approximately 20 times as many beta particles to do the same damage as a given amount of alpha particles. This is to be expected, because alpha particles are larger and deposit a greater amount of energy in the target tissue.

Prefixes
Pico: 1×10^{-12}
  1/1,000,000,000,000
Nano: 1×10^{-9}
  1/1,000,000,000
Micro: 1×10^{-6}
  1/1,000,000
Milli: 1×10^{-3}
  1/1,000

Q is called the relative biological effectiveness, sometimes referred to as RBE.

Table 7.3		Physiological Effects of a Single Dose of Radiation
Dose (rem)	**Dose (Sv)**	**Likely effect**
0–25	0–0.25	No observable effect
25–50	0.25–0.5	White blood cell count decreases slightly
50–100	0.5–1	Significant drop in white blood cell count, lesions
100–200	1–2	Nausea, vomiting, loss of hair
200–500	2–5	Hemorrhaging, ulcers, possible death
>500	>5	Death

The scientific community in the United States, as well as in most of the world, no longer uses the rem. Instead, the **sievert (Sv)** is employed, which is equal to 100 rem. Alternatively, 1 rem equals 0.0100 Sv. The likely effects of a single dose of radiation at various levels are given in Table 7.3. The rem is a smaller unit than the sievert, so the values for the same dose of radiation are higher in rems than in sieverts.

Because most doses of radiation are significantly less than one sievert (or one rem), smaller units such as microsieverts (μSv) and millirems (mrem) are employed.

1 microsievert is 0.1 millirem.

$$1 \text{ microsievert} = 1/1{,}000{,}000 \text{ of a sievert} = 1 \times 10^{-6} \text{ Sv} = 1 \text{ μSv}$$

$$1 \text{ millirem} = 1/1000 \text{ of a rem} = 1 \times 10^{-3} \text{ rem} = 1 \text{ mrem}$$

Table 7.4 uses microsieverts to report the radiation exposure associated with various activities and lifestyle factors. This information is the basis for estimating your personal annual radiation exposure in Your Turn 7.21. Once you have completed this exercise, you can check Table 7.4 to see how your exposure compares with that of the average American.

Your Turn 7.21   Your Personal Radiation Exposure

Use Table 7.4 to estimate the approximate radiation dosage you receive each year.

Nearly 3000 μSv or about 4/5 of the 3600 μSv absorbed per year by a typical U.S. resident comes from natural background sources—mostly radon, cosmic rays, soil, and rock—and not from human-related sources. Of the 670 μSv of human-produced radiation absorbed annually, about 600 μSv is attributable to medical procedures such as diagnostic X-rays. As is evident from Table 7.4, the radiation emitted by a properly operating nuclear power plant is negligible compared with normal background radiation, including the natural radiation of your own body. For example, about 0.01% of all the potassium ions (K^+) that are essential to your internal biochemistry are K-40, a radioactive isotope. These K-40 ions give off about 200 μSv per year, approximately 1000 times more radioactivity than that received as a result of living within 20 miles of a nuclear power plant. In fact, although bananas are rich in K^+, a steady diet of them will not increase your personal radioactivity because K^+ is not retained.

Thousands of K-40 nuclei undergo radioactive decay in your body each second.

To further put things in perspective, note that the immediate physiological effects of radiation exposure are generally not observable below a single dose of 0.25 Sv (250,000 μSv; see Table 7.3). This is nearly 70 times the average annual exposure. However, there is still uncertainty about the long-term effects of low doses of radiation. The assumption is usually made that there is no threshold below which no damage occurs. However, the effects of low doses are so small and the time span so great that scientists have not been able to make reliable measurements. Moreover, tests with animals are not always reliable because there is considerable species-to-species variation in the effect of radiation.

Table 7.4	Your Annual Radiation Dose	
Source of Radiation		**(μSv/yr)**
1. Location of your town or city		
a. Cosmic radiation at sea level (U.S. average 260 μSv*)		260
b. Additional dose if you are above sea level		___
1000 m (3300 ft) add 100 μSv		
2000 m (6600 ft) add 300 μSv		
3000 m (9900 ft) add 900 μSv		
2. House construction		
Building materials contain tiny amounts of radioisotopes.		
Brick (700 μSv); wood (300 μSv); concrete (70 μSv)		___
3. Ground		
Radiation from rocks and soil (U.S. average)		260
4. Food, water, and air (U.S. average)		400
5. Fallout from nuclear weapons testing (U.S. average)		40
6. Medical and dental X-rays		
a. Chest X-ray (100 μS)		___
b. Gastrointestinal tract X-ray (5000 μSv)		___
c. Dental X-rays (100 μSv each visit)		___
d. Other X-rays (estimate)		___
7. Jet travel (exposure to cosmic radiation)		
A 5-hour flight at 30,000 ft is 30 μSv		___
8. Other		
Live within 50 miles of a nuclear plant site, add 0.09 μSv		___
Live within 50 miles of a coal-fired power plant, add 0.3 μSv		___
Use a computer terminal, add 1 μS		___
Go through X-ray check stations at airports, add 0.02 μSv		___
Smoke 1.5 packs of cigarettes a day, add 13,000 μSv		___
Your Total Annual Dose of Radiation		___
U.S. annual average = 3600 μSv		

*Based on the "BEIR Report III," National Academy of Sciences, Committee on Biological Effects of Ionizing Radiation 1987. *The Effects on Populations of Exposure to Low Levels of Ionizing Radiation.* Washington, DC: National Academy of Sciences.

Other models are possible. For example, low doses of a harmful substance (such as radiation) may actually be beneficial. This idea is called hormesis. End-of-chapter question #56 allows you to further investigate hormesis.

The issue then is how to extrapolate the known high-dose data to low doses. Two dose-response models are illustrated in Figure 7.19. The first, the more conservative of the two, is the **linear, nonthreshold model.** This assumes a linear relationship between the adverse effects and the radiation dose, with radiation being harmful at all doses, even low ones. Thus, if the adverse effect is getting cancer, doubling the radiation dose doubles the incidence of cancer and tripling it causes three times as much. No cellular repair of the damage caused by radiation is assumed to take place, even at low doses. Setting no threshold for the effects of radiation continues to be controversial. Dr. John Boice Jr., the director of the International Epidemiology Institute in Maryland, notes that the controversy relates to just how low the effects are: "Are they really low or are they really, really, low?" However, an exhaustive report (263 pages!) compiled in 2002 by the National Council on Radiation Protection and Measurement stated that there is "no conclusive evidence" on which to reject the linear, nonthreshold model, noting that "it may never be possible to prove or disprove the validity."

The model represented by Figure 7.19(b) makes a different assumption about low doses: they give no observable adverse effects because cellular repair takes place. Thus, until a threshold is reached where the damage cannot be repaired, the response curve does not rise. By this model, low doses of radiation are safe.

Currently, the linear, nonthreshold model is being used by federal agencies such as the Environmental Protection Agency (EPA) in setting exposure standards. Stephen

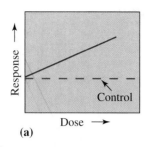
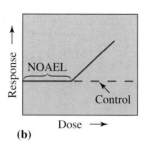

(a)   **(b)**

Figure 7.19

Two dose-response curves for radiation.
(**a**) linear, nonthreshold model. (**b**) threshold model.

Note: NOAEL means no observed adverse effect level.

Source: *Environmental Science & Technology,* "Redrawing the Dose-Response Curve," Volume 38, No. 5, March 1, 2004, page 90A.

Page, the director of EPA's Office of Radiation and Indoor Air, notes that with the non-threshold model "the risk from radiation is within the allowable range from toxic chemicals, 1 in 10,000 to 1 in a million chances of developing cancer." But by being more conservative, this model also is more costly to implement.

Consider This 7.22   **Radiation Dose Response**

You have just read about two models for the dose-response curve for low level radiation (see Figure 7.19). The linear, nonthreshold model represents the more stringent hypothesis used by federal agencies in setting exposure limits. The ramifications of adopting a specific model are both biological and economical. The more stringent linear dose model requires stricter limits on workers' acceptable radiation dose limit than the no observed adverse effect model. By using the linear model, are we being "better safe than sorry" or are we wasting a lot of money protecting ourselves from an emotional issue without looking at the science behind it?

   As a nuclear technician who must operate under the stricter federal limits for radiation safety at a hospital, write a letter to an interested friend giving your position on this issue and the reasons for it.

7.9   How Long Will Nuclear Waste Products Remain Radioactive?

It is hard to exaggerate the problems of nuclear waste. The fission products formed in nuclear reactors can indeed have dangerously high levels of radioactivity. These high levels will persist for thousands of years, as the radioactive decay process cannot be speeded up. The fact that most of us experience considerably more radioactivity from natural sources than from artificial ones should not lull us into a false sense of security. Nuclear waste disposal presents formidable problems.

   In part, the problem arises because one radioactive isotope may decay into a series of other radioisotopes. For example, the U-238 remaining in the spent fuel pellets decays to Th-234, which in turn yields Pa-234 (see Figure 7.17) and next U-234, and then Th-230, and then radium and radon . . . and so forth. Each must be contained in turn, until a stable isotope of lead is reached.

   A particularly significant consideration in the storage and disposal of radioactive waste is the *rate* at which radioactivity declines. Depending on the particular isotope, the decline can occur very rapidly over a short time or very slowly over a much longer period. The rate of decay is typically reported in terms of the **half-life,** the time required

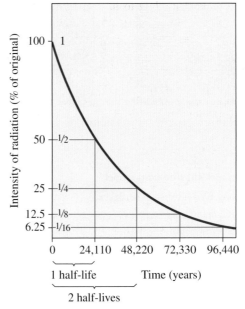

Figure 7.20

Radioactive decay of Pu-239.

See nuclear equations 7.7 and 7.8
for the production of plutonium.

for the level of radioactivity to fall to one half of its value. For example, plutonium-239, the alpha-emitting isotope formed in uranium reactors, has a half-life of 24,110 years. This means that it will take 24,110 years for the radiation intensity of any sample of Pu-239 waste to drop to one-half its value. At the end of a second half-life of another 24,110 years, the radiation will be one fourth the original level. And in three half-lives (a total of 72,330 years), the level will be one eighth of the original (Figure 7.20).

The half-life of a particular isotope is a constant and is independent of the physical or chemical form in which the element is found. Moreover, the rate of radioactive decay is essentially unaltered by changes in temperature and pressure. But, when various radioisotopes are compared, their half-lives are found to range from milliseconds to millennia. For example, the half-life of uranium-238 (see equation 7.9) is 4.5 billion years. Coincidentally, this is approximately the age of the oldest rocks on Earth, a determination made by measuring their uranium content. By contrast, other radioisotopes have much shorter half-lives (Table 7.5). Note that different isotopes of the same element have different half-lives.

Table 7.5	Half-Lives for Selected Isotopes
Radioisotope	**Half-life**
Uranium-238	4.5×10^9 years
Potassium-40	1.3×10^9 years
Plutonium-239	24,110 years
Carbon-14	5715 years
Cesium-137	30.2 years
Strontium-90	29.1 years
Thorium-234	24.1 days
Radon-222	3.82 days
Iodine-131	8.04 days
Plutonium-231	8.5 minutes
Polonium-214	0.00016 seconds

Each radioisotope decays according to its own clock. We can use half-life values to do calculations based on this clock. For example, plutonium-239, discovered in 1941, has a long half-life of about 24,000 years. Over a dozen other isotopes of plutonium have been characterized, but only very recently in 1999 did researchers identify a sample of Pu-231 which has a much shorter half-life—a matter of mere minutes. These researchers had to work fast! For example, once Pu-231 was generated, what percent of a sample would remain after about 25 minutes? In contrast, what percent of uranium-238 would remain after 25 minutes? The half-lives of both isotopes are listed in Table 7.5.

To answer the first question about Pu-231, recognize that 25 minutes is about three half-lives (8.5 minutes × 3). After one half-life, 50% of the sample has decayed and 50% remains. After two half-lives, 75% of the sample has decayed and 25% remains. And after three half-lives, 87.5% has decayed and 12.5% remains. Note that this question could have been phrased as "After 25.5 minutes (exactly three half lives), what percent remains?" However, when estimating radioactivity, it is helpful to be able to make a quick estimate. Answering the second question is easier. With such a short period of time, a mere instant in the span of a billion years, essentially all of the uranium-238 would remain.

Carola Laue and Darleane Hoffman, working with a team at Lawrence Berkeley National Laboratory, characterized plutonium-231.

Your Turn 7.23 | Tritium

Hydrogen-3 (tritium) is sometimes formed in the primary coolant water of a nuclear reactor. Tritium is a beta emitter with a half-life of 12.3 years. After how many years will only about 12% of the original tritium remain?

Your Turn 7.24 | Radon

Radon-222 is a radioactive gas produced from the decay of radium, a radioisotope naturally present in many rocks. Its half-life is in Table 7.5.

a. Where did the radium in rocks come from? *Hint:* See Figure 7.17.
b. Radon activity is usually measured in picocuries (pCi). Suppose that the radioactivity from Rn-222 in your basement was measured as 16 pCi (which would be high). If no additional radon entered the basement, how much time would pass before the radiation level fell to 0.50 pCi? *Hint:* Note that in dropping from 16 to 1 pCi, the radiation level is reduced by half four times: 16 to 8 to 4 to 2 to 1. This corresponds to four half-lives, each 3.82 days.
c. Why is it incorrect to assume that no more radon will enter your basement?

Short-lived radioisotopes can be useful for the diagnosis and treatment of illnesses. For example, small amounts of iodine-131 (half-life about 8 days) in the form of potassium iodide is used to treat hyperthyroidism in persons with Graves' disease. In this procedure, the orally administered radioactive iodide ions concentrate in the overactive thyroid gland, fully or partially destroying it (Figure 7.21). In most patients, thyroxin, the iodine-containing hormone normally secreted by the thyroid, must then be supplemented with a synthetic substitute.

On the other hand, radioisotopes of iodine can be quite dangerous. Because they are a product of nuclear fission, radioactive iodine can enter the biosphere in the chemical form of iodide ion in the case of either an atomic blast or a reactor accident. When taken up by the thyroid gland in larger amounts, radioactive iodine can cause thyroid cancer. The incidence of thyroid cancer among children in the Chernobyl radioactive fallout area seems significantly higher than normal, and iodine-131 has been implicated (Figure 7.22).

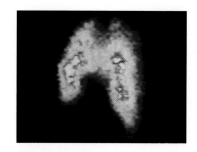

Figure 7.21
A thyroid image produced with I-131. The yellow and red regions show the area of the thyroid gland in which radioactive iodine has concentrated.

See end-of-chapter question #57 to learn more about taking potassium iodide tablets in case of exposure to radioactive iodine.

Figure 7.22
A child being checked for thyroid function at a clinic north of Minsk, near the Chernobyl nuclear power plant.

Carbon-14 dating was used to establish the age of the famous Shroud of Turin.

| **Your Turn 7.25** | **"Iodine"** |

When people speak of radioactive iodine, they may be referring to iodine in the chemical form of an iodine atom, an iodine molecule, or an iodide ion. Write Lewis dot structures to distinguish among these three chemical species. Which chemical form is used in medical imaging?

Strontium-90 is another particularly dangerous isotope found as a product of nuclear fission. It entered the biosphere in the 1940s and 1950s from the fallout from the atmospheric testing of nuclear weapons. Strontium ions are chemically similar to calcium ions; both elements are in Group 2A of the periodic table. Hence, strontium (Sr^{2+}), like Ca^{2+}, concentrates in milk and bone. Once ingested, the radioactive strontium poses a lifelong threat because of its 29.1-year half-life. Like I-131, Sr-90 is among the fission products produced in nuclear reactors, and persons living near Chernobyl were exposed to harmful levels of both.

On a cheerier note, we turn to carbon-14, a well-known radioisotope used for carbon-14 dating. With its half-life of 5715 years, carbon-14 decays to nitrogen-14 through the process of beta decay. Carbon-14 is naturally occurring because it is produced in our upper atmosphere by the interaction of cosmic rays. This radioactive carbon gets incorporated into carbon dioxide molecules and diffuses down into the troposphere where we live and breathe. Our atmospheric carbon dioxide contains a constant steady-state ratio of one radioactive carbon-14 atom for every 10^{12} atoms of nonradioactive carbon-12. Living plants and animals incorporate the isotopes in that same ratio. However, when the organism dies, exchange of CO_2 with the environment ceases. Thus, no new carbon is introduced to replace the C-14 converted to nitrogen-14 by beta decay. As a consequence, the concentration of C-14 decreases with time, dropping by half every 5715 years.

In the 1950s, W. F. Libby first recognized this decrease by experimentally measuring the C-14/C-12 ratio in a sample. The ratio provided an estimate of when the organism died. Human remains and many human artifacts contain carbon, and fortunately, the rate of decay of C-14 is a convenient one for measuring human activities. Charcoal from prehistoric caves, ancient papyri, mummified human remains, and suspected art forgeries have all revealed their ages by this technique. The C-14 technique provides ages that agree to within 10% of those obtained from historical records, thus validating the legitimacy of the radiocarbon dating technique.

7.10 How Will We Dispose of Waste from Nuclear Plants?

The experience of more than 50 years suggests that the answer to this vitally important question is, "slowly and with difficulty." Whether that is the correct answer is another matter. In a June 1997 *Physics Today* article, John Ahearne, past chair of the U.S. Nuclear Regulatory Commission, reminds us that, "... Like death and taxes, radioactive waste is with us—it cannot be wished away...."

High-level radioactive waste (HLW) has high levels of radioactivity and, because of the long half-lives of the radioisotopes involved, requires essentially permanent isolation from the biosphere. It consists of the radioactive materials that result from the reprocessing of spent nuclear fuel (SNF) and comes in a variety of chemical forms, including ones that are highly acidic or basic. It also contains heavy metals that are toxic. Thus, HLW is sometimes labeled a "mixed waste" in that it is hazardous *both* because of the chemicals it contains *and* from their radioactivity. Furthermore, this waste also presents a security risk because it contains fissionable plutonium that could be extracted and used to construct nuclear weapons.

Federal statutes define HLW by its source rather than by its chemical and nuclear characteristics. For example, the waste created when fuel is reprocessed to produce plutonium for military uses or that from commercial nuclear power plants is HLW. HLW also includes several other highly radioactive materials that require permanent isolation. Approximately 99% of the volume of HLW in the United States originated in the nuclear weapons industry. In 1996, the U.S. Department of Energy reported that the accumulated high-energy radioactive wastes generated by the Department of Defense occupied a total volume of approximately 350,000 m^3 with a radioactivity of about 900 million curies. This volume corresponds to nine football fields covered to a depth of 30 feet. Military waste is in the form of solutions, suspensions, slurries, and salt cake stored in barrels, bins, and tanks. In comparison, each commercial nuclear reactor each year typically produces about 30 tons of spent fuel, a much smaller amount.

Spent nuclear fuel (SNF), is regulated as HLW. Radioactive spent fuel is an unavoidable by-product of nuclear reactors. After about three or four years of use, the U-235 concentration in the fuel rods of a nuclear power reactor, initially at 3–5%, drops to the point where it is no longer effective in sustaining the fission process. Approximately one fourth to one third of the fuel rods are replaced annually, on a rotating schedule. Nevertheless, the spent fuel rods are still "hot"—both in temperature and radioactivity. They contain various isotopes of uranium plus plutonium-239 formed by the capture of neutrons by U-238, and a wide variety of fission products such as iodine-131, cesium-137, and strontium-90. Remotely controlled machinery, operated by workers protected by heavy shielding, removes the spent rods from the reactor and replaces them with new fuel rods.

The spent rods are transferred to on-site deep pools where they are cooled by water containing a neutron absorber. As of January 1, 1997, over 34,000 metric tons of SNF have been discharged from the nation's commercial reactor sites. This figure does not include the spent waste from defense reactors or nuclear submarines, and is estimated to reach 52,000 metric tons by 2005. Most of this waste is currently under water in storage pools on site at the nuclear plant where it was used (Figure 7.23). For example, the storage facility at Seabrook is a 34-foot deep steel-lined concrete pool in a secure building. It has the capacity to hold up to 25 years' worth of nuclear waste.

The on-site storage at nuclear power plants of high-level radioactive waste for 25 years is hardly ideal. John Ahearne points out that "Almost all of the [spent fuel] waste is currently being stored at the sites where it was generated, in facilities that were not built for long-term storage." The initial plans, begun in the 1950s and early 1960s, had been to reprocess the spent fuel to extract plutonium and uranium from it and recycle these elements as nuclear fuel to produce additional energy. Storage capacity for spent fuel rods on site was designed with such reprocessing in mind. However, only one of several planned reprocessing plants actually went into operation, and then only briefly (1967–1975). Thus, reprocessing never was capable of keeping up with the rate of spent fuel production, about 2000 tons every year. In 1977, then President Jimmy Carter, a nuclear engineer, declared a moratorium on commercial nuclear fuel reprocessing that

> Unlike uranium, all isotopes of plutonium can be used to construct nuclear weapons.

> Spent nuclear fuel was mentioned earlier in Section 7.6.

Figure 7.23
Spent cooling rods from a nuclear power reactor in a cooling chamber.

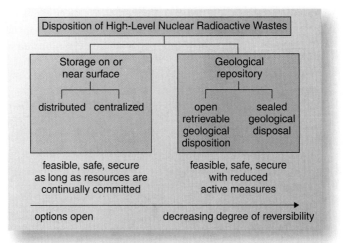

Figure 7.24

Methods of high-level nuclear waste deposition.

Source: *Disposition of High-Level Wastes and Spent Nuclear Fuel,* National Academy Press, 2000.

Each of the 103 U.S. commercial nuclear reactors produces about 20 tons of spent fuel annually.

continues to this day. Long-term geological storage of HLW, first proposed in 1957 by the National Academy of Sciences, became the alternative option. It is estimated that, by 2010, the country will face a total disposal problem of more than 100,000 tons of military and civilian high-level nuclear waste. The year is significant, because it is the earliest date when a permanent underground repository can possibly open. Given past and recent experience, 2010 seems an unrealistic goal.

Two feasible options for the storage of HLW are now under consideration: monitored storage on or near the surface, and storage in geological repositories deep underground. These differ in a key variable: *active management* (Figure 7.24). In surface storage, human societies over thousands of years must commit resources to maintain the integrity of the wastes. In geological repository storage, the wastes may be accessible and retrievable (although less easily) or sealed "forever," requiring minimal human vigilance. In a report published in 2000 by the National Academies of Science, the latter option of deep underground storage was favored, noting that it was not prudent to assume that future societies on Earth would be able to maintain surface storage facilities.

Yet, no long-term HLW storage facility of any type currently exists in the United States (or in any other country). The absence of a long-term repository is becoming a significant impediment to the use of nuclear power. In fact, in the 1970s, some states passed laws prohibiting the construction of any new nuclear power plants until the federal government demonstrated that radioactive wastes could be disposed of safely and permanently. The Department of Energy contracted with electrical utility companies to begin accepting spent fuel elements for underground long-term storage in 1998, but had still not met this deadline in 2004.

Progress in preparing a national underground disposal site for HLW continues to be painfully slow. A site must be found that is suitable for storage that will remain isolated from the ground water for tens of thousands of years. One concept is to carve out a chamber at least 1000 ft below ground, 1000 ft above the water table, in an appropriate rock formation. Geological formations such as salt, basalt, tuff, granite, and shale have all been considered. Salt domes, which are formations composed entirely of salt, are particularly attractive because they are very stable, extremely dry, and self-sealing if cracks should appear. Granite and basalt always contain cracks, but they have a great capacity to chemically absorb most wastes.

In such a chamber, HLW would be stored for at least 10,000 years to allow the high levels of radioactivity to decrease significantly. A sobering sample time-line for storage is shown in Table 7.6. The time needed is determined by the half-lives and amounts of all the radioisotopes. For example, beta-emitting fission products such as Sr-90 and Cs-137 with relatively short half-lives need to be isolated for over 300 years,

Table 7.6	Sample Time-Line for HLW Underground Repository
Year	**Event**
2010	Construction on underground storage site begins.
2015	Waste storage begins.
2040	Loading ends.
2065	Waste packages retrievable until this time.
2320	Repository sealed by this year.
3000	Most dangerous radioactive substances have decayed to stable products. First waste package is assumed to fail because of manufacturing defects.
12010	End of regulatory period of 10,000 years. Radioactive exposure of farmers in nearby valley is predicted to be 0.007 μSv/year, an insignificant amount.
312010	Radioactive exposure for nearby farmers predicted to reach 250 μSv/yr, a dose that concerns regulators.
622010	Peak radioactive exposure for farmers predicted at 850 μSv/year.

Source: From *The New York Times, Science Times,* August 10, 1999. Reprinted with permission of The New York Times.

that is, about 10 half-lives (see Table 7.5). This would allow over 99% of the strontium-90 and cesium-137 to decay. Other radioisotopes such as plutonium with longer half-lives require isolation for much longer periods.

> After 10 half-lives, the radioactivity of a sample drops essentially to background level.

Most plans call for encasing the spent fuel elements in ceramic or glass, packing the product in metal canisters, and burying them deep in the Earth in a designated repository. A method called vitrification has been developed to contain reprocessed defense wastes, including Pu-239, for future geological burial. The wastes are mixed with finely ground glass and melted to about 1150 °C. The molten glass and wastes are then poured into stainless steel canisters, cooled, and capped for on-site storage until development of a long-term underground repository (Figure 7.25). More than one million pounds of waste have been treated this way.

Consider This 7.26 Nuclear Waste Warning Markers

The Department of Energy recently asked 13 experts to design a system of markers to be installed near an underground nuclear waste repository in New Mexico, warning future generations of the existence of nuclear waste. The markers must last for at least 10,000 years (more than four times the age of the pyramids of Egypt), the end of the regulatory period. The message on the markers must be intelligible to Earthlings of the future. Try your hand at designing these warning markers, keeping in mind the changes that have occurred in *Homo sapiens* during the past 10,000 years and those that might occur in the next 10 millennia.

Figure 7.25

Encapsulating reprocessed HLW in glass canisters (vitrification).

To establish deep geological storage of HLW, the federal government must deal with state legislatures and tribal government whose land rights are involved. Currently, Yucca Mountain (Figure 7.26) in Nevada is the leading candidate for HLW storage by the Department of Energy. But it is not certain that the Yucca Mountain depository will ever become operational. The selection of Yucca Mountain as the site for study of underground burial of HLW required overcoming difficult political barriers as well as technical ones. The 1982 Nuclear Waste Policy Act projected that the initial geological repository would be constructed in time to receive high-level nuclear waste by 1998. Obviously this year has come and gone. In 1987, the Nuclear Waste Policy Amendments Act designated Yucca Mountain as the sole site to be studied as an underground long-term, high-level nuclear waste repository. To fulfill the requirements of the 1982 Act, the Department of Energy (DOE) contracted with nuclear utilities to begin accepting spent fuel beginning in 1998. To date, utility companies have paid over $14 billion to

(a)

(b)

Figure 7.26

(a) Map of Yucca Mountain and state of Nevada

(b) Yucca Mountain, looking south into the desert

Source: (a) Department of Energy

fund the development of a repository to do so. According to a recent court decision, this responsibility remains with DOE, even though no such facility is available now.

The political battles continue. In early 2002, President George W. Bush sent a letter to both houses of Congress stating "I consider the Yucca Mountain site qualified for application for a construction authorization for a repository. Therefore, I recommend the Yucca Mountain site for this purpose." His recommendation was based on an earlier one to him from the Secretary of Energy, which in turn was based on thousands of pages of documentation. But in April, 2002, Nevada Governor Kenny C. Guinn countered with an official Notice of Disapproval to the Senate. His accompanying statement proclaimed "As a matter of science and the law, and in the interests of state comity and sound national policy, Yucca Mountain should not be developed as a high-level nuclear waste repository." In July 2002, the Senate gave final approval for the site at Yucca Mountain, essentially overriding the wishes of the State of Nevada.

However, the approval of a site is not the end of the process. Next, as the main player, the DOE must submit an application to the Nuclear Regulatory Commission (NRC) for a license to begin construction. The job of the Office of Nuclear Material Safety and Safeguards at the NRC is to evaluate the technical information on which this license is based. In April 2004, this NRC office issued a memo that the technical information being provided had been improved, but was still inadequate, perhaps delaying the process further. If and when an application is approved, extensive hearings and legal review then follow. Site standards must be met before the repository can accept waste, and these standards are not met by the estimated radiation releases listed in Table 7.6. Only if the NRC concludes that the repository will function safely as required by the technical demands, will it authorize construction of the site.

Meanwhile, tunnels are being dug 1400 feet beneath Yucca Mountain, Nevada (see Figure 7.26) to determine its adequacy for deep, long-term storage of HLW. The DOE has already spent an estimated $54 billion on the project. If completed, the site will be the largest radioactive storage facility in the world, with a capacity of 70,000 metric tons of spent fuel and 8000 tons of high-level military waste. As shown in Table 7.6, roughly 25 years will be required just to transport the waste to the Nevada site, at a rate of 20 shipments per day.

Congress again got involved with HLW disposal by proposing legislation to create the Nuclear Waste Policy Act of 1997. Under this legislation, an *interim* above-ground storage site would have been created for 40,000 tons of spent fuel at the Nevada Test Site, adjacent to the unfinished Yucca Mountain site. Spent fuel was to be stored at the interim site, operational by January 2002, in metal canisters inside concrete bunkers until a permanent repository is ready. In discussing the bill, Nevada Representative Jim Gibbons complained that, "The people in support of this bill are the ones who have nuclear waste in their districts and want to get it out—get it from wherever it is into the state of Nevada." Nevada has no commercial nuclear reactors, and Nevada officials are concerned that a permanent site might never be approved, thereby leaving Nevada stuck with the temporary facility indefinitely.

Congressional opponents of the bill have dubbed it the "mobile Chernobyl" bill. They are concerned that the spent fuel high-level waste would be transported by rail and highway through 43 states within half a mile of 50 million Americans before it reached the proposed Nevada Test Site interim repository. Speaker of the House Dennis Hastert of Illinois, a supporter, declared that the bill "assures that another 15 years will not pass before the federal government lives up to its responsibility of accepting spent fuel." His position is not surprising given that Illinois has more commercial nuclear reactors (11) than any other state. Former President Clinton vetoed the bill, forbidding delivery of nuclear waste to Yucca Mountain before the underground repository was completed.

Since the terrorist attacks of September 11, 2001, politicians have revisited the issue of having HLW stored around the country, rather than in a central and presumably more secure location. Currently, many perceive the HLW sitting in temporary storage facilities around the country as more dangerous. Accordingly, the 2004 spending bill proposed by President George W. Bush set a high priority in completing the Yucca Mountain project. Coalitions in Nevada continue to resist the waste dump in their state; in fact, the mood in Nevada has been termed as "outright defiance." The mayor of Las Vegas, Oscar Goodman, stated in 2003 "If it comes by rail, the only rail goes right through the heart of my city. And I guarantee you one thing: as long as I'm the mayor, it ain't coming through."

 Consider This 7.27   **Yucca Mountain**

What is the current status of using Yucca Mountain as a HLW long-term repository? What position are people in Nevada taking? Use the resources of the Web to find out. Be sure to cite your sources.

Some members of Congress have argued that a decision on a 10,000-year storage facility should be postponed and a 100-year storage facility should be developed. Congress (or its successor) may still be debating the issue 24,110 years from now, when the plutonium-239 completes its *first* half-life. Other disposal methods seem even less promising. Disposal in deep-sea clay sediments, under 3000–5000 m of water was investigated. Proposals to bury the radioactive waste under the Antarctic ice sheet or to rocket it into space have largely been discredited. But one thing is sure: Whatever disposal methods are ultimately adopted, they must be effective over an extremely long time.

Consider This 7.28   **Storage of Nuclear Waste in Developing Countries**

Some industrialized nations have proposed a novel method of foreign aid, with strings attached. The producers of nuclear power would ship their radioactive wastes to developing countries and pay them to store it. If you were the president of a developing country that was considering such an arrangement, what issues would you need to consider? Prepare a position paper in outline form listing the reasons for and against such an offer of foreign aid, and then reach a decision based on the relative importance of the factors in your list.

7.11 What Is Low-Level Radioactive Waste?

Nearly 90% of the volume of all nuclear waste is low-level, rather than high-level. **Low-level radioactive waste (LLW)** is waste contaminated with smaller quantities of radioactive materials than HLW, and specifically excludes spent nuclear fuel. LLW waste includes a wide range of materials. Some, such as laboratory clothing, gloves, and cleaning tools from medical procedures using radioisotopes, and from discarded smoke detectors, have radioactivity levels that can be quite low. Other types of LLW have higher levels of radioactivity, such as the wastes generated by nuclear fuel fabrication facilities, the manufacture of radioactive pharmaceuticals, mining, and research facilities. Approximately twice the volume of LLW comes from military sources as from commercial and medical sources. It is estimated the United States will have 4.5 million cubic meters of low-level nuclear waste by the year 2030.

Since low-level nuclear waste is far less radioactive, it is disposed of differently. For example, certain types of LLW are put into sealed canisters and buried in lined trenches 10 m deep (Figure 7.27). Low-level military nuclear waste is disposed at federally owned sites maintained by the Department of Energy. Nonmilitary LLW disposal is the responsibility of the state where it is generated.

Although low-level waste poses significantly less danger than high-level waste, the "not-in-my-backyard" syndrome operates with low-level nuclear waste as well. The very idea of radioactive waste, even if low-level, is sufficient to generate considerable opposition to proposed sites for LLW. Congress expected that states, through the Low Level Radioactive Waste Policy Act of 1980, would form regional compacts by which compact members would send their low-level waste to a disposal site in one state. Such compacts have not been successful, including failed efforts in Illinois and in New York, each costing $55 million over eight years.

Currently, two commercial LLW disposal sites are in operation, one at Barnwell, South Carolina, and the other at Richland, Washington. About 70% of the waste shipped to the South Carolina site is from other states. The Washington state site accepts the other 30%, but only from a limited number of states, all nearby except Alaska and Hawaii. A third site in Clive, Utah, also accepts some commercial and Department of Energy low-level waste in addition to other types of waste. Thirty-five years ago, six commercial LLW sites were in operation. Two were closed in the 1970s, a third in 1986, and a fourth in 1992. In 2001, the National Academy of Sciences issued a report expressing concern, stating that efforts to open any new sites were "deadlocked" politically. There is an ongoing need to safely store the low-level radioactive wastes generated across the nation by hospitals, nuclear manufacturers, and research laboratories. Lack of access to such disposal sites could compromise the ability of such facilities to function.

Figure 7.27
Burial of low-level nuclear waste.

7.12 Nuclear Power Worldwide

Globally, 16–17% of the electricity produced and consumed is generated in about 440 nuclear power plants (and these figures are slowly rising). If this amount of energy were to be replaced, it would require the entire annual coal production of the United States or the former Soviet Union. Thus, the international reliance on nuclear energy is relatively small but still significant. From Figure 7.28, you can see that the United States has the largest number of nuclear reactors and is by far the largest generator of electricity from nuclear power. But, as of 2004, nearly a third of the U.S. nuclear units are over 30 years old. Furthermore, as noted in the opening section of this chapter, no new reactors are currently under construction.

The numbers of reactors and total power output, however, do not tell the complete story. A more interesting measure is the percentage of electrical power a country obtains from fission reactors, as graphed in Figure 7.28. On a percentage basis, France leads the world in nuclear power. As of 2004, it has 59 operational nuclear power plants that generate 78% of the electricity used in France. The Swiss also have a high nuclear dependency, producing 40% of their electricity with only five reactors. Most of the countries that generate over 40% of their electricity from nuclear power plants are in Western Europe; Hungary, and South Korea are exceptions.

> In August, 2004, tweny-six nuclear reactors were under construction throughout the world.

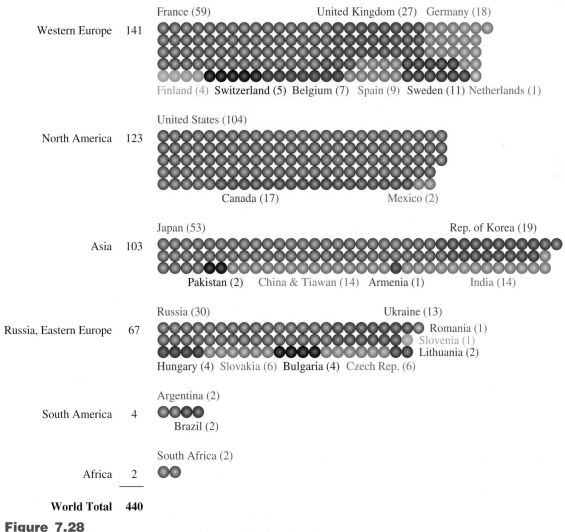

Figure 7.28

Number of reactors in operation worldwide, as of January 31, 2004.

Source: http://www.iaea.org/programmes/a2/index.html, International Atomic Energy Agency

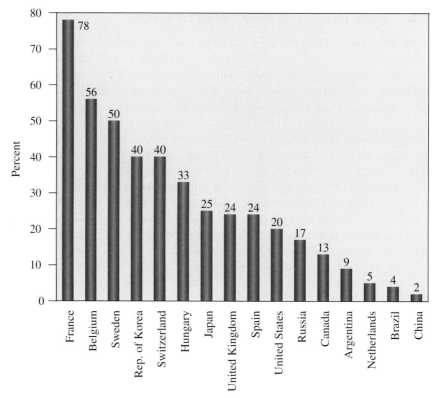

Figure 7.29

Percent of electrical power generated by nuclear power reactors in selected countries, 2003.

Source: World Nuclear Association. Taken from www.world-nuclear.org/info/reactors.htm

Also noteworthy are the countries *not* included in Figure 7.28. Thus far, only indus-trialized nations have major commercial development of nuclear fission. As of 2004, Mexico and Pakistan each had two operating reactors, and Romania and Slovenia each had one. India obtains less than 4% of its electricity from its 14 reactors, but con-struction has begun on 6 new plants. China now has nine operating nuclear power reactors, with two more planned. Reactors are now under construction in North Korea, Iran, and in Ukraine. Ironically, in spite of the fact that much of the world's uranium comes from Africa, most nations on that continent use no nuclear energy.

Consider This 7.29   Worldwide Nuclear Power

a.   From the data in Figure 7.29, which countries are the top three producers of elec-tricity from nuclear power in terms of megawatts of power? In general, how would you characterize these countries in terms of their location and energy use?

b.   Name three countries missing from Figure 7.29. In general, how would you char-acterize these countries?

c.   How does this list compare with the "top emitters" for carbon dioxide referenced in Consider This 3.35? Explain any similarities or differences.

Consider This 7.30   Nuclear Neighborhood

On its Web site, the Nuclear Regulatory Commission (NRC) provides a map of all reactors now licensed to operate in the United States. Identify the three nuclear reactors nearest to where you live. How many years have these reactors been in operation? Who owns and operates them? You can find a direct link to the NRC at the *Online Learning Center.*

7.13 Living with Nuclear Power: What Are the Risks and Benefits?

From the previous section, it is obvious that countries differ markedly in the extent to which they use nuclear power to generate electricity. What is *not* so obvious are the reasons for this variability. Some nations have fiscal problems so severe that it is difficult or impossible for them to fund the construction or expansion of nuclear power facilities. For others, an adequate supply of relatively cheap electricity is available from water power, fossil fuels, or other sources. Therefore, there is little need for nuclear energy. Nuclear power provides a means for some countries to gain a greater independence from needing to import fossil fuels. In still other countries, such as France, a conscious choice has been made to use nuclear energy to produce the bulk of electrical power to reduce dependency on imported fossil fuel. And, just the opposite conclusion has been reached by other nations. In Sweden, which currently obtains 39% of its electricity from fission, a referendum has called for the halt of nuclear power generation by 2010.

Regardless of whether a country contains many nuclear-powered electrical generators or only a few, associated risks and benefits must be weighed. Such risk–benefit analyses are never easy, though in a sense we do it every day. We commonly regard risk as the probability of being injured or losing something, but there are many types of risk. They can be voluntary, such as those associated with wind surfing or bungee jumping, or involuntary, such as inhaling someone else's cigarette smoke. When we drive a car we control the risks (at least to some extent), but we have no control over the increased risk of radiation exposure at high altitudes or of a commercial plane crash. Counterbalancing risks are benefits such as the improvement of health, increased personal comfort or satisfaction, saving money, or reducing fatalities. Everyday living inevitably involves risks and their related benefits: crossing a street, riding a motorcycle or in a car, cooking a meal or eating one, and even the simple act of getting up in the morning. Because there is some element of risk in everything we do, we almost automatically make judgments about what level of risk we consider acceptable. Most people do not intentionally put themselves at high risk, even when the potential benefit is also high, such as going into a burning building to save a child. On the other hand, there is an alarming increase in the number of people who expect "zero risk" in whatever they do or whatever surrounds them, although it is impossible to achieve. *There is no such thing as zero risk.*

In the case of nuclear energy, we are dealing with social benefits in relation to technological risks, but we must not make the mistake of only considering the risks and benefits that relate directly to fission. We must also weigh the risks associated with the alternatives—especially the coal-fired power plants that nuclear reactors are designed to replace. Recall, for example, the estimate in Chapter 4 that over 100,000 workers have been killed in American coal mines since 1900, many prior to the 1950s when higher safety standards were instituted. Table 7.7 summarizes the risk of fatalities from the annual operation of a 100-megawatt power plant using either coal or nuclear power. The conclusion is that, at least for the hazards identified here, the risks associated with energy produced by nuclear power are considerably less than those of coal-burning plants.

Paradoxically, coal-fired power plants release more radioactivity on a daily basis than nuclear plants. In addition to C, H, O, N, and S, coal contains other elements, including uranium and thorium, as impurities. According to W. Alex Gabbard, a physicist at the Oak Ridge National Laboratory, trace quantities of uranium in coal can be as high as 10 ppm and the amount of thorium is usually more than twice that of uranium. Gabbard has estimated that in 1982, power plants in the United States burned a total of 616 million tons of coal and released 801 tons of uranium and 1971 tons of thorium into the environment. In fact, the quantity of uranium emitted by the coal-fired plants exceeded the mass of uranium consumed in nuclear plants. He predicted that if in the year 2040 the U.S. burns 2516 million tons of coal, over 145,000 tons of uranium will be released, noting that about 1000 tons of these will be uranium-235.

The 19th century poet William Wordsworth spoke of technological risks and benefits as ". . . Weighing the mischief with the promised gain . . ." He was speaking, in this case, about the railroad, a technology new in his time.

Table 7.7	**Risks from Coal and Nuclear-Powered Electricity Generators**	
Hazard Type	**Coal**	**Nuclear**
Routine occupational hazards	Coal-mining accidents and black-lung disease constitute a uniquely high risk.	Risks from sources not involving radioactivity dominate.
Deaths*	2.7	0.3–0.6
Routine population hazards	Air pollution produces relatively high, though uncertain, risk of respiratory injury. Significant transportation risks.	Low-level radioactive emissions are more benign than the corresponding risks from coal. Significant transportation risks incompletely evaluated.
Deaths*	1.2–50	0.03
Catastrophic hazards (excluding occupational)	Acute air pollution episodes with hundreds of deaths are not uncommon. Long-term climatic change, induced by CO_2, is conceivable.	Risks of reactor accidents are small compared with other quantified catastrophic risks. The problem lies in as yet unquantified risks for reactors and the remain-der remainder of the fuel cycle.
Deaths*	0.5	0.04
General environmental degradation	Strip mining and acid runoff; acid rainfall with possible effect on nitrogen cycle, atmospheric ozone; eventual need for strip mining on a large scale.	Long-term contamination with radioactivity.

Source: Modified from *Perilous Progress: Managing the Hazards of Technology,* by Robert W. Kates, Ed., 1985, Westview Press, Boulder, Colorado.

*Deaths are the number expected per year for a 100-megawatt power plant. In all cases, 6000 man-days lost are assumed to equal one death.

Coal-fired power plants also produce huge amounts of carbon dioxide, a waste prod-uct of combustion for which currently we have no large-scale technology for remediation. Annually, one 1000-megawatt coal-fired electric power plant releases about 4.5 million tons of CO_2 into the atmosphere. In a year, such a plant also generates about 3.5 million cubic feet of waste ash, a substantial volume. By comparison, a 1000-megawatt nuclear power reactor produces about 70 cubic ft of high-level waste per year. Thus, the risks and benefits for power stack up differently, depending on how this power is generated.

Your Turn 7.31   Uranium Revisited

a.   If 1000 of the tons of uranium released in the year 2040 were U-235, which iso-tope would be found in the remaining tons of uranium?

b.   In coal, why is thorium found together with uranium?

As we have noted, nuclear energy carries tremendous emotional overtones. In part, these stem from mystery, misunderstandings, and mushroom clouds. The risks of radiation and the possibility of a major disaster, however remote, loom large in human consciousness. The accidents at Three Mile Island and Chernobyl, though hardly equivalent, have made the public wary. We have limited trust in technology and perhaps even less in people. We are apprehensive about human error in the design, construction, and management of nuclear power plants. After all, human errors and technicians' responses to them were the weak points in the prescribed safety procedures that caused the accidents at Three Mile Island and Chernobyl.

Consider This 7.32 | **Informed Citizens**

a. What should you know to be an informed citizen about nuclear power plants? Make a list of questions (at least five) that would be important to ask about a specific nuclear reactor.

b. Check out the specifications of a particular reactor at the Web site provided by the U.S. Nuclear Regulatory Commission (NRC). Choose any reactor in the country you wish. Does the information provided answer the questions you posed? Comment on what you would like to know that you were unable to find. If others in your class selected different reactors, you may wish to compare notes.

7.14　What Is the Future for Nuclear Power?

This final question is, in many ways, the most difficult one posed in this chapter. The answer remains uncertain. Early in the chapter, a provocative quote from Professor Gregory Choppin indicated that a revival of nuclear energy will occur because we have no other option. An accompanying statement by Jeff Johnson points out that the public remains divided on the issues of nuclear energy, as noted by others.

This question of the future of nuclear power does not stand in isolation from other international issues. At present, the world community is actively seeking ways to reduce greenhouse gases. Because heat from the fission of U-235 produces steam used to generate electricity, a nuclear power plant releases no carbon dioxide, a major greenhouse gas. Nuclear power also has been touted as a way to reduce acid rain, because nuclear fission releases no acidic oxides of sulfur and nitrogen. Recall that the Seabrook plant generates electricity at a rate of 1160 megawatts (1160 million joules per second) or 1×10^{14} joules per day. Approximately 10,000 tons of coal would have to be burned in a conventional power plant to generate this daily energy output. Burning this quantity of coal could easily release 300 tons of SO_2 and perhaps 100 tons of NO_x. Whether the risks associated with nuclear power outweigh those of global warming and acid rain is a difficult question for which there are no clear-cut answers, in spite of extended study and debate. Proponents line up on each side of the argument.

In August 1988, during a summer of record heat, 15 U.S. senators cosponsored a bill to fund research to combat the enhanced greenhouse effect by developing carbon dioxide-free energy sources, including safer and more cost-effective nuclear power plants of standardized design. Alan Crane, an energy policy specialist, appeared before a House subcommittee hearing and spoke of global warming and nuclear energy risks in these terms:

> *"There is significant, though not yet quantifiable, risk that the resulting climate changes will wreak devastating changes in agricultural production throughout the world, among other problems. Such changes could lead to the death of far more people and cause far greater environmental damage than any nuclear reactor accident, and appear to be considerably more likely."*

Two tons of U-235 (40–60 tons of enriched uranium fuel) can fuel a 1000-megawatt–producing commercial nuclear reactor for about one and a half years. To produce this amount of electricity, a coal-fired plant would use the coal carried by a train with 200 cars each carrying 15 tons of coal *every day* for the same one and a half years.

Others conclude that nuclear power can reduce global warming only slightly and suggest that it would be much less expensive to invest in enhanced energy efficiency. Bill Keepin and Gregory Kats of the Rocky Mountain Institute have estimated that to reduce carbon dioxide emissions significantly through the use of nuclear power would require the completion of a new nuclear plant every 2 days for the next 38 years. Oak Ridge National Laboratory staff members reported that before any massive replacement of fossil fuel with nuclear power could occur, several new techniques would have to be developed. These include commercial-scale recycling of nuclear fuels, breeder reactors to extend existing fuels, and possibly uranium recovery from seawater.

Hans Blix, former director general of the International Atomic Energy Agency, in describing the international dimensions of the problem, noted that developing nations will not likely build nuclear plants in the near future: "If nuclear power is to be relied on to alleviate our burdening of the atmosphere with carbon dioxide, it is therefore to the industrialized countries that we must first look. They are in the position to use these advanced technologies—and they are also the greatest emitters of carbon dioxide."

The original nuclear era, synonymous with the growth of nuclear power in the United States, began in the early 1960s and lasted until 1979. It fell victim to stabilized demand for electricity, which was brought on by enormous oil price hikes in 1975 and the Three Mile Island incident in 1979. As a consequence of that accident, the required number of nuclear plant personnel and their training requirements grew significantly. In addition, the mandatory retrofitting of existing nuclear facilities to enhance their safety added significantly to the cost of an already capital-intensive industry. As a result, the electrical power industry became understandably reluctant to invest further in nuclear facilities.

A second nuclear era may occur. Some experts believe that such a rebirth is possible through the development of smaller, more efficiently designed reactors in the 600-megawatt range, rather than the current 1000–1200-megawatt facilities. Unlike the many different designs used to build the current reactors, these new reactors would be of a standardized, easily replicated design, have a longer operational lifetime (60 years versus the current 30), and be demonstrably safer and more economical in operation. James Lake, past president of the American Nuclear Society and laboratory director at a DOE nuclear energy laboratory, says "The energy crisis has shined a spotlight on us [nuclear power]. . . . There are 438 reactors worldwide, and people are thinking there should be 4000 in the next 20 years." The NRC certified three new nuclear reactor designs in the 1990s. Under new NRC rules, if a reactor with such a design were used, a company could bring a new nuclear power plant on line in just 5–6 years, rather than the 8–10 years needed to construct the plant for a yet-to-be-certified reactor.

But in all of this, the unsolved problem of the safe disposal of radioactive waste remains perhaps the greatest impediment. Nuclear engineers William Kastenberg and Luca Gratton conclude their *Physics Today* article (June 1997) with the sobering thought: "For a high-level waste depository of the type proposed for Yucca Mountain, it is clear that natural processes will eventually redistribute the waste materials. Present design efforts are directed toward ensuring that, at worst, the degraded waste configurations will eventually resemble stable, natural ore deposits, preferably for periods exceeding the lifetimes of the more hazardous radionuclides. Perhaps that's the best we can hope for."

 Consider This 7.33   **Risks and Benefits**

We have examined nuclear fission as a source of electrical power in some detail. Now we ask you to list the risks and benefits associated with currently operating nuclear fission reactors. Using this list, take a stand on the question of the future use of nuclear fission-powered plants. Write an editorial for a local newspaper proposing your view of a viable 20-year national policy on the issues.

Consider This 7.34   Second Opinion Survey

Now that you are near the end of your study of nuclear power, return to the personal opinion survey of Consider This 7.1 and answer the questions one more time.

    After completing the survey for a second time, compare your answers in the second survey with those from the first. Are there any striking differences in your opinions of nuclear power between the first and second surveys? If so, which of your opinions about nuclear power changed the most? What was responsible for this shift?

C O N C L U S I O N

Nearly 50 years have passed since the first commercial nuclear power plant began producing electricity in the United States. The glittering promise of boundless, unmetered electricity, drawn from the nuclei of uranium atoms, has proved illusory. But the needs of our nation and our world for safe, abundant, and inexpensive energy are far greater today than they were in 1957. So scientists and engineers continue their atomic quest. Where the search will lead is uncertain, but it is clear that people and politics will have a major say in ultimately making the decision. Reason, together with a regard for those who will inhabit our planet in both the near and far future, must govern our actions.

Chapter Summary

Having studied this chapter, you should be able to:

- Tell how nuclear fission occurs (7.2)
- Write balanced nuclear equations for alpha and beta decay, and for nuclear fission (7.2)
- Use mathematical relationships to calculate the amount of energy produced by a fission reaction (7.2)
- Compare and contrast how electricity is produced by a conventional power plant with how it is produced by a nuclear power plant (7.3)
- Summarize the reasons a nuclear power reactor cannot undergo a nuclear explosion (7.5)
- Develop a personal radiation dose inventory and describe the biological effects of nuclear radiations (7.8)
- Understand and apply the concept of half-life to the use of radioisotopes, radiocarbon dating techniques, and nuclear waste storage (7.9)
- Relate the issues surrounding the use of nuclear power in this country and abroad (7.12)
- Describe the issues associated with the production and storage of high-level radioactive waste, including spent fuel (7.10)

- Take an informed stand on the storage of high-level radioactive wastes (7.10)
- Read and hear news articles on nuclear power and nuclear waste issues with confidence in your ability to interpret the accuracy and efficacy of such reports (7.10–7.11)
- Summarize the nature of low-level radioactive waste and its storage (7.11)
- Report on the use of nuclear power for electricity generation globally and the reasons why several countries have very high percentages of electrical power production from nuclear reactors compared with the United States (7.12)
- Describe the risks and benefits of the use of nuclear power (7.13)
- Take an informed stand on the use of nuclear power for electricity production (7.14)
- Discuss why the 1950s and 1960s promise of abundant and cheap nuclear energy was not realized in this country (7.14)
- Outline the factors that will allow or oppose the growth of nuclear energy in the next decade (7.14)

Questions

Emphasizing Essentials

1. $E = mc^2$ is one of the most famous equations of the 20th century. What does each of the letters in the equation represent?

2. What is the difference between the chemical symbol N and representations such as ^{14}N or ^{15}N?

3. a. How many protons does an atom of $^{239}_{94}$Pu contain?

   b. What element contains one more proton than uranium? Two more protons?

   c. How many protons does radon-222 contain?

4. For each of these nuclei, find the number of neutrons.

   a. C-14 (a radioisotope of carbon)

   b. C-12 (a stable isotope of carbon)

   c. H-3 (tritium, a radioisotope of hydrogen)

   d. Tc-99 (a radioisotope used in medicine)

5. Describe all the ways you can think of that carbon atoms can differ from each other. Then describe all the ways in which carbon atoms differ from uranium atoms. For the latter, there are at least three ways.

6. a. Boron (atomic number 5) has two common stable isotopes: B-10 and B-11. Given that the periodic table lists the average atomic mass as 10.81, which isotope must be more abundant? Explain your reasoning.

   b. The average atomic mass of chlorine is 35.453. Must this mean that there are just two stable isotopes, Cl-35 and Cl-36, or are there other possibilities? Explain your answer.

7. Give an example of any nuclear equation and of any chemical equation. In what ways are the two equations alike? Different?

8. This nuclear equation represents a plutonium target being hit by an alpha particle. Show that the sum of the subscripts on the left is equal to the sum of the subscripts on the right. Then do the same for the superscripts.

$$^{239}_{94}\text{Pu} + {}^{4}_{2}\text{He} \longrightarrow [{}^{243}_{96}\text{Cm}] \longrightarrow {}^{242}_{96}\text{Cm} + {}^{1}_{0}\text{n}$$

9. For the nuclear equation shown in the previous question,

   a. From where do you think the ${}^{4}_{2}\text{He}$ came?

   b. ${}^{1}_{0}\text{n}$ is a product. What does this represent?

   c. Why is curium-243 written in square brackets?

10. Write nuclear equations to represent each of these.

    a. Hydrogen-2 (better known as deuterium) can undergo nuclear fusion. For example, two H-2 nuclei can fuse to form another nucleus and a neutron.

    b. A U-238 target, when bombarded with N-14, produces another nucleus and five neutrons.

    c. Pu-239 is hit with a neutron to form Ce-146, another nucleus, and three neutrons.

11. When 4.00 g of hydrogen nuclei undergoes fusion to form helium in the Sun, the change in mass is 0.0265 g and energy is released. Use Einstein's equation, $\Delta E = \Delta mc^2$, to calculate the energy equivalent of this change in mass.

12. Under conditions like those on the Sun, hydrogen can fuse with helium to form lithium, which in turn can form different isotopes of helium and of hydrogen.

$$^{2}_{1}\text{H} + {}^{3}_{2}\text{He} \longrightarrow [{}^{5}_{3}\text{Li}] \longrightarrow {}^{4}_{2}\text{He} + {}^{1}_{1}\text{H}$$
$$\text{2.01345 g} \quad \text{3.01493 g} \qquad\qquad \text{4.00150 g} \quad \text{1.00728 g}$$

    a. What is the mass difference between a mole of the reactant and the product isotopes? The mass for a mole is given below each isotope.

    b. How much energy (in joules) released in this reaction?

13. This schematic diagram represents the reactor core of a nuclear power plant.

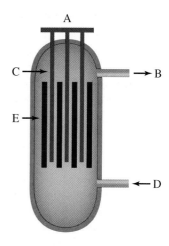

Match each letter with one of these terms.

   fuel rods

   cooling water into the core

   cooling water out of the core

   control rod assembly

   control rods

14. Identify the segments of the nuclear power plant diagrammed in Figure 7.7 that contain radioactive materials and those that do not.

15. Explain the difference between the primary coolant and the secondary coolant. The secondary coolant is not housed in the containment dome. Why not?

16. One important distinction between the Chernobyl reactors and those in the United States is that those in Chernobyl used graphite as a moderator to slow neutrons whereas U.S. reactors use water. In terms of safety, give two reasons why water is a better choice.

17. Consider the uranium fuel pellets used for commercial nuclear power plants.

    a. Which isotopes of uranium occur naturally in uranium ore?

    b. Why is it necessary to enrich the uranium before manufacturing the fuel pellets?

    c. The fuel pellets are enriched only to a few percent, rather than to 80–90%. Name three reasons why.

    d. It is not possible to separate the isotopes of uranium by chemical means. Why not?

    e. Describe one way in which uranium isotopes can be separated.

18. Nuclear fission can occur through multiple pathways. For the fission of uranium-235 induced by a neutron, write a nuclear equation to form:

    a. bromine-87, lanthanum-146, and more neutrons.

    b. a nucleus with 56 protons, a second with a total of 94 neutrons and protons, and 2 additional neutrons.

19. Pu-239 is most hazardous when particles of it are inhaled. Explain the reasons behind this observation.

20. What do α, ^4_2He, and $^4_2\text{He}^{2+}$ all represent? How do these representations differ?

21. Radioactive decay is accompanied by a change in the mass number, a change in the atomic number, a change in both, or a change in neither. For the following types of radioactive decay, which change(s) do you expect?

    a. alpha emission

    b. beta emission

    c. gamma emission

22. Write nuclear equations for each of these.

    a. Iodine-131 undergoes beta decay. This radioisotope concentrates in the thyroid gland and can be used for medical imaging.

    b. Thallium-206 undergoes beta decay. This radioisotope is used for medical imaging as well, especially for the heart muscle.

    c. Mo-98 is bombarded with a neutron to produce Tc-99 (used widely in medical imaging) and another particle.

23. In a fashion similar to U-238 (see Figure 7.17), U-235 goes through a series of alpha and beta decays before reaching a stable isotope. For practice, write the first six, which will bring you to an isotope of radon. In order, the steps in the full radioactive decay series are α, β, α, β, α, α, α, β, α, β, α, ending in stable Pb-207. Some steps have accompanying γ radiation, but you may omit this.

24. Given that the average U.S. citizen receives 3600 μSv of radiation exposure per year, use the data in Table 7.4 to calculate the percentage of radiation exposure the average U.S. citizen receives from each of these sources.

    a. food, water, and air

    b. a dental X-ray twice a year

    c. the nuclear power industry

25. What percent of a radioactive isotope would remain after two half-lives, four half-lives, and six half-lives? What percent would have decayed after each period?

26. Suppose somebody tells you that a radioisotope is "gone" after about seven half-lives. Critique this statement, explaining both why it could be a reasonable assumption and why it might not be.

27. Estimate the half-life of X from this graph.

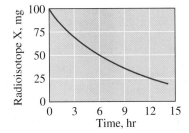

Concentrating on Concepts

28. In Consider This 7.1, you were asked to answer several questions about nuclear power. Extend this survey by asking the same questions of someone at least one generation older than you and someone still in high school. What similarities and differences did you find in their answers compared with your opinions?

29. a. Why were citizens' groups in Massachusetts concerned about construction of the Seabrook Nuclear Power Plant in New Hampshire? To get started, you might want to consult a road atlas to find the location of Seabrook.

    b. What aspects of the site chosen for the Seabrook plant were advantages in the minds of the designers and builders of the Seabrook plant, but were disadvantages in the minds of some protesting citizens' groups?

30. The Seabrook power plant at full capacity uses only a few pounds of uranium to generate 1160 megawatts of power, which is equivalent to 1.16×10^9 J/s. To produce the same amount of energy would take about two million gallons of oil or about 10,000 tons of coal in a conventional power plant. What is the fundamental difference in the way that energy is produced in the Seabrook plant, compared with conventional power plants?

31. Considering that Einstein had proposed the equation $E=mc^2$ over 30 years earlier, why were Otto Hahn and Fritz Strassmann puzzled when, in 1938, they discovered the element barium among the products formed when uranium was bombarded with neutrons?

32. In a chemical reaction, it is often said that matter is conserved. Why is it incorrect to say that mass is conserved in a nuclear reaction?

33. If you look at nuclear equations in sources other than this textbook, you may find that the subscripts have been omitted. For example, you may see an equation for a fission reaction written this way.

    $$^{235}\text{U} + {}^1\text{n} \longrightarrow [^{236}\text{U}] \longrightarrow {}^{87}\text{Br} + {}^{146}\text{La} + 3\,{}^1\text{n}$$

    a. How do you know what the subscripts should be? Why can they be omitted?

    b. Why are the superscripts *not* omitted?

34. More than two decades have passed since the incident at Three Mile Island. Use the Web to answer these questions.

    a. Is Reactor 2, the site of the problem in 1979, back on line producing electricity?

    b. How has the nuclear waste from the accident been treated?

35. What are the similarities between the incidents at Chernobyl and at Three-Mile Island? What are the differences?

36. Not all the neutrons in a nuclear reactor produce fission reactions. Name some other possibilities as to what might happen to these neutrons.

37. **a.** Is depleted uranium (DU) still radioactive? Explain your answer.

   b. Is spent nuclear fuel (SNF) still radioactive? Explain your answer.

38. It is generally believed that terrorists would be more likely to construct a nuclear bomb using Pu-239 reclaimed from breeder reactors than using U-235. Use your knowledge of chemistry to offer reasons for this.

39. A Web site describing an X-ray procedure reports, "Despite its negative connotations, people are exposed to more radiation on a daily basis than they may realize. For example, infrared radiation is released whenever there is extreme heat. The sun generates ultraviolet radiation, and a little exposure to it results in a tan. In addition, the body contains naturally radioactive elements." Critique this explanation.

40. What does the term *decommission* mean, as in "decommissioning a nuclear power plant"? What are the technical challenges involved? You might want to start by learning more about the decommissioning of the Yankee Rowe facility (see Table 7.1). The resources of the Web can help you answer this question.

41. **a.** What are the characteristics of high-level radioactive waste? Name at least two processes that produce it.

   b. Two options are being considered for the storage of HLW: monitored storage on the surface, and storage deep underground. Cite the advantages and disadvantages of both.

   c. Explain how low-level waste (LLW) differs from HLW.

Exploring Extensions

42. Einstein's equation, $\Delta E = \Delta mc^2$, also applies to chemical changes as well as to nuclear reactions. An important chemical change studied in Chapter 4 was the combustion of methane, which releases 50.1 kJ of energy for each gram of methane burned.

   a. What mass loss corresponds to the release of 50.1 kJ of energy?

   b. To produce the same amount of energy, what is the ratio of the mass of methane burned in a chemical reaction to the mass loss converted into energy according to the equation $\Delta E = \Delta mc^2$?

   c. Use your results in parts **a.** and **b.** to comment on why Einstein's equation, although correct for both chemical and nuclear changes, is usually only applied to nuclear changes.

43. Lise Meitner and Marie Curie were both pioneers in developing an understanding of atomic nuclei. You likely have heard of Marie Curie and her work, but may not have heard of Lise Meitner. How are these two women related in time and in their scientific work?

44. Gallium consists of just two stable isotopes: Ga-69 and Ga-71.

   a. If the atomic mass of elemental gallium is 69.72, which isotope is present in a larger percentage?

   b. If Ga-69 has a mass of 68.9257 and Ga-71 has a mass of 70.9249, what percentage of each isotope is present?

45. Fluorine only has one naturally occurring radioisotope, F-19. If fluorine also occurred in nature as F-18, would this necessarily complicate the separation of $^{238}UF_6$ and $^{235}UF_6$? Explain.

46. Alchemists in the Middle Ages dreamed of converting base metals, such as lead, into precious metals—gold and silver. Why did they never succeed? Has the situation changed since then? Explain your answer.

47. In question #12, the energy of a mole of H-2 joining with He-3 was calculated. What is the ratio of the energy released in this nuclear reaction to the energy released in the combustion of a mole of hydrogen gas? *Hint:* This value can be calculated using bond energies from Table 4.2.

48. Californium, element number 98, was first synthesized by bombarding an element with alpha particles. The products were californium-245 and a neutron. What was the target isotope used in this nuclear synthesis?

49. Consider this representation of a Geiger–Müller counter (also called a Geiger counter), a device commonly used to detect ionizing radiation. The probe contains a gas under low pressure.

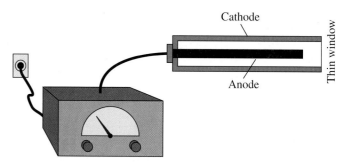

   a. How does radiation enter the Geiger–Müller counter?

   b. Why does this device only detect radiation that is capable of ionizing the gas contained in the probe?

   c. What are other methods for detecting the presence of ionizing radiation?

50. A stockpile of approximately 50 metric tons of plutonium exists in the United States as a result of disassembling warheads from the nuclear arms race. What should the fate of this plutonium be? *Hint:* A search for "plutonium disposal" on the Web will bring up references. Try also including United States and DOE as search terms.

   a. Some propose that the plutonium be sent to local nuclear power plants to "burn" as fissionable fuel. What are the advantages and disadvantages of such a course of action?

b. Others propose that it be stored permanently in a repository. Again, list the advantages and disadvantages.

51. Advertisements for Swiss Army watches stress their use of tritium. One ad states that the "... hands and numerals are illuminated by self-powered tritium gas, 10 times brighter than ordinary luminous dials...." Another advertisement boasts that the "... tritium hands and markers glow brightly making checking your time a breeze, even at night...." Evaluate these statements and, after doing some Web research, discuss what form the tritium is in these watches, and what its role is.

52. The amount of exposure from medical X-rays varies considerably, depending on the procedure. For example, dental X-rays may be taken as "bitewings," full mouth, or panoramic, with each involving different amounts of radiation. Pick a particular type of medical X-ray and research the exposure to radiation involved. Report your data in both rems and sieverts (using mrem, mSv, or μSv as appropriate).

53. MRI, or magnetic resonance imaging, is an important tool for some types of medical diagnoses.

a. What is the scientific basis for this technique?

b. What information can an MRI give a physician that cannot be obtained through direct examination of a patient?

c. This MRI method used to be called NMR, nuclear magnetic resonance. Why do you think the name was changed?

54. Make a time line of nuclear history, putting at least a dozen dates on your line. For example, start with Roentgen's discovery of radioactivity in 1896. Other candidates include Chernobyl, Hiroshima, the opening of the first commercial reactor, discovery of various medical isotopes, use of uranium glazes in Fiesta ware, and the nuclear test ban treaty.

55. The Tennessee Valley Authority's nuclear reactor, Browns Ferry 1, is licensed, but has not operated since 1985. After repairs, it is scheduled to reopen by 2007. Find out the status of this reactor. Will it go (or has it gone) on line, thus increasing the number of commercially operating nuclear reactors from 103 to 104?

56. The hormesis phenomenon, mentioned in a margin note in Section 7.8, is that toxic substances in small amounts can increase one's resistance to the same substance in large amounts. Analogous to the dose-response curves of Figure 7.19, the following figure indicates the zone of hormesis where the curve dips below the control line into a therapeutic region. Use the resources of the Web to investigate hormesis. Prepare a summary of your findings. One starting point is an article, "Is Radiation Good for You?" at the science Web site The Why? Files. *Hint:* As in Figure 7.19, NOAEL means no observed adverse effect level.

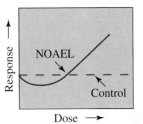

Source: *Environmental Science & Technology*, "Redrawing the Dose-Response Curve," Volume 38, No. 5, page 90A March 1, 2004.

57. Taking potassium iodide tablets can protect your thyroid from exposure to radioactive iodine, thus reducing your risk of thyroid cancer.

a. By what mechanism does potassium iodide protect you?

b. How long does the protection last?

c. Are the tablets expensive? *Hint:* the FDA Web site is a good source of information for parts **a.** and **b.**

58. According to Table 7.4, smoking 1.5 packs of cigarettes a day adds 13,000 μSv to your annual radiation dose.

a. Which radioactive isotopes are responsible for this dose in cigarette smoke?

b. A nonsmoker, living with a 1.5 pack a day smoker, may receive the equivalent of 12 chest X-rays per year as a result of the second-hand smoke. How many μSv would this add to the nonsmoker's annual dose?

Energy from
Electron Transfer

Millions of cell phone users depend on compact, lightweight, long-lasting, and rechargeable batteries.

Portable fuel cells can power computers, making field work a reality for users.

The battery-powered Segway human transporter is useful for personal transportation.

Over 100 cleaner vehicles of the present and future gathered at the 2003 Michelin Challenge event in Sonoma, California. Electric, hybrid, and hydrogen-powered cars, SUVs, trucks, and buses are shown. Solar-powered vehicles are also part of the future energy mix for transportation.

As these photographs illustrate, consumers make personal choices about the energy they use. In all cases, the energy is produced by the transfer of electrons. The title of this chapter reflects the importance of this fundamental process. Our way of life is heavily dependent on the convenience of personal power sources, such as batteries. The portable electronics market's demand for compact, mobile, and lightweight energy sources has been a major factor in developing advanced battery technology. But are all batteries the same? What is the difference between a rechargeable battery and one that must be discarded after it "runs down"? Transportation is turning to new technology as well, with batteries, fuel cells, and gasoline–electric hybrid power sources all gaining acceptance as technology develops. But what are fuel cells and how can they be used to transfer electrons and generate power? Why are some cars called zero-emissions vehicles (ZEVs) and are they really as advertised? What are the advantages of hybrid technology that combines combustion of gasoline with other ways to transfer electrons to produce energy? Will hydrogen, with its potential to provide clean, reliable, and affordable energy, be the answer to all of our energy needs? Will the promise of obtaining endless power from our Sun's radiation be realized? Each of these questions about energy from electron transfer will be considered in this chapter.

Consider This 8.1   **Take a Battery Inventory**

To help understand your personal dependence on electron transfer in consumer products, make a list of the things you own or use that run on batteries.

a. Which items on your list use a battery as the main source of energy and which as a backup source?

b. What type of battery is used in each case?

c. Which batteries are rechargeable and which are not?

d. How do you dispose of batteries that are not rechargeable?

This is not the first time in this text that we have discussed our ever-increasing need for energy. For example, Chapter 4 emphasized the energy obtained from burning fossil fuels to produce electrical energy and also to power transportation. Combustion processes also involve the transfer of electrons, but the widespread use of fossil-fuel combustion gave us reason to study them in a separate chapter. Chapter 7 focused on energy released by splitting fissionable atomic nuclei, a process that does not involve the transfer of electrons. Centralized power plants, whether fueled by coal or fission, distribute electricity regionally through vast power networks to offices, classrooms, and residences. But supplies of both fossil and fission fuels are limited, and when they are gone, they are gone forever. Moreover, these fuels also create huge environmental costs. Combustion of coal and petroleum releases vast quantities of carbon dioxide, contributing to global warming. Sulfur dioxide and nitrogen oxides are also released, leading to degradation of air quality and increases in acid precipitation. Nuclear fission also has its critics for its perceived safety risks and for not solving the problem of disposing of high-level nuclear wastes. The conclusion seems obvious. If our species is to continue to inhabit this planet and achieve the quality of life we desire, we must develop and depend on other sources of energy.

8.1 Electrons, Cells, and Batteries: The Basics

The famous inventor Thomas Edison was convinced in the late-19th century that batteries were doomed to failure. He branded such devices as ". . . a sensation, a mechanism for swindling the public by stock companies, . . ." He went on to say that, although such batteries were scientifically all right, their commercial success would be ". . . as

Edison's view might have been clouded by his ownership of a large, municipal electrical power company.

absolute a failure as one can imagine." He clearly was better at inventing than at judging the future success of batteries! Why are batteries important in our times? As you realize, if you are wearing a battery-powered watch, carrying a cellular phone, or using a laptop computer, you depend on batteries every day.

A **battery** is a system for the direct conversion of chemical energy to electrical energy. Batteries are found everywhere in today's society because they are convenient, transportable sources of stored energy. Batteries are also big business in the United States, with consumers expected to spend over $100 billion on electronics in 2004 according to the Consumer Electronics Association. Many of these products require batteries, spurring continued growth in that industry as well.

Although the term *battery* is in common use, a standard flashlight "battery" is more correctly called a cell, an electrochemical cell, or a galvanic cell. A **galvanic cell** is a device that converts the energy released in a spontaneous chemical reaction into electrical energy. A collection of several galvanic cells wired together constitutes a true battery, such as the battery in your automobile. A galvanic cell is the opposite of an **electrolytic cell,** one in which electrical energy is converted to chemical energy.

All galvanic cells produce useful energy from reactions involving the transfer of electrons from one substance to another. The transfer of electrons involves two changes, each represented by a separate equation. An equation for a **half-reaction** is a type of chemical equation that shows the electrons either lost or gained. The **oxidation half-reaction** shows the reactant that loses electrons, and the electrons appear on the product side of the equation. The **reduction half-reaction** shows the reactant that gains electrons, and the electrons appear on the reactant side of the equation. Two half-reactions—one oxidation and the other reduction—are always involved.

Oxidation = Loss of electrons
Reduction = Gain of electrons

Let us start by looking at a simplified example of the reaction that takes place in a nickel–cadmium (NiCad) battery:

$$\text{Oxidation half-reaction: } Cd \longrightarrow Cd^{2+} + 2\,e^{-} \qquad [8.1]$$

$$\text{Reduction half-reaction: } 2\,Ni^{3+} + 2\,e^{-} \longrightarrow 2\,Ni^{2+} \qquad [8.2]$$

$$\text{Overall cell reaction: } 2\,Ni^{3+} + Cd \longrightarrow 2\,Ni^{2+} + Cd^{2+} \qquad [8.3]$$

In this case, two electrons are given off, or "lost," in the oxidation half-reaction (equation 8.1). The number of electrons given off during oxidation must equal the number of electrons gained through reduction for the overall equation to balance. For this reason, the coefficient "2" appears in the reduction half-reaction (equation 8.2). Electrons do *not* appear in the overall reaction shown in equation 8.3, obtained by adding together equations 8.1 and 8.2.

Your Turn 8.2   **Electrons in Half-Reactions**

Which of these represents oxidation half-reactions? Which represent reduction half-reactions? What is the basis for your decision?

a. $Al^{3+} + 3\,e^{-} \longrightarrow Al$

b. $Zn \longrightarrow Zn^{2+} + 2\,e^{-}$

c. $Mn^{7+} + 3\,e^{-} \longrightarrow Mn^{4+}$

d. $2\,H_2O \longrightarrow 4\,H^{+} + O_2 + 4\,e^{-}$

e. $2\,H^{+} + 2\,e^{-} \longrightarrow H_2$

The transfer of electrons through an external circuit produces **electricity,** the flow of electrons from one region to another that is driven by a difference in potential energy. The reaction provides the energy needed to drive a cordless razor, a power tool, or countless other battery-operated devices. The chemical species oxidized in the cell and the species reduced must be connected in a way such that the electrons released during oxidation are transferred to the species being reduced. To enable this transfer, **electrodes,** electrical conductors placed in the cell as sites for chemical reaction, are used. Different processes take place at each electrode, and they are given different specific names. The **anode** is the electrode where oxidation takes place. The **cathode** receives the electrons sent from the anode through the external circuit. At the cathode, the electrons are used in the reduction half-reaction. The electrical energy is the result of the spontaneous reaction that occurs in the cell. The resulting **voltage,** the difference in electrochemical potential between the two electrodes, is expressed in units called volts (V). The greater the difference in potential between two electrodes, the higher the voltage will be and the greater the energy associated with the transfer. In the case of a NiCad cell, the maximum difference in electrochemical potential is measured as 1.46 V under specified conditions. Several cells are connected together in series to produce greater potential differences necessary to power larger devices (Figure 8.1).

The complete reaction in a NiCad battery is actually just a bit more complicated than represented in equations 8.1–8.3. The cell contains a water-based paste of either NaOH or KOH as an electrolyte. The chemical structure of both nickel and cadmium change as the reaction proceeds. Cadmium metal, the anode, is oxidized to Cd^{2+} ions; simultaneously, Ni^{3+} ions (in hydrated NiO(OH)) on a nickel cathode are reduced to Ni^{2+} ions (in $Ni(OH)_2$).

Anode reaction: (Oxidation half-reaction)
$$Cd(s) + 2\,OH^-(aq) \longrightarrow Cd(OH)_2(s) + 2\,e^- \qquad [8.4]$$

Cathode reaction: (Reduction half-reaction):
$$2\,NiO(OH)(s) + 2\,H_2O(l) + 2\,e^- \longrightarrow 2\,Ni(OH)_2(s) + 2\,OH^-(aq) \qquad [8.5]$$

Overall cell reaction:
$$Cd(s) + 2\,NiO(OH)(s) + 2\,H_2O(l) \longrightarrow 2\,Ni(OH)_2(s) + Cd(OH)_2(s) \qquad [8.6]$$

These three equations show exactly the same transfer of electrons represented in equations 8.1–8.3, but now all the different states and chemical forms are indicated. Figure 8.2 illustrates how the galvanic cell can be engineered.

Two types of electrodes: **anode** and **cathode**
Two types of ions: **anion** and **cation** (See Section 5.7)

The unit honors the Italian physicist Alessandro Volta (1745–1827). He is credited with inventing the first electrochemical battery in 1800.

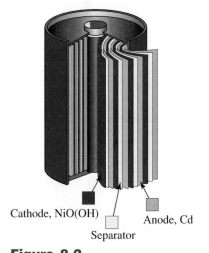

Figure 8.1

A portable power drill, such as this Black & Decker model, usually comes with two NiCad battery packs.

Your Turn 8.3 — Checking Balance and Charge

a. Consider equations 8.1–8.6 from the standpoint of number and type of atoms. Is each equation balanced?

b. Consider equations 8.1–8.6 from the standpoint of charge. Is the total charge zero for both reactants and products in these balanced half-reactions? Does it have to be?

c. What is the quick way to tell an overall cell reaction from a half-reaction?

The NiCad battery is rechargeable. This is possible because the starting materials that undergo oxidation and reduction are solids, as are the products. The solids formed by the forward (discharging) reaction cling to a stainless steel grid within the battery. Thus, they are still available and permit the reaction to be reversed during the recharging process. No gases are produced during either the recharging or the discharging, so the battery can be totally sealed. Transfer of electrons takes place during both the discharging and recharging processes, just in opposite directions. Equation 8.7 indicates this reversible process.

$$Cd(s) + 2\,NiO(OH)(s) + 2\,H_2O(l) \underset{\text{recharging}}{\overset{\text{discharging}}{\rightleftharpoons}} 2\,Ni(OH)_2(s) + Cd(OH)_2(s) \quad [8.7]$$

To construct a commercially useful galvanic cell, more than just the electrochemical potential must be considered. Successful cells must be of reasonable cost, last a reasonable length of time, and be safe to use and discard or recharge. In some applications, the size and weight of the cell is of paramount importance. Solids, pastes, gels, or thick slurries act as electrolytes to carry ions between electrodes. Aqueous

Cathode, NiO(OH)   Separator   Anode, Cd

Figure 8.2

A NiCad galvanic cell in which Cd is oxidized at the anode, and Ni^{3+} is reduced at the cathode.

Electrolytes were defined in Section 5.7.

solutions are often too hazardous to use in commercial cells because of the possibility for leakage. However, aqueous solutions are often used in simple laboratory cells. Electrons are transferred through the external circuit from anode to cathode, but positive and negative ions are often transported internally through a conducting salt bridge to complete the circuit. Ask if you can set up the galvanic cell in Your Turn 8.4 in the laboratory, or if you can see it as a demonstration in class.

Your Turn 8.4   **A Laboratory Galvanic Cell**

As this galvanic cell operates, a reddish coating of impure copper metal begins to appear on the surface of the copper cathode. This is the overall equation for the reaction.

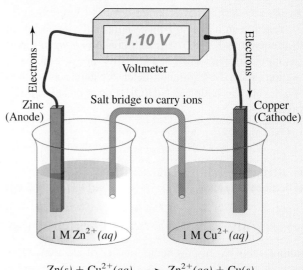

$$Zn(s) + Cu^{2+}(aq) \longrightarrow Zn^{2+}(aq) + Cu(s)$$

a. Write the oxidation half-reaction taking place at the anode.
b. Write the reduction half-reaction taking place at the cathode.
c. This laboratory galvanic cell is not rechargeable. Why not?

8.2 Some Common Batteries

Almost everyone has used an alkaline battery such as one of those shown in Figure 8.3. How does the electron transfer work in this case? Figure 8.4 details the operation of a typical alkaline battery.

These are the half-reactions for the alkaline cell illustrated in Figure 8.4.

Anode reaction: (Oxidation half-reaction)
$$Zn(s) + 2\,OH^-(aq) \longrightarrow Zn(OH)_2(s) + 2\,e^- \qquad [8.8]$$

Cathode reaction: (Reduction half-reaction)
$$2\,MnO_2(s) + H_2O(l) + 2\,e^- \longrightarrow Mn_2O_3(s) + 2\,OH^-(aq) \qquad [8.9]$$

The overall cell reaction is the sum of the two half-reactions.
$$Zn(s) + 2\,MnO_2(s) + H_2O(l) \longrightarrow Zn(OH)_2(s) + Mn_2O_3(s) \qquad [8.10]$$

The cell voltage depends primarily on which elements and compounds participate in the reaction. The voltage does not depend on factors such as the overall size of the cell, the amount of material it contains, or the size of the electrodes. Thus all alkaline cells, from the tiny AAA size to the large D cells, have the same voltage, 1.54 V. The oxidation half-reaction at the anode sends electrons through the external circuit to power your flashlight, radio, or other device, before returning the electrons to the cathode to participate in the reduction half-reaction. The **current,** or the rate of electron flow

Alkaline (basic) solutions were discussed in Section 6.4.

Figure 8.3
These AAA to D alkaline cells all produce 1.54 V, but the larger cells can sustain the transfer of electrons through the external circuit for a longer time.

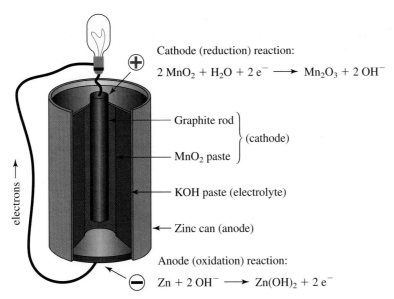

Cathode (reduction) reaction:

$$2\,MnO_2 + H_2O + 2\,e^- \longrightarrow Mn_2O_3 + 2\,OH^-$$

Graphite rod ⎫
                ⎬ (cathode)
MnO$_2$ paste ⎭

KOH paste (electrolyte)

Zinc can (anode)

Anode (oxidation) reaction:

$$Zn + 2\,OH^- \longrightarrow Zn(OH)_2 + 2\,e^-$$

electrons →

Figure 8.4

Diagram of an alkaline cell (battery).

through the external circuit, is measured in amperes (amps, A) or very likely in milliamps (mA) for small cells. Larger cells contain more materials and can sustain the transfer of electrons over a longer period, useful for many applications.

Many different galvanic cells have been developed, each for a specific purpose. Examples are listed in Table 8.1, along with their voltages, whether they are rechargeable, and their common uses.

> Current is measured in amperes (amps, A) to honor André Ampère (1775–1836). He was a largely self-taught French mathematician who devoted himself to the study of electricity and magnetism.

Consider This 8.5   Cell Phone Batteries

Many cell phones use lithium ion batteries. Use the resources of the Web to find out more about this type of battery by searching for "lithium battery chemistry."

a. Why is a lithium ion battery suited for use in portable devices?

b. What materials form the anode and the cathode of a lithium ion battery?

c. How does the lithium ion battery differ from a lithium–iodine battery?

d. What other types of batteries are used in cell phones? What are the advantages and disadvantages of each compared with lithium ion batteries?

Table 8.1	Some Common Galvanic Cells		
Type	**Voltage**	**Rechargeable?**	**Examples of Uses**
Alkaline	1.54	No	Flashlights, small appliances
Lithium–iodine	2.8	No	Camera batteries, pacemakers
Lithium ion	3.7	Yes	Laptop computers, cell phones, digital music players
Lead–acid (storage battery)	2.0	Yes	Automobiles
Nickel-cadmium (NiCd)	1.25	Yes	Consumer electronics
Nickel-metal hydride (NiMH)	1.25	Yes	Replacing NiCad for many uses; hybrid vehicles
Mercury	1.3	No	Formerly widely used in cameras, other appliances

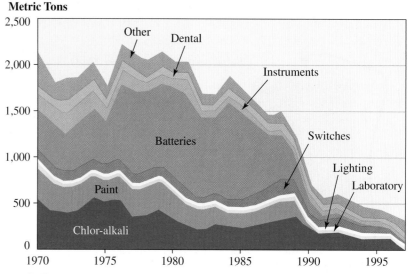

Figure 8.5

Total mercury use in the United States, 1970–1998. Mercury use for the manufacture of batteries has declined sharply in recent years.

Source: *U.S. Geological Survey Minerals Yearbook.* Taken from *Chemical & Engineering News,* February 5, 2001, p. 22.

Compact, long-lasting batteries play a special role in medical applications. The widespread use of cardiac pacemakers has been due, in large part, to the improvements made in the batteries used to power them, rather than in the pacemakers themselves. All lithium batteries take advantage of the low density of lithium metal to make a light-weight battery, but lithium–iodine cells are so reliable and long-lived that they are often the battery of choice for this application. A lithium–iodine pacemaker battery implanted in the chest can last as long as 10 years before it needs to be replaced. Persons with pacemakers are advised to avoid electromagnetic fields that can interfere with the operation of the device, such as holding a cell phone close to the location of the pacemaker or going through an airport security arch.

Because mercury batteries can be made very small, they once were used widely in watches, camera equipment, hearing aids, calculators, and other devices that use transistors and integrated circuits that do not require large currents. Unfortunately, the toxicity of mercury (Chapter 5) makes the disposal of these cells a hazard. Burning trash may exacerbate and extend the problem by releasing mercury vapor into the atmosphere. Development of safer battery alternatives and the need to recycle many types of batteries led to passage of the Mercury-Containing and Rechargeable Battery Management Act (the Battery Act) in 1996. The act mandated the phaseout of mercury in batteries and represents a major step forward in the effort to facilitate recycling nickel–cadmium and certain small sealed lead-acid rechargeable batteries. The change in mercury use, including for batteries, is shown in Figure 8.5.

Consider This 8.6   The Federal Battery Act and You

Explore the provisions of the Battery Act by going to the EPA Web site or using the link provided at the *Online Learning Center.*

a. Why was it necessary at the federal level to regulate battery manufacture, use by consumers, and disposal of batteries?

b. What types of batteries are regulated by the Battery Act?

c. What are some hazards associated with the improper disposal of batteries?

d. What do you personally do with your "dead" batteries? Explain the options available to you in your community.

Most batteries convert chemical energy into electrical energy with an efficiency of about 90%. This can be compared with the much lower efficiencies that typically characterize the conversion of heat to work (30–40%) in electricity-generating plants. However, it is important to remember that considerable energy is required to manufacture galvanic cells. Metals and minerals must be mined and processed, and the various components manufactured and assembled. Moreover, a battery has a finite life. Even rechargeable batteries will eventually fail and have to be replaced. Sooner or later, the chemical reaction in the battery will be complete, the voltage will drop below usable levels, and electrons will no longer flow. At this point, the battery is "dead" and ready for disposal.

8.3 Lead-Acid (Storage) Batteries

The best known rechargeable battery is the lead-acid battery. Lead-acid batteries are called **storage batteries** because they "store" electrical energy. Until very recently, such batteries were used in every automobile. The lead-acid storage battery is a true battery because it consists of six cells, each generating 2.0 V, for a total of 12.0 V. The overall cell reaction is given by equation 8.11.

$$\text{Pb}(s) + \text{PbO}_2(s) + 2\,\text{H}_2\text{SO}_4(aq) \underset{\text{recharging}}{\overset{\text{discharging}}{\rightleftharpoons}} 2\,\text{PbSO}_4(s) + 2\,\text{H}_2\text{O}(l) \qquad [8.11]$$

$$\text{lead}\quad\text{lead dioxide}\quad\text{sulfuric acid}\qquad\qquad\text{lead sulfate}\qquad\text{water}$$

The anode is made of metallic lead and the cathode of lead dioxide, PbO_2. The electrolyte is a concentrated sulfuric acid solution (Figure 8.6). Although the weight of the lead and the corrosive properties of the acid are disadvantages, the lead-acid storage battery is dependable and long-lasting.

The key to the success of the lead-acid storage battery is the fact that reaction 8.11 is reversible. As it spontaneously proceeds in the direction indicated by the arrow to the right, the reaction produces the energy necessary to power a car's starter, headlights, and various devices. But, as the reaction proceeds, the battery "discharges." The electrical demands of a modern car are so great that in a short time, most of the reactants would be converted to products, significantly reducing the voltage and the current. To counter this, the battery is attached to a generator, or alternator, turned by the engine. The alternator generates direct current electricity, which is run back through the battery.

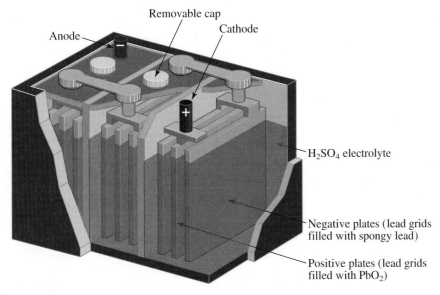

Removable cap

Anode

Cathode

H_2SO_4 electrolyte

Negative plates (lead grids filled with spongy lead)

Positive plates (lead grids filled with PbO_2)

Modern lead-acid batteries are permanently sealed. They do not require addition of water and therefore are not designed with a removable cap.

Figure 8.6
Cutaway view of a lead-acid storage battery.

This input of energy reverses the reaction, represented by the arrow to the left in equation 8.11, and recharges the battery. In a high-quality lead-acid storage battery, this process of discharging and recharging can go on over a period of five years or more.

In environments where the emissions from internal combustion engines cannot be tolerated, sealed lead–acid storage batteries provide the only source of energy for locomotion. Thus, forklifts in warehouses, passenger carts in airport terminals, golf carts, and wheelchairs are typically powered by lead–acid storage batteries. Because of the dependability of such batteries, they are sometimes used in conjunction with wind turbine electrical generators. The generator charges the batteries when the wind is blowing, and the batteries provide electricity when the wind stops.

Your Turn 8.7   **Another Look at the Storage Battery**

Examine equation 8.11 for the reaction that takes place in a lead-acid storage battery while it is discharging.

a. Which material is being oxidized?
b. Which material is being reduced?
c. Write the chemical equation for the *recharging* reaction. What material is being oxidized? Which is reduced?

A group of Israeli chemists led by Doron Aurbach and his colleagues at Bar-Ilan University have been working on the development of a more environmentally friendly storage battery. Their strategy has been to replace lead with magnesium, a metal that is far safer, lighter in weight, and less expensive. The new magnesium-acid battery, like the lead–acid storage battery, is rechargeable. It produces up to 1.3 V but the chemists hope to improve that to 1.7 V in the near future with further refinements to the materials. Magnesium is a far more reactive metal than lead, providing some challenges in preventing unwanted side reactions between magnesium and the electrolytes used in the battery. One of the first applications envisioned for the new type of storage battery will be to provide uninterrupted power to computer networks during power outages. This battery may also become useful to power gas/electric hybrid vehicles.

8.4 Fuel Cells: The Basics

Suppose someone were to suggest a way to combine H_2 and O_2 to form H_2O without the hazards of combustion. Suppose that this individual also claimed that the reaction could be carried out with no direct contact between the hydrogen and the oxygen. The Sceptical Chymist might well dismiss such assertions as sheer nonsense—an outright impossibility. And yet, sometimes what appears to be completely contrary to common sense can happen in the natural world. The operation of a fuel cell is a case in point. A **fuel cell** is a galvanic cell that produces electricity by converting the chemical energy of a fuel directly into electricity without burning the fuel. Fuel cells are sometimes called "flow" batteries, because both fuel and oxidizer must constantly flow into the cell to continue the chemical reaction. After we consider this important type of battery, we can better understand new approaches to powering cars and other types of vehicles.

William Grove, an English physicist, invented fuel cells in 1839. However, they remained a mere curiosity until the advent of the U.S. space program in the 1960s. The Space Shuttle, for example, carried three sets of 32 cells fueled with hydrogen, but never "burned" it. Instead, the hydrogen was used in the fuel cell to produce electricity for the shuttle.

A fuel cell functions somewhat like a conventional flashlight battery. But, unlike batteries, fuel cells use a constant external supply of fuel (such as hydrogen) and oxidant (such as the oxygen in air). Therefore, they produce electricity as long as fuel

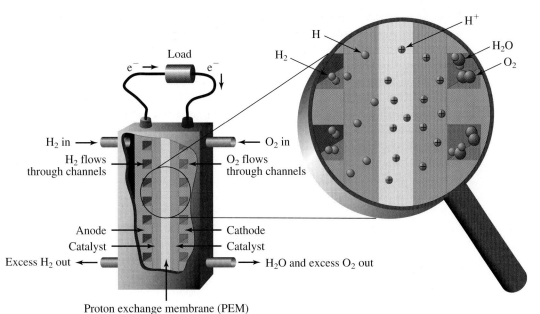

Figure 8.7
A PEM fuel cell in which H_2 and O_2 are combined without combustion.

 Figures Alive! Visit the *Online Learning Center* to learn more about the chemistry of a PEM fuel cell.

and oxidant are provided and consequently do not need to be recharged in the same sense as traditional batteries. In a fuel cell, the chemical reaction is physically divided into two parts, each of which occurs in a different region of the cell. One region behaves as the anode, where oxidation of the fuel will take place. The other region behaves as the cathode, where reduction of the oxidant takes place.

The material that separates the anode from the cathode in a fuel cell serves the same purpose as the electrolyte in a traditional galvanic cell. Changing the electrolyte in a fuel cell changes its properties and therefore its usefulness. The most common electrolytes are aqueous solutions of phosphoric acid (H_3PO_4) and of potassium hydroxide (KOH), an alkaline solution. These types are currently being used by NASA. But acidic and alkaline electrolyte solutions are corrosive and require that the liquid be fully contained, adding to the complexity and cost of the fuel cell.

Hydrogen fuel cells developed more recently use a solid polymer to separate the reactants and to act as the electrolyte. The polymer electrolyte membrane, also called a proton exchange membrane (PEM), is permeable to H^+ ions and is coated on both sides with a platinum-based catalyst. These electrolytes can operate at reasonably low temperatures, typically from 70 °C to 90 °C, and can start transferring electrons to provide electrical power quickly. As a result, PEM fuel cells are currently very popular among automakers for new fuel cell prototype vehicles and for personal consumer applications. A typical design is shown in Figure 8.7.

Polymers are discussed in Chapter 9.

An H^+ ion is a proton, the simplest possible cation. See Section 5.7.

Your Turn 8.8   **Revisiting the PEM fuel cell**

Use Figures Alive! at the *Online Learning Center* to answer these questions.

a. How is a fuel cell different from other batteries discussed earlier in this chapter?
b. Can a PEM fuel cell be recharged ? Why or why not?
c. Why is the combination of H_2 and O_2 in a fuel cell not classified as combustion?

The attraction for a proton and polar water molecules was discussed in Section 5.5.

One drawback to PEM fuel cells is that they must remain constantly moist. The species that moves through the membrane is not just a proton, but a water molecule plus a proton, called the **hydronium ion, H_3O^+.** Therefore, water is required for the proton transfer. Water produced at the cathode is recycled back to the anode to maintain the required moisture. If the membrane dries out, the PEM fuel cell abruptly will stop working. Because constant moisture levels are essential, PEM fuel cells are extremely sensitive to high temperatures. If the temperature rises above 100 °C, the water may evaporate from the electrolyte and the anode.

PEM fuel cells are currently the best option for automobiles, but they require expensive catalysts and complex construction techniques to be reliable over the long term. Scientists are now working on developing solid acid electrolyte fuel cells. An advantage of these electrolyte materials is that they transport "bare" protons through their structure, reducing the need for constant moisture levels present in PEM fuel cells. Solid electrolytes are impermeable to gases and liquids, thereby serving to keep the fuel and oxidizer separated. Furthermore, they can be used with various fuels, not just hydrogen gas. For example, they can be used in fuel cell systems in which methanol is the source of hydrogen gas.

In all fuel cells, two chemical changes always occur. One is oxidation, which means a loss of electrons. The other change is reduction, which means the gain of electrons. Typically, hydrogen gas (H_2) is the fuel used in conjunction with oxygen. The oxidation and reduction half-reactions are represented by equations 8.12 and 8.13. As a molecule of hydrogen passes through the membrane, it is oxidized and loses two electrons to form two H^+ ions.

Anode reaction: (Oxidation half-reaction)

$$H_2(g) \longrightarrow 2\,H^+(aq) + 2\,e^- \qquad [8.12]$$

These H^+ ions flow through the proton exchange membrane and combine with oxygen (O_2) and two electrons to form water.

Cathode reaction: (Reduction half-reaction)

$$^{1/2}\,O_2(g) + 2\,H^+(aq) + 2\,e^- \longrightarrow H_2O(l) \qquad [8.13]$$

The overall cell reaction is the sum of the two half-reactions.

$$H_2(g) + {}^{1/2}\,O_2(g) + \cancel{2\,H^+(aq)} + \cancel{2\,e^-} \longrightarrow \cancel{2\,H^+(aq)} + H_2O(l) + \cancel{2\,e^-} \quad [8.14]$$

The electrons lost by H_2 through oxidation (equation 8.12) are gained by O_2 in the reduction process (equation 8.13). The 2 e^- and 2 H^+ that appear on each side of the arrow in equation 8.14 can be cancelled to give the net equation shown in equation 8.15.

This same equation appears in Chapters 1, 4, 5, and 7, an indication of its significance.

$$H_2(g) + {}^{1/2}\,O_2(g) \longrightarrow H_2O(l) \qquad [8.15]$$

Thus, in a fuel cell, there is a transfer of electrons from H_2 to O_2. Oxidation cannot occur alone; that would be rather like one hand clapping. Oxidation (electron loss) must always be paired with reduction (electron gain). The net reaction is a combination of the oxidation and the reduction half-reactions, taking into account gain and loss of electrons.

To produce electricity, the two half-reactions of the fuel cell must be connected in such a way that the electrons released during the oxidation reaction are *transferred* to the reduction reaction (see Figure 8.7). As with other types of batteries, electrodes provide the sites for chemical reaction. Oxidation occurs at the anode, and reduction takes place at the cathode. The electrical energy produced is the result of the spontaneous reaction that occurs in the fuel cell. Hydrogen and oxygen gases that have not been consumed can be recycled into the cell. The electrons flowing from the anode to the cathode of a fuel cell move through an external circuit to do work, which is the whole point of the device. On the Space Shuttle, these electrons are used to illuminate bulbs, power small motors, and operate computers.

The reaction shown in equation 8.15 is clearly the equation for the burning of hydrogen with oxygen. But in a fuel cell, it is "burning" without a flame and with

Table 8.2		Comparison of Combustion with Fuel Cell Technology		
Process	**Fuel***	**Oxidant**	**Products**	**Other Considerations**
Combustion	H_2	O_2 from air	H_2O, heat, light, and sound	Rapid process, flame present, lower efficiency, most useful for producing heat
Fuel cell	H_2	O_2 from air	H_2O, electricity, some heat	Slower process, no flame, quiet, higher efficiency, most useful for generating electricity

*Compounds containing hydrogen, such as natural gas or alcohols, can be used as fuels. Since these compounds contain carbon as well, CO or CO_2 (or both) are released as products.

relatively little heat and no light being produced. Electrons are transferred, producing an electric current. Water is the only chemical product if hydrogen is the fuel. There are no greenhouse gases, no air pollutants, nor spent nuclear fuel to contend with. These are among the reasons that hydrogen fuel cells are considered a more environmentally friendly way to produce electricity than is burning fossil fuel or fissioning uranium atoms. A comparison of the combustion of H_2 with O_2 and fuel cell technology is given in Table 8.2.

The net reaction represented by equation 8.15 releases 286 kJ of energy per mole of water formed. But instead of liberating most of this energy in the form of heat, the fuel cell converts 45–55% or more of it into electrical energy. This direct production of electricity eliminates the inefficiencies associated with using heat to do work to produce electricity. Internal combustion engines are only 20–30% efficient in deriving energy from fossil fuels. Moreover, a fuel cell does not "run down" or require recharging. It keeps functioning as long as hydrogen and oxygen are supplied. The fuel can be hydrogen gas, or hydrogen formed from the onboard decomposition of liquid methanol, CH_3OH, producing hydrogen for fuel as needed. The catalytic decomposition of methanol into hydrogen and carbon dioxide is an example of a **reforming process** (Figure 8.8). Reforming processes also have been developed to allow gasoline or even diesel fuel to be used as the source of hydrogen.

Economical supply and safe storage of hydrogen gas are two of the issues that have prevented full commercialization of fuel cells until now.

Remember that it *requires* energy to separate hydrogen gas from compounds containing hydrogen.

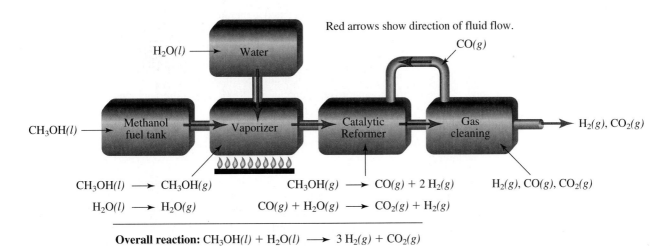

Figure 8.8

Hydrogen obtained from methanol by a reforming process.

Solid oxide fuel cells appear to be the most promising technology for small electric power plants. A common combination used has been zirconium oxide (ZrO_2) with calcium oxide (CaO). The oxide crystal lattice is coated on both sides with specialized porous electrode materials. This design of fuel cell operates at higher temperatures than PEM fuel cells, typically in the range of 700 °C to 1000 °C. They have efficiencies around 50%, and are reliable over long periods. Solid oxide fuel cells are "fuel-flexible" because they are tolerant to impurities and can at least partially internally reform hydrocarbon fuels. The waste thermal energy from a solid oxide fuel cell can be used to spin a gas turbine, generating more electricity. Such hybrid schemes are expected to have efficiencies approaching 70% as further improvements take place.

Consider This 8.9   **Other Fuel Cell Technologies**

a. Use the resources of the Web to learn about the current status, advantages, and disadvantages of solid oxide fuel cell technology.
b. Are the applications for PEM fuel cells and solid oxide fuel cells likely to be the same? Why or why not?
c. Are other alternative fuel cell technologies currently being proposed for future development? If so, what are the prospects for the near future that these will become successful alternatives to PEM and solid oxide fuel cells?

8.5   Fuel Cells: Large and Small

The U.S. space program accelerated the development of fuel cell technology in the past. More recently interest has been renewed in using fuel cells to generate power at stationary locations that may or may not be connected to a power grid. The energy may be used locally or for standby or backup power on the grid. This practice is called **distributed generation,** placing power-generating modules of 30 MW or less near the end user. Worldwide, over 150 demonstration fuel cell plants have been installed for power generation. Nearly 75% of these are in Japan, about 15% are in North America, and 9% are in Europe. The largest fuel cell assembly in the United States supplies electrical power to 1000 homes in Santa Clara, California. The health maintenance organization giant Kaiser Permanente has installed fuel cell units to furnish electricity in three of its California hospitals.

Ballard Power Systems is a world leader in developing, manufacturing, and marketing PEM fuel cells for generating electricity. Their most recent partnership is with Nippon Petroleum in Yokohama, Japan. They are jointly developing and testing a kerosene-fueled 1-kW stationary fuel cell generator to be used for the Japanese residential market. About 26% of residential energy requirements in Japan are met by burning kerosene, but with a high environmental cost. Generating electricity in this fuel cell system will be another example of Ballard's approach to developing zero-emissions energy. Ballard Power Systems built a successful 60-kW prototype stationary fuel cell power generator in Vancouver, Canada, in 2000. This company has partners in the United States, Germany, Switzerland, and Japan for a series of field trials of 250-kW stationary power generators that started in 1999. The trials continue with the goal of obtaining performance data in real-world operations before moving ahead with further development.

Sceptical Chymist 8.10   **"Zero" Emissions from Fuel Cells**

Hydrogen–oxygen fuel cells are considered to be environmentally friendly "zero-emissions" power sources. Does the Sceptical Chymist agree with this statement? What if the source of hydrogen were methanol or natural gas, largely methane? Are the same emissions expected, and are they still just as environmentally friendly?

Engineers are now developing fuel cells to provide portable power cleanly and quietly for consumer products. One day, fuel cells may be routinely powering your lawn mower and your laptop computer; a prototype fuel cell now exists small enough to power a cell phone. Many predict that development of these far smaller fuel cells, sometimes called **microcells,** will become reality before larger-scale applications such as electrical generation and powering automobiles. In 2003, the Japanese manufacturer Nippon Electric Company (NEC) announced the introduction of a methanol fuel cell-powered laptop computer that can run for 5 hours, longer than on today's lithium ion batteries. Allied Business Intelligence, a New York market research company that tracks new technology, predicts some 200–500 million microcells might be sold annually by 2011, generating as much as $5 billion in revenue. Microcells could replace batteries, delivering many more hours of usable power.

More robust portable power supplies could allow professionals in government, military, and industry to increase their mobility and efficiency using human transporters (Figure 8.9). Personal-sized fuel cells could help meet increasing consumer demands for sophisticated color displays, wireless access to the Internet, tablet computers, multifunction cell phones, and MP3 players capable of storing and playing more than 10,000 songs. There is even an entrant in the technological challenge to produce the smallest fuel cell. Researchers at the University of Texas, Austin, have developed an enzyme-based fuel cell that can power small silicon-based microelectronics.

Unlike batteries, fuel cells cannot deliver large bursts of power. Because of this, some portable power designs combine both battery and fuel cell technology.

Figure 8.9
Fuel cell-powered Segway human transporter.

 Consider This 8.11   **Military Electrons**

Fuel cells find many applications in the military. To find the military fuel cell demonstration site nearest you, go to the Web site of the Department of Defense (DoD) on fuel cells. Check out several of the demonstration sites across the country to learn (a) the different uses of these fuel cells and (b) any cost savings reported.

Consider This 8.12   **Batteries—Functions and Uses**

The type of battery used for a specific application depends on the best match between technical capability and intended function. Based on what you have learned to this point, consider each of these uses and discuss the criteria that will determine which battery is best for the intended use: (a) TV remote controls; (b) heart pacemakers; (c) deep-space probes; (d) cellular phones; (e) automobiles.

We turn next to how fuel cells, batteries, and combinations of energy sources dependent on the transfer of electrons can be used to power automobiles and other forms of transportation.

8.6 Alternative Energy Sources for Transportation

Because PEM fuel cells are compact, light, and no longer require hazardous electrolytes such as potassium hydroxide, they are prime candidates for use in electric vehicles. However, problems in developing safe and efficient storage of hydrogen fuel have limited fuel cell use for vehicles. In addition, bulky storage containers have somewhat limited a vehicle's driving range. Nevertheless, as shown in Figure 8.10, hydrogen fuel can be used with stacked PEM fuel cells to power a car by electron transfer.

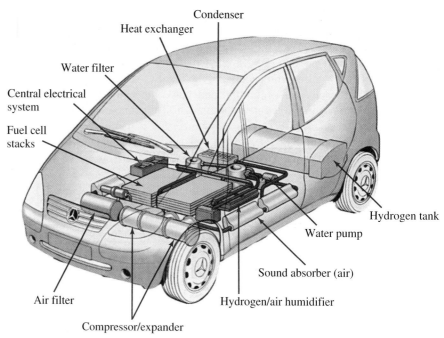

Figure 8.10

The electrochemical engine for vehicles.

Source: Reprinted with permission of George Retseck Illustration.

The overall process of obtaining H_2 from CH_3OH was shown in Figure 8.8.

Another option is to use methanol (CH_3OH) as the source of hydrogen in PEM fuel cells. The onboard reformer converts the hydrogen in the compound methanol into H_2O. A different catalyst, an alloy made from four metals, replaces the platinum-based one used with hydrogen-only fuel cells, which cannot be used with methanol and water. At the anode of a methanol fuel cell, a 3% solution of methanol and water reacts to produce carbon dioxide, hydrogen ions, and electrons.

Anode reaction: (Oxidation half-reaction)
$$H_2O(l) + CH_3OH(aq) \longrightarrow CO_2(g) + 6\,H^+(aq) + 6\,e^- \qquad [8.16]$$

Air or oxygen is blown into the cathode compartment where the oxygen picks up electrons and reacts with hydrogen ions to produce water.

Cathode reaction: (Reduction half-reaction)
$$^3/_2\,O_2(g) + 6\,H^+(aq) + 6\,e^- \longrightarrow 3\,H_2O(l) \qquad [8.17]$$

Here is the net reaction.

$$CH_3OH(aq) + ^3/_2\,O_2(g) \longrightarrow CO_2(g) + 2\,H_2O(l) \qquad [8.18]$$

As with hydrogen fuel cells, the electricity produced by methanol fuel cells is used to power electric motors for the vehicle. Equation 8.18 reveals that an electric vehicle powered by methanol–oxygen fuel cell, unlike one powered by a hydrogen–oxygen fuel cell, is not strictly speaking a **ZEV,** a zero-emissions vehicle. One of the products is the greenhouse gas CO_2. However, the amount of CO_2 generated per unit of useful energy is lower than that emitted in the direct combustion of the fuel, because the efficiency is higher. The operating temperature is lower as well, which lessens the formation of nitrogen monoxide, such as is produced from internal combustion engines. In addition, methanol is a renewable fuel, unlike gasoline. Another plus is that fuel cells have no moving parts, so the "engine" of these vehicles should require less repair and last longer than conventional ones. According to Dr. Halpert, Program Manager for Batteries and Fuel Cells at the Jet Propulsion Laboratory, "This [methanol] fuel cell may well become the power source of choice for energy-efficient, nonpolluting electric vehicles."

Figure 8.11
A methanol-based fuel cell bus at Georgetown University. The Federal Transit Administration funded the bus.

Consider This 8.13 **Fuel Cells in Your Future?**

Many different types of fuel cells are under development. One promising type is the propane–oxygen fuel cell. Here is the equation for the chemical reaction that takes place.

$$C_3H_8(g) + 5\,O_2(g) \longrightarrow 3\,CO_2(g) + 4\,H_2O(l)$$

a. Identify the chemical that undergoes oxidation and the one that undergoes reduction.

b. Unlike batteries, fuel cells do not store chemical energy. Explain the significance of this statement for the future of fuel cells.

c. List reasons why fuel cells have not been the energy source of choice in the past, but may become viable in the future.

Electric vehicles powered by fuel cells are not a far-fetched idea. Buses carrying up to 60 passengers have been powered by PEM fuel cells in Chicago, IL, and at Georgetown University in Washington, D.C. (Figure 8.11).

In November 2000, DaimlerChrysler unveiled NECAR 5 (New Electric Car), an electric vehicle that operates on methanol fuel cells (Figure 8.12). This Mercedes-Benz

Figure 8.12
The NECAR 5, a Mercedes-Benz A-class car by DaimlerChrysler, operates using methanol fuel cells.

A-class car is the technological successor to a series of electric vehicles developed since 1994. NECAR 5 averages about 25 miles per gallon of methanol and has a 250-mile range without refueling, close to that of more conventional vehicles. It can reach speeds of more than 90 mph. In 2002, NECAR 5 established a long-distance record, driving across the United States from San Francisco to Washington, D.C. Dr. Ferdinand Panik with DaimlerChrysler remarked: "In the end, I believe the fuel cell can be done for the same price as the piston engine, or lower. And I believe it can let the owner travel 50% farther for the fuel used, with an engine that will be truly maintenance-free."

Whether Panik's prediction comes true remains to be seen. Reliable predictions of the actual costs of automobiles powered by fuel cells are difficult to obtain, but the prices of the cell components are declining. The fuel cell manufacturer, Ballard Power Systems, predicts that competitively priced, zero-emissions vehicles will start to show up in automobile showrooms for the 2004–2005 model year. To make the prediction a reality, Ballard Power Systems has formed a worldwide alliance with DaimlerChrysler and Ford Motor Company to become the largest commercial producer of fuel-cell-powered electric drive trains and components for cars, trucks, and buses. The alliance is gearing up to hit their target of mass-producing 100,000 fuel-cell-powered electric vehicles by 2004–2005. "The beauty of fuel-cell vehicles is that they are pollution-free and energy efficient, and we can make the fuel right here in America," said Paul Lehman, a fuel-cell researcher at Humboldt State University (California). "In electric cars, fuel cells offer important advantages over batteries: They have greater range, and they take minutes to refuel—not hours to recharge."

Consider This 8.14   Fuel-Cell Cars—Today's Reality?

Use the Web and inquiries to local car dealers to explore whether fuel-cell cars are a reality for consumers in your area.

a.   Can you buy a methanol fuel cell car such as DaimlerChrysler's NECAR today? If so, give the particulars of performance and price. If not, explain why not.

b.   Another alternative is a hydrogen fuel cell car from Honda, the EV-Plus. Can you buy this car today in your area? If so, give the particulars of performance and price. If not, explain why not.

Arthur D. Little, an energy-consulting firm, has announced the development of a prototype fuel cell that converts gasoline to hydrogen. In the prototype fuel cell, gasoline vapor is converted into hydrogen and carbon monoxide. The carbon monoxide, in contact with a special catalyst, is then reacted with steam to produce carbon dioxide and additional hydrogen. The hydrogen can then be used to make electricity, as in a conventional fuel cell. As a fuel to produce hydrogen for fuel cells, gasoline would have the advantage of using a preexisting fuel distribution system, a feature not yet available for distributing hydrogen directly. Still in development, the gasoline fuel cell is not likely to be available commercially until later this decade.

In contrast with other automobile makers planning to produce pollution-free vehicles, General Motors decided to take a different approach. The first all-electric passenger car to be available in limited quantities for leasing was GM's electric vehicle, the Saturn EV-1. Introduced in 1997 after 10 years of research and development, this zero-emissions electric vehicle represented a transformation in the way in which the world considers alternatives to fossil fuels as energy sources. Twenty-six lead storage batteries, weighing a total of about 1100 lb, were at the energetic heart of the 137-horsepower EV-1. This car was developed in response to legislation enacted in California and subsequently in New York and Massachusetts. In an effort to remedy what has been called the worst air pollution in the United States, a tentative plan had been adopted that would require all cars in the Los Angeles basin to be converted to

Figure 8.13

The General Motors Saturn EV-1 did not require a tailpipe.

Figure 8.14

The GEM Neighborhood Electric Vehicle has a top speed of 25 mph and a range of up to 30 miles per charge.

electric power or other clean fuel by 2007. Given current technology at the time these proposals were passed, regulators felt that the only way to achieve this goal was to rely on lead storage batteries.

The all-battery driven EV-1 cars had the obvious advantage of not releasing any carbon dioxide or carbon monoxide, any oxides of sulfur or nitrogen, or any unburned hydrocarbons—they did not even have a tailpipe (Figure 8.13). Such cars required no gasoline or other fuel. Unfortunately, zero emissions on the road does not translate to a similar absence of pollution elsewhere. Batteries lost power at low temperatures and had to be recharged at frequent intervals, a process requiring electricity. As you know, electrical power generation is notoriously inefficient; less than half of the energy released from burning fuel is converted into electricity. Furthermore, power plants that burn fossil fuels for conversion to electrical energy release sulfur dioxide, nitrogen oxides, and carbon dioxide. In fact, calculations indicate that the SO_2 and NO_x emitted from power plants generating the electricity to keep a fleet of battery-powered cars operational would exceed the amount of these two gases that would be released by the gasoline-powered cars that would have been replaced! The overall CO_2 emission does decline when electric cars are substituted for internal combustion automobiles, but by less than 50%.

California has since relaxed the 1990 law mandating that 10% of all cars sold between 2003 and 2008 be pollution-free ZEVs. Emerging hybrid technologies discussed in the next section allow automakers to produce partial ZEVs that meet the current state and federal standards, are more efficient, and certainly less costly to produce. All production of EV-1 cars ended in 2000, and the leases are being withdrawn. All cars had to be returned to GM by August 2004. The technology lives on, however, in the street-legal GEM vehicle (Figure 8.14). This is called a Neighborhood Electric Vehicle (NEV) and is produced by Global Electric Motorcars, a company of DaimlerChrysler based in Fargo, North Dakota. There are currently over 28,000 vehicles that have been sold nationally since 1998. Electric vehicles such as GEMs have many niche uses for consumers, businesses, and even the military.

Consider This 8.15   **Revolt of EV-1 Loyalists**

EV-1 cars have a group of very loyal fans who are not taking the demise of their ZEVs lightly. Many lessees have offered to buy their EV-1 vehicles outright, even if they will not receive any further technical support from GM. In a very public display of their frustration, over 100 celebrities, engineers, and fans gathered at Hollywood Forever Cemetery to hold a mock funeral for the EV-1 in July 2003.

(continued on p. 374)

Consider This 8.15   Revolt of EV-1 Loyalists (*continued*)

a.   What do you think originally motivated the EV-1 lessees to choose the EV-1 vehicles for their transportation needs?

b.   Why are the EV-1 lessees now unhappy with GM? What are their specific concerns?

c.   Would it meet the concerns of the EV-1 lessees if GM offered to give each of them a GEM NEV or its equivalent? Why or why not?

d.   What will happen to the EV-1 vehicles after they are returned to GM?

8.7   Hybrid Vehicles

The name Prius comes from the Latin "to go before".

Concerns about energy supplies and environmental effects drove the development of ZEVs. However, fuel-cell-powered cars are still in the future for the average consumer, and all-battery powered cars have not proven practical. Some compromise was needed so that consumers could enjoy the convenience and range of a gasoline-powered car combined with the environmental advantages of an electric vehicle. Toyota and Honda have led the way in developing practical **hybrid cars** that combine conventional gasoline engines with battery technology. The Toyota Prius (Figure 8.15), a hybrid about the size of a Toyota Corolla, was first available in Japan in 1997 and soon thereafter in the United States.

Consider This 8.16   NiMH Batteries

Toyota's successful Prius gas–electric hybrid car uses nickel metal hydride (NiMH) batteries. The auto company states that the batteries should not have to be either recharged or replaced during the normal lifetime of the car.

a.   Identify the oxidation half-reaction, the reduction half-reaction, and the overall chemical change that takes place in a NiMH battery.

b.   What features of NiMH batteries make them superior to lead-acid batteries that were used in EV-1 vehicles?

c.   Why are NiMH batteries replacing NiCd and lithium ion batteries for uses such as cell phones and portable computers?

Hybrids, unlike conventional gasoline-powered cars, deliver better mileage in city driving than at highway speeds.

With a 1.5-L gasoline engine sitting side-by-side with nickel–metal hydride batteries, an electric motor, and an electric generator, the Prius needs no recharging. It consumes only about half the gasoline, emits 50% less carbon dioxide and far less nitrogen oxides than a conventional car, while delivering 52 miles per gallon of gasoline around town and 45 out on the highway. The electric motor draws power from the batteries to get the car moving or to power it at low speeds. Using a process called regenerative braking, the kinetic energy of the car is transferred to the generator, which charges the batteries during deceleration and braking. The gasoline engine assists the electric motor during normal driving, with the batteries boosting power when extra acceleration is needed. The attractive $18,000 price tag was set artificially low to stimulate sales in Japan, which have been brisk enough for demand to outpace supplies. The Prius became available in the United States in 2000. In the same year, Honda introduced a slightly smaller hybrid passenger car, the Insight, that gets 70–80 miles per gallon overall. Newer, more powerful passenger car hybrids are now available from both companies and several others are rapidly entering the field. The new midsize Toyota Prius, introduced in the United States in 2003, is rated at 60 mpg in the city and 51 mpg on the highway.

(a)

(b)

Figure 8.15

(a) The Toyota Prius gas–electric hybrid automobile.

(b) The "engine" of the Prius. The NiMH batteries are not shown because they are under the back seat and trunk areas.

A midsize Prius with hybrid technology was introduced in 2004. It is advertised as an ATPZEV, advanced technology partial zero-emissions vehicle.

Consider This 8.17 **Hybrid Cars**

Visit Web sites for at least two different companies to find the latest information on their hybrid vehicles. Search for "hybrid car" or "advanced vehicles". Check the *Online Learning Center* for links to several sites describing hybrid vehicles.

a. What does the manufacturer say about them?

b. What do the owners or lessees of these cars say about each type of hybrid car?

c. Which hybrid cars are available in your region of the United States? What is the waiting period for a hybrid car?

d. Design either a poster or a radio announcement that would help market a hybrid car.

Automobile industry leaders agree that there will be no mass market for alternative energy vehicles—battery-powered, hybrid cars or ones using fuel cells—unless the performance and price of such vehicles match those of conventional models. Jack Smith, former CEO of General Motors, said: "People are too practical. There's a certain element who will fall in love with new technology, but technology won't survive unless it's cost effective." The common strategy for companies now is to provide more than one option for fuel-efficient, low emission vehicles, then allow consumers to choose the best option for their personal transportation needs.

The best selling vehicles in the United States right now are not passenger cars. SUVs and trucks are the hot items, and they can employ hybrid technology proven successful in cars. Reluctant consumers have to be convinced that their desire for muscle and size will not prevent them from enjoying the fuel savings and environmental benefits possible with hybrid technology. Ford had its hybrid SUV called Escape in showrooms in summer 2004. Toyota introduced the 2005 Highlander hybrid gas–electric SUV at the North American International Auto Show in Detroit in January 2004. The Chevrolet Tahoe and GMC Yukon for 2007 will use gas–electric hybrid technology, improving fuel economy by 30% over current models. GM also will offer the 2008 Chevy Silverado and GMC Sierra full-size pickups with gas–electric technology, promising to not reduce the load-carrying or trailer-towing capabilities of these vehicles. Many other vehicles currently are on the drawing board, including high-performance cars from Lexus and Subaru.

Consider This 8.18   Delivering with Hybrids and Fuel Cells

a. FedEx executives announced in May 2003 that they planned to switch 20 of their medium-sized trucks to energy-saving, environmentally friendly hybrid vehicles. If the initial switch is successful, they will extend this change over their entire fleet of 30,000 trucks. The new trucks will be diesel–electric hybrids, not gas–electric hybrids. What reasons did FedEx cite for this choice of hybrid technology?

b. UPS joined with DaimlerChrysler to produce at least one delivery truck powered by a hydrogen fuel cell by the end of 2004. EPA agreed to supply a hydrogen refueling station, located in Ann Arbor, Michigan, for the project. What reasons did UPS cite for this choice of technology?

> The name Hy-Wire refers to the hydrogen fuel cell and the wires connecting the electronics and computers that control the car.

Experimenting with the conversion of one truck to use a hydrogen fuel cell is certainly a fine idea, and so is GM's development of the prototype hydrogen car, called the Hy-Wire. President George W. Bush was photographed in 2003 with this concept car, just after announcing that $1.2 billion would go into researching hydrogen gas as a replacement for gasoline in a program called "Freedom Fuel." Could it be that hydrogen can provide a way to enjoy better environmental quality while ensuring energy independence from foreign oil? Before going ahead with a hydrogen-based economy, we need to take a more careful look at this essential element, hydrogen. Where will we get it, and how will we build the infrastructure to safely produce, store, and distribute hydrogen? We turn to these questions in the next section.

8.8   Splitting Water: Fact or Fantasy?

In Jules Verne's 1874 novel, *Mysterious Island,* a shipwrecked engineer speculates about the energy resource that will be used when the world's coal supply has been used up. "Water," the engineer declares, "I believe that water will one day be employed as fuel, that hydrogen and oxygen which constitute it, used singly or together, will furnish an inexhaustible source of heat and light."

Is this simply science fiction, or is it energetically and economically feasible to break water into its component elements? Can hydrogen really serve as a useful fuel? And what does this have to do with advanced vehicles and light from the Sun? To answer these questions and assess the credibility of the claim by Verne's engineer, a Sceptical Chymist needs to examine the energetics of this reaction.

$$2\,H_2(g) + O_2(g) \longrightarrow 2\,H_2O(l) \qquad [8.19]$$

Experiment shows that the reaction, as written, gives off 572 kJ when 2 mol of liquid water is formed from the combination of 2 mol of hydrogen and 1 mol of oxygen. Burning of 1 mol of H_2 will yield half of this, or 1 mol of H_2O and $1/2(572)$ of energy. Here is the chemical equation.

$$H_2(g) + 1/2\,O_2(g) \longrightarrow H_2O(l) + 286\ \text{kJ} \qquad [8.20]$$

Sceptical Chymist 8.19   **Checking Bond Energies**

Use the bond energy values in Table 4.2 to check the energy released by the reaction in equation 8.20. Are the values the same? Why or why not? To be convincing, clearly show your reasoning.

Because energy is *evolved,* the energy change in this combustion reaction is −286 kJ for each mole of H_2 burned to form liquid water. This is equivalent to releasing 143 kJ per gram of H. In comparison, the heat of combustion of coal is 30 kJ/g, octane (a major component in gasoline) is 46 kJ/g, and methane (natural gas) is 54 kJ/g when the products of combustion are $CO_2(g)$ and $H_2O(l)$. Clearly, hydrogen has the capability of being a powerful energy source. In fact, on a per-gram basis, hydrogen has the highest heat of combustion of any known substance. The extraordinary energy per gram of hydrogen when it burns raises a tantalizing prospect—the practical use of hydrogen as a fuel to power motor vehicles that would produce only water vapor. No pollutants such as the carbon monoxide and nitrogen oxides would form, unlike what happens when burning fossil fuels.

Before using hydrogen gas in automobiles can become common practice, we must overcome several challenges. For example, how could we obtain a sufficient supply of hydrogen if it were to be used for fueling motor vehicles? On the one hand, things look promising because hydrogen is the most plentiful element in the universe. Over 93% of all atoms are hydrogen atoms. Although hydrogen is not nearly that abundant on Earth, there is still an immense supply of the element. But, on the other hand, essentially all of it is tied up in compounds. Hydrogen is too reactive to exist for long in its diatomic form, H_2, in the presence of the other elements and compounds that make up the atmosphere and Earth's crust. Therefore, to obtain hydrogen for use as a fuel, we must extract hydrogen from hydrogen-containing compounds, which requires energy.

Your Turn 8.20   **Back to Basics**

Calculate the number of moles and grams of H_2 that would have to be burned to yield an American's daily energy share of 260,000 kcal. *Hint:* 1 kcal = 4.18 kJ.

Answer
3800 mol, 7600 g

Hydrogen gas is traditionally generated in the laboratory by the action of certain acids on certain metals. Perhaps the most common combination is sulfuric acid and zinc.

$$H_2SO_4(aq) + Zn(s) \longrightarrow ZnSO_4(aq) + H_2(g) \qquad [8.21]$$

Although convenient as a small-scale source of hydrogen, the reaction of a metal with an acid is far too expensive to scale up for industrial applications. Therefore, we turn back to the most abundant earthly source of hydrogen—water.

Because the formation of 1 mol of water from hydrogen and oxygen releases 286 kJ (equation 8.20), an identical quantity of energy must be absorbed to reverse the reaction to produce hydrogen. Figure 8.16 summarizes the processes.

All that is needed to bring about this reaction is a source of 286 kJ. The most convenient method of decomposing water into hydrogen and oxygen is by **electrolysis,** the process of passing a direct current of electricity of sufficient voltage through water to decompose it into H_2 and O_2 (Figure 8.17). When water is electrolyzed, the volume of hydrogen generated is twice that of oxygen. This suggests

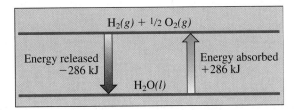

Figure 8.16
Energy differences in the hydrogen–oxygen–water system.

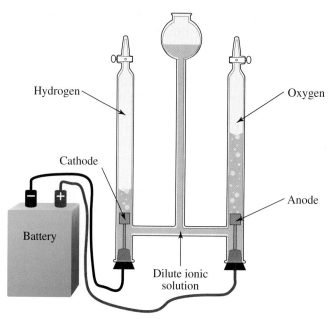

Figure 8.17
Electrolysis of water.

that a water molecule contains twice as many hydrogen atoms as oxygen atoms, testimony to the formula H_2O.

$$286 \text{ kJ} + H_2O(l) \longrightarrow H_2(g) + 1/2\ O_2(g) \qquad [8.22]$$

Of course, the question remains: How will the electricity for large-scale electrolysis be generated? Most electricity in the United States is produced by burning fossil fuels in conventional power plants. If we only had to deal with the first law of thermodynamics, the best we could possibly achieve would be to burn an amount of fossil fuel equal in energy content to the hydrogen produced in electrolysis. But recall from Chapter 4 that we must also deal with the consequences of the second law of thermodynamics. Because of the inherent and inescapable inefficiency associated with transforming heat into work, the maximum possible efficiency of an electrical power plant is 63%. When we add the additional energy losses caused by friction, incomplete heat transfer, and transmission over power lines, it would require at least twice as much energy to produce the hydrogen than we could obtain from its combustion. This is comparable to buying eggs for 10 cents each and selling them for five, which is no way to do business.

Considerations such as these lead some observers to say that the pollution from producing hydrogen could offset the benefits. Furthermore, most methods of generating electricity have a variety of negative environmental effects. At one time it was thought that the "cheap" extra electricity from nuclear fission could be used to produce hydrogen to fuel the economy, but that energy utopia has hardly been realized. It should be apparent, therefore, that electricity generated from fossil fuels or nuclear fission is not a currently practical method to split water to produce hydrogen for use as a fuel.

A second possibility is to use heat energy to decompose water. One process used to produce commercial hydrogen does, in fact, use heat. You have already encountered it, in Chapter 4, in the discussion of substitutes for fossil fuels. Hot steam is passed over coke (essentially pure carbon) at 800 °C.

The mixture of H_2 and CO is called water gas. See equation 4.10.

$$131 \text{ kJ} + H_2O(g) + C(s) \longrightarrow H_2(g) + CO(g) \qquad [8.23]$$

The mixture of hydrogen and carbon monoxide produced can be burned directly. It can serve as the starting material for the synthesis of hydrocarbon fuels and other important compounds, or the hydrogen can be separated from the mixture and used as needed.

Reaction 8.23 is being studied in an effort to find catalysts that will make it possible to carry out the reaction at lower temperatures.

Simply heating water to decompose it thermally into H_2 and O_2 is not commercially promising. To obtain reasonable yields of hydrogen and oxygen, temperatures of over 5000 °C would be required. The attainment of such temperatures is not only extremely difficult, but it also would consume enormous amounts of energy—at least as much as would be released when the hydrogen was burned. Thus, we have again reached a point where we would be investing a great deal of time, effort, money, and energy to generate a quantity of hydrogen that would, at best, return only as much energy as we invested. In practice, a good deal less energy would result.

Methane, the major component of natural gas, is currently the chief source of hydrogen. Hydrogen can be formed from the endothermic reaction of methane with steam.

$$165 \text{ kJ} + CH_4(g) + 2 H_2O(g) \longrightarrow 4 H_2(g) + CO_2(g) \qquad [8.24]$$

Researchers continue to find ways to increase the efficiency of this method of producing hydrogen.

Your Turn 8.21   Back to Bond Energies

a. Use the average bond energy values in Table 4.2 to check the energy required by the reaction in equation 8.24. Show your work clearly.
b. Is this reaction endothermic or exothermic?
c. Did your calculated value match the 165 kJ given in the equation? Why or why not?

8.9 The Hydrogen Economy

If and when we were to find feasible methods for producing hydrogen cheaply and in large quantities, we still would face significant problems in storing and transporting it. Although H_2 has a high energy content per gram, it occupies a very large volume—about 12 L (a bit over 12 qt) per gram at normal atmospheric pressure and room temperature. If H_2 were to be stored and transported in its gaseous state, large, heavy-walled metal cylinders would be required, thus eliminating much of the advantage of the favorable energy–mass ratio. To save space, gases are typically converted to the liquid state under high pressure, as in the case of "bottled gas" or "liquid propane." But hydrogen must be cooled to −253 °C before it liquefies. Keeping it in liquid form requires maintaining the gas at low temperature and therefore involves high costs.

Attempts have been made to circumvent these problems by storing H_2 in other forms. One involves absorbing gaseous H_2 into activated carbon, a solid that can hold a great deal of H_2 at moderate pressures. The carbon can be heated to release H_2 just as it is needed for reaction.

A different approach involves reacting the H_2 with certain metals to produce compounds called hydrides, which are relatively stable solids with a reasonably high storage capacity. For example, when 10 L (slightly over 10 qt) of H_2 gas at 25 °C and 1 atm of pressure react with Li metal, the lithium hydride that forms occupies a mere 4.3 mL. That is slightly less than a teaspoon.

$$Li(s) + 1/2 H_2(g) \longrightarrow LiH(s) \qquad [8.25]$$

When such hydrides react with H_2O, they produce H_2.

$$LiH(s) + H_2O(l) \longrightarrow H_2(g) + LiOH(aq) \qquad [8.26]$$

The product H_2 can then be reacted in the usual fashion.

Lithium has a 1+ charge in LiH. Thus, hydrogen has an unusual 1− charge in this compound.

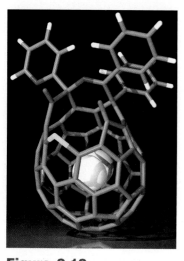

Figure 8.18

An open-cage fullerene, serving as a nanocontainer, is filled with a H_2 molecule, shown in white. Also: red = oxygen, yellow = sulfur, and blue = nitrogen.

Allotropes were defined in Section 2.1.

Nanotechnology was defined in Chapter 1 and will be discussed further in Chapter 9.

Your Turn 8.22   Active Metal Hydrides

a. Sodium, another member of Group 1A, will form a metallic hydride in a reaction similar to that of lithium. Write a chemical equation showing the reaction of Na with H_2.

b. Sodium hydride will react with water to release H_2. What is the other product of this reaction?

c. In Consider This 8.16, you researched nickel metal hydride (NiMH) batteries used in hybrid cars. Is the role of NiMH to release H_2 through a reaction similar to that in equation 8.26? Explain.

Prototype vehicles that operate on this principle have been built and used in a number of locations. Clearly, such approaches would greatly improve the safety and convenience of handling H_2 and perhaps would be the decisive factor in determining the extent of its acceptance as a fuel. As we discussed earlier in this chapter, a hydrogen-fueled car would produce only water vapor and none of the carbon monoxide or nitrogen oxides emitted from using gasoline-fueled internal combustion engines.

A more recent suggestion is to store hydrogen in a compound derived from an allotropic form of carbon, C_{60}. The class of compounds based on C_{60} is called **fullerenes,** and the particular one shown in Figure 8.18 contains an open "cage" with S, N, and O atoms nearby. Scientists looking for efficient ways to utilize fullerenes cages to trap either metal atoms or gases have made great strides in just a few years of research. In 2003, Japanese scientists at Kyoto University achieved 100% encapsulation of H_2 over 8 hours. The H_2 is released slowly when a solution containing H_2 in the open-cage fullerene is slowly warmed above 160 °C. This development is yet another example of the growth of nanotechnology over the last few years.

Even if we were to manage to solve all of the production, storage, and transport problems just identified, we must consider how best to extract energy from our hydrogen. The most obvious way would be to burn it in power plants, vehicles, and homes. A stream of pure hydrogen burns smoothly, quietly, and safely in air, delivering 143 kJ per gram and forming only nonpolluting water as an end product. But when hydrogen is mixed directly with oxygen, a spark can be sufficient to produce a devastating explosion, limiting the usefulness of hydrogen as a direct fuel. However, the promise of using hydrogen in fuel cells is creating renewed interest in the hydrogen economy.

In April 2004, the U.S. Department of Energy awarded $350 million in grants to more than 130 research institutions and companies, including the Big Three U.S. automakers. Combined with private funding, this is a $575 million effort to help remove some of the current obstacles to developing cleaner-burning technology for widespread use. The major goal is to put hydrogen-fueled vehicles on the road by 2015.

Meeting the energy needs of our modern world will not be possible with only one technology. In the long run, pollution-free transportation will benefit from a mix of improved fuel cells, new battery technologies, and changes in public attitudes toward utilization and conservation of energy. However, one more source must be considered, one that may well play an increasingly important role in our energy mix. In the next section you will see this is yet another example of electron transfer.

 Consider This 8.23 Iceland's Hydrogen Economy

The small country of Iceland is taking bold steps to be the first to cut its ties to fossil fuels. Their plan, first announced in February 1999, is to demonstrate that it can produce, store, and distribute hydrogen as a means to power both public and private transportation.

a. Who are the partners in this venture?

b. What has been its first tangible outcome?

c. Do you think the lessons learned in Iceland will be relevant for the United States? Explain your reasoning.

8.10 Photovoltaics: Plugging in the Sun

Every hour of every day, enough energy from the Sun reaches Earth to meet the world's energy demand for a whole year. But, currently, less than 0.5% of the power generated in the United States comes directly from the Sun. This statistic will need to change, according to Nobel Laureate Richard E. Smalley, professor of physics, chemistry, and astronomy at Rice University. Dr. Smalley predicts that nanotechnology will provide many new solutions to problems currently limiting the use of alternative technologies. He sees solar energy as part of the solution to the top problem humanity will confront over the next 50 years—which is increasing the supply of energy necessary to sustain the expected 8–10 billion people living on Earth by 2050.

All life on Earth depends on the Sun's energy. We harness that energy, converting it to food and fuel. Looking back through time, the fossil fuels we burn today once depended on solar energy that enabled plant material to grow. Now we burn those fuels for energy, much of it in the form of electricity. Electricity supplies about 35% of U.S. energy needs, and two thirds of that electricity is used in residential and commercial buildings. Indirect conversion of solar energy into electricity is not new. The challenges are to convert solar radiation *directly* into electricity in a practical and economically feasible way, enabling more widespread use of this available energy source.

Photovoltaic cells (solar cells) convert radiant energy directly to electrical energy, without the intermediary of hydrogen or some other fuel. Photovoltaic (PV) technologies consider research results on photovoltaic cells and develop practical energy sources that help to meet our energy needs. It takes only a few PV cells to produce enough electricity for your calculator or digital watch. If more power is required, PV cells can be connected together to form arrays capable of generating electricity for large-scale use. PV devices have already demonstrated their practical utility for satellites, highway signs, security and safety lighting (Figure 8.19), navigational buoys, and automobile recharging stations, just to mention a few common uses.

PV cells are made from a class of materials called **semiconductors,** materials that do not normally conduct electricity well, but can do so under certain conditions. Semiconductors are solids with structures closely resembling metals. We know that the structure of metals, such as copper or aluminum, enables them to be good conductors of electricity. **Metallic bonding** can be described with the "electron sea" model, in which outer (valence) electrons are shared among all the atoms in the substance. There is a highly regular array of positively charged nuclei, surrounded by a "sea" of electrons. The outermost electron on each metal atom is loosely attracted to its nucleus. When a large number of atoms form a tightly packed array, the valence electrons can act like a liquid that is spread over all the nuclei. The solid is held together by the mutual attraction of the metal cations for the mobile, highly delocalized electrons.

It is easy to induce electrons to move in this metallic electron sea, since the bonding of an electron to any single individual nucleus is relatively weak. The motion of electrons is what constitutes a current, what we think of as conducting electricity. In semiconductors, the bonding is very similar to that in metals. The nuclei are arranged in regular arrays. However, the electrons around each nucleus have a stronger attraction to the nuclei that they surround. Therefore, before the electrons are free to move around and conduct electricity, energy must be added.

The element silicon was one of the first semiconducting materials developed for use in PV cells. A crystal of silicon consists of an array of silicon atoms, each bonded to four other by means of shared pairs of electrons (Figure 8.20a). These shared electrons are normally fixed in the bonds and unable to move through the crystal. Consequently, silicon is not a very good electrical conductor under ordinary circumstances. However, if a valence electron absorbs sufficient energy, it can be excited and released from its bonding positions (Figure 8.20b). Once freed, the electron can move throughout the crystal lattice, making the silicon an electrical conductor.

For a PV cell to generate electricity, light must induce such a movement of electrons in the cell. This electron movement depends on the interaction of matter and

Figure 8.19
Photovoltaic (solar) cells are used to improve security, enhance safety, and direct pedestrians and vehicles.

The term *photovoltaic* reminds you that radiant energy (*photo*) is converted to electricity (*voltaic*).

Using solar cells rather than batteries in navigational buoys saves the U.S. Coast Guard an estimated $6 million annually through reduced maintenance and repair.

Conductivity in aqueous ionic solution is discussed in Section 5.7.

Many of the businesses that developed semiconductors were clustered in California's "Silicon Valley," which took its nickname from the element.

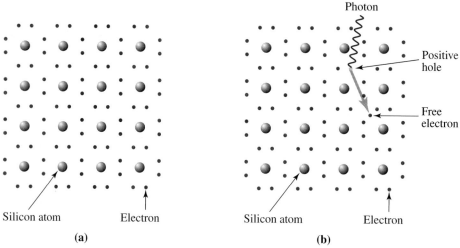

Figure 8.20
(a) Schematic of bonding in silicon.
(b) Photon-induced release of a bonding electron in a silicon semiconductor.

photons of radiant energy—a topic already treated in considerable detail in Chapters 2 and 3. In those chapters, we pointed out that the portion of the Sun's radiation reaching Earth's surface is mainly in the visible and infrared regions of the spectrum, with a maximum intensity near a wavelength of 500 nm. Light of this wavelength is in the visible range and has energy of about 3.6×10^{-19} J per photon, corresponding to 220 kJ per mole of photons. The energy required for silicon to release an electron from a bond is 1.8×10^{-19} J per photon, which is equivalent to radiation with a wavelength of 1100 nm. Visible light has a wavelength range of 400–700 nm. Recall that the shorter the wavelength of radiation, the greater the energy per photon. Therefore, photons of visible sunlight have more than enough energy to excite electrons in silicon semiconductors.

Your Turn 8.24 **Confirming Calculations**

a. Using relationships found in Section 2.5, show that the statements just made about energy per photon and energy per mole of photons for 500-nm wavelength visible light are correct.
b. Repeat the energy calculations for light of a wavelength of 400 nm and of 700 nm. Then comment on the ability of each wavelength to release electrons from silicon.

In order to make PV cells that can capture and store energy from sunlight, it is important that the electric currents generated have certain predictable and controllable properties. Therefore, it became necessary to fabricate the PV cells not from a single pure substance, but from a combination of materials with different characteristics. A very common method of adjusting the properties of a pure semiconductor material is through **"doping,"** a process of intentionally adding small amounts of other elements to pure silicon. The doping materials are chosen for their ability to facilitate the transfer of electrons. For example, about 1 ppm of gallium (Ga) or arsenic (As) is often introduced into the silicon. These two elements and others from the same periodic groups are used because their atoms differ from silicon by a single outer electron. Silicon has four electrons in its outer energy level, gallium has three, and arsenic has five. Thus, when an atom of As is introduced in place of Si in the silicon lattice, an extra electron is added. The replacement of a Si atom with a Ga atom means that the crystal is now one electron "short."

Ga is in Group 3A.
Si is in Group 4A.
As is in Group 5A.

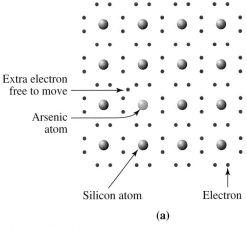

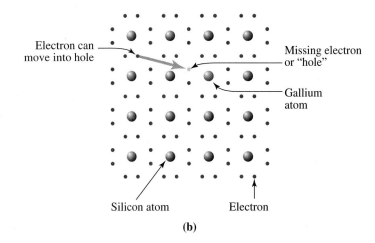

Figure 8.21

(a) An arsenic-doped *n*-type silicon semiconductor.
(b) A gallium-doped *p*-type silicon semiconductor.

The extra electrons in arsenic-doped silicon are not confined to bonds between atoms. Rather, electrons move easily through the lattice, increasing the electrical conductivity of the material over that of pure silicon. Silicon doped in this manner is called an ***n*-type semiconductor** in which there are freely moving negative charges, the electrons. On the other hand, for each silicon atom replaced with a gallium ion, an electronic vacancy, or "hole," is introduced into what is normally a two-electron bond. When an electron moves into this vacancy, a new hole appears where the mobile electron formerly was located and the positive charge has developed in a new location. Gallium-doped silicon is therefore a ***p*-type semiconductor** in which there are freely moving positive charges, or holes. Negatively charged electrons and positively charged holes move in opposite directions. Figure 8.21 illustrates both *n*- and *p*-type semiconductors. Both types of doping increase the conductivity of the silicon because less energy is needed to get extra electrons or holes moving. This means that photons of lower energy (longer wavelength) can induce electron release and transport in doped crystals.

Your Turn 8.25	**Other Doping Materials**

Some solar cell designs use phosphorus and boron to dope silicon crystals.

a. Which will form a *n*-type semiconductor? Explain your reasoning.
b. Which will form a *p*-type semiconductor? Explain your reasoning.

"Sandwiches" of *n*- and *p*-type semiconductors are used in transistors and other miniaturized electronic devices that have revolutionized communication and computing. Similar sandwich structures are central to the direct conversion of sunlight to electricity. A photovoltaic cell typically includes sheets of *n*- and *p*-type silicon in close contact (Figure 8.22). The *n*-type semiconductor is rich in electrons and the *p*-type is rich in positive holes. When they are placed in contact, electrons tend to diffuse from the *n*-region into the *p*-region. Likewise, the positive holes tend to be displaced from the *p*-region to the *n*-region. This generates a voltage or potential difference at the junction between the semiconductors. The voltage difference accelerates the electrons released when sunlight strikes the doped silicon. If the two layers are connected by a conducting wire, electrons flow through the external circuit from the *n*-semiconductor, where their concentration is higher, to the *p*-semiconductor, where it is lower.

Not only does the use of a *p-n* junction facilitate the conduction of electricity, but also it ensures that the current will flow in a specific direction through the PV cell.

An individual solar cell produces at most about 0.5 V.

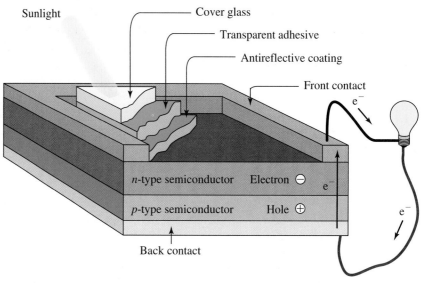

Figure 8.22

Schematic diagram of a photovoltaic (solar) cell.

Figure 8.23

A technician is taking measurements on ultrapure silicon ingots.

The transfer of electrons generates a direct current of electricity that can be intercepted to do essentially all the things electricity does including being stored in batteries for later use. As long as the cell is exposed to light, the current will continue to flow, powered only by solar energy.

The fabrication of photovoltaic cells is not without some significant problems. The first is that although silicon is the second most abundant element in Earth's crust, it is found combined as silicon dioxide, SiO_2. You know this material by its common name, sand, or more correctly as quartz sand. The good news is that the starting material from which silicon is extracted is cheap and abundant. The not so good news is that processes to extract and purify silicon are *not* inexpensive. Many of the early PV cell designs require ultrapure 99.999% silicon (Figure 8.23).

A second complication is that the direct conversion of sunlight into electricity is not very efficient. A photovoltaic cell could, in principle, transform into electricity up to 31% of the radiant energy to which it is sensitive. But some of the radiant energy is reflected or absorbed by the material making up the cell. Typically a commercial solar cell now has an efficiency of only 15%, but even this is a significant increase from the first solar cells built in the 1950s, which had efficiencies of less than 4%. In Chapter 4, we lamented the 63% maximum efficiency of converting heat to work in a conventional power plant. It might seem that we should be even more distressed at the lower limits that can be achieved by photovoltaics. Remember, however, that the first use of solar cells was to provide electricity in NASA spacecraft. For that application, the intensity of radiation was so high that low efficiency was not a serious limitation and costs were not of paramount concern. For commercial use on Earth, costs and efficiency are issues. Our Sun is an essentially unlimited energy source, and converting it to electricity even inefficiently is free from many of the environmental problems associated with burning fossil fuels or with storage of spent fuel from nuclear fission. These considerations add impetus to research and development of solar cells.

In addition to making an effort to improve the performance of silicon semiconductors by doping, scientists have been searching for other substances that exhibit the same or better semiconductor properties. Among promising substitutes are germanium, an element found in the same group of the periodic table as silicon, and compounds in which the elements have the same number of outer electrons as Sn or Ge. Included in this latter list are gallium arsenide (GaAs), indium arsenide (InAs), cadmium selenide (CdSe), and cadmium telluride (CdTe). There are also some very new combinations of indium, gallium, and nitrogen. Some of these new semiconductors have enhanced the efficiency of radiation-to-electricity conversion and made possible photovoltaic cells that are responsive to wider regions of the spectrum. Siemens Solar Industries has developed a copper indium selenide semiconducting thin film that has significant advantages over amorphous silicon and cadmium telluride.

Your Turn 8.26 Two-Element Semiconductors

Using the periodic table as a guide, show why the electron arrangement and bonding in GaAs and CdSe would be very similar to that in pure Ge.

Replacing crystalline silicon with the noncrystalline form of the element is another approach to commercial viability. Photons are more efficiently absorbed by less highly ordered Si atoms, a phenomenon that permits reducing the thickness of the silicon semiconductor to 1/60th or more of its former value. The cost of materials is thus significantly reduced. Other researchers are developing multilayer solar cells. By alternating thin layers of *p*-type and *n*-type doped silicon, each electron has only a short distance to travel to reach the next *p-n* junction. This lowers the internal resistance within the cell and raises the efficiency of the cells at an increasingly competitive price. Figure 8.24 gives a sense of just how thin these layers actually are. Smaller quantities of low-grade silicon are required and the production process can become highly automated.

Long-range prospects for photovoltaic solar energy are encouraging. Its cost is decreasing while the cost of electricity generated from fossil fuels is increasing. Limited and uncertain supply and the expense of pollution controls are driving the cost of fossil-fuel electricity still higher. Given this situation and the continued improvements in the performance and decreases in the cost of solar cells, electricity from photovoltaic systems could become even more competitive with that from fossil fuels early

> Maximum theoretically predicted efficiencies increase dramatically for stacks of *p*- and *n*-junctions. Efficiencies jump to 50% for 2 junctions, 56% for 3 junctions, and way up to 72% with 36 junctions.

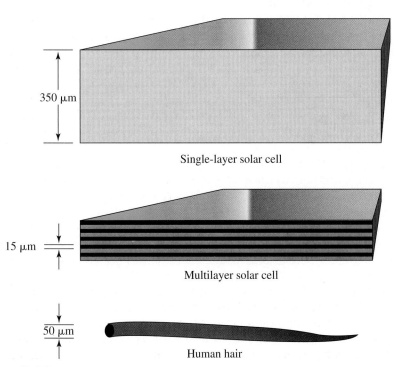

350 μm

Single-layer solar cell

15 μm

Multilayer solar cell

50 μm

Human hair

Figure 8.24

A comparison of the relative thickness of a solar cell layer, either in a single or multilayer cell, to the diameter of an average human hair. *Note:* 1 μm = 10^{-6} m.

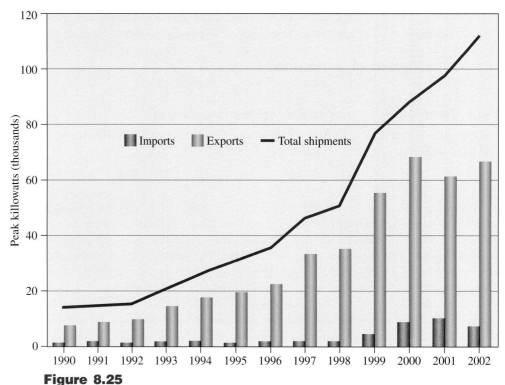

Figure 8.25

The growth of world photovoltaics. Import, export, and total shipments of photovoltaic cells, 1990–2002. *Note*: The total includes all import and export shipments and may include imported photovoltaics that subsequently were shipped to domestic or foreign customers. Source: www.eia.doe.gov/fuelrenewable.html

in the 21st century. As a result of these and other advances in photovoltaic technology, the cost of producing electricity in this manner has dropped dramatically from $3 per kilowatt-hour in 1974 to 20–30 cents per kilowatt-hour in 2003. Solar energy can be produced even more cheaply by vast solar arrays in desert areas, says Avi Brenmiller, chairman of Solel Solar Systems, an Israeli firm that built a giant solar thermal plant in California's Mojave Desert. There, the costs are estimated to be about 10 cents per kilowatt-hour.

Although Israel is a largely desert country, it does not use much solar power within its borders. Conventional fossil-fuel power is still far less expensive.

"World solar markets are growing at 10 times the rate of the oil industry, which has expanded only 1.4% per year since 1990," wrote Christopher Flavin and Molly O'Meara of the Worldwatch Institute. Most major energy companies, such as Shell, Amoco, and British Petroleum have invested heavily in the solar business, now valued at about $2.7 billion in annual sales worldwide; only 25% of those sales, however, are in the United States. Photovoltaic power has grown at an average rate of 16% per year since 1990 (Figure 8.25). Even so, solar power still represents less than 1% of global power supplies.

The largest solar installation in the United States, located in Carrisa Plains, California, was built by ARCO Solar, Inc. together with Pacific Gas and Electric Company. It generates about 7 MW at peak power. Although this generating capacity is small compared with that of fossil fuel, nuclear, and hydroelectric plants, much larger solar installations are expected in the future. A 200-MW plant, which could provide household power for 300,000 people, could be erected on a square mile of land at relatively modest capital expense and minimal maintenance. At currently attainable levels of operating efficiency, all the electricity needs of the United States could be supplied by a photovoltaic generating station covering an area of 85 miles by 85 miles, roughly the area of New Jersey.

Photovoltaic technology is already in use in a number of countries. From small, remote villages in developing countries to upscale suburbs in Japan and the United

Figure 8.26

This array of 1600 photovoltaic cells in Sacramento, California, produces 2 MW (2 million watts) of electricity.

States, 500,000 homeowners worldwide use solar cells to generate their own electricity. Aided by generous tax credits, over 23,000 homes in Japan have rooftop solar units, installed to overcome the high cost of electricity. In sunny Sacramento, California, the Municipal Utility District has constructed an array of 1600 photovoltaic cells that produces 2 MW (2×10^6 W) of electricity, sufficient to serve 660 homes (Figure 8.26). An additional 420 homes have rooftop photovoltaic systems. The homeowners sell back excess electricity to the utility company. In a touch of irony, the photovoltaic site is adjacent to a now-closed nuclear power plant. The Sacramento Municipal Utility District voted to close its nuclear reactors in favor of using photovoltaics and other "cleaner" energy technologies.

Shell Solar and the German company GEOSOL are building the world's largest solar power station near Leipzig, Germany (Figure 8.27). When completed, it will have some 33,500 solar modules with a total output of 5 MW. This will be enough power

Figure 8.27

This computer-simulated image shows the planned array of 33,500 solar modules being built near Leipzig, Germany. When completed, it will be the largest solar power station in the world.

Figure 8.28
Two model solar Habitat for Humanity homes built in Tennessee by Oak Ridge National Laboratories (ORNL) in partnership with the Million Solar Roofs program. Each has a 2-kW solar electric system.

to meet the electricity demand for 1800 households. This solar power project will save about 3700 metric tons of CO_2 emissions from entering the atmosphere, another important consideration. The first phase of the project was completed in July 2004.

The Million Solar Roofs (MSR) program plans to install a million rooftop photovoltaic systems in the United States by 2010. The U.S. program started in 1997, and uses a combination of tax credits with private, local, and state partnerships to help meet the goal. The initiative employs two types of solar technology. Photovoltaics produce needed electricity from sunlight and solar thermal systems produce heat that can be used for domestic hot water, space heating, or even heating swimming pools. Competitive grants totaling more than $1.6 million were received by 35 partnerships in 2003. Habitat for Humanity (Figure 8.28) has been one of the most active partners, helping to promote energy-efficient homes and increased use of solar energy in many different regions of the country. Another example of a MSR partnership is in San Diego, California. Devastating fires in October of 2003 caused many residents to lose their homes. "Rebuild a Greener San Diego" is designed to encourage energy efficiency and the use of photovoltaics in the replacement homes.

 Consider This 8.27 **Million Solar Roofs Partnerships**

Explore the MSR initiative and pick a project in the United States that you find particularly interesting.

a. Which project did you pick and where is it located?
b. Who are the partners in this project?
c. What are its major purposes?
d. When was this project funded? Is it completed at this time?
e. Is there a MSR partnership in your community? If so, what is it? If not, how does one start such a partnership?

More than a third of Earth's population is not hooked into an electrical network because of the costs associated with constructing and maintaining equipment, and supplying the fuel to generate the electricity. Because photovoltaic installations are essentially maintenance-free and can be used almost anywhere, they are particularly attractive for electrical generation in remote regions. For example, the highway traffic lights in certain parts of Alaska, far from power lines, operate on solar energy. A similar, but more significant application of photovoltaic cells may be to bring electricity to isolated

Energy costs are high in remote areas, such as in Alaska, or in developing nations. Thus, electricity from photovoltaics is economically competitive in these areas.

villages in developing countries. In recent years, more than 200,000 solar lighting units have been installed in residential units in Colombia, the Dominican Republic, Mexico, Sri Lanka, South Africa, China, and India.

An exemplary application of the use of photovoltaic technology is in Indonesia, an archipelago of more than 13,000 islands. About 70% of all households there do not have access to electrical lines. Therefore, installing photovoltaic cells is a attractive alternative. In the village of Leback, solar electric units have been installed in 500 homes. Electricity has also been supplied to public buildings (Figure 8.29), shops, streetlights, and a satellite antenna system for the 11 public television units. Before the photovoltaic systems, Leback villagers used kerosene for lighting and batteries for radios. Kerosene costs $6–12 per month, depending on availability. Under a loan–purchase agreement, villagers pay $4–5 per month for their home solar electric system.

The transfer of electrons in solar cells can even power cars. Sunrayce USA has developed into an 11-day, 2300-mile race sponsored by the U.S. Department of Energy (DOE), its National Renewable Energy Laboratory (NREL), Terion Communications, and several other private corporations. The race is a popular activity for engineering students. Student teams design, build, test, and drive cars powered by photovoltaic cells and battery packs (Figure 8.30). Because of their avant garde designs, the cars are not ready to be put on the roads for everyday use. Australia's SunRace 21 C is an equally demanding 10-day, 2300-km race for solar and electric vehicles. The 2001 race went from Adelaide through Melbourne and Canberra, finishing in front of the famous Opera House in Sydney. The Ultimate Solar Car Challenge in 2002 was a 4000-km transcontinental race from Perth on the west coast to Sydney on the east coast. This race celebrated the 20th anniversary of the world's first solar car journey. Both countries regularly hold solar bike racing events and the United States has an intercollegiate solar boat competition called Solar Splash. In addition, a remotely piloted airplane powered only by battery-charging amorphous (noncrystalline) silicon solar cells has flown more than 2400 miles in less than 120 hours.

Figure 8.29
Solar (photovoltaic) cells bring electricity to remote areas.

Your Turn 8.28   Checking the Races

a. Which solar vehicle race was longer—the U.S. Sunrayce 2001 (2300 miles) or the Australian SunRace 21.C (2300 km)? Explain.

b. Which solar vehicle race was longer—the U.S. Sunrayce 2001 (2300 miles) or the Australian Ultimate Solar Car Challenge (4000 km)? Again, explain.

Figure 8.30
The Solar Phantom V, built by students at Rose-Hulman Institute of Technology, Terre Haute, Indiana, is shown driving along Route 66 from Albuquerque to Gallup, New Mexico, in preparation for the 2001 Sunrayce.

Although prodigious amounts of sunshine hit the Earth daily, the rays do not strike any specific spot on our planet for 24 hours a day, 365 days a year. This means that the electricity generated by photovoltaic cells during the day must be stored using batteries for use at night. Nevertheless, the direct conversion of sunlight to electricity has many advantages. In addition to freeing us from our dependence on fossil fuels, an economy based on solar electricity would reduce the environmental damage of extracting and transporting these fuels. Furthermore, it would help to lower the levels of air pollutants such as sulfur oxides and nitrogen oxides. It would also help avert the dangers of global warming by decreasing the amount of carbon dioxide released into the atmosphere. Fossil fuels will certainly remain the preferred form of energy for certain applications. On balance, however, the future looks sunny for solar-based energy. Indeed, the only better source of energy might be if some modern Prometheus could steal a part of the Sun and bring it down to Earth.

Prometheus was a Greek mythic figure who stole fire from the gods.

CONCLUSION

Fossil fuels, the Sun's ancient investments on Earth, are fast disappearing, and we must seek alternatives for tomorrow. We look to many different forms of electron transfer for the future. Cells and batteries provide a flow of electrons that can be captured for useful energy. Fuel cells are one of the most successful new strategies and may become a major energy source for future electrical, personal power use, and transportation needs. Advances in research and changes in global economies may make it fiscally and energetically feasible to use solar radiation to extract hydrogen from water or some other hydrogen source. The hydrogen can either be burned directly as a clean fuel or combined with oxygen in a fuel cell that generates electricity rather than heat. Photovoltaic cells convert sunlight directly into electricity. There is no need for intermediate steps in which heat energy is transformed first into mechanical and then into electrical energy, losing energy along the way.

But the laws of thermodynamics and human nature are such that these transformations will not occur spontaneously. Energy alternatives cannot be developed without hard work and the investment of intellect, time, and money. Yet, in the United States, the amount of effort and money devoted to research on new energy sources sometimes appears to be directly proportional to the cost of oil. When international crises drive up the price of petroleum or when regional energy shortages occur, there is a sudden flurry of official interest in energy conservation and the development of alternative technologies. When oil supplies are plentiful and prices are low at the gasoline pump, few seem to care about preparing for the time when fossil fuels will be depleted or they become much too polluting or too expensive to burn. Someday, maybe teams of chemists, physicists, and engineers will even succeed in simulating the Sun by controlling the Sun's nuclear reactions here on Earth as an affordable energy source, not a mere laboratory curiosity. But, until then, we need to establish national and personal priorities, and act on them. We have been the beneficiaries of a bountiful nature, but in turn, we have an obligation to ensure energy sources for unborn generations.

Chapter Summary

Having studied this chapter, you should be able to:

- Discuss the principles governing the transfer of electrons in galvanic cells, including the processes of oxidation and reduction (8.1)

- Describe the design, operation, applications, and advantages of several different types of batteries (8.1–8.3)

- Describe the design, operation, applications, and advantages of fuel cells (8.4, 8.5)

- Compare and contrast the principles, advantages, and challenges of producing and using battery-powered, fuel-cell-powered, and hybrid vehicles (8.6, 8.7)
- Explain the energetics of producing hydrogen and using it as a fuel (8.8)
- Discuss issues related to developing a hydrogen economy (8.9)

- Describe the principles governing the operation of photo-voltaic (solar) cells and their current and future uses (8.10)
- Express informed opinions about the future development of all types of electron transfer technology for producing electrical energy on personal, regional, national, and global scales (8.1–8.10)

Questions

Emphasizing Essentials

1. **a.** Define oxidation.

   b. Define reduction.

   c. Why must these processes take place together?

2. Which of these half-reactions represent oxidation and which reduction? Explain your reasoning.

   a. $Fe \longrightarrow Fe^{2+} + 2\,e^-$

   b. $Ni^{4+} + 2\,e^- \longrightarrow Ni^{2+}$

   c. $2\,H_2O + 2\,e^- \longrightarrow H_2 + 2\,OH^-$

3. Consider this galvanic cell. A coating of impure silver metal begins to appear on the surface of the silver electrode as the cell discharges.

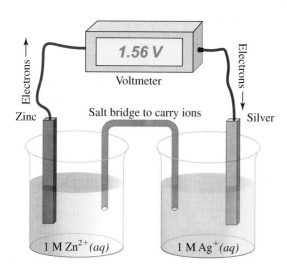

1.56 V

Voltmeter

Electrons →

← Electrons

Zinc

Salt bridge to carry ions

Silver

$1\,M\,Zn^{2+}(aq)$

$1\,M\,Ag^+(aq)$

   a. Identify the anode in this cell.

   b. Write the oxidation half-reaction.

   c. Identify the cathode.

   d. Write the reduction half-reaction.

4. Explain the difference between a rechargeable battery and one that must be discarded. Contrast a NiCad battery with an alkaline battery in your explanation.

5. Is there a difference between a galvanic cell and an electrochemical cell? Explain, giving examples to support your answer.

6. In the lithium–iodine cell, Li is oxidized to Li^+; I_2 is reduced to $2\,I^-$.

   a. Write equations for the two half-reactions that take place in this cell, labeling the oxidation half-reaction and the reduction half-reaction.

   b. Write an equation for the overall reaction in this cell.

   c. Identify the half-reaction that occurs at the anode and the half-reaction that occurs at the cathode.

7. Two common units associated with electricity are volts and amps. What does each unit measure?

8. Explain the significance of the title of this chapter, "Energy from Electron Transfer."

9. **a.** Is the voltage from a tiny AAA-size alkaline cell the same as that from a large D alkaline cell? Why or why not?

   b. Will both batteries sustain the flow of electrons for the same time? Why or why not?

10. The mercury battery has been used extensively in medicine and the electronics industries. Its overall reaction can be represented by this equation.

$$HgO(l) + Zn(s) \longrightarrow ZnO(s) + Hg(l)$$

   a. Write the oxidation half-reaction.

   b. Write the reduction half-reaction.

11. Figure 8.5 shows how the use of mercury has changed since 1970. What was the most common use in 1970? 1980? 1990? What are some reasons for the observed changes in mercury use?

12. These are the *incomplete* equations for the half-reactions in a lead storage battery. These half-reactions do not show the electrons either lost or gained.

$$Pb(s) + SO_4{}^{2-}(aq) \longrightarrow PbSO_4(s)$$

$$PbO_2(s) + 4\,H^+(aq) + SO_4{}^{2-}(aq) \longrightarrow PbSO_4(s) + 2\,H_2O(l)$$

   a. Balance both equations with respect to charge by adding electrons on either side of the equations, as needed.

   b. Which half-reaction represents oxidation and which represents reduction?

   c. One of the electrodes is made of lead, the other of lead dioxide. Which is the anode and which is the cathode?

13. a. What is the function of the electrolyte in a galvanic cell?

    b. What is the electrolyte in an alkaline cell?

    c. What is the electrolyte in a lead-acid storage battery?

14. What is the difference between a storage battery and a fuel cell?

15. Is the conversion of $O_2(g)$ to $H_2O(l)$ in a fuel cell an example of oxidation or reduction? Use electron loss or gain to support your answer.

16. Consider this diagram of a hydrogen–oxygen fuel cell used in earlier space missions.

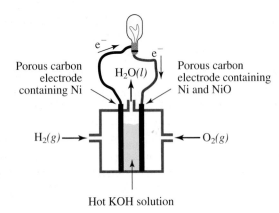

a. How does the reaction between hydrogen and oxygen in a fuel cell differ from the combustion of hydrogen with oxygen?

b. Write the half-reaction that takes place at the anode in this fuel cell.

c. Write the half-reaction that takes place at the cathode in this fuel cell.

17. What is a PEM fuel cell and how does it differ from the fuel cell represented in question #16?

18. In addition to the studies on hydrogen fuel cells, experiments are being done on fuel cells using methane as a fuel. Balance the given oxidation and reduction half-reactions and write the overall equation for a methane-based fuel cell.

Oxidation half-reaction:

$$\_\_ CH_4 + \_\_ OH^- \longrightarrow \_\_ CO_2 + \_\_ H_2O + \_\_ e^-$$

Reduction half-reaction:

$$\_\_ O_2 + \_\_ H_2O + \_\_ e^- \longrightarrow \_\_ OH^-$$

19. The U.S. Department of Transportation (DOT) prohibits passengers from carrying flammable fluids aboard aircraft. Explain how this might affect the development of microfuel cells for use in consumer electronics such as portable computers.

20. What is meant by the term *hybrid car?*

21. a. What is meant by the acronym ZEV? By the acronym NEV?

    b. Is a hybrid car a ZEV? Explain your reasoning.

    c. Is a fuel-cell-powered car a ZEV? Explain your reasoning.

22. Why will GM's EV-1, an all lead-acid battery-powered car, no longer be produced?

23. a. How are equations 8.19 and 8.20 the same and how are they different?

    b. How will the energy released in the reaction shown in equation 8.19 compare with the energy released in the reaction represented by equation 8.20? Explain your reasoning.

24. Given that 286 kJ of energy are released per mole of H_2 burned, how much energy will be released when 370 kg of H_2 is used?

25. a. Use bond energies from Table 4.2 to calculate the energy released when 1 mol of hydrogen burns.

    b. Compare your result with the stated value of 286 kJ. Account for any difference.

26. a. Potassium, a Group 1A metal, reacts with H_2 to form a hydride. Write the chemical equation for the reaction.

    b. Potassium hydride reacts with water to release H_2 and form potassium hydroxide. Write the chemical equation for the reaction.

    c. Offer a possible reason that potassium is not used to store H_2 for use in fuel cells.

27. Every year, 5.6×10^{21} kJ of energy comes to Earth from the Sun. Why can't this energy be used to meet all of our energy needs?

28. This *unbalanced* equation represents the last step in the production of pure silicon for use in solar cells.

$$\_\_ Mg(s) + \_\_ SiCl_4(l) \longrightarrow \_\_ MgCl_2(l) + \_\_ Si(s)$$

a. How many electrons are transferred per atom of pure silicon formed?

b. Is the production of pure silicon an oxidation or a reduction reaction? Why do you think so?

29. The symbol • represents an electron and the symbol ⬤ represents a silicon atom. Does this diagram represent a gallium-doped *p*-type silicon semiconductor, or does it represent an arsenic-doped *n*-type silicon semiconductor? Explain your answer.

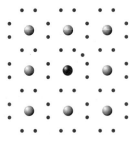

30. **a.** What is the Million Roofs program?

    b. Who are the partners in this program?

    c. What does this program hope to achieve?

Concentrating on Concepts

31. You have seen several examples of oxidation–reduction reactions in this chapter and identified the exchange of electrons taking place. Now examine these equations and decide which are oxidation–reduction reactions and which are not. Explain your decisions.

Equation 1:

$$Zn(s) + 2\ MnO_2(s) + H_2O(l) \longrightarrow Zn(OH)_2(s) + Mn_2O_3(s)$$

Equation 2:

$$HCl(aq) + NaOH(aq) \longrightarrow NaCl(aq) + H_2O(l)$$

Equation 3:

$$CH_4(g) + 2\ O_2(g) \longrightarrow CO_2(g) + 2\ H_2O(g)$$

32. Identify the type of battery commonly used in each of these consumer electronic products. Assume none uses solar cells.

    a. battery-powered watch

    b. MP-3 player

    c. digital camera

    d. hand-held calculator

33. Ag-Zn batteries are replacing lead-acid batteries in small airplanes, such as Cessna-172s.

    a. Why are these batteries, although more expensive, preferable to the lead–acid batteries used previously?

    b. What is the oxidation half-reaction? What is the reduction half-reaction?

34. The battery of a cell phone discharges when the phone is in use. A manufacturer, while testing a new "power boost" system, reported these data.

Time, min.sec	Voltage, V
0.00	6.56
1.00	6.31
2.00	6.24
3.00	6.18
4.00	6.12
5.00	6.07
6.35	6.03
8.35	6.00
11.05	5.90
13.50	5.80
16.00	5.70
16.50	5.60

    a. Prepare a graph of these data.

    b. The manufacturer's goal was to retain 90% of its initial voltage after 15 minutes of continuous use. Has that goal been achieved? Justify your answer using your graph.

35. The text describes a prototype fuel cell that converts gasoline to hydrogen and carbon monoxide. The carbon monoxide, in contact with a catalyst, then reacts with steam to produce carbon dioxide and more hydrogen.

    a. Write a set of reactions that describe this prototype fuel cell, using octane (C_8H_{18}) to represent the hydrocarbons in gasoline.

    b. When is this fuel cell expected to be commercially available?

    c. Speculate as to the future economic success of this prototype fuel cell.

36. Fuel cells were invented in 1839, but never developed into practical devices for producing electrical energy until the U.S. space program in the 1960s. What advantages did fuel cells have over previous power sources?

37. Hydrogen, H_2, and methane, CH_4, can each be used with oxygen in a fuel cell. Hydrogen and methane also can be burned directly. Which has greater heat content when burned, 1.00 g of H_2 or 1.00 g of CH_4? *Hint:* Write the balanced chemical equation for each reaction and use the bond energies in Table 4.2 to help answer this question.

38. Why are electric cars powered by lead-acid storage batteries alone only a short-term solution to the problem of air pollution emissions from automobiles? Outline your reasoning.

39. Assuming that hybrid cars are available in your area, what questions would you ask the car dealer before deciding to buy or lease a hybrid? Which of these questions do you consider most important? Offer reasons for your choices.

40. **a.** GM's Saturn EV-1 had to be plugged in to have its batteries recharged. Explain why.

    b. You never need to plug in Toyota's gasoline–battery hybrid car to recharge the batteries. Explain why not.

41. Prepare a list of the environmental costs and benefits associated with hybrid vehicles. Compare that list with the environmental costs and benefits of vehicles powered by gasoline. On balance, which energy source do you favor, and why?

42. William C. Ford, Jr., the chief executive officer of Ford Motor Company, is quoted as saying that going "totally green" with zero-emissions vehicles will be a real challenge. Regular drivers won't buy high-tech clean cars, Ford admits, until the industry has a "no-trade-off" vehicle widely available. What do you think he means by a no-trade-off vehicle? Do you think he is justified in this opinion?

43. **a.** What is meant by "the hydrogen economy"?

    b. Even if methods for producing hydrogen cheaply and in large quantities were to become available, what problems would still remain for the hydrogen economy?

44. Although hydrogen gas can be produced by the electrolysis of water, this reaction is usually not carried out on a large scale. Suggest a reason for this fact.

45. Consider this diagram of two water molecules in the liquid state.

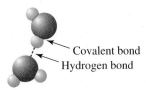

Covalent bond
Hydrogen bond

a. What bonds are broken when water boils? Are these intermolecular or intramolecular bonds? (*Hint*: See Chapter 5 for definitions.)

b. What bonds are broken when water is electrolyzed? Are these intermolecular or intramolecular bonds?

46. The cost of electricity generated by solar thermal power plants currently is greater than that of electricity produced by burning fossil fuels. Given this economic fact, suggest some strategies that might be used to promote the use of environmentally cleaner electricity.

47. Name some of the current applications of photovoltaic cells *other* than the production of electricity in remote areas?

Exploring Extensions

48. The aluminum-air battery is being explored for use in automobiles. In this battery, aluminum metal undergoes oxidation to Al^{3+} and forms $Al(OH)_3$. Oxygen from the air undergoes reduction to OH^- ions.

a. Write equations for the oxidation and reduction half-reactions. Use H_2O as needed to balance the number of hydrogen atoms present, and add electrons as needed to balance the charge.

b. Add the half-reactions to obtain the equation for the overall reaction in this cell.

c. Specify which half-reaction occurs at the anode and which occurs at the cathode in the battery.

d. What are the benefits of the widespread use of the aluminum–air battery? What are some of the limitations? Write a brief summary of your findings.

e. What is the current state of development of this battery? Is it in use in any vehicles at the present time? What is its projected future use?

49. An iron-based "superbattery" is a promising alternative for delivering more power with fewer environmental effects than alkaline batteries. Find out how the superbattery is designed and its state of commercial acceptance.

50. Although Alessandro Volta is credited with the invention of the first electric battery in 1800, some feel this is a reinvention. Research the "Baghdad Battery" to evaluate the merit of this claim.

51. Consider these three sources of light: a candle, a battery-powered flashlight, and an electric light bulb. For each source, provide:

a. The origin of the light

b. The immediate source of the energy that appears as light

c. The original source of the energy that appears as light. *Hint:* Trace this back stepwise as far as possible.

d. The end-products and by-products of using each

e. The environmental costs associated with each

f. The advantages and disadvantages of each light source

52. Hydrogen gas can be prepared in the lab by reacting metallic sodium with water, as shown in this equation.

$$2\,Na(s) + 2\,H_2O(l) \longrightarrow H_2(g) + 2\,NaOH(aq)$$

a. Calculate the grams of sodium needed to produce 1.0 mol of hydrogen gas.

b. Calculate the grams of sodium needed to produce sufficient hydrogen to meet an American's daily energy requirement of 1.1×10^6 kJ.

c. If the price of sodium were \$94/kg, what would be the cost of producing 1.0 mol of hydrogen? Assume the cost of water is negligible.

53. a. Using hydrogen as a fuel has both advantages and disadvantages. Set up parallel lists for the advantages and the disadvantages of using hydrogen as the fuel for transportation and for producing electricity.

b. Do you advocate the use of hydrogen as a fuel for transportation or for the production of electricity? Explain your position in a short article for your student newspaper.

54. If all of today's technology presently based on fossil fuel combustion were replaced by H_2–O_2 fuel cells, significantly more H_2O would be released into the environment. Is this effect a concern? Find out what other effects might be anticipated from switching to a hydrogen economy.

55. The text refers to fossil fuels as the "... Sun's ancient investment on Earth." How would you interpret this statement to a friend who is not enrolled in your course?

56. The text describes projections made by Ballard Power Systems, in worldwide alliances with DaimlerChrysler and Ford Motor Company, to become the largest commercial producer of fuel-cell-powered electric drive trains and components for cars, trucks, and buses. The alliance plans to produce 100,000 fuel-cell-powered electric vehicles by 2004–2005. What is the progress toward that goal? See if you can find whether they've revised their predictions. Write a brief report, gearing it toward an investor who wants to evaluate the future growth of the company.

57. Richard E. Smalley is quoted in the text as predicting that nanotechnology will help provide many new solutions to problems currently limiting the use of alternative technology. Dr. Smalley is identified as a Nobel laureate. Learn more about the research for

which he was so honored. Prepare a short oral report or a poster communicating your findings to the class.

58. Figure 8.18 shows an open-cage fullerene serving as a nanocontainer. Have other similar structures been identified to encapsulate H_2 gas?

59. Figure 8.26 shows an array of photovoltaic cells installed by the Municipal Utility District in Sacramento, California. Where else in the United States or in the world is there a comparable array? Use the Web to learn of other large-scale photovoltaic cell installations. What factors help to influence this approach, one that uses a centralized array rather than using individual rooftop solar units?

60. The Million Roofs program is so little publicized that most people in the United States are unaware of its existence. Design a poster to explain and promote the program to the general public.

9

The World of Plastics and Polymers

Mountain biking makes extensive use of polymers, both synthetic and natural ones.

"Hey, I'm out of here this weekend. I sure need a break from studying and exams. I've got a date with this mountain bike before the cold weather really sets in. But I don't even care if it's cold or rainy. Check out the new biking gear I just bought. In case I need it, I have my Thinsulate-lined Gore-Tex jacket that will keep me dry. Do you like my C-Tech polyester microfiber jersey and these new Porelle Drys socks that will keep me warm? I've got Spandura tights that are 18 times tougher than my older nylon/Lycra ones. And I can stash the rest of my gear in my lightweight nylon backpack with polyurethane padded straps for comfort. Like I said, I'm outta here. It will feel great to get away from the synthetic world and commune with Ma Nature for a change. Catch ya later."

It should be obvious that our biking student missed the point. She cannot get away from synthetic polymers—especially not when going biking, hiking, or camping; or anywhere else for that matter. Polymers are all around us. They are present in synthetic forms, such as rubber and nylon. They are also in and around us in natural forms, such as in proteins, and in cellulose, the structural material in plants. Synthetic polymers are the focus of this chapter.

Consider This 9.1 **Skiing Gear Analysis**

Polymers and other synthetic materials have revolutionized sports equipment in recent years. Such technology has created lighter, stronger, and more responsive materials to match recreational needs. Look more closely at the snowboarder shown in Figure 9.1. Identify the parts of his equipment and outer clothing that are created by chemists, and describe the properties of these materials that make them well suited for their intended uses.

Consider This 9.2 **Polymers All Around Us**

Information regarding the effect of polymers on our lives is available from the American Plastics Council. A direct link is available on the *Online Learning Center.* Select two areas where polymers are used. Identify the polymers, their properties, and their uses.

Figure 9.1
An example of using polymers to play.

At this moment, you probably are wearing or carrying at least a half dozen materials that did not exist 70 years ago or perhaps 10 years ago. Your shoes alone may consist of six or more different kinds of polymers: the sole, the trim, the foam padding, the sock liner, the upper, the laces, and even the lace tips. Very likely at least some of your clothing contains synthetic fibers. The pen that you take notes with is made largely of polymers. Your calculator has a polymer case and so does your cell phone and computer. And the CDs and DVDs you play are made of polymers, as are their cases. Several kinds of polymers also are essential components of your car, which contains nearly 300 lb of them.

Many of these polymers came from a single raw material: petroleum. But the planet's supply of petroleum is limited and nonrenewable, a fact that creates a serious dilemma. As you know, most petroleum is burned as fuel (Section 4.8). Only about 3% is reserved as a chemical feedstock for manufacturing polymers and other important chemicals. This 3% is essential to current methods of manufacturing the polymers that have reshaped modern life.

Therefore, we return to the dilemma posed earlier: To burn or not to burn. Should petroleum be converted to gasoline, kerosene, and fuel oil—substances that we burn? Or should more of it be diverted for use as a raw material for the synthesis of polymers? Currently these synthetic materials cannot be made from any other source. And, given their overwhelming presence all around us, it is clear that we have come to depend on them. What are the risks and benefits—the economic and social trade-offs—in using oil as a source of energy rather than as a source of synthetic products? If a greater share of petroleum is not reserved for nonfuel uses, the age of plastics may be very short. Very soon, chemistry may provide a solution to our dependence on nonrenewable resources. In principle, polymers can be made from any carbon-containing starting material. Crude oil is simply the most convenient and the most economical.

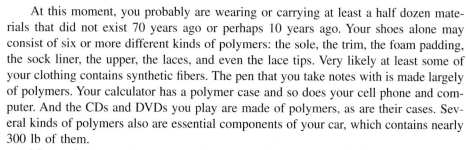
Chemical companies are beginning to demonstrate that it is possible to convert renewable biological materials such as wood, cotton fibers, straw, starch, and sugar into new polymers. This work has implications for land use, crop productivity, the environment, and no doubt much more. To transform biomass into synthetic polymers, new methods and new technologies are being developed, but more

Chapter 4, *Energy, Chemistry and Society,* and Chapter 6, *Neutralizing the Threat of Acid Rain,* both have information relevant for the "to burn or not to burn" question.

research is still needed. For example, Cargill Dow recently introduced "NatureWork PLA," a plastic made from corn glucose that can be used for clothing, packaging of foods, and even the plastic parts of automobiles. The DuPont chemical company recently introduced a family of polymers called Sorona, derived from corn-based chemicals. DuPont is manufacturing Sorona in a chemical plant that can be converted from petroleum-based production to bio-based production "when process economics and market demand justify the transition." This underscores our point that the cost of the research and the manufacturing for bio-based products is still much higher than for petroleum-based ones. Moreover, the demand for finished products will have a large impact on the speed with which the transition to bio-based processes occurs.

Consider This 9.3   **Bio-Based Synthetic Polymers**

Search on the Web for NatureWork from Cargill Dow, Sorona from DuPont, or some other bio-based polymers. What are the various source materials for the polymer you selected? What are some of the uses of the polymer?

To understand the complexities surrounding the sources and uses of polymers, we first must understand their structure and how they are synthesized. We will focus on this next and then return to the issues of polymer use in our society.

9.1   Polymers: The Long, Long Chain

You are likely familiar with the common or brand names of many different polymers: rayon, nylon, Lycra, polyurethane, Teflon, Saran, Styrofoam, and Formica to list only a few. These seemingly very different materials all are synthetic polymers. What they have in common is evident at the molecular level. All **polymers** are large molecules made up of long chains of atoms covalently bonded together. **Monomers** (from *mono* meaning "one" and *meros* meaning "unit") are small molecules used to synthesize the polymeric chain. Each monomer is a single link of the chain. The polymers (*poly* means "many") can be formed from the same type of monomer or from a combination of monomers (Figure 9.2).

These polymer molecules can be very long indeed. Sometimes they are referred to as **macromolecules** because they involve thousands of atoms, and their molecular masses can reach over a million.

Synthetic polymers are often referred to by the word *plastic*. We often apply the term *plastic* to many different materials with a broad range of properties and applications. According to a standard dictionary definition, *plastic* is an adjective meaning "capable of being molded" or a noun referring to something that is capable of being molded. More specifically, the *Merriam-Webster Collegiate Dictionary,* 11th edition, mentions "any of numerous organic synthetic or processed materials that are mostly thermoplastic or thermosetting polymers of high molecular weight and that can be

> Microfiber is not a new type of polymer. It is simply a polymer made into very fine (small-diameter) threads. The fine threads can be made into fabrics that have desirable properties, such as insulating capacity, water repellency, and breathability, among many others.

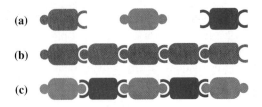

Figure 9.2
(a) Representations of three different types of monomers.
(b) Representation of a polymer formed from one type of monomer.
(c) Representation of a polymer formed from two different types of monomers.

molded, cast, extruded, drawn, or laminated into objects, films, or filaments." As it turns out, some metals have properties that can be referred to as plastic because they can be "cast, extruded, and drawn." Therefore, the word *plastic* has many applications beyond that of describing some synthetic polymers. We will generally use the word *polymer* in this chapter. Because the focus of this chapter is primarily on synthetic polymers, we may sometimes also refer to these as plastics.

Keep in mind that chemists did not invent polymers. Polymeric materials were here long before chemists, and chemists also are made up of many polymers. Natural polymers are found in plants as well as in animals. For example, wood, wool, cotton, starch, natural rubber, skin, hair—even some minerals, such as asbestos and quartz—are polymers. Polymeric molecules give strength to an oak tree, delicacy to a spider's web, softness to goose down, and flexibility to a blade of grass.

When chemists began trying to synthesize polymers, they were hoping to copy some of the properties of natural polymers. Indeed, many synthetic polymers were originally created as substitutes for expensive or rare naturally occurring materials or to improve on natural polymers. As an increasing number of new polymers were developed, the variety of their properties and uses expanded dramatically. For example, polymers have become increasingly important in automobile manufacturing because they are considerably less dense than structural metals, making cars more fuel-efficient. Plastic packaging reduces weight, eliminates breakage, and helps save fuel during shipping. Polymeric construction materials have replaced wood in some applications, and plastic pipes substitute effectively for ones made from lead, iron, or copper.

As we've already seen, recreation has been revolutionized by the introduction of synthetic polymers. Football is played on artificial turf by players wearing polymeric helmets, padding, and pants. Tennis balls, tennis rackets, and strings are all made from synthetic polymers. In-line skates are made almost entirely of synthetic polymers. Carbon fibers embedded in plastic resins provide the strength and flexibility required in trail bikes, fishing rods, golf-club shafts, and sailboat hulls and sails. Ice skaters and hockey players can skate without ice on rinks of Teflon or high-density polyethylene. Most modern canoes are made of Kevlar, ABS, or polyethylene synthetic polymers, not birchbark, wood, or aluminum as they once were.

> Density was discussed in Section 5.6. Steel has an approximate density of 8.0 g/cm^3. Polymers generally have densities between 1 and 2 g/cm^3. This means that an automobile body built from polymers rather than steel would weigh only one eighth to one quarter as much and would require less energy (and therefore less fuel) to move.

Consider This 9.4   Polymers for Fun

Choose a favorite activity—on the land, water, or even in the air. Do a Web search to find a company that manufactures or sells equipment for this activity. Which polymers can you find mentioned on this Web site? Make a table of their names, the equipment they are used in, and any desirable properties cited as advertising points. *Hint:* Many polymer names start with *poly,* such as polyester or polypropylene. Other polymers have trade names, such as Gore-Tex, Orlon, or Styrofoam. Still other polymers are coatings and resins and may be mentioned as epoxides or acrylics.

Consider This 9.5   Personal Polymers

a. A Teflon ear bone, fallopian tube, or heart valve? A Gore-Tex facial implant or hernia repair? Some polymers are biocompatible and now used to replace or repair body parts. List four properties that would be desirable for polymers used *within* the human body.

b. Other polymers may be used outside your body, but in close contact with it. For example, no surgeon is needed for you to use your contact lenses—you insert, remove, clean, and store them yourself. From which polymers are contact lenses made? What properties are desirable in these materials? Either a call to an optometrist or a search on the Web may provide some answers.

9.2 The "Big Six": Theme and Variations

Polymer manufacturing has a considerable economic impact. Overall, the U.S. plastics industry and its suppliers employ about 2.2 million workers, about 2 percent of the U.S. workforce. In 2002, nearly $393 billion in total annual shipments from total plastics activity were recorded, despite economic factors unfavorable to growth. In fact, the production of polymers has long since eclipsed that of metals. Since 1976, the United States has manufactured a larger volume of synthetic polymers than the volume of steel, copper, and aluminum combined. Six polymeric materials account for about 76% of all the plastics used in the United States, and these are referred to as the "Big Six." A good way to become familiar with polymers is by collecting various types of polymer-based objects and making some observations.

Consider This 9.6   Plastics You Use

Gather a variety of polymeric items from your apartment or residence hall: plastic bags, soda bottles, CD cases, Styrofoam cups, nylon backpacks, or whatever happens to be at hand. Make a list of the objects and note the properties of the polymers. Include color, transparency, flexibility, elasticity, hardness, and other properties that could be used to classify and identify the plastics. Try to draw conclusions about which objects are made from the same material. *Hint:* Table 9.1 will be of some help.

You have no doubt discovered from the Consider This 9.6 activity (or from prior experience) that polymers exhibit a wide range of properties. We can illustrate this with a few objects that might well be found in your room or residence. Your backpack is water-resistant, lightweight, and sturdy, chosen from a myriad of styles and colors, possibly even your school's colors. The plastic that makes up most soft-drink and water bottles is transparent and has moderate hardness and flexibility. A cup made of Styrofoam is white, opaque, light, soft, and easily deformed and torn, but it is an excellent heat insulator. A CD "jewel case" is hard, brittle, and almost glasslike in its transparency. The CD inside the case is very strong, yet delicate; its surface can be scratched easily. Videotape is shiny, opaque, dark, flexible, and durable. A plastic milk bottle is somewhat opaque or at least translucent, is impermeable to liquid, and it can be deformed somewhat. Finally, in our brief survey, a typical grocery plastic bag is light, flexible, and easily stretched. But it is sometimes difficult to correlate these properties unambiguously with the chemical composition of the polymers. If you were trying to sort for recycling the items described in this paragraph, you might find it hard to do so. In fact, it may be surprising that objects as different as a foam coffee cup and a CD case are made of the same plastic—polystyrene. The videotape and soda bottle are composed primarily of polyethylene terephthalate, and the plastic bag and the milk bottle are both polyethylene.

Today, more than 60,000 synthetic polymers are known. Most have been developed for special purposes, ranging from frying pan coatings to resins for restoring antiques. Yet the bulk of the plastics we regularly encounter are the Big Six: low-density polyethylene (LDPE), high-density polyethylene (HDPE), polypropylene (PP), polystyrene (PS), polyvinyl chloride (PVC), and polyethylene terephthalate (PET or PETE). All are ultimately derived from petroleum. Over 40 million tons of these six polymers are made annually in the United States.

Table 9.1 summarizes information about the Big Six. Six monomers are used to make six different polymers, but perhaps not in the way you would expect. The first two polymers use the same monomer. The next three each use a different single monomer. And the last polymer is built from not one, but two different monomers. In any case, the monomer is the key to understanding the polymer.

Table 9.1	The Big Six (Including Identifying Code of the Polymers)		
Polymer	**Monomer**	**Properties of Polymer**	**Uses of Polymer**
Polyethylene (LDPE) △4 LDPE	Ethylene $C=C$ (H, H / H, H)	Opaque, soft, flexible, impermeable to water vapor, unreactive toward acids and bases, absorbs oils and softens, melts at 100 °C–120 °C, does not become brittle until −100 °C, oxidizes on exposure to sunlight, subject to cracking	Plastic bags, toys, electrical insulation, bubble wrap
Polyethylene (HDPE) △2 HDPE	Ethylene $C=C$ (H, H / H, H)	Similar to LDPE, more opaque, denser, mechanically tougher, more crystalline and rigid	Milk, juice, and water jugs, stiff plastic bags and containers
Polyvinyl chloride △3 V	Vinyl chloride $C=C$ (H, H / H, Cl)	Rigid, thermoplastic, impervious to oils and most organic materials, transparent, high impact strength	Plumbing pipe, garden hose, shower curtains, blister packs
Polystyrene △6 PS	Styrene $C=C$ (H, H / H, phenyl)	Glassy, sparkling clarity, rigid, brittle, easily fabricated, upper temperature limit of 90 °C, soluble in many organic materials	Styrofoam insulation, inexpensive furniture, drinking glasses
Polypropylene △5 PP	Propylene $C=C$ (H, H / H, CH_3)	Opaque, high melting point (160 °C–170 °C), high tensile strength and rigidity, lowest density commercial plastic, impermeable to liquids and gases, smooth surface with high luster	Battery cases, indoor-outdoor carpeting, bottle caps, auto trim
Polyethylene terephthalate △1 PETE, or PET	Ethylene glycol $HO-CH_2CH_2-OH$ Terephthalic acid (structure: benzene ring with two $COOH$ / $C=O$, HO, OH groups)	Transparent, high impact strength, impervious to acid and atmospheric gases, not subject to stretching, most costly of the six	Soft-drink bottles, clothing, audio- and videotapes, film backing

Note: The structures of the first five monomers differ only by the atoms shown in blue.

To understand different monomers, we first need to know about some of the common groups of atoms that these monomers may contain. These common groups are known as **functional groups.** They are distinctive arrangements of atoms that impart characteristic chemical properties to the molecules that contain them. One example of a functional group is the —OH group. It is called the hydroxyl group and, among other places, it is found in alcohol molecules. Another example is the —COOH group, known as the carboxylic acid group. It is found in all organic acids. The actual arrangement of atoms in this group is:

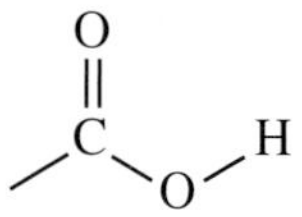

$$
\begin{array}{c}
\text{O} \\
\parallel \\
{-}\text{C} \\
\diagdown \text{O} {-} \text{H}
\end{array}
$$

Think of this group as a unit. Although part of the carboxylic acid group is an —OH group, since it also has the C=O part, we call it a carboxylic acid group.

When the H atom in the carboxylic acid is replaced by a carbon-containing group, the result is an ester. Its general structure is:

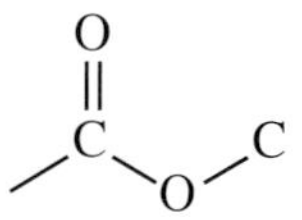

$$
\begin{array}{c}
\text{O} \\
\parallel \\
{-}\text{C} \\
\diagdown \text{O} {-} \text{C}
\end{array}
$$

A fourth example is the amine group or —NH$_2$. This functional group has a nitrogen atom bonded to two hydrogen atoms.

The largest individual functional group we will discuss in this chapter is the phenyl group, —C$_6$H$_5$. This group consists of six carbon atoms, each at a corner of a hexagon. Five of the six carbon atoms are bonded to hydrogen atoms; the sixth carbon atom is linked to some other atom. Typically the symbols for the carbon and hydrogen atoms in the phenyl group are omitted, and the entire —C$_6$H$_5$ structure is represented by a hexagon with a circle inside (see Figure 9.3).

The structure in Figure 9.3a indicates that bonding in the —C$_6$H$_5$ ring can be viewed as consisting of three carbon-to-carbon single bonds and three carbon-to-carbon double bonds, alternating around the hexagon. Experiments have shown, however, that the electrons are uniformly distributed around the ring, with all the carbon–carbon bonds of equal strength and length. We will use the shorthand symbol of a hexagon with a circle inside (Figure 9.3b) to indicate the uniformity of these six bonds.

These functional groups and others will be discussed in greater detail in the following chapter. This set will be sufficient for you to understand the basics of polymers and is summarized in Table 9.2.

With these functional groups in mind, we can now return to our discussion of the monomers found in Table 9.1. Molecules of ethylene, vinyl chloride, styrene, and propylene are similar in that they each contain two carbon atoms connected by a C-to-C double bond, C=C. This double bond is a recurring theme in our polymers. In ethylene, two hydrogen atoms are attached to each of the double-bonded carbon atoms, H$_2$C=CH$_2$. But, in vinyl chloride, propylene, and styrene molecules, one of the hydrogen atoms has been replaced with something else. These replacements provide the variations on the theme of the C-to-C double bond. In the case of vinyl chloride, hydrogen is replaced by a chlorine atom. A methyl group substitutes for a hydrogen atom in propylene. In styrene, the substituent is a phenyl group.

These replacements in the monomers create variety in the polymers formed from them. Moreover, the substituents give chemists greater latitude in designing polymers for particular uses, as shown in Table 9.1. As we shall see in the next section, the first

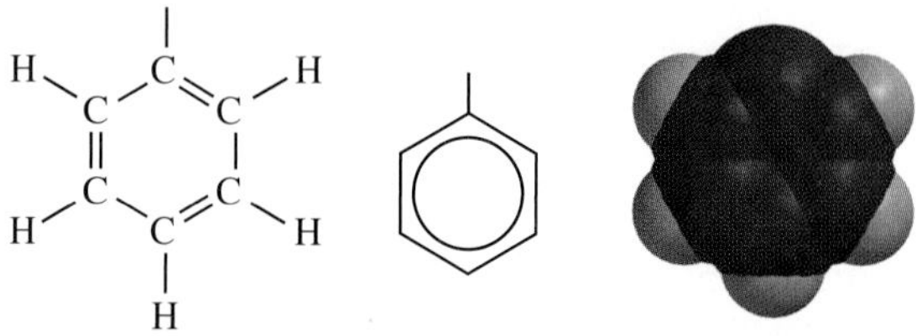

The phenyl group is one of the most common structures in organic chemistry and will be discussed further in Chapter 10.

The methyl group, —CH$_3$, is not usually classed as a functional group because it is not the primary reactive site in a molecule.

(a)   **(b)**   **(c)**

Figure 9.3

Various representations of the phenyl group. These include **(a)** the line drawing, **(b)** the shorthand representation and **(c)** the space-filling molecular model.

Table 9.2	Some Groups Found in Monomers	
Name	**Formula**	**Structure**
methyl	—CH₃	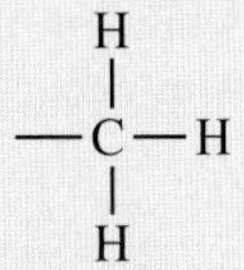
hydroxyl	—OH	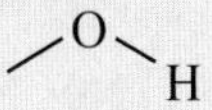
carboxylic acid	—COOH	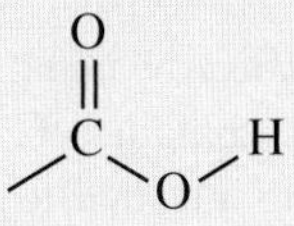
ester	—COOC—	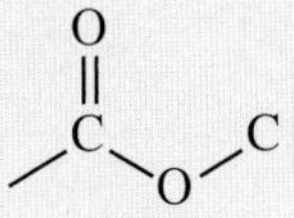
amine	—NH₂	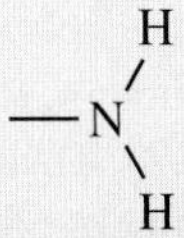
phenyl	—C₆H₅	

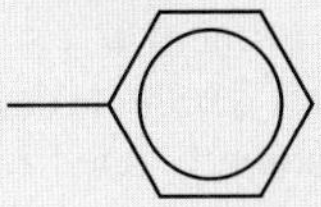

five of the Big Six are made by addition polymerization (see Section 9.3) from monomers containing C-to-C double bonds. The sixth polymer, polyethylene terephthalate, is a special case among the Big Six; it is made by a condensation polymerization process using two different monomers. We will return to polyethylene terephthalate, but only after spending more time with the other five members of this sextet.

Table 9.1 also lists some of the more important properties of these six polymers. They are all **thermoplastic;** that is, they can be melted and shaped, and all tend to be flexible. Materials that are formed from three of them, low- and high-density polyethylene and polypropylene, have various microscopic regions in which the molecules are arranged differently. In some of these regions, the molecules have a very orderly and repeating pattern, like one would find in a crystalline solid. In these **crystalline regions,** the long polymer molecules are arranged neatly and tightly in a regular pattern. In the other regions, the polymer is **amorphous,** which means the molecules are found in a very random arrangement, and their packing is much looser. Because of their structural regularity, the crystalline regions impart toughness and resistance to abrasion, and they make polypropylene and high-density polyethylene opaque. The amorphous regions promote flexibility. Materials formed from the other three polymers—polyethylene terephthalate, polystyrene, and polyvinyl chloride—are amorphous in structure. The range of properties among different polymers means that they are differently suited for specific applications. The next Consider This gives you an opportunity to match polymers with their properties and uses.

PET or PETE is used to abbreviate polyethylene terephthalate.

Consider This 9.7   **Uses of the Big Six Polymers**

For each of these uses, specify the desirable properties of a plastic and, using the information in Table 9.1, suggest the most suitable polymer or polymers.

a.   a bread bag
b.   a soft-drink bottle
c.   "bubble" wrap
d.   bottle caps
e.   outdoor lawn furniture
f.   a water bottle to carry on a bicycle

Whatever use is made of them, the six major polymers also generally have small amounts of other materials added to them. Because all six are colorless, coloring agents often are introduced. **Plasticizers,** substances that improve the flexibility of the polymer, are commonly added, as are a variety of other substances that enhance the performance and durability of the material. Indeed, the smell associated with certain polymers (and interiors of new cars) is sometimes due to plasticizers.

9.3 Addition Polymerization: Adding Up the Monomers

We already noted that the monomers ethylene, vinyl chloride, styrene, and propylene each contain a C-to-C double bond, three hydrogen atoms, and one variable atom or group attached to these two carbon atoms. The general structure can be written as:

$$\begin{array}{cc} H & H \\ \diagdown & \diagup \\ C \!\!=\!\! C \\ \diagup & \diagdown \\ H & R \end{array}$$

In this structure, R is a hydrogen atom, a chlorine atom, or a group of atoms (see Tables 9.1 and 9.2). These four monomers form polymers by a process called **addition polymerization.** In addition polymers, the monomers simply add to the growing polymer chain in such a way that the product contains all the atoms of the starting material. No other products are formed, and no atoms are eliminated. Thus, ethylene molecules become bonded together to form polyethylene. The two carbon atoms in ethylene are linked with a double bond that is capable of reacting with another ethylene molecule. In the process, the C-to-C double bonds are converted to C-to-C single bonds, and the polymer contains only C-to-C single bonds with hydrogen atoms attached to the carbon atoms.

> Sometimes an X is used to indicate a halide atom, such as Cl or F. R is reserved for hydrocarbon groups or the H atom.

The production of polyethylene begins with the joining of two ethylene molecules to form a four-carbon chain (equation 9.1).

$$2\ \begin{array}{cc} H & H \\ \diagdown & \diagup \\ C\!\!=\!\!C \\ \diagup & \diagdown \\ H & H \end{array} \longrightarrow \ \begin{array}{cccc} H & H & H & H \\ | & | & | & | \\ -C-C-C-C- \\ | & | & | & | \\ H & H & H & H \end{array} \qquad [9.1]$$

An additional ethylene molecule joins by adding to the growing chain (equation 9.2).

$$\begin{array}{cccc} H & H & H & H \\ | & | & | & | \\ -C-C-C-C- \\ | & | & | & | \\ H & H & H & H \end{array} + \begin{array}{cc} H & H \\ \diagdown & \diagup \\ C\!\!=\!\!C \\ \diagup & \diagdown \\ H & H \end{array} \longrightarrow \begin{array}{cccccc} H & H & H & H & H & H \\ | & | & | & | & | & | \\ -C-C-C-C-C-C- \\ | & | & | & | & | & | \\ H & H & H & H & H & H \end{array} \quad [9.2]$$

By continuing to add ethylene molecules, the chain grows rapidly to form a long polymer containing n monomeric units (equation 9.3).

$$n\ \begin{array}{cc} H & H \\ \diagdown & \diagup \\ C\!\!=\!\!C \\ \diagup & \diagdown \\ H & H \end{array} \longrightarrow \ \left[\!\!\begin{array}{cc} H & H \\ | & | \\ C-C \\ | & | \\ H & H \end{array}\!\!\right]_n \qquad [9.3]$$

The numerical value of n and hence the length of the chain varies with the reaction conditions. Often n is in the hundreds or thousands. During the manufacturing process, n will be adjusted in order to select specific properties for the polymer. Moreover, n can vary even within a single sample because a typical synthetic polymer is a mixture of individual polymer molecules of varying length and mass. The value of n will specify the length and, therefore, the molecular mass of the polymer. Molecular masses of polyethylene are generally between 10,000 and 100,000 g/mol. In every case, however, the carbon atoms are attached to one another by single bonds, and the hydrogen atoms are bonded to the carbon atoms. Thus, a polyethylene molecule is a macromolecular version of a hydrocarbon molecule, such as those in petroleum.

Your Turn 9.8	**Polyethylene Structural Formula**

Draw the structural formula of a polyethylene chain containing eight ethylene units.

Ethylene and polyethylene are made up only of carbon and hydrogen atoms. Therefore, the structure is always the same: a long chain of C-to-C single bonds with hydrogen atoms attached to every carbon. But the vinyl chloride molecule, $H_2C=CHCl$, has a single chlorine atom. This introduces an opportunity for variability in the structure of polyvinyl chloride (PVC).

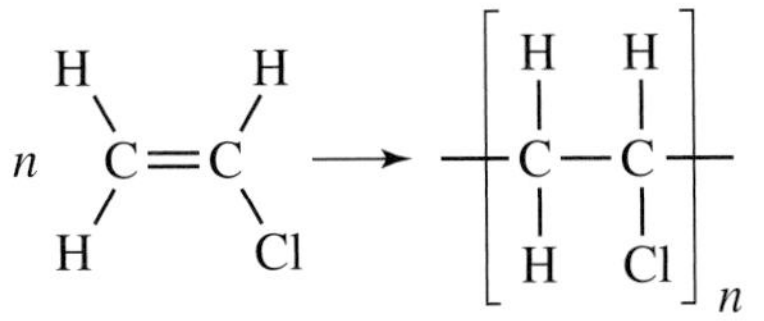

$$n \quad \underset{H}{\overset{H}{C}}=\underset{Cl}{\overset{H}{C}} \longrightarrow \left[\begin{array}{cc} \overset{H}{\underset{H}{C}} & \overset{H}{\underset{Cl}{C}} \end{array} \right]_n \qquad [9.4]$$

The chlorine atom creates asymmetry in the molecule. As an example, let us arbitrarily think of the carbon atom bearing two hydrogens (CH_2) as the "head" of a vinyl chloride molecule and the chlorinated carbon atom (CHCl) as its "tail." Because of the chlorine atom, when vinyl chloride molecules add to one another to form polyvinyl chloride, the molecules can be oriented in three possible arrangements: repeating head-to-tail, alternating head-to-head and tail-to-tail, and a random distribution of heads and tails (Figure 9.4).

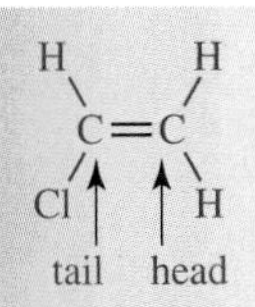

In the repeating head-to-tail structure, chlorine atoms are on alternate carbons. In a head-to-head/tail-to-tail arrangement of PVC, chlorine atoms are on adjacent carbons. In the random polymer, an irregular mixture of the previous two types occurs. In each case, the properties are somewhat different. Controlling monomer orientation within the chain is one method used to influence polymer properties. The head-to-tail arrangement

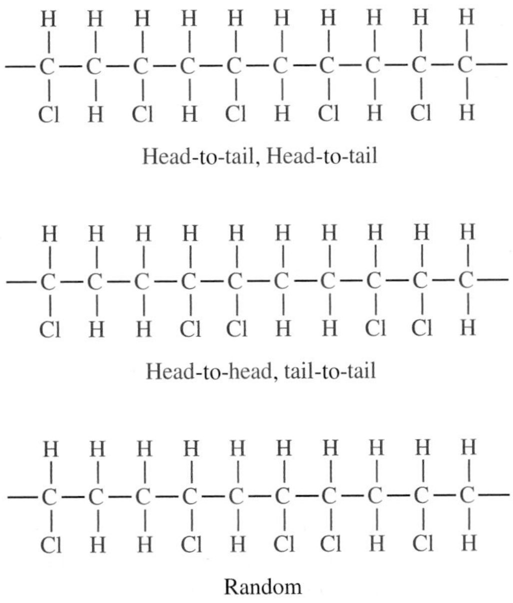

Head-to-tail, Head-to-tail

Head-to-head, tail-to-tail

Random

Figure 9.4

Three possible arrangements of monomer units in PVC.

Plasticizers are used to control flexibility in polymers. See Section 9.2.

is the usual product for polyvinyl chloride. The orientation of monomer units in the chain is one of the factors that can affect the flexibility of PVC. Stiff PVC finds use in drain and sewer pipes, credit cards, house siding, toys, furniture, and various automobile parts. The flexible version is familiar in wall coverings, upholstery, shower curtains, garden hoses, and insulation for electrical wiring.

Polypropylene (PP) is also formed by addition polymerization, using propylene monomers (see Table 9.1). A particularly useful form of polypropylene has the monomeric units bonded in a repeating head-to-tail fashion. This regularity imparts a high degree of crystallinity and makes the polymer strong, tough, and able to withstand high temperatures. Its uses reflect these properties. Polypropylene is found in indoor–outdoor carpeting, videocassette cases, and cold-weather underwear. Strength and chemical resistance make polypropylene a good choice for applications in which structural ruggedness is required, such as in indoor–outdoor carpeting.

The familiar white foam hot beverage cup is the most common example of polystyrene (PS). The styrene monomer has the phenyl group in place of a hydrogen atom on one of the carbon atoms in the C-to-C double bond (equation 9.5). Styrene usually polymerizes to polystyrene with the head-to-tail arrangement. The by-now-familiar type of addition polymerization equation applies; here n equals about 5000.

$$[9.5]$$

We noted earlier that the hard, brittle, transparent jewel cases for CDs are chemically almost identical to light, white, opaque foam coffee cups; both are polystyrene. Such Styrofoam cups are made by expansion molding. Polystyrene beads containing 4–7% of a low-boiling liquid are placed in a mold and heated using steam or hot air. The heat causes the liquid to vaporize, and the expansion of the gas also expands the polymer (similar to the baking of bread). The expanded particles are fused together into the shape determined by the mold. Because it contains so many bubbles, this plastic foam is not only light, but it also is an excellent thermal insulator.

Chlorofluorocarbons were at one time used as foaming agents. Concern over the involvement of CFCs in the destruction of stratospheric ozone (see Chapter 2) led to their replacement in 1990. Gaseous pentane (C_5H_{12}) and carbon dioxide are now frequently used for this purpose. The hard, transparent version of polystyrene is made by molding the melted polymer without the foaming agent. It is used to fabricate wall tile, window moldings, and radio and television cabinets, in addition to CD cases.

The Dow Chemical Company developed a new process that uses pure carbon dioxide as a blowing (foaming) agent to produce Styrofoam for packaging material. Using the Dow 100% CO_2 technology eliminates the use of 3.5 million pounds of CFC-12 (see Section 2.9) or HCFC-22 (Section 2.12) as blowing agents. The CO_2 used in the process is a by-product from existing commercial and natural sources, such as ammonia plants and natural gas wells. Thus, it does not contribute additional CO_2 to global warming. Because it is nonflammable, carbon dioxide is preferred as a blowing agent over pentane, which is flammable.

Consider This 9.9   *Polypropylene—A Tough Plastic*

Polypropylene, one of the Big Six, is used to construct a number of items in which toughness counts. It may not be as familiar to you as polyethylene and PET, because many polypropylene items are not marked with a recycling symbol (and are not collected at curbside). Search the Web to identify half a dozen items manufactured from polypropylene.

Your Turn 9.10   Polystyrene Structural Formula

Draw the structural formula of a polystyrene chain containing five styrene units arranged in the repeating head-to-tail arrangement. See Table 9.1 for the structural formula of styrene. Why do you think this arrangement is favored rather than the head-to-head arrangement?

Your Turn 9.11   Teflon

The monomer used to form Teflon is tetrafluoroethylene, $F_2C{=}CF_2$.

a. Write the structural formula of a polytetrafluoroethylene chain containing eight tetrafluoroethylene units.
b. Why is a repeating head-to-tail arrangement not possible for this polymer?

9.4 Polyethylene: A Closer Look at the Most Common Plastic

During manufacturing, various strategies are possible for modifying polymer chains to create differing properties of the material. We will use polyethylene as an example. Polyethylene has a wide variety of uses, which suggests a similarly wide range of properties for this single polymer. Yet, as we have seen in the previous section, all polyethylene is made from the same starting material—ethylene, $H_2C{=}CH_2$. How can this one material form polymers that can be used in so many different ways?

Ethylene is a compound extracted from petroleum (see Section 4.8). At ordinary temperatures and pressures, ethylene is a gas. However, in the 1930s, it was discovered that by using a special catalyst to initiate the reaction, individual ethylene molecules could be made to react with one another to form a polymer. The polymerization of ethylene involves a rearrangement of its electrons. The reaction is initiated by a catalyst that is a free radical with one unpaired electron. In Figure 9.5, a representation of the polymerization of polyethylene, the free radical is represented by R· (the dot indicates an unpaired electron). The radical reacts easily with a $H_2C{=}CH_2$ molecule. One of the two bonds between the carbon atoms in ethylene breaks, and one of the electrons from that bond pairs with the unpaired electron of the radical to form a covalent

Roy Plunkett, a DuPont chemist, discovered Teflon while experimenting with gaseous tetrafluoroethylene. Plunkett was curious enough about the solid that formed accidentally to study it sufficiently to realize that he had made a previously unknown polymer. This exemplifies Louis Pasteur's maxim that "Chance favors only the prepared mind."

Ethylene is also known as ethene.

Curiously, ethylene acts as a natural plant hormone that ripens fruit. Some fruits (such as tomatoes) emit ethylene, which is why putting tomatoes in a paper bag can help to ripen them.

Free radicals are very reactive species, as you may recall from the role of Cl· atoms in stratospheric ozone destruction (Section 2.9).

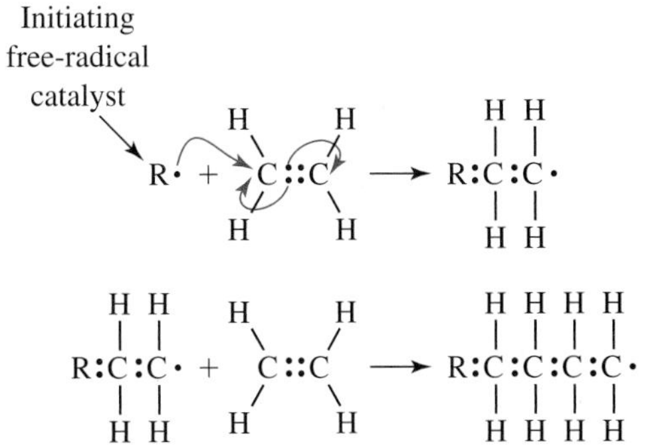

Figure 9.5
The polymerization of ethylene.

Figures Alive! Visit the *Online Learning Center* to see an interactive and animated version of this figure. Look for the Figures Alive! icon in this chapter as a guide to related activities.

The tetrahedral geometry around each carbon atom makes the carbon atoms in polyethylene align in a zigzag arrangement rather than a straight chain.

bond (see Figure 9.5). The result, •CH$_2$CH$_2$R, is also a free radical because it carries an unpaired electron left over from the broken C-to-C double bond. Therefore, •CH$_2$CH$_2$R can react with another ethylene molecule that bonds to the carbon atom with the unpaired electron at the reactive, growing end of the polymer. As the chain grows, C-to-C double bonds in monomers are converted to C-to-C single bonds in the polymer. This process is repeated many times over in many chains at the same time. Occasionally, the free radical ends of two polymers join to form a bond and stop the chain growth. The process can also be stopped by adding specific compounds to the reaction that will "cap" the radical and terminate the polymer chain. The result of all this chemistry is that gaseous ethylene is converted to solid polyethylene.

Many of the properties of polyethylene are related to the presence of these long chains of polymer molecules. Relatively speaking, they are very long indeed. If a polyethylene molecule were as wide as a piece of spaghetti, the molecular chain could be as much as half a mile long. To continue the analogy, in the polyethylene used to make plastic bags, these chains are arranged somewhat like spaghetti on a plate. The strands are jumbled up and not very well aligned, although in some regions the molecular chains are parallel. Moreover, the polyethylene chains, like spaghetti strands, are not bonded to one another.

In Chapter 5 (Section 5.6) you learned about hydrogen bonding, a force *between* water molecules in the liquid phase. The hydrogen bond is not a true covalent bond. Rather, the hydrogen bond is known as an **intermolecular attractive force,** an attraction between two molecules or ions resulting from the interactions of the electron clouds and nuclei. These attractive forces are different from the covalent bonds that exist *within* each molecule in which electrons are truly shared between two atoms. Intermolecular attractive forces are much weaker than covalent bonds, but they still have an effect on the behavior of a group of atoms. In water, these forces help to keep the molecules very close together, sliding past and around one another as the liquid flows, unlike in the gas phase where the molecules are far apart and only occasionally bump into one another.

Polymeric materials also have a type of intermolecular attractive force that keeps them close to one another and makes the material solid, rather than a gas. The type of attraction that exists between strands of polyethylene is a direct result of the molecule being very long. Because the chains are so long, they contain many atoms and there are many electrons along the chain of the molecule. These electrons can have attractions to the atoms on other, nearby polyethylene strands. The result is a bit like friction, or like the attraction between the two halves of Velcro. The bigger the surface area of the Velcro strip, the better it will hold. The intermolecular forces holding polyethylene together are called **dispersion forces,** and they are attractions between molecules that result from a distortion of the electron cloud that causes an uneven distribution of the negative charge. The bigger a molecule is, the easier it will be to distort its electron cloud. Therefore, dispersion forces are stronger in larger molecules and can be quite significant in very large molecules such as polymers.

Evidence of the molecular arrangement of polyethylene can be obtained by doing a little experiment. Cut a strip from a heavy-duty transparent polyethylene bag, grab the two ends of the strip, and pull. A fairly strong pull is required to start the plastic stretching, but once it begins, less force is needed to keep it going. The length of the plastic strip increases dramatically as the width and thickness decrease (Figure 9.6a). A little shoulder forms on the wider part of the strip and a narrow neck almost seems to flow from it in a process called "necking." Unlike the stretching of a rubber band, the necking effect is not reversible, and eventually the plastic thins to the point where it tears.

Figure 9.6b is a representation of the necking of polyethylene from a molecular point of view. As the strip narrows and necks down, the previously mixed-up molecular chains move. They shift, slide, and align parallel to one another and to the direction of the pulling force. In some plastics, such stretching or "cold drawing" is carried out as part of the manufacturing process to alter the three-dimensional arrangement of the chains in the solid. Of course, as the force and stretching continue, the polymer eventually reaches a point at which the strands can no longer realign, and the plastic breaks. Paper, a natural polymeric material, tears when pulled because the strands (fibers) in paper are rigidly held in place and are not free to slip like the long molecules in polyethylene.

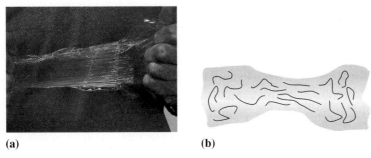

(a)   **(b)**

Figure 9.6
(a) A plastic bag stretched until it "necks." (b) Molecular rearrangement as polyethylene is stretched.

Consider This 9.12 **Necking Polyethylene**

Necking changes the properties of polyethylene.

a. Does necking affect the number of monomer units, n, in the average polymer? Explain your reasoning.
b. Does necking affect the bonding between the monomer units within the polymer? Explain your reasoning.

Another strategy to control the molecular structure and physical properties of polymers is to regulate the branching of the polymer chain. This approach is used to produce the two general types of polyethylene: high-density polyethylene (HDPE) and low-density polyethylene (LDPE) (Figure 9.7).

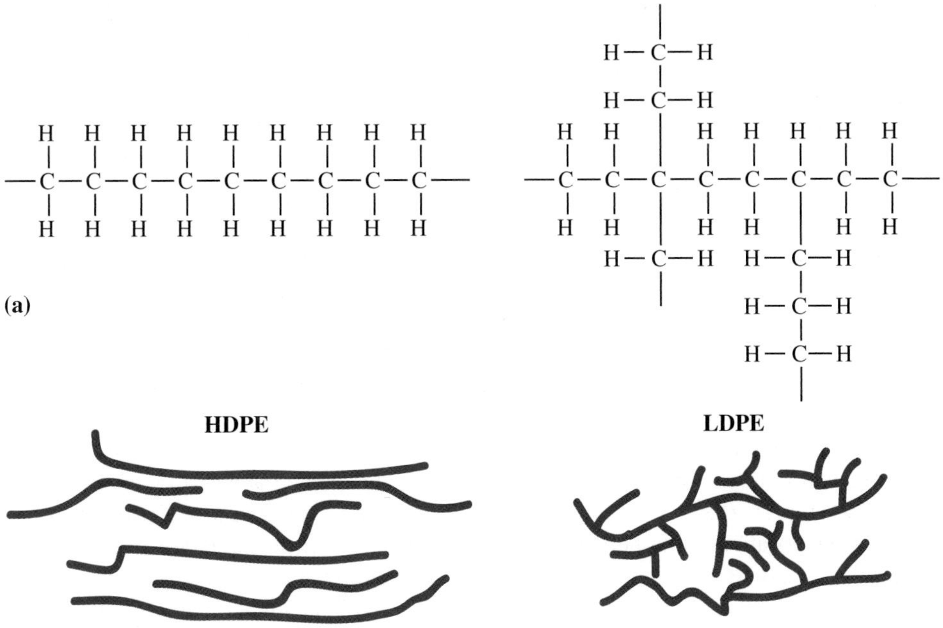

(a)

HDPE

LDPE

(b)

Figure 9.7
(a) Detail of bonding in high-density (linear) polyethylene (HDPE) and low-density (branched) polyethylene (LDPE). (b) Pictorial representation of high-density (linear) polyethylene and low-density (branched) polyethylene.

The version found in the plastic bags used for supermarket fruits and vegetables is LDPE. This version is soft, stretchy, transparent, and not very strong. This low-density form was the first type of polyethylene to be manufactured. Study of its structure reveals that the molecules consist of about 500 monomeric units and that the central polymer chain has many side branches, like the limbs radiating from a central tree trunk.

About 20 years after the discovery of LDPE, chemists were able to adjust reaction conditions to prevent branching and make HDPE. In their Nobel Prize-winning research, Karl Ziegler (1898–1973) and Giulio Natta (1903–1979) developed new catalysts that enabled them to make linear (unbranched) polyethylene chains consisting of about 10,000 monomer units. Having no side branches, these long chains can be arranged parallel to one another (see Figure 9.7). The structure of HDPE is thus more like a regular crystal than the irregular tangle of the polymer chains in LDPE. The more highly ordered structure of HDPE gives it greater density, rigidity, strength, and a higher melting point than LDPE. Furthermore, the high-density form is opaque; the low-density form tends to be transparent.

Consider This 9.13   **High- and Low-Density Polyethylene**

Use the structures of HDPE and LDPE found in Figure 9.7 to explain why the density of HDPE is greater than that of LDPE.

The differences in properties of high- and low-density polyethylene give rise to different applications. HDPE is used to make toys, containers, stiff or "crinkly" plastic bags, and heavy-duty pipes. One new use of HDPE was spurred by the AIDS epidemic. A surgeon who cuts her or his skin during an operation on an HIV-positive patient runs the risk of acquiring the human immunodeficiency virus through contact with the patient's blood. Allied-Signal Corporation has produced a linear polyethylene fiber called Spectra that can be fabricated into liners for surgical gloves. Spectra gloves are said to have 15 times more cut resistance than medium-weight leather work gloves, but they are so thin that a surgeon can retain a keen sense of touch. A sharp scalpel can be drawn across the glove with no damage to the fabric or the hand inside. Such strength is in marked contrast to the properties of the common LDPE plastic grocery bag.

Consider This 9.14   **Shopping for Polymers**

The Web site Macrogalleria is an excellent place to learn about polymers. At Level One of the site is a "shopping mall" featuring a host of different polymers. Use the link at the *Online Learning Center* to find at least six products made of LDPE and HDPE other than those mentioned in this chapter. Make your selections from two different "stores" at the "mall" and find two uses for both LDPE and for HDPE.

It would be a mistake, though, to conclude that polyethylene is restricted to the extremes represented by highly branched or strictly linear forms. By modifying the extent and location of branching in LDPE, its properties can be varied from soft and wax-like (coatings on paper milk cartons) to stretchy (plastic food wrap). Because of its linearity and close packing, HDPE is sufficiently rigid to be used for plastic milk bottles. Unfortunately, consumers are sometimes unaware of the consequences of such structural tinkering. For example, the higher melting point of HDPE (130 °C) permits plasticware made from it to be washed in automatic dishwashers. But objects made of LDPE, with a melting point of 120 °C, will melt in dishwashers.

Finally, we should note that one of the first and most important uses of polyethylene was a consequence of the fact that it is a good electrical insulator. During World

The plastic is melted by the high temperature of the dishwasher's heating element, not the hot water.

War II, polyethylene was used by the Allies as insulation to coat electrical cables in aircraft radar installations. Sir Robert Watt, who discovered radar, described polyethylene's critical importance in these words.

> "The availability of polythene [polyethylene] transformed the design, production, installation, and maintenance problems of airborne radar from the almost insoluble to the comfortably manageable . . . A whole range of aerial and feeder designs otherwise unattainable was made possible, a whole crop of intolerable air maintenance problems was removed. And so polythene played an indispensable part in the long series of victories in the air, on the sea, and on land, which were made possible by radar."

> Quoted by J. C. Swallow in "The History of Polythene" from
> *Polythene—The Technology and Uses of Ethylene Polymers* (2nd ed.)
> A. Renfrew (ed) London: Iliffe and Sons, 1960.

Consider This 9.15   **Yet Another Type of Polyethylene**

In addition to LDPE and HDPE, polyethylene is manufactured in other forms. Use the Web to find an alternative form of polyethylene. What are the properties and uses of this form of the polymer?

9.5 Condensation Polymers: Bonding by Elimination

Unlike the other five polymers of the Big Six described earlier, polyethylene terephthalate (PET or PETE) is not formed by an addition reaction. Rather, it is produced via a condensation reaction. Many polymers are formed by condensation reactions. Natural ones include cellulose, starch, wool, silk, and proteins; synthetics are nylon, Dacron, Kevlar, ABS, and Lexan.

In **condensation polymerization,** monomer units join by eliminating (splitting out) a small molecule, often water. Thus, a condensation polymerization has two products: the polymer itself plus the small molecules split out during the polymer's formation. Polyethylene terephthalate is a **copolymer,** a combination of two or more different monomers—ethylene glycol and terephthalic acid (see Table 9.1). Ethylene glycol, $HOCH_2CH_2OH$, is a dialcohol; it has an —OH on each carbon atom in the molecule. A molecule of terephthalic acid, $HOOCC_6H_4COOH$, has a —COOH group at each end. Because terephthalic acid has two organic acid groups per molecule, it is a diacid. The two different monomers condense as shown in equation 9.6.

ABS is a condensation polymer built from *a*crylonitrile, *b*utadiene, and *s*tyrene monomers.

Ethylene glycol is the primary ingredient in automobile antifreeze.

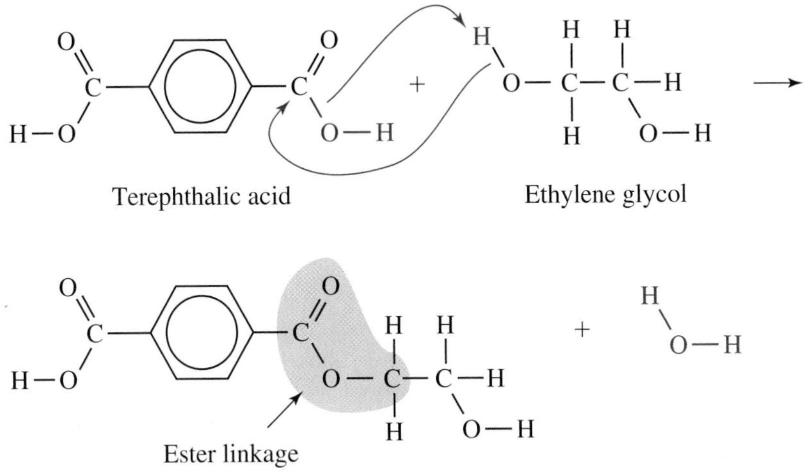

Terephthalic acid   Ethylene glycol

[9.6]

In equation 9.6, the benzene ring is represented by a hexagon, in which two of the H atoms on benzene have been replaced by bonds to other C atoms.

Ester linkage

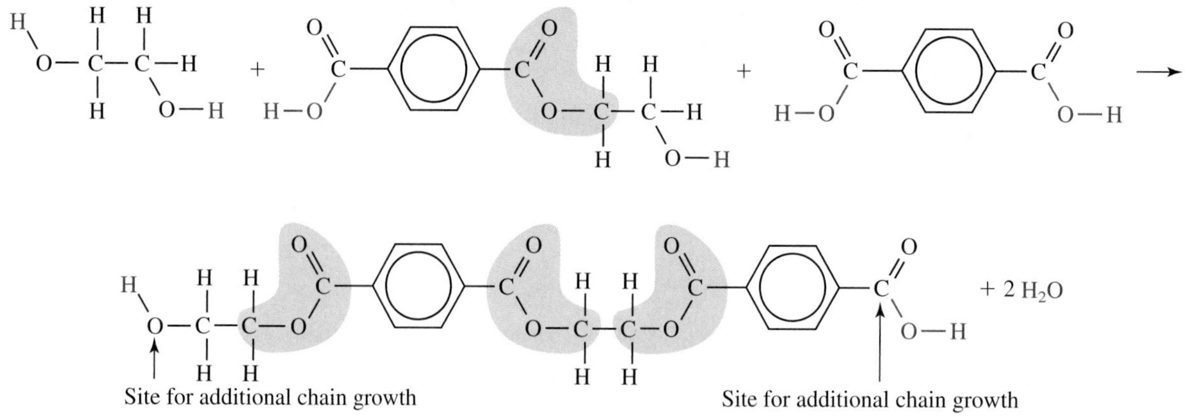

Figure 9.8

A growing PET polymer chain. Ester linkages are highlighted in blue. Note that no hydrogen atoms are included in the ester linkage.

Each time a monomer reacts, the —OH of a carboxylic acid group and the H in the —OH of an alcohol group react to form HOH, a water molecule. The water is given off as polyethylene terephthalate forms. The remaining portions of the alcohol and the acid join by forming an ester linkage. The condensation polymerization reaction of ethylene glycol with terephthalic acid is represented by equation 9.6; the ester linkage is highlighted in blue.

Note that the molecule produced in equation 9.6 and Figure 9.8 still has a —COOH group on one end and an —OH on the other. The carboxylic acid group can react with an alcohol group of another ethylene glycol molecule; likewise, the alcohol group of the growing polymer can react with a carboxylic acid group of another terephthalic acid molecule. This process, represented in Figure 9.8, occurs many times over to yield a long polymeric chain of polyethylene terephthalate.

Consider This 9.16   **Can All Organic Acids and Alcohols Form Polyesters?**

You have seen that terephthalic acid and ethylene glycol are capable of forming a polyester. Consider two other possible monomers: acetic acid and ethyl alcohol (ethanol):

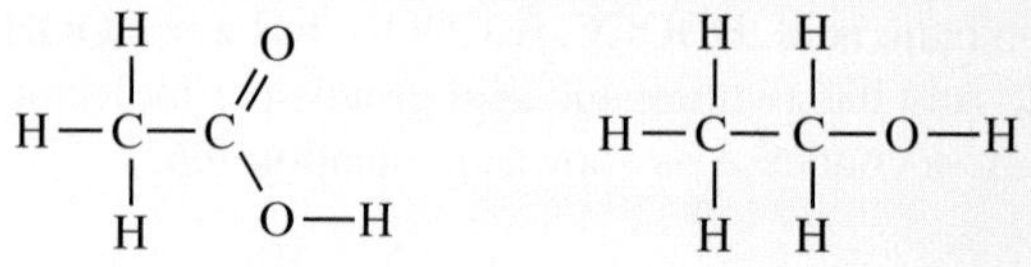

Acetic acid   Ethyl alcohol

a.   Can these two substances form an ester?
b.   Can they form a polyester? Explain your reasoning.

Polyethylene terephthalate (PET) is classified as a polyester. Since their introduction, polyester fibers have found many uses in fabrics and clothing. The polymer is perhaps most familiar under the trade name Dacron. This polyester is frequently mixed with cotton, wool, or other natural polymers to make fabrics, but it has many other uses. Indeed, over 7.5 billion pounds of thermoplastic polyesters is produced annually in the United States. The most common use for this plastic is in soft-drink bottles because PET is semirigid, colorless, and gas-tight. Narrow, thin-film ribbons of it (under the trade name Mylar) are coated with metal oxides and magnetized to make audiotapes and videotapes. Dacron tubing is used surgically to replace damaged blood vessels, and artificial hearts contain parts made of PET. Photographic and X-ray film are made from PET, and containers are made from it for medical supplies to be sterilized by irradiation.

Consider This 9.17 New Combinations

Polyethylene naphthalate (PEN) is a polymer widely used for bar code labels. In both PET and PEN, the alcohol monomer is ethylene glycol, but the organic acid monomers differ slightly. This is the structural formula of the organic acid monomer used to produce PEN.

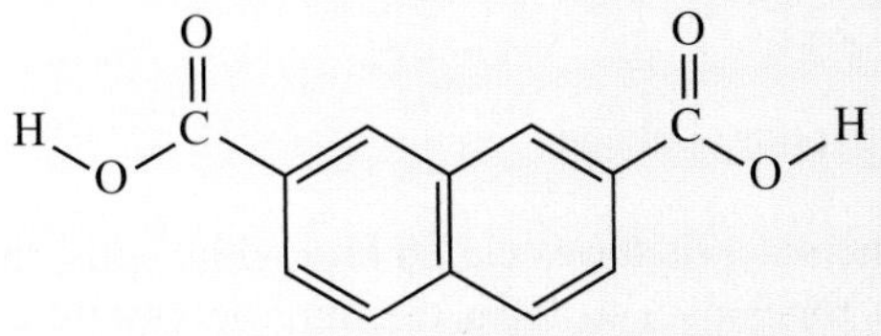

Napthalic acid

Draw structural formulas to show the reaction of two molecules of naphthalic acid with two molecules of ethylene glycol.

9.6 Polyamides: Natural and Nylon

No discussion of condensation polymerization can be complete without including proteins, one of the most important classes of natural polymers, and nylon, a synthetic substitute that brilliantly duplicates some of the properties of a natural polymer: silk.

A wide variety of proteins make up our skin, hair, muscle, and enzymes. All of these biological macromolecules are **polyamides,** polymers of amino acids. An **amino acid** molecule contains an amine group ($-NH_2$) as well as a carboxylic acid group ($-COOH$). This is the general formula for an amino acid:

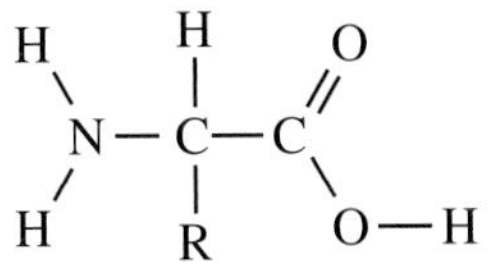

In Section 9.3 we used R to indicate any functional group. For amino acids, R represents one of 20 specific different groups, resulting in 20 different naturally occurring amino acids that are found in most proteins. These amino acids differ in the identity of their R groups. In some amino acids, R consists of carbon and hydrogen atoms, as in alanine, where R is a methyl group, $-CH_3$. In others, R may include additional atoms, such as oxygen, nitrogen or sulfur. Some R groups have acidic properties; others are basic.

Amino acids are the monomers of proteins. Analogous to polyester formation, the OH in a $-COOH$ group of one amino acid reacts with the H in the $-NH_2$ group of another in a condensation reaction to eliminate a H_2O molecule. A **peptide bond,** which consists of a $-C=O$ functional group joined with a $-NH$ functional group, thereby attaching the remaining portions of the two amino acids. The reaction is represented

See Sections 11.6 and 12.3 for more about amino acids and proteins.

by equation 9.7, in which the peptide bond is highlighted in blue. One amino acid contains the substituent R, and the other contains the substituent R′.

$$[9.7]$$

Amide (peptide) linkage

In the sophisticated chemical factories called biological cells, this condensation reaction is repeated many times over to form the long polymeric chains called proteins. Given the fact that there are 20 different naturally occurring amino acid building blocks, a great variety of proteins can be synthesized. Some proteins are made up of combinations containing hundreds of these 20 amino acids, whereas others contain only a few of them.

Chemists are often well advised to attempt to replicate the chemistry of nature. What we have learned about the structures of natural polymers informs production of synthetic polymers. For example, in the 1930s, a brilliant chemist working for the DuPont Company, Wallace Carothers (1896–1937) was studying a variety of polymerization reactions, including the formation of peptide bonds (Figure 9.9). Instead of using amino acids, Carothers tried combining adipic acid, $HOOC(CH_2)_4COOH$, and hexamethylenediamine, $H_2N(CH_2)_6NH_2$ (also known as 1,6-hexanediamine). Note from equation 9.8 that a molecule of adipic acid has an acid group on both ends, and the hexamethylenediamine molecule has an amine group on each end. As in the case of protein synthesis, the acid and amine groups react to eliminate water and form peptide bonds. But in this instance, the polymer consisted of alternating adipic acid and hexamethylenediamine monomer units. This polymer is known as nylon.

Figure 9.9

Wallace Carothers, the inventor of nylon.

Adipic acid   Hexamethylenediamine   First step in making nylon

Site for additional chain growth   Site for additional chain growth

$$[9.8]$$

DuPont executives decided the new polymer had promise, especially after company scientists learned to draw it into thin filaments. These filaments were strong and smooth and very much like the protein spun by silkworms. Therefore, nylon was first introduced to the world as a substitute for silk. The world greeted it with bare legs and open pocketbooks. Four million pairs of nylon stockings were sold in New York City on May 15, 1940, the first day they became available (Figure 9.10). But, in spite of consumer passion for "nylons," the civilian supply soon dried up, as the polymer was diverted from hosiery to parachutes, ropes, clothing, and hundreds of other wartime uses. By the time World War II ended in 1945, nylon had repeatedly demonstrated that it was superior to silk in strength, stability, and rot resistance.

Figure 9.10
Customers eagerly lined up to buy nylon stockings in 1940, when they were first available commercially.

Today this polymer, in its many modifications, continues to find wide applications in clothing, sportswear, camping equipment, the workroom, the kitchen, and the laboratory.

Consider This 9.18   Silk

Find the structure of the silk protein by doing a search on the Web with the terms *silk* and *protein structure*. Identify the differences and similarities in the chemical structure between the silk protein and nylon.

Your Turn 9.19   Kevlar

Kevlar is a condensation polymer used to make bulletproof vests. It is made from terephthalic acid and phenylenediamine (structures given here).

Terephthalic acid   Phenylenediamine

Use these structural formulas to draw a segment of a Kevlar molecule containing three units of terephthalic acid and three phenylenediamine units.

9.7 Plastics: Where From and Where To?

We began this chapter by considering the fact that polymers are derived from crude oil, the source of fossil fuels. Given the world's dwindling supply of crude oil, we should pay close attention to the raw materials incorporated into plastics. Equally important, and a direct result of the constraints of the law of conservation of matter,

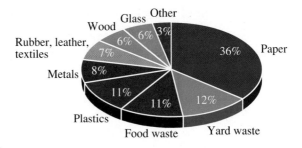

Figure 9.11

What's in your garbage? Composition of municipal solid waste.

Source: U.S. Environmental Protection Agency, 2001.

we need to consider the disposal or recycling of polymeric materials after their use. Indeed, there seems to be a good deal more concern about where plastics go than where they come from.

Consider This 9.20   **How Much Plastic Do You Throw Away?**

Keep a journal of all the plastic and plastic-coated products you throw away (not recycle) in one week. Include plastic packaging from food and other products that you purchase. Keep the journal handy because you will be asked to review it in a later activity (Consider This 9.31).

Sceptical Chymist 9.21   **All Those Disposable Diapers**

True or false? Disposable diapers are the main items in municipal landfills. Check out the link at the *Online Learning Center*.

Much of the plastic we use eventually ends up in a landfill, along with lots of other types of municipal and domestic solid wastes, in the usual "out of sight, out of mind" approach. As a nation, we daily discard enough trash to fill two Superdomes. Of course that's just a ballpark figure, but it corresponds to about 3 tons of trash annually per family. The EPA estimates that about 56% of all municipal solid waste now finds its way into landfills, 30% is recycled, and 14% is incinerated. Figure 9.11 provides information about the contents of a typical landfill given in terms of percent by weight. However, it is the volume of the buried materials, not only their weight, that causes landfills to reach their capacity.

Consider This 9.22 **Absorbent Properties**

What is the polymer in disposable diapers that allows them to absorb so much liquid?

You will note from Figure 9.11 that only 11% of municipal solid waste is plastic, about 30% of which comes from packaging materials. But the largest percentage of municipal solid waste (36%) is paper and paper products. This raises a question that is often discussed: Which constitutes the lesser environmental burden, paper or plastic? One issue to consider is that 1000 plastic bags weigh 17 lb and have a volume of 1219 in.^{3} (about 2/3 of a cubic foot). The same number of paper grocery bags weighs 122 lb and takes up 8085 in.^{3} (about 4.5 ft^3). Figure 9.12 shows the volume relationship to scale.

Consider This 9.23　Plastic versus Paper

When you are in a supermarket, which do you usually request, plastic or paper grocery bags? List the advantages and disadvantages of each and decide which is preferable. What factors, other than volume of municipal wastes, should be considered with respect to this issue? Are there alternatives other than just paper or plastic bags from the supermarket? Prepare a letter to the editor of your local newspaper clearly stating your choice of plastic vs. paper and explaining your reasons for it.

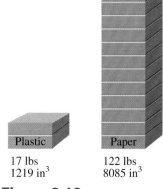

Plastic　　Paper

17 lbs　　122 lbs
1219 in^3　　8085 in^3

Figure 9.12

Relative volumes occupied by 1000 plastic bags and by 1000 paper bags.

Every year, more than 100 billion pounds of plastic is produced in the United States—about 350 lb for every woman, man, and child. Much of this ultimately finds its way into landfills. Given this huge quantity, there is little consolation in the fact that landfills contain considerably more paper than plastic. The reduction of the amount of all wastes, including plastics going into landfills, remains a high priority. Five strategies for reducing the reliance on landfills suggest themselves: incineration, biodegradation, reuse, recycling, and source reduction. In the paragraphs that follow, we will examine each of these approaches and attempt to weigh the relative merits.

Because the Big Six and most other polymers are composed primarily of carbon and hydrogen, incineration is an excellent way to dispose of used plastics. Indeed, a recent study in Germany led to the conclusion that burning waste plastic does less damage to the environment than any other method of disposal. The chief products of combustion are carbon dioxide, water, and a good deal of energy. In fact, pound for pound, plastics have a higher energy content than coal. Although plastics account for only 11% of the weight of municipal solid waste, they have approximately 30% of its energy content. The German study found that the greater the percentage of plastic in the refuse burned in a garbage incinerator, the more efficient the burning, the greater the quantity of energy released, and the lower the emission of airborne pollutants. It has been estimated that incineration can decrease the volume of plastic headed for landfills by as much as 90%.

But incineration of plastics is not without some drawbacks. The repeated message of Chapters 1–4, that burning does not destroy matter, applies here as well. Effluent gases produced by combustion may be "out of sight," but they had best not be "out of mind." Burning plastics produces CO_2, a greenhouse gas. Of special concern in incineration are chlorine-containing polymers such as polyvinyl chloride that release hydrogen chloride during combustion. Because HCl dissolves in water to form hydrochloric acid, such smokestack exhaust could make a serious contribution to acid rain. Chlorine-containing plastics can sometimes produce phosgene gas ($COCl_2$), which is very toxic. Moreover, some plastic products have inks containing heavy metals such as lead and cadmium. These toxic elements concentrate in the ash left after incineration and thus contribute to a secondary disposal problem. On balance, however, if carefully monitored and controlled, incineration can lead to a large reduction in plastic waste, generate much-needed energy, and have little negative impact on the environment.

Phosgene was used as a chemical warfare agent during World War II.

Your Turn 9.24　Complete Combustion Products

When polypropylene burns completely, it produces just carbon dioxide and water. Write a balanced chemical equation for the combustion of polypropylene. Assume an average chain length of 2500 monomers. *Hint:* Check the molecular formula of propylene in Table 9.1.

Another strategy for disposing of plastic wastes is to enlist bacteria to do the job—in other words, to employ biodegradation. The problem is that bacteria and fungi do not find most plastics very appetizing. These microorganisms lack the enzymes necessary to break down plastics. However, because bacteria and fungi evolved in our natural environment, they possess enzymes to break down naturally occurring polymers into simpler molecules. Many strains of bacteria use cellulose from plants or proteins from plants and animals as their primary energy sources. You have already encountered several instances of such processes in this text. In Chapter 3 you read about the release of methane by cattle. Actually, the methane is produced when bacteria decompose cellulose in the cow's rumen. In the same chapter, we also mentioned that methane is generated by natural decomposition of organic material in landfills, another result of bacterial activity.

Scientists are now engineering biodegradability into some synthetic polymers. Certain bonds or groups are introduced into the molecules to make them susceptible to fungal or bacterial attack, or to decomposition by moisture. Recently, research scientists at DuPont developed a biodegradable polymer called Biomax that decomposes in about eight weeks in a landfill. This new polymer is a close chemical relative of PET. Biomax uses other monomers in conjunction with those conventionally used to prepare PET (ethylene glycol and terephthalic acid). When polymerized, these co-monomers create sites in the polymer chains that are susceptible to degradation by water. Once the moisture does its job of breaking the polymer into smaller chains, naturally occurring microorganisms feed off the smaller chains, converting them to CO_2 and water. Biomax could be used in a variety of applications such as lawn bags, bottles, liners of disposable diapers, disposable eating utensils, and cups.

Achieving biodegradability in polymers raises some concerns, as an EPA report cautions.

> "Before the application of these technologies can be promoted, the uncertainties surrounding degradable plastics must be addressed. First, the effect of different environmental settings on the performance (e.g., degradation rate) of degradables is not well understood. Second, the environmental products or residues of degrading plastics and the environmental impact of degradables on plastic recycling is unclear."

Part of the difficulty is that even natural polymers do not decompose as completely in landfills as was suggested earlier in this chapter. Modern waste disposal facilities are covered and lined to prevent leaching of waste and waste by-products into the surrounding ground. Landfill linings and coverings create anaerobic (oxygen-free) conditions that impede bacterial and fungal action. As a result, many supposedly biodegradable substances decompose slowly or not at all when buried. Excavation of old landfills have found 37-year-old newspapers that are still readable (Figure 9.13) and 5-year-old hot dogs that, while hardly edible, are at least recognizable.

Figure 9.13
Some buried wastes can remain intact for a long time. This newspaper from 1952 was excavated 37 years later and could clearly be read.

Consider This 9.25 Landfill Liners

A Web search for *landfills* can bring up high-quality information about garbage. If you browse through these sites, you will see that some controversy surrounds the plastic materials used to construct the liner. Search for *landfill liners* and check out the plastics involved.

a. Which polymers are now used for liners?
b. What are their drawbacks?
c. Are there new polymers that offer more desirable properties? Cite the author and the URL of the Web sites, as well as when they were last updated.

Reusing a plastic product is one way to divert it from a landfill. Although not all plastics are directly reusable, many are. Plastic bottles can be reused by cleaning them and filling them with the same substance (water, shampoo, etc.), or used in other ways. A recent study reported that 80% of Americans reuse plastic products, such as food storage containers and refillable bottles.

In the United States, nearly 50% of certain plastic parts from damaged or discarded cars are repaired and reused. In a bold move to foster reuse on a large scale, new cars in Germany must be designed and built so that when the life of the car is over, its plastic parts can be removed easily and used to build other automobiles, thus creating automobiles that renew themselves. Through reusing rather discarding, far less petroleum is needed to make new plastics in the first place. The savings in this case can be substantial considering that the use of polymers in automotive applications has risen dramatically in the last 30 years. In 1973, about 4% of the weight of an average vehicle came from plastics. By 1998 that fraction rose to 7.5%, with 244 lb of polymeric materials contributing to the weight of a typical vehicle. Estimates show that in five years the average weight of polymers in an automobile will be almost 11% of the vehicle weight as smaller, lighter automobiles are developed.

Consider This 9.26 Plastics in an Automobile

Use the links in the *Online Learning Center* to find the kinds of plastics in an automobile and the uses of them. In particular, find this information for

a. the exterior of the car
b. the interior of the car
c. the fuel system
d. the engine

9.8 Recycling of Plastics

Given the problems associated with landfill disposal of natural and synthetic polymers, attention has logically turned to *recycling* both. Although recycling polymers does not literally dispose of them as does incineration or biodegradation, it helps to reduce the amount of new plastic entering the waste stream. However, in contrast to incineration, recycling polymers requires an input of energy. Furthermore, if the waste plastic is dirty or of low quality, more energy is needed to recycle it than to produce a comparable quantity of new virgin plastic.

Although the rate of recycling grew throughout the 1990s, it has now stabilized at about 23% for all plastics. The United States still lags far behind other developed nations in the percentage of plastics recycled. Extensive recycling has been an integral part of everyday life in Germany for over 10 years, leading to an extensive commercial recycling network. In 2000, Germany collected over 30 million tons of garbage. That total included over 5 million tons of recyclable plastics used in packaging, an average of 65 kg per person.

In the United States, more than 200 million milk and water jugs have been recycled to convert their high-density polyethylene into a fiber. The fiber is then made

200 million milk jugs joined end-to-end would form a chain long enough to reach nearly twice around the Earth.

directly into Tyvek, a material used in such broad applications as sports clothing, durable mailing envelopes, and insulating wrap for new buildings.

Consider This 9.27   Recycling Plastics

Go to the American Plastics Council Web site to find the latest about plastics recycling.

a.   What types of plastic make up the bulk of recycled plastics?
b.   What percent of PET (PETE) bottles are now being recycled?
c.   Use the links in the *Online Learning Center* to find information about the major opportunities and obstacles for recycling programs that collect all types of plastic bottles, not just those made of PET and HDPE.

Figure 9.14
Activewear made from recycled PET.

Increasing quantities of "post-consumer" plastics are being used in the United States. In 2002, total post-consumer plastic bottle recycling reached an all time record high of 1.6 billion pounds. Residents of more than 20,000 American communities, serving 63% of the population, have access to plastics recycling programs. Although the recycling rate for all plastic bottles in the United States has stabilized, the recycling rate for PET in the United States is declining and is now under 20%. A major reason is that nearly 35% of bottles collected for recycling in 2002 within the United States were shipped to China. Lower labor costs there allow production of post-consumer products at a more competitive price.

Polyethylene terephthalate soft-drink bottles are particularly easy to melt and reuse. About 800 million pounds of PET plastic containers was recycled in the United States in 2002. Much of recycled PET was converted into polyester fabrics, including carpeting, T-shirts, the popular "fleece" used for jackets and pullovers (Figure 9.14), and the fabric uppers in jogging shoes. Five recycled 2-L bottles can be converted into a T-shirt or the insulation for a ski jacket; it takes just about 450 such bottles to make polyester carpeting for a 9 × 12 ft room.

The chemist Nathaniel Wyeth (Figure 9.15) used his creativity to develop the plastic soda bottle in contrast with his artist brother Andrew Wyeth, who expresses his creativity on canvas. Nathaniel Wyeth had an even bigger vision than T-shirts or rugs from recycled PET. In *ChemMatters* magazine (October 1994) he was quoted as having said: "One of my dreams is that we're going to be able to melt the returned bottles down, mix them with reinforcing fibers, and make car bodies out of them. Then, once the car has served its purpose, rather than put it in the junk pile, melt the car down and make bottles out of it."

Another major recycling initiative involves national supermarket chains recycling their grocery bags. In fact, if you read the labels on plastic materials, you will increasingly find them made of a mixture of virgin and recycled (post-consumer) plastics. Some of the use of post-consumer plastics is now mandated by law. For example, since 1995, all packaging used in California has been required to contain 25% recycled material.

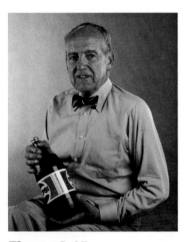

Figure 9.15
Nathaniel Wyeth (1911–1990), with his invention of the PET soda bottle still in use today.

Consider This 9.28   Polymers in Your Computer

Twenty years ago, recycling personal computers was not a concern because there weren't enough of them around to matter. These days, given the number of monitors, keyboards, and "mice" in circulation, there is good reason to keep these out of the landfill.

a.   What polymers are used in making your computer?
b.   Is it possible to recycle the polymers in your computer? Use the Web to find out.
c.   What is being done currently to recycle plastics from computers? You might want to search for "computers" and "recycling" to get started. Which plastic-containing computer supplies and accessories should have recycling programs?

However, simply collecting plastics to be recycled is not enough. For recycling to be successful and self-sustaining, a number of factors must be coordinated. They involve not only science and technology, but also economics and sometimes politics as well. True recycling involves a closed loop (Figure 9.16) in which plastics are collected and sorted, then converted into products that consumers buy, use, and later recycle. First of all, a dependable supply of used plastic must be consistently available at designated locations. This creates the formidable task of collecting the discarded objects, along with the associated job of sorting and separating the various polymers. The codes that appear on plastic objects (see Table 9.1) are provided to help facilitate this process. But, because of the large volume of material to be sorted, automated sorting methods are being developed. The plastics industry has spent more than $1 billion nationally on recycling research and developing environmentally responsible and sustainable plastic recycling programs.

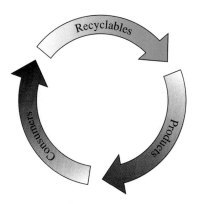

Figure 9.16

True recycling involves a never-ending loop.

Source: Reprinted by permission of NAPCOR (National Association for PET Container Resources).

Your Turn 9.29 Separating Polymers

Plastics vary in density. When placed in a liquid, a plastic floats if its density is less than that of the liquid, and it sinks if it is more dense than the liquid.

The densities (g/mL) of four common plastics are:

Plastic	Density (g/mL)
High-density polyethylene	0.95–0.97
Polyethylene terephthalate	1.38–1.39
Polypropylene	0.90–0.91
Polyvinyl chloride	1.18–1.65

The densities of six liquids at the same temperature are:

Liquid	Density (g/mL)
Methanol	0.79
An ethanol/water mixture	0.92
A different ethanol/water mixture	0.94
Water	1.00
Saturated solution of $MgCl_2$	1.34
Saturated solution of $ZnCl_2$	2.01

Using the densities of the plastics and of the liquids given, develop a procedure by which the individual plastics can be separated from a mixture containing pieces of all four plastics.

Almost any polymer that is not extensively cross-linked can be melted. However, the first thing that must be done is to sort the plastics into different types. If the supply of waste is homogeneous, that is, if it contains only one type of plastic, the molten polymer can be used directly in the manufacturing of new products. Alternatively, it can be solidified, pelletized, and stored for future use. However, when mixtures of various polymers are melted, the product tends to be darkly colored with varying properties, depending on the nature of the mixture. Although this reprocessed material does not have outstanding working properties, it is good enough for general lower grade uses such as parking lot bumpers, disposable plastic flower pots, and cheap plastic lumber. Such mixed material is obviously not as valuable as the pure, homogeneous recycled polymer. Hence, this underscores the importance of sorting plastics. For similar reasons, manufacturers prefer to use only a single polymer in a product to avoid having to separate the various polymers.

A variant on this method of reprocessing plastics is to decompose the polymers into simpler molecules, in some cases back to the actual original monomers. Fanciful as Nathaniel Wyeth's bottles-to-cars-to-bottles idea might seem, it is nonetheless a reality. DuPont chemists won a Presidential Green Chemistry Challenge Award for improving on Wyeth's dream. They developed a proprietary process for treating post-consumer PET that, like pulling beads apart from a necklace, unlocks (depolymerizes) the polymer back to its original monomers. These can then be reused to make new PET for other products. Mary Johnson, a DuPont employee, says "Because these monomers retain their original properties, they can be reused over and over again in any first-quality application. A popcorn bag can become an overhead transparency, then a polyester peanut butter jar, then a snack food wrapper, then a roll of polyester film, then a popcorn bag again."

The essential final step in recycling plastics is marketing. But a company or a city would be well advised to determine or create the demand for recycled polymers before completing all the other steps. Without a product and buyers, recycling programs are doomed to fail. In fact, recycling laws in a number of cities have not been implemented and enforced because one of the links in this polymeric chain of supply, collecting, sorting, processing, manufacturing, and marketing is missing. Without all these, the system will not work, unless it is heavily subsidized. More municipalities are becoming willing to provide the necessary funds as the market for post-consumer plastics grows stronger and prices increase enough to justify significant recycling activity.

Consider This 9.30   Responsible Consumerism

Unless consumers buy products made from recycled plastics, manufacturers will have little financial incentive to produce such products, and this may threaten the viability of plastics recycling. Use the Web to find at least half a dozen products from recycled plastics, other than those mentioned in the textbook. Choose products that show the range of types of plastics and their use. Identify the polymer(s) used in each product.

The remaining option for dealing with plastics, source reduction—using less material in the first place, and thus generating less waste—appears to be simplest and most direct. Simply decrease the quantity of plastics produced and used. The advantages are many: resources would be conserved, pollution would be reduced, and potentially toxic materials would be minimized. Examples of source reduction are the improvements made in plastic technologies that have reduced the amounts of plastic needed to make high-volume products; the 2-L soda bottle now uses 25% less plastic than when it was introduced in 1975, and the 1-gal milk jug weighs 30% less than a decade ago.

Yet source reduction, a seemingly innocuous option, is far more complicated than it appears. The problem is that something else is generally used to replace the plastic, and this substitution can be fraught with hidden pitfalls. In making choices between alternative materials, the decisions must be informed by the source and nature of chemical feedstocks (reactants), the method of manufacturing, waste products produced during manufacturing and their disposal, and many other factors. Energy costs as well as economic costs must be taken into account. How much energy must be expended in the entire life cycle of a product from raw material to final disposal?

Obviously, the identification and weighing of all of the possible variables is a complex and difficult task. But, when the job is done properly, one sometimes discovers that attempts to reduce the amount of plastic waste by substitution for the plastic may actually increase the overall amount of waste and the associated negative environmental impact. For example, 2 lb of plastic is enough to make containers able to hold about 8 gal of juice, soft drinks, or water. To hold that same amount of beverage would require 3 lb of aluminum, 8 lb of steel, or 27 lb of glass.

As another, more detailed example, consider the replacement of a plastic cup with one made of paper. Each occupies about the same volume in a landfill, where both will probably remain undecomposed for a long time. The "conventional wisdom" of public

Table 9.3	Paper versus Styrofoam Cups	
Item	**Paper Cup**	**Styrofoam Cup**
Per cup		
Raw materials		
Wood and bark, g	33	0
Petroleum, g	4.1	3.2
Finished mass, g	10	11.25
Cost	2.5 times that of plastic	1
Per million grams of material		
Utilities		
Steam, kg	9,000–12,000	5,000
Power, 10^9 J	3.5	0.4–0.6
Cooling water, m^3	50	154
Water effluent		
Volume, m^3	50–190	0.5–2
Suspended solids, kg	35–60	Trace
Air emissions, kg	7–22	35–50
Recycling potential		
Primary user	Possible	Easy
After use	Low	High
Ultimate disposal		
Heat recovery (million J/kg)	20	40
Mass in landfill, g	10.1	1.5
Biodegradable	Slowly, if at all	No

opinion has it that paper cups are more environmentally friendly than Styrofoam ones. But a detailed analysis of the issue made by Martin Hocking of the University of British Columbia counters that position. Hocking did a cradle-to-grave type of life cycle analysis in which he considered all aspects of the production and disposal of the two types of cups (Table 9.3). Hocking's conclusion is that paper cups are not as environmentally friendly as commonly thought. The paper cups require more raw material and consume as much petroleum as that used to produce the Styrofoam cups. The harsh nature of the chemicals used, the large volume of water required, and the nature of effluents generated into the air and water in paper making are far greater than those affiliated with producing polystyrene cups. Both types of cups are not very biodegradable in sealed landfills, so paper offers no significant advantage there. The Styrofoam cups are easier to reuse and recycle, and are about as easy to incinerate as paper ones.

Consider This 9.31   Reuse and Alternatives

To complete the Consider This 9.20, you kept a journal of all the plastic and plastic-coated products you discarded in a week. Review that list and indicate ways you could reuse those discarded products or suggest alternatives for the plastic in the product.

Of course, the best method of source reduction is not to replace plastics, but to do without them or their substitutes whenever possible. Among many letters that readers sent to newspapers about the paper versus plastic debate, one in particular, written over a decade ago in the *Syracuse Herald Journal,* reminds us of the larger issues involved:

> "There is a danger in this grocery bag controversy of losing sight of issues of greater importance. One of these is the matter of legitimate, responsible use of resources. Plastics are made from one of the most precious resources, one which cannot be renewed or replaced. In many respects we should regard it as more precious than gold or diamond. There are products essential to human health and well-being which can be made only from petroleum. There are also nonessential, wasteful uses of this priceless commodity. Where did we get the idea that it is our right to waste millions of barrels of oil each year exceeding the speed limit? Who said we're justified in manufacturing and using plastic items like shopping bags, burger boxes, and disposable diapers which are instant garbage? How did we get hooked on the consumer habits that are destroying not only a level of comfort we take for granted, but the very air and water we need to survive?"

There is no single, best solution to the problems posed by plastic waste and solid waste in general. Incineration, biodegradation, reuse, recycling, and source reduction all provide benefits and all have associated costs. Therefore, it is likely that the most effective response will be the development of an integrated waste management system that will employ all four of these strategies. The goal of such an integrated system would be to match the methods to the composition of the waste stream, thus optimizing efficiency, conserving energy and material, and minimizing cost and environmental damage.

CONCLUSION

Synthetic polymers are at the very center of modern living, yet their existence depends on a precious resource that is disappearing rapidly—"more precious than gold or diamond" according to the letter quoted earlier. We have come to not only depend on synthetic polymers, but in many cases to take them for granted to the point of being wasteful. Once more we come to an issue of lifestyle. Over the past nearly 70 years, chemists have created an amazing array of polymers and plastics—new materials that have made our lives more comfortable and more convenient. Many of these plastics represent a significant improvement over the natural polymers they replace. Furthermore, many products we use today would be impossible without synthetic polymers and plastics. There would be no CDs or DVDs, no cell phones, no videotape, no kidney dialysis apparatus, and no heart–lung machines or artificial hearts. We have become dependent on synthetic polymers, and it would be difficult if not impossible to abandon their use. The chemical industry has given consumers what they want. But there now appears to be rather more of it than we would like or perhaps can deal with responsibly—mountains of soft-drink bottles and miles of plastic bags. We must learn to cope with this glut of stuff while saving matter and energy for tomorrow. To create a new world of plastics and polymers will require the intelligence and efforts of policy planners, legislators, economists, manufacturers, consumers, and of course, chemists.

Chapter Summary

Having studied this chapter, you should be able to:

- Understand the nature of plastics and polymers, their typical properties and molecular structures (9.1)
- Describe typical uses for the Big Six polymers (9.2)
- Be able to name and draw several functional groups (9.2)
- Understand the molecular mechanism of addition polymerization (9.3) and condensation polymerization (9.5)

- Recognize the chemical composition and molecular structure of the Big Six polymers:
  Low-density polyethylene (LDPE) and high-density polyethylene (HDPE) (9.4)
  Polyvinyl chloride (PVC) (9.3)
  Polystyrene (PS) (9.3)
  Polypropylene (PP) (9.3)
  Polyethylene terephthalate (PET) (9.5)

- Tell how amino acids and proteins are related chemically (9.6)
- Use structural formulas to write the chemical equation for the synthesis of nylon (9.6)
- Identify sources of materials for manufacturing plastics (9.7)

- Relate the technical, economic, and political issues in methods for disposing of waste plastic: incineration, biodegradation, reuse, recycling, and source reduction (9.7 and 9.8)

Questions

Emphasizing Essentials

1. This chapter deals with plastics and polymers. Do these terms mean the same thing?

2. Why are polymers sometimes referred to as macromolecules?

3. Assume you have two different monomers and you want to form a polymer with three different sections to it (a "trimer"). How many different trimers can you form? *Hint:* Be sure to consider all the possible arrangements, including a trimer made of only one type of monomer.

4. Some people object to polymers, saying that they are not "natural." Give examples of two natural polymers.

5. **a.** What functional group is attached to the double-bonded carbon atom in the styrene monomer? *Hint:* see Table 9.1.

   b. What functional group is attached to the carbon atoms in the ethylene glycol monomer? *Hint:* see Table 9.1.

6. What is the major reason that plastics are substituted for metal in automobiles?

7. Glucose from corn is the source of some new bio-based polymer materials. What is the chemical structure of glucose?

8. Consider these data:

Year	U.S. population, millions	Plastics produced in the U.S., billions of pounds
1977	220	34
1997	269	89
2003	290	107

   a. How many pounds of plastic were produced in the United States in 2003?

   b. How many pounds of plastic were produced per person in the United States in 2003?

   c. What is the percent change in the total number of pounds of plastic produced per person between 1977 and 2003?

   d. What is the percent change in the number of pounds of plastic produced per person between 1997 and 2003?

9. Do you expect the heat of combustion of polyethylene, as reported in kJ/g, to be more similar to that of hydrogen, coal, or octane, C_8H_{18}? Explain your prediction.

10. Circle and identify all the functional groups in this molecule:

11. Equations 9.1 and 9.2 show the polymerization of ethylene monomers to form a small segment of polyethylene. Use the bond energies of Table 4.2 to calculate the energy change during the reaction in equation 9.1. Is the reaction endothermic or exothermic?

12. How would your result from question 11 differ if, rather than using ethylene as the monomer, tetrafluoroethylene were used as the monomer, forming a small segment of the Teflon polymer, polytetrafluoroethylene? This is the structure of the monomer.

13. **a.** Determine the number of CH_2CH_2 monomeric units, *n*, in one molecule of polyethylene with a molar mass of 40,000 g.

   b. What is the total number of carbon atoms in this molecule?

14. **a.** What is the molar mass of the phenyl functional group?

   b. What is the percentage difference in the molar mass of the styrene and ethylene monomers?

15. Explain the role of a free radical in the polymerization of ethylene.

16. This is a representation of a small segment of polyethylene. The hydrogen atoms are omitted for the sake of clarity.

Are the C-to-C bond angles in polyethylene all 180°? Explain your reasoning.

17. This is a representation for the formation of polyvinyl chloride from vinyl chloride. As this process takes place at the molecular level, how does the Cl-to-C-to-H bond angle change?

$$n \quad \overset{H}{\underset{H}{\diagdown}} C = C \overset{H}{\underset{Cl}{\diagup}} \quad \longrightarrow \quad \left[\begin{matrix} H & H \\ | & | \\ C - C \\ | & | \\ H & Cl \end{matrix} \right]_n$$

18. Describe how each of these strategies would be expected to affect the properties of polyethylene and give an atomic/molecular level explanation for each effect.

    a. increasing the length of the polymer chain

    b. aligning the polymer chains with one another

    c. increasing the degree of branching in the polymer chain

19. Both bottles are made of polyethylene. How do the two bottles differ at the molecular level?

20. The manufacturers of some plastic household containers say that it is fine to place the item in the dishwasher, particularly on the top shelf. Others do not claim that their products are dishwasher-safe.

    a. Why is it recommended that dishwasher-safe plastics be put only on the top shelf?

    b. Assume that you have lost the information sheet that originally accompanied the dishwasher in your kitchen. Also, assume that you want to use the dishwasher as much as possible, rather than washing plastic containers by hand. What general properties of the plastic containers should help to guide you in avoiding problems in the dishwasher?

21. Do all the Big Six polymers have a common structural feature (see Table 9.1)? If so, identify that feature. If not, identify which of the six do share a common structural feature.

22. Table 9.1 lists features of the Big Six polymers.

    a. Which are the most flexible?

    b. Which do not have crystalline regions?

    c. Which are soluble in organic materials?

    d. Which is the most costly?

23. a. The structure for styrene is given in Table 9.1. Rewrite that structure showing all of the atoms present.

    b. What is the molecular formula for styrene?

    c. What is the molar mass of a polystyrene molecule consisting of 5000 monomers?

24. Vinyl chloride can join in several different orientations to form polyvinyl chloride. Several different arrangements are shown in Figure 9.4. Which of those arrangements is shown here?

$$\begin{matrix} Cl & H & H & Cl & Cl & H \\ | & | & | & | & | & | \\ - C - C - C - C - C - C - \\ | & | & | & | & | & | \\ H & H & H & H & H & H \end{matrix}$$

25. Butadiene, $H_2C{=}CH \quad CH{=}CH_2$, is polymerized to make buna rubber.

    a. Write an equation representing this process.

    b. Is this an example of addition or condensation polymerization?

26. The Dow Chemical Company has developed a new process that uses CO_2 as the blowing, or foaming, agent to produce Styrofoam packaging material. What compound does CO_2 likely replace in the process, and why is this substitution environmentally beneficial?

27. Kevlar is a type of nylon called an *aramid* that contains rings similar to those found in benzene. Because of its great mechanical strength, Kevlar is used in radial tires and in bulletproof vests. It was discussed in Your Turn 9.19 and the structures for the two monomers, terephthalic acid and phenylenediamine, were given there. What linkage between monomers forms the resulting polymer?

28. Polyacrylonitrile is a polymer made from the monomer acrylonitrile, CH_2CHCN.

    a. Draw a Lewis structure of this monomer.

    b. Polyacrylonitrile is used in making Acrilan fibers used widely in rugs and upholstery fabric. What danger do rugs or upholstery made of this polymer create in the case of house fires?

29. Section 9.7 has this heading: "Plastics: Where From and Where To?" Answer both questions posed in this heading, concentrating on the major source and major means of disposal.

Concentrating on Concepts

30. What are the relationships among these terms: natural, synthetic, polymers, nylon, protein? Show these relationships by writing an outline with enough information to show how the terms are related.

31. You were asked in Consider This 9.20 to keep a journal of all the plastic and plastic-coated products you *throw away* in one week. Now consider all of the plastic items that you *recycle* in one week. Are there any from your first list of items thrown away that could be on your second list? Why or why not?

32. Celluloid was the first commercial plastic, developed in response to the need to replace ivory for billiard balls and piano keys.

    a. Speculate on what specific properties were required for celluloid that allowed it to be a suitable substitute for ivory in these products.

    b. Bakelite was an early plastic. Was it also developed in response to a specific need? Explain your reasoning.

33. What is the chemical structure of the monomer in DuPont's Sorona polymer?

34. This graph shows U.S. production of plastics from 1977 through 2003.

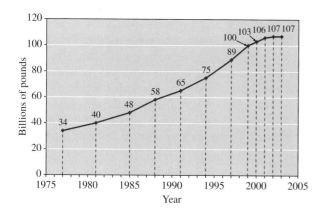

    a. What is the approximate increase in plastic production for any 5-year period before 2003?

    b. How many years were required for plastics production in 1977 to double?

    c. Redraw the graph as a bar graph showing the relationship of the year to the pounds of plastics produced. Discuss whether the bar graph is easier to use than the line graph to establish the doubling time.

    d. Suggest some factors that may have contributed to changes in production from 2001 to 2003.

35. The properties of a plastic are a consequence of more than just its chemical composition. What are some of the other features of polymer chains that have an influence on the properties of the polymer formed?

36. Consider the polymerization of 1000 ethylene molecules to form a large segment of polyethylene.

$$1000 \; CH_2{=}CH_2 \longrightarrow +CH_2CH_2{+}_{1000}$$

    a. Calculate the energy change during this reaction. *Hint:* Use Table 4.2 of bond energies.

    b. Should heat be supplied or must heat be removed from the polymerization vessel to carry out this reaction? Explain.

37. Isoprene is the monomer that forms natural rubber. It has the structure shown here:

$$\underset{1}{CH_2}{=}\underset{2}{C}\;{-}\;\underset{3}{CH}{=}\underset{4}{CH_2}$$
$$\qquad\;\; |$$
$$\qquad CH_3$$

The carbon atoms are numbered 1 through 4 for clarity. When isoprene monomers join in an addition polymerization to form polyisoprene (natural rubber), the polymer has a C-to-C double bond between carbon atoms 2 and 3. How does this double bond form? *Hint:* Each double bond has four electrons in it. When a new single bond is formed between two monomers, that single bond only needs two electrons in it, one from each of the monomers that are joined by that new bond.

38. Catalysts are used to help control the average molar mass of polyethylene, an important strategy to control polymer chain length. During World War II, low-pressure polyethylene production used varying mixtures of triethylaluminum, $Al(C_2H_5)_3$, and titanium tetrachloride, $TiCl_4$, as a catalyst. Here are some data showing how the molar ratio of the two components of the catalyst affects the average molar mass of the polymer produced.

Moles $Al(C_2H_5)_3$	Moles $TiCl_4$	Average Molar Mass of Polymer, g
12	1	272,000
6	1	292,000
3	1	298,000
1	1	284,000
0.63	1	160,000
0.53	1	40,000
0.50	1	21,000
0.20	1	31,000

    a. Prepare a graph to show how the molar mass of the polymer varies with the mole ratio of $Al(C_2H_5)_3/TiCl_4$.

    b. What conclusion can be drawn about the relationship between the molar mass of the polymer and the mole ratio of $Al(C_2H_5)_3/TiCl_4$?

    c. Use the graph to predict the molar mass of the polymer if an 8:1 ratio of $Al(C_2H_5)_3$ to $TiCl_4$ were used.

    d. What ratio of $Al(C_2H_5)_3$ to $TiCl_4$ would be used to produce a polymer with a molar mass of 200,000?

    e. Can this graph be used to predict the molar mass of a polymer if either pure $Al(C_2H_5)_3$ or pure $TiCl_4$ were used as the catalyst? Explain.

39. When you try to stretch a piece of plastic bag, the length of the piece of plastic being pulled increases dramatically and the thickness decreases. Does the same thing happen when you pull on a piece of paper? Why or why not? Explain on a molecular level.

40. Consider Spectra, Allied-Signal Corporation's HDPE fiber, used as liners for surgical gloves. Although the

Spectra liner has a very high resistance to being cut, the polymer allows a surgeon to maintain a delicate sense of touch. The interesting thing is that Spectra is *linear* HDPE, which is usually associated with being rigid and not very flexible.

a. Suggest a reason why branched LDPE cannot be used in this application.

b. Offer a molecular level reason for why linear HDPE is successful in this application.

41. One limitation of the Big Six is the relatively low temperatures at which they melt, 90–170 °C (see Table 9.1). Suggest ways to raise the upper temperature limits while maintaining the other desirable properties of these substances.

42. All the Big Six polymers are insoluble in water, but some of them dissolve or at least soften in hydrocarbons or in chlorinated hydrocarbons (Table 9.1). Use your knowledge of molecular structure and solubility concepts to explain this behavior.

43. When Styrofoam packing peanuts are immersed in acetone (the primary component in many nail-polish removers), they dissolve. After the acetone is allowed to evaporate, a solid remains. The solid is Styrofoam, but it is very solid and dense. Explain what happened. *Hint:* Remember that Styrofoam is made with foaming agents.

44. Vinyl chloride can join in several different orientations to form polyvinyl chloride (see Figure 9.4). Do these two structures represent the same possible arrangement?

$$\begin{array}{cccccc} Cl & H & Cl & H & Cl & H \\ | & | & | & | & | & | \\ -C & -C & -C & -C & -C & -C- \\ | & | & | & | & | & | \\ H & H & H & H & H & H \end{array}$$

and

$$\begin{array}{cccccc} H & H & Cl & H & H & H \\ | & | & | & | & | & | \\ -C & -C & -C & -C & -C & -C- \\ | & | & | & | & | & | \\ Cl & H & H & H & Cl & H \end{array}$$

Explain your answer by identifying the orientation in each arrangement.

45. What structural features must a monomer possess to undergo addition polymerization? Explain, giving an example.

46. What structural features must a monomer possess to undergo condensation polymerization? Explain, giving an example.

47. Plastics are widely used in packaging. Check the recycling code on 10 containers to identify the plastic used for each of them (see Table 9.1). How many of these containers are made from addition polymers? From condensation polymers?

Exploring Extensions

48. "Buckminsterfullerene," a form of elemental carbon, is shaped somewhat like a soccer ball. Each carbon is located at a corner where two 6-membered rings and one 5-membered ring come together. Locate a structural drawing of buckminsterfullerene on the Web and find a corner of the drawing that fits this description.

49. Dr. Richard Smalley, Dr. Harry Kroto, and Dr. Robert Curl, Jr. won the 1996 Nobel Prize in chemistry. Research their contributions and write a short report to describe why they won the Nobel Prize.

50. What is the difference in the material used in "hard" and "soft" contact lenses? How do the differences in properties affect the ease of wearing of contact lenses?

51. Two terms that have been added to the vocabulary of plastics are "virgin plastic" and "post-consumer waste" plastics. What do these terms imply about the plastics being used?

52. Free-radical peroxides promote the polymerization of ethylene into polyethylene. They also play a key role in tropospheric smog formation. Use the Web to learn more about how the peroxides promote ethylene polymerization and how peroxides are involved with photochemical smog formation in the troposphere. Write a brief report comparing the types of peroxides important with each of these cases. Give references for the Web information.

53. **a.** What is the structure of the monomers used in SBR synthetic rubber?

b. How do natural and synthetic rubber differ, and how are they alike?

54. Synthetic rubber is usually formed through addition polymerization. An important exception is silicone rubber, which is made by the condensation polymerization of dimethylsilanediol. This is a representation of the reaction.

$$n \ HO-\underset{\underset{CH_3}{|}}{\overset{\overset{CH_3}{|}}{Si}}-OH \longrightarrow \left[O-\underset{\underset{CH_3}{|}}{\overset{\overset{CH_3}{|}}{Si}}-O \right]_n + n \ H_2O$$

a. Predict some of the properties of this polymer. Explain the basis for your predictions.

b. Silly Putty is a popular form of silicone rubber. What are some of the properties of Silly Putty?

55. **a.** Name some other functional groups that were not discussed in this chapter.

b. Find the structure of the molecule acetone. What functional group is in this molecule?

56. Who first synthesized Kevlar? What was the background and academic training of these scientists?

Was the potential for using this polymer in radial tires immediately understood? What are other applications of Kevlar? Write a short report on the results of your findings, giving references either to books or Web information.

57. This is the structural formula for Dacron, a condensation polyester:

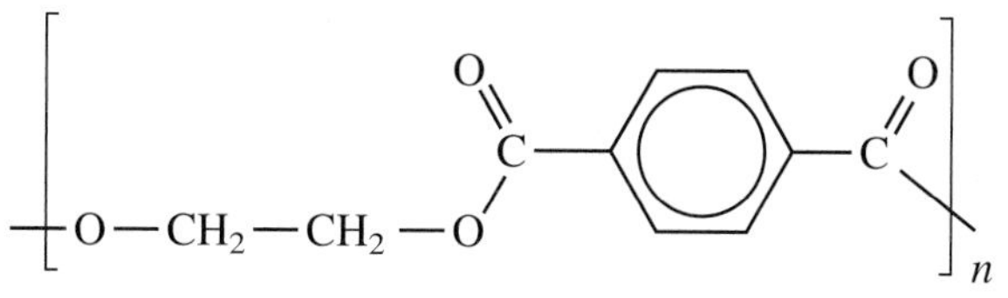

Dacron is formed by the reaction of a dialcohol and a dicarboxylic acid. Write the structural formulas for the alcohol and the acid monomers used to produce Dacron.

58. Cotton, rubber, silk, and wool are natural polymers. Consult other sources to identify the monomer in each of these polymers; specify which are addition polymers and which are condensation polymers.

59. How does your college's or university's community dispose of plastics? Of all the strategies for disposing plastics described in this chapter, which are used? How are the alternatives presented to the people in the community? Find out the current practices in your community, and then offer some suggestions for improving them.

60. What are some alternative strategies you can engage in to avoid the paper versus plastic dilemma with grocery bags or coffee cups?

Manipulating Molecules and Designing Drugs

Medications range from prescription drugs...

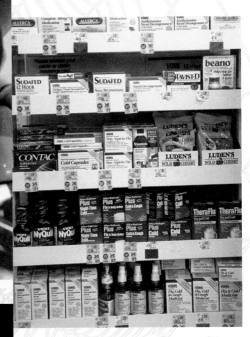

...to over-the-counter medicines...

...to herbal alternatives...

...to illicit drugs.

"Medicine is not only a science; it is also an art. It does not consist of compounding pills and plasters; it deals with the very processes of life, which must be understood before they are guided."

Paracelsus

"The outrage of the anti-abortionists was understandable because RU-486 promises to decentralize the provision of abortion to a woman's bedroom, which can neither be bombed nor picketed."

Carl Djerassi

"I don't do drugs, I am drugs."

Salvador Dali

Drugs. This word elicits hope, relief, fear, intrigue, outrage, or maybe simply disdain. These quotes only hint at the many reactions possible. Pharmaceuticals (drugs) are substances intended to prevent, moderate, or cure illnesses. Medicinal chemistry is the science that deals with the discovery or design of new therapeutic chemicals and their development into useful medicines.

Modern pharmacology has its origins in folklore, and the history of medicine is full of herbal and folk remedies. The use of herbs, roots, berries, and barks for relief from illness can be traced to antiquity as illustrated in documents recorded by ancient Chinese, Indian, and Near East civilizations. The Rig-Veda (compiled in India between 4500 and 1600 B.C.), one of the oldest repositories of human learning, refers to the use of medicinal plants. The Chinese Emperor Shen Nung prepared a book of herbs over 5000 years ago. In it, he described a plant called *Ma Huang* (now called *Ephedra sinica*), used as a heart stimulant. This plant contains ephedrine, a drug we will look at later in this chapter.

More recently, chemists have designed, synthesized, and characterized a vast array of prescription and over-the-counter drugs. Today, drugs help patients regulate their blood sugar, blood pressure, cholesterol, and allergies. They help AIDS patients stay alive while scientists search for a cure. Effective anticancer drugs and powerful analgesics now exist. Other drugs can even manage mental disorders that once were thought to be untreatable.

On the other hand, people have long taken drugs for the purpose of altering their perceptions and moods. The famed philosopher Nietzsche said that no art could exist without intoxication. Many writers and artists have found that drugs act as a creative and destructive force in their lives and work. People abuse drugs primarily because of the promise of instant relief or pleasure and the possibility of heightened awareness. It is a common misconception to assume that today's problem with drug abuse is a fairly recent phenomenon. The reality is that human history has been marked with drug use and abuse.

In discussing drugs, we will consider these questions: Where do the ideas and resources to develop drugs come from? What is the process by which pharmaceuticals make it to market? Why does a drug have a certain effect, and which features of its molecular structure contribute to the biological activity? What are the merits and pitfalls of herbal medicines, and are naturally occurring drugs "safer" than synthetic drugs? Which drugs are most commonly abused?

Consider This 10.1 Today's Drugs

a. Consider the modern pharmaceuticals prescribed in the United States today. List what you think are the top five most frequently prescribed drugs.

b. Consider the most commonly abused drugs in use today. List what you think are the top five most frequently abused drugs.

c. Share your lists with a small group of students. Do some of the drugs that you or others listed appear in both parts **a** and **b**?

10.1 A Classic Wonder Drug

In the fourth century B.C., Hippocrates, perhaps the most famous physician of all time, described a "tea" made by boiling willow bark in water. The concoction was said to be effective against fevers. Over the centuries, that folk remedy, common to many different cultures, ultimately led to the synthesis of a true "wonder drug," one that has aided millions of people.

One of the first systematic investigators of willow bark (Figure 10.1) was Edmund Stone, an English clergyman. His report to the Royal Society (1763) set the stage for a series of further chemical and medical investigations. Chemists were subsequently able to isolate small amounts of yellow, needle-shaped crystals of a substance from the willow bark extract. Because the tree species was *Salix alba,* this new substance was named salicin. Experiments showed that salicin could be chemically separated into two compounds. Clinical tests provided evidence that only one of these components reduced fevers and inflammation. It was also demonstrated that the active component was converted to an acid in the body. Unfortunately, the clinical testing revealed some troubling side effects. The active component not only had a very unpleasant taste, but also its acidity led to acute stomach irritation in some individuals.

The active acid was used as a treatment for pain, fever, and inflammation. But because of its serious side effects, chemists set out to modify the structure of the active acid to form a related compound that still would be effective, but without the undesirable taste or stomach distress. The first modification attempt took a very simple approach. The acid was neutralized with a base, either sodium hydroxide or calcium hydroxide, to form a salt of the acid. It turned out that the resulting salts had fewer side effects than the parent compound. Based on this finding, chemists correctly concluded that the acidic part of the molecule was responsible for the undesirable properties. Consequently, the next step was to seek a structural modification that would lessen the acidity of the compound without destroying its medicinal effectiveness.

One of the chemists working on the problem was Felix Hoffmann, an employee of a major German chemical firm. Hoffmann's motivation was more than just scientific curiosity or assigned work. His father regularly took the acidic compound as treatment for arthritis. It worked, but he suffered nausea. The younger Hoffmann succeeded in converting the original compound into a different substance, a solid that reverted back

Acid–base neutralization reactions are discussed in Section 6.3.

Figure 10.1

The white willow tree, *Salix alba,* source of a miracle drug.

Table 10.1	Side Effects of the "Wonder Drug"	
Symptoms	**Frequency**	**Severity***
Drowsiness	Rare	4
Rash, hives, itch	Rare	3
Diminished vision	Rare	3
Ringing in the ears	Common	5
Nausea, vomiting, abdominal pain	Common	2
Heartburn	Common	4
Black or bloody vomit	Rare	1
Black stool	Rare	2
Blood in the urine	Rare	1
Jaundice	Rare	3
Anaphylaxis (severe allergic reaction)	Rare	1
Unexplained fever	Rare	2
Shortness of breath	Rare	3

Source: H. W. Griffith, *The Complete Guide to Prescription and Non-Prescription Drugs,* 1983, HP Books, Tucson, Arizona.

*The severity scale ranges from 1, life-threatening, seek emergency treatment immediately to 5, continue the medication and tell the physician at the next visit.

to the active acid once it was in the body. This molecular modification greatly reduced nausea and other adverse reactions; a new drug had been discovered (1898).

Extensive hospital testing of Hoffmann's compound began along with simultaneous preparation for its large-scale manufacture by a well-known pharmaceutical company. The new drug itself could not be patented because it was already described in the chemical literature. However, the company hoped to recoup its investment by patenting the manufacturing process. Clinical trials showed the drug to be nonaddicting and relatively nontoxic. Its toxicity by ingestion is classified as low, but 20–30 g ingested at one time may be lethal. At the suggested dose of 325–650 mg (0.325–0.650 g) every 4 hours, it is a remarkably effective <u>antipyretic</u> (fever-reducing), <u>analgesic</u> (antipain medication), and <u>antiinflammatory agent.</u> Data from clinical tests uncovered the side effects noted in Table 10.1. The drug was also found to increase blood clotting time and to cause at least some small, almost always medically insignificant, amounts of stomach bleeding in about 70% of users.

Consider This 10.2   Miracle Drug

In the United States, the final step for approval of a drug is the submission of all of its clinical test results to the Food and Drug Administration (FDA) for a license to market the product.

a. If you were an FDA panel member presented with the information in Table 10.1, Side Effects of the "Wonder Drug," would you vote to approve this drug that treats pain, fever, and inflammation?

b. If approved, should this drug be released as an over-the-counter drug, or should its availability be restricted as a prescription drug? Write a one-page report defending your position.

Perhaps you have already guessed the identity of the miracle drug related to willow bark tea. Its chemical name, 2-(acetyloxy)-benzoic acid or (more commonly) acetylsalicylic acid, may not help much. But the power of advertising is such that, had we revealed that the firm that originally marketed the drug was the Bayer division of

I. G. Farben, we would have let the tablet out of the bottle. The compound in question is the world's most widely used drug, even a century after its discovery. Americans annually consume nearly 80 billion tablets of this miracle medicine. You know it as aspirin.

Admittedly, we have compressed the time somewhat. Most of the development, testing, and design of aspirin occurred in the 18th and 19th centuries. Stone's letter to the Royal Society was written in 1763, and Felix Hoffmann's modification of salicylic acid to yield aspirin was done in 1898. Furthermore, the clinical testing of aspirin was somewhat less systematic than our account implies. But the basic facts and the steps that led to aspirin's full development are essentially correct. We must also add one more very important fact. Aspirin did not have to receive drug approval before being put on the market; no such certifying process was in place at the time. Had such approval based on clinical test results been necessary, it is quite likely that aspirin would have been available only on a prescription basis.

Consider This 10.3   **What Should a Drug Be Like?**

Make a list of the properties you think a drug should have. Then compare your list with those of your classmates. Note similarities and differences.

a.   Are any items missing from your list that you now think you should include?

b.   Are any items present on your list that you now think you should delete?

10.2   The Study of Carbon-Containing Molecules

One of carbon's interesting properties is its ability to form a wide variety of molecules, especially ones containing bonds linking multiple carbon atoms. This element is so widely distributed throughout nature that the largest subdiscipline of chemistry, **organic chemistry,** is devoted to the study of carbon compounds. The name *organic* is historical and suggests a biological origin for the substances under investigation, but this is not necessarily true. In practice, most organic chemists confine themselves to compounds in which carbon is combined with a relatively small number of other elements: hydrogen, oxygen, nitrogen, sulfur, chlorine, phosphorus, and bromine. Even with this restriction, over 12 million of the 23 million total known compounds are considered organic. The chemical behavior (i.e., properties and reactivity) of organic compounds enables us to organize them into a relatively small number of categories. As a result, in this chapter we concentrate on only a few and stress their important roles within the functions of living things.

To identify a specific organic compound from among the myriad of possibilities, the compound must be named. Chemists use a formal set of nomenclature rules established by an international committee so each of the 12 million compounds can be uniquely named. However, many of these compounds have been known for a long time by common names such as alcohol, sugar, and morphine. When a headache strikes, even chemists do not call out for 2-(acetyloxy)-benzoic acid; they simply say "Give me some aspirin!" Likewise, prescriptions specify penicillin-N rather than 6[(5-amino-5-carboxy-1-oxopentyl)amino]-3,3-dimethyl-7-oxopentyl-4-thia-1-azabicyclo[3.2.0] heptane-2-carboxylic acid. Mouthfuls like this are the cause of great merriment to those who like to satirize chemists. Nonetheless, chemical names are important and unambiguous to those who know the system. You can rest easy because in this chapter, we will use common names in almost all cases.

An incredible variety of organic compounds exists because of the remarkable ability of carbon atoms to bond in multiple ways both to other carbon atoms and to atoms of other elements. To better understand such possibilities, we need a few basic rules for bonding in organic molecules. You used one of these in Chapter 2, the octet rule. When

The octet rule is discussed in Section 2.3.

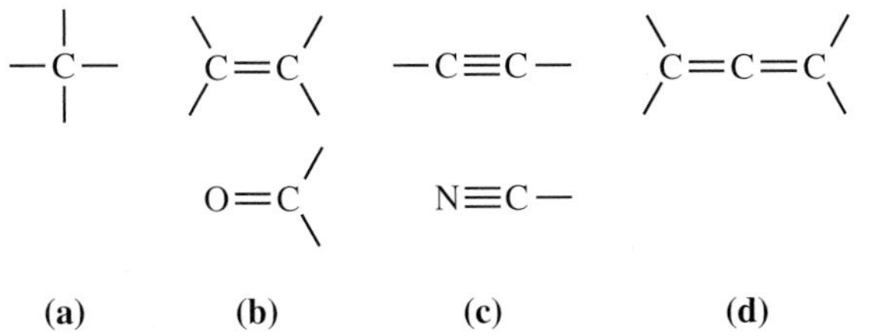

Figure 10.2

Common bonding arrangements for carbon.

bonded, each carbon atom has a share in eight electrons, an octet. Eight electrons can be arranged to form four bonds, with a pair of shared electrons in each covalent bond. The most common configurations for these four bonds around a carbon atom are (a) four single bonds, (b) two single bonds and one double bond, (c) one single bond and one triple bond, or (d) two double bonds. These arrangements are illustrated in Figure 10.2.

Other elements exhibit different bonding behavior in organic compounds. A hydrogen atom is always attached to another atom by a single covalent bond. An oxygen atom typically attaches either with two single bonds (to two different atoms) or one double bond (to a single atom). A nitrogen atom commonly forms three single bonds (to three different atoms), but also can form either a triple bond (to one other atom), or a single and a double bond.

Your Turn 10.4   **Satisfying the Octet Rule**

Examine each carbon in Figure 10.2. Is the octet rule followed?

Chemical formulas such as C_4H_{10} indicate the kinds and numbers of atoms present in a molecule, but do not show how the atoms are arranged or connected. To get that higher level of detail, **structural formulas** are used that show the atoms and their arrangement with respect to one another in a molecule. Here is the structural formula for normal butane, or *n*-butane (C_4H_{10}), a hydrocarbon fuel used in cigarette lighters and camp stoves.

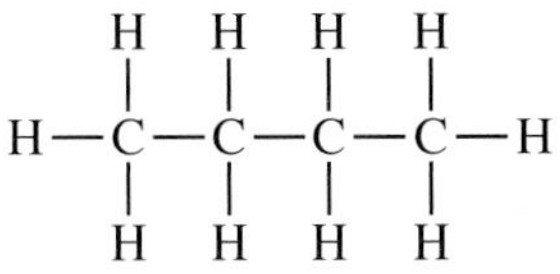

A drawback to writing structural formulas, at least in a textbook, is that they take up considerable space. To convey the same information in a format that is easily typeset into a single line, we use **condensed structural formulas** where carbon-to-hydrogen bonds are not drawn out explicitly, but simply understood to be single bonds. Here are condensed structural formulas for *n*-butane.

$$CH_3 - CH_2 - CH_2 - CH_3 \quad \text{or} \quad CH_3CH_2CH_2CH_3$$

Note that the carbons are bonded directly to other carbon atoms, and that the hydrogen atoms do not intervene in the chain. Rather, two or three hydrogens are attached to each carbon atom, depending on its position in the molecule.

The same number and kinds of atoms can be arranged in different ways, helping to explain why there are so many different organic compounds. **Isomers** are molecules with the same chemical formula (same number and kinds of atoms), but with different structures and properties. You already encountered isomers in the discussion of octane, C_8H_{18}, in Chapter 4.

Here we illustrate isomers with C_4H_{10}. One way to arrange these atoms is in a chain to form *n*-butane. Another arrangement is possible in which the four carbon atoms

Although *n*-butane could be drawn as $H_3C - CH_2 - CH_2 - CH_3$, for ease we often write $CH_3 - CH_2 - CH_2 - CH_3$.

Isomers are also described in Section 4.9.

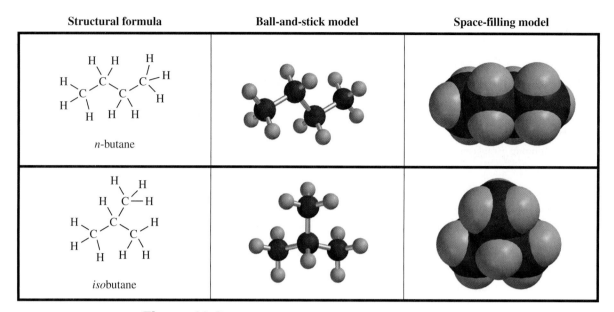

Structural formula	Ball-and-stick model	Space-filling model

n-butane

*iso*butane

Figure 10.3
Various depictions of the isomers *n*-butane and *iso*butane.

are not all in a line. This other isomer is known as *iso*butane. The linear *n*-butane is shown for comparison, now represented in a more realistic zigzag form.

n-butane   *iso*butane

The chemical formulas of these two isomers are the same; the way the atoms are connected is different. Note that the central carbon in *iso*butane has three carbons connected to it, and all of the other carbons have one other carbon (and three hydrogens) connected to them. Rotating this representation doesn't change how the atoms are connected.

Just like its linear isomer, *iso*butane can be written using a condensed structural formula.

$$CH_3-CH-CH_3 \quad \text{or} \quad CH_3CH(CH_3)CH_3 \quad \text{or} \quad CH_3CH(CH_3)_2$$

with CH_3 above the central CH.

Here, the parentheses around the CH_3 groups indicate that they are attached to the carbon to their left. Note that the CH_3 attached to the central CH carbon atom introduces a "branch" into the molecule.

Figure 10.3 shows three depictions of *n*-butane and *iso*butane. The first column shows the simple structural formula and the second a ball-and-stick model. In the third column we see space-filling models that present a more realistic view of the molecular shape.

Only two isomers of C_4H_{10} exist. As the number of atoms in a hydrocarbon increases, so does the number of possible isomers. Thus, C_8H_{18} has 18 isomers and

Your Turn 10.5　　　Isomers of C_5H_{12}

The formula C_5H_{12} represents three isomers. Draw a structural formula and a condensed structural formula for each.

Table 10.2 Molecular Representations

Compound	Chemical Formula	Structural Formula	Line-Angle Drawing
n-butane	C_4H_{10}	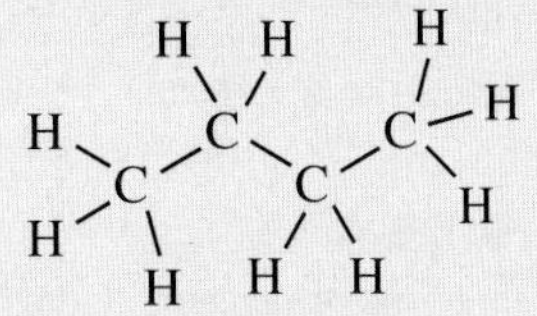	
*iso*butane	C_4H_{10}	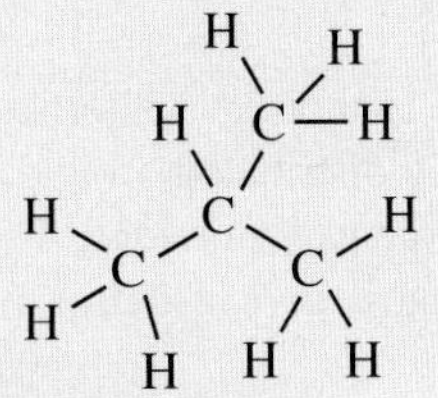	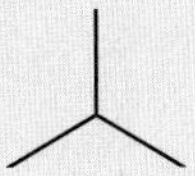
n-hexane	C_6H_{14}	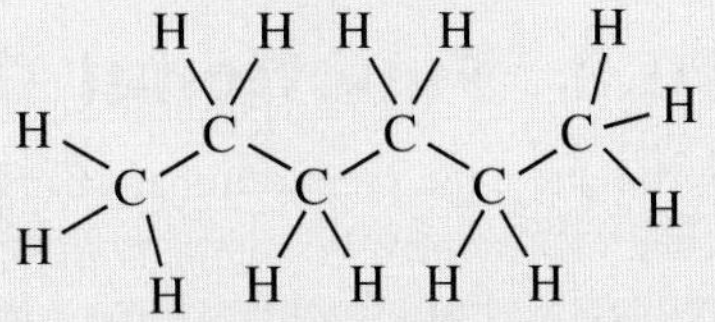	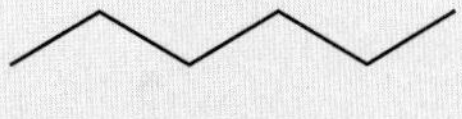
cyclohexane	C_6H_{12}	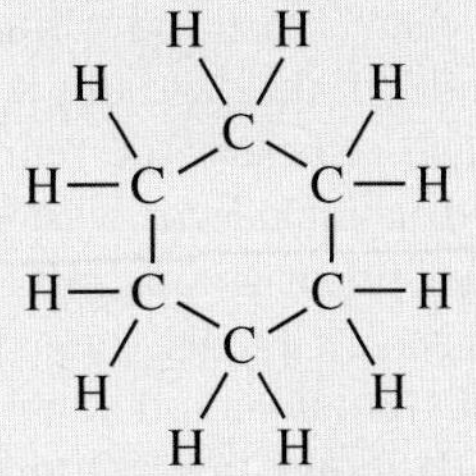	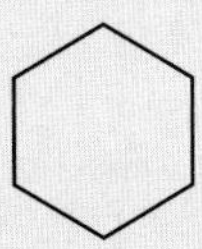

$C_{10}H_{22}$ has 75. Given a chemical formula, no simple calculation can be performed to obtain the number of isomers.

Chemists routinely use another form of structural formula called a **line-angle drawing.** This is a simplified version of a structural formula that is most useful for representing larger molecules. A line-angle drawing focuses on the backbone of carbon atoms. One carbon atom is assumed to occupy each vertex position. Any line extending from the backbone signifies another carbon atom (actually a —CH_3 group), unless the symbol for another element is given. Hydrogen atoms are not indicated in the line-angle drawing. Line-angle drawings for *n*-butane, *iso*butane, and two other simple molecules are shown in Table 10.2.

Your Turn 10.6   Structural Isomers

a. Are *n*-butane and *iso*butane isomers? Why or why not?
b. Are *n*-hexane and cyclohexane isomers? Why or why not?

Many molecules, including aspirin, have carbon atoms arranged in a ring. For example, examine the structure of cyclohexane, C_6H_{12}, in Table 10.2. Its ring has six carbons, and rings most commonly contain five or six carbon atoms. In aspirin, however, the six-membered ring is based on benzene, C_6H_6, rather than on cyclohexane. The structural formula for benzene is shown in Figure 10.4(a).

Structure (b) in Figure 10.4 is a line-angle drawing for benzene. Note the alternating single and double bonds between the adjacent carbon atoms. But the actual structure of benzene, based on experimental evidence, is flat with all carbon-to-carbon bonds having the same length. Given that C-to-C single bonds are longer than C-to-C double bonds,

A carbon-to-carbon single bond length is 0.154 nm, a double bond length is 0.134 nm, and the bond lengths in benzene are 0.139 nm.

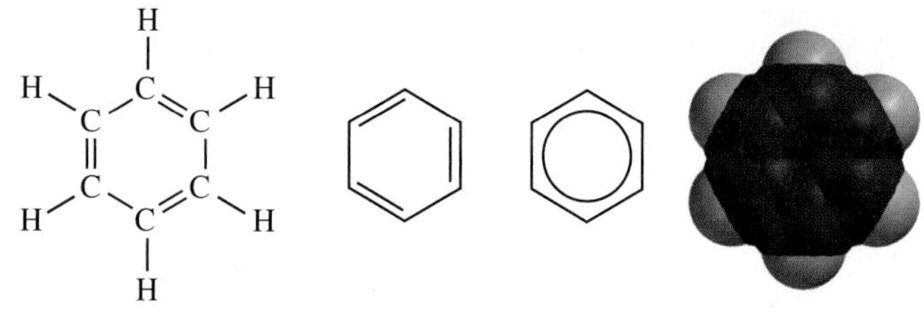

(a)   **(b)**   **(c)**   **(d)**

Figure 10.4
Representations of benzene, C_6H_6.

The uniform distribution of electrons in the benzene ring, shown in (c) and (d), helps us to visualize what is meant by resonance (Section 2.3).

benzene cannot have alternating single and double bonds. Rather, the electrons must be uniformly distributed around the ring. The circle within the hexagon in structure (c) is an effort to convey this idea. This same hexagonal structure is found in the $-C_6H_5$ phenyl group that is part of many molecules, including styrene and polystyrene (Chapter 9).

10.3 Functional Groups

Fortunately, the great number of organic compounds is somewhat simplified by the existence of a relatively small number of functional groups that appear with considerable frequency. **Functional groups** are distinctive arrangements of groups of atoms that impart characteristic physical and chemical properties to the molecules that contain them. Indeed, these groups are so important that we often show them in structural formulas and represent the remainder of the molecule with an "R". The R is generally assumed to include at least one carbon or hydrogen atom. You already encountered some functional groups in Chapter 9. The generic formula for an alcohol is ROH, as in methanol, CH_3OH, (an alcohol derived from degradation of wood) and ethanol, CH_3CH_2OH (alcohol derived from fermentation of grains and sugar). The presence of the $-OH$ group attached to a carbon makes the compound an alcohol.

Section 9.5 discusses alcohols in polymerization reactions.

An alcohol has an $-OH$ group *covalently* bound to the rest of the molecule. This is different from the hydroxide ion, OH^-, which is *ionically* bonded to a cation.

Similarly, acidic properties are conferred by a carboxylic acid group, commonly

written as [structure] , $-COOH$ or $-CO_2H$. In aqueous solution, an H^+ ion (a proton) is transferred from the $-COOH$ group to an H_2O molecule to form a hydronium ion, H_3O^+. We represent an organic acid with the general formula RCOOH, or RCO_2H. In acetic acid (CH_3COOH), the acid in vinegar, the R group is $-CH_3$, a methyl group.

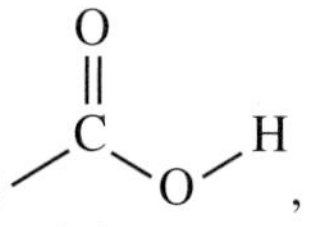

Table 10.3 lists eight functional groups found in drugs and other organic compounds. Each functional group is characteristic of an important class of compounds. Table 10.3 includes the name, structural formula, and condensed structural formula of an example of each.

Your Turn 10.7   Line-Angle Drawings

For each of these condensed structural formulas, make a line-angle drawing. Name the functional group found in each one.

a. $CH_3CH_2CH_2COCH_3$
b. $CH_3CH_2CH(CH_3)CH_2OH$
c. $CH_3CH(NH_2)CH_2CH_3$
d. $CH_3COOCH_2CH_3$
e. CH_3CH_2CHO

Answers

a. 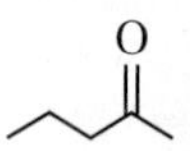 (a ketone)   b. 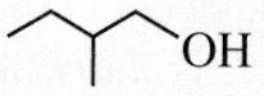 (an alcohol)

Table 10.3		Some Important Organic Functional Groups		
			Specific Examples	
Functional Group	**Generic Formula**	**Name**	**Structural Formula**	**Condensed Structural Formula**
alcohol	(generic structure)	ethanol (ethyl alcohol)	(structure)	CH_3CH_2OH
ether	(generic structure)	dimethyl ether	(structure)	$CH_3—O—CH_3$ or CH_3OCH_3
aldehyde	(generic structure)	propanal	(structure)	$CH_3CH_2—\overset{O}{\overset{\|}{C}}—H$ or CH_3CH_2CHO
ketone	(generic structure)	2-propanone (dimethyl ketone, acetone)	(structure)	$CH_3—\overset{O}{\overset{\|}{C}}—CH_3$ or CH_3COCH_3
carboxylic acid	(generic structure)	ethanoic acid (acetic acid)	(structure)	$CH_3—\overset{O}{\overset{\|}{C}}—OH$ or CH_3CO_2H or CH_3COOH
ester	(generic structure)	methyl ethanoate (methyl acetate)	(structure)	$CH_3—\overset{O}{\overset{\|}{C}}—OCH_3$ or CH_3COOCH_3
amine	(generic structure)	ethylamine	(structure)	$CH_3CH_2NH_2$
amide	(generic structure)	propanamide	(structure)	$CH_3CH_2—\overset{O}{\overset{\|}{C}}—NH_2$ or $CH_3CH_2CONH_2$

The presence and properties of functional groups are responsible for the action of all drugs. Aspirin has three functional groups, shown in Figure 10.5. You will recognize that the green area encloses a benzene ring. Its presence makes aspirin soluble in fatty compounds that are important cell membrane components. The other two

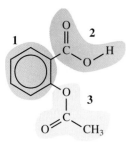

Figure 10.5

Structural formula of aspirin.

functional groups are responsible for the drug activity. You have just been reminded that the —COOH group indicates an organic acid (blue area). The remaining functional group (yellow area) is an ester. An ester may be formed by reacting an acid and an alcohol; a water molecule is eliminated in the process.

Felix Hoffmann prepared aspirin by modifying the structure of salicylic acid. But note that he did not modify the carboxylic acid group on the molecule. Salicylic acid also contains an —OH group that Hoffmann reacted with acetic acid via equation 10.1. The product was an ester of acetic acid and salicylic acid, which accounts for one of aspirin's names: acetylsalicylic acid.

$$[10.1]$$

Formation of an ester is shown for condensation polymers in Section 9.5.

Because aspirin retains the —COOH group of the original salicylic acid, it still has some of the undesirable acidic properties of the parent compound. However, the ester group (yellow area in Figure 10.5) reduces the strength of the acid group and makes the compound more palatable and less irritating to the stomach lining. Once aspirin is ingested and reaches the site of its action, reaction 10.1 is reversed. The ester splits into acetic acid and salicylic acid, and the latter compound exerts its antipyretic (fever-reducing) and analgesic (pain-reducing) properties.

Your Turn 10.8   Ester Formation

Draw structural formulas for the esters that form when these acid and alcohol pairs react.

a. CH_3CH_2OH +

b. + CH_3OH $\xrightarrow{H^+}$

The concept of *like dissolves like* is introduced in Section 5.9.

Functional groups can play a role in the solubility of a compound, an important consideration in the uptake, rate of reaction, and residence time of drugs in the body. The general solubility rule, "like dissolves like," applies in the body as well as in the test tube. When that rule was introduced in Chapter 5, a distinction was made between polar and nonpolar molecules. A polar molecule has a nonsymmetrical distribution of electric charge. This means that a partial negative charge builds up on some part (or parts) of the molecule, while other regions of the molecule bear a partial positive charge. Water is an excellent example of a polar molecule. Relatively speaking, the oxygen atom is slightly negatively charged and the hydrogen atoms are slightly positive. Because the molecule is bent, it has a nonsymmetrical charge distribution. This is represented in Figure 10.6; the δ^+ and δ^- symbols represent partial charges. Functional groups containing oxygen and nitrogen atoms (for example, —OH, —COOH, and —NH$_2$) usually increase the polarity of a molecule. This in turn enhances its solubility in a polar substance such as water, which is advantageous for drug molecules.

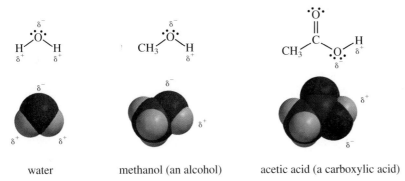

Figure 10.6

Examples of polar molecules.

By contrast, compounds whose molecules do not contain such functional groups, but consist primarily or exclusively of carbon and hydrogen atoms, are typically nonpolar. A hydrocarbon such as octane, C_8H_{18}, is a good example; this compound is insoluble in water. However, it does dissolve in nonpolar solvents that are structurally similar to it. For the same reason, drugs with significant nonpolar character tend to accumulate in cell membranes and fatty tissues that are largely hydrocarbon and nonpolar.

Another method for improving the water solubility of a drug is to convert it to its salt form. For instance, many drugs contain nitrogen and are considered to be bases, that is, they will accept a proton (H^+) from an acid. When a molecule containing nitrogen possessing a lone pair of electrons is treated with an acid like HCl or H_2SO_4, the nitrogen becomes protonated (it accepts an H^+ from the acid). As a result, the nitrogen becomes positively charged, and this charge is balanced by the chloride or sulfate ion, which is negatively charged. Prior to being protonated, the compound is electronically neutral and is said to be in its freebase form. A freebase is a nitrogen-containing molecule in which the nitrogen is in possession of its lone pair of electrons.

Consider the drug pseudoephedrine, a common decongestant used in over-the-counter remedies for the common cold:

pseudoephedrine (freebase)   pseudoephedrine hydrochloride salt

The nitrogen of the amino group in pseudoephedrine is protonated when treated with hydrochloric acid. The resulting positive charge on the nitrogen is balanced by the chloride ion, and pseudoephedrine is thus converted to its hydrochloride salt (an ionic compound). The salt form of pseudoephedrine is preferable as a drug because it is more stable, has less of an odor, and is water-soluble. An estimated half of all drug molecules used in medicine are administered as salts that improve their water-solubility and stability, which in turn increase their shelf life. Conversion of a salt back into its freebase form may be accomplished by treating the salt with a base like NaOH.

Drugs with similar physiological properties often have similar molecular structures and include some of the same functional groups. Of the approximately 40 alternatives to aspirin that have been produced, ibuprofen and acetaminophen are the most familiar. Figure 10.7 gives the structural formulas of the three leading analgesics. All are based on a benzene ring with two substituents, an atom or functional group that has been substituted for a hydrogen atom, but these substituents differ. In Your Turn 10.9, you have an opportunity to identify the structural similarities and differences of these analgesics.

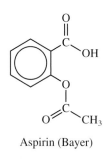
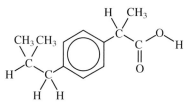
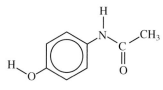

Aspirin (Bayer)   Ibuprofen (Advil)   Acetaminophen (Tylenol)

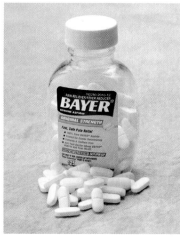
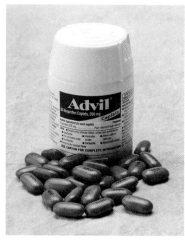
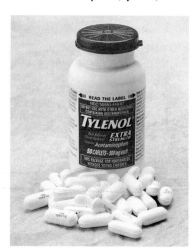

Figure 10.7
Structural formulas and samples of some common analgesics.

Your Turn 10.9   **Common Structural Features of Analgesics**

Look at the structural formulas in Figure 10.7. Identify the structural features and functional groups that aspirin, ibuprofen, and acetaminophen have in common.

The current commercial method for producing ibuprofen is a stunning application of green chemistry. Previous methods of ibuprofen production required six steps, used large amounts of solvents, and generated significant quantities of waste. By using a catalyst that also serves as a solvent, BHC Company, a 1997 Presidential Green Chemistry Challenge Award winner, makes ibuprofen in just three steps with a minimum of solvents and waste. In the BHC process, virtually all the reactants are converted to ibuprofen or another usable by-product; any unreacted starting materials are recovered and recycled. Nearly 8 million pounds of ibuprofen, enough to make 18 billion 200-mg pills, is produced annually in Bishop, Texas, at the BHC facility, built specifically for the commercial production of the drug.

10.4 How Aspirin Works: Relating Molecular Structure to Activity

To understand the action of aspirin, it is necessary to know something about the body's chemical communication system. We normally think of internal communication as consisting of electrical impulses traveling along a network of nerves. This is certainly true for the system that triggers movement, breathing, heartbeats, and reflex actions. Most of the body's messages, however, are conveyed not by electrical impulses, but through chemical processes. In fact, your very first communication with your mother was a chemical signal saying "I'm here; better get your body ready for me." It is much more efficient to release chemical messengers into the bloodstream where they can be circulated to appropriate body cells, than to "hardwire" each individual cell with nerve endings.

These chemical messengers are called **hormones,** and they are produced by the body's endocrine glands. Figure 10.8 is a representation of such chemical communication. Hormones encompass a wide range of functions and a similarly wide range of chemical composition and structure. Thyroxine, a hormone secreted by the thyroid gland, is essential for regulating metabolism. The ability of the body to use glucose (blood sugar) for energy depends on insulin. This hormone, a small protein of only 51 polymerized amino acids, is secreted by the pancreas. Persons who suffer from diabetes are often required to take daily injections of insulin. Yet another well-known hormone is adrenaline (epinephrine), a small molecule that prepares the body to "fight or flee" in the face of danger. And the hormonal messages that are so compelling in adolescents are carried by steroids, a sexy set of molecules that we will visit in a few pages.

Aspirin and other drugs that are physiologically active, but not antiinfectious agents, are almost always involved in altering the chemical communication system of the body. A significant problem is that this system is very complex, allowing many compounds to be used to send more than one message simultaneously. The wide range of aspirin's therapeutic properties, as well as its side effects, are clear evidence that the drug is involved in several chemical communication systems. It works in the brain to reduce fever, it relieves inflammation in muscles and joints, and it apparently decreases the chances of stroke and heart attack. It may even lessen the likelihood of colon, stomach, and rectal cancer.

In large measure, the versatility of aspirin and similar nonsteroidal antiinflammatory drugs (NSAIDs) is related to their remarkable ability to block the actions of other molecules. Research on the activity of aspirin indicates that one of its modes of action involves blocking cyclooxygenase (COX) enzymes. **Enzymes** are proteins that act as biochemical catalysts, influencing the rates of chemical reactions. Most enzymes speed up reactions and channel them so that only one product (or a set of related products) is formed. In the case of cyclooxygenases, the reaction is the synthesis of a series of hormone-like compounds called prostaglandins. Prostaglandins cause a variety of effects. They produce fever and swelling, increase sensitivity of pain receptors, inhibit blood vessel dilation, regulate the production of acid and mucus in the stomach, and assist kidney functions. By preventing prostaglandin production, aspirin reduces fever and swelling. It also suppresses pain receptors and so functions as a painkiller. Because the benzene ring conveys high fat solubility, aspirin is also taken up into cell membranes. In certain specialized cells, the drug blocks the transmission of chemical signals that trigger inflammation. This process also appears to be related to aspirin's effectiveness as a pain reliever.

The NSAIDs exhibit these same properties in varying degrees. For example, because acetaminophen blocks cyclooxygenase (COX) enzymes, but does not affect the specialized cells, it reduces fever but has little antiinflammatory action. On the other hand, ibuprofen is a better enzyme blocker and specialized cell inhibitor. Consequently, ibuprofen is both a better pain reliever and fever reducer than aspirin. Ibuprofen has fewer functional groups than aspirin, which may be the reason why ibuprofen has fewer side effects. With fewer polar functional groups, ibuprofen is more lipid-soluble than aspirin. Its antiinflammatory activity is five to 50 times that of aspirin.

All these structurally related antiinflammatory drugs appear to affect the way cell membranes respond to stimuli. Research has shown that this is yet another possible mode of action for aspirin and its chemical relatives. On the other hand, aspirin is unique among these three compounds in its ability to inhibit blood clotting. This property has led to the suggestion that low regular doses of aspirin can help prevent strokes or heart attacks. Of course, these anticoagulation characteristics also mean that aspirin is not the painkiller of choice for surgical patients or those suffering from ulcers. That is why "more hospitals use Tylenol." You already read that some people experience stomach irritation when they take aspirin. Another drawback of the drug is that in rare cases in the presence of certain viruses, it can trigger a sometimes fatal response known as Reye's syndrome, particularly in children under the age of 15. Furthermore, in some patients, aspirin can trigger acute episodes of asthma.

Section 12.6 describes the use of genetic engineering to obtain human insulin from bacteria.

You encountered catalysts in several other contexts, including automobile emissions control (Section 1.11), petroleum refining (Section 4.8), and addition polymerization (Section 9.3).

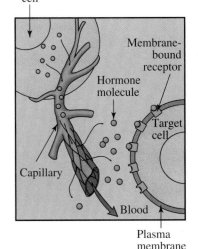

Figure 10.8

Chemical communication in the body. Hormone molecules travel from the cell where they are made, through the bloodstream, to the target cell.

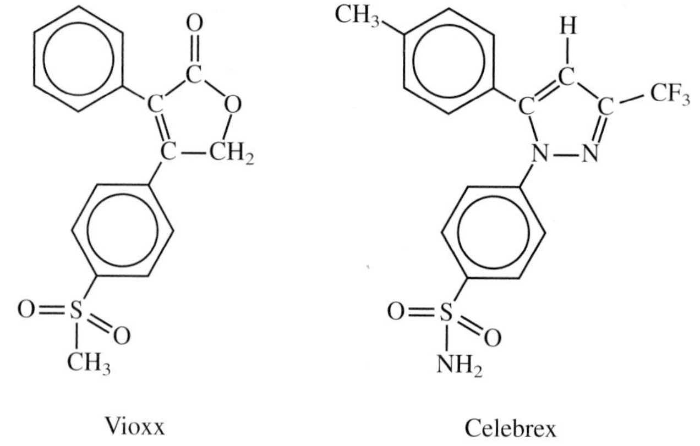

Vioxx   Celebrex

Figure 10.9

Two COX-2 inhibitors.

Recently, scientists have been able to better understand how aspirin, ibuprofen, and acetaminophen affect the two COX enzymes. These drugs block one of the enzymes, COX-2, that makes prostaglandins associated with inflammation, pain, and fever, thereby reducing these symptoms. But the drugs also inhibit the other enzyme in the pair, COX-1, that primarily makes hormones that maintain proper kidney function and keep the stomach lining intact. Thus, the drugs are not sufficiently selective to affect COX-2 without shutting down COX-1 as well.

By determining the crystal structure of COX-2 in 1992, researchers were then guided in making nearly a dozen new candidate drugs that block COX-2 alone. This work resulted in the emergence of a new class of medicines in the late 1990s called COX-2 inhibitors. Two wildly popular and heavily prescribed COX-2 inhibitors were Vioxx and Celebrex, shown in Figure 10.9. These new "superaspirins" were touted as being safer and more effective than currently available NSAIDS. The premise is that since they do not act on the COX-1 enzymes, there should be fewer gastrointestinal (GI) side effects. The reality is that they have not been proven to be more effective or safer than aspirin or ibuprofen. With regard to GI side effects, in February 2001, the FDA's Arthritis Advisory Committee concluded that Celebrex "did not demonstrate statistical superiority to NSAIDs." In September 2004 Merck & Co., the maker of Vioxx, withdrew the drug from the market because of confirmed increased risk for heart attack and stroke. Despite the uncertainty in their safety and effectiveness, both of these drugs have been huge sellers, with 2003 sales figures of $2.5 billion for Vioxx and $2.9 billion for Celebrex.

If there is uncertainty regarding the effectiveness or safety of these drugs, then why are they so successful at the cash register? It may be in part due to an emerging marketing tool used by pharmaceutical companies: direct-to-consumer advertising. Flip through most any popular magazine or watch an hour of prime-time television and you will find a number of advertisements for prescription drugs. Combine this relatively new form of drug marketing with the pharmaceutical industry's very powerful system of marketing drugs to physicians and medicines of uncertain effectiveness can become today's "blockbuster" drugs.

Consider This 10.10   **Patient Knows Best?**

Direct-to-consumer advertising of a prescription drug has proved to be a successful marketing tool for pharmaceutical companies. Twenty percent of consumers say that advertisements prompted them to call or visit their doctor to discuss the drug, according to PharmTrends, a patient-level syndicated tracking study of consumer behavior by market research organization Ipsos-NPD. Make a list of the pros and cons of this type of marketing from both the patient and from the physician point of view.

A few final comments about NSAIDS seem appropriate. Because it is a specific chemical compound, aspirin is aspirin—acetylsalicylic acid, regardless of its manufacturer. Indeed, about 70% of all acetylsalicylic acid produced in the United States is made by a single manufacturer. But, although all aspirin molecules are identical, not all aspirin tablets are the same. The commercial products are mixtures of various components, including inert fillers and bonding agents that hold the tablet together. Buffered aspirin tablets also include weak bases that counteract the natural acidity of the aspirin. Some coated aspirins keep the tablet intact until it leaves the stomach and enters the intestine. These differences in formulation can influence the rate of uptake of the drug and hence, how fast it acts, and the extent of stomach irritation. Furthermore, although standards for quality control are high, it is conceivable that individual lots of aspirin may vary slightly in purity. Aspirin also decomposes with time, and the smell of vinegar can signify that such a process has begun. Fortunately, none of this poses a significant threat to health, and the benefits of aspirin far outweigh the risks for the great majority of people.

Consider This 10.11 Aspirin in a Large Bottle

A friend who suffers from heart disease has been told by the doctor to take one aspirin tablet a day. To save money, your friend often buys the large 300-tablet bottle of aspirin. You, on the other hand, rarely take aspirin, but cannot pass up a good bargain. You also buy the large bottle.

a. Why is the "giant economy size" bottle of aspirin not as good a deal for you as it is for your friend?

b. What chemical evidence supports your opinion?

10.5 Modern Drug Design

The evolution of "willow bark tea" to aspirin and further modifications to this painkiller's structure to enhance its beneficial effects and decrease its side effects represent stages in historical drug design. Penicillin is another example of a miracle drug whose origin lies in "natural" sources. Its story includes an accidental discovery by the British bacteriologist Alexander Fleming in 1928. Fleming's curiosity was aroused by the chance observation that in a container of bacterial colonies, the area contaminated by the mold *Penicillium notatum* was free of bacteria (Figure 10.10). The reduced growth of bacteria

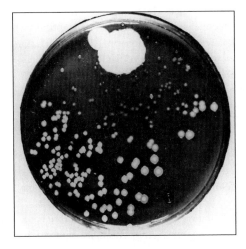

Figure 10.10
Photograph of the original culture plate of the fungus *penicillium notatum*. This image was photographed by Sir Fleming for his 1929 paper on penicillin. The large white area at 12 o'clock is the mold *penicillium notatum;* the smaller white spots are areas of bacterial growth.

in the vicinity of the mold was evidence that a compound produced by the mold inhibited the growth of bacteria. He correctly concluded that the mold gave off a substance that inhibited bacterial growth, and he named this biologically active material penicillin.

A careful reconstruction has indicated that a series of critical, but fortuitous events had to occur for the discovery to be made. Spores from the mold, part of an experiment in a nearby lab, drifted into Fleming's laboratory and accidentally contaminated some Petri dishes containing *Staphylococcus* (bacteria) growing on a nutrient medium. Then came a series of chance incidents involving poor laboratory housekeeping, a vacation, and the effects of weather. Fleming fortunately noticed the dishes in which the *Staphylococcus* had been killed. Using former experience, he correctly interpreted the phenomenon, recognizing that the unknown substance being given off by the *Penicillium* was a potential antibacterial agent for the treatment of infection. "The story of penicillin," Fleming wrote, "has a certain romance in it and helps to illustrate the amount of chance, of fortune, of fate, or destiny, call it what you will, in anybody's career." But of course, the discovery would not have happened without Fleming's powers of observation and insight in response to the unexpected. The episode admirably illustrates the often misquoted maxim of the great French scientist, Louis Pasteur: "In the fields of observation, chance favors only the prepared mind." Most versions of this famous aphorism neglect the "only." It was only because Fleming's mind was prepared that he was able to capitalize on this chain of unlikely events.

Taking penicillin from the Petri dish to the pharmacy was not easy. The first step was a systematic effort to isolate the active agent produced by *Penicillium notatum*. Once identified, the substance had to be separated, purified, and concentrated by new, sophisticated techniques. Also, the efficacy of penicillin in treating humans had to be demonstrated. World War II gave increased impetus to this research and to the development of new methods for preparing large quantities of penicillin. Because the scientists were successful in doing so, thousands of lives were saved during the war, and millions since then.

The discovery of penicillin may have been serendipitous, but the next several "generations" of antibiotics in this class have involved systematic and careful research. Small changes are designed into a drug and its efficacy examined and tested. Just as in the case of NSAIDs, beneficial effects are optimized while side effects are decreased. More than 10 different penicillins are currently in clinical use: penicillin G (the original discovered by Fleming and the form that causes an allergic reaction in about 20% of the population), ampicillin, oxacillin, cloxacillin, penicillin O, and amoxicillin (the pink, bubble gum-flavored concoction you might have been given as a child). Amoxicillin is still available in capsule form too; it is commonly prescribed for being effective against a broad spectrum of bacteria and is usually well tolerated.

Bacteria develop resistance to penicillin by secreting an enzyme that destroys the penicillin structure before it can act. The newer penicillins differ in the organisms against which they are most effective and in the drug's susceptibility to the enzyme. Closely related to the penicillins are the cephalosporins (like cephalexin [Keflex]) that are particularly effective against resistant strains of bacteria. Careful research on structural modifications has led to other effective medicines like cyclosporine, a major drug to prevent tissue rejection that has revolutionized the success of organ transplant surgery.

So how do chemists know which structural features are important to a drug's function? The modern approach to chemotherapy and drug design probably began early in the 20th century with Paul Ehrlich's search for an arsenic compound that would cure syphilis without doing serious damage to the patient. His quest was for a "magic bullet," a drug that would affect only the diseased site and nothing else. He systematically varied the structure of many arsenic compounds, simultaneously testing each new compound for activity and toxicity on experimental animals. He finally achieved success with arsphenamine (Salvarsan 606), so named because it was the six hundred sixth compound investigated. Since then, medicinal chemists have adopted Ehrlich's strategy of carefully relating chemical structure and drug activity.

Drugs can be broadly classified into two groups: those that produce a physiological response in the body and those that inhibit the growth of substances that cause infections. You already learned that aspirin falls in the first group. So do synthetic hormones and psychologically active drugs. These compounds typically initiate or block a chemical action that generates a cellular response, such as a nerve impulse or the synthesis of a protein. Antibiotics exemplify drugs that prevent the reproduction of foreign invaders. They do so by inhibiting an essential chemical process in the infecting organism. Thus, they are particularly effective against bacteria.

Although drugs vary in their versatility, many of them act only against particular diseases or infections. This specificity is consistent with the relationship that exists between the chemical structure of a drug and its therapeutic properties. Both the general shape of the molecule and the nature and location of its functional groups are important factors in determining its physiological efficacy. This correlation between form and function can be explained in terms of the interaction between biologically important molecules. Although many of these molecules are very large, consisting of hundreds of atoms, each molecule often contains a relatively small active site or receptor site that is of crucial importance in the biochemical function of the molecule. A drug is often designed to either initiate or inhibit this function by interacting with the receptor site.

An example is provided by a receptor site that controls whether a cell membrane is permeable to certain chemicals. In effect, such a site acts as a lock on a cellular door. The key to this lock may be a hormone or drug molecule. The drug or hormone bonds to the receptor site, opening or closing a channel through the cell membrane. Whether the channel is open or closed can significantly influence the chemistry that occurs in the cell. In fact, under some circumstances, the cell may be killed, which may or may not be beneficial to the organism.

This lock-and-key analogy is often used to describe the interaction of drugs and receptor sites. Just as specific keys fit only specific locks, a molecular match between a drug and its receptor site is required for physiological function. The process is illustrated in Figure 10.11. If a perfect lock-and-key match were required in the body, it would mean that each of the millions of physiological functions would have a unique receptor site and a specific molecular segment to fit it. Simple logic suggests that such rigid demands would not promote cellular efficiency. Consequently, the lock-and-key model, although a good starting point that works in a limited number of cases, must be modified.

Using another analogy, a receptor site is like a size-9 right footprint in the sand. Only one foot will fit it exactly, and many feet (all left feet and any right feet much larger or smaller than size 9) will not fit perfectly. But many other right feet can fit into the print reasonably well. So it is with receptor sites and the molecules or functional groups that bind to them. Some active sites can accommodate a variety of molecules including drugs. Indeed, the way most drugs function is by replacing a normal protein, hormone, or other substance in the invading organism. The general term **substrate** refers to these substances whose reactions are catalyzed by an enzyme. In this substrate inhibition model of enzyme activity, the presence of the drug molecule thus prevents the enzyme from carrying out its required chemistry. As a result, the growth of an invading bacterium is inhibited, or the synthesis of a particular molecule is turned off (Figure 10.12).

Generally speaking, the drug that best fits the receptor site has the highest therapeutic activity. In some cases, however, a drug molecule does not need to fit the receptor site particularly well. The bonding of functional groups of the drug to the receptor site may even alter the shape of the drug, the site, or both. Often what counts is for the drug to have functional groups of the proper polarity in the right places. Therefore, one important strategy in designing drugs is to determine its **pharmacophore,** the three-dimensional arrangement of atoms or groups of atoms responsible for the biological activity of a drug molecule. Medicinal chemists then synthesize a molecule having that specific active portion, but with a much simpler, nonactive remainder. These researchers custom design the molecule to meet the requirements of the receptor site. In effect, they design feet to fit footprints.

An outstanding example of this approach is provided by opiate drugs such as morphine. Morphine, a complex molecule, is difficult to synthesize. However, the pharmacophore responsible for opiate activity has been identified and is highlighted in Figure 10.13. The flat benzene ring fits into a corresponding flat area of the receptor, and the nitrogen atom binds the drug molecule to the site. Incorporating this particular portion

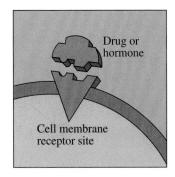

Figure 10.11

Lock-and-key model of biological interactions.

The lock-and-key analogy was first proposed in 1894 by Emil Fischer, a famous biochemist.

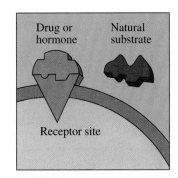

Figure 10.12

Drug molecule displacing a natural substrate on receptor site.

The term *pharmacophore* was originally described by Paul Ehrlich more than 100 years ago. Ehrlich was a German physician and biochemist who won a Nobel Prize in medicine in 1908 for his work on immunization.

into other less complex molecules, such as meperidine (Demerol), creates opiate activity. Demerol is much less addictive than morphine but also less potent.

Consider This 10.12 3-D Drugs

See for yourself how drug molecules appear in three dimensions by visiting the Three-Dimensional Drug Structure Data Bank at the National Institutes of Health (NIH). A direct link is provided at the *Online Learning Center,* or you can locate the site by searching for "Center for Molecular Modeling" or "NIH."

a. Select several drugs and examine their three-dimensional structure. How do these computer representations differ from the structural formulas of drugs shown in this chapter?

b. What advantages do the computer representations have over two-dimensional drawings? What are their limitations compared with "real" molecules? Are there any disadvantages?

The discovery that only certain functional groups are responsible for the therapeutic properties of pharmaceutical molecules was an important breakthrough. Sophisticated computer graphics are now used to model potential drugs and receptor sites. Thanks to these representations with their three-dimensional character, medicinal chemists can "see" how drugs interact with a receptor site. Computers can then be used to search for compounds that have structures similar to that of an active drug. Chemists can also modify structures in computer models and visualize how the new compounds will function.

Such techniques help to minimize the time it takes to prepare a so-called **lead compound,** a drug (or a modified version of that drug) that shows high promise for becoming an approved drug. Combinatorial chemistry is a recent development that accelerates the creation of lead compounds by using the fact that organic molecules contain functional portions, that is, "pieces" of a molecule responsible for the chemical property of the molecule, and nonfunctional portions. For example, an 8 × 12 array of small wells can be used to examine 96 combinations of structural features for their potential as drugs. Each row in one direction contains 8 functional group variations in a target molecule; each row in the perpendicular direction contains 12 different functional groups or variations in another portion of the same target molecule. In other words, in a given row, one part of the molecule remains unchanged while the second site is changed. Each of the 96 wells is examined to see whether any of these structural variants show the sought-after chemical or biomedical activity. If activity is observed in a well, those wells are analyzed for their chemical composition.

The process can be repeated several times, each time seeking to determine when reactions have formed potential drugs. Unpromising reactions can be screened out quickly. From the continuing candidates, the company can develop a so-called library

The term *lead compound,* pronounced differently, could also refer to a compound containing the element Pb.

A student at the University of Mississippi removing sample wells from a combinatorial synthesis instrument.

  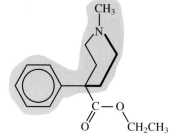

Morphine　　　　　　Active area　　　　　　Demerol

Figure 10.13

Molecular structures of morphine and Demerol. The highlighted "active areas" or pharmacophores are the portions of the molecule that interact with the receptor.

of molecular diversity for a huge array of synthesized compounds, any of which might become a lead compound in the search for a drug. Used in conjunction with computers, combinatorial chemistry can minimize the trial-and-error aspects and expense, thus speeding up drug design and development (Figure 10.14). Using traditional methods, a medicinal chemist could prepare perhaps four lead compounds per month at an estimated cost of $7000 each. With combinatorial chemical methods, the chemist can prepare nearly 3300 compounds in that same time for only about $12 each.

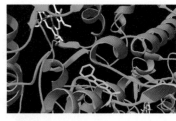

Figure 10.14
Computer modeling of the arthritis drug flurbiprofen as it binds at a protein receptor site. A search of a combinatorial library found many molecules that would be expected to bind at the same receptor site.

10.6   Left- and Right-Handed Molecules

Drug design is further complicated when drug–receptor interaction involves a common, but subtle phenomenon called optical isomerism or chirality. **Chiral,** or **optical, isomers** have the same chemical formula, but they differ in their three-dimensional molecular structure and their interaction with polarized light, hence the name. Chirality most frequently arises when four different atoms or groups of atoms are attached to a carbon atom. A compound having such a carbon atom can exist in two different molecular forms that are nonsuperimposable mirror images of each other. These are chiral isomers, also called optical isomers.

> Polarized light waves move in a single plane; nonpolarized light waves move in many planes.

   Nonsuperimposable mirror images should be familiar to you. You carry two of them around with you all the time—your hands. If you hold them palms up, you can recognize them as mirror images. For example, the thumb is on the left side of the left hand and on the right side of the right hand. Your left hand looks like the reflection of your right hand in a mirror. But your two hands are not identical. Figure 10.15

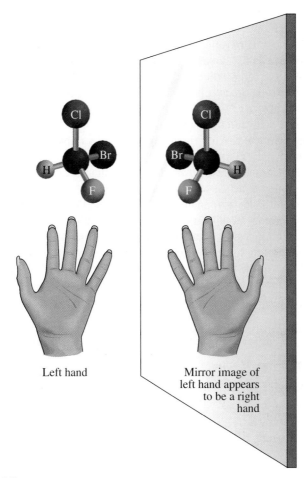

Left hand

Mirror image of left hand appears to be a right hand

Figure 10.15
Mirror images of molecular models and hands. In the molecule CHClFBr, each bond connects a different atom to carbon, forming a molecule with a tetrahedron shape.

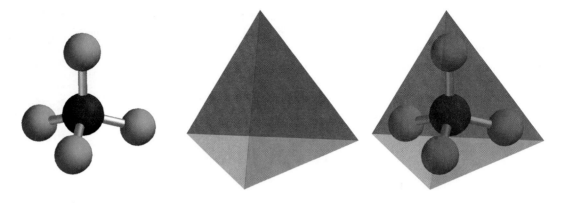

Tetrahedral molecule   Tetrahedron   Tetrahedral molecule fitted inside

Figure 10.16

A tetrahedron has four equilateral triangular faces.

illustrates this relationship for both hands and molecules. Note that the four atoms or groups of atoms bonded to the central carbon atom are in a tetrahedral arrangement (Figure 10.16). The positions of these four atoms correspond to the corners of a three-dimensional figure with equal triangular faces. The "handedness" of these molecules gives rise to the term *chirality,* from the Greek word for hand.

Sugars and amino acids are discussed in Sections 11.2 and 11.6.

Many biologically important molecules, including sugars and amino acids, exhibit chirality. This is significant because, although most chemical and physical properties of a pair of optical isomers are very nearly identical, their biological behavior can be profoundly different. Generally, the explanation for this difference is related to the necessity of a good molecular fit between a molecule and its receptor site. Maybe Lewis Carroll's Alice had some inkling of this when, in *Through the Looking Glass,* she remarked to her cat, "Perhaps looking-glass milk isn't good to drink."

You can illustrate this relationship between chirality and biological activity by taking things in your own hands. Your right hand fits only a right-handed glove, not a left-handed one. Similarly, a right-handed drug molecule fits only a receptor site that complements and accommodates it. Any drug containing a carbon atom with four different atoms or groups attached to it will exist in chiral isomers, only one of which usually fits into a particular asymmetrical receptor site (Figure 10.17).

The extreme molecular specificity created by chirality makes the medicinal chemist's job more complex. A drug molecule must include the appropriate functional groups, and these groups must be arranged in the biologically active configuration.

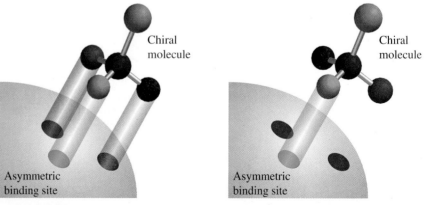

Chiral molecule

Chiral molecule

Asymmetric binding site

Asymmetric binding site

Figure 10.17

A chiral molecule binding (left) or not binding (right) to an asymmetrical site.

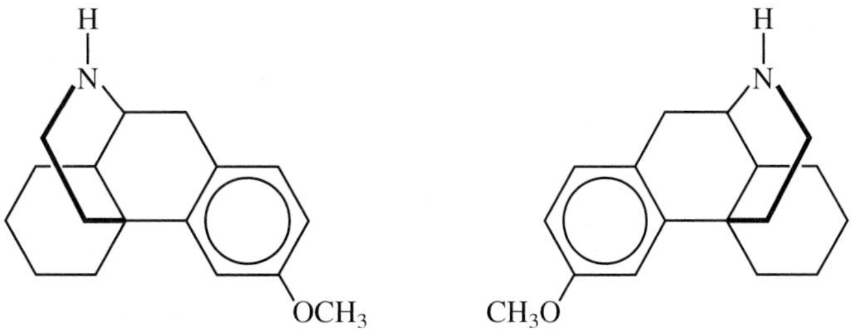

Levomethorphan   Dextromethorphan

Figure 10.18
Levo- and dextromethorphan.

The darker lines indicate that these bonds are in front of the rings.

Often the "right" and "left" optical isomers are produced simultaneously. Such a situation results in a **racemic mixture** consisting of equal amounts of each optical isomer. But frequently only one optical isomer is pharmaceutically active. For example, many opiate drugs exist as optical isomers, only one of which may have opiate activity. In Figure 10.18, levomethorphan, the left-handed (levo, or L) isomer of methorphan, is an addictive opiate. On the other hand, its right-handed (dextro, or D) mirror image is a nonaddictive cough suppressant. This permits the use of dextromethorphan in many over-the-counter cough remedies, but the right-handed (dextro) isomer must either be synthesized in pure dextro form or separated from a mixture with its levo isomer.

Many other drugs exhibit chirality and are active only in one of the isomeric forms. This is true for some antibiotics and hormones, and for certain drugs used to treat a wide range of conditions, including inflammation, cardiovascular disease, central nervous system disorders, cancer, high cholesterol levels, and attention deficit disorder. Among the widely used chiral drugs are ibuprofen, the antirejection drug cyclosporine used in organ transplants, and the lipid-reducing drug atorvastatin (Lipitor). Ibuprofen is sold as a racemic mixture of D- and L-isomers (Figure 10.19). L-Ibuprofen is a pain reliever whereas D-ibuprofen is not. However, in the body the D-form is converted to the L-isomer. Therefore, it is likely that someone taking ibuprofen is just as well off taking the racemic mixture rather than the more expensive L-ibuprofen. On the other hand, naproxen, a common pain reliever, is one example of many in which one isomer is preferred, even required. One form of naproxen relieves pain; the other causes liver damage.

Consequently, drug companies have active research programs designed to create chirally "pure" drugs, those having only the beneficial isomer of a drug in pure form. Although making the proper, single isomer might seem like an exercise of interest only to chemists, it is big business. Nine out of ten of the most successful prescription drugs sold worldwide are single isomer drugs, with total sales of $48.3 billion in 2003. These include the chiral "blockbusters", drugs with global sales over one billion dollars per year, of Lipitor, Zocor, Nexium, Plavix, and Zoloft. At least one major drug company predicts that 80% of its sales within the next five years will be from specific chiral isomer drugs. Worldwide sales of all chirally pure drugs were $161 billion in 2003 and are predicted to reach $200 billion by 2008, a very good return for knowing how to produce molecules of the correct chirality.

The vitamin E sold in stores is generally a racemic mixture of D- and L-isomers. The D-isomer is the physiologically active one that can be purchased in pure form at a significantly higher price.

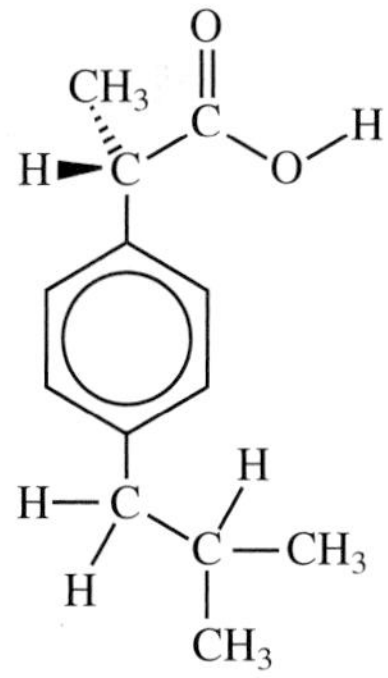

Figure 10.19
D-Ibuprofen.

William Knowles, Barry Sharpless, and Ryoji Noyori shared the 2001 Nobel Prize in chemistry for their research that developed new catalytic methods for synthesizing chiral drugs.

Your Turn 10.13   **Examining Ibuprofen**

Carefully examine the structural formula for D-ibuprofen given in Figure 10.19.

a. Which is the chiral carbon atom?
b. Identify all of the functional groups present.
c. Draw the structural formula of L-ibuprofen.

10.7 Steroids: Cholesterol, Sex Hormones, and More

Consider certain cellular components, contraceptives, muscle-mass enhancers, and abortive agents. What do they have chemically in common? The surprising answer is that they are all steroids, a family of compounds that arguably best illustrates the relationship of form and function. Certainly no other group of chemicals is more controversial than steroids, because their uses range from contraception to vanity promoters. The naturally occurring members of this ubiquitous group of substances include structural cell components, metabolic regulators, and the hormones responsible for secondary sexual characteristics and reproduction. Among the synthetic steroids are drugs for birth control, abortion, and bodybuilding.

In spite of the tremendous range of physiological functions represented in Table 10.4, all steroids are built on the same molecular skeleton. Thus, these compounds also provide a marvelous example of the economy with which living systems use and reuse certain fundamental structural units for many different purposes. The body synthesizes the many large molecules necessary for life by combining smaller molecular fragments. Once such a process is established, the same fundamental biochemical reactions are used to incorporate these molecular fragments into a variety of complex compounds. This wonderfully efficient process is rather like having a standardized house plan that can be reproduced readily—a unit that gains individuality by changes in the types of windows and doors or by the interior decorations.

The common characteristic of steroids is a molecular framework (nucleus) consisting of 17 carbon atoms arranged in four rings—"three rooms and a garage" if you like (six-membered rings are the "rooms"; the five-membered ring the "garage"). This steroid nucleus is illustrated here.

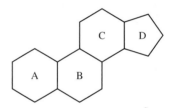

Recall that in such a representation (a line-angle drawing), carbon atoms are assumed to occupy the vertices of the rings, but they are not explicitly drawn. The 3 six-membered carbon rings of the steroid nucleus are designated A, B, and C, and the five-membered ring is designated D. Although the steroid nucleus is drawn flat, it has a three-dimensional shape. The dozens of natural and synthetic steroids are all variations on this theme. They differ only slightly in structural detail, but can differ profoundly in physiological function. Extra carbon atoms or functional groups at critical positions on the rings are responsible for this variation.

Table 10.4	**Steroid Functions**
Function	**Examples**
Regulation of secondary sexual characteristics	Estradiol (an estrogen); testosterone (an androgen)
Reproduction and control of the reproductive cycle	Progesterone and the gestagens
Regulation of metabolism	Cortisol; cortisone derivatives
Digestion of fat	Cholic acid; bile salts
Cell membrane component	Cholesterol

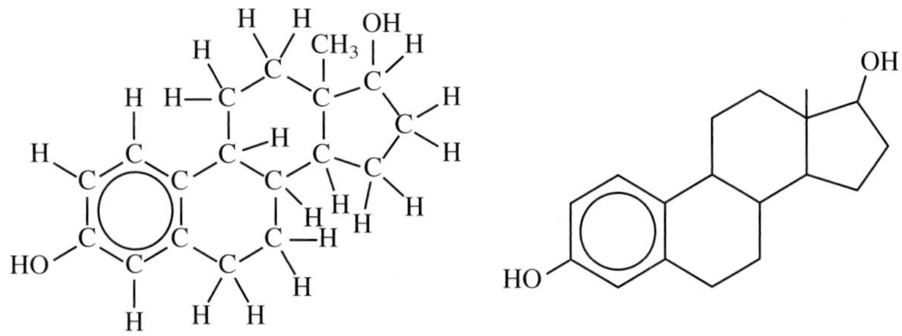

Figure 10.20

Representations of estradiol.

The steroid estradiol is a female sex hormone and is shown in Figure 10.20. The figure on the left includes all the atoms in the molecule; the one on the right gives the skeletal representation using a line-angle drawing.

This same system is used in Figure 10.21 to represent the molecular structures of six vitally important steroids. The shaded areas enclose the regions in which structural variations occur. Careful examination of the figure indicates that some very subtle molecular differences can result in profoundly altered properties. For example, the only differences between a molecule of estradiol and one of testosterone are associated with ring A. The female sex hormone has no localized double bonds in the ring and an attached —OH group; the male hormone has one double bond in the A ring, an =O in place of the —OH, and a —CH₃ group (represented by the vertical line where the A and B rings come together). It is, of course, naive to suggest that the differences between men and women are all due to a carbon atom and a few hydrogen atoms, but it is tempting.

In this chapter we concentrate on only a small number of the many steroid compounds. We begin with cholesterol, the most abundant steroid in the body and probably the best known. The average-sized adult has about half a pound of cholesterol in his or her body. Cholesterol is a starting point for the production of steroid-related hormones and a major component of cell membranes. Because their shape is relatively long, flat, and rigid, cholesterol molecules help to enhance the firmness of cell membranes. Although cholesterol is essential for human life, too much of the compound in the blood may lead to the buildup of plaque, fatty deposits in the blood vessels. This plaque restricts blood flow and can lead to a stroke or heart attack. Therefore, people are advised to regulate their dietary intake of cholesterol, which is found in milk, butter, cheese, egg yolks, and other foods rich in animal fats. But one must keep in mind that some "cholesterol-free" foods can nevertheless contribute to the build-up of cholesterol in the body, where it is synthesized from fatty acids of animal or vegetable origin. A diet rich in "saturated fats" (those with no carbon-to-carbon double bonds) is particularly likely to lead to elevated serum cholesterol.

More information about dietary cholesterol appears in Section 11.5.

Your Turn 10.14   **Structural Similarities of Steroids**

a. Carefully examine the structural formulas given in Figure 10.21. Identify the similarities in each of these pairs.
   1. estradiol and testosterone
   2. estradiol and progesterone
   3. cholic acid and cholesterol
b. Write molecular formulas for any three of the steroids in Figure 10.21.

The role of cholesterol in the body is relatively passive, but steroid hormones are involved in a tremendous range of physiologically vital processes, including such popular pastimes as digestion and reproduction. Because of the importance of these functions, medicinal chemists have, over the past 50 years, synthesized many derivatives of naturally occurring steroid hormones. These drugs, developed to mimic or inhibit the

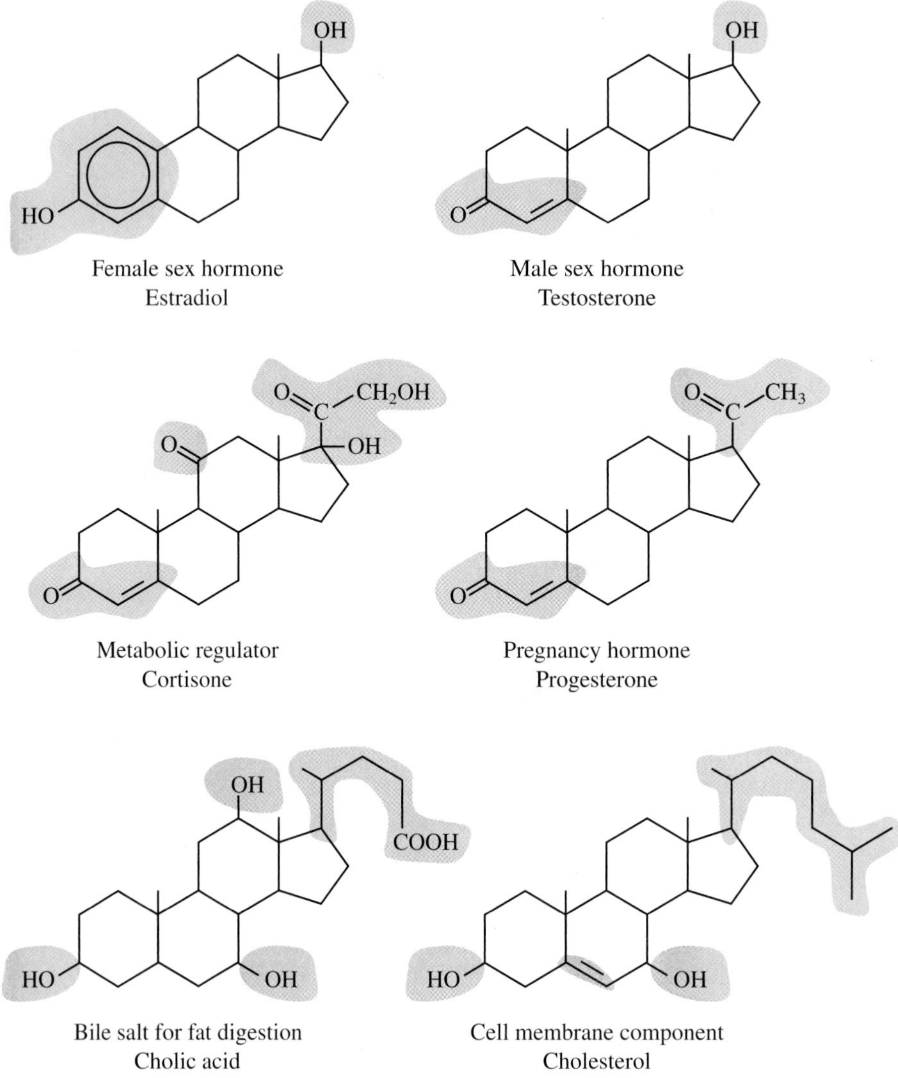

Figure 10.21

Molecular structures of some important steroids.

Figures Alive! Visit the *Online Learning Center* to learn more about these molecules and their three-dimensional shapes.

activities of the hormones in the body, have been variously described as miracle drugs, killer compounds, or sleazy therapeutic agents. Perhaps more than any other type of pharmaceutical, steroid-related drugs are involved with social and ethical issues. These issues include birth control, abortion, dieting, bodybuilding, drug abuse, and drug testing. We begin by looking at drugs related to sex hormones.

10.8 "The Pill"

Sex hormones are the chemical agents that determine the secondary sex characteristics of individuals. **Estrogens** are female sex hormones; **androgens** are male sex hormones. All males have a low concentration of female sex hormones, and all females have low levels of male sex hormones. However, androgens predominate in males and estrogens in females.

Because of their importance, androgens and estrogens were the first steroidal hormones studied in great detail. When this work was just beginning in the 1930s, techniques for determining molecular structure were in their infancy. A sample of several milligrams of the pure substance was required—much more than is needed today. Because sex hormones occur only in very small quantities, Herculean efforts were required to obtain sufficient amounts for the early chemical studies. For example, one ton of bull testicles was processed to yield just 5 mg of testosterone, and four tons of pig ovaries provided only 12 mg of estrone, a precursor of estradiol. A **precursor** is a molecule that can be converted directly to another molecule. Fortunately, improved technology and instrumentation allow modern chemists to determine molecular structures with samples weighing only a fraction of a milligram. After the molecular structures of the sex hormones were determined, work could proceed on the synthesis of drugs of similar structure. These efforts ultimately led to the creation of "the Pill," the oral contraceptive that has had such a profound effect on modern society.

As with many drugs, birth control drugs came about through molecular modifications, in this case, changing substituents selectively on the steroid nucleus. Interestingly, the initial motivation of the research that ultimately led to oral contraceptives was the enhancement of fertility in women who found it difficult to conceive. When fertilization occurs, the hormone progesterone is released, carrying a number of chemical messages. Some of these messages help prepare the uterus for the implantation of the embryo. Others block the release of pituitary hormones that stimulate ovulation. The reason for this is clear. Ovulation during pregnancy could lead to very serious complications. In 1956, Gregory Pincus and Min Chueh Chang (researchers at the Worcester Foundation in Massachusetts) and John Rock (an eminent gynecologist) injected progesterone into patients to block ovulation and stimulate body changes related to pregnancy. Their hope was that when the therapy was discontinued, a kind of rebound would occur and ovulation would be stimulated. Such a response, now known as the "Rock rebound," does take place and fertility increases.

Unfortunately, progesterone was expensive and not very effective when administered orally. It also caused some serious side effects in a small percentage of patients. Therefore, chemists working in a number of pharmaceutical firms set out to develop a synthetic analog for progesterone that could be taken orally, would reversibly suppress ovulation, and would have few side effects. The ultimate goal of these efforts soon became the inhibition of fertility, not its enhancement. In the mid-1950s, Frank Colton, a chemist at G. D. Searle, synthesized norethynodrel. The molecular structure of this compound (Figure 10.22) shows some subtle but significant differences from progesterone, notably the replacement of —COCH$_3$ on the D ring with —OH and —C≡CH. As a consequence of these changes, the norethynodrel molecule is tightly held on a receptor site, which prevents its rapid breakdown by the liver and permits its oral administration. Norethynodrel became the active ingredient in Enovid, the first commercially available oral conceptive, which was approved for sale in 1960.

The drug's availability had an immediate and substantial impact. In 1962, 1.2 million women in the United States used the "Pill," then containing 150 micrograms (μg) of estrogen and approximately 10 mg of progestin, a close chemical relative of progesterone. Since the development of the original birth control pill, further molecular modifications (many of them minor) have led to decreased dosage and minimized side effects. Currently, over 16 million users ingest one of about 50 commercial varieties of oral contraceptives with

Prepackaged contraceptive pills. The label carries a warning that while contraceptives are intended to prevent pregnancy, they do not protect against HIV infection (AIDS) and other sexually transmitted diseases.

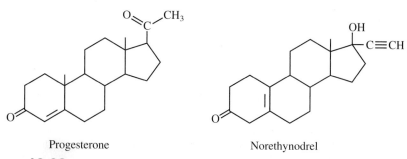

          Progesterone   Norethynodrel

Figure 10.22
Molecular structures of progesterone and norethynodrel.

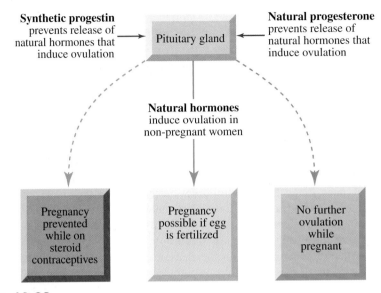

Figure 10.23
Action of a steroid contraceptive.

just 20–35 μg of estrogen and approximately 1 mg of progestin or the "minipill" that just contains the progestin. About 8 out of 10 women in the United States have used the birth control pill at some point in their lives, and 4 out of 10 American women between the ages of 18 to 29 currently take the Pill. The current low effective dosages are evidence of how just scant amounts of the hormones act as major chemical messengers with profound effects. Recent advances include an implant (Norplant) composed of six small plastic rods surgically placed under the skin of the upper arm for up to five years, during which time the rods slowly release progestin into the body. Another development is "the shot" (Depo-Provera), a highly effective progestin injection given by a physician every three months, although infertility may last up to a year thereafter.

The mechanism for the action of steroid-based contraceptives is diagrammed in the simple schematic of Figure 10.23. In effect, the drug "fools" the female reproductive system by mimicking the action of progesterone in true pregnancy. Birth control steroids, being progesterone-like molecules, send a chemical message that is similar to the message carried by progesterone. Because pregnancy is simulated, ovulation is inhibited. In effect, the message this time is not "Hey, Mom, I'm here!" but rather "Hey, you think I'm here, but I'm not!"

In the fall of 2000, a University of Edinburgh study reported that researchers found a drug, desogestrel, a synthetic hormone that suppresses daily sperm production while maintaining normal levels of testosterone in the body. Some minor side effects occur, including mood swings, weight gain, and increased appetite, some of the same side effects felt by women who take contraceptive pills. After discontinuing the dose of desogestrel, sperm concentrations in all of the men in the study returned to prestudy levels within 16 weeks. Dr. Christina Wang, professor at Harbor-UCLA Research & Education Institute is currently testing an implantable male contraceptive in which testosterone and progestin are used to turn off sperm production. Perhaps gender equity in contraception will soon be achieved!

Consider This 10.15   **Oral Contraceptives**

The advent of the first commercially available birth control pill in 1960 was coincident with a cultural and sexual revolution. Direct control of fertility was put into the hands of women, the sex most directly affected by the birth of children. With this new control came added responsibility. Suggest some of the scientific reasons why chemical control of reproduction was developed first for female rather than for male reproductive systems. Suggest some cultural reasons why it might be so.

10.9 Emergency Contraception and the "Abortion Pill"

Controversy still surrounds the synthetic steroidal hormones that control fertility by inhibiting ovulation. But a far more controversial approach to birth control is to take a drug after ovulation has occurred. Such drugs fall into two categories. The first are Emergency Contraceptive Pills (ECPs or "the morning-after pills"), actually a large dose of ordinary oral contraceptives taken after intercourse. Like other hormonal methods, these function to suppress ovulation and to make the uterus inhospitable to a fertilized ovum. ECPs were first used in the 1960s for rape victims, but the FDA has approved their emergency use when a woman has had noncontraceptive intercourse within the previous 72 hours. They only can be obtained from a physician.

The second category came into legal use in the United States in September 2000, when the Food and Drug Administration approved RU-486 (Figure 10.24) for marketing under the name Mifeprex. Since its discovery was announced in France in 1982, RU-486, otherwise known as "the French abortion pill," has been used by over 400,000 women in France, Sweden, the United Kingdom, and China. The drug RU-486 was extensively tested and found to be 96% reliable. Moreover, it appears to have a relatively low incidence of serious side effects. Mifeprex, which is actually two pills (mifepristone and misoprostol) taken under a doctor's supervision over two days, has proven to be very effective in ending pregnancies up to the fiftieth day following conception.

> China manufactures its own version of RU-486, a copy of the Rousell–Uclaf product first synthesized in France.

Your Turn 10.16   A Closer Look at RU-486

What is the molecular formula for RU-486? Compare the structure of RU-486 with that of progesterone.

Roussel–Uclaf, which holds the drug's patent, has sold RU-486 in Europe since 1988. Although abortion is legal in the United States, it is certainly not universally accepted by the American public and the medical community. Because of threats of boycotts and protests, Roussel–Uclaf did not introduce RU-486 into this country until the company had arranged to donate the rights to the drug to the Population Council, a nonprofit research organization. The Council began clinical testing of RU-486 in October 1994, administering it at a dozen clinics and hospitals in various parts of the country. In March 1996, after completion of its clinical trials, the Population Council requested approval of RU-486 from the FDA. The FDA final approval took almost four years because of the enormous pressure from both political and medical organizations.

RU-486 and progesterone have very similar molecular structures. Thus, RU-486 is able to act as an **antagonist,** a drug that fits into a receptor site but does not have the customary drug effect, for progesterone. RU-486 thus blocks the action of the female hormone progesterone that is necessary for initiating and sustaining pregnancy. By

> Extending the lock-and-key analogy, an antagonist is a key that fits into a lock but is unable to open it.

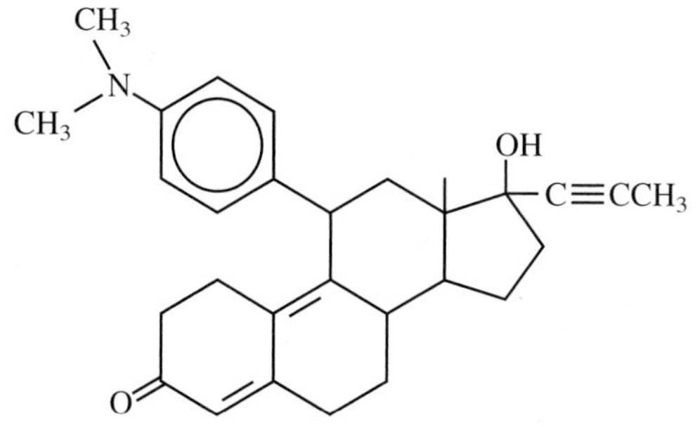

Figure 10.24
Molecular structure of RU-486.

occupying the progesterone receptor site, it changes the shape of the site somewhat with the result that no pregnancy protein production signal is sent. When there is no progesterone present, the uterine lining breaks down and bleeding occurs, resulting in the termination of pregnancy.

Regarding the safety of RU-486, Danco Laboratories (the U.S. producer of the drug) estimates that 200,000 women in the United States and more than 1 million women worldwide have used this drug since it was first invented. But in September 2003, an 18-year-old American woman who had concealed her pregnancy died of complications one week after she began taking RU-486, the abortion pill. Although the exact cause of death is unknown, it is surmised that she had died after a massive infection caused by fragments of the fetus left inside her uterus caused her to go into septic shock. This tragedy will no doubt lead to further investigations of the safety of RU-486 and will reignite the opposition of antiabortion groups to the use of this controversial drug.

An alternative method for drug-induced abortion has been developed in the United States that could make the use of RU-486 in the United States a moot point. In August 1995, gynecologists announced that two widely available prescription drugs used in combination had been shown to have results very similar to those associated with mifepristone for nonsurgical abortions no more than seven weeks after conception. The medical protocol is much like that followed with RU-486, but the initial drug administered is methotrexate. Unlike RU-486, methotrexate is not a progesterone mimic. Rather, it works as an abortifacient (induces miscarriages) by blocking a B vitamin called folic acid required for normal cell growth and division. Hence, methotrexate inhibits the development of the embryo and placenta. As with RU-486, the expulsion of the fetus is induced by misoprostol.

What makes the use of methotrexate and misoprostol for this purpose particularly interesting is the fact that each had been already approved by the FDA, although for different uses. Methotrexate has been used for some time in large doses to treat some cancers and in small doses for rheumatoid arthritis and psoriasis. Misoprostol is prescribed to protect the stomach linings of people who require daily doses of powerful antiinflammatory drugs or antiulcer medications. Once a drug has FDA approval, a licensed physician can use it for any purpose, including "off-label" uses not originally specified or intended. Thus, no exhaustive approval process was required before these drugs could be employed in combination as abortive agents.

A debate regarding the safety of off-label drugs is brewing in the medical community while the use of unapproved prescriptions is soaring. A report published by the KnightRidder/Tribune organization indicated that between 1997 and 2003 the number of prescriptions written for conditions *not* approved by the FDA increased by 96%. In addition to the steroids mentioned, other off-label drugs and the conditions for which they were prescribed include antiseizure medications for weight loss, antipsychotics for insomnia, and antidepressants for attention deficit hyperactivity disorder.

Consider This 10.17   "Off-Label" Drugs

Additional information can be found at the FDA Web site about off-label uses of drugs. Drugs such as methotrexate and misoprostol can be prescribed for off-label uses (also called unapproved, unlabeled, or extra-label uses).

a. In general, does prescribing medications to be used in this way strike you as reasonable for physicians? Explain your reasoning. Compare your opinion with that of the FDA.

b. Search for "off label." From the lengthy list of hits, pick two that interest you and summarize their contents.

c. Did what you learned from this exercise strengthen or modify your opinion of off-label uses? Explain.

10.10 Anabolic Steroids and Designer Steroids

Anabolic steroids are synthetic steroidal hormones used to stimulate muscle and bone growth. Like birth control drugs, anabolic steroids are controversial. Moreover, like some of their contraceptive chemical cousins, anabolic steroids also were created for quite a different purpose than their ultimate use. These steroids were developed initially to help patients suffering from wasting illnesses to regain muscle tissue. Ironically, their use has now become perverted by the strong who seek to become even stronger.

It long has been known that testosterone promotes muscle growth as well as the development of male secondary sexual characteristics. Drug companies sought to pursue this avenue to produce a testosterone-like drug that would stimulate muscle growth in debilitated patients, such as those recovering from long-term illness. The intent was to modify the testosterone molecule in such a way that its analog would have the desired effects on muscle development without serious negative side effects. Anabolic steroids were the result. These drugs are available legally only by prescription.

Since the 1950s, some athletes have been taking anabolic steroids to build muscle and boost their athletic performance. Studies show that, over time, anabolic steroids will have an impact on a person's health including increased risk of heart attack, strokes, and liver problems. Other undesirable body changes also may occur including breast development and genital shrinking in men, masculinization of the body in women, and acne and hair loss in both sexes.

Nonmedical use of anabolic steroids is illegal and banned by most major sports organizations. Despite this and the health risks, some athletes persist in taking them, believing that these substances provide a competitive advantage. Therefore procedures and laboratories for screening athletes for illegal drug use had to be created. The Olympic Analytical Laboratory at the University of California, Los Angeles, is one of about 30 testing labs worldwide that is accredited by the International Olympic Committee (IOC) and the only one in the United States accredited by the IOC and the World Anti-Doping Agency (WADA). The United States Anti-Doping Agency (USADA) and WADA are independent, nonprofit organizations created in 2000 that govern drug testing and provide drug education for athletes.

Scientists look for evidence of drug use by analyzing compounds extracted from urine using a technique called mass spectrometry (MS). This technique breaks up the molecules and sorts the resulting fragments by mass. Steroids contain functional groups with characteristic masses and can therefore be identified.

In 2003, an athletic coach sent the USADA a syringe containing a trace amount of a substance suspected to be a new, "undetectable" steroid used by some athletes. This was forwarded to the Olympic Analytical Laboratory where it was subjected to mass spectrometry experiments. Under typical MS experimental conditions, the suspected steroid appeared to be unstable, yielding none of the tell-tale functional groups present in currently banned steroids. Other techniques were employed and the substance was identified as the anabolic steroid tetrahydrogestrinone (THG, Figure 10.25) that was not in the data libraries of banned steroids. This substance is structurally related to two other anabolic steroids, gestrinone and trenbolone (see Figure 10.25). Although purveyors of THG may have represented it as a dietary supplement, it does not meet the dietary supplement definition. Rather, it is a purely synthetic "designer" steroid derived by simple chemical modification from another anabolic steroid that is explicitly banned by the USADA.

See Section 10.14 for a definition of dietary supplements.

The Olympic Analytical Laboratory then developed a rapid process to screen urine samples for THG based on its characteristic chemical features. Urine samples collected at the 2003 U.S. Outdoor Track & Field Championships were retested for THG. Four U.S. athletes and one U.K. athlete tested positive. The National Football League, Major League Baseball, and Major League Soccer have banned the use of THG.

In October 2003, the FDA released a statement on THG disclosing that the steroid is an unapproved new drug and that the agency was working with other federal law enforcement agencies to aggressively prosecute those who manufacture, distribute, or

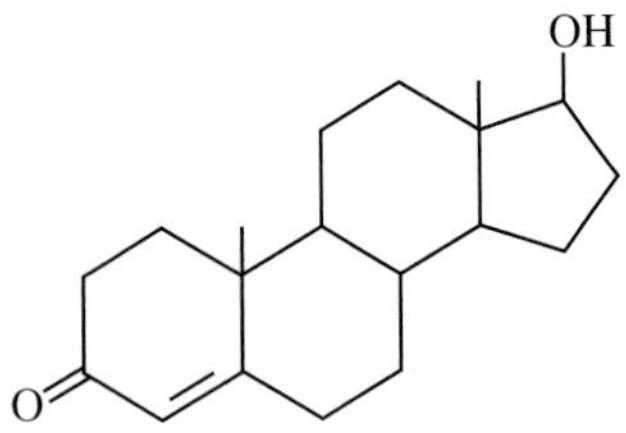

Testosterone

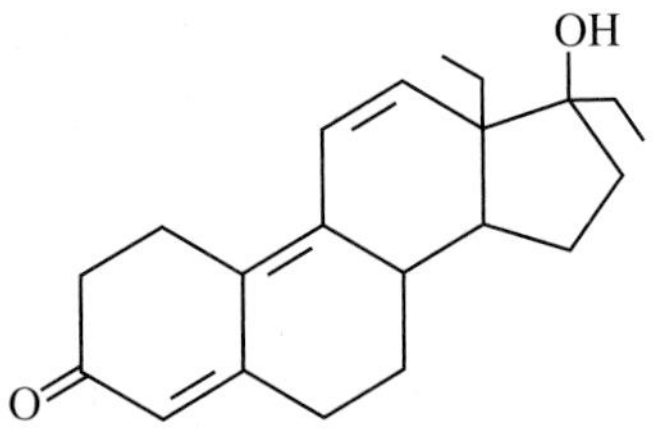

Tetrahydrogestrinone

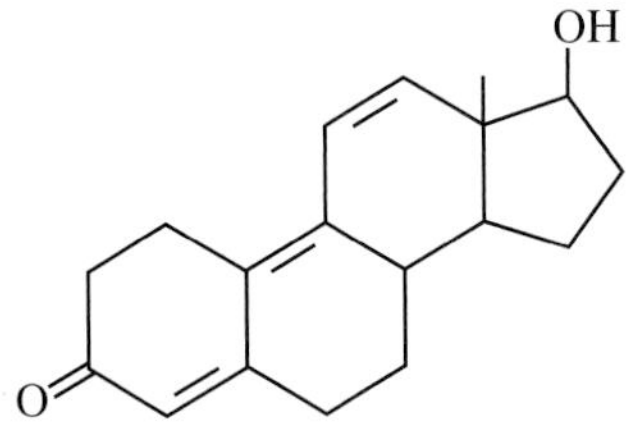

Trenbolone

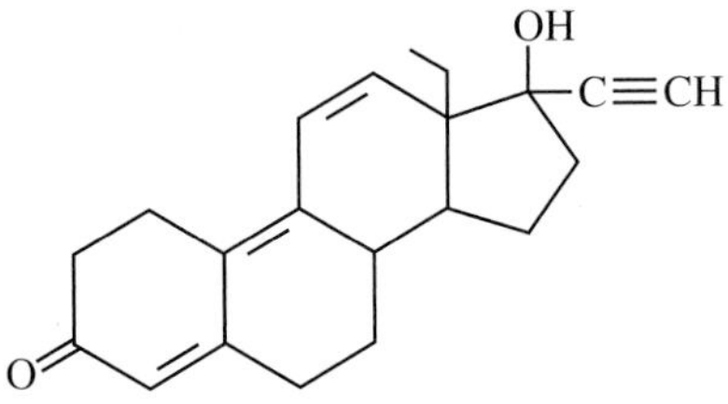

Gestrinone

Figure 10.25

Testosterone and some anabolic steroids.

market THG. Terry Madden, the Chief Executive Officer of the USADA expressed extreme concern about THG by decrying "a conspiracy involving chemists, coaches and certain athletes using what they developed to be undetectable designer steroids to defraud their fellow competitors and the American and world public who pay to attend sports events."

On a brighter note, in January of 2004, the International Olympic Committee reported that samples taken from athletes at the 2002 Salt Lake City Winter Olympics were retested for THG and all were negative. This suggests that the steroid was used on a limited scale. As this book went to press, THG had not been reported at the 2004 Athens Summer Olympics either, reinforcing this conclusion.

10.11   Drug Testing and Approval

Many drugs and medicines were once available without a prescription. Before the Food and Drug Administration regulations existed, virtually anything could be sold as a cure. In fact alcohol, cocaine, and opium were included in some early products without notification to users. The Food, Drug, and Cosmetic Act of 1938 and its amended versions in 1951 and 1962 defined prescription drugs as ones that could be habit forming, toxic, or unsafe for use except under medical supervision. Virtually everything else is available for sale over the counter.

Prescription drugs are manufactured on a colossal scale to meet patient demands. In 2002, over 3 billion prescriptions were filled in the United States, accounting for about $183 billion at the cash register. But the pathway for a new drug from a laboratory to a pharmacy shelf is long and complicated. All proposed new drugs, whether extracted from natural materials or synthesized in the laboratory, are subjected to an exacting series of tests before they obtain FDA approval. Current law requires evidence that the drugs are safe as well as effective before approval is granted.

From discovery to approval, the development of a new drug takes, on average, nearly 12 years and about $500 million—over three times the cost of a decade ago. The expenses are principally for the various stages of drug testing, probably the most complicated and thorough premarketing process ever developed for any product.

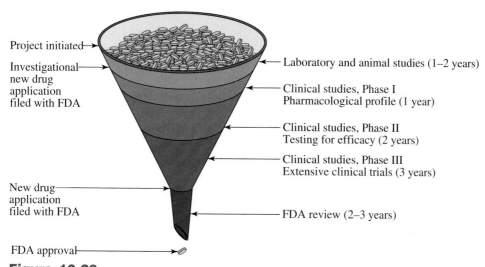

Figure 10.26

Schematic of the drug approval process in the United States.

Although the number of pills getting through the funnel of Figure 10.26 gets progressively smaller with time, the diagram does not begin to convey the high rejection rate of proposed drugs. Currently, the odds of getting a candidate drug from identification to approval are 1 in 10,000. For every 10,000 trial compounds that begin the process, 20 make it to the level of animal studies, half that many get clearance for use in clinical testing with humans, and finally one gets FDA approval.

Examples already encountered in this chapter suggest the long process of chemical hide-and-seek that often precedes the identification of a compound with therapeutic properties. Once the promising candidates have been identified, they are subject to *in vitro* studies, those carried out in laboratory flasks. Simultaneously, a wide range of activity is undertaken by the pharmaceutical company. Chemists and chemical engineers investigate whether the compound can be produced in large volume with consistent quality control. Chemists and pharmacists together carry out studies of the most effective way to formulate the drug for administration—as capsules, pills, injection, syrup, or perhaps something more unusual such as a nasal spray, skin patch, or implant. Chemical stability and shelf life are evaluated. Economists, accountants, patent attorneys, and market analysts conduct research on the likelihood of deriving a profit from the product.

Only a small fraction of compounds survive this scrutiny to move on to animal testing. Such *in vivo* tests are designed to determine the drug's efficacy, safety, dosage, and side effects. At this stage pharmacologists typically determine the drug's mode of action, how it is metabolized, and its rate of absorption and excretion. The tests are carefully controlled, requiring the collection of very specific kinds of data. For example, drugs are evaluated for their short- and long-term effects on particular organs (such as the liver or kidneys) and on more general systems (such as the nervous or reproductive system). Perhaps the most controversial toxicity testing involves the determination of the lethal dose-50 (LD_{50}), the minimum dose that kills 50% of the test animals.

In vitro means "in glass."

In vivo means "in life."

Consider This 10.18   **Animals and Drug Testing**

Animal rights groups often target the LD_{50} standard as an example of callous indifference to animal welfare. Other groups argue that standards such as LD_{50} are necessary to ensure drug safety and effectiveness. Take a position on the issue and write an article for your college newspaper.

Results of animal tests must be submitted to the FDA for evaluation before permission is granted to proceed to the next stage—clinical testing of the drug on humans. In addition, approval must be obtained from local agencies and authorities such as a hospital's ethics panel or medical board. The FDA must establish whether the drug is effective and safe before it can be sold to the public. What goes on the label regarding use, side effects, and warnings must also be determined.

Typically, clinical studies involve the three phases identified in Figure 10.26. First there is phase I, developing a pharmacological profile. Then comes phase II that tests the efficacy of the drug. Phase III is carrying out the actual clinical tests. Most of the safety tests of phase I are done with healthy volunteers, who are given single and repeat doses of the drug in various amounts. It is also at this stage that researchers look for interactions with other drugs. Double-blind placebo tests are administered to small patient groups in phase II to test the drug's effectiveness on patients having the condition that the drug is designed to affect. In this protocol, neither the patient nor the physician knows which patients are receiving the drug and which are receiving a placebo, an inactive imitation that looks like the "real thing." Such tests are designed to eliminate bias from the interpretation of the results.

Long-term toxicity studies are also initiated during phase II. The clinical trials are expanded in phase III, while manufacturing processes are scaled up and tests are carried out on the stability of the drug. The entire process often requires six years or more. In the early 1980s, an average of 30 clinical trials were done on drugs that were ultimately approved. Over the past 20 years, that number has risen to over 70, the trials have become more complex, and the number of patients treated per trial (about 4000) has more than doubled.

Large-scale clinical trials are desirable because a large pool will more likely include a wide range of subjects. Variety is important because the drug in question may have markedly different effects on the young and the old, on men and women, on pregnant or lactating women, on infants, nursing infants, and unborn infants, and on persons suffering from diabetes, poor circulation, kidney problems, high blood pressure, heart conditions, or a host of other maladies.

Once clinical trials have been completed successfully (typically by only 10 drugs out of an original pool of 10,000 compounds), the test data are submitted to the FDA as part of a new drug application. This document can easily exceed 3500 pages. On review, the agency may require the repetition of experiments or the inclusion of new ones, thus adding years to the approval process. Of the drugs submitted to clinical testing, only about 1 in 10 is finally approved.

Consider This 10.19   **Double-Blind Testing**

Double-blind protocols have other uses than testing for the effectiveness of a drug. For example, physicians may use a double-blind test for diagnosing food allergies. In such tests, the physician administers a series of foods and placebos in disguised form. The test substances are labeled in code known only to a third person.

a. Why do you think double-blind tests for the effectiveness of a drug or for establishing a food allergy are necessary?

b. Compared with the single-blind tests in which only the patient is unaware of the drug or food being administered, how do double-blind tests affect the reliability of the information gained?

Once it receives the FDA's permission, a drug can be sold in the United States. Nevertheless, it still remains under scrutiny, monitored through reports from physicians. Drugs are removed from the market if serious problems occur. Some side effects show up only when large numbers of users are involved. A new drug application, for example, typically includes safety data on several hundred to several thousand patients. An adverse event occurring in 1 in 15,000 or even 1 in 1000 users, could be missed in

Table 10.5	Recently Recalled Prescription Drugs		
Drug generic (Brand name)	**Use/Treatment**	**Year Removed**	**Reason**
fenfluramine (Pondimin)	Weight loss	1997	Heart valve abnormalities
dexfenfluramine (Redux)	Weight loss	1997	Heart valve abnormalities
mibefradil (Posicor)	Blood pressure, chest pains	1998	Dangerous reactions with 25 other medications
bromfenac (Duract)	Pain reliever	1998	Severe liver damage
terfenadine (Seldane)	Antihistamine	1998	Abnormal heart rhythms
aztemizole (Hismanal)	Antiallergies	1999	Cardiac arrest
quinine sulfate (Raxar)	Antibiotic	1999	Liver toxicity
cisapride (Propulsid)	Heartburn	2000	Cardiac arrest
methylprednisolone (Rezulin)	Adult diabetes	2000	Liver failure
alosetron (Lotronex)	Irritable bowel syndrome	2001	Severe bleeding of the colon
cerivastatin (Baycol)	Lipid-lowering	2001	Severe muscle damage
rofecoxib (Vioxx)	Arthritis	2004	Cardiac arrest, stroke

clinical trials, but it could pose a serious safety problem when the drug is used by many times that number of patients. Such an example is terfenadine (Seldane), a second-generation antihistamine for treating seasonal allergies without causing drowsiness. A small, but statistically significant number of patients who took Seldane along with certain antibiotics or antifungal medicines developed abnormal heart rhythms. In 1998, the FDA removed Seldane from the approved list.

A number of other recalls were for drugs on which the public had come to rely. Some of these are noted in Table 10.5. The spate of withdrawals has encouraged some to question the FDA's procedures.

The lengthy process for drug testing and approval is not without controversy. Most people probably favor thorough screening of any proposed drug, but the price of such protection is high. The most obvious costs are monetary. Bringing a new drug to market is incredibly expensive, and the numbers of new drug approvals are not keeping pace with rising research and development costs (Figure 10.27). Of course much of the expense is passed on to the consumer. For example, in a hospital a single dose of

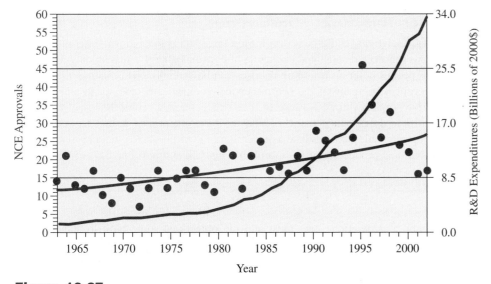

Figure 10.27

Drug development statistics: New certified approvals (NCE or new chemical entity) and development costs (R&D expenditures).

Source: Tufts CSDD Approved NCE Database, PhRMA, 2004.

tirofiban (Aggrastat), a medication that dramatically increases the likelihood of surviving a heart attack (and eliminates subsequent health costs), has a price tag of $1100. Such prices have driven the costs of medical care and medical insurance to astronomical levels. One issue in the debate over health care reform is who will pay for the research and development that ultimately leads to new medication.

Consider This 10.20   Who Should Pay?

Who should pay for Aggrastat, a life-prolonging drug: individuals, medical insurance, pharmaceutical companies, or the government? Take a stand and defend your position.

But more than money is at stake. In some cases, the costs of the protracted drug approval process may be human lives. When a patient is suffering from an almost certainly fatal disease such as some cancers, the risk–benefit equation changes. When there is nothing to lose, people are willing to take great risks, including imperfectly tested drugs. Some mortally ill patients have smuggled drugs from countries where the approval process is less stringent than in the United States. Some have grasped at the straw of largely unproven remedies. And, within the system, some advocates have urged that the FDA approval process be short-circuited to permit the use of experimental drugs on patients who have no other options.

Within the limits of its legal responsibilities to balance benefits with risks, the FDA has responded appropriately. Ten years ago, an FDA review for a new drug required nearly three years, whereas today it is less than a year. A new "fast-track" system has been instituted for priority drugs—those that address life-threatening ailments or new drug therapies for conditions that had no such therapies. The fast-track policy promises to have priority drugs, if found to be acceptable, approved within six months of application. Action on nonpriority drugs is to be taken within 10 months, down from the initial target of 12 months.

People suffering from rare diseases may not be able to purchase appropriate medication at any price because it may not exist. There is a significant financial disincentive for a pharmaceutical company to invest heavily in developing "orphan drugs" that will be used by only a small fraction of the population.

Consider This 10.21   Orphan Drugs

Antibiotics, analgesics, and other drugs have a large and profitable market around the world. However, development and marketing of essential drugs needed by only a small number of people suffering from rare diseases can be an enormous economic drain on a pharmaceutical company. If the pharmaceutical companies decide not to make and market these "orphan drugs" because of their low economic return, how will people who need these drugs obtain them? Should the government step in and require successful drug companies to contribute a percentage of their profits to a fund for research, development, and production of these orphan drugs? Take a position and outline your reasons.

To such considerations, one must add the objections some have to standard test protocols. Animal rights advocates are highly critical of the use of any animal subjects in drug screening. The sacrificing of test animals in establishing LD_{50} values is especially controversial. For others, the generally accepted methods of human testing are at issue. The argument is that because the drug may have some benefit and will probably do no significant harm, it is unethical to withhold it from a control population. Some terminally ill AIDS patients, by mixing and sharing test drugs and placebos, have refused to cooperate with double-blind clinical studies.

Consider This 10.22　The High Price of Drugs

Other issues arise from prescription drug manufacturing and sale outside of the United States. In some cases American prescription drugs are sold in foreign countries, often at prices that are much lower than those in the United States. As a result, the press has reported that senior citizens travel to Canada or Mexico to buy their prescription drugs at reduced prices. The situation with several successful drugs used in treating AIDS patients is another issue. Brazil has decided to disregard patents on the anti-AIDS drug nelfinavir (Viracept) and allow its manufacture in government laboratories. Pick one of these two issues and compile a list of arguments supporting each side. Which side of your chosen issue do you support?

Consider This 10.23　Safety and Standards on the Fast Track

The credibility of the FDA's fast-track drug testing program was called into doubt when 11 drugs released under the program were withdrawn within a period of 5 years (1997–2001). For example, some would argue that the recall of the pain killer Duract or heartburn medicine Propulsid show that adequate testing had not taken place on these drugs. Others could counter that standards of testing have not changed, only the speed with which data are evaluated, and that no tests can produce data that are 100% reliable in humans. Search the FDA and other Web sites to gather information about Duract or Propulsid. Use what you learn to reach a decision as to whether you think fast-track drug testing is overall a benefit or a risk to the health of Americans.

10.12　Brand Name or Generic Prescriptions

Enter customer. The pharmacist may ask "Brand name or generic?" This scenario is played out daily in thousands of pharmacies across the country. How is the person to decide? For millions of Americans, the cheaper generic version can mean the difference between getting the necessary medication and not being able to afford it, although not all approved drugs are available in generic form.

The two forms can be differentiated rather simply. A pioneer drug is the first version of a drug that is marketed under a brand name, such as Valium, an antianxiety drug. A **generic drug** is chemically equivalent to the pioneer drug, but cannot be marketed until the patent protection on the pioneer drug has run out after 20 years. The lower priced drug is commonly marketed under its generic name, in this case diazepam instead of Valium. The 20-year patent protection on the pioneer drug begins when it is patented, not when first put on the market. In cases requiring a long preapproval time, the actual marketing period can be relatively short, even less than six years. In such a situation, a drug company has very little time to recapture its research and development costs before a generic competitor can be manufactured. Almost 80% of generic drugs are produced by brand-name firms (Figure 10.28). Like pioneer drugs, generic drugs must also be approved by the FDA.

In 1984, Congress passed the Drug Price Competition and Patent Restoration Act that greatly expanded the number of drugs eligible for generic status. This act eliminated the need for generics to duplicate the efficacy and safety testing done on counterpart pioneer drugs. Doing so saves drug manufacturers considerable time and money. The FDA also issued specific guidelines for a generic drug's comparability to the pioneer drug. By FDA mandate, the generic and pioneer versions must be bioequivalent in dosage form, safety, strength, route of administration, quality, performance characteristics, and intended use. In other words, it must deliver the same amount of active ingredient into a patient's bloodstream at the same rate.

Health insurance companies and the FDA suggest that policyholders choose generic rather than brand-name drugs when possible, for obvious economic reasons. According

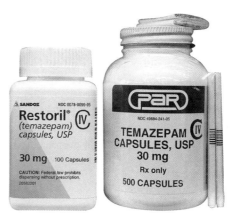

Figure 10.28

A brand name drug (Restoril) and its generic counterpart (temazepam).

to the Congressional Budget Office, generic drugs saved consumers an estimated $8–10 billion a year at retail pharmacies. Additional billions of dollars are saved when hospitals use generics. The concern for health care costs, along with the graying of baby boomers, will likely accelerate the use of generics, as will patents that expire on additional brand name drugs, making their generic versions possible.

10.13   Over-the-Counter Drugs

Over-the-counter (OTC) drugs allow people to relieve many annoying symptoms and cure some ailments without the need to see a physician. Nonprescription medications now account for about 60% of all medications used in the United States and may be used to treat or cure about 400 ailments. In accordance with laws and regulations prevailing in this country, drugs including OTC drugs are subjected to an intensive, extensive, and expensive screening process before they can be approved for sale and public use. Just as in the case of prescription drugs, the ultimate question to be answered with over-the-counter drugs is, "Do the benefits outweigh the risks?" The answer to this question for an OTC drug depends on whether a consumer is using it properly. Therefore, the FDA and pharmaceutical manufacturers must try to balance OTC safety and efficacy. More than 80 therapeutic categories of OTC drugs exist, ranging from acne products to weight control products. Table 10.6 contains several major categories of OTC products and their chief components.

Earlier we described OTC pain relievers and NSAIDs and their mode of action. Their side effects include increased stomach bleeding (aspirin), gastrointestinal upset (aspirin, ibuprofen), aggravating asthma (aspirin), and kidney (acetaminophen) or liver (acetaminophen) damage at high dosage or with chronic use.

More than 100 viruses are responsible for the misery from the common cold. A whole host of cold remedies, most with multiple components, are designed to help the sufferer. Decongestants reduce swelling when viruses invade the mucous membranes, but the adverse effects include nervousness and insomnia. Nasal sprays relieve the swollen nasal tissues, but their use beyond a three-day limit often leads to a rebound effect, or a return of the runny nose.

Antihistamines relieve the runny nose and sneezing associated with allergies, but cause drowsiness and often light-headedness. Because they induce drowsiness, it is not surprising that approved sleep aids are often antihistamines. However, children often experience insomnia and hyperactivity after taking them.

Coughing is a natural way to rid the lungs of excess secretions. Expectorants make the phlegm thinner and therefore easier to cough up, while suppressants provide relief and restful sleep. The presence of both in most cough remedy preparations seems

Table 10.6	**Examples of Over-the-Counter Drugs**

Analgesics and Antiinflammatory Drugs

aspirin
ibuprofen (Advil)
naproxen (Aleve)
acetaminophen (Tylenol)

Antacids and Indigestion Aids

aluminum and magnesium salts (Maalox)
calcium carbonate (Tums)
calcium and magnesium salts (Rolaids)

Agents to Block Acid Formation

cimetidine (Tagamet)
famotidine (Pepcid)
nizatidine (Axid)

Motion Sickness Drugs

dimenhydrinate (Dramamine or Travelin)
cyclizine (Marezine)
diphenhydramine (Benylin)
meclizine hydrochloride (Bonine)

Cough Remedies

expectorant
   guaifenesin
cough suppressants
   codeine (only in some states)
   dextromethorphan

Cold Remedies

antihistamines
   brompheniramine
   chlorpheniramine
   diphenhydramine

Decongestants

pseudoephedrine
phenylpropanolamine (PPA, now banned)
phenylephrine

Sleep Aids

diphenhydramine (Benylin)
doxylamine (Unisom)

senseless. Codeine and dextromethorphan have equally good cough-suppressing potential, but the former has a reputation for being habit-forming, a property that limits its over-the-counter availability in some states.

Diet aids once contained phenylpropanolamine (PPA) or benzocaine, a topical painkiller that apparently deadens the sensitivity of the taste buds. PPA, also a decongestant, has an unimpressive effect in weight loss, and has an efficacy limited to three to four months. In the early fall of 2000, the FDA started advising people to stop taking over-the-counter cold medicines or appetite suppressants that contain PPA, citing a possible risk of stroke. The regulatory agency also asked drug manufacturers to discontinue use of this ingredient in other products. The FDA banned PPA in over-the-counter products in November of 2000.

Heartburn, indigestion, and "acid" stomach are targets of antacids and related drugs. Antacids are basic compounds containing aluminum, magnesium, or calcium hydroxides (or combinations of them) that neutralize excess stomach acid. The popularity of the calcium-containing alternatives has grown with the promotion of the need for younger and older adults to maintain a regular supply of dietary calcium to prevent degradation of bones (osteoporosis).

Your Turn 10.24	Antacids

Give the chemical formulas for these commonly used antacids.

a. aluminum hydroxide
b. magnesium hydroxide
c. calcium hydroxide

A second type of antacid uses a completely different mode of activity. Histamine-2 blockers prevent the formation of excess acid. You may be familiar with the television commercials that point out the immediate activity of the compounds that neutralize excess stomach acid versus the long-term relief that comes from those that block acid production. The latter group contains brands like Axid and Pepcid that represent drugs that went from creation in the prescription stage to OTC status in about 20 years.

The self-care revolution of the last several decades has encouraged the availability of safe and effective OTC drugs and has provided additional pressure for the reclassification of many prescription drugs to OTC status. According to the Consumer Healthcare Products Association, about 80 ingredients or reduced dosages of drugs have made the OTC switch since 1976. Recent prescription to OTC switches include ranitidine (Zantac, acid reducer, 1995), minoxidil (Rogaine, for hair growth, 1996), cimetidine (Tagamet, acid reducer, 1999), loratidine (Claritin, antihistamine, 2002), and omeprazole (Prilosec, acid reducer, 2003). Right now, more than 700 OTC products have ingredients that were once only available by prescription. Additional pressure to have the switch occur comes from the health insurance industry. Changing widely used prescription drugs to OTC status greatly diminishes the insurance companies' share of payments. The conditions for which the drugs are prescribed must be common, non-life-threatening, and self-diagnosable by the average consumer. The FDA can change the status of an OTC drug back to prescription status if significant safety problems are uncovered.

From the drug manufacturer's standpoint, changing the status of a product from prescription to OTC often allows the manufacturer to market the product for several more years without generic competition. Sales volume also increases when a product is reclassified to OTC status. For the consumer, out-of-pocket expense may actually increase when going with OTC therapeutics because few third-party health insurance payers provide reimbursement for OTC products. But in September 2003, the U.S. Treasury Department and the IRS announced that OTC drugs can now be paid for with pre-tax dollars through health care flexible-spending accounts. "Flexible Spending Accounts are an important tool in helping people meet their health care costs," stated Treasury Secretary John Snow. "Since many prescription drugs have moved to the over-the-counter market, this action today makes paying for them a little bit easier to swallow."

For purposes of itemizing medical expenses on tax returns, the cost of such OTC drugs continues to be nondeductible. In addition, the cost of dietary supplements (explored in the following section) that are merely beneficial to the employee's health are not excluded from income.

10.14 Herbal Medicine

A growing number of people worldwide are using herbal products for preventive and therapeutic purposes. Herbal remedies and folk medicines abound in most cultures. This should not be surprising or astounding; nature is a very good chemist. Some (but not all) compounds found in plants and simple organisms are likely to have positive physiological effects in humans. U.S. sales of such popular herbal remedies as ginkgo biloba, St. John's wort, echinacea, ginseng, garlic, and kava kava have steadily risen over the past decade to $4 billion in 2003. Table 10.7 lists some herbs and plants and reasons for ingesting them.

The Herb Research Foundation reports data from over 2000 patients in 23 clinical studies that have consistently found that a preparation of St. John's wort, a plant, is just as effective against mild to moderate depression as standard antidepressant drugs. In April 2001, however, a study published in the *Journal of the American Medical Association* reported that St. John's wort is ineffective against severe depression (Figure 10.29). The herb worked no better than the placebo in over 200 adults diagnosed with severe depression. The study was funded partly by the National Institutes of Mental Health and partly by Pfizer Incorporated, which makes sertraline (Zoloft), the most commonly prescribed antidepression drug in the United States, with $2.5 billion is sales in the

Table 10.7	Common Herbs and Why They Are Used
Herb or Plant	**Symptoms to be Relieved**
Valerian, passion flower	Anxiety
Licorice, wild cherry bark, thyme	Coughs
Echinacea, garlic, goldenseal root	Colds, flu
St. John's wort	Depression
Chamomile, peppermint, ginger	Nausea, digestion problems
Valerian, passion flower, hops, lemon balm	Insomnia
Ginkgo biloba	Memory loss
Valerian, passion flower, kava kava, Siberian ginseng	Stress, tension

year 2002. Dr. Richard Skelton from Vanderbilt University, a coauthor of the new study, recommended further studies on the use of the herb for mild depression, saying "I would like to see people with mild depression studied, and see if it works in those folks. If it works, that would be great." He recommended against using the herbal medicine until further studies are done.

Consider ephedra (Figure 10.30), a naturally occurring substance derived from the Chinese herbal *ma huang* as well as from other plant sources. Although ephedra has long been used to treat certain respiratory symptoms in traditional Chinese medicine, in recent years it has been heavily promoted and used for the purposes of aiding weight loss, enhancing sports performance, and increasing energy.

Ephedra contains six amphetamine-like alkaloids including ephedrine and pseudoephedrine. Ephedrine, the main constituent, is a bronchodilator (opens the airways) and stimulates the sympathetic nervous system. It has valuable antispasmodic properties, acting on the air passages by relieving swellings of the mucous membrane. Pseudoephedrine (Sudafed) is a nasal decongestant and has less stimulating effect on the heart and blood pressure.

In their synthetic form, these drugs were regulated as OTC drugs and used as a decongestant for the short-term treatment of runny nose, asthma, bronchitis, and

 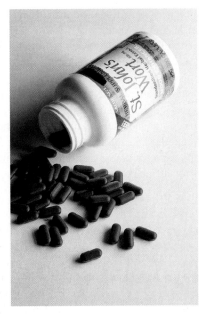

Figure 10.29

St. John's Wort plant *(Hypericum perforatum)* and an extract in tablet form.

Figure 10.30
Ephedra's source and a common formulation.

allergic reactions by opening the air passages in the lungs. Figure 10.31 shows three related structures; methamphetamine is presented so you can see the structural similarities between ephedra drugs and this potent stimulant. Ephedra does not contain methamphetamine.

Dietary supplements that contained ephedra, which made headlines in 2003 because of the deaths of well-known athletes, were linked to rare but serious health consequences. Side effects reported by ephedra users included nausea and vomiting, psychiatric disturbances such as agitation and anxiety, high blood pressure, irregular heartbeat, and, more rarely, seizures, heart attack, stroke, and even death.

At present, no evidence supports the claim that ephedra enhances athletic performance. And only preliminary evidence suggests that ephedra aids in modest, temporary weight loss. However, evidence does indicate that ephedra is associated with an increased risk of side effects, possibly even fatal ones. Effective in 2003, the International Olympic Committee, the National Football League, the National Collegiate Athletic Association, minor league baseball, and the U.S. Armed Forces all banned the use of ephedra.

The FDA recently reviewed whether ephedra-containing products should be on the market. And in February 2003, the Department of Health and Human Services announced a number of actions to address concerns about ephedra's safety. First, warning letters were sent to dozens of ephedra manufacturers challenging them to remove unproven claims or substantiate those claims, with a particular focus on athletic performance enhancement claims. Given the limited evidence on ephedra's benefits, the FDA and the Federal Trade Commission decided to assess whether further enforcement actions were warranted against other manufacturers. Second, a new, mandatory warning label for all marketed ephedra products was proposed. It made it clear to users, via a

A solid wedge indicates a group coming out towards you. A dashed wedge indicates that the group is pointing away from you. These conventions were first introduced in Section 3.3.

Ephedrine   Pseudoephedrine   Methamphetamine

Figure 10.31
Chemical structures of ephedrine and two related drugs.

black-box warning on the front of the product, as well as additional information in the product labeling, that serious adverse events and death have been reported after using ephedra, and that risks of adverse events are particularly high with strenuous exercise or use of stimulants including caffeine. Third, the FDA solicited comments from health professionals, the supplement industry, and the general public on any additional data on ephedra's safety, so that they could acquire the most complete picture possible of the product's risks.

As a result of this investigation, in December 2003 the FDA issued a consumer alert on the safety of dietary supplements containing ephedra. The alert advised consumers to immediately stop buying and using ephedra products. Finally, in February 2004, the FDA published a final rule stating that dietary supplements containing ephedra present an unreasonable risk of illness or injury. The rule effectively banned the sale of these products, which took effect 60 days after its publication.

The St. John's wort and ephedra examples point to several issues concerning herbal medicines. Psychiatrists report that many of their patients with depression have tried St. John's wort before coming for medical help. One estimate suggests that over a million people in the United States alone have tried it or are using St. John's wort. Irrespective of the accuracy of the estimate, large numbers of Americans are apparently using herbal medicines under circumstances with little or no medical supervision.

Herbal remedies are only loosely regulated by the FDA. The Dietary Supplement Health and Education Act (DSHEA) of 1994 changed their classification from "food or drug" to "dietary supplement." **Dietary supplements** by definition include vitamins, minerals, amino acids, enzymes, and herbs and other botanicals. Many dietary supplements have shown no adverse effects and in fact have proven to be beneficial to good health. Others, like ephedra, have been shown to be problematic.

Under the DSHEA, the FDA does not review dietary supplements for safety and effectiveness before they are marketed. Rather, the law allows the FDA to prohibit sale of a dietary supplement if it "presents a significant or unreasonable risk of injury." When the FDA seeks to take regulatory action against a supplement (such as in the ephedra case), the burden of proof for establishing harm falls on the government. Unlike prescription or OTC drugs, there is no assessment of purity of preparations or concentrations of active ingredients, set amounts, or delivery protocols. Furthermore, there are no requirements for studies of interactions among herbal medicines or between them and traditional medicines. Since the manufacturers of herbal remedies are not required to submit proof of safety and efficacy to the FDA before marketing, information regarding the interactions between herbal remedies and other drugs are largely unknown.

In the spring of 2001, representatives from the American Society of Anesthesiologists reported concerns that patients undergoing surgery may risk unexpected bleeding when they take certain herbs within two weeks before surgery. To this point, no scientific studies linking the bleeding and a specific herb have been published. According to Dr. John Neeldt, president of the American Society of Anesthesiologists, the familiar question "Are you taking any medications?" should be augmented with, "Are you taking any herbal remedies?"

Consider This 10.25　 **Does Natural Mean Safer?**

The legal standard of "significant or unreasonable risk" implies a risk–benefit calculation based on the best available scientific evidence. This suggests that the FDA must determine if a product's known or supposed risks outweigh any known or suspected benefits, based on the available scientific evidence. This must be done in light of the claims the product makes and with the understanding that the product is being sold directly to consumers without medical supervision.

When deciding to take any medication, a risk–benefit analysis is always involved, even if the consumer is unaware of it. One element of such an analysis is the *perceived* risk. Do you think that the general population perceives naturally occurring drugs as safer than synthetic ones?

Consider This 10.26   A Female Aphrodisiac

In 2001, Niagara, a dietary supplement that is a caffeinated, nonalcoholic energy drink made with several South American herbs touted itself as an aphrodisiac that boosts the female sex drive. Its similar name, Niagara, prompted Pfizer, the manufacturer of sildenafil (Viagra), to sue the maker of the dietary supplement. Make a list of the arguments that the producers of Viagra and Niagara might make during a court appearance.

10.15   Drugs of Abuse

Before concluding our foray into the world of medicinal drugs, an examination of the abuse of drugs is warranted. People have taken nonmedically indicated drugs for a variety of reasons since the beginning of history.

According to the Substance Abuse and Mental Health Services Administration (SAMHSA), in 2002 an estimated 19.5 million Americans, or 8.3% of the population age 12 or older, were current illicit drug users. Current means use of an illicit drug during the month prior to the survey interview. Notable statistics from the SAMHSA survey include the following.

- Marijuana is the most commonly used illicit drug. Of the 14.6 million users of marijuana in 2002, about one third, or 4.8 million persons, used it on 20 or more days in the past month.

- In 2002, an estimated 2.0 million persons (0.9%) were current cocaine users, 567,000 of whom used crack. Hallucinogens were used by 1.2 million persons, including 676,000 users of ecstasy. There were an estimated 166,000 current heroin users.

- An estimated 6.2 million persons, or 2.6% of the population age 12 or older, were current users of psychotherapeutic drugs taken nonmedically. An estimated 4.4 million used pain relievers, 1.8 million used tranquilizers, 1.2 million used stimulants, and 0.4 million used sedatives.

- In 2002, approximately 1.9 million persons age 12 or older had used oxycodone (OxyContin, a narcotic analgesic) nonmedically at least once in their lifetime.

- Among youths age 12–17, 11.6% were current illicit drug users. The rate of use was highest among young adults (18–25 years) at 20.2%. Among adults age 26 or older, 5.8% reported current illicit drug use.

- In 2002, an estimated 11.0 million persons reported driving under the influence of an illicit drug during the past year.

- An estimated 120 million Americans age 12 or older reported being current drinkers of alcohol in the 2002 survey (51.0%). About 54 million (22.9%) participated in binge drinking at least once in the 30 days prior to the survey, and 15.9 million (6.7%) were heavy drinkers.

- An estimated 71.5 million Americans (30.4% of the population age 12 or older) reported current use of a tobacco product in 2002.

In this section we will examine a few of the most commonly abused drugs.

Consider This 10.27   Which Drug Is Most Harmful?

Of all the drugs that humans abuse, which do you think causes the most harm? That is, can you point a finger at one drug that is the most disruptive in the home and workplace, causing more health problems and death than all of the others?

Table 10.8	Drug Schedules	
CLASS	Description	Examples
SCHEDULE I	Drug has no current accepted medical use. Drug has a high potential for abuse.	heroin LSD marijuana MDMA (Ecstasy)
SCHEDULE II	Drug has current accepted medical use. Drug has high potential for abuse.	oxycodone (OxyContin) hydromorphone (Dilaudid) methadone cocaine certain amphetamines and barbiturates
SCHEDULE III	Drug has current accepted medical use. Drug has medium potential for abuse.	hydrocodone and acetaminophen (Vicodin) acetaminophen with codeine (Tylenol with codeine)
SCHEDULE IV	Drug has current accepted medical use. Drug has low potential for abuse.	alprazolam (Xanax) propoxyphene and acetaminophen (Darvocet) diazepam (Valium)
SCHEDULE V	Drug has accepted medical use. Drug has lowest potential for abuse.	diphenoxylate and atropine (Lomotil) promethazine (Phenergan)

10.15.1 Drug Schedules

In 1970, the Comprehensive Drug Abuse Prevention and Control Act was passed into law. Title II of this law, the Controlled Substances Act, is the legal foundation of narcotics enforcement in the United States. The Controlled Substance Act regulates the manufacture and distribution of drugs, and places all drugs into one of five schedules. The criteria for the schedules and examples of each are shown in Table 10.8.

Consider This 10.28 **Low Potential for Abuse?**

The sedatives diazepam (Valium) and alprazolam (Xanax) are currently listed as schedule IV drugs, indicating that they have a low potential for abuse. Search the Web for sites that discuss addiction to these powerful sedative drugs. Do you agree with the current scheduling?

10.15.2 Marijuana

Marijuana is the most commonly used illicit drug in the United States Some slang names for marijuana include pot, weed, grass, Mary Jane, and chronic. It is a mixture of the dried leaves, stems, seeds, and flowers of the hemp plant, *Cannabis sativa*. The drug is usually smoked and occasionally eaten. The first known record of marijuana use dates to ca. 2737 B.C. by the Chinese Emperor Shen Nung, who prescribed use of the plant for the treatment of malaria, gas pains, and absent-mindedness.

Basically, hemp is the plant whose botanical name is *Cannabis sativa*. Other plants are called hemp, but *Cannabis* hemp is the most useful of these plants. Hemp is any

Cannabis sativa means useful (*sativa*) hemp (*cannabis*).

durable plant that has been used since prehistory for many purposes. Fiber is its most well known product, and the word *hemp* can mean the rope or twine made from the hemp plant, as well as just the stalk of the plant that produced it. The major psychoactive drug in *Cannabis sativa* is concentrated in the leaves and flowers of the hemp plant, so one cannot get high from smoking hemp rope or wearing clothing woven from hemp fibers.

The Chinese routinely used marijuana for fibers used in clothing as well as for medicinal purposes. Much later, hemp was planted in the Jamestown area in 1611 for the purpose of making rope, but there is no recorded evidence of its medicinal use by the early settlers. George Washington kept a field of hemp at Mt. Vernon, and it is believed that he used the plant for both rope and medicine.

Extracts of marijuana were employed by physicians in the early 1800s for a tonic and euphoriant. But in 1937 the Marijuana Tax Act prohibited its use as an intoxicant and its medical use was regulated as national concern of its use emerged. The Marijuana Tax Act required anyone producing, distributing, or using marijuana for medical purposes to register and pay a tax that effectively prohibited nonmedical use of the drug. Although the act did not make medical use of marijuana illegal, it did make it expensive and inconvenient.

In 1942, marijuana was removed from the U.S. Pharmacopoeia, the government's official compendium of medicines, because it was believed to be a harmful and addictive drug that caused psychoses, mental deterioration, and violent behavior. The current legal status of marijuana was established in 1970 with the passage of the Controlled Substances Act.

The major psychoactive chemical in marijuana is Δ^9-tetrahydrocannabinol, or THC (Figure 10.32). The concentration of THC varies depending on how the hemp is grown: temperature, amount of sunlight, and soil moisture and fertility. High-potency varieties of marijuana are grown across the United States, with THC levels as high as 7%. In hashish, or hash, the dried and pressed flowers and resin of the plant, THC concentrations can be as high as 12%.

When someone smokes marijuana, THC rapidly passes from the lungs into the bloodstream, which carries the chemical to organs throughout the body, including the brain. In the brain, THC connects to specific sites called cannabinoid receptors on nerve cells and influences the activity of those cells. Some brain areas have many cannabinoid receptors; others have few or none. Many cannabinoid receptors are found in the parts of the brain that influence pleasure, memory, thought, concentration, sensory and time perception, and coordinated movement. THC leaves the blood rapidly through metabolism and uptake into the tissues. The chemical may remain stored in body fat for long periods; research has indicated that a single dose can take up to 30 days for complete elimination. The short-term effects of marijuana use can include problems with memory and learning, distorted perception, difficulty in thinking and problem solving, loss of coordination, decreased blood pressure, and increased heart rate.

Medicinal marijuana may be indicated for treatment of nausea, glaucoma, pain management, and appetite stimulation. Such treatment may be a last resort when all other medications have failed, such as with the unrelenting nausea and vomiting that may accompany weeks of chemotherapy in treating diseases such as leukemia and

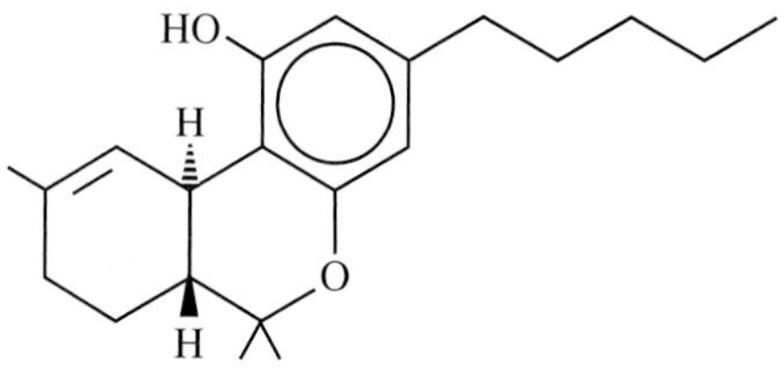

Figure 10.32

Molecular structure of THC.

AIDS. Clinical studies on the usefulness of marijuana are difficult to conduct as many barriers discourage researchers. For example, the scarcity of funding combined with complicated regulations enforced by both federal and state agencies make research in this area daunting.

The current debate over the medical use of marijuana is basically a debate over the value of its medicinal properties relative to the risk posed by its use. The debate is colored by complex moral and social judgments that underlie current drug control policy in the United States. The 1996 California referendum known as Proposition 215 allowed seriously ill Californians to obtain and use marijuana for medical purposes without criminal prosecution or sanction, with the stipulation that a physician's recommendation is required. A similar referendum was passed in Arizona the same year. In November 1998, Arizona voters passed a second referendum that allowed physicians to prescribe marijuana as medicine, but this is still at odds with federal law. By the summer of 1998, eight states (California, Connecticut, Louisiana, New Hampshire, Ohio, Vermont, Virginia, and Wisconsin) had laws that permit physicians to prescribe marijuana for medical purposes. These too are in conflict with federal laws.

10.15.3 MDMA (Ecstasy)

During the 1980s a unique psychoactive substance called MDMA emerged. The chemical name for this compound is 3,4-methylenedioxy-*n*-methylamphetamine, also known as Adam, Ecstasy, or XTC. Ecstasy gained national attention as the drug of choice at club parties, called "raves." It is still a widely used and abused drug taken recreationally, often by young people. The drug shares structural similarities to both mescaline, a hallucinogen, and to amphetamines, a family of stimulants. Figure 10.33 shows the similarities of MDMA and methamphetamine.

Although the drug is often referred to as a hallucinogen, the effects of MDMA are different from psychedelic drugs like LSD. MDMA does not seem to bring about perceptual distortions like hallucinogens. Sometimes it is called a "love drug" as the user is endowed with enhanced empathy for others while experiencing its effects. In fact, the drug was used in psychotherapy until 1985, when it was made illegal by the DEA (Drug Enforcement Agency). Drug users take MDMA orally, experiencing effects lasting approximately 4–6 h. The typical dose is between one and two tablets, with each containing approximately 60–120 mg of MDMA. As with most illegal drugs, these tablets contain other substances and adulterants, so it is impossible for the user to know how much MDMA is being ingested.

MDMA works in the brain by increasing the activity levels of at least three neurotransmitters: serotonin, dopamine, and norepinepherine. Much like other amphetamines, MDMA causes these neurotransmitters to be released from their storage sites in neurons, increasing brain activity. Compared with the potent stimulant methamphetamine, MDMA triggers a larger increase in serotonin and a smaller increase in dopamine. Serotonin is a major neurotransmitter involved in regulating mood, sleep, pain, emotion, and appetite, as well as other behaviors. By releasing large amounts of serotonin and also interfering with its synthesis, MDMA leads to a significant depletion of this important neurotransmitter. As a result, it takes the human brain a

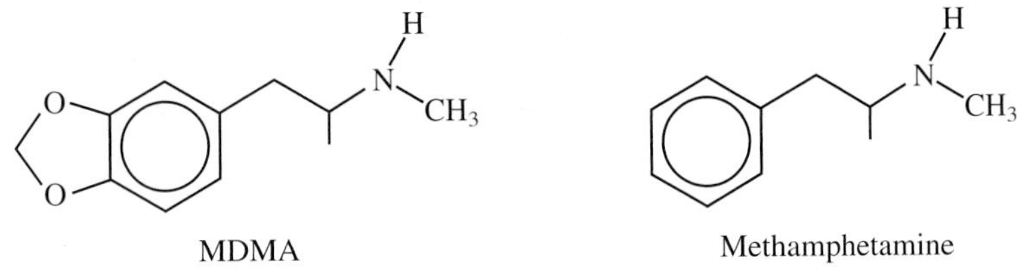

MDMA   Methamphetamine

Figure 10.33

Molecular structure of psychoactive drugs.

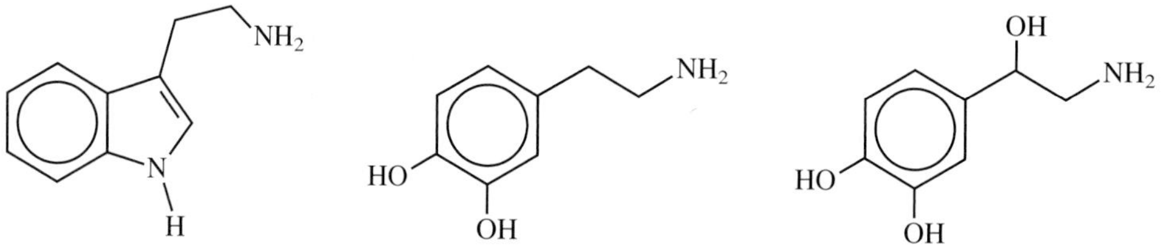

Sertonin　　　　Dopamine　　　　Norepinephrine (noradrenaline)

Figure 10.34

Molecular structures of neurotransmitters.

significant amount of time to rebuild the store of serotonin needed to perform important physiological and psychological functions. This has been shown to cause impairment in visual and verbal memory. The structures of these neurotransmitters are shown in Figure 10.34.

10.15.4　OxyContin—Hillbilly Heroin?

OxyContin (Figure 10.35) is the tradename for the drug oxycodone hydrochloride, a morphine-like narcotic. Other analgesics like Percocet and Percodan contain the same chemical, but in much smaller amounts (5–7.5 mg/tablet). OxyContin tablets contain up to 80 mg of oxycodone with a time release mechanism that allows the drug to be delivered slowly over a longer period.

The drug is supposed to be taken orally, but abusers of OxyContin are able to get around the time release mechanism by crushing the tablet and either snorting it or dissolving it in water and injecting. The effect is similar to that achieved from using heroin. Obviously, absorbing this much of the powerful narcotic can have dramatic consequences.

Federal authorities monitor emergency room visits caused by overdoses of narcotic pain relievers. In 1995, before OxyContin arrived on the market, 45,254 emergency room visits were linked to the abuse of narcotic painkillers. In 2002, that number hit 119,185, representing a 163% increase, according to the federal Substance Abuse and Mental Health Services Administration. At least 1.9 million Americans have admitted to taking OxyContin illegally at least once, according to the federal Drug Enforcement Agency (DEA). And from 1996 to 1999, deaths linked to the abuse of OxyContin jumped to 268 from 51, according to the DEA.

> Oxycodone is a Schedule II drug under the Controlled Substances Act because of its high propensity to cause dependence and abuse.

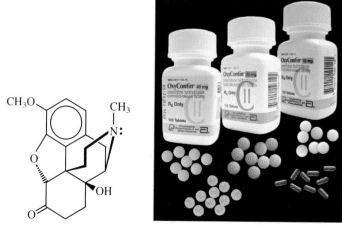

Figure 10.35

Oxycodone molecular structure and OxyContin pills.

Source: Photo © 2004, Publishers Group, www.streetdrugs.org.

The problem was first observed in rural areas of Kentucky, Virginia, West Virginia, and Maine; hence the slang names "hillbilly heroin" and "poor man's heroin." But it has since spread to other areas in the United States, and in March 2002, an 18-year-old female became the first UK fatality attributed to OxyContin.

Purdue Pharma, the manufacturer of the product, has come under fire recently for allegedly turning a blind eye to the mounting reports of abuse of the drug. As of 2003 more than 200 lawsuits were pending against Purdue Pharma for overpromoting the prescription painkiller while playing down its addictive side. In an effort to educate health care providers about these risks, the drug manufacturer issued a warning in the form of a "Dear Healthcare Professional" letter. The company's other efforts have included discontinuing the most powerful pill, a 160-mg form of the drug; stamping pills from Mexico and Canada to help authorities trace illicit supplies; passing out tamper-proof prescription pads to doctors; and instituting educational programs aimed at doctors as well as potential abusers, with a particular focus in the Appalachian region.

Your Turn 10.29   **Oxycodone Formulation**

The oxycodone in an OxyContin pill is not a freebase, but rather is formulated as a salt. What acid is used to form this salt? Draw a structural formula for the drug in its salt form.

Consider This 10.30   **Who Is Right? Who Is Wrong?**

Use the Web to investigate several sites that discuss the abuse of OxyContin. Also find sites that describe the legal woes that Purdue Pharma has encountered. Do you think that the drug manufacturer bears some responsibility in this matter? Write an essay taking one side or another or present your stance in a class discussion.

Morphine, oxycodone, hydrocodone (found in Vicodin, Lortab, and Lorcet), and codeine belong to a class of drugs known as opiates, and they all have the same mode of action. These drugs bind to a chemical receptor called *mu*, that interrupts the transmission of pain in the spinal cord. Opiates also stimulate areas of the brain involved in pleasure, called the reward, or endorphin, pathways. In a normal brain, the chemical dopamine crosses between brain cells in the reward pathway, producing pleasurable feelings. Opiates stimulate the release of higher levels of dopamine, strengthening reward signals, and producing intense euphoria. Repeated use of opiates causes the brain of a user to become accustomed to an overstimulated reward pathway. This in turn brings about the phenomenon of tolerance—greater amounts of opiates are required to achieve the euphoria the user once experienced.

The drugs just described are but a few of today's commonly abused substances. There are many more, including alcohol which is possibly the most damaging of all drugs when taken to excess. Just like with prescribed medicines, the choice to take illicit drugs involves a risk–benefit analysis. All drug users should have this basic information before making this choice.

CONCLUSION

Molecular modifications by chemists have created a vast new pharmacopoeia of wonder drugs that have significantly increased the number and quality of our days. Thanks to penicillin, sulfa drugs, and more recent antibiotics, the great majority of bacterial infections are easily controlled. Dreaded killers such as typhoid, cholera, and pneumonia have been largely eliminated—at least in wealthy, industrialized societies. Synthetic steroids, used for birth control and many other reasons, have transformed society. And the humble aspirin tablet has been supplanted by newer NSAIDs, which in turn are being replaced by a new generation of innovatively designed "superaspirins."

But no drug can be completely safe, and almost any drug can be misused. These issues become the focus when the FDA is asked to change the status of a drug from prescription to over-the-counter. Taking a medication is a conscious choice between the benefits derived from the drug and the risks associated with its side effects and limits of safety. Because most drugs have very wide, carefully established margins of safety, their benefits far outweigh their risks for the general population. For some drugs, however, the trade-off between effectiveness and safety involves a different balance. A drug with severe side effects may be the only treatment available for a life-threatening disease. Someone suffering from AIDS or advanced, inoperable cancer understandably has a different perspective on drug risks and benefits than a person with a cold. And the impersonal anonymity of averages takes on new meaning at the bedside of a loved one. That some countries like Brazil have ignored patents on anti-AIDS drugs is the first evidence of decisions based on the moral dilemmas. Herbal and alternative medicines raise new questions. Who is responsible for defining their efficacy, monitoring their purity, and developing contraindications to their use with other drugs? When chemistry is applied to medicine, science must be guided by morality, and reason must be tempered with compassion.

Chapter Summary

Having studied this chapter, you should be able to:

- Describe the discovery, development, and physiological properties of aspirin (10.1)

- Understand bonding in carbon-containing (organic) compounds (10.2)

- Apply the concept of isomerism to organic compounds (10.2)

- Convert molecular formulas to structural formulas, condensed structural formulas, and line-angle drawings (10.2)

- Recognize functional groups and the classes of organic compounds that contain them; draw structural formulas for organic molecules containing various functional groups (10.3)

- Understand that functional groups may be chemically modified to change a molecule's properties; predict the products of ester formation reactions; describe how amines may be converted to their salt forms (10.3)

- Relate the molecular structure of aspirin to other analgesics (10.3)

- Understand the mode of action of aspirin and other analgesics (10.4)

- Describe the discovery of penicillin (10.5)

- Describe the lock-and-key mechanism of drug action (10.5)

- Understand differences in molecular structure between chiral (optical) isomers (10.6)

- Recognize the structure of the steroid nucleus (10.7)

- Consider the chief functions of steroids, and some specific examples: sex hormones (testosterone, progesterone), cell-membrane components (cholesterol) (10.7)

- Understand how birth control pills inhibit ovulation (10.8)

- Understand the mechanisms by which RU-486 and methotrexate induce abortion (10.9)

- Discuss the ethical issues in the use of steroids for birth control and abortion (10.8–10.9)

- Understand the nature and use of anabolic steroids and designer steroids (10.10)

- Discuss the procedure for drug testing and approval and the associated benefits and costs (10.11)

- Understand the similarities and differences between brand name and generic drugs (10.12)

- Identify some of the over-the-counter drug categories and their uses (10.13)

- Understand the process of a drug going from prescription to OTC (10.13)

- Describe some of the potential benefits and risks of herbal medicine (10.14)

- Describe the scheduling of prescription drugs (10.15)

- Discuss the use of marijuana, MDMA, and oxycodone in terms of their physiological and social effects (10.15)

Questions

Emphasizing Essentials

1. a. What is the intended effect of an antipyretic drug?

    b. What is the intended effect of an analgesic drug?

    c. What is the intended effect of an antiinflammatory drug?

    d. Can one drug exhibit all of these effects?

2. The field of chemistry has many subdisciplines.

    a. What do organic chemists study?

    b. How does this differ from what biochemists study?

3. Write condensed structural formulas and line-angle drawings for the three isomers of C_5H_{12} assigned in Your Turn 10.5: Isomers of C_5H_{12}.

4. Write the structural formula and line-angle drawings for each different isomer of C_6H_{14}. *Hint:* Be sure the bonding really is different, not just a different paper-and-pencil representation in two dimensions of the same structure.

5. Consider the isomers of C_4H_{10}. How many different isomers could be formed by replacing a single hydrogen atom with an —OH group; that is, how many different alcohols have the formula C_4H_9OH? Draw the structural formula for each.

6. For each compound, identify the functional group present.

    a. $CH_3 \diagup O \diagdown CH_3$ b. $CH_3CH_2 \diagup \overset{\overset{O}{\|}}{C} \diagdown \overset{H}{O}$

    c. $CH_3CH_2 \diagup \overset{\overset{O}{\|}}{C} \diagdown CH_3$ d. $CH_3CH_2 \diagup \overset{\overset{O}{\|}}{C} \diagdown NH_2$

    e. $CH_3CH_2 \diagup \overset{\overset{O}{\|}}{C} \diagdown OCH_3$

7. Which of these can contain only one carbon atom?

    a. an alcohol   d. an ester

    b. an aldehyde   e. an ether

    c. a carboxylic acid f. a ketone

    Draw a structural formula for each one-carbon molecule. Then explain why the others must contain more than one carbon atom.

8. For each of these, identify the functional group.

    a. CH_3CH_2—OH

    b. $CH_3CH_2 \diagup \overset{\overset{O}{\|}}{C} \diagdown H$

    c. $CH_3CH_2 \diagup \overset{\overset{O}{\|}}{C} \diagdown OCH_3$

Then, draw an isomer that contains a different functional group.

9. Histamine causes runny noses, red eyes, and other symptoms in allergy sufferers. Here is its structural formula.

    a. What is the molecular formula of this compound?

    b. Circle the amine functional groups in histamine.

    c. Which part (or parts) of the molecule do you think make the compound water-soluble?

10. Figure 10.7 shows a somewhat condensed structural formula for acetaminophen, the active ingredient in Tylenol.

    a. Draw the complete structural formula for acetaminophen, showing all atoms and all bonds.

    b. What is the molecular formula for this compound?

    c. Children's Tylenol is a flavored aqueous solution of acetaminophen. Predict what part or parts of the molecule make acetaminophen water-soluble.

11. Identify the functional groups in each of these.

    a. Barbital (a sedative)

    b. Penicillin-G

    c. Amyl dimethylaminobenzoate (an ingredient in sunscreens)

12. Ibuprofen is relatively insoluble in water but readily soluble in most organic solvents. Explain this solubility behavior based on its structural formula found in Figures 10.7 and 10.19.

13. Here is the structural formula for diazepam, the sedative found in Valium.

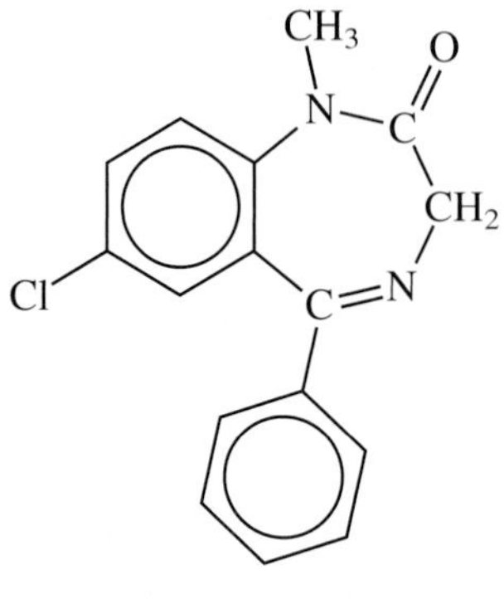

Judging from its structure, do you expect it to be more soluble in fats or in aqueous solutions? Why?

14. Draw the structure of each ester that forms when acetic acid reacts with these alcohols:

$CH_3CH_2CH_2OH$ $(CH_3)_2CHOH$ $(CH_3)_3COH$

a. *n*-propanol   **b.** isopropanol   **c.** *t*-butanol

15. Interpret this sentence by giving the meaning of each acronym and explaining the effect. "NSAIDs have an effect on COX enzymes."

16. Compare the physiological effects of aspirin with those of acetaminophen and ibuprofen. Relate differences to the nature of each compound at the molecular and cellular levels.

17. Would aspirin be more active if it were to interact with prostaglandins directly, rather than by blocking the activity of COX enzymes? Explain your reasoning.

18. What are "superaspirins?" How do they differ from regular aspirin and other NSAIDs?

19. The text states that 80 billion tablets of aspirin a year are consumed in the United States If the average tablet contains 500 mg of aspirin, how many pounds of aspirin does this consumption represent?

20. Identify the functional groups in morphine and meperidine (Demerol), using the structural formulas found in Figure 10.13. Can these molecules be assigned to a particular class of compound (i.e., an alcohol, ketone, or amine)? Why or why not?

21. What is meant by the term *pharmacophore?*

22. Sulfanilamide is the simplest sulfa drug, a type of antibiotic. It appears to act against bacteria by replacing *para*-aminobenzoic acid, an essential nutrient for bacteria, with sulfanilamide. Use these structural formulas to explain why this substitution is likely to occur.

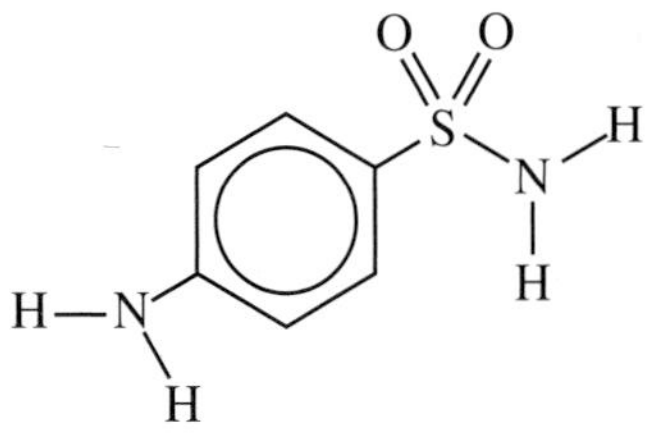

Sulfanilamide

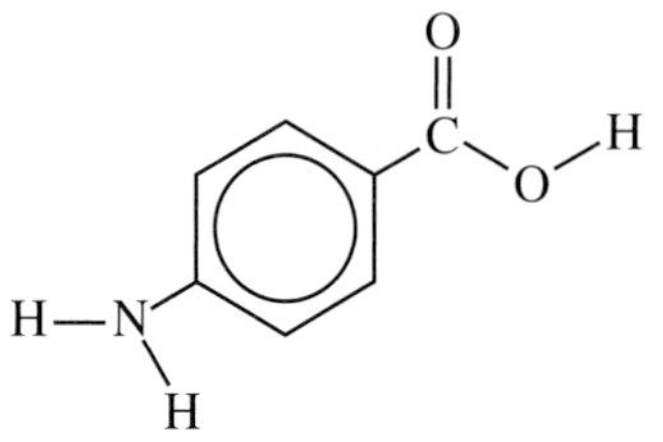

Para-aminobenzoic acid

23. Which of these compounds can exist in chiral forms?

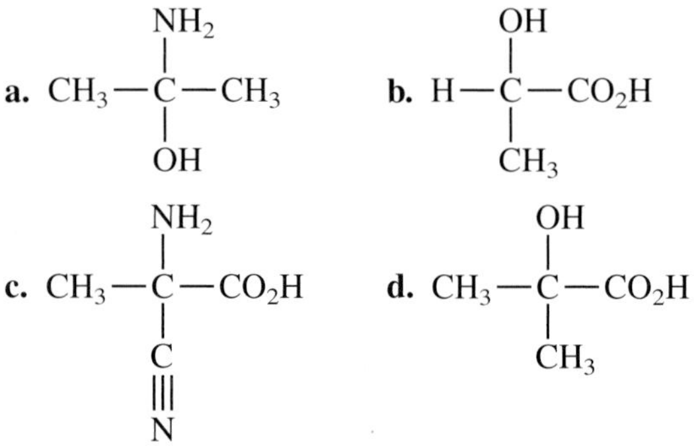

24. Which of these compounds can exist in chiral forms?

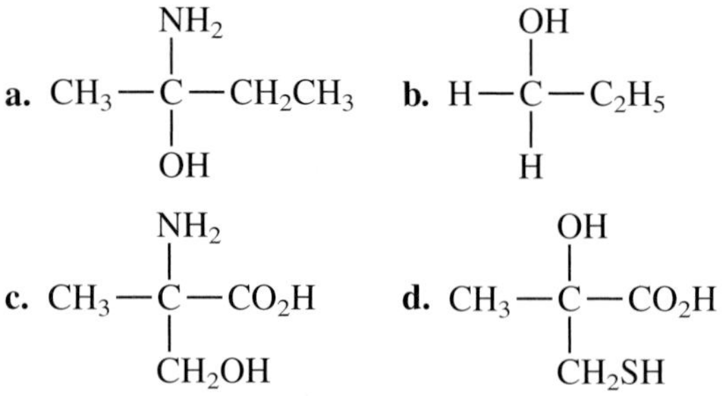

25. Methamphetamine hydrochloride is a powerful stimulant that also goes by the street names "crystal," "crank," or "meth." This structural formula shows its salt form.

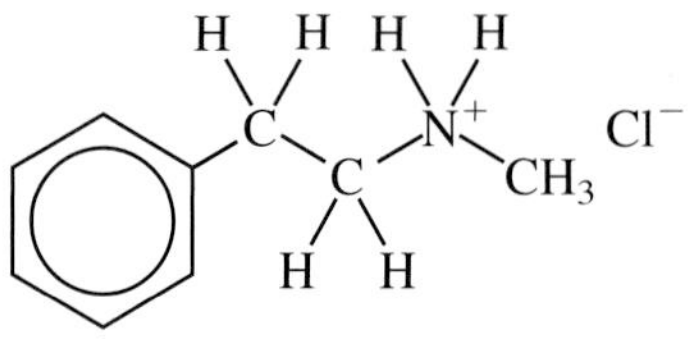

The freebase form of this drug, called "ice," is also abused. What does the term *freebase* mean and how might the drug be converted to this form?

26. Use the structural formulas in Figure 10.21 to answer these questions.

a. Identify the functional groups in cortisone.

b. Suggest a reason why cholic acid is more soluble in water than cholesterol.

27. Molecules as diverse as cholesterol, sex hormones, and cortisone all contain common structural elements. Draw the structural formula that shows the common structural elements.

Concentrating on Concepts

28. The text states that some remedies based on the medications of earlier cultures contain chemicals that are effective against disease, others are ineffective but harmless, and still others are potentially harmful. How might it be determined into which of these three categories a recently discovered substance fits?

29. Draw structural formulas for each of these molecules and determine the number and type of bonds (single, double, or triple) used by each carbon atom.

a. H_3CCN (acetonitrile, used to make a type of plastic)

b. $H_2NC(O)NH_2$ (urea, an important fertilizer)

c. C_6H_5COOH (benzoic acid, a food preservative)

30. Carbon usually forms four covalent bonds, nitrogen usually three, oxygen usually two, and hydrogen only one bond. Use this information to draw structural formulas for:

a. A compound that contains one carbon atom, one nitrogen atom, and as many hydrogen atoms as needed.

b. A compound that contains one carbon atom, one oxygen atom, and as many hydrogen atoms as needed.

31. In Your Turn 10.5, you were asked to draw structural formulas for each of the three isomers of C_5H_{12}. One student

submitted this set of isomers, with a note saying that six isomers had been found (the hydrogens have been omitted for clarity). Help this student see why some of the answers are incorrect.

32. Styrene, $C_6H_5CH{=}CH_2$, contains a ring similar to benzene, C_6H_6. One hydrogen has been replaced by the side chain $-CH{=}CH_2$. Draw structural formulas to show that this molecule, like benzene, has resonance structures.

33. Aspirin is a specific compound, so what justifies the claims for the superiority of one brand of aspirin tablets over another?

34. Figure 10.8 represents chemical communication within the body. Write a paragraph explaining what this figure means to you in helping to explain chemical communication.

35. Consider this statement. "Drugs can be broadly classed into two groups: those that produce a physiological response in the body and those that inhibit the growth of substances that cause infections." In which class does each of these drugs fall?

a. aspirin   **c.** antibiotics   **e.** amphetamine

b. superaspirin   **d.** hormones   **f.** penicillin

36. Consider the structure of morphine in Figure 10.13. Codeine, another strong analgesic with narcotic action, has a very similar structure in which the $-OH$ group attached to the six-membered ring is replaced by an $-OCH_3$ group.

a. Draw the structural formula for codeine and label its functional groups.

b. The analgesic action of codeine is only about 20% as effective as morphine. However, codeine is less addictive than morphine. Is this enough evidence to conclude that replacement of $-OH$ groups with $-OCH_3$ groups in this class of drugs will always change the properties in this way? Why or why not?

37. Dopamine is found naturally in the brain. The drug L-dopa is found to be effective against the tremors and muscular rigidity associated with Parkinson's disease. Identify the chiral carbon in L-dopa, and comment on why L-dopa is effective whereas D-dopa is not.

L-dopa

38. Vitamin E is often sold as a racemic mixture of the D- and L-isomers. Use the Web to find answers to these questions.

a. Which is the more physiologically active isomer?

b. How does the cost of the racemic mixture compare with the price of the pure, physiologically active isomer?

Isomer 1

Isomer 2

Isomer 3

Isomer 4

Isomer 5

Isomer 6

39. Consider the fact that L-methorphan is an addictive opiate, but D-methorphan is safe enough to be sold in many over-the-counter cough remedies. From a molecular point of view, how is this possible?

40. Figure 10.23 diagrams the action of a steroid contraceptive. Study this diagram and then explain it in your own words as if you were speaking to an interested friend.

41. Why are many projects to isolate or synthesize new drugs started in this country, but few actually receive FDA approval for general use?

42. Until the early 19th century, it was believed that organic compounds had some sort of "life force" and could only be produced by living organisms. This was the basis of a concept called vitalism. This view was dispelled in the late 1800s. Use the Web to find out what changed that opinion.

Exploring Extensions

43. One avenue for successful drug discovery is to use the initial drug as a prototype for the development of other similar compounds called analogs. The text states that cyclosporine, a major antirejection drug used in organ transplant surgery, is an example of a drug discovered in this way. Research the discovery of this drug to verify this statement. Write a brief report describing your findings, citing your sources.

44. Dorothy Crowfoot Hodgkin first determined the structure of a naturally occurring penicillin compound. What was her background that prepared her to make this discovery? Write a short report on the results of your findings, citing your sources.

45. Before the cyclic structure of benzene was determined (Figure 10.4), there was a great deal of controversy about how the atoms in this compound were arranged.
   a. Count outer electrons for C and H in C_6H_6. Then draw the structural formula for a possible linear isomer.
   b. Give the condensed structural formula for your answer in part **a**.
   c. Compare your structure with those drawn by classmates. Are they all the same? Why or why not?

46. Antihistamines are widely used drugs for treating allergies caused by reactions to histamine compounds. This class of drug competes with histamine, occupying receptor sites on cells normally occupied by histamine. Here is the structure for a particular antihistamine.

a. What is the molecular formula for this compound?
b. What similarities do you see between this structure and that of histamine (shown in question 9) that would allow the antihistamine to compete with histamine?

47. Over the next few years, the FDA may consider deregulating more than a dozen drugs, nearly as many as have already been approved for over-the-counter sales during the past decade. The products that have led this trend have been the widely advertised drugs for heartburn.
   a. Which questions need to be answered before a drug is deregulated?
   b. Will these questions change if you are considering this need from the viewpoint of the FDA, a pharmaceutical company, or as a consumer?

48. Find out more about the new process for the manufacture of ibuprofen that won the 1997 Presidential Green Chemistry Challenge Award. How does this process differ from the earlier one for manufacturing ibuprofen? Write a brief report on your research, giving your references.

49. Testosterone and estrone were first isolated from animal tissue. One ton of bull testicles was needed to obtain 5 mg of testosterone and 4 tons of pig ovaries was processed to yield 12 mg of estrone.
   a. Assuming complete isolation of the hormones was achieved, calculate the mass percentage of each steroid in the original tissue.
   b. Explain why the calculated result very likely is incorrect.

50. Herbal remedies are prominently displayed in supermarkets, drug stores, and discount stores.
   a. What influences your decision to buy one of these remedies?
   b. Choose a remedy and carefully examine its label for information about the active ingredients, inert ingredients, anticipated side effects, the suggested dosage, and the cost per dose.
   c. How confident are you that the safety and efficacy of these remedies are ensured? Explain your answer.

51. Habitrol was the most successful smoking-cessation prescription until the introduction of bupropion (Zyban) in 1997, which now has over 50% of the market. How are these two approaches to drug therapy different? What is the current market share of each drug?

52. Drug approval laws in other nations are not the same as those in the United States. Choose a country and find out how its drug approval process works compared with the process in the United States. Construct relative time lines that reveal what steps must be taken and approximately the length of time each step may require. Also, find out if specific drugs are available in that country that are not available in the United States and comment on what factors may be influencing the policies of each country.

53. Over-the-counter drugs allow consumers to treat a myriad of symptoms and ailments. An advantage is that the user can purchase and administer the treatment without the effort or the expense of consulting a physician. To provide a wider margin of safety for these circumstances, the OTC versions of drugs are often administered in lower doses. See if this is true by looking at information for prescription and OTC versions of painkillers like Motrin and heartburn treatments like Axid and Pepcid. Report on your findings.

54. Herbal or alternative medicines are not regulated in the same way as prescription or OTC medicines. In particular, the issues of concern are identification and quantification of the active ingredient, quality control in manufacture, and side effects when the herbal remedy is used in conjunction with another alternative or prescription medicine. Look for evidence from herbal supplement manufacturers that address these issues, and write a report documenting your findings, giving your references.

55. Danco Laboratories, the U.S. company that produces RU-486 (Mifeprex), has reported data demonstrating the increasing use of Mifeprex in the United States:

- Sales in the United States were 45% higher for 2002 compared with 2001.

- As of January 2003, more than 130,000 women had chosen Mifeprex for their nonsurgical abortion since product availability.

And consider the claim by Eric Schaff, chair of the National Abortion Federation (which promotes nonsurgical abortion), that aspirin causes more deaths than RU-486. Use the Web to find information to support or refute this claim.

56. The antibiotic ciprofloxacin hydrochloride (Cipro) treats bacterial infections in many different parts of the body. This drug made headlines in 2001 for use in patients who had been exposed to the inhaled form of anthrax. Use the Web or another source to obtain the structure of Cipro. Draw its structure and identify the functional groups.

57. Some of the most effective anticancer drugs in use today are isolated from natural sources. Find three of these drugs and also state the source from which each was isolated.

58. In 2003, a series of spot examinations of mail shipments of foreign drugs to U.S. consumers conducted by the FDA and U.S. Customs and Border Protection revealed that these shipments often contain unapproved or counterfeit drugs that pose serious safety problems. Although many drugs obtained from foreign sources purport, and may even appear to be, the same as FDA-approved medications, these examinations showed that many are of unknown quality or origin. Of the 1153 imported drug products examined, the overwhelming majority, 1019 (88%), were illegal because they contained unapproved drugs. Many of these imported drugs could pose clear safety problems. Use the FDA Web site to determine which drugs were most commonly counterfeited and their countries of origin.

59. Thalidomide was first marketed in Europe in the late 1950s. It was used as a sleeping pill and to treat morning sickness during pregnancy. At that time it was not known to cause any adverse effects. By the late 1960s however, the drug was banned after it was found to be a teratogen, causing deformed limbs in the children of women who took it early in pregnancy. Use the Web for information to write a short paper that describes the optical isomers of thalidomide, why the FDA never approved thalidomide for use in the United States and for what purpose the FDA has recently approved the use of thalidomide.

60. Conventional acne treatments are maintenance therapies. Antibiotics, such as tetracycline and erythromycin, are not expected to result in long-term improvement once they are stopped. Some individuals may not respond to any of the conventional medications for acne. A possible solution for those patients is the vitamin A derivative known as isotretinoin (Accutane). However, Accutane is by no means a frontline therapy. Using the Web, find the serious side effects that may accompany the use of Accutane.

Nutrition: Food for Thought

Nutrition has become a national issue—in medicine and health, in business and advertising, and in our daily lives. From television advertisements to the neighborhood newsstand, there is an abundance of information about using diet to lose weight and prevent disease. Often, the information seems contradictory.

"Obesity is near to overtaking smoking as the No. 1 cause of death in the United States . . . and other research shows that its adverse health effects could soon wipe out many recent improvements in health. . . ."

Centers for Disease Control and Prevention Reports,
(*March 3, 2004*)

"Your choice of diet can influence your long-term health prospects more than any other action you might take."

Dr. C. Everett Koop (1988),
Former U.S. Surgeon General

"The current USDA dietary pyramid misses an enormous opportunity for improving the health of Americans. . . . It's clear that we need to rebuild the pyramid from the ground up. Every American deserves it."

Walter C. Willett, M.D.,
Professor of Epidemiology and Nutrition,
Harvard University

With the high standard of living and an amazing capacity to produce food in the United States, it is ironic that the leading cause of death among Americans may soon become obesity and related adverse health effects. In March 2004 the Centers for Disease Control and Prevention reported that poor diet and physical inactivity caused 400,000 deaths, or 16.6% of the total annual mortality. How can it be that we are eating ourselves to death in this country?

To understand this "malfunction" in a land of plenty, we have to begin with a look at the materials of which our bodies are composed and the purposes for which we consume food items. But first we might ask, "Who tells us what to eat?" The U.S. Department of Agriculture (USDA) has promoted various guidelines for eating with a series of food charts over the last 50 years. A recently developed and highly promoted guide is the food pyramid. The base of the pyramid represents the most recommended food types and largest amounts; as we move to the top, the food groups change and the recommended portions decrease. In recent years, health researchers and others have recommended alternative food pyramids. Some even argue that the USDA food pyramid has contributed to the growth of obesity.

Before discussing "a nutrition system gone haywire," we start with the categories of food—the nature and structures of carbohydrates, the family of fats (and controversial cholesterol), proteins, vitamins and minerals. Besides being the fodder of broadcast advertising, "low-carb" diets are now the target of an increasing number of scientific studies. We also know that "too many calories" is part of the dilemma leading to an increase in obesity, and this is a tip-off that our discussion should include energy. We will consider both the energy content of food and also how the body might handle the various structural forms that store the energy from food. We visit some of those issues and the often-confusing set of packaging labels like "low fat," "fat-free," and "organic." The chapter ends with a review of some old and new methods of food preservation and an examination of a more global perspective on nutrition.

11.1 You Are What You Eat

Whether we eat delicious culinary creations elegantly prepared to celebrate some memorable social occasion or some junk food gobbled on the run, all of us eat because food provides the four fundamental types of materials required to keep our bodies functioning. These materials are water, energy sources, raw materials, and metabolic regulators.

Water serves as both a reactant and a product in metabolic reactions, as a coolant and thermal regulator, and as a solvent for the countless substances that are essential for life. Our bodies are approximately 60% water, but H_2O cannot be burned in the body or elsewhere. Therefore, we need food as a source of energy to power processes as diverse as muscle action, brain and nerve impulses, and the movement of molecules and ions in suitable ways at appropriate times and places. We also eat because raw

Many of these important properties of water are discussed in Chapter 5.

1 dietary calorie =
1 Calorie =
1 kcal = 1000 calories

Consider This 11.1   Calories and Carbohydrates

The composition of two types of tortilla chips is given here—one a regular brand and the second promoting itself as a "low-carb" variety. The percent of daily values given are based on a 2000-Calorie diet.

	Regular tortilla chips		Low-carb tortilla chips	
	Amount	% Daily Value	Amount	% Daily Value
Serving Size	1 oz (28 g)		1 oz (28 g)	
Calories	142		150	
Calories from fat	61		70	
Total Fat	7 g	10	8 g	12
Saturated fat	1.4 g	7	1 g	5
Cholesterol	0 mg	0	0 mg	0
Sodium	150 mg	7	190 mg	9
Total Carbohydrates	18 g	5	8 g	2
Dietary Fiber	1.8 g	8	4 g	18
Protein	2 g		12 g	

a. Does there appear to be a significant difference in the carbohydrates in the two kinds of tortilla chips or does this appear to be a case of deceptive advertising?

b. What other food group categories are impacted by changes in carbohydrate content?

c. Some specialty foods manipulate the "taste" of their products by increasing the salt and fat content. Is there any evidence of that being practiced here?

materials are needed for the syntheses of new bone, blood, enzymes, muscles, hair, and for the replacement and repair of cellular materials. And finally, food supplies some of the enzymes and hormones that function as regulators to control the biochemical reactions associated with metabolism and all other vital processes.

Eating properly is a matter of consuming the proper foods, not simply a case of eating sufficient amounts of food. It is possible to consume food regularly, even to the point of being overweight, and still be malnourished. The meaning of this commonly used term is important. **Malnutrition** is caused by a diet lacking in the proper mix of nutrients, even though the energy content of the food eaten may be adequate. Malnutrition contrasts with **undernourishment,** in which the daily caloric intake is insufficient to meet the metabolic needs of a person. Although there are malnourished and undernourished people in this country, hunger is typically not a consideration in a nation where nearly 55% of the population is up to 40% over their ideal weights, each consuming nearly 1000 excess dietary Calories (kilocalories) per day. One result of such excess is that about 50% of adult women and 25% of adult men annually attempt to practice girth control by dieting, generally with very limited long-term success.

Sceptical Chymist 11.2   A Lifetime of Food

During a lifetime, you will eat a truly prodigious amount of food, estimated to be about 700 times your adult body weight. This statement is itself quite an extraordinary assertion. Do calculations to check that the statement is in the ballpark. State all of your assumptions clearly.

Hint: Start by assuming a life span of approximately 78 years and that your present weight is your adult weight. Estimate the weight of food eaten daily at present, and use these data to project your lifetime consumption of food.

Table 11.1	Percentage of Water, Carbohydrates, Fats, and Proteins			
Food	**Water**	**Carbohydrates**	**Fats**	**Proteins**
White bread	37	48	4	8
2% Milk	89	5	2	3
Chocolate chip cookies	3	69	23	4
Peanut butter	1	19	50	25
Steak (3 oz sirloin)	57	0	15	28
Fish (3 oz tuna)	63	0	2	30
Black beans (cooked)	66	23	<1	9

Source: U.S Department of Agriculture, Agricultural Research Service, *Home and Garden Bulletin* 72.

In many foods, the contents are conveyed by food labels, that prominently display the amounts of carbohydrates, fats, and proteins (Figure 11.1). These are the macronutrients that provide essentially all of the energy and most of the raw material for repair and synthesis. Sodium and potassium ions (not the metals) are present in much lower concentrations, but these ions of metallic elements are essential for the proper electrolyte balance in the body. A number of other minerals and an alphabet soup of vitamins are listed in terms of the percent of recommended daily requirements supplied by a single serving of the product. It should be self-evident that all these substances, whether naturally occurring or added during processing, are chemicals. Unfortunately, this fundamental fact is apparently lost on those who pursue the impossible dream of a "chemical-free" diet. *All* food is inescapably and intrinsically chemical, even food claiming to be "organic" or "natural."

Table 11.1 indicates the mass percentages (grams of component per 100 g of food item) of water, carbohydrates, fats, and proteins in several familiar foods. The table reveals that for this particular selection of foods, the variation in composition is considerable. But in every case, these four components account for almost all of the matter present. The percentage of water ranges from a high of 89% in 2% milk to a low of 1% in peanut butter. Peanut butter is comparable to steak and fish in percent of protein and also leads these foods in fat content. Chocolate chip cookies have the highest percentage of carbohydrate because of high sugar and refined flour content.

Now compare Table 11.1 with Figure 11.2, which presents similar data for the human body. It is not surprising that the human body is composed of the same basic types of molecules that make up the food we eat. We are wetter and fatter than bread and contain more protein than milk; we are more like steak than chocolate chip cookies. From the data of Figure 11.2, we can calculate that a 150-lb person consists of 89 lb of water (150 lb × 60 lb water/100 lb body) and 30 lb of fat. The remaining 30 lb is almost all composed of various proteins and carbohydrates plus the calcium and phosphorus in the

Nutrition Facts

Serving Size 1 cup (228g)
Servings Per Container 2

Amount Per Serving	
Calories 250	Calories from Fat 110

	% Daily Value*
Total Fat 12g	**18%**
Saturated Fat 3g	**15%**
Cholesterol 30mg	**10%**
Sodium 470mg	**20%**
Total Carbohydrate 31g	**10%**
Dietary Fiber 0g	**0%**
Sugar 5g	
Protein 5g	

Vitamin A	4%
Vitamin C	2%
Calcium	20%
Iron	4%

* Percent Daily Values are based on a 2,000 calorie diet. Your Daily Values may be higher or lower depending on your calorie needs:

	Calories:	2,000	2,500
Total Fat	Less than	65g	80g
Sat Fat	Less than	20g	25g
Cholesterol	Less than	30mg	300mg
Sodium	Less than	2,400mg	2,400mg
Total Carbohydrates		300g	375g
Dietary Fiber		25g	30g

Figure 11.1

A nutrition facts label from a package of nuts.

Vitamins and minerals are discussed in Sections 11.8 and 11.9.

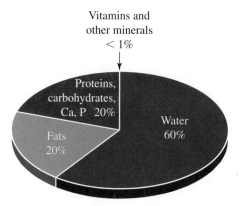

Figure 11.2

Composition of the human body.

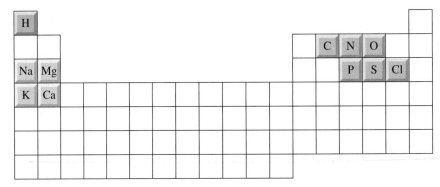

Figure 11.3
Periodic table indicating the 11 major elements of the human body.

bones. The other minerals and the vitamins together total less than 1 lb. This indicates that a little bit of each of them goes a long way, a point discussed in Section 11.9.

Of the nearly 90 naturally occurring elements, 11 make up over 99% of the mass of your body. Figure 11.3 is a periodic table showing only these elements: hydrogen, carbon, nitrogen, oxygen, phosphorus, sulfur, chlorine, sodium, magnesium, potassium, and calcium. The first seven are nonmetals, the latter four are metals. Hydrogen, carbon, nitrogen, and oxygen are the "building-block" elements used to construct body cells and tissues; the other seven elements are the macronutrients.

Table 11.2 lists the mass percentages (grams of element/100 g body weight) of these 11 elements in the human body and gives their relative atomic abundances. Oxygen is the most abundant element when measured in grams per 100 g body weight, but hydrogen is the most plentiful element in terms of actual number of atoms present.

Your Turn 11.3   **Abundance of Atoms in Mass or Number**

Table 11.2 gives both the mass of the major elements per 100 g of body weight and the relative abundance in the number of atoms per million atoms in the body.

a. How is relative abundance calculated from mass percent (grams/100 g body weight)?

b. Why is oxygen the most abundant element when measured in grams per 100 g of body weight, but only the second most abundant when measured in terms of relative abundance per million atoms in the body? *Hint:* Consider the atomic masses of each element.

Table 11.2	**Major Elements of the Human Body**		
Element	**Symbol**	**Grams per 100 g of body weight**	**Relative abundance per million atoms in body**
Oxygen	O	64.6	255,000
Carbon	C	18.0	94,500
Hydrogen	H	10.0	630,000
Nitrogen	N	3.1	13,500
Calcium	Ca	1.9	3,100
Phosphorus	P	1.1	2,200
Chlorine	Cl	0.40	570
Potassium	K	0.36	580
Sulfur	S	0.25	490
Sodium	Na	0.11	300
Magnesium	Mg	0.03	130

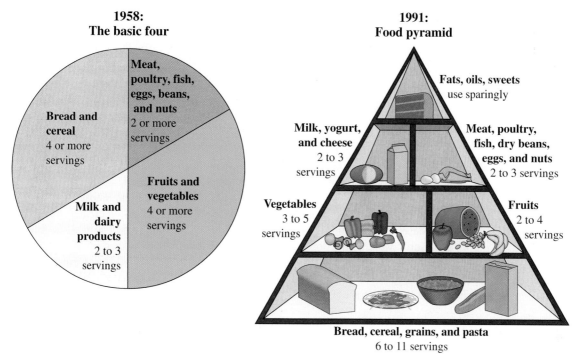

1958:
The basic four

1991:
Food pyramid

Figure 11.4
The USDA basic four food groups and the food pyramid.

Source: United States Department of Agriculture.

The preponderance of oxygen, carbon, hydrogen, and nitrogen is fully consistent with the composition of the major chemical components of the human body. Hydrogen and oxygen are, of course, the elements of water. Moreover, along with carbon atoms, they constitute all carbohydrates and all fats. Finally, these three elements plus nitrogen are found in all proteins. Thus, nature uses very simple units—aggregates of oxygen, carbon, hydrogen, and nitrogen atoms—in a myriad of elegantly functional combinations to produce the major constituents of a healthy body and a healthful diet.

The current recommendations for a healthful diet, approved in 1991 by the USDA, are represented by the food pyramid of Figure 11.4. The pyramid incorporates the traditional basic four food groups of the 1958 pie chart, but with different emphases. Look at the 1958 basic-four pie chart and the new food pyramid. The pyramid reorganizes the food categories, separating some that are lumped together in the 1958 chart, and redefines some of the serving sizes. Of particular interest, the pyramid includes a category not found on the pie chart, namely "fats, oils, and sweets." The category of "fruits and vegetables" found on the pie chart is separated into two categories, "vegetables" and "fruits," on the pyramid. The pyramid calls for eating habits that increase the proportions of foods at the base of the pyramid and decrease those near or at the top. This food pyramid urges us to eat significantly more bread, cereal, rice, and pasta than was implied by the 1958 guide, and more fruits and vegetables. Simultaneously, it urges us to reduce the percentage of fats, oils, and sweets in our daily diet.

Consider This 11.4   **The USDA Food Pyramid**

Politics entered the picture when the USDA initially released the food pyramid (see Figure 11.4). The meat and dairy industries pressured the USDA to delay public dissemination of the pyramid. Some nutritionists claim that the USDA, which is obligated to both promote and regulate agricultural products, has a conflict of interest between this responsibility and its obligation to promote the health of American citizens. Imagine that you were an administrator of the American National Cattlemen's Beef Association or the American National Cattlewomen, Inc.

(continued on p. 490)

Consider This 11.4 The USDA Food Pyramid (*continued*)

a. How would you reconcile your concern for your members' livelihood, which depends on the quantity and price of the beef sold, and your interest in preserving their health?

b. Would you support or oppose the USDA's food pyramid, which encourages eating more grains and less red meat? Give reasons for your answer.

c. Do you think the USDA should be the agency to make this determination? Draft a letter to a friend stating and defending your position on this issue.

In 1998 an expert panel at the National Institutes of Health (NIH) developed new guidelines for body weight based on a body mass index (BMI). Under the new guidelines, people with a BMI of less than 25.0 have a healthy weight. Those whose BMI is 25.0–29.9 are defined as overweight; adults with a BMI of 30 or greater are classified as obese. The NIH guidelines have a BMI lower limit of 18.5. The BMI is defined as a person's weight in kilograms (1 kg = 2.2 lb) divided by her or his height, in meters, squared, that is, kg/m^2 (1 m = 39.37 in.; $1\ m^2 = 1.55 \times 10^3\ in.^2$). For example, a person who is 5 ft 4 in. should weigh no more than 145 lb to have a BMI below 25. These are the steps in the calculation.

$$145\ lb \times \frac{1\ kg}{2.2\ lb} = 65.9\ kg$$

$$5\ ft\ 4\ in. = 64\ in.$$

$$(64\ in.)^2 = 4.09 \times 10^3\ in.^2$$

$$4.09 \times 10^3\ in.^2 \times \frac{1\ m^2}{1.55 \times 10^3\ in.^2} = 2.64\ m^2$$

$$BMI = \frac{65\ kg}{2.64\ m^2} = \frac{24.9\ kg}{1\ m^2}$$

The new guidelines apply best to average people; exceptions are possible. For example, slightly muscled folks with small bones could have too much body fat and yet have a BMI below 25. On the other hand, persons who are naturally large-boned and heavily muscled may not have excess body fat, yet can have BMIs over 25.

Your Turn 11.5　　　Body Mass Index

a. Calculate your BMI.

b. Explain the BMI to another person and calculate that person's BMI.

The goal of the USDA food pyramid is to give Americans advice for healthful eating, and because there have been criticisms of the USDA pyramid, others have taken it upon themselves to suggest alternatives. Figure 11.5 is the Healthy Eating Pyramid that has been proposed by nutrition experts from the Harvard School of Public Health. Its authors suggest that it is based on the best scientific knowledge about the links between diet and health and that it is independent of business interests from the food industry. This new pyramid is designed to fix the flaws in the USDA version and to offer sound options so that people can make better choices. The Healthy Eating Pyramid sits on a foundation of exercise and weight control. Suggested carbohydrates are whole grains. Against current trends in nutrition but consistent with research evidence and common eating habits, fats (especially unsaturated ones) appear near the foundation.

There is still plenty of controversy surrounding the ideal balance of the macronutrients and the best sources of these compounds. As nutrition experts turn up new information the recommendations will change, but there is no question that a diet that promotes health requires water, carbohydrates, fats, and proteins. We therefore turn now to a consideration of each of these macronutrients—their chemical composition, molecular structure, properties, and sources.

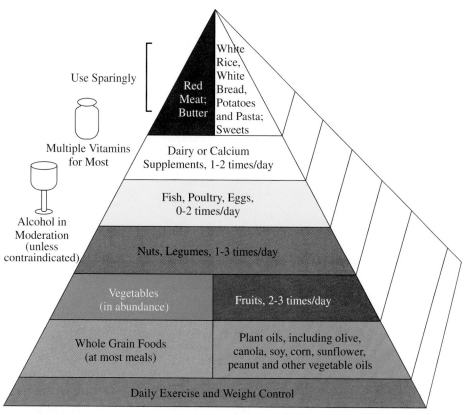

Figure 11.5
The Healthy Eating Pyramid from the Harvard School of Public Health.

Source: http://www.hsph.harvard.edu/nutritionsource/pyramids.html.

Consider This 11.6 **Healthy Eating Pyramid**

Search the Web to find complete details about the Healthy Eating Pyramid. Make a list of the food types and features of the USDA Food Pyramid and Healthy Eating Pyramid and compare them item by item. What are the similarities? What are the significant differences? Do you find that your lifestyle conforms with either of them?

Consider This 11.7 **Other Food Pyramids**

The Asian, Latin, Mediterranean, and vegetarian diets are favored by some health professionals in preference to the USDA Food Pyramid.

a. Search the Web to find out the details of two of these diets and compare them with the Food Pyramid.
b. What are the major differences in the diets?
c. Find out why the alternative diets are favored by some health professionals.

11.2 Carbohydrates—Sweet and Starchy

The best known dietary carbohydrates are sugars and starch. **Carbohydrates** are compounds containing carbon, hydrogen, and oxygen, the last two elements in the same 2:1 atomic ratio as found in water. Glucose, for example, has the formula $C_6H_{12}O_6$. This

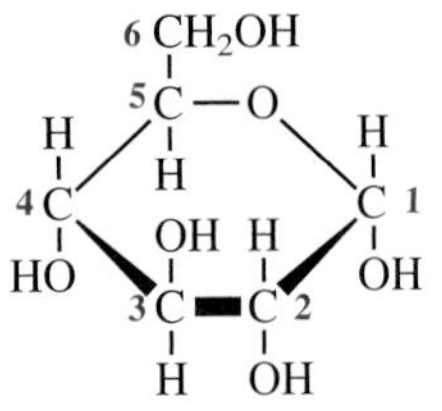

α-Glucose,
a monosaccharide

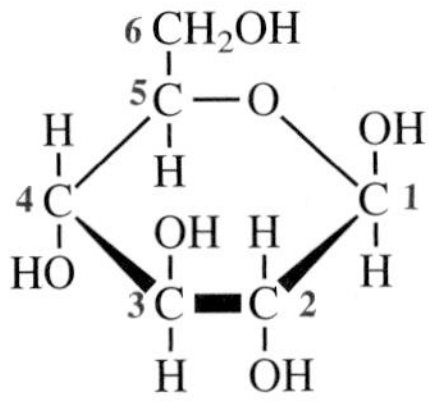

β-Glucose,
a monosaccharide

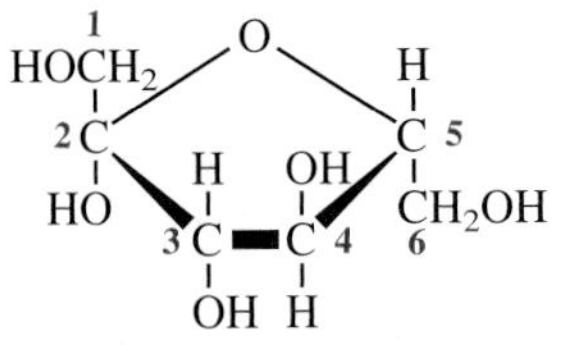

β-Fructose,
a monosaccharide

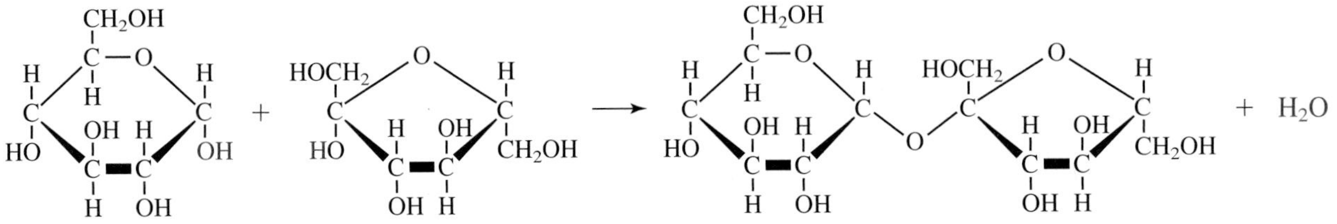

α-Glucose   β-Fructose   Sucrose

Figure 11.6
Molecular structures of some sugars.

composition gives rise to the name, "carbohydrate," which implies "carbon plus water." But the hydrogen and oxygen atoms are not bonded together to form water molecules. Rather, carbohydrate molecules are built of rings containing carbon atoms and an oxygen atom (Figure 11.6). The hydrogen atoms and —OH groups are attached to the carbon atoms. This general arrangement provides many opportunities for differences in molecular structure. Thus, 32 distinct isomers (including chiral isomers) have the formula $C_6H_{12}O_6$. The isomers differ slightly in their properties, including intensity of sweetness.

The simplest sugars are **monosaccharides,** or "single sugars," such as fructose and glucose, both $C_6H_{12}O_6$. Figure 11.6 indicates that molecules of these compounds include a single ring consisting of four or five carbon atoms and one oxygen atom. The best way to consider these two-dimensional representations of a three-dimensional structure is to imagine that the ring is perpendicular to the plane of the paper, with the bold-print edges facing you. The H atoms and —OH groups are thus either above or below the plane of the ring. This results in two forms of glucose: alpha (α) and beta (β). In α–glucose, the —OH group on carbon 1 is on the opposite side of the ring from the —CH₂OH group attached to carbon 5. As shown in Figure 11.6, the β–glucose form has the two groups on the *same* side of the ring. This is also the case for β-fructose, with the —OH on carbon 2 and the —CH₂OH group at carbon 5 being on the same side of the ring.

Ordinary table sugar, sucrose, is an example of a **disaccharide,** a "double sugar" formed by joining two monosaccharide units. In a sucrose molecule, an α-glucose and a β-fructose unit are connected by a C—O—C linkage created when an H and an OH are split out from the monosaccharides to form a water molecule. This is a condensation reaction similar to the ones for polymerization as described in Section 9.5. Water produced in this way is used for other reactions in the body. This reaction and the structure of the $C_{12}H_{22}O_{11}$ sucrose molecule are also shown in Figure 11.6.

The linking of monosaccharide molecules by this reaction is by no means restricted to the formation of disaccharides. Some of the most common and abundant carbohydrates are **polysaccharides,** polymers made up of thousands of glucose units. As the name implies, these macromolecules consist of "many sugar units." Polysaccharides form when monosaccharide monomer units join into a chain through splitting out a water molecule each time a monomer unit is incorporated into the chain.

Starch, cellulose, and glycogen are three familiar examples of polysaccharides, sometimes called complex carbohydrates. Starch is the primary carbohydrate component of several foods such as potatoes. Cellulose is the primary fibrous component in the cell walls of plants, and therefore a primary component of paper. Glycogen is the

Chirality and optical isomerism are discussed in Section 10.6.

The ending -ose is typical for sugars.

polysaccharide form in which carbohydrates are stored in our bodies. Glycogen has a molecular structure similar to that of starch. The chains of glucose units in glycogen, however, are longer and more branched than those in starch. Glycogen is vitally important because it is a storehouse of energy in molecular form. It accumulates in muscles and especially in the liver, where it is available as a quick source of internal energy.

A healthful diet derives more of its carbohydrates from complex carbohydrates than from simple sugars, such as mono- and disaccharides. Enzymes in our saliva initiate the process of breaking down the long polysaccharide chains into glucose molecules, an important first step in metabolism. But the body also synthesizes glucose from a variety of precursors, including other sugars.

Your Turn 11.8   **Tasty Crackers**

Sustained chewing of an unsweetened, unsalted cracker results in a sweet taste. What is the molecular explanation for this phenomenon?

Like starch, cellulose is a polymer of glucose, but these two polysaccharides behave very differently in the body. Humans are able to digest starch by breaking it down into individual glucose units; on the other hand, we cannot digest cellulose. Consequently, we depend on starchy foods such as potatoes or pasta as carbohydrate sources rather than literally devouring toothpicks or textbooks. The reason for this is a subtle difference in how the glucose units are joined in starch and in cellulose. In the alpha (α) linkage in starch, the bonds connecting the glucose units have a particular orientation, whereas the beta (β) linkage between glucose units in cellulose is in a different orientation (Figure 11.7).

Our enzymes, and the enzymes of many mammals, are unable to catalyze the breaking of beta linkages in cellulose. Consequently, we can't dine on grass or trees. Cows, goats, sheep, and other ruminants manage to break down cellulose with a little help. Their digestive tracts contain bacteria that decompose cellulose into glucose monomers. The animals' own metabolic systems then take over. Similarly, the fact that termites contain cellulose-hungry bacteria means that wooden structures are sometimes

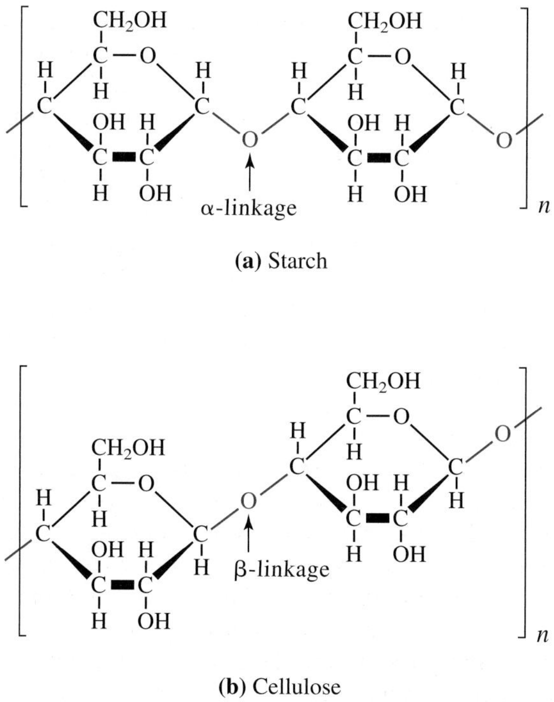

(a) Starch

(b) Cellulose

Figure 11.7

The bonding between glucose units in **(a)** starch and **(b)** cellulose.

See Section 3.8 for more
about methane and other
greenhouse gases.

at risk. And, as you have already read in Chapter 3, the methane released by termites may be contributing to global warming.

Lactose intolerance, a common metabolic dysfunction, is somewhat related to the difference in digestibility between starch and cellulose. This condition is shared by about 80% of the world's population. Although most Northern Europeans, Scandinavians, and people of similar ethnic background eat milk, cheese, and ice cream with no ill effects, they are the exception. The great majority of people who are lactose-intolerant have difficulty digesting dairy products. Consumption of these foods is often followed by diarrhea and excess gas. The symptoms result from the inability to break down lactose (milk sugar) into its component monosaccharides, glucose and galactose. The linkage between the two monosaccharides in lactose is a beta form, similar to that in cellulose. People who are lactose-intolerant have a lack of or a low concentration of lact*ase*, the enzyme that catalyzes the breaking of this bond. In such individuals, the intact lactose is instead fermented by our own intestinal bacteria. This process generates carbon dioxide and hydrogen gases, and lactic acid, the principal cause of the diarrhea. Given milk's importance for growing bones and teeth, it is significant that infants of all ethnic groups generally produce sufficient lactase to digest a milk-rich diet. But, as we age, this production decreases. By adulthood, most people in the world do not have enough of the enzyme to accommodate a diet heavy in dairy products.

-ase is generally used as the ending
for enzymes that act on sugars
(ending in –ose).

Your Turn 11.9   Differences in Molecular Structure

Speculate why the slight difference in molecular structure between starch and cellulose is enough to make the latter polysaccharide indigestible to human beings. *Hint:* Section 10.5 can be helpful.

Consider This 11.10   Lactose Intolerance—A Closer Look

Search the Web search for "lactose intolerance" to answer these questions. Reading the cartons in your cupboard or at a nearby store can also help you milk the labels for some information.

a. Over-the-counter digestive aids allow you to increase your intake of dairy products. How do these work? What are their advantages and disadvantages?

b. Even with digestive aids, you may risk not getting enough calcium, an essential mineral that you will learn more about in Section 11.11. What other foods can you eat to obtain enough calcium in your diet?

c. Sometimes lactose turns up in foods in which you least expect it, such as bread. Although lactose may not be listed on the label, you will see ingredients such as whey, milk products, nonfat dry milk, or dry milk solids—all of which contain lactose. Find three other nondairy foods that you may have to watch out for if you are lactose-intolerant.

d. Look at the labels of dairy products whose containers make claims such as "lactose-free milk." What has been added to these products so that they can be consumed by people who are lactose-intolerant?

11.3 Fats and Oils: Part of the Lipid Family

Everyone knows the properties of fats from personal experience. They are greasy, slippery, soft, low-melting solids that are not soluble in water. Butter, cheese, cream, whole milk, and certain meats and fish are loaded with them. All of these products are of animal origins. But margarine and some shortenings are evidence that fats can also be of vegetable origin. Oils, such as those obtained from olives, corn, or nuts, exhibit many of the properties of animal-based fats, but in liquid form. The fact that these properties are also shared by petroleum-based oils and greases suggests a chemical and structural similarity. But there are some important differences. You are well aware that petroleum is made up almost exclusively of hydrocarbons. In the molecules of these compounds, carbon atoms are bonded to one another (often in chains) and to hydrogen atoms. Hydrocarbon molecules are nonpolar, and hence they do not mix well with water or other polar substances.

The inference that edible fats and oils must also be nonpolar is fully justified. The molecules of fats of animal and vegetable origin also include long hydrocarbon chains. But biological fats are a bit more structurally complex than their petroleum-based chemical cousins. Of particular significance is the fact that edible fats and oils always contain some oxygen. Most of these compounds are classified chemically as triglycerides. **Fats** are triglycerides that are solid at room temperature, whereas **oils** are triglycerides that are liquid at room temperature. Whether solid or liquid, triglycerides belong to the **lipid** family, a class of compounds that includes cholesterol and other steroids and molecules of some complex compounds, such as lipoproteins that contain fatty segments (see Chapter 10).

To a chemist, the term *triglyceride* indicates a particular kind of lipid. A **triglyceride** is formally defined as an ester of three fatty acid molecules and one glycerol molecule. The overwhelming majority of fatty acids in the body, almost 95%, are transported and stored in the form of triglycerides. To see the connection, it is necessary to consider the molecular structures of fatty acids and glycerol.

Naturally occurring **fatty acid** molecules are characterized by two structural features: a long hydrocarbon chain generally containing an even number of carbon atoms (typically 12 to 24) including a carboxylic acid group (—COOH) at the end of the chain. This functional group is what puts the *acid* in fatty acid because —COOH can release a hydrogen ion (H^+). The long hydrocarbon chains, on the other hand, give fats most of their characteristic properties. Stearic acid, $C_{17}H_{35}COOH$, is a fatty acid found in animal fats. Its structural formula and condensed chemical formula are given here.

$$CH_3CH_2CH_2CH_2CH_2CH_2CH_2CH_2CH_2CH_2CH_2CH_2CH_2CH_2CH_2CH_2CH_2 - C \begin{smallmatrix} O \\ \\ OH \end{smallmatrix}$$

Stearic acid, a fatty acid

$$CH_3(CH_2)_{16}COOH$$

Condensed chemical formula for stearic acid

Glycerol, or glycerine as it is commonly called, is a sticky, syrupy liquid that is sometimes added to soaps and hand lotions. The following structural formula indicates that a molecule of this compound includes three —OH groups, which classifies it as an alcohol.

$$\begin{array}{ccc} H & H & H \\ | & | & | \\ H-C-C-C-H \\ | & | & | \\ OH & OH & OH \end{array}$$

Glycerol

The formation of a triglyceride can be represented by a word equation:

$$3 \text{ fatty acids} + \text{glycerol} \longrightarrow \text{triglyceride} + 3 \text{ water} \qquad [11.1]$$

In this process, as in the formation of polysaccharides, smaller units join to form more complex molecules by splitting out water molecules. The presence of three fatty acid units in the final product molecule makes it a *tri*glyceride. In fatty acids and in glycerol we have the acid and alcohol functional groups, respectively, required to form an ester (Section 9.5). In fact, each of the three —OH groups in a glycerol molecule can react with a fatty acid molecule. Thus, equation 11.2 represents the combination of three stearic acid molecules with a glycerol molecule to form a triester or triglyceride.

Hydrocarbons in petroleum were discussed in Section 4.8. Reasons why nonpolar substances do not dissolve in water were examined in Section 5.12.

organic acid + alcohol ⟶ ester + water

This group is characteristic of an ester:

$$\begin{smallmatrix} O \\ \| \\ C \\ O \end{smallmatrix} C$$

The ester group was discussed in Sections 9.2 and 10.3.

$$3 \quad CH_3(CH_2)_{16}-C\begin{smallmatrix}O\\ \diagup\diagdown\\ OH\end{smallmatrix} \quad + \quad \begin{matrix}H\\|\\HO-C-H\\|\\HO-C-H\\|\\HO-C-H\\|\\H\end{matrix} \quad \longrightarrow \quad \begin{matrix}H\\|\\CH_3(CH_2)_{16}COO-C-H\\|\\CH_3(CH_2)_{16}COO-C-H\\|\\CH_3(CH_2)_{16}COO-C-H\\|\\H\end{matrix} \quad + \quad 3\ H_2O \qquad [11.2]$$

Esterification from acids and alcohols to make polyesters was described in Section 9.5.

This is the process involved in the formation of most animal and vegetable fats and oils. Variety is built in by having up to three different fatty acids incorporated into the same triglyceride rather than just one, as in the stearic acid example given.

11.4 Saturated and Unsaturated Fats and Oils

Figures Alive! Visit the *Online Learning Center* to learn more about fats and fatty acids.

Animal and vegetable fats and oils exhibit considerable variety. A chief reason for this diversity is that not all fatty acids are identical. As we already noted, they vary in the number of carbon atoms and therefore in the length of the hydrocarbon chain. Additionally, fatty acids can contain one or more C-to-C double bonds and can differ in where these double bonds are located in the molecule. A fatty acid is said to be **saturated** if the hydrocarbon chain contains only C-to-C single bonds between the carbon atoms and no C-to-C double bonds. In this case, the hydrocarbon contains the maximum number of H atoms on the C chain that its structure can accommodate. This is the case with stearic acid. The fatty acid is **unsaturated** if the molecule contains one or more C-to-C double bonds between carbon atoms. Oleic acid, with one C-to-C double bond per molecule, is classified as monounsaturated. Those fatty acids containing more than one C-to-C double bond per molecule are called polyunsaturated. Linoleic acid, which contains two C-to-C double bonds per molecule, and linolenic acid with three C-to-C double bonds per molecule, are polyunsaturated. Note that each of these three different unsaturated fatty acids contains the same number of carbon atoms, 18.

$$CH_3(CH_2)_{16}COOH$$
Stearic acid, a saturated fatty acid

$$CH_3-(CH_2)_7-CH=CH-(CH_2)_7-COOH$$
Oleic acid, a monounsaturated fatty acid

$$CH_3-(CH_2)_4-CH=CH-CH_2-CH=CH-(CH_2)_7-COOH$$
Linoleic acid, a polyunsaturated fatty acid

$$CH_3-CH_2-CH=CH-CH_2-CH=CH-CH_2-CH=CH-(CH_2)_7-COOH$$
Linolenic acid, a polyunsaturated fatty acid

The three fatty acids in a single triglyceride molecule can all be identical, two can be the same and the third can be different, or all three can be different. Moreover, the fatty acids in a triglyceride molecule can exhibit varying degrees of unsaturation. The fatty acids that a given fat or oil contains govern its extent of unsaturation.

Stearic acid is a solid at body temperature, whereas oleic and linoleic acids are liquids.

The physical properties of fats also depend on their fatty acid content. Table 11.3 indicates that there are trends within a given family of fatty acids. In saturated fats, for example, the melting points increase as the number of carbon atoms per molecule (and the molecular mass) increase. On the other hand, in a series of fatty acids with a similar number of carbon atoms, increasing the number of C-to-C double bonds decreases the melting point. Thus, when the melting points of the 18-carbon fatty acids are compared, saturated stearic acid (no C-to-C double bonds per molecule) is found to melt at 70 °C, oleic acid (one C-to-C double bond per molecule) melts at 16 °C, and linoleic acid (two C-to-C double bonds per molecule) melts at −5 °C. These trends are carried over to the triglycerides containing the fatty acids. This explains why fats

Table 11.3	Melting Points of Some Fatty Acids	
Name	**Carbon Atoms per Molecule**	**Melting Point, °C**
Saturated fatty acids		
Capric	10	32
Lauric	12	44
Myristic	14	54
Palmitic	16	63
Stearic	18	70
Unsaturated fatty acids		
Oleic (1 double bond/molecule)	18	16
Linoleic (2 double bonds/molecule)	18	−5
Linolenic (3 double bonds/molecule)	18	−11

rich in saturated fatty acids are solids at room or body temperature, whereas highly unsaturated ones are liquids.

Figure 11.8 gives evidence of this generalization. The bar graphs present the composition of various dietary fats and oils in terms of saturated, polyunsaturated, and monounsaturated components. Typically, these naturally occurring lipids are mixtures of various triglycerides. In general, solid or semisolid animal fats, such as lard and beef tallow, are high in saturated fats. In contrast, olive, safflower, and other vegetable oils consist mostly of unsaturated triglycerides. However, the figure reveals that some surprising differences in the composition of oils. For example, palm and coconut oil contain much more saturated fat than corn and canola oil. Ironically, the coconut oil used in some nondairy creamers is 92% saturated fat, far more than the percentage found in the cream it replaces. In fact, coconut oil contains more saturated fats than pure butterfat. Concern over the high degree of saturation in coconut and palm oil accounts for the statement sometimes printed on food labels: "Contains no tropical oils."

Normal body temperature is 37 °C; room temperature is approximately 20 °C.

Coconut oil has been used for making popcorn at movie concession stands and for cooking French fries at fast-food establishments.

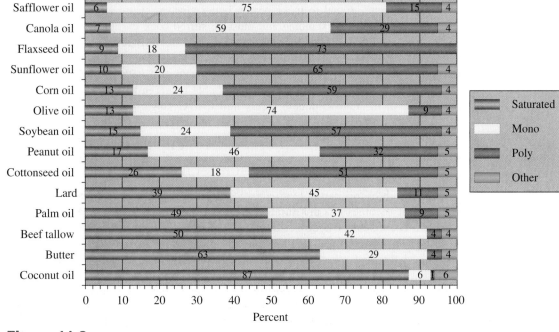

Figure 11.8

Saturated and unsaturated fats.

Source: *Nutrition Action Healthletter*, July/August 2002, www.cspinet.org.

Your Turn 11.11   Fatty Acid Composition

Using Figure 11.8 and information from this section, identify the predominant fatty acids likely to be present in each of these.

a.   canola oil
b.   olive oil
c.   lard

Consider This 11.12   Omega-3 Fatty Acids

What are they? They are often in the news and are recommended by nutrition experts. Search the Web, or other sources, to find out:

a.   What are omega-3 fatty acids?
b.   What foods are good sources of omega-3s?
c.   What do research studies say are the health benefits of omega-3s?

Consider This 11.13   Trans Fatty Acids and Saturated Ones

Unsaturated fatty acids can be either in a cis or trans orientation. There is controversy about the role of trans fatty acids in the diet. Use the Web to answer these questions. Write a news article about your findings.

a.   What is the structural difference between cis and trans fatty acids?
b.   What is the controversy regarding dietary trans fatty acids?
c.   Have studies been completed that resolve the controversy? If so, cite them.

Some food labels also reveal that not all the fats and oils in our diet are consumed in their naturally occurring molecular forms. Unless you eat "natural" peanut butter, the jar on your shelf probably is labeled something like, "oil modified by partial hydrogenation." Peanuts ground to make peanut butter always release a quantity of peanut oil. This oil separates on standing and must be stirred back into the solid before use. The oil extracted from peanuts is rich in mono- and polyunsaturated fats, and thus is a liquid. It can be treated chemically and converted into a semisolid that does not separate from peanut butter. This is done by reacting the oil with hydrogen gas over a metallic catalyst. The hydrogen adds to the double bonds in the oil, converting some, but not all, of the C-to-C double bonds into C-to-C single bonds, such as with the linoleic acid in peanut oil.

$$CH_3(CH_2)_4{-}CH{=}CH{-}CH_2{-}CH{=}CH{-}(CH_2)_7COOH + H_2 \longrightarrow$$

$$CH_3(CH_2)_4{-}CH_2{-}CH_2{-}CH_2{-}CH{=}CH{-}(CH_2)_7COOH \quad [11.3]$$

As a result of this partial hydrogenation, the number of double bonds in the lipid decreases, it becomes less unsaturated, and it is transformed from an oil into a semisolid fat. The extent of hydrogenation can be carefully controlled to yield products of desired unsaturation and resultant melting point, softness, and spreadability. Such customized fats and oils are in many products, including margarines, cookies, and candy bars.

In our preoccupation with dietary fat, it is important to realize that fats often enhance our enjoyment of food. They improve "mouth feel" and intensify certain flavors. Of prime significance, however, is the fact that fats are essential for life. They are the most concentrated source of energy in the body (Section 11.10), and they provide insulation that retains body heat and cushions internal organs. Moreover, triglycerides and other lipids, including cholesterol, are the primary components of cell membranes and nerve

Cholesterol is discussed in Section 10.7.

sheaths. Although "fathead" is hardly a compliment, in fact our brains are rich in lipids. Because fats play many important roles in our bodies, a variety of triglycerides are required to make them, incorporating a wide range of fatty acids—saturated, monounsaturated, and polyunsaturated. Fortunately, our bodies can synthesize almost all of the necessary fatty acids from the starting materials provided by a normal diet. The exceptions are linoleic and linolenic acids. These two essential fatty acids must be obtained directly from the foods we eat; our body cannot produce them. Generally this does not create a problem because linoleic and linolenic acids are found in many foods including plant oils, fish, and leafy vegetables.

Consider This 11.14   **Spreadables and Fat Content**

This table lists the fat content for Crisco (a partially hydrogenated shortening) and three soft, butter substitutes.

	Crisco	Brummel & Brown	I Can't Believe It's Not Butter	Benecol
Total fat (g)	12	10	10	9
Saturated	3	2	2	1
Monounsaturated	6	3	2.5	4
Polyunsaturated	3	2	4.5	3

The three butter substitutes do have less fat than butter, as advertised. All the butter substitutes are partially hydrogenated. Notice that the sum of the saturated, monounsaturated, and polyunsaturated fats does not equal the mass of total fat in the three products. The difference is the amount of trans fats, so called "stealth" fats. Thus, Brummel & Brown contains 10 g − 7 g = 3 g of trans fats per serving.

a. Calculate the grams of trans fats in each of the other two butter substitutes. Of the three butter substitutes, which one has the highest percentage of trans fats?

b. What percentage of the total fat in Crisco is comprised of monounsaturated fats? How does this percentage compare with the percentage of monounsaturated fats in the three butter substitutes?

c. Conduct a minisurvey of butter and margarine products in a supermarket in your area. List the total fat content and the mass of polyunsaturated, saturated, monounsaturated, and trans fats for examples of each of these five products: butter stick; regular margarine stick; regular margarine tub; "light" margarine stick; and "light" margarine tub.

> Trans fats arise from partial hydogenation of vegetable oils. Scientific studies show that they raise the level of triglycerides and LDL (or bad) cholesterol. The latter is discussed in the next section. In December of 2003, a bill was passed in which the FDA will require foods to include trans-fat information on the labels by 2006.

11.5 Controversial Cholesterol

Although dietary fat is an essential part of a balanced diet, the fact remains that many Americans consume too much of it, and too much of the wrong kind. Fats provide about 40% of the calories in the average American diet. Health care specialists recommend that this value should be 30% or less. Much of the concern and controversy regarding cardiac health problems is focused on cholesterol, one of the steroids introduced in Section 10.7.

Cholesterol has drawn heavy media attention. By the end of the 1980s, pooled data from a wide variety of studies led to the conclusion that high serum (blood) cholesterol levels *appear* to predict the potential for a stroke or heart attack. Although an irrefutable direct linkage has not yet been demonstrated, the data seem compelling. Waxy deposits of excess cholesterol (plaque) cause arteries to narrow and harden, which may elevate blood pressure and increase the risk of heart disease and stroke (Figure 11.9).

Figure 11.9

Cross sections of a healthy artery (top) and an artery clogged with atherosclerotic plaque (bottom).

Cholesterol falls into the lipid class of molecules, but it has a very different structure from the fats and oils that we have been discussing until now. Cholesterol is a four-ring structure with a hydrocarbon chain attached to the five-carbon ring (Figure 11.10). This four-ring structure is similar to that found in other hormones such as testosterone and progesterone. This class of compounds is known as *steroids* (see Section 10.7). Cholesterol is the most abundant steroid in the human body, being both synthesized internally and absorbed from food.

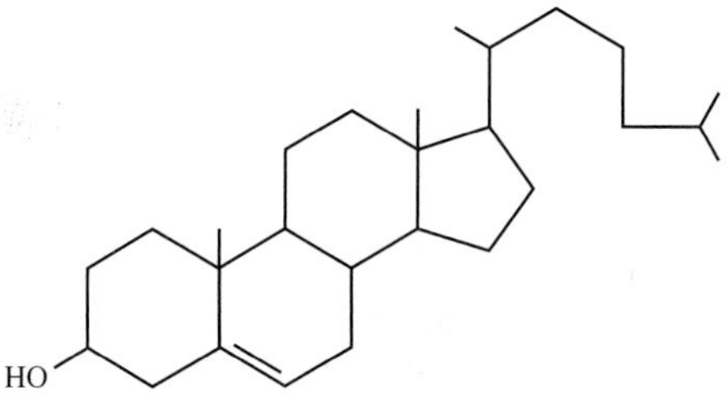

Figure 11.10

The structure of cholesterol.

Today there is general agreement that elevated blood cholesterol levels are associated with atherosclerosis, although consensus has not quite been reached on what constitutes dangerously high concentrations. Many medical researchers and the American Heart Association (AHA) consider values equal to or greater than 240 mg cholesterol per 100 mL of blood as the critical point for medical intervention. Cholesterol concentrations of 200–239 mg/100 mL are considered borderline high. According to the AHA, individuals with a total cholesterol of 240 mg/100 mL, in general, have twice the risk of heart attack than those who have a cholesterol level of 200 mg/100 mL.

One obvious response to elevated serum cholesterol is to restrict consumption of cholesterol. The American Heart Association recommends a maximum intake of 300 mg of cholesterol per day. This means cutting back on animal fats, which are rich in the compound. Included are fatty red meats, cream, butter, and cheese. Egg yolks are particularly rich in cholesterol, each yolk containing on average a whopping 213 mg. In contrast, egg whites contain no cholesterol, nor do fruits, vegetables, or vegetable oils.

Consider This 11.15 Cholesterol Content of Various Foods

An egg yolk contains significant amounts of cholesterol, whereas foods that are not derived directly from animals such as fruits and vegetables have no cholesterol. Use the Web to find the cholesterol content of various foods. List two foods that are high in cholesterol and two foods (not fruits or vegetables) that have much lower cholesterol levels.

But restricting dietary cholesterol is only one part of a two-part situation. It is possible that even a strict vegetarian with negligible cholesterol intake might have elevated serum cholesterol. This is because most of the body's cholesterol does not come from the diet directly, but is synthesized by the body. About 1 g of cholesterol is synthesized daily in the liver to maintain the minimum concentration required for use in cell membranes and to produce estrogen, testosterone, and other steroid hormones. The liver produces cholesterol principally from dietary saturated fats. Consequently, a high intake of saturated fats can result in a high concentration of cholesterol. Although cutting down on cholesterol consumption is an important step in lowering serum cholesterol, reducing the amount of saturated fats in the diet may be even more significant. The AHA recommends that only 8–10% of total Calories should come from such fats, and to especially limit the intake of those with certain numbers of carbon atoms per molecule: 12 (lauric acid), 14 (myristic acid), and 16 (palmitic acid).

Your Turn 11.16   A Fast Food Meal

The composition of a fast food meal is given here. Do calculations to determine whether the meal eaten meets the guideline that only 8–10% of total calories should come from saturated fats. *Hint:* Each gram of fat contains 9 Calories.

	Cheeseburger	French Fries	Shake
Calories	330	540	360
Calories from fat	130	230	80
Total fat (g)	14	26	9
Saturated fat (g)	6	4.5	6
Cholesterol (mg)	45	0	40
Sodium (mg)	830	350	250
Carbohydrates (g)	38	68	60
Sugars (g)	7	0	54
Proteins (g)	15	8	11

Perhaps the most important factor influencing serum cholesterol is genetics. This may explain why some people seem to eat fatty foods without suffering from heart disease, while others who carefully watch their diets are afflicted with it. Because we are not in complete control of our genes (at least not yet), physicians urge us to do what we can to lower our serum cholesterol. Reducing dietary cholesterol and fatty acids, exercising regularly, decreasing weight and stress, and eating certain types of dietary fiber appear to be important for maintaining good health.

Two compounds present in the body help carry cholesterol and triglycerides through the bloodstream: HDL (high-density lipoprotein) and LDL (low-density lipoprotein). These may help prevent the build-up of plaque in the arteries, and their concentration in the blood is affected by eating and exercise habits. The HDLs are the "good" lipoproteins, more effective in transporting cholesterol than LDLs. The AHA recommends a concentration of greater than 35 mg HDL/100 mL of blood and an LDL value of less than 130 mg/100 mL. It appears that people with high values for the LDL/HDL concentration ratio are particularly susceptible to heart disease. Evidence suggests that regular exercise (20–30 minutes at least three times per week) increases HDL at the expense of LDL, and also serves to burn calories, a process necessary to prevent weight gain. Several studies suggest that having one alcoholic drink per day also raises the HDL level and reduces risk of heart disease. But alcohol has its own risks, so moderation is clearly important.

Consider This 11.17   Current AHA Recommendations

Have the American Heart Association's recommended values for cholesterol, HDL, and LDL changed from those reported in this section? Check the current recommendations at the AHA Web site. If there are any differences, explain why.

Sceptical Chymist 11.18   Olean, A "Fake" Fat

Americans eat about 22 lb of salty snack foods per capita every year. Olean, a nonfattening, nonmetabolizable fat developed by the Procter & Gamble Company, was approved by the FDA in 1996 for use in salty snack foods such as potato chips and tortilla chips. In spite of having FDA approval, Olean remains controversial, with supporters and detractors. Use the Web to locate the Olean Web site as well as other sites that present contrasting viewpoints.

(continued on p. 502)

Sceptical Chymist 11.18   Olean, A "Fake" Fat (*continued*)

a. How does Olean work? Why is it not digested?

b. Why is the use of Olean in snack foods controversial?

c. In August 2003, the FDA removed the requirement for a warning label for products containing Olean. Why did the FDA take this action? Give examples of opposing viewpoints regarding the use of the Olean warning labels.

d. What is your decision about using products containing Olean? Explain your reasoning.

The campaign to lower blood cholesterol is not without cautions. For example, when researchers worldwide investigated blood cholesterol concentrations in relation to all premature deaths, not just those from heart disease, they discovered a very interesting result. As seen in Figure 11.11, the graphs of the results were a slightly U-shaped curve for men and a flat, declining line for women. In other words, men were at greater risk at both high and low cholesterol levels, whereas women showed no such signs, even at high cholesterol levels. Men with cholesterol levels greater than 240 mg/100 mL tended to die prematurely of heart disease while men with cholesterol levels lower than 160 mg/100 mL died prematurely from cancer and from respiratory and digestive diseases. Thus, although low-fat diets appear to be correlated to reduced incidence of heart disease, they may be connected to increased susceptibility to other conditions. In response to these findings, the American College of Physicians now suggests that "cholesterol reduction is certainly worthwhile for those at high, short-term risk of coronary heart disease but of 'much smaller or . . . uncertain' benefit for everyone else."

Sceptical Chymist 11.19   Dietary Fat

As you just read, the role of diet and cholesterol in heart disease and other health conditions is controversial. Ron Krauss, Chair of the American Heart Association Dietary Committee, suggests that it is "scientifically naive to expect that a single dietary regime can be beneficial for everybody: The 'goodness' or 'badness' of anything as complex as dietary fats will ultimately depend on the context of the individual."

a. Why does Krauss think that it is "scientifically naive to expect that a single dietary regime can be beneficial for everybody"?

b. Explain his sentence "The 'goodness' or 'badness' of anything as complex as dietary fat and its subtypes will ultimately depend on the context of the individual."

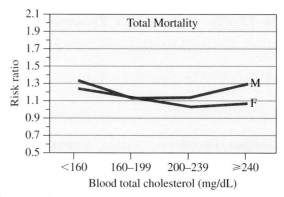

Figure 11.11
Risk of death and blood cholesterol levels.

Source: *Science,* Vol. 291, 30 March 2001, p. 3543.

11.6 Proteins: First Among Equals

The word *protein* derives from *protos,* Greek for "first." The name is misleading. Life depends on the interaction of thousands of chemicals, and to assign primary importance to any single compound or class of compounds is simplistic. Nevertheless, proteins are an essential part of every living cell. They are also major components in hair, skin, and muscle; and they transport oxygen, nutrients, and minerals through the bloodstream. Many of the hormones that act as chemical messengers are proteins, as are all the enzymes that catalyze the chemistry of life.

Proteins are polyamides or polypeptides, polymers made up of amino acid monomers. The great majority of proteins are made from various combinations of the 20 different naturally occurring amino acids. Molecules of all these amino acids share a common structural pattern. Four chemical species are attached to a carbon atom: (1) a carboxylic acid group, —COOH; (2) an amine group, —NH_2; (3) a hydrogen atom, —H; and (4) a side chain designated R in the following structure.

Amino acids and proteins were mentioned briefly in Section 9.6.

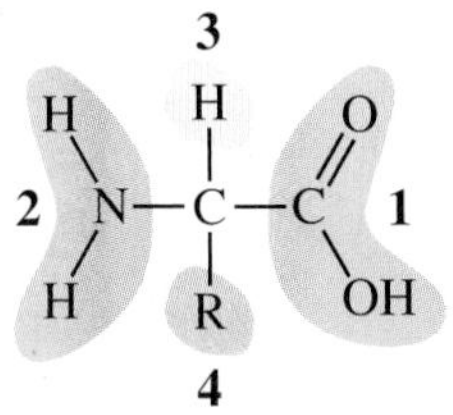

Variations in the R side-chain group differentiate the individual amino acids (Figure 11.12). In glycine, the simplest amino acid, the R is a hydrogen atom. In alanine, R is a —CH_3 group; in aspartic acid (found in asparagus), it is —CH_2COOH; and in phenylalanine, it is a group with the formula —$CH_2(C_6H_5)$. Here, C_6H_5 designates the hexagonal phenyl ring first introduced in Chapter 9 (Section 9.2). Note the structural relationship between alanine and phenylalanine.

Two of the 20 naturally occurring amino acids have R groups that bear a second —COOH functional group, three have R groups containing amine groups, and two others contain sulfur atoms. Because all amino acids except glycine involve four different units bonded to a central carbon atom, they all exhibit chirality. All the naturally occurring amino acids that are incorporated into proteins are in the left-handed isomeric form.

See Section 10.6 for a discussion of chirality.

Combining amino acids to form proteins depends on the presence of the two characteristic functional groups that give this family of compounds its name: the amine group and the acid group. Equation 11.4 represents the reaction of glycine with alanine to form a dipeptide, a compound formed from two amino acids. Here, the acidic —COOH group of a glycine molecule reacts with the —NH_2 group of alanine, and an H_2O molecule is eliminated. In the process, the two amino acids become linked by a peptide bond

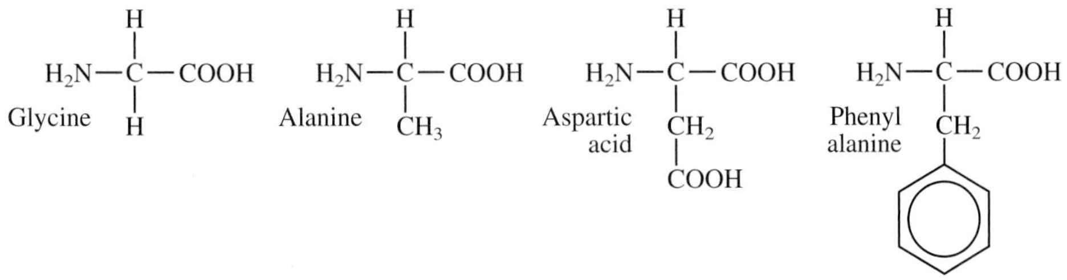

Figure 11.12

Examples of amino acids with different side chains.

The peptide bond that forms when two amino acids react is discussed in Section 9.6.

(indicated in the box). Once incorporated into the peptide chain, the amino acids are known as **amino acid residues.**

$$\underset{\text{Glycine}}{\overset{\displaystyle \underset{|}{\overset{|}{\text{H}}}\overset{\text{H}}{\underset{|}{\text{N}}}\!-\!\underset{\underset{\text{H}\ \ \text{H}}{|}}{\text{C}}\!-\!\overset{\displaystyle \overset{\text{O}}{\|}}{\text{C}}\!-\!\text{O}\!-\!\text{H}}{}} \quad + \quad \underset{\text{Alanine}}{\overset{\displaystyle \text{H}\text{N}\!-\!\text{C}\!-\!\text{C}\!-\!\text{O}\!-\!\text{H}}{}} \quad \longrightarrow \quad \underset{\text{Dipeptide}}{\overset{\displaystyle \text{Dipeptide}}{}} \quad + \quad \underset{\text{Water}}{\text{H}_2\text{O}} \qquad [11.4]$$

The reaction in equation 11.4 is another example of condensation polymerization, already encountered in the formation of polysaccharides (Section 11.2) and some synthetic polymers (Section 9.5). The reaction is equivalent to the reaction in equation 9.7.

Because each amino acid bears an amine group and an acid group, there are two ways the amino acids can join. Hence, two dipeptides are possible. We illustrate the options with simple block diagrams for the amino acids. The first case is that of equation 11.4 in which glycine acts as the acid and alanine as the amine.

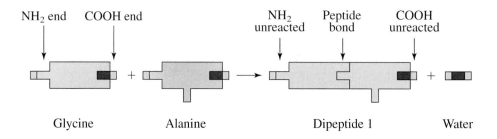

In the second case, the amino acids reverse roles; alanine provides the —COOH and glycine the —NH$_2$ for the reaction.

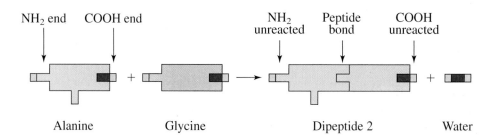

Examination of the molecular structures of the two dipeptides indicates that they are not identical. In dipeptide 1, the unreacted amine group is on the glycine residue and the unreacted acid group is on the alanine residue; in dipeptide 2, the —NH$_2$ is on the alanine residue and the —COOH on the glycine residue.

The point of all this is that the order of amino acid residues in a peptide makes a difference in its properties. The particular protein formed depends not only on what amino acids are present, but also on their sequence in the protein chain. Assembling the correct amino acid sequence to make a particular protein is like putting letters in a word; if they are in a different order, a completely new meaning results. Thus, a tripeptide consisting of three different amino acids is like a three-letter word containing the letters *a*, *e*, and *t*. There are six possible combinations of these letters. Three of them—*ate, eat,* and *tea*—form recognizable English words; the other three—*aet, eta,* and *tae*—do not. Similarly, some sequences of amino acids may be biological nonsense.

Putting the amino acids of a protein into their proper order is like assembling a train correctly by placing each car in the right sequence.

Still restricting ourselves to three-letter words and only the letters *a, e,* and *t,* but allowing the duplication of letters, we can make perfectly good words such as *tee* and *tat,* and lots of meaningless combinations such as *aaa* and *tte.* There are, in fact, a total of 27 possibilities, including the 6 identified earlier. Just as many words use letters more than once, most proteins contain specific amino acids incorporated more than once. More information about the structure and synthesis of proteins is included in Chapter 12.

Your Turn 11.20 **Making Tripeptides**

You can see from the block diagrams in this section that one glycine (Gly) and one alanine (Ala) molecule can combine to form two dipeptides: GlyAla and AlaGly. If one permits multiple use of each of the two amino acids, two other dipeptides are possible: GlyGly and AlaAla. Thus, a total of four different dipeptides can be made from two amino acids if each amino acid can be used more than once. Eight different tripeptides can be made from supplies of two different amino acids, assuming that each amino acid can be used once, twice, three times, or not at all. Use the symbols Gly and Ala to write down representations of the amino acid sequence in all eight of these tripeptides. *Hint:* Start with GlyGlyGly.

Consider This 11.21 **3-D Amino Acids**

Structural features of amino acids are more readily apparent if you look at their three-dimensional representations. At the *Online Learning Center* you can view and rotate the molecular structures for several different amino acids.

a. How is the three-dimensional structure of glycine different from the two-dimensional structure shown in your text?

b. Glycine is the simplest amino acid. It contains only two functional groups (—NH_2 and —COOH) and only the elements C, H, O, and N. Browse through a Web-based collection of amino acids and then describe two ways in which their structures are more complex than glycine's.

c. In leucine, what four different groups are bonded to a central carbon atom? Is this molecule optically active? Explain your answer.

11.7 Good Nutrition and Alternative Diets: Getting Enough Protein

Dietary protein requirements are usually expressed in terms of grams of protein per kilogram of body weight per day, and they vary with age, size, and energy demand. Each day, infants require 1.8 g protein/kg of body mass, middle-school children about 1.0 g/kg, and adults 0.8 g/kg. Therefore, a 20-lb (9-kg) child needs 16 g of protein daily to provide the raw materials for body growth and development. A 165-lb (75-kg) adult requires 60 g each day to maintain proper physiological function.

The body does not normally store a reserve supply of protein, so foods containing these nutrients must be eaten every day. As the principal source of nitrogen for the body, proteins are constantly being broken down and reconstructed. A healthy adult on a balanced diet is in nitrogen balance, excreting as much nitrogen (primarily as urea in the urine) as she or he ingests. Growing children, pregnant women, and persons recovering from long-term debilitating illness or burns have a positive nitrogen balance. This means that they consume more nitrogen than they excrete because they are using the element to synthesize additional protein. A negative nitrogen balance exists when more protein is being decomposed than is being made. This occurs in starvation, when the energy needs of the body are unmet from the diet, and muscle is metabolized to maintain physiological functions. In effect, the body feeds on itself.

Table 11.4	The Essential Amino Acids	
Histidine	Lysine	Threonine
Isoleucine	Methionine	Tryptophan
Leucine	Phenylalanine	Valine

Another cause of a negative nitrogen balance may be a diet that does not include enough of the essential amino acids. Of the 20 natural amino acids that make up our proteins, we can synthesize 11 from simpler molecules, but 9 must be ingested directly. If any of the nine essential amino acids identified in Table 11.4 are missing from the diet, many important proteins cannot be produced in the body in sufficient quantity. The result can be severe malnutrition.

Good nutrition thus requires protein in sufficient quantity and suitable quality. Beef, fish, poultry, and other meats contain all the essential amino acids in approximately the same proportions found in the human body. Therefore, meat is termed a complete protein. However, most people of the world depend on grains and other vegetable crops rather than meat as their major sources of protein. If such a diet is not sufficiently diversified, some essential amino acids may be lacking. For example, Mexican and Latin American diets are rich in corn and corn products, a protein source that is *incomplete* because corn is low in tryptophan, an essential amino acid. A person may eat enough corn to meet the total protein requirement, but still be malnourished because of insufficient tryptophan.

Fortunately for millions of vegetarians, a reliance on vegetable protein does not necessarily doom one to malnutrition. The trick is to apply a principle nutritionists call **complementarity,** combining foods that complement one another's essential amino acid content so that the total diet provides a complete supply of amino acids. You do this, likely unknowingly, every time you eat a peanut butter sandwich. Bread is deficient in lysine and isoleucine, but peanut butter supplies these amino acids. On the other hand, peanut butter is low in methionine, a compound provided by bread. The traditional diets of many countries meet protein requirements through nutritional complementarity. In Latin America, beans are used to complement corn tortillas; soy foods are eaten with rice in parts of Southeast Asia and Japan. People in the Middle East combine bulgur wheat with chickpeas or eat hummus, a sauce of sesame seeds, and chickpeas, with pita bread. In India, lentils and yogurt are eaten with unleavened bread.

Consider This 11.22   Vegetarian Complementarity

Use the Web to find at least two additional examples of complementarity.

a.   What essential amino acids are involved in the combination?
b.   Do these combinations involve common foods of that country (such as peanut butter and bread in the United States)?

Livestock, especially beef cattle, also benefit from complementarity. They are fed a variety of grains with a complete set of amino acids to incorporate ultimately into steaks and hamburger. However, the second law of thermodynamics applies to beef cattle as well as to everything else. There is a loss of efficiency with each step of energy transfer, whether in electrical power plants or in cells during metabolism. Cattle are notoriously inefficient in converting the energy in their feed into meat on the hoof. It takes about 7 lb of grain to produce 1 lb of beef. Put into human terms, the 1.75 lb of grain used to produce a "quarter-pounder" can provide two days of food for someone on a vegetarian diet. Other animals are more efficient than cattle in converting grain to meat. Hogs require 6 lb of grain per pound of meat, turkeys need 4 lb, and chickens even less, only 3 lb. It is obviously much more efficient to get food energy directly from grains, rather than through secondary or tertiary sources further along the food chain. On the other hand, one should keep in mind that pasture land used to graze cattle is often unsuitable for growing crops. Moreover, much of the food consumed by animals would be indigestible or unpalatable to humans.

A postscript to the protein story is provided by the unusual case of aspartame, a sweet dipeptide. Because of the great American preoccupation and battle with excess Calories and excess pounds, artificial sweeteners have become a billion-dollar business. Gram for gram, these compounds are much sweeter than sugar, but they have little if any nutritive value. Hence, they are nonfattening. The principal use (75%) of artificial sweeteners is in soft drinks. Currently the most widely used artificial sweetener is aspartame, the principal ingredient in NutraSweet and Equal. Somewhat surprisingly, the compound is related to proteins. Aspartame is a dipeptide made from aspartic acid and a slightly modified phenylalanine. The molecular structure of aspartame is given in Figure 11.13.

Alone, neither of the two amino acids in aspartame tastes sweet. Yet, the compound that results from their chemical combination is about 200 times sweeter than sucrose. The fact that sucrose and aspartame are chemically and structurally very different invites speculation about the molecular features that convey sweetness. But this digression will be long enough without taking up the issue of sweetness. For whatever reason, aspartame is sufficiently sweet to be used by millions of people worldwide. A few cases of adverse side effects have been attributed to aspartame, but exhaustive reviews have failed to show an unequivocal and direct connection between the symptoms and the sweetener. For the vast majority of consumers, aspartame is a safe alternative to sugar. One group of people, however, definitely should not use aspartame. The warning on packets of artificial sweeteners and products containing aspartame is explicit: "Phenylketonurics: Contains Phenylalanine" (Figure 11.14).

This is a case where one person's meat is another person's poison. Phenylalanine is an essential amino acid converted in the body to tyrosine, another amino acid. Individuals with phenylketonuria, a genetically transmitted disease, lack the enzyme that catalyzes this transformation. Consequently, the conversion of dietary phenylalanine to tyrosine is blocked and the phenylalanine concentration rises. To compensate for the elevated phenylalanine, the body converts it to phenylpyruvic acid, excreting large quantities of this acid in the urine. Phenylpyruvic acid is termed a "keto" acid because of its molecular structure; hence, the disease is known as phenyl*keto*nuria or PKU. People with the disease are called phenylketonurics.

Excess phenylpyruvic acid causes severe mental retardation. Therefore, the urine of newborn babies is tested for this compound, using special test paper placed in the diaper. Infants diagnosed with PKU must be placed on a diet severely limited in phenylalanine. This means avoiding excess phenylalanine from milk, meats, and other sources rich in protein. Commercial food products are available for such diets, their composition adjusted to the age of the user. Because phenylalanine is an essential amino acid, a minimum amount of it must still be available, even in phenylketonurics. Supplemental tyrosine may also be needed to compensate for the absence of the normal conversion of phenylalanine to tyrosine. A phenylalanine-restricted diet is recommended for phenylketonurics at least through adolescence. Adult phenylketonurics must also limit their phenylalanine intake, and hence curtail their use of aspartame.

> Americans drink an annual average of about 53 gallons of soft drinks per person, up nearly 50% from the amount in 1985.

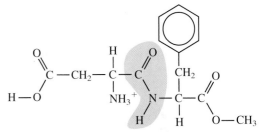

Figure 11.13
The molecular structure of aspartame.

> The ketone is a functional group that is discussed in Section 10.3.

Figure 11.14
Warning: "Phenylketonurics: Contains Phenylalanine" on a diet beverage can.

 Consider This 11.23 **Searching for Sweetness**

Aspartame received FDA approval in 1981, but the search for artificial sweeteners has continued. Use the Web to find out if any artificial sweeteners have received FDA approval since 1981.

a. Identify the artificial sweetener(s) and its (their) intended uses.
b. Describe the chemical composition of any approved sweetener(s).

11.8 Vitamins: The Other Essentials

Your daily diet should supply an adequate number of Calories, but Calories alone are not enough. You have already read about the essential fatty acids and amino acids that must be ingested for good health, and you are well aware that a balanced diet must also provide certain vitamins and minerals. Unfortunately, many popular foods that are high in sugars and fats lack these essential micronutrients. It is thus possible that a person can be overfed with excess Calories but malnourished through a diet lacking adequate vitamins and minerals.

A detailed understanding of the role of vitamins and minerals is of relatively recent origin. Over the ages, humans learned that if certain foods were lacking, illness often resulted, but the correlation between diet and health was often accidental and anecdotal. More systematic studies began early in the 20th century, with the discovery of "Vitamine B_1" (thiamine). The particular designation, B_1, was the label on the test tube in which the sample was collected. The general term *vitamin* was chosen because the compound, which is *vit*al for life, is chemically classed as an *amine*. The final "e" disappeared with the discovery that not all vitamins are amines. Today, **vitamins** are defined by their properties: they are essential in the diet, although required in very small amounts; they all are organic molecules with a wide range of physiological functions; and they generally are not used as a source of energy, although some of them help break down macronutrients.

Vitamins are often classified on the basis of solubilities; they either dissolve in water or in fat. Vitamins A, D, E, and K are all nonpolar molecules and therefore dissolve in fat but not in water. For example, the molecular structure of vitamin A, shown below, is based almost exclusively on carbon and hydrogen atoms. Thus, it is similar to the hydrocarbons derived from petroleum. Vitamins that are not fat-soluble are soluble in water because these polar molecules contain several —OH groups, which form hydrogen bonds with water molecules. Vitamin C is a case in point.

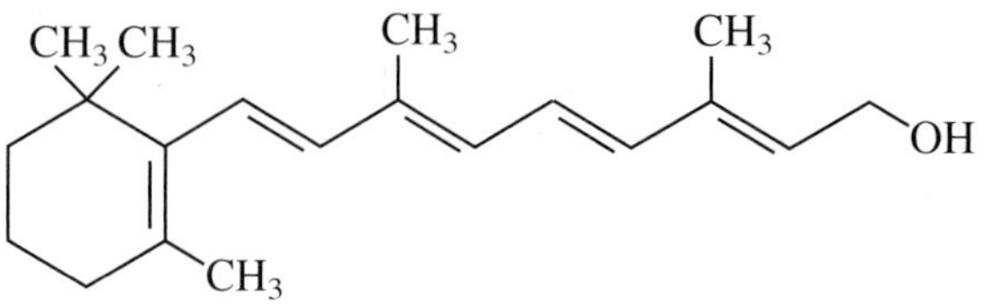

> The relationship betweeen molecular structure and solubility was discussed in Section 5.12. Hydrogen bonds were discussed in Section 5.8.

Vitamin A, a fat-soluble vitamin 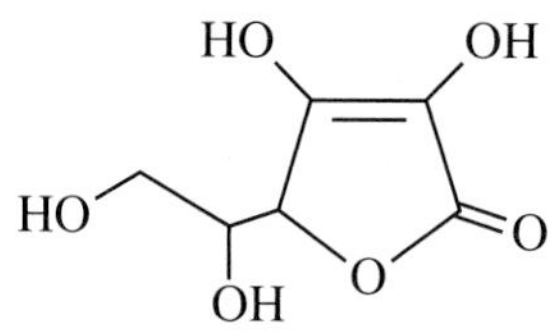   Vitamin C, a water-soluble vitamin

Your Turn 11.24   **Vitamin Structures and Solubility**

Examine the molecular structures given for vitamin A and vitamin C. Use solubility and molecular structure relationships described in Section 5.12 under the general principle that "like dissolves like" to explain why vitamin A is soluble in fat but vitamin C is soluble in water.

These solubility differences among vitamins have significant implications for nutrition and health. Because of their fat-solubility, vitamins A, D, E, and K are stored in cells rich in lipids, where they are available on biological demand. This means that the fat-soluble vitamins need not be taken daily. It also means that these vitamins can build up to toxic levels if taken far in excess of normal requirements. For example, high doses of vitamin A can result in fatigue, headache, dizziness, blurred vision, dry skin, nausea, and liver damage. Vitamin D toxicity occurs at just four to five times its Recommended Daily Allowance (RDA), making vitamin D the most toxic vitamin. Cardiac and kidney damage can result. Such high levels of the vitamin are reached using vitamin supplements, not through a normal diet.

Water-soluble vitamins, by contrast, are not generally stored; any unused excess is excreted in urine. Thus, they must be consumed frequently and in small doses.

Unfortunately, when taken in extremely large doses, even water-soluble vitamins can accumulate at toxic levels, although such cases are rare. For example, there are reports that vitamin B_6, taken at 10–30 times the recommended dose per day for extended periods, results in nerve damage, including paralysis. Even higher doses of vitamin B_6 supplements, up to 1000 times the recommended dosage, have been consumed to alleviate the symptoms of premenstrual syndrome (PMS), again causing abnormal neurological symptoms. For most people, a balanced diet should provide all the necessary vitamins and minerals in appropriate amounts, making vitamin supplements unnecessary.

Even a brief review of various essential vitamins and minerals is well beyond the scope of this book, but a few observations might be of interest. For example, niacin, or nicotinic acid, illustrates the way in which vitamins, especially members of the B family, act as coenzymes. **Coenzymes** are generally small molecules that work in conjunction with enzymes to enhance the enzyme's activity. Niacin plays an essential role in energy transfer during glucose and fat metabolism. The synthesis of niacin in the body requires the essential amino acid tryptophan. Thus, a diet deficient in tryptophan may lead to niacin deficiency. Such a deficiency causes pellagra, a condition involving a darkening and flaking of the skin, as well as behavioral aberrations.

Vitamin C (ascorbic acid) must also be supplied in the diet, typically via citrus fruits and green vegetables. An insufficient supply of the vitamin leads to scurvy, a disease in which collagen, an important structural protein, is broken down. The link between citrus fruits and scurvy was discovered more than 200 years ago when it was found that feeding British sailors limes or lime juice on long sea voyages prevented the disease. Ascorbic acid is also required for the uptake, use, and storage of iron, important in the prevention of anemia. The claims that high doses of vitamin C can prevent colds and ward off certain cancers remain largely unsubstantiated.

> This practice also led to British sailors being called "limeys."

The last vitamin in this brief overview is vitamin E, important in the maintenance of cell membranes and as protection against high concentrations of oxygen, such as those that occur in the lungs. Vitamin E is so widely distributed in foods that it is difficult to create a diet deficient in it, although people who eat very little fat may need supplements. Vitamin E deficiency in humans has been linked with nocturnal cramping in the calves and fibrocystic breast disease.

Consider This 11.25 Megadosing Vitamin C

In the past, there have been claims that megadoses of vitamin C prevent the common cold and may be effective against certain types of cancers. Use the Web to find evidence to either support or refute the claims about the efficacy of megadoses of vitamin C.

a. What dose of vitamin C constitutes a megadose?
b. Where was the research conducted?
c. What were the results of the research?

11.9 Minerals: Macro and Micro

An adequate supply of a number of minerals (ions or inorganic compounds) is also essential for good health. Table 11.2 listed calcium, phosphorus, chlorine, potassium, sulfur, sodium, and magnesium among the major elements in the body. These seven macrominerals, although not nearly as abundant as oxygen, carbon, hydrogen, or nitrogen, are nevertheless necessary for life. Macrominerals are needed daily in our diet in amounts greater than 100 mg (0.100 g) or are present in the body in amounts greater than 0.01% of body weight. The adult RDAs for these macrominerals typically range from 1 to 2 g. The body requires lesser amounts of iron, copper, zinc, and fluorine—the so-called microminerals. Trace minerals, including iodine, selenium, vanadium, chromium, manganese, cobalt, nickel, molybdenum, and tin are usually measured in micrograms (1×10^{-6} g). Arsenic, cadmium, and even lead, which are generally classified as toxic, are actually needed by the body, but in very small amounts. Although

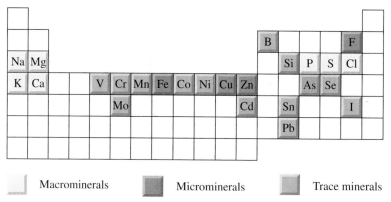

Figure 11.15

Periodic table indicating macrominerals, microminerals, and some trace minerals necessary for human life. Boron, silicon, arsenic, lead, and cadmium are essential in animals and likely essential in humans.

the total amount of trace elements in the body is only about 25–30 g, their slight amounts belie the disproportionate importance they have in good health.

Only the essential minerals are shown in the periodic table in Figure 11.15. The metallic elements exist in the body as cations, for example, Ca^{2+} (calcium), Mg^{2+} (magnesium), K^+ (potassium), and Na^+ (sodium). The nonmetals typically are present as anions, thus chlorine is found as Cl^- and phosphorus appears in the phosphate ion, PO_4^{3-}.

The physiological functions of minerals are widely diverse. Calcium is the most abundant mineral in the body. Along with phosphorus and smaller amounts of fluorine, it is a major constituent of bones and teeth. Blood clotting, muscle contraction, and transmission of nerve impulses also require Ca^{2+} ions.

Sodium is also essential for life, but not in the relatively excessive amounts supplied by the diets of most Americans. Physicians recommend a maximum of 1.2 g (1200 mg) of sodium (Na^+) per day. This corresponds to 3 g of salt (NaCl) and is twice the estimated minimum requirement. Most Americans exceed the recommended daily sodium intake, sometimes by three- to fourfold. The major culprits are processed foods and fast foods, which are very heavily salted for flavor. The major concern with excess dietary sodium is its correlation with high blood pressure (hypertension).

> The Latin word for salt is *sal*. Salt was so highly valued in Roman times that soldiers were paid in *sal*, thereby forming the root for the modern word *salary*.

Your Turn 11.26   **Sodium in Your Diet**

The average U.S. daily diet contains about 9 g of NaCl, which is about 3.6 g Na^+. Make a list of how much sodium is in some of the typical foods you eat. Some common examples are: processed macaroni and cheese, 1086 mg Na^+ per serving; white bread slice, 114 mg Na^+ per serving. Recall that 1 g = 1000 mg.

Oranges, bananas, tomatoes, and potatoes help supply the recommended daily requirement of 2 g of potassium, a mineral that is essential for the transmission of nerve impulses and intracellular enzyme activity. Potassium and sodium are elements in the first column of the periodic table. Neutral atoms of the two elements each have one electron in the outer level. These electrons are readily lost to form K^+ and Na^+ ions, the forms found in the human body. Because sodium and potassium ions have similar chemical properties, their physiological functions are also closely related. In intracellular fluid (the liquid within cells), the concentration of potassium ions is considerably greater than that of sodium ions. The reverse situation holds in the lymph and blood serum outside the cells. There the concentration of potassium ions is low and that of sodium ions is high. The relative concentrations of K^+ and Na^+ are especially important for the rhythmic beating of the heart. Individuals who take diuretics to control high blood pressure may also take potassium supplements to replace potassium excreted in the urine. However, such supplements should be taken only under a physician's directions because of the potential danger that they could dramatically alter the potassium–sodium balance and lead to cardiac complications.

In most instances, the microminerals and trace elements have very specific biological functions and may be incorporated in only a relatively small number of biomolecules. Iron, for example, is an essential part of hemoglobin, the protein that transports oxygen in the blood, and of myoglobin, which is used for temporary oxygen storage in muscle. A hemoglobin molecule has four Fe^{2+} ions, each of which binds reversibly to an O_2 molecule. Insufficient iron in the diet causes iron deficiency anemia, a condition in which the red blood cells are low in hemoglobin and correspondingly can carry less oxygen. Symptoms include fatigue, listlessness, and decreased resistance to infection. Iron deficiency anemia is a major problem in developed as well as developing nations. Iron deficiencies have been estimated as high as 20% in the United States, particularly among postpuberty women. On the other hand, too much iron in the diet can cause gastrointestinal distress and contribute to cirrhosis of the liver. Children have been fatally poisoned by ingesting iron supplement tablets. To be utilized by the body, iron must be absorbed as Fe^{2+} ions, not in the Fe^{3+} form or simply as elemental iron. Therefore, it is a bit surprising that some iron-fortified cereals contain metallic iron dust that can be removed by a magnet from a slurry of the cereal and water. Foods naturally rich in the iron that is usable by the body include liver and spinach.

Iodine is another element with a specific biological function. Most of the body's iodine is concentrated in the thyroid gland, where it is incorporated into thyroxine, a hormone that regulates metabolism. Excess thyroxine is associated with hyperthyroidism or Graves' disease, in which basal metabolism is accelerated to an unhealthy level, rather like a racing engine. On the other hand, a thyroxine deficiency, sometimes caused by insufficient dietary iodine, slows metabolism and results in tiredness and listlessness. Both hyper- and hypothyroidism can lead to goiter, an enlargement of the thyroid gland. One way to help prevent goiter is by consuming adequate amounts of iodine. Seafood is a rich source of the element, but it is also provided by iodized salt, normal sodium chloride to which 0.02% of potassium iodide (KI) has been added. The tendency of the thyroid gland to concentrate iodine is key to the use of radioactive iodine-131 as a treatment for an overactive thyroid and to the risks of accidental exposure to this isotope.

Consider This 11.27 **Getting Well Using Radioactive Iodine**

Hyperthyroid individuals suffer from an overactive thyroid and accelerated metabolism. Use the Web to find out how radioactive iodine-131 is used to treat hyperthyroidism.

a. What role does I-131 play in the treatment?
b. Isn't radioactivity bad for the patient? Explain.
c. What further treatment does the patient need after the I-131 is used?

11.10 Energy from the Metabolism of Food

If you need a refresher on chemical energetics, see Section 4.4.

All the energy needed to run the complex chemical, mechanical, and electrical system called the human body comes from carbohydrates, fats, and proteins. This energy initially arrives on Earth in the form of sunlight, which is absorbed by green plants during photosynthesis. Under the influence of a catalyst called chlorophyll, carbon dioxide and water are combined to form glucose. In the process, the Sun's energy is stored in chemical bonds of the sugar.

$$\text{Energy (from sunshine)} + 6\,CO_2 + 6\,H_2O \longrightarrow C_6H_{12}O_6 + 6\,O_2 \qquad [11.5]$$

During metabolism, the photosynthetic process is reversed, the food is converted into simpler substances and the stored energy is released.

$$C_6H_{12}O_6 + 6\,O_2 \longrightarrow 6\,CO_2 + 6\,H_2O + \text{Energy (from metabolism)} \qquad [11.6]$$

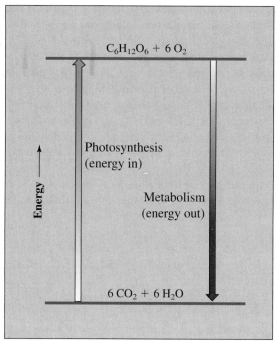

Figure 11.16
Energy from photosynthesis and metabolism.

The breaking of chemical bonds in glucose and oxygen molecules requires the absorption of energy. But, more energy is released in the exothermic reactions in which carbon dioxide and water are formed. Thus, there is a net release of energy. The energy balance between equations 11.5 and 11.6 is schematically represented in Figure 11.16.

Energy provided by food we eat is used to drive the chemical reactions that constitute the processes of life. The most obvious example of an energy-requiring process is muscular motion, including the beating of the heart. But most of the energy released by metabolism goes to maintain differences in ionic concentrations across cell membranes. The natural tendency is for diffusion to move substances from regions of higher concentration to those of lower concentration. Energy is required to prevent this from happening. The proper concentration differences that are essential for nerve action and other physiological functions are maintained at great energetic expense. In short, spontaneous reactions furnish the energy for nonspontaneous reactions to occur. As an analogy, consider an automobile storage battery. The battery produces electrical energy spontaneously because of chemical reactions in the battery. These spontaneous processes provide energy that can be used to permit nonspontaneous processes to take place, for example, starting the car or making the headlights and horn work.

In addition to having a supply of sufficient energy, the body must have some way of regulating the rate at which the energy is released. Without such control, wild temperature fluctuations and high inefficiency could result. Again, the automobile provides an analogy. Dropping a lighted match into the fuel tank would burn all the gasoline (and the car as well). This is a drastic and not particularly effective way to move a car. Under normal operating conditions, just enough fuel is delivered to the ignition system to supply the automobile with the energy it needs without raising the temperature of the car and its occupants beyond reason. In this way, by releasing a little energy at a time, the efficiency of the process is enhanced. So it is with the body. The conversion of foods ultimately into carbon dioxide and water occurs over many small steps, each one involving enzymes, enzyme regulators, and hormones. As a result, energy is released gradually, as needed, and body temperature is maintained within normal limits.

The chief sources of this energy in a well-balanced diet are carbohydrates and fats. When metabolized, *carbohydrates* provide about 4 kcal/g and *fats* release about 9 kcal/g. Although proteins, like carbohydrates, yield about 4 Cal/g, they are not a major energy

Each heartbeat uses one joule of energy.

Osmosis was discussed in Section 5.17.

1 dietary calorie = 1 Cal
            = 1 kcal
            = 1000 calories.

source, but rather a store of molecular parts for building skin, muscles, tendons, ligaments, blood, and enzymes. A kilocalorie (kcal) is identical to a dietary Calorie, written with a capital C. Package labels and nutritional tables that state how much energy we get from our meals typically use Calories, as in "one chocolate chip cookie provides 50 Calories." We will do the same. Keep in mind that whatever units are used, on a gram-for-gram basis, fats provide about 2.5 times as much energy as carbohydrates.

The reason for this dramatic difference is implicit in the chemical composition of these two types of material. Compare the formula of a fatty acid, lauric acid, $C_{12}H_{24}O_2$, with that of sucrose (table sugar), $C_{12}H_{22}O_{11}$. Both compounds have the same number of carbon atoms per molecule and very nearly the same number of hydrogen atoms. When the fatty acid or the sugar burns, its carbon and hydrogen atoms combine with added oxygen to form CO_2 and H_2O, respectively. But more oxygen is required to burn a gram of lauric acid, $C_{12}H_{24}O_2$, than a gram of sucrose, $C_{12}H_{22}O_{11}$. This is evident from the equations for the two reactions.

$$C_{12}H_{24}O_2 + 17\,O_2 \longrightarrow 12\,CO_2 + 12\,H_2O \qquad [11.7]$$

$$C_{12}H_{22}O_{11} + 12\,O_2 \longrightarrow 12\,CO_2 + 11\,H_2O \qquad [11.8]$$

In the language of chemistry, the sugar is already more "oxygenated" or more "oxidized" than the fatty acid. There are more CH bonds in the fatty acid to "burn" to CO_2 and H_2O and release more energy than from sucrose. Hence, sucrose is chemically and energetically "closer" to the end products of CO_2 and H_2O and needs less oxygen to form them. Therefore, when 1 g of sucrose is burned, 3.8 Calories is released, compared with 8.8 Cal/g of lauric acid.

The energetics involving oxygenated fuels was discussed in Chapter 4.

The fact that fats are such concentrated energy sources means that it is easy to get an unhealthy percentage of our daily Calories from fats. The problem is nicely illustrated by Your Turn 11.28 and Sceptical Chymist 11.29. In accordance with the 1991 USDA food pyramid, nutritionists and the American Heart Association advise that no more than 30% of your caloric intake should come from fats, and 55–60% should be derived from carbohydrates, especially polysaccharides. The remainder, 10% or less, should be contributed by protein.

Your Turn 11.28   Calories from Fat

Bagel chips are an alternative to potato chips. The label on one popular brand of bagel chips lists 130 Calories per serving, with 35 Calories from fat. The total fat is listed as 4 g per serving, with 1 g being saturated fat, 1 g being polyunsaturated fat, and 2 g being monounsaturated fat.

a. What percentage of daily value of calories based on a 2000-Cal diet is provided by one servings of these chips?
b. Potato chips have 150 Cal in a 28-g serving, 10 g of which is fat. Are bagel chips a more healthful alternative to potato chips in your opinion? Give reasons for your answer.

Sceptical Chymist 11.29   Low-Fat Cheese

A popular brand of low-fat shredded cheddar cheese advertises that it provides 1.5 g of fat with 15 Cal from total fat per serving. There are 50 Cal per serving and of the total fat, 1.0 g is saturated fat. A serving is defined as 1/4 cup or 28 g. Is this a "low-fat" cheese? Defend your decision with some calculations. Remember that the dietary recommendation is that no more than 30% of Calories should come from fat.

So how much energy or how many Calories does a person need? The answer is vague: "It depends." The number of Calories your diet should supply each day depends on your level of exercise or activity, the state of your health, your sex, age, body size, and a few other factors. You can probably find the category that best describes you in

Table 11.5	Recommended Daily Energy Intake (United States)				
Age (yrs)	Avg. weight (kg)	Avg. weight (lb)	Avg. height (in.)	Avg. Cal/kg	Avg. Cal/day
0.5–1.0	9	20	28	98	850
4–6	20	44	44	90	1800
7–10	28	62	52	70	2000
Males					
15–18	66	145	69	45	3000
19–24	72	160	70	40	2900
25–50	79	174	70	37	2900
51+	77	170	68	30	2300
Females					
15–18	55	120	64	40	2200
19–24	58	128	65	38	2200
25–50	63	138	64	36	2200
51+	65	143	63	30	1900

Table 11.5, which summarizes the daily food energy intakes that have been recommended for Americans. Most men require more Calories per day than women do, but the difference is not totally due to differences in body weight. The column showing the recommended number of Calories per kilogram of body weight indicates that the magnitude of this indicator decreases with age. Growing children need a proportionally large energy intake to fuel their high level of activity and provide raw material for building muscle and bone. Therefore, children are particularly susceptible to undernourishment and malnutrition. Indeed, mortality rates among infants and young children are disproportionately high in famine-stricken countries.

Your basal metabolism rate is approximately 1 Cal/kg body mass per hour.

The **basal metabolism rate (BMR)** is the minimum amount of energy required daily to support basic body functions. The BMR is sufficient to keep the heart beating, the lungs pumping, the brain active, the blood circulating, all major organs working, and body temperature at 37 °C. This corresponds to approximately one Calorie per kilogram (2.2 lb) of body weight per hour, although it varies with size and age. The BMR is experimentally determined in a resting state, and the quantity of energy used in digestion is eliminated by having the subject fast for 12 hours before the measurement is made. To put this on a personalized basis, consider a 20-year old female weighing 55 kg (121 lb). If her body has a minimum requirement of 1 Cal/(kg·h), her daily basal metabolism rate will be 1 Cal/(kg·h) × 55 kg × 24 h/day or about 1300 Cal/day. According to Table 11.5, the recommended daily energy intake for a woman of this age and weight is 2200 Cal. This means that 59% of the energy derived from this food goes just to keep her body systems going.

$$\frac{1300 \text{ Cal}}{2200 \text{ Cal}} \times 100 = 59\%$$

Where the rest of it goes depends on what she does. The law of conservation of energy decrees that the energy must go somewhere. If she "burns off" the extra Calories in exercise and activity, none will be stored as added fat and glycogen. But, if the excess energy is not expended, it will accumulate in chemical form. Putting it more crassly, "those who indulge, bulge," unless they work and play hard.

Some indication of how hard and how long we have to work or play to use up dietary Calories is given in Tables 11.6 and 11.7. The former reports the energy expenditures for various activities as a function of body weight. Table 11.7 quantifies exercise in readily recognizable units such as hamburgers, potato chips, and beer. Of course, by combining the information in this section with the information in earlier parts of this chapter about the *types* of nutrients in food, it should be clear that a healthful diet cannot be achieved simply by consuming the correct number of calories. A 2000-Cal diet of only potato chips and beer would leave a person malnourished. Proper nutrition is not simply a matter of how much, but also of what kind of food a person consumes.

Table 11.6	Energy Expenditure (Cal/min) for Various Activities in Relation to Body Mass (lb)			
Activity	**120 lb (Cal/min)**	**150 lb (Cal/min)**	**180 lb (Cal/min)**	**200 lb (Cal/min)**
Running (7 min/mile)	12	15	18	20
Basketball (vigorous)	10	13	15	17
Rollerblading (12 mph)	10	12	13	14
Jogging (10 min/mile)	9	11	14	15
Swimming (fast)	9	11	13	14
Aerobics (vigorous)	7	9	11	12
Bicycling (11 mph)	6	7	9	10
Golf (carrying clubs)	5	6	7	8
Volleyball	5	6	7	8
Walking (20 min/mile)	3	4	5	6
Studying/reading	1.3	1.7	1.9	2.0

Table 11.7	How Much Exercise Must I Do If I Eat This Cookie? Calories and Minutes of Exercise for a 150-lb Person		
Food	**Calories**	**Walk (min)**	**Run (min)**
Apple	125	31	8
Beer (regular) 8 oz	100	25	7
Chocolate chip cookie	50	12	3
Hamburger	350	88	23
Ice cream, 4 oz	175	44	12
Pizza, cheese, 1 slice	180	45	12
Potato chips, 1 oz	108	27	7

Your Turn 11.30   Jogging Off the Calories

A 150-lb person consumes a meal consisting of two hamburgers, 3 oz of potato chips, 8 oz of ice cream, and a 12-oz beer. Calculate the number of Calories in the meal and the number of minutes the person would have to run to "work off" the meal.

Answer
1524 Cal, 102 min running (from Table 11.6)

Your Turn 11.31   Calculate Your Energy Needs

First calculate your BMR. Then select from Table 11.6 the activities you do in a typical day. Calculate the supplemental energy you need for these activities. Then add your BMR and your supplemental energy needs to determine the total number of Calories you require per day. How does this result compare with the recommended energy intake for your age and sex (see Table 11.5).

11.11 **Quality versus Quantity: Diet Fads**

Despite great strides in improving healthcare and modern medicine in the last century, a new health crisis has emerged. Obesity is now at virtually epidemic proportions in the United States, as one of the opening quotes of this chapter stated. In the last 30 years scientific research on nutrition has made tremendous strides. However, with advertising and the mass media, people seem more frustrated than ever in trying to understand this advice. One day the "experts" say one thing; the next they seem to say the opposite. Because of an emphasis on short but often sensational news pieces, the media generally only reports the results of single studies. Often these are only reported when they are at odds with the currently accepted standards. But nutritional studies are very complex and the scientific results cannot be summarized into simple sound-bites while still being fully explained. Because human beings are complex biochemical machines, a great deal of care must be taken to understand the limitations and true implications of any single data set.

So how can we judge for ourselves? Recently, nutrition experts and diet-book authors have urged the government to test some of the weight-loss plans. As a result, we can expect even more data in the next several years. In many cases, there are a host of studies about a single, particular feature of diet. For example, numerous studies have confirmed that vitamin E acts as an antioxidant. When many research studies are carried out, with different groups of people and under different conditions, and they show a consensus, then the results can begin to provide a basis for peoples' behavior.

Large studies done on humans rather than mice, rats, and monkeys are more reliable. But all human-based studies have certain limitations based on the population used and on how the data are gathered. Furthermore, they are generally difficult and costly to undertake, and many years are required to get results. For example, heart disease and other chronic ailments take many decades to develop, and markers like narrowing of the arteries don't always develop into the disease.

There are two major approaches to nutritional and health studies: case–control and cohort studies. In **case–control studies** a group of people with an outcome (i.e., the health problem) are compared with a group without it. The former group who are ill often recall previous behavior differently than those without the problem. **Cohort studies** follow large groups of people over long periods of time. Two of the largest and longest running cohort studies of diet are the Harvard-based Nurses' Health Study and Health Professionals Follow-Up Study, with over 90,000 women and 50,000 men, respectively, being followed over several decades. These studies have provided a great deal of information that has often led to suggestions for dietary changes (including the development of the USDA food pyramid), and have sometimes fueled diet trends.

To understand the proliferation of diets, it is useful to begin with some history. In the 1960s experiments based on controlled-feeding of particular food items to participants for several weeks, showed that saturated fat increased cholesterol levels. But these studies also showed that polyunsaturated fats—found in vegetables and fish—reduced cholesterol. Advice in the following decades was to replace saturated fats with unsaturated ones rather than reducing total fat. The "sat fat is bad" movement led to the greatly expanded use of vegetable oils. Unfortunately, as was noted in Section 11.4, the partial hydrogenation of these oils to create products such as margarine gave rise to trans fats, a category of fat that may prove to be worse than saturated forms.

In addition "carbs are good" seemed to accompany "sat fat is bad," in part because animal protein sources are associated with saturated fats, whereas grain-based proteins are not. But the overindulgence on carbohydrates appears to have had its own undesirable consequences. Controlled feeding studies in the early 1990s showed that when a person replaces saturated fat with a caloric equivalent of carbohydrates, the total cholesterol, LDL, and HDL levels all fall. Because the LDL to HDL ratio does not change, a person's risk of heart disease is reduced only slightly. If a person switches from unsaturated fats to an equal caloric portion of carbohydrates, the HDL falls but the LDL actually rises, making the cholesterol ratio even worse. In addition the carbohydrates boost the blood levels of triglycerides, whose high levels are associated with higher risk of heart disease.

The large cohort Harvard studies have shown that a participant's risk of heart disease was strongly influenced by the type of dietary fat consumed. Eating trans fat increases the risk substantially while saturated fat increases it slightly. Unsaturated fats decrease the risk. Therefore, the total fat intake alone is not associated with heart disease risk. Furthermore, epidemiological studies have shown little evidence that total fat or specific fats affect the risks of several cancers. The rise in obesity has been blamed on fat but American consumption of calories from fat has decreased since the 1980s while the rate of obesity still continues to grow. All of this would seem to imply, then, that fat is not the culprit it was once believed to be. Or, at least, that the solution to American dietary woes is not quite so clear-cut.

So what about the "carbs?" Diet books that address the issue of carbohydrates have spanned the spectrum from banning all carbs to a "good carb versus bad carb" differentiation. The claim is that "bad" carbohydrates cause a quick rise in blood sugar, followed by a spike in blood insulin level and a weight gain caused by making the body store fat or making the person feel hungry again from the low blood sugar. Insulin is a hormone secreted by the pancreas that allows the cells in your body to absorb and store sugar that is in the blood. The sugar that is not immediately burned for energy is converted to and stored as fat in your cells. Furthermore, glucagon is also a hormone secreted by the pancreas that, essentially, has the opposite effect of insulin: it promotes the use of stored glucose in cells. The release of glucagon will decrease after a "glucose spike," such as would result from eating "bad carbs," but it is known to increase after consumption of proteins. Therefore, many of the new "low-carb" diets promote the consumption of increased amounts of protein as a way to use up stored calories. The long-term health effects of this approach have yet to be seen.

But the metabolism and classification of carbohydrates is more complicated than just defining them as "good" or "bad." In the early 1980s, a group of Canadian researchers defined the glycemic index, a scale to measure the rapidity and degree with which a fixed quantity of food increases blood sugar. But a particular food item may not even have a definitive value—the glycemic index GI depends on how the food was processed, stored, ripened, cut, or cooked. For example, glucose has a GI of 100, baked potatoes 93, carrots 49, pasta 39, and peanuts 14. Bread is high-GI whether whole wheat or white since it is made from finely ground flour. Pasta is low-GI whether wheat or white depending on whether it is thick (lower GI) or thin (higher GI). Rice ranges from high-GI (instant white) to low-GI (converted) with brown in the middle. Sugars range from glucose (high) to fructose (low GI) with sucrose (table sugar) in the middle. Data from the Nurses' Health Study suggest that a high dietary glycemic load from refined carbohydrates increases the risk of chronic heart disease, and an increased intake of whole grains may protect against it. Just as in the case of fats, the type of carbohydrates may be far more important than the total amounts. And in the battle between fats and carbs, a study of severely obese patients showed that those on a low-carbohydrate diet lost more weight than those on a low-fat diet (5.8 vs. 8.6 kg) and had greater decreases in triglyceride levels. Insulin sensitivity also improved in the low-carbohydrate subjects.

Clearly, nutrition and dieting are complex subjects. However, research in this area is increasing and is contributing to our understanding of these issues. What is interesting is that, in the United States the concern is on how to keep people from dying of obesity. But, historically, and in many parts of the world today, the concern has been on how to keep people from dying of starvation. This highlights, once again, the distinction between undernourishment and malnourishment. We turn next to the issue of hunger on our planet.

Consider This 11.32 Comparing Diets

The growth in girth of Americans has been accompanied by a proliferation of diet plans and books. Among others, healthful living has been promoted by the American Heart Association, the Atkins (phase I and phase II), the South Beach, Dean Ornish (very low fat), and Weight Watchers diets among others.

a. Pick two diets. Use the resources of the Web to compare and contrast the features. How close are either of these to your lifestyle and diet?

b. What are the scientific arguments behind the diets you have selected?

c. Use a Web-based search engine to find additional diets that appear to be significantly different from those above.

11.12 Feeding a Hungry World

According to the Food and Agricultural Organization of the United Nations (FAO) 2000 report, more than 800 million people worldwide were undernourished during 1998–1999, 99% of them in underdeveloped countries. Although the number of under-nourished people in 37 developing countries fell by 100 million, principally in India and China, another 59 developing nations suffered an increase, totaling 60 million indi-viduals. Translating these sterile statistics into terms of human misery means that nearly one in six people in the developing countries is undernourished, evidence of the mag-nitude and tenacity of hunger in the world (Figure 11.17).

This dietary discrepancy is evident from Table 11.8, which reports the average dietary energy supplies (DES) in Calories available per person per day for different global regions. Although the DES has generally increased since 1979, in each of the three time intervals listed, the average DES for developed nations was significantly greater than that for developing countries. In 1986–1988, the average daily individual energy intake in the developed countries was 44% higher than that in less industrially advanced nations. DES values for North America are conspicuously high—well above the recommended daily energy intake (see Table 11.5). Another way to compare the data of Table 11.8 is by cal-culating the DES in various parts of the world as a percentage of the North American level. Again for 1997–1999, the DES for Africa was only 73% that for North America, the Far East was 79% the North American value, and Latin America was 84%.

To meet growing populations, global grain and cereal production has almost doubled over the past quarter century, and supplies of vegetables, fruits, milk, meat, and fish have also increased. It is estimated that at current production levels there is enough food to provide for 6.1 billion people—which is less than today's world population of 6.36 bil-lion. Nevertheless, many in the world go hungry; over 800 million of our fellow human beings are undernourished, mostly women and children. Approximately 24,000 people die daily of undernourishment. Clearly, the world's food supply is not equally distributed among its inhabitants. Piles of corn and wheat rot in the American Midwest or on docks around the world, while half a billion men, women, and children go to bed hungry.

Economic, political, and social, as well as agricultural reasons account for this inequity. Some are geographic fates—certain areas of the world simply do not have enough arable land and adequate soil to produce sufficient food for their people. This is especially true in areas in which population growth is exploding. In some areas, pro-longed droughts reduce crop yields, and episodic floods in other locales wash away crops. The fertilizers needed to supplement mineral-depleted soils may not be available because of cost. Furthermore, animals for working the fields, to say nothing of trac-tors, may be too expensive for farmers in some regions.

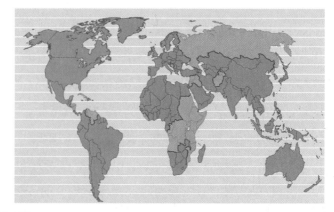

Figure 11.17

Countries (in salmon color) experiencing food supply shortfalls and requiring exceptional assistance.

Source: Reprinted with permission of The Food and Agriculture Organization of the United Nations, from *The State of Food and Agriculture,* 2000.

Table 11.8	Average Dietary Energy Supplies (DES) in Calories Per Person Per Day		
Region	**1979–1981**	**1986–1988**	**1997–1999**
Developed countries			
Averages	3300	3389	3220
North America	3487	3626	3300
Western Europe	3371	3445	3140
Developing countries			
Averages	2317	2352	2627
Africa	2148	2119	2415
Latin America	2675	2732	2770
Near East	2794	2914	…
Far East	2185	2220	2620
World			
Averages	2587	2671	…

Source: Data from *The State of Food and Agriculture,* United States, New York, Rome: 1990 and 2000; Food and Agriculture Organization of the United Nations, United Nations, New York, 1991 and 1999.

A nation that cannot feed itself must import food, which means that it must have something to sell. Consequently, economics and international trade dictate the nutrition equation. Under these conditions, some countries face the difficult choice between supporting domestic agriculture to achieve self-sufficiency or investing in manufacturing goods for export to establish a favorable trade balance. And, civil strife in some countries, created by political and military actions, blocks the flow of food and other agricultural products.

The disparity in food production is particularly great between developed and developing countries. The developed nations, including the United States and Canada, supply over 50% of the world's food, but have only 20% of its population. Food production per capita is one of the most meaningful ways of looking at the data, because it corrects for differences in population. In general, overall per capita food production has increased more rapidly in developed countries over the past two decades than it has in most developing countries, even though the latter started at much lower levels. One mitigating factor is the higher rate of population growth in developing regions. In many developing countries, food production has generally not kept pace with growing populations. The exception has been Asia, where the rate of increase in per capita food production has been greater than that even in the developed world. In stark contrast, sub-Saharan Africa has experienced a long-term, continuing decline in per capita food production. One major reason for the increase in Asian crop yields has been greater use of fertilizers and pesticides. The application of both has often been criticized as being harmful to the environment, but the fact remains that millions have been saved from starvation, thanks to fertilizers and pesticides.

Urea, $(NH_2)_2CO$, is a major fertilizer used worldwide. It provides nitrogen to soil by decomposing to ammonia and carbon dioxide when acted on by urease, an enzyme in soils. The ammonia is then taken up by plants.

$$(NH_2)_2CO + H_2O \xrightarrow{\text{urease}} 2\,NH_3 + CO_2 \qquad [11.9]$$

However, the efficiency of urea as a fertilizer is typically reduced because of the direct loss of ammonia through evaporation, in excess of 30%, before it can be taken up by plant roots. To overcome this inefficiency, the IMC-Agrico Company, a 1997 entrant in the Presidential Green Chemistry Challenge Awards program, developed AGROTRAIN, a formulation containing a compound that is converted into a urease inhibitor. Spread on a field, the AGROTRAIN-linked product produces the urease inhibitor, which reduces the rate at which urease decomposes urea so that ammonia is released more slowly and efficiently. The higher efficiency is important, especially in no-till applications, an environmentally friendly approach in which there is little or no disturbance of topsoil. This method reduces soil erosion and requires much less energy for application of the fertilizer.

An even more striking, and generally less controversial, contribution to world agriculture has been the Green Revolution. A fundamental component of this enterprise has been the development of high-yield grains, principally wheat, rice, and corn, that were genetically modified to grow best in particular regions. These new varieties mature faster, permitting more harvests per year, so that the same amount of cultivated land can produce more crops. Since 1960, the Green Revolution has helped world grain harvests to more than double. Billions of people in India, Asia, and Africa have benefited from the practice. But the Green Revolution is neither a panacea nor the ultimate answer. In spite of its successes, the Green Revolution is not universally applicable, and has not been without costs. It works best in areas where water for irrigation is abundant; where money is available for supplemental fertilizers such as ammonia, urea, or nitrates; and where technological understanding and application exist.

Researchers estimate that, within 20 years, global demand for the world's three most important crops—rice, maize (a type of corn), and wheat—will have increased by 40%, simply to keep pace with global food requirements. Genetic engineering and other applications of biotechnology now hold out promise for a second Green Revolution to meet such demands. In the next chapter, we turn to more closely examine genetic engineering—its methods, accomplishments, and limitations.

11.13 Food Preservation

Over the centuries, people have used a variety of methods to preserve foods. Heavily salting foods or storing them in concentrated sugar syrups were the traditional methods used before modern refrigeration. These two methods create a salt or sugar concentration very much greater than that in any contaminating microorganisms such as bacteria, yeasts, and molds. At such high concentrations of salt or sugar, osmosis causes water to leave the cells of the microorganisms, killing them by rupturing their cell membranes. These methods are no longer commonly used, largely because they greatly alter the taste of the food. Heat is also used to kill microorganisms, as is done in home canning or pasteurization. Modern refrigeration retards, but does not ultimately prevent, spoilage.

> Osmosis was discussed in Section 5.17.

To supplement these methods, substances are added to foods to reduce spoilage and extend useful shelf life. As increasing numbers of consumers turn to packaged foods for meals rather than cooking "from scratch," adequate shelf life of packaged products assumes greater importance. Such products carry warning labels like "Best if used by (date)" or ones that give a specific expiration date.

> Refrigeration lowers the temperature so that the rates of reactions in offending microorganisms in foods are slowed, thus retarding spoilage.

Antioxidants are one type of such additives, compounds that prevent packaged, processed foods from becoming rancid due to oxidation of oil or fats, which form harmful free radicals. If you examine the label of any such processed foods (dry cereals, potato chips, and other so-called junk food snacks), you are likely to see the letters BHT or BHA. These stand for *b*utylated *h*ydroxy*t*oluene and *b*utylated *h*ydroxy*a*nisole, respectively, the two most common antioxidant food additives.

BHT   BHA

Antioxidants such as BHT and BHA act by preventing the build-up of free radicals, which are molecular fragments formed when fats and oils react with oxygen from the air in the food package.

fat (or oil) + oxygen $\longrightarrow$ free radicals + other products

Free radicals were discussed in Sections 2.11 and 9.4 in relation to ozone depletion and polymerization, respectively.

A **free radical** is an atom or molecule with an unpaired electron (designated by a single dot in a Lewis-dot diagram of a molecule), which makes the species highly reactive. BHT, BHA, and other antioxidants scavenge the unpaired electron from the free radical to form a stable radical species. This prevents further oxidation of the fat, thus preventing rancidity.

Lewis dot diagrams were discussed in Section 2.3.

A different method of food preservation uses radiation. Food, of course, is irradiated in a number of circumstances for different purposes. The outcome of that irradiation is a matter of what part of the electromagnetic spectrum is used (Section 2.4). It should be rather obvious that radiation is essential for food production; visible light from the Sun drives photosynthesis to give us fruits and vegetables. We use longer wavelength IR radiation from a stove, or even longer wavelength microwaves to cook food or warm up leftovers. Food preservation through **irradiation** is an entirely different case because it uses short-wavelength, high-energy gamma radiation to kill microorganisms. Such radiation is *ionizing* radiation, in contrast to that from visible, infrared, or microwave radiation, which are nonionizing.

Classified as a food additive by Congress in 1958, food irradiation was approved by the FDA in 1963. It has been used to preserve food for astronauts to take with them as they travel in space. The method is used in more than 30 countries and is especially prevalent in Europe, Mexico, and Canada. It has the enthusiastic endorsement of the Food and Agricultural Organization of the United Nations. Irradiated foods even have their own international logo (Figure 11.18). Yet irradiated foods are controversial, especially in the United States. Why the controversy?

Figure 11.18
The international label for irradiated food.

The irradiation procedure is relatively simple in the 160 such facilities worldwide where it is done. The material to be irradiated is placed on a conveyer belt that moves past a tight beam of high-energy gamma radiation generated by a cobalt-60 or cesium-137 source. The source and the irradiation facility are enclosed and shielded so that extraneous radiation does not escape. Over 40 different foods have been approved internationally for preservation by irradiation. Yet, only a small number of irradiated foods have been approved by the United States, including potatoes and strawberries for domestic consumption, and fish, shrimp, and grapefruit for export (Figure 11.19). In 1999, the USDA announced its allowance of irradiation of red meat—beef, pork, and lamb—as a way to reduce or eliminate disease-causing organisms. Irradiation of poultry was approved in 1992.

Figure 11.19
Strawberries preserved by irradiation.

Those opposed to food irradiation question whether irradiated foods are safe to eat. The effectiveness as well as the need for this technology are questioned, including the proliferation of the radioactive materials used as gamma ray sources for a possibly unneeded commercial application. Critics also express concern that food irradiation will be used to cover up improper food-handling processes. Even so, critics and proponents agree that irradiating foods to preserve them does not make them radioactive beyond the normal background radiation that all foods naturally possess (Section 7.8).

The most serious charge brought by critics concerns the formation of possible "unique radiolytic products" (URPs) generated by gamma radiation breaking chemical bonds. Recall from Section 2.4 that, with its short wavelength, gamma radiation has sufficient energy to break chemical bonds and create free radicals or ions. Gamma radiation has much more energy per photon than microwave, infrared, visible, or even ultraviolet radiation. Most foods contain a high percentage of water, and gamma radiation is absorbed by water to form extremely small amounts of irradiation products. These then react with other components of food to form stable products. It must be kept in mind that cooking food also causes chemical changes in the food, changes that are many times greater than those from gamma irradiation. Nearly five decades of research suggests that the by-products of irradiation are the same chemical substances formed by conventional cooking or other preservation methods. Studies based on animal feeding research have repeatedly demonstrated no toxic effects from irradiated foods. The World Health Organization, the

Food and Agricultural Organization of the United Nations, and the U.S. FDA all have concluded that food irradiation is safe when proper procedures and practices are used.

The need for food preservation is serious and should be kept in perspective. Worldwide, food spoilage and contamination is a significant problem, claiming up to 50% of a food crop in some parts of the world, including many developing nations. Closer to home, we are not immune to such contamination in our food supply. Outbreaks of food poisoning in the United States occur periodically from ground beef and chicken tainted with bacteria, commonly *Salmonella,* due to inadequate treatment in processing plants. Food contaminated with this organism has been linked to about 4000 deaths in the United States alone. The symptoms of food poisoning—abdominal pain, diarrhea, nausea, and vomiting—mimic those of short-term gastroenteritis (stomach flu). Therefore, food poisoning is often mistaken as being caused by flu. It is estimated that almost half of the raw chicken sold in the United States is contaminated with *Salmonella.* With proper handling and cooking, even chicken that is contaminated with bacteria can be prepared so as to prevent illness. Irradiation of chicken meat would lower the threat of accidental poisoning by *Salmonella.*

Food irradiation can be viewed in terms of risk versus benefit: Does the benefit outweigh the risk? Some would respond that irradiating strawberries to keep them fresh for a few days longer is a misuse or misapplication of a suspect technology. People are not likely to become ill or die from eating strawberries that are a bit past their peak. On the other hand, trichinosis, a serious disease, can occur by eating pork contaminated with the *Trichinella spiralis* parasite. Low-level gamma irradiation of pork kills the parasite, making the pork safe to eat. Although irradiated beef, pork, and chicken have been approved by the FDA, firms that process these meats are wary that consumers will not buy these irradiated products. This is in spite of the fact that in countries where humans have consumed irradiated foods for years, including poultry and seafood, no adverse effects have been observed. Apparently the U.S. chicken processing companies feel that the risks (and costs) do not outweigh the benefits (at least to them).

Refrigerated chicken has a shelf life of three days; after gamma irradiation, chicken can have a three-week refrigerated shelf life.

Consider This 11.33   Food Irradiation . . . Thanks or No Thanks?

Food irradiation remains controversial. The Web site of the Foundation for Food Irradiation Education suggested that the Web provides "a unique opportunity to communicate the facts about food irradiation to journalists, educators, food company executives and the general public." Indeed, the Web can link a host of constituents with differing viewpoints on a topic such as food irradiation. Use the Web to prepare a position paper on whether food should be irradiated. The paper can be written from the standpoint of a food company executive, a manufacturer of irradiation equipment, a government official, or a consumer activist. Be sure to cite your sources. Later, you may wish to join with others to stage a class debate about the issues involved.

CONCLUSION

This chapter began with the threat of obesity as an epidemic and the anxiety about losing weight and the current focus on low-carbohydrate foods. It ends with the spectre of trying to find enough food worldwide to feed the growing population and ways to store and protect that food. Even though our individual tastes vary, our biological needs are much the same. We need carbohydrates and fats as our primary energy source; fats for cell membranes, synthesis, and lubrication; proteins to build muscle and create the enzymes that catalyze the wonderful chemistry of life; and vitamins and minerals to help make that chemistry happen. Nutrition, like water quality, is a global issue that affects the health of all human beings, regardless of where they live. People with too much to eat, like most Americans, seem preoccupied with food, although generally with too little regard for what they eat. The hungry and the starving think of little else beyond how to feed themselves. Chemistry is only part of the solution to one of the great challenges of our time—how to meet all individual dietary needs, regardless of region or wealth.

Chapter Summary

Having studied this chapter, you should be able to:

- Recognize the frequency and regional occurrence of malnutrition and undernourishment (11.1)
- Understand the physiological functions of food (11.1)
- Describe the distribution of water, carbohydrates, fats, and proteins in the human body and some typical foods (11.1)
- Identify the major chemical elements found in the human body (11.1)
- Recognize and use the chemical composition and molecular structure of carbohydrates (11.2)
- Differentiate among the structures and properties of sugars, starch, and cellulose (11.2)
- Describe the symptoms and cause of lactose intolerance (11.2)
- Recognize and use the chemical composition and molecular structure of fats and oils or triglycerides (11.3)
- Identify sources of saturated and unsaturated fats and their significance in the diet (11.4)
- Differentiate among saturated, monounsaturated, and polyunsaturated fatty acids and fats (11.4)
- Discuss sources of cholesterol and its significance in the diet (11.5)
- Give the general molecular structure of amino acids (11.6)
- Identify and use the chemical composition and molecular structure of proteins (11.6)
- Discuss the importance of essential amino acids and their dietary significance (11.7)

- Describe the symptoms and cause of phenylketonuria (11.7)
- Discuss the effects of selected vitamins on human health (11.8)
- Differentiate chemically between fat-soluble and water-soluble vitamins (11.8)
- Describe the effects of selected minerals on human health (11.9)
- Discuss the necessity of macrominerals, microminerals, and trace minerals for human health (11.9)
- Explain carbohydrates, fats, and proteins as energy sources (11.10)
- Discuss typical recommended daily energy intakes (11.10)
- Relate energy expenditures in various activities (11.10)
- Identify and use basal metabolism rate (BMR) (11.10)
- Discuss the relationship between chronic heart disease and consumption of amounts and types of fats and carbohydrates (11.11)
- Discuss the glycemic index and its relationship to carbohydrate metabolism (11.11)
- Describe various strategies for feeding the world's growing population (11.12)
- Differentiate among international variations in dietary energy supplies (11.12)
- Discuss various methods of food preservation, including the advantages and disadvantages of food irradiation (11.13)

Questions

Emphasizing Essentials

1. Food provides four fundamental types of materials to keep our bodies functioning. What are those types of materials?

2. Is it possible for a person to be malnourished even when eating a sufficient number of Calories every day to meet metabolic needs? Explain.

3. **a.** What are macronutrients and what role do they play in keeping us healthy?

   b. Name the three major classes of macronutrients.

4. Consider this chart.

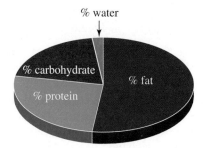

% water

Based on the relative percentages of protein, carbohydrate, water, and fat given, is this graph more likely a rep-

resentation of steak, peanut butter, or chocolate chip cookies? Justify your choice based on the relative percentages of the components shown.

5. Answer each of these questions about the common foods shown in Table 11.1.

   a. Identify the top three foods that are good sources of carbohydrates and arrange them in order of decreasing percentage of carbohydrates.

   b. Identify the top three foods that are good sources of protein and arrange them in order of decreasing percentage of protein.

   c. Which of these foods should be avoided if you are controlling dietary intake of fat? Identify the top three and arrange them in order of decreasing percentage of fat.

6. Answer each of these questions about the common foods shown in Table 11.1.

   a. Which food has the highest protein-to-fat ratio? Calculate that ratio.

   b. Which food has the highest fat-to-protein ratio? Calculate that ratio.

7. An 18-oz steak is the manager's special at a local restaurant. Use the information in Table 11.1 to calculate the number of ounces of protein, of fat, and of water that the customer eating this entire steak would consume.

8. Water is not considered as a macronutrient, but it clearly is essential in maintaining health. What are some of the roles that water plays in our bodies? *Hint:* You may want to refer to Chapter 5.

9. Examine the data in Table 11.2 and explain why hydrogen ranks first in atomic abundance in the human body, but third behind oxygen and carbon in terms of mass percent.

10. Use the information in Table 11.2 to answer these questions.

    a. What is the ratio of the relative abundance of potassium to sodium in the human body?

    b. What is the ratio of grams of potassium to grams of sodium in the human body?

    c. Are these elements included in the composition of the human body shown in Figure 11.2? Why or why not?

11. a. Consider the composition of the human body shown in Figure 11.2. What are the principal elements that make up water, proteins, carbohydrates, and fats? Which elements are in common among these major components of the human body?

    b. Compare your answers for part **a** with the relative abundance of the elements in the body given in Table 11.2. Is there a correlation? Explain the correlation between your lists and the relative abundances of the elements in the body.

12. This figure is a schematic diagram of the food pyramid. What foods are in each section of the pyramid, and how many servings of each should you consume daily?

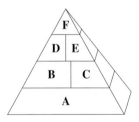

13. A large piece of sausage pizza would contain several food groups. Identify each group and name the part of the pizza that is responsible for representing that particular food group.

14. Fructose, $C_6H_{12}O_6$, is a carbohydrate.

    a. Rewrite the formula for fructose to emphasize the original meaning of the term *carbohydrate*.

    b. Draw a structural formula for one of the isomers of fructose.

    c. Do you expect the different isomers of fructose to all have the same sweetness? Explain why or why not.

15. Fructose and glucose both have the formula, $C_6H_{12}O_6$. How do their structural formulas differ?

16. State what is meant by each term, and give an example of a substance that fits that term.

    a. monosaccharide   b. disaccharide

    c. polysaccharide

17. What problems can arise from regularly consuming excess dietary servings of carbohydrates?

18. What is "Splenda?" How is it chemically similar to and different from sucrose?

19. Use the lock-and-key model discussed in Section 10.5 to offer a possible explanation why individuals who suffer from lactose intolerance can digest other sugars such as sucrose and maltose, but not lactose.

20. a. What are the similarities between fats and oils?

    b. What are the differences between fats and oils?

21. From the entries in Figure 11.8, identify the fat or oil with the highest percentage of each type of fat.

    a. polyunsaturated fat   c. total unsaturated fat

    b. monounsaturated fat   d. saturated fat

22. The label of a popular brand of soft margarine lists "partially hydrogenated" soybean oil as an ingredient. What does "partially hydrogenated" mean? Why does the label not simply say soybean oil, rather than partially hydrogenated soybean oil?

23. The text describes substitutes that have been developed for fat (Olean) and sugar (NutraSweet). Why have there not been attempts to develop a comparable substitute for protein?

24. What is the nutritional significance of the elements shaded on this periodic table?

25. Why is it safer to take large doses of vitamin C than it is to take large doses of vitamin D?

26. What is the difference between case–control studies and cohort studies with respect to gathering data about health and nutrition?

27. What is the meaning and role of the glycemic index GI?

28. Use the information in Figure 11.17 to answer these questions.

    a. Which areas of the world are experiencing food supply shortfalls and require exceptional assistance?

    b. Give reasons why these areas do not include North America.

Concentrating on Concepts

29. Explain to a friend why it is impossible to go on a highly advertised "all organic, chemical-free" diet.

30. Your friend wants to cut food costs and has learned that peanut butter is a good protein source. What additional information should your friend consider before making the decision to make peanut butter the major dietary protein source? *Hint:* There is relevant information in Table 11.1.

31. **a.** What percentage of the total number of elements in the periodic table is utilized by the human body to produce proteins, carbohydrates, and fats?

    b. What relationship is there between the type of bonds these elements can form and what makes them so prevalent in the human body?

32. When the USDA made the decision to change dietary recommendations, it also changed the way the information was visually displayed to consumers. Rather than using a restructured pie chart, the food pyramid was introduced. Is the food pyramid a better communication tool for dietary recommendations than the pie chart used previously? Why or why not?

33. According to one USDA study, nearly 40% of the food that the average American eats each day consists of milk or dairy products. Would such a diet be possible and still meet the guidelines of the food pyramid?

34. An alternative Healthy Eating Pyramid appears as Figure 11.5. Compare the features of it to the USDA Food Pyramid. Why would exercise be part of a food pyramid?

35. In people who exhibit lactose intolerance, the enzyme lactase that normally catalyzes the lactose breakdown, is either missing or is present at levels too low to support normal enzymatic activity. How does this inability to break down lactose parallel our ability to metabolize starch, but not cellulose?

36. For each statement, indicate whether it is always true, may be true, or cannot be true. Justify your answers by explaining your reasoning.

    a. Plant oils are lower in saturated fat than are animal fats.

    b. Lard is more healthful than butterfat.

    c. There is no need to include fats in our diets because our bodies can manufacture fats from other substances we eat.

37. Experimental evidence suggests that some physiological effects of saturated fats, compared with unsaturated fats, may be caused by differences in folding or wrapping of the molecules. The hydrocarbon chains in saturated fatty acids can fold or wrap more tightly than those of unsaturated or polyunsaturated fatty acids.

    a. Explain why saturated fatty acid molecules are able to fold more tightly than molecules of unsaturated or polyunsaturated fatty acids. *Hint:* If you have a model set available, make suitable molecular models to help you see the effect single or double bonds can have on the ease of folding.

    b. Explain why the extent of molecular folding influences the melting points of stearic, oleic, linoleic, and linolenic acids. See Table 11.3 for melting point values.

38. Some people prefer to use nondairy creamer rather than real cream or milk. Some, but not all nondairy creamers, use coconut oil derivatives to replace the butterfat in cream. Should a person trying to reduce dietary saturated fats by using nondairy creamer use nondairy creamers such as these? Why or why not?

39. The reaction of free radicals and oxidizing agents with unsaturated and polyunsaturated fats in the body has been suggested as a cause of premature aging. What is the chemical basis for this assertion? *Hint:* You might find it helpful to consider the mechanism of addition polymerization in Chapter 9.

40. Why is it more difficult for a person to control her or his cholesterol level than to control her or his fat intake? What steps are effective in minimizing cholesterol in the blood?

41. **a.** Which are the "good" lipoproteins: LDL or HDL?

    b. What function do the "good" lipoproteins perform?

    c. Does a person with an LDL reading of 100 mg/100 mL and an HDL reading of 150 mg/100 mL meet current healthy heart guidelines from the AHA?

42. Consider this information in the table below about the sugar content of different food products.

Food Product	Sugar	Calories	Serving Size
Altoids, peppermint	2 g	10	3 pieces (2 g)
Ginger snaps	9 g	120	4 cookies (28 g)
Critic's Choice Tomato Ketchup	3 g	15	1 tbsp (13 g)
Del Monte Pineapple Cup	13 g	50	Individual cup (113 g)
Dr Pepper soft drink	40 g	150	1.5 cups
French Vanilla Coffee-Mate	5 g	40	1 tbsp (15 mL)
Hostess Twinkies	14 g	150	1.5 oz
LifeSavers, WintOGreen	15 g	60	4 mints (16 g)
Tropicana Home Style Orange Juice	22 g	110	8 oz (1 cup)
Snickers bar	29 g	200	2.1 oz
Sunkist orange soda	52 g	190	1.5 cups
Wheatables crackers	4 g	130	13 crackers (29 g)

a. Examine this list of 12 food products. Which item has the highest ratio of grams of sugar to the number of Calories (g sugar/Cal) in one serving?

b. Does the sugar content of any of these foods surprise you? Explain your response.

c. Do you expect that the specific sugars in Dr Pepper are the same sugars found in Sunkist orange soda? In cranberry juice? In the pineapple cup? Why or why not?

d. The complete label for WintOGreen Lifesavers shows 16 g of total carbohydrates per serving, 15 g of which is sugars. What type of compounds do you think accounts for the other 1 g of carbohydrates?

43. Consider the structure for riboflavin, one of the B vitamins found in leafy green vegetables, milk, and eggs.

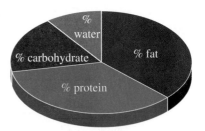

Why is it somewhat safer to take large doses of vitamin B than it is to take large doses of vitamin D?

44. American diets depend heavily on bread and other wheat products. A slice of whole wheat bread (36 g) contains approximately 1.5 g of fat (with 0 g saturated fat), 17 g of carbohydrate (with about 1 g of sugar), and 3 g of protein.

a. Calculate the total Calorie content in a slice of this bread.

b. Calculate the percent Calories from fat.

c. Do you consider bread a highly nutritious food? Explain your reasoning.

45. Use the resources of the Web to find the latest scientific studies concerning the effectiveness of some of the so-called fad diets. Does it appear the ultralow fat Ornish diet or low-carb Atkins diet is proving to be better? Possible criteria for judging could include greater weight loss or larger numbers of people affected.

46. What is your opinion about food preservation by irradiation? Are there some cases in which you feel irradiation is justified as a way to ensure better quality food to the consumer? Explain your position and be prepared to defend it.

Exploring Extensions

47. The second Green Revolution may come from the advances provided by transgenic foods. What is meant by this statement? Is the second Green Revolution likely to be more controversial if transgenic grains are concerned? Explain why or why not.

48. How has the proportion of undernourished people in different areas of the world changed over the past 30 years? How have the *total numbers* of undernourished people changed during that time? Focus on any one region of the world and find the necessary information to speak to these two points. Then devise a visual way to represent these data.

49. Compare these two pie charts for the percentage of macronutrients in soybeans and wheat.

a. Use these charts to help explain why the World Health Organization has helped develop several soy-based, rather than wheat-based, food products for distribution in parts of the world where protein deficiency is a major problem.

b. Suggest some cultural reasons why soy might be preferable to wheat for some areas of the world.

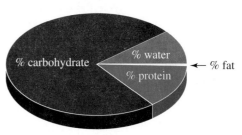

Soybeans

Wheat

50. The Sceptical Chymist finds the statement that the composition of the human body is ". . . roughly similar to the stuff we stuff into it" an idea hard to believe, but is willing to try to justify this statement, at least for the macronutrients. Compare the information found in Table 11.1 and Figure 11.2. Does it give you adequate information to decide whether the ". . . roughly similar to the stuff we stuff into it" statement is reasonable, assuming you eat only the foods shown in Table 11.1? Why or why not?

51. How does the elemental composition of the human body compare with the elemental composition of Earth's crust? How does the human body's composition compare with the elemental composition of the universe? Table 11.2 gives the values for the human body. Research the composition of Earth's crust and that of the universe, listing your references. Then comment on the comparative values for the first five elements listed in order of mass abundance in each of the three circumstances—the human body, Earth's crust, and the universe.

52. The food guide pyramid gives a range of servings for each food group. What factors do you think determine the number of servings you should eat?

53. 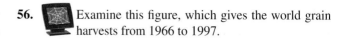 The food guide pyramid gives a range of servings for each food group. To use this information, the consumer must know what constitutes a reasonable serving size. Investigate what constitutes reasonable serving sizes for one of the food groups, and then prepare a poster with your results to share with others who are investigating the reasonable serving sizes for other food groups. Were you surprised by any of the serving sizes? Which ones?

54. Not everyone considers milk nature's "perfect food." Compare and contrast the viewpoints of the dairy industry with groups that work against the dairy industry. What are some of the specific benefits attributed to milk, and what are some of the reasons that milk has been called "nature's not-so-perfect food"?

55. Here is the label information from a popular brand of canned chicken noodle soup.

Serving Size: 1/2 cup (4 oz; 120 g)

Servings per container: about 2.5

Amount per serving

Calories 75   Calories from Fat 25

	Amount	**% Daily Value**
Total fat	2.5 g	4
Saturated fat	1.5 g	8
Cholesterol	20 mg	7
Sodium	970 mg	40
Total carbohydrates	9 g	3
Dietary fiber	1 g	4
Sugars	1 g	
Protein	4 g	
Vitamin A		15
Vitamin C		2
Calcium		2
Iron		4

a. Analyze this information to see if the soup conforms to dietary recommendations of the AHA.

b. Is the serving size recommended on the label adequate? Explain.

c. What effect would changing the serving size have on your answer to part **a**?

56. Examine this figure, which gives the world grain harvests from 1966 to 1997.

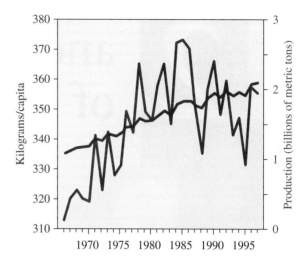

a. Write a paragraph summarizing the information displayed by the graph.

b. Has this information changed since the last year shown in this graph? Explain.

57. Every month, a certain consumer-advocate organization presents an "Unnatural Living Award" to a person, product, or institution that demonstrates an unnatural ability to provide an unnatural product to the American people. What are the criteria by which you would make your nomination for this award? What do you consider would be a good candidate to receive this award? Explain your reasons for suggesting this candidate.

Genetic Engineering and the Chemistry of Heredity

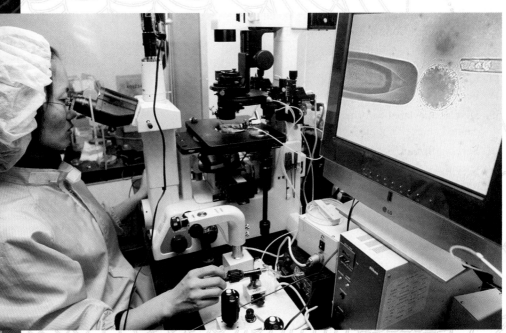

A researcher in the human embryo cloning project at Seoul National University manipulates a cell shown on the monitor, while watching through a microscope.

"Our research team has successfully culled stem cells from a cloned human embryo. . . . The result of our research proves it is possible scientifically for human cloning, and we are likely to revive the controversy over human cloning"

Hwang Yoon-Young announcing that South Korean researchers
had 'cloned human embryos.' CNN,
February 12, 2004

"I think the big question is: If you make this kind of thing in a dish, have you created a human life? . . . Can you make something that people have strong moral views about in terms of destroying it, in order to benefit other people?"

Arthur Kaplan, medical ethicist,
University of Pennsylvania, CNN,
February 12, 2004

Thus, in 2004, the controversy of human cloning reentered the public view. A team of South Korean scientists led by Dr. Hwang Woo Suk of Seoul National University published a paper in the journal *Science* describing a detailed process of how to create human embryos by cloning. The scientists were not intending to make babies but instead were practicing what is known as therapeutic cloning, a process ultimately intended to produce treatments for a multitude of illnesses including diabetes, spinal cord injuries, and Parkinson's disease.

The chemistry of genetics begins with a report of human cloning. Ironically, 2004 marked the 51st anniversary of the unraveling of the structure of DNA, deoxyribonucleic acid, and the 4th anniversary of the first complete draft of the human genome, the molecular set of codes that defines how we are put together. At the same time, the members of the European Union and other countries worldwide debate the merits of banning transgenic or bioengineered food, naturally grown products that contain genetic material from other species that scientists have introduced in order to improve some trait.

What genetic material is manipulated in bioengineered food? How is genetic information passed along from one generation to the next? Chemistry provides a description of the code contained on the long strands of DNA. The "backbone" of the DNA strands is a polymer, a mind-boggling pattern of alternating deoxyribose sugar and inorganic phosphate groups, with an important structural mission but little or no specific genetic information. So-called bases, nitrogen-containing cyclic molecules, are the variables in the code, and only four different ones in DNA (and another one unique to ribonucleic acid RNA) define the structural components in the entirety of a human being and for that matter all animals and plants.

After looking at the structural components of the chemical code, we will examine how DNA replicates and how it defines protein structure and synthesis. After reflecting on the magnitude of the accomplishment within the human genome project, we begin to examine human-made changes in DNA or so-called **recombinant DNA.** From this we look at the ability to engineer new drugs and vaccines, diagnoses via DNA, and uses of genetic fingerprinting. At this stage it becomes possible to understand the process of how genetic material can be mixed between species, leading to transgenic foods, and finally to cloning.

Consider This 12.1   Fiftieth Anniversary of DNA

The 50th anniversary of the "discovery" of DNA really marked the stage at which its structure was elucidated. Use the Web to define the critical stages of development of our understanding about DNA structure and function by creating a timeline. The benchmarks are often recognized by awards like the Nobel Prizes in physiology or medicine and chemistry.

12.1 The Chemistry of Heredity

Each second the human body hosts thousands of chemical reactions involving an even greater number of chemicals. Some compounds are decomposed and others are synthesized; energy is released, transformed, and used; and chemical signals are transferred and processed. But, in spite of the dazzling complexity of these processes, the last half-century has seen a phenomenal increase in our knowledge of the chemistry of life. Much biological research has refocused on molecules rather than on cells or organisms. This "molecular" research has led to an understanding of the very basis of life itself.

The practical manifestations of this intellectual achievement are manifold. In 1900, average life expectancy at birth in the United States was less than 50 years; today it is almost 77 years (Figure 12.1). Reasons for this dramatic increase include better nutrition; improved sanitation; advances in public health; more accurate medical diagnoses; new medical procedures; and numerous new medicines, drugs, and vaccines. Chemistry has contributed to all these innovations. Biotechnology and molecular engineering are integral parts of the latest revolution in health care. There seems little doubt that genetic engineering will profoundly affect human life in the 21st century.

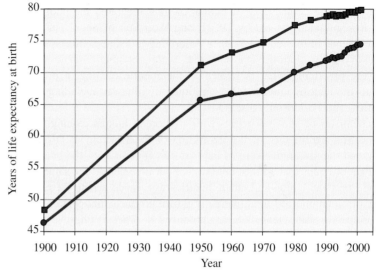

Figure 12.1

Average life expectancy at birth in the United States since 1900 for females (red) and males (blue).

Source: Centers for Disease Control and Prevention, National Center for Health Statistics, http://www.cdc.gov/nchs/fastats/lifexpec.htm

Your Turn 12.2 **Increases in Life Expectancy**

a. Use Figure 12.1 to determine the percent increase in average human life expectancy at birth for women and for men in the United States between 1900 and 1950, and between 1950 and 2001.

b. Speculate as to why the average life expectancy of women and men differ.

| Human red blood cells do not have nuclei. |

To begin from a personal perspective, you contain about 10 million million (10×10^{12}) cells that have a nucleus. Each of these cell nuclei contains a complete set of the genetic instructions that make you what you are—at least biologically. This information is organized into **chromosomes,** 46 compact structures of intertwined molecules of DNA in a human cell nucleus, and approximately 30,000 **genes,** components that convey one or more hereditary traits. This is the **human genome,** the totality of human hereditary information in molecular form.

Your special template of life is written in a molecular code on a tightly coiled thread, one invisible to the unaided eye. This thread is **deoxyribonucleic acid (DNA),** the molecule that carries genetic information in all species. Unraveled, the DNA in *each* of your cells is about 2 m (roughly 2 yd) long. If all of the DNA in all 10 million of your cells were placed end to end, the resulting ribbon would stretch from here to the Sun and back more than 60 times! But as you will soon discover, this astronomical figure is far from the most astounding feature of this amazing molecule.

Sceptical Chymist 12.3 **Stretching DNA**

Sometimes authors get carried away with their rhetoric. Check the correctness of the claim that the DNA in an adult human being would stretch from Earth to the Sun over 60 times.

Hints: It is approximately 93 million miles from Earth to the Sun. Other necessary information is in the preceding paragraphs, and the unit conversion factors are in Appendix 1.

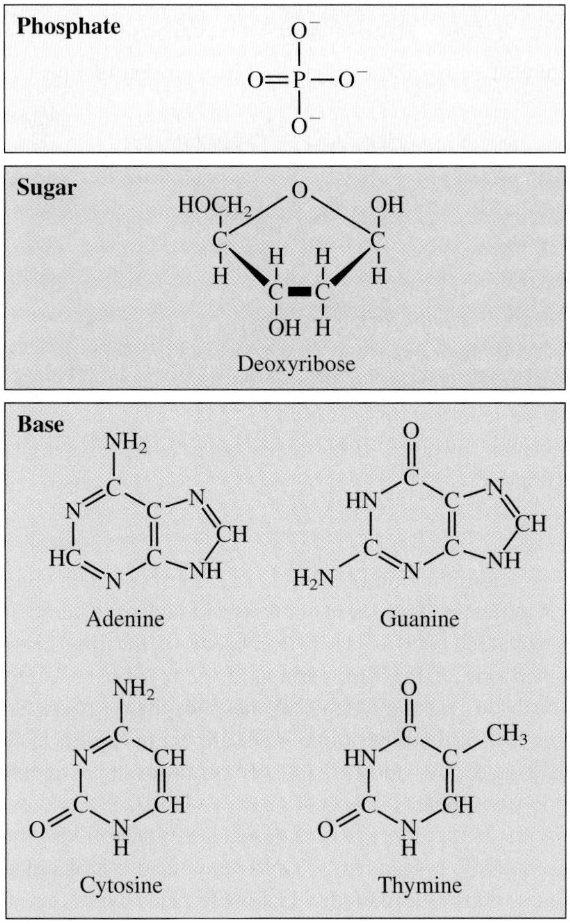

Figure 12.2
The components of deoxyribonucleic acid DNA.

The molecular structure of deoxyribonucleic acid dictates the way DNA encodes genetic information. A strand of DNA consists of fundamental chemical units, repeated thousands of times. Each of the units is composed of three parts: nitrogen-containing bases, the sugar deoxyribose, and phosphate groups. All are illustrated in Figure 12.2. Two of the bases, adenine (symbolized by **A**) and guanine (**G**), are built on a six-membered ring fused to a five-membered ring. Carbon and nitrogen atoms make up the rings. Cytosine (**C**) and thymine (**T**) each contain six-membered molecular rings consisting of four carbon atoms and two nitrogen atoms. These four compounds are bases because they react with water to produce basic solutions. H^+ ions (protons) are transferred from H_2O molecules to nitrogen atoms of the nitrogen-containing bases, creating OH^- ions in solution, forming a basic (alkaline) solution from the resulting hydroxide ions.

$$H_2O(l) + \text{N-base}(aq) \rightleftharpoons {}^+\text{HN-base}(aq) + OH^-(aq) \qquad [12.1]$$

Deoxyribose is a monosaccharide (a "single" sugar) with the formula $C_5H_{10}O_4$. Figure 12.2 reveals that the deoxyribose molecule contains a five-membered ring formed by four carbon atoms and one oxygen atom. The "deoxy" of deoxyribose means that a hydroxyl (—OH) group in ribose has been replaced by a hydrogen atom. That replacement in the sugar structure occurs at the ring carbon shown with two hydrogen atoms. The phosphate group can be represented as PO_4^{3-}, but, depending on the pH, an H^+ ion can be attached to one or more of the O atoms. The form of phosphate in which each of three oxygen atoms has a proton attached is H_3PO_4, phosphoric acid. The ionizable hydrogen atoms on the phosphate groups make nucleic acids acidic.

H^+ and OH^- ions are discussed in Sections 6.1 and 6.2. The double arrows indicate that reactants are not completely converted to products.

Sugars are discussed in Section 11.2.

Consider This 12.4   Ribose and Deoxyribose

Compare the structure of deoxyribose (Figure 12.2) with that of ribose:

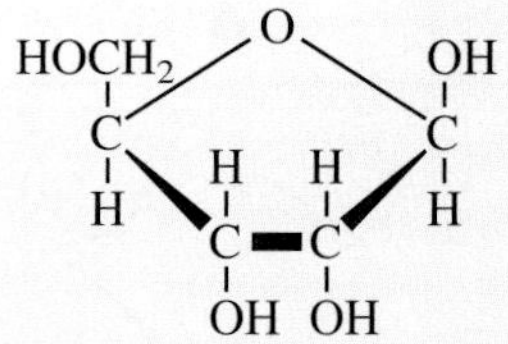

$$HOCH_2-\overset{\displaystyle O}{\underset{\displaystyle}{C}}\cdots OH$$

a. Write the molecular formula for both sugars.
b. In Section 11.2 carbohydrates were shown to have the formula $C_nH_{2n}O_n$. Do both ribose and deoxyribose follow this pattern?
c. Which atoms have lone pairs of electrons?
d. Which atoms can be involved in hydrogen bonds? *Hint:* Hydrogen bonding was defined in Section 5.8.

A **nucleotide** is a combination of a base, a deoxyribose molecule, and a phosphate group. Figure 12.3 indicates how these units are linked in a nucleotide called adenosine phosphate. A covalent bond exists between one of the ring nitrogen atoms of the adenine molecule and one of the ring carbons in deoxyribose. Another covalent bond connects the deoxyribose sugar molecule to the phosphate group. Similar nucleotides can be formed using any of the other three bases shown in Figure 12.2. Each nucleotide consists of a base joined covalently to the deoxyribose sugar and with the sugar, in turn, attached to the phosphate.

A typical DNA molecule consists of thousands of nucleotides covalently bonded in a long chain. Consequently, a segment of DNA may have a molecular mass in the millions. The phosphate groups link the individual nucleotides. Note that in Figure 12.3, one —OH group on the deoxyribose ring remains unreacted. The phosphate group of another nucleotide can react with this —OH group, forming and eliminating a H_2O molecule, and connecting the two nucleotides. Figure 12.4 shows an example in which four nucleotides are linked in this manner to form a segment of DNA. This alternating chain of deoxyribose–phosphate–deoxyribose–phosphate units runs the length of the nucleic acid molecule. Attached to each of the deoxyribose rings is one of the four possible bases.

The formation of the bond between phosphate and deoxyribose in which water is eliminated is another example of a condensation polymer (Section 9.5).

The specific bases and their sequence in a strand of DNA turn out to have great significance. Some of the early clues to the structure of DNA and the mechanism by which it conveys genetic information came as a result of the research of Erwin Chargaff in the 1940s and 1950s. Chargaff and his coworkers were able to determine the percentage of the four bases present in DNA from a variety of species. They found that the relative amounts of the bases in a DNA sample are identical for all members of the same species. Moreover, these percentages are independent of the age, nutritional state, or environment of the organism studied. For example, according to Chargaff's data, the DNA from all

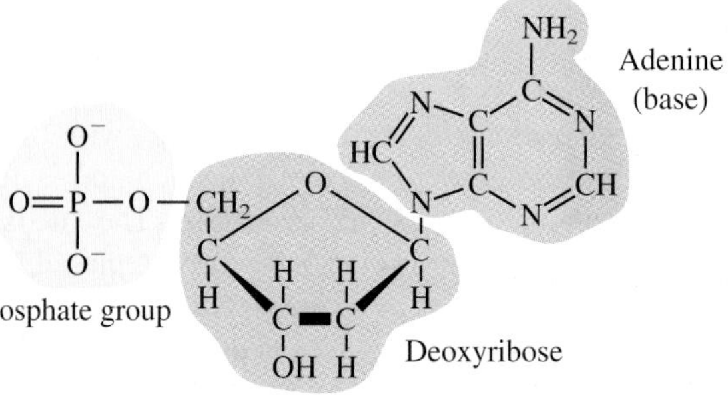

Figure 12.3

The molecular structure of a nucleotide, adenosine phosphate, is composed of adenine, a nitrogen base (green), deoxyribose (blue), and a phosphate group (yellow).

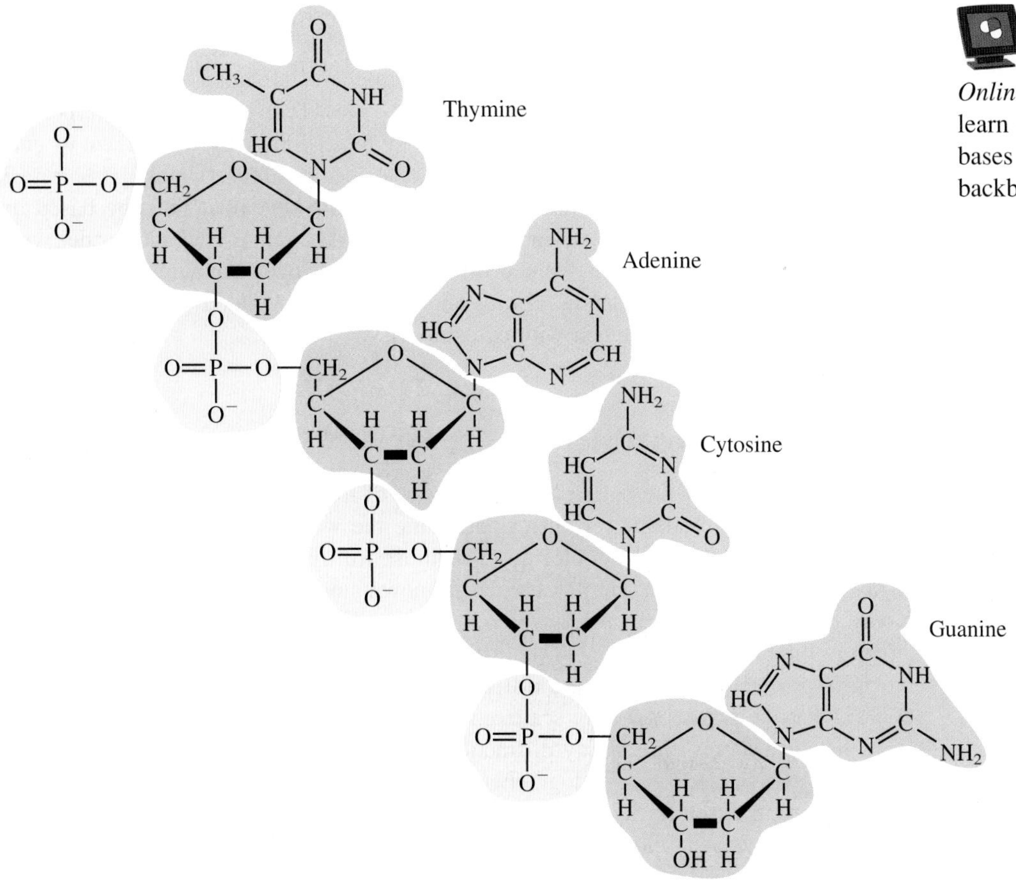

Figures Alive! Visit the *Online Learning Center* to learn more about the four bases and the sugar-phosphate backbone.

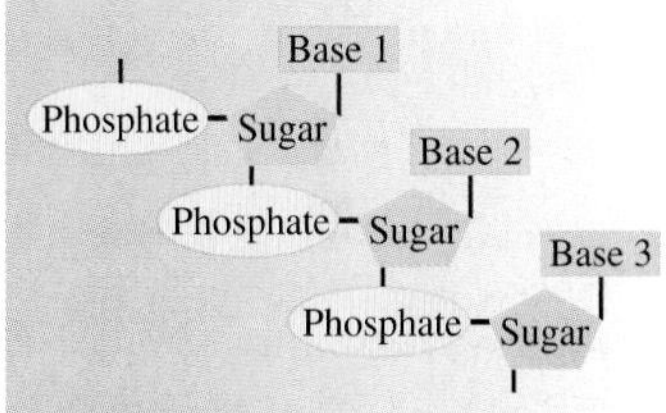

Figure 12.4

A segment of DNA. A phosphate group connects one deoxyribose to an adjacent one. Each of the four bases, thymine (T), adenine (A), cytosine (C), or guanine (G), is attached to a deoxyribose sugar.

DNA can be represented as a polymer (see Chapter 9) of nucleotide monomers or repeating units. The schematic shows the sugar–phosphate–sugar–phosphate chain with different bases attached to each of the sugars, deoxyribose.

members of our species, *Homo sapiens,* contains 31.0% adenine, 31.5% thymine, 19.1% guanine, and 18.4% cytosine. Table 12.1 also contains such findings for other species. Humans, fruit flies, and bacteria do not seem to have very much in common, and it is perhaps reassuring that the mix of the four bases is quite different in the three species. But it turns out that the more closely related the species are, the more similar the base composition of their DNA. This observation certainly suggests that the base composition of the nucleic acid must have something to do with inherited characteristics.

Table 12.1	The Percent Base Compositions of DNA for Various Species			
Species	**Adenine**	**Thymine**	**Guanine**	**Cytosine**
Homo sapiens (human)	31.0	31.5	19.1	18.4
Drosophila melanogaster (fruit fly)	27.3	27.6	22.5	22.5
Zea mays (corn)	25.6	25.3	24.5	24.6
Neurospora crassa (mold)	23.0	23.3	27.1	26.6
Escherichia coli (bacterium)	24.6	24.3	25.5	25.6
Bacillus subtilis (bacterium)	28.4	29.0	21.0	21.6

Source: From I. Edward Alcamo, *DNA Technology: The Awesome Skill.* © 1996 The McGraw-Hill Companies, Inc. All rights reserved. Reprinted with permission.

Note that the percentages of adenine and thymine are consistently similar, as are the percentages of cytosine and guanine.

A more careful examination of Table 12.1 discloses that the DNA of *Homo sapiens* and the DNA of *Escherichia coli* (*E. coli*), a species of bacteria that inhabits the human intestine, exhibit a very important common characteristic. They obey the same compositional regularity, now called **Chargaff's rules.** In every species, the percent of adenine almost exactly equals the percent of thymine. Similarly, the percent of guanine is essentially identical to the percent of cytosine. Put more simply: %A = %T and %G = %C. Such a correlation can hardly be coincidental; it must be based on biochemical form and function at the molecular level. As soon as Chargaff's rules were communicated, the conclusion seemed obvious: The nitrogen-containing DNA bases somehow come in pairs. Adenine always appears to be associated with thymine, and guanine is consistently matched with cytosine.

12.2   The Double Helix of DNA

Not obvious was how the paired bases were part of the overall molecular structure of DNA. Therefore, scientists set out to determine the way in which nucleotides were incorporated into the DNA molecule. The most fruitful experimental strategy was X-ray diffraction, a technique that had been known since early in the 20th century. In **X-ray diffraction,** a beam of X-rays is directed at a crystal. The X-rays strike the atoms in the crystal, interact with their electrons, and bounce off the atoms. Stated a bit more precisely, the X-rays are diffracted, or scattered, by the atoms. The crucial point is that the X-rays are only scattered at certain angles that are related to the distance between atoms. The X-ray diffraction pattern of a DNA fiber was obtained in late 1952 by the British crystallographer Rosalind Franklin (Figure 12.5).

Two scientists, James D. Watson, a 24-year-old American, and Francis H. C. Crick, a Cambridge University biophysicist, were responsible for combining a collection of data, including Franklin's X-ray information. Watson and Crick (Figure 12.6) concluded that a pattern in Franklin's diffraction photograph was consistent with a repeating helical arrangement of atoms, similar to a loosely coiled spring. Moreover, the X-ray photographs contained evidence of a regular repeat distance of 0.34 nm within a DNA molecule.

With these clues, Crick and Watson set out to create a molecular model. A major breakthrough was the recognition that the adenine and thymine portions of the molecule fit together almost perfectly, like pieces in a jigsaw puzzle. Moreover, these two bases can be linked by two hydrogen bonds (Figure 12.7). Similarly, cytosine and guanine align by forming three hydrogen bonds. Adenine and thymine are said to be **complementary**

X-rays have short wavelengths and high energies.

---

1 nm = 1 × 10⁻⁹ m

Hydrogen bonds are not covalent chemical bonds, but relatively weak interactions between molecules like water or DNA bases. See Section 5.9.

Figure 12.5

Rosalind Franklin, whose work contributed significantly to elucidating the DNA structure.

Figure 12.6

James Watson and Francis H. C. Crick, codiscoverers of the double-helical structure of DNA.

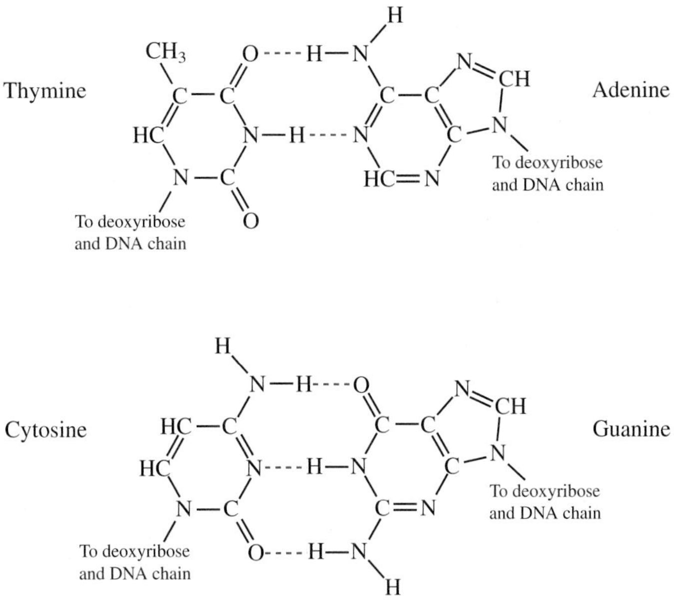

Figure 12.7

Base pairing of adenine with thymine and cytosine with guanine in DNA. Chemical bonds are drawn as solid lines, and the intermolecular attractions due to hydrogen bonds are shown as dashed lines.

bases, as are cytosine and guanine, since they are capable of forming a hydrogen-bonded base pair. This base pairing is the molecular basis underlying Chargaff's rules: A pairs with T and C pairs with G.

In the model of DNA developed by Watson and Crick, the hydrogen bonds between the complementary bases help hold together two strands of a **double helix.** Perhaps an even better metaphor is a spiral staircase. The steps of this molecular staircase are the bases, always pairing A with T and C with G. One member of each base pair belongs to one strand of a helix, the other to the matching, complementary helical strand, thus creating a double helix. Recall that the bases are connected to the deoxyribose rings, which in turn are linked by phosphate groups. Thus, the deoxyribose and phosphate units are, in effect, the stair rails to which the steps are attached (Figure 12.8). Watson and Crick concluded that the base pairs are parallel to each other, perpendicular to the axis of the DNA fiber, and separated by 0.34 nm, the repeat distance calculated from the diffraction pattern. In addition, Franklin's results also suggested another repeat distance of 3.4 nm. Watson and Crick took this to be the length of a complete helical turn consisting of 10 base pairs.

Your Turn 12.5   Complementary Base Sequences

Identify the base sequences that are complementary to each of these sequences.

a. ATACCTGC
b. GATCCTA

Answers
a. TATGGACG   b. CTAGGAT

Your Turn 12.6   From Bases to Distance

The distance between base pairs in a molecule of DNA is 0.34 nm.

a. Calculate the length (in centimeters) of the shortest human chromosome, which consists of 50,000,000 base pairs.

(continued on p. 536)

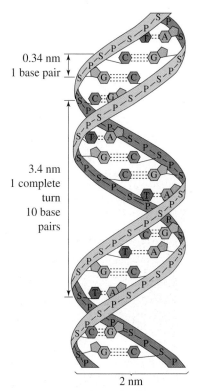

Figure 12.8

A model of DNA with
P = phosphate group;
S = the sugar, deoxyribose;
A = adenine; T = thymine;
C = cytosine; G = guanine. The alternating sugar and phosphate groups are represented on the two twisting ribbons. The four bases are attached to this backbone and paired A to T, C to G.

Your Turn 12.6   From Bases to Distance (*continued*)

b. Mark off that length on your paper. If this is the length of the unstretched DNA molecule, what does this imply about the organization of DNA in the chromosome?

Answer

a.
$$\frac{0.34 \text{ nm}}{\text{base pair}} \times \frac{1 \text{ m}}{10^9 \text{ nm}} \times \frac{10^2 \text{ cm}}{1 \text{ m}} \times \frac{50,000,000 \text{ base pairs}}{\text{chromosome}} = \frac{1.7 \text{ cm}}{\text{chromosome}}$$

b. This is a small distance; 1.7 cm is about two-thirds of an inch. The best way that this many base pairs could fit into such a small space would be to have them tightly packed in a spiral or folded.

Your Turn 12.7   The Length of DNA

The DNA in each human cell consists of 3 billion base pairs. Calculate the length of this DNA. Does this length agree with that given in Section 12.1? Why or why not?

Watson and Crick's research paper, "Molecular Structure of Nucleic Acids: A Structure for Deoxyribose Nucleic Acid," appeared in the scientific journal *Nature* on April 24, 1953. It is only one page long, and the Watson–Crick paper is written with the customary passionless detachment of contemporary scientific prose. Even the most significant statement in the communication is delivered with typical British understatement: "It has not escaped our notice that the specific pairing we have postulated immediately suggests a possible copying mechanism for the genetic material."

In 1968, James Watson published a highly personal account of the research that led to the determination of the DNA structure. This book, entitled *The Double Helix*, is recommended to anyone who still doubts that scientists are flesh and blood. In the book, Watson candidly describes the process of scientific discovery: the competition and ambition, the lucky guesses and the blind alleys. His account does not always conform to a textbook definition of the scientific method, but then again neither does scientific research. Two decades later, Crick followed with *What Mad Pursuit*, his own reminiscences of those heady days in Cambridge.

It is noteworthy that Watson and Crick were not experts in the field of genetics when they began their research on DNA. Moreover, they did few experiments themselves. Instead, they drew on the work of experts such as Erwin Chargaff, Rosalind Franklin and her crystallographer colleague Maurice Wilkins, and the American chemist Linus Pauling. At the time, all these scientists were better known and more highly regarded than Francis Crick and his young American collaborator. But Watson and Crick seem to have brought a fresh point of view to the problem of DNA structure, hence they saw what more experienced and better-informed scientists missed. Some rightly might argue that Rosalind Franklin's very significant contribution of crystallographic data was not sufficiently acknowledged, either in 1953 or in *The Double Helix*. But few would quarrel with the decision to award Watson, Crick, and Wilkins the 1962

Both crystallographer Maurice Wilkins and Francis Crick, the co-discoverer of the DNA structure died in 2004.

 Consider This 12.8 Discovering Rosalind Franklin

In 1958, Rosalind Franklin died an untimely death at age 37. Thus, she did not live long enough to add her own account to the history that was written (and rewritten) about the discovery of DNA. Because her work was long minimized or ignored, some historians now assert that both DNA and Rosalind Franklin were discovered. To set the record straight, several excellent biographies of Franklin are now available. You can find reviews of these books as well as other accounts of her life on the Web. What were her contributions to the structure of DNA? Why was her work not given full credit during her time? What questions would you ask if you could interview her?

Nobel Prize in physiology or medicine. Unfortunately, Franklin had died of ovarian cancer by that time, and the Nobel Prize is never awarded to anyone posthumously.

The history of science is full of examples of how a single discovery can release a flood of related research. So it was with the discovery of the structure of DNA. Scientists immediately set out to discover the molecular details of how DNA is replicated, how it encodes genetic information, and how that information is translated into physiological characteristics. **Replication,** the process by which copies of DNA are made, is now well understood and is diagramed in Figure 12.9.

Your Turn 12.9   Chargaff's Rules

What role did knowledge of Chargaff's rules play in the discovery of the DNA double helix?

Consider This 12.10   Checking All Bases

The structural features of a DNA molecule are more apparent if you can look at them in 3-D. At the *Online Learning Center,* you can view and rotate several different molecular structures.

a. How is a 3-D structure of DNA shown at that site similar to the one shown in the text? How is it different?

b. Look carefully at the structures of the four bases that make up DNA. What are their common structural features? How do they differ?

c. Look again at the large DNA molecule. Can you find the bases? How are they aligned?

Before a cell divides, the double helix partially and rapidly unwinds at a rate of about 10,000 turns per minute. This results in a region of separated complementary single strands of DNA, as pictured in the middle portion of Figure 12.9. Individual nucleotides in the cell are selectively hydrogen-bonded to these two single strands that serve as templates: A to T, T to A, C to G, and G to C. Held in these positions, the nucleotides are bonded together (polymerized) by the action of an enzyme. Every minute, about 90,000 nucleotides are added to the growing chain. By this mechanism, each strand of the original DNA generates a complementary copy of itself. The original template strand and its newly synthesized complement coil about each other to form a new double helix, a daughter molecule identical to the first. Similarly, the other separated strand of the original molecule twines around its new partner, another new daughter molecule. Thus, where there was only one double helix, there are now two, represented at the bottom of Figure 12.9. As the nucleus splits and the cell divides into two daughter cells, one complete set of chromosomes is incorporated into each of them. This process is repeated again and again, so that each of the 2×10^{12} nucleated cells in a newborn baby contains all the genetic information first assembled from parental DNA when the sperm combined with the ovum.

12.3 Cracking the Chemical Code

The discovery of the molecular code in which the genetic information is written is arguably history's most amazing example of cryptography. Key to the code is the sequence of bases in the DNA. The 3 billion base pairs repeated in every human cell provide the blueprint for producing one human being. Although these specifications are carried in DNA, they are expressed in proteins. Proteins are everywhere in the body: in skin, muscle, hair, blood, and the thousands of enzymes that regulate the chemistry of life. It follows that, by directing the synthesis of proteins, DNA can dictate the characteristics of the organism.

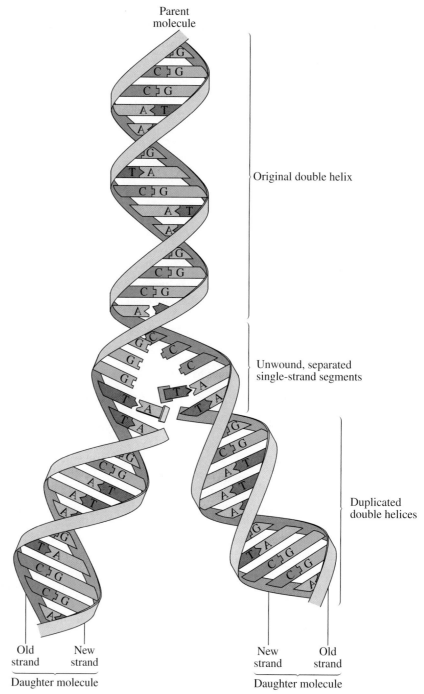

Parent
molecule

Original double helix

Unwound, separated
single-strand segments

Duplicated
double helices

Old	New		New	Old
strand	strand		strand	strand
Daughter molecule			Daughter molecule	

Figure 12.9

Diagram of DNA replication. The original DNA double helix (top portion of figure) partially unwinds and the two complementary portions separate (middle). Each of the strands serves as a template for the synthesis of a complementary strand (bottom). The result is two complete and identical DNA molecules.

Proteins are large molecules formed by the combination of individual amino acids. The 20 amino acids that commonly occur in proteins can be represented by this general formula.

$$
\begin{array}{c}
H \qquad\ H \qquad O \\
\diagdown \qquad | \qquad \diagup\!\!\!\diagdown \\
N\!-\!C\!-\!C \\
\diagup \qquad | \qquad\quad \diagdown \\
H \qquad\ R \qquad\ OH
\end{array}
$$

For more information about amino acids and proteins, see Section 11.6.

The amine or amino group is —NH$_2$, the acid group is —COOH, and R represents a side chain that is different in each of the 20 amino acids. When the amino acids combine, the —COOH group of one of them reacts with the —NH$_2$ group of another, forming what is known as a peptide bond and eliminating an H$_2$O molecule. A **protein** is therefore a long chain of amino acid residues, as these structural units are called once they have been joined together.

The biochemists who set out to decipher the genetic code concentrated on translating base language into amino acid language. They assumed that somehow the order of bases in DNA determines the order of amino acids in a protein. The hypothesis that the code is related to the sequence of base pairs is the only reasonable one. The phosphate and deoxyribose units are identical in all DNA. Therefore, they could not supply the variability in the DNA structure to account for the individuality among species, and variation is essential in genetic material. Only the base pairs provide the opportunity for variability in the structure of DNA.

It was obvious at the outset that the code could not be a simple one-to-one correlation between bases and amino acids. There are only four bases in DNA. If each base corresponded to an individual amino acid, DNA could encode for only four amino acids. But 20 amino acids appear in our proteins. Therefore, the DNA code must consist of at least 20 distinct code "words," each word representing a different amino acid. And the words must be made up of only four letters—A, T, C, and G—or, more accurately, the bases corresponding to those letters.

Some simple statistics can help us determine the minimum length of these code words. To find out how many words of a given length can be made from an alphabet of known size, one raises the number of letters available to a power n, corresponding to the number of letters per word.

$$\text{words} = (\text{letters})^n$$

Thus, using four letters to make two-letter words generates 4^2, or 16, different two-letter words. Similarly, DNA bases read in pairs (akin to two letters per word) could code for only 16 amino acids. This vocabulary is too limited to provide a unique representation for each of the 20 amino acids. So we repeat the calculation, this time assuming that the code is based on three sequential base pairs or, if you prefer, that we are dealing with three-letter words. Now the number of different triplet-base combinations is 4^3, or $4 \times 4 \times 4 = 64$. This system provides more than enough capacity to do the job.

Your Turn 12.11 **Quadruplet-Base Code**

Suppose that the DNA code used four sequential base pairs instead of a triplet-base code. How many different four-base sequences would result?

Answer

$4 \times 4 \times 4 \times 4 = 4^4 = 256$ different four-base sequences

---

Obviously, more than mathematical reasoning was required to prove the molecular basis of genetics. Once again, Francis Crick was a leader in this research. His work clearly established that the genetic code is written in groupings of three DNA bases, called **codons**. And today, thanks to other scientists, this triplet-base code has been cracked, and specific amino acids have been related to particular codons.

The process by which the three-base sequence results in an amino acid selection involves several intermediate stages. During those steps, the DNA base sequence is ultimately transferred to messenger- or m-RNA, the molecule literally used as the template for the amino acids in the protein. Thus, the codon refers to the three bases in both DNA and RNA.

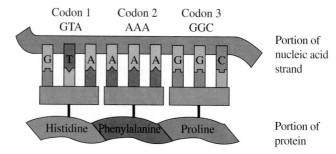

Figure 12.10

A nine-base nucleic acid sequence showing three codons.

The markings on the Rosetta Stone were in several languages and helped to decipher Egyptian hieroglyphics.

No Rosetta Stone was available to aid these scientists in their efforts at translation. Instead, they relied on elegant and imaginative experiments that ultimately yielded a genetic dictionary. If you were to use the letters A, T, C, and G in a game of Scrabble, you could generate 64 different three-letter combinations. A few, CAT, TAG, and ACT, for example, make sense. Most are like AGC, TCT, and GGG and are meaningless—at least in English. Nature does far better than that; 61 of the 64 possible triplet codons specify amino acids. Thus, the codon sequence GTA in a DNA molecule signals that a molecule of the amino acid histidine should be incorporated into the protein, AAA codes for phenylalanine, and GGC stands for proline. The three-base sequences that do not correspond to amino acids are signals to start or stop the synthesis of the protein chain. An example of a nine-base nucleic acid segment and how it codes for three amino acids is shown in Figure 12.10.

Because there are more codons than amino acids, the code has redundancy. Some amino acids are represented by more than one codon. Leucine, serine, and arginine have six codons each. On the other hand, tryptophan and methionine are each represented by only a single codon. Significantly, the code is identical in all living things. The instructions to make Albert Einstein, bacteria, or trees are written in the same molecular language.

The amount of information carried by your deoxyribonucleic acid is truly phenomenal. The DNA in each of your cell nuclei consists of approximately 1 billion (1×10^9) triplet codons. You have just read that each triplet is at least potentially capable of encoding 1 of the 20 amino acids found in human protein. If each codon could be assigned a letter of the English alphabet, rather than an amino acid, your DNA could encode 1×10^9 letters or about 2×10^8 five-letter words. These words would fill 400,000 pages of 500 words each, or 1000 volumes of 400 pages each. And you carry that library in 2 m of a helical thread, invisible to all but scientists peering with electron microscopes. The miniaturization of this information to the molecular level puts to shame the most sophisticated supercomputers.

Unfortunately, scientific fact interferes a bit with the hyperbole of the previous paragraph. Scientists have determined that less than 2% of human DNA actually constitutes unique gene sequences. The human genome contains multiple copies of some genes that code for frequently used proteins. Moreover, many copies of DNA sequences are too short to function as genes. For example, millions of copies of sequences consist of only 5–10 base pairs. But the presence of this "junk" DNA in no way detracts from the wonder of molecular genetics or from its challenge. It merely gives scientists something else to study.

Your Turn 12.12   Duplicate Codons

Suggest some advantages of a genetic code in which several codons represent the same amino acid.

Your Turn 12.13   DNA Unique to Humans

The human genome contains about 3 billion base pairs, but only about 2% of this DNA consists of unique genes. The number of genes is estimated at 30,000. Use this information to calculate the average number of base pairs per gene.

Answer

$$\frac{3 \times 10^9 \text{ base pairs}}{\text{human genone}} \times \frac{2 \text{ unique genes}}{100 \text{ base pairs}} \times \frac{1 \text{ human genome}}{3 \times 10^4 \text{ genes}} \times \frac{1 \text{ base pair}}{1 \text{ unique gene}} = 2 \times 10^3 \frac{\text{base pairs}}{\text{gene}}$$

12.4 Protein Structure and Synthesis

The mechanism by which DNA directs protein synthesis is known in great detail—too much detail for this text. For our purposes, it is sufficient to recognize that the transfer of information and matter is extremely complicated. Given this complexity, it is amazing that errors in protein synthesis are very rare. Consider, for example, chymotrypsin. This protein, an enzyme that catalyzes the digestion of other proteins, consists of 243 amino acid residues. Chymotrypsin is just one of the proteins that can be made from 20 different amino acids by using 243 amino acid residues. Statistically, 20^{243} different protein molecules could be formed. Expressed relative to the more familiar base 10, this number corresponds to 1.4×10^{316}, a number larger than the estimated number of atoms in the universe! Each member of this immense group of molecules has its own unique **primary structure,** the identity and sequence of the amino acids present. One and only one primary structure is the biologically correct form of chymotrypsin with the desired enzymatic properties. The fact that the body unfailingly (or almost unfailingly) synthesizes this particular protein out of 1.4×10^{316} possibilities is evidence of a molecular blueprint and a cellular assembly line of almost incomprehensible specificity and accuracy. And, of course, similar considerations apply to each of the proteins in the entire organism.

After the amino acids are strung together in the correct sequence, the resulting protein chain should be able to twist and turn into an infinite number of shapes. Surprisingly, it does not. Rather, the protein molecule assumes a characteristic shape that is the result of structural features but can also be influenced by variables such as temperature and pH. Once again, X-ray diffraction provides a means of determining this structure. For chymotrypsin, the result is pictured in Figure 12.11. What looks like a jumble of ribbon is the carefully ordered backbone of the protein molecule. The figure shows helical segments and parallel chains that constitute the intermediate level of molecular organization, called the **secondary structure.** The overall shape or conformation of the molecule is termed its **tertiary structure.** Evidence suggests that the three-dimensional conformation of a protein molecule is stabilized by the interaction of various functional groups. Hydrogen bonds are particularly important in stabilizing secondary structural subunits that occur in many proteins.

The catalytic activity of chymotrypsin and any other enzyme is evidence of the reliable regularity of the tertiary structure of the protein. For an enzyme to carry out its chemistry, functional groups on certain amino acid residues must come close enough to form an active site. The **active site** is the region of the enzyme molecule where its catalytic effect occurs. Sometimes, the amino acids involved are adjacent; in other cases, they are widely separated in the protein chain, but close together in the tertiary structure. Figure 12.11 indicates that the active site in chymotrypsin consists of three amino acids that would be far apart if the protein were unwound. These groups help hold the **substrate,** the molecule or molecules whose reaction is catalyzed by the enzyme. In the case of chymotrypsin, the active site of the enzyme catalyzes the breaking of peptide bonds in the substrate, another protein. In some other enzymes, the active site promotes the formation of chemical bonds. In all cases, the orientation of the active site and the conformation of the rest of the enzyme molecule are of critical importance. The fact that a newly synthesized protein molecule automatically assumes the enzymatically active

All enzymes are proteins, but not all proteins are enzymes.

The lock-and-key model described in Section 10.5 can explain the mode of action of many enzymes.

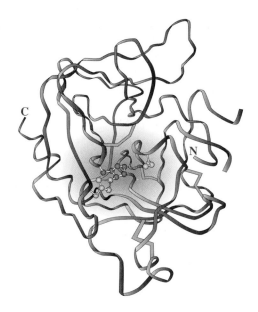

Figure 12.11

Tertiary structure of the enzyme, chymotrypsin. The "ribbon" portion represents the polymerized amino acid chain; the central colorized portion is the active site at which the enzymatic chemistry takes place. The two ends are marked C for the carboxylic acid group and N for the amine group.

shape almost suggests that the chain of amino acid residues possesses some sort of molecular memory. In fact, the favored tertiary structure is the most energetically stable conformation.

A subtle change in the primary structure of a protein can have a profound effect on its properties. A much-studied example is provided by hemoglobin, the blood protein that transports oxygen, and the condition called *sickle cell anemia*. When an individual with a genetic tendency toward sickle-cell disease is subjected to conditions that involve high oxygen demand, some red blood cells distort into rigid sickle or crescent shapes (Figure 12.12). Because these cells lose their normal deformability, they cannot pass through tiny openings in the spleen and other organs. Some of the sickled cells are destroyed and anemia results. Other sickled cells can clog organs so badly that the blood supply to them is reduced.

The property of sickling has been traced to a minor change in the amino acid composition of human hemoglobin. A hemoglobin molecule has 574 amino acid residues. The only difference between normal hemoglobin and hemoglobin S in persons with the sickle cell trait is in two of these amino acids. In hemoglobin S, two of the residues that should be glutamic acid are replaced with valine. Apparently this substitution causes the hemoglobin to convert to the abnormal form at low oxygen concentration.

Figure 12.12

Scanning electron micrographs of normal red blood cells (left) and red blood cells showing the effect of sickle cell disease (right).

Sickle cell anemia is hereditary; the error in the amino acid sequence reflects a corresponding error in a DNA codon. Normally, mutations detrimental to a species are eliminated by natural selection. Perhaps the sickle cell trait has survived because it may also convey some benefit. A clue to what the benefit might be comes from studying the carriers of the gene for hemoglobin S. The gene is most common in people native to Africa and other tropical and subtropical regions and in their descendants. The fact that these are also areas with the highest incidence of malaria has led to speculation that an individual whose hemoglobin has a tendency to sickle may be protected against malaria. Specific mechanisms have been proposed to account for this protection. If the hypothesis is correct, it is an interesting example of how a genetic trait that originally had survival advantage can become a detriment in a different environment. Of course, the fact that sickle cell anemia is a genetic disease at least raises the possibility that genetic engineering may some day eliminate it.

When scientists want to know the differences in amino acid order between proteins like hemoglobin and hemoglobin S, they have two alternatives. They can take samples of each protein and painstakingly determine the order of the amino acids. This process often involves using enzymes to clip the whole protein into smaller pieces and then removing one amino acid at a time from one end of each piece and identifying it. The second means of determining amino acid order, another brute force approach, is to determine the sequence of the bases in DNA that code for that protein. Although not intended to address the sequence in a specific protein, the following section describes how the sequence of bases for all genetic material has been mapped.

12.5 The Human Genome Project

The **Human Genome Project** is the effort to map all the genes in the human organism. After more than a decade of research, on June 26, 2000, scientists announced that a rough draft of the project to decode the genetic makeup of humans had been completed. The goal, to determine the sequence of all 3 billion base pairs in the entire genome, was completed for the approximately 30,000 genes found on the 46 human chromosomes. The Human Genome Project was supported by the U.S. National Institutes of Health (NIH), the Wellcome Trust (a philanthropic organization based in London), and by Celera Genomics, a private company in Maryland. The completion of the goal was announced in a joint statement by NIH and Celera representatives. Upon the completion of this first phase, this massive enterprise was described as "... the most important, wondrous map ever produced by humankind ..." by then-President Bill Clinton and as "... the outstanding achievement not only of our lifetime but perhaps in the history of mankind ..." by Dr. Michael Dexter of the Wellcome Trust. It is anticipated that it will take many years to discover the traits the genome conveys.

The Human Genome Project began in 1989, with James Watson of DNA fame as its first director. At an estimated $3 billion, the project cost about $1 per base pair. Instrumental in accomplishing so much so quickly was the high level of cooperation among international teams of researchers, and the competition to be the first to complete the project between publicly funded efforts (NIH) and corporate research (Celera).

Sometimes scientists do research for the same reason that mountain climbers scale a mountain: "Because it's there." But the Human Genome Project involves more than the spirit of adventure or the thrill of discovery. The more we know about our genetic makeup, the more likely we will be to diagnose and cure disease, understand human development, trace our evolutionary roots, and recreate our family tree. And one hopes these discoveries will be utilized for the benefit of our species and others.

You may wonder whose DNA has been selected for this unprecedented scrutiny. In fact, most of the DNA analyzed in the Human Genome Project came from members of over 60 multigenerational French families whose lineage is well documented. But it does not really matter; any one of us could have served as a DNA donor and as a representative of *Homo sapiens*. In spite of our apparent differences and our long history of disputes based on those differences, the DNA of all humans is remarkably similar.

It differs from individual to individual by about 0.1% of the base sequences. Within that tiny fraction resides our genetic uniqueness. Biology and chemistry provide irrefutable evidence of a lesson we as nations and as individuals have been slow to learn, a lesson stated by former President Clinton: "The most important fact of life on this Earth is our common humanity."

Determining the sequence of the DNA base pairs was difficult and time-consuming. The smallest human chromosome contains 50,000,000 base pairs. Given those numbers, researchers had to develop automated base sequencers that were both fast and accurate. The accuracy is very important, because a single missed base will throw off all the subsequent base assignments, just as a skipped buttonhole is transmitted down the length of a shirt. The target error rate is 1 in 10,000 bases, a 99.99% level of accuracy. Celera Genomics and NIH researchers used different methods. The data released in February 2001 enabled some comparisons between the two groups. By spring 2003, the "finished" draft covered 99% of the genome and only 300 gaps of the 300,000 in the rough draft remain.

Groups of scientists across the world argued for free access to the information as it was discovered. Unfortunately, the spirit of cooperation was already strained as the project progressed. Celera Genomics funded its efforts by patenting genes it identified and selling access to some of its data. The first patent application created a controversy that involved both scientists and the broader public. Because NIH funded, supervised, and completed much of the research, NIH applied for some of the first patents. Spokespersons for NIH argued that by securing patents on some of the genes, they would be protecting the public funds invested in the project. Critics responded that because government sources provided most of the funds, the results should belong to the public. Others reasoned that free and open exchange of information is essential in such a complex international project, and that patents would inhibit this communication. James Watson called the idea of patenting genes "sheer lunacy." In 1992, the U.S. Patent Office ruled that gene fragments could not be patented unless they had some known function. The patent problem remains unresolved.

Despite the many potential benefits of the Human Genome Project, the enterprise is not without its critics. Some point out that a genetic map ("being caught with your genes down") represents the ultimate invasion of privacy. Information about an individual's genetic makeup might be used by insurance companies to discriminate against those with a hereditary tendency toward certain diseases. The information might be used by businesses to refuse to hire people who may be genetically at risk, or by ruthless governments to identify the "genetically inferior." Indeed, shortly before the June 2000 human genome completion announcements, a CNN–Time magazine poll reported that 46% of the respondents said they expected harmful results from the endeavor while 40% expected benefits; 41% thought that the project is morally wrong compared with 47% who disagreed with that position. Nevertheless, a majority of respondents, 61%, reported that they would like to know if they were predisposed to developing a genetic disease, compared with 35% who would not want to know.

Research continues to refine the initial data and to produce the genetic map for other species. Comparison of such data across species is likely to generate information about fundamental, biochemical functions in living organisms. The even longer task remains—turning the base pair sequences into useful information regarding the specific details of their cellular functions.

Consider This 12.14   The Human Genome Project

In 2000, the Human Genome Project completed the sequence for all of the base pairs at least to the level of a "rough draft." What were some of the problems with the early version? Are there any ways to know if there were mistakes? Since 2001 scientists have completed the genetic maps of various species—the thale cress plant (a weed), the bacterium that causes cholera, and more recently a large series of microbes. Are any of these accomplishments useful to the sequence data for humans?

12.6 Recombinant DNA

Mythology is full of fanciful creatures: the sphinx with the head of a woman and the body of a lion; the griffin, which is half-lion and half-eagle; and the chimera with a lion's head, a goat's body, and a serpent's tail. In 1973, two American scientists, Herbert Boyer and Stanley Cohen, created another hybrid. They introduced a gene for manufacturing a protein from the African clawed toad into a common bacterium, *E. coli.* Upon replication, the *E. coli* bacteria produced the toad's protein. To be sure, humans have been manipulating the gene pool for thousands of years. We have created mules by crossbreeding horses and donkeys, dogs as diverse as Chihuahuas and Saint Bernards, and fruits and vegetables that never existed in nature. All these were done by selective breeding. Boyer and Cohen created their chemical fantasy in laboratory glassware through the manipulations of genetic engineering.

A statue of a griffin.

To illustrate the technique, consider a real response to a very real need. Insulin is a small protein consisting of 51 amino acids. It is produced by the pancreas and influences many metabolic processes. Most familiar is its role in reducing the level of glucose in blood by promoting the entry of that sugar into muscle and fat cells. People who suffer from a common type of diabetes have an insufficient supply of insulin, and hence elevated levels of blood sugar. Left untreated, the disease can result in poor blood circulation, especially to the arms and legs, amputations, blindness, kidney failure, and early death. But, diabetes can be controlled by diet, exercise, and insulin injections.

Before 1982, all insulin used by diabetics was isolated from the pancreas glands of cows and pigs, collected in slaughterhouses. It turns out that the insulin produced by cattle and hogs is not identical to human insulin. Bovine insulin differs from the human hormone in three out of 51 amino acids; porcine (pig) and human insulins differ in only one. These differences are slight, but sufficient to undermine the effectiveness of bovine and porcine insulin in some human diabetics. For many years, there seemed to be no hope of obtaining enough human insulin to meet the need. Although insulin has been synthesized in the laboratory, the process is far too complex for industrial adaptation. However, since 1982 the lowly bacterium, *E. coli,* has been tricked into making human insulin.

This unlikely bit of interspecies cooperation is a consequence of using recombinant DNA techniques to introduce the gene for human insulin into this simple organism (Figure 12.13). Bacteria contain rings of DNA called **plasmids.** These rings can be removed and cut open by the action of special enzymes. Meanwhile, the gene for human insulin is either prepared synthetically or isolated from human tissue. This human DNA is inserted into the plasmid ring by other enzymes. The result is interspecies recombinant DNA. These modified plasmids are called **vectors** since they are used to carry the DNA back into the bacterial "host." Once inside the cell, the biochemistry of the bacterium takes over (see Figure 12.13). Every 20 minutes, the *E. coli* population doubles, soon producing millions of copies or clones of the "guest" (human) DNA.

Clones are a collection of cells or molecules identical to an original cell or molecule. It is possible to harvest the cloned DNA, but in the insulin example we are interested in a supply of the protein, not its gene. Therefore, the bacteria are allowed to synthesize the proteins encoded in the recombinant DNA. Although the recombinant *E. coli* has no use for human insulin, it generates it in sufficient quantities to harvest, purify, and distribute to diabetics. Today, the cost of human insulin produced by bacteria is less than that of insulin isolated from animal pancreas. Currently, more than 3 million Americans use genetically engineered insulin to treat their diabetes.

Figure 12.13 is a representation of the recombinant DNA techniques just described. The actual operations are a good deal more complicated than the figure suggests, and many details have been omitted. A variety of vectors and host organisms have been used in molecular engineering. Other bacterial species are sometimes used instead of *E. coli,* and yeasts and fungi are often employed because all of these species

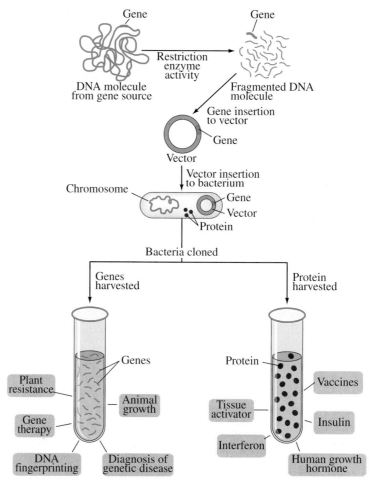

Figure 12.13

A representation of genetic engineering.

reproduce very quickly. DNA with specific properties is fragmented and the appropriate portion introduced into a vector that carries it into a host organism. Cloning of the host, a bacterium in this case, gives rise to two categories of end products: cloned genes and proteins.

Consider This 12.15   Sources of Insulin

If a person must take insulin daily, how would that person determine the source of the insulin? Does the source make a difference in how the insulin acts in the body? Use the resources of the Web or consult a pharmacist to answer these questions.

Another success for genetic engineering has been the synthesis of human growth hormone (HGH). This protein, produced by the pituitary gland, stimulates body growth by promoting protein synthesis and the use of fat as an energy source. Children with insufficient HGH fail to reach normal size. If the condition is diagnosed early, injections of the hormone over 8–10 years can prevent this form of dwarfism. Formerly, a year's HGH therapy for one person required the pituitary glands from about 80 human cadavers. That source is no longer used, thanks to the production of HGH in bacteria. However, the cost of treatment, even with cloned HGH, can be as high as $20,000–$50,000 per year.

12.7 Engineering New Drugs and Vaccines

You have just read about two examples of replacement therapy, in which an insufficient natural supply of an essential protein is augmented with a genetically engineered supplement. Similar biochemical methods are also being used to create new drugs or larger supplies of already known drugs. The gene coding for the drug is introduced into a host organism, which then synthesizes the desired product. This is currently one of the most rapidly growing applications of recombinant DNA technology.

Other products of biotechnology appear to be effective against viruses. **Viruses** are simple, infectious, almost living biochemical species. We say "almost living" because viruses do not have the necessary biochemical machinery to carry out metabolism or to reproduce by themselves. A typical virus consists of nucleic acid and protein; many viruses are essentially inert. When it invades a cell, however, the virus takes charge, forcing the cell to make more viral nucleic acid and protein. In effect, the virus does naturally what genetic engineers accomplish with recombinant DNA techniques. Because they are so simple, viruses are notoriously difficult to combat or defend against. Thus, some forms of pneumonia, "strep," or other bacterial infections, though potentially far more dangerous than a common cold, are much easier to treat than a cold, which is caused by a virus. Only bacterial infections can be destroyed by one of a host of antibiotics.

Genetically engineered interferons may help change all that. Interferons are nature's way of providing protection against viruses. Three main classes of naturally occurring interferons have been identified (alpha, beta, gamma) and a minor one (omega). All are proteins, and 13 genes have been currently identified that code for interferon-alpha. Interferons also contain carbohydrate portions. Although the mechanism by which these molecules operate against viral invasion is not fully understood, they have been exploited to fight infections. Thus far, genetically engineered interferons show promise against hepatitis B, herpes zoster (shingles), a type of multiple sclerosis, and a variety of cancerous malignancies including some forms of leukemia, malignant melanoma, multiple myeloma, carcinoid tumors, lymphoma, and certain kidney cancers. The use of an interferon nasal spray to control the common cold may still be years away because of the high costs and biochemical complexity; the common cold is thought to be caused by over 200 viruses.

For the treatment of many diseases, the ultimate goal is not just the development of a drug to treat it, but the creation of a vaccine to prevent getting the disease. A **vaccine** is preparation of killed microorganism, attenuated living organisms, or fully active organisms that produce or increase immunity to a particular disease. Vaccines work by mobilizing the body's own defense mechanisms. The idea is to expose the body to a molecule or organism closely related to the virus or bacterium that causes the disease. The immune system responds to this stimulus by generating antibodies against it. These **antibodies,** or protein "markers," remain in the body and offer protection against subsequent infection by the virus or bacterium. Of course, the vaccine should not itself cause the disease. Therefore, vaccines are typically made from bacteria or viruses that have been killed or weakened or from fragments or subunits of the unwelcome invaders. The latter approach is preferable, because there is no danger that the disease will be transmitted in the process of vaccination. Fortunately, it is here that genetic engineering is most promising. The DNA encoding for a characteristic but non-infectious part of a virus—for example, its protein coat—can be introduced into plasmids. The bacteria consequently produce this particular protein. The protein is then isolated, concentrated, and used as a vaccine that carries essentially no risk of infection. This technique has been employed to synthesize a vaccine against hepatitis B. Because hepatitis is transmitted by blood, health care professionals are often vaccinated against the disease. If and when a vaccine is developed against HIV, it may be through a similar technology. Antiviral vaccines now exist for influenza A and RSV (respiratory syncytial virus, a virus that manifests itself as the common cold in adults but which can be deadly to infants and adults with compromised immune systems). Antiviral drugs are available for a series of infections including herpes, chicken pox, and forms of meningitis.

Consider This 12.16 Interferons

Being commercially available, interferons now can be used more extensively in trials against a variety of viral infections and other diseases. Use the Web to explore the progress made in using interferons against the HIV virus.

12.8 Diagnosis Through DNA

Until very recently, the most sensitive methods of diagnosing disease were based on detecting enzymes or antibodies in an infected organism. Antibodies are generated by a host organism in response to a microbial invasion. As a result, an infection is often well established before a positive test can be obtained. Fortunately, genetic engineering has enabled diagnosticians to identify the DNA of the infectious agent, even at early stages and at a low concentration.

Such sensitivity is possible as the result of two important techniques: DNA probes and the polymerase chain reaction. **DNA probes** are DNA segments selected or engineered so that they are complementary to some segment of the infecting viral or bacterial DNA (the target). The probes are single-stranded DNA, with from 10 to over 10,000 bases, and they are usually labeled with a radioisotope. These radioactive probe molecules are introduced into a sample of biological fluid or cytoplasm suspected of containing an infectious agent. The test is carried out at a temperature and pH at which the DNA double helix of the infectious agent separates into single strands. If the probe encounters a strand with a complementary segment of bases, hydrogen bonds form between A and T and C and G, and the probe sticks to the target molecule. Because the probe is radioactive, it can easily be traced.

> DNA probes were used to identify *Helicobacter pylori* as the causative agent of gastric ulcers and also as a likely cause of stomach cancer. Radioisotopes are discussed in Sections 7.7–7.9.

Successful early diagnosis involves detecting the infecting virus or bacterium before the disease is well established. Even the most sensitive DNA probes will not work if the concentration of the invading DNA is too low. Here, the **polymerase chain reaction (PCR)** has proven to be of great utility. This technology makes it possible to start with a single segment of DNA and make millions or even billions of copies of it in a few hours. The PCR process, which won its inventor, Kary Mullis, the Nobel Prize in chemistry in 1993, is diagrammed in Figure 12.14.

A researcher starts with a sample, denoted as a double-stranded DNA at the top of the figure. In step (a), the sample is heated to unwind and separate the two complementary strands, cooled, and combined with short segments of "primer" DNA. These **primers** are synthetic, single-stranded nucleotides that bracket and identify the section of the DNA to be copied. In step (b), the enzyme polymerase and an ample quantity of the four free nucleotides bind the separated DNA strands. The polymerase enzyme starts at the location of the primer and directs the formation of a new strand using the free nucleotides, thereby creating a complementary copy of the single strand. In this way (step c), a copy is made of each of the two original strands. This sequence of steps yields four strands derived from the original double strand. Repeating the process causes all four strands to be reproduced. The overall yield of DNA molecules is 2^n, where n is the number of times that this process is repeated. One can see that the yield of identical copies of the original double strand grows very quickly. This should be obvious by considering that $2^{10} = 1.024 \times 10^3$, $2^{20} = 1.048 \times 10^6$, and $2^{100} = 1.27 \times 10^{30}$ molecules.

PCR is very rapid; only 1–2 minutes is needed for a complete cycle. Thus, 100 cycles would require just 2–3 hours. But, the reaction would run out of starting materials long before then. A little arithmetic (Sceptical Chymist 12.17) shows that 1.27×10^{30} double-stranded DNA molecules, each of 100 base pairs, would weigh almost 140,000 tons!

Sceptical Chymist 12.17   **140,000 Tons of Base**

Check the assertion that 1.27×10^{30} double-stranded DNA molecules of 100 base pairs each would weigh almost 140,000 tons. *Hint:* Start by using Avogadro's number to determine how many moles of DNA are represented by this large number of base pairs. Assume an average mass of 300 g per mole of nucleotide.

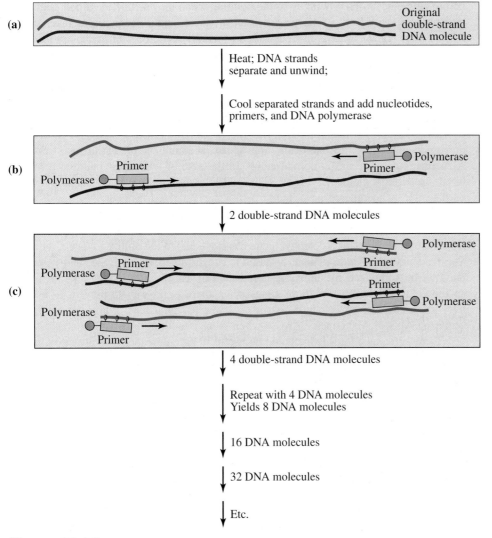

Figure 12.14

Diagram of the polymerase chain reaction.

Your Turn 12.18   **Using PCR to Copy DNA**

A technician starts with a single DNA molecule. How many cycles of PCR must take place to produce these numbers of DNA molecules?

a.   16 DNA molecules

b.   256 DNA molecules

c. 1.0×10^6 DNA molecules

Answer

a.   4 cycles ($1 \longrightarrow 2 \longrightarrow 4 \longrightarrow 8 \longrightarrow 16$)

    PCR technology has proven indispensable in any procedure where a small sample of DNA must be dramatically amplified to get a sufficiently large supply for subsequent studies. One such set of applications is in developing "DNA fingerprints" for use in criminal cases, in studying archeological remains, and, of course, in diagnosing disease. In the last instance, the DNA in specimens thought to contain infecting or defective DNA is multiplied by PCR so that DNA probes can be used.

    The early diagnosis of HIV infection is one of the achievements of this new technology. The DNA of human immunodeficiency virus can be detected several weeks before antibodies to HIV build up to the point at which they can be identified. Similarly, recombinant DNA methods have been used to speed up the diagnosis of tuberculosis. DNA probes and PCR have also been used to identify a number of hereditary diseases. Defective genes have been identified for cystic fibrosis, Huntington's disease, some forms of Alzheimer's disease, and amyotrophic lateral sclerosis (ALS), better known as Lou Gehrig's disease, after the great New York Yankee first baseman who died of ALS in 1942. Scientists identified the altered gene responsible for the disease in 1993. As in sickle cell anemia, the mutation is slight and subtle. A single amino acid is altered in superoxide dismutase, an enzyme that eliminates free radicals, highly reactive chemical species with unpaired electrons. Free radicals appear to accumulate in Parkinson's disease, Alzheimer's disease, and in normal aging. In the case of ALS, their build-up results in the destruction of motor nerves. ALS gives little early warning. Early diagnosis, based on detection of the altered gene, might permit preventative treatment.

The "might" in the previous sentence indicates a major problem in the diagnosis of hereditary diseases and defects. In many cases, the ability of modern science to respond to the effects of a defective gene has not equaled the ability to detect the gene. The tendency to develop genetic diseases is programmed in the DNA, and it may or may not be stimulated by infection. One can well ask the advantage of knowing that an individual carries one of these genes when there is no way to prevent or even to treat its deleterious effects. Is it helpful to know that you have a high probability of acquiring ALS or Alzheimer's disease when there is nothing you can do about it?

To be sure, such knowledge could be useful in case new therapies are developed in the future. Moreover, some carriers of inheritable diseases choose not to have children or they choose to use in vitro fertilization and genetic screening. A married couple, both of whom were carriers of the gene for cystic fibrosis, recently took the latter approach. Five ova taken from the woman were fertilized with her husband's sperm, and the resulting embryos were analyzed for the cystic fibrosis gene at the eight-cell stage. An embryo with no or only one copy of the defective gene, neither of which would result in the disease, was implanted into the woman's uterus, and a healthy baby was born.

Genetic screening is more frequently used on embryos and fetuses that are further developed. If there is a likelihood of the parents passing on a defective gene, they sometimes request **amniocentesis,** a procedure in which a sample of the amniotic fluid is withdrawn from the mother's uterus. This fluid contains fetal cells that are then analyzed for their genetic makeup. By this means, the defective genes for Down syndrome and other hereditary conditions can be detected. If they are present, the parents face a difficult decision with a significant ethical component: whether to abort the fetus. Such difficult choices could be avoided if science were to develop ways of actually changing the DNA of the fetus or even of children and adults. We now consider this prospect.

A revolutionary approach to fight or prevent disease involves an attempt to correct the basic problem by altering the genetic makeup of some of a patient's own cells. **Gene therapy** has the goal of supplying cells with normal copies of missing or flawed genes. Medical researchers estimate that about 4000 known genetic disorders in addition to cancers, heart disease, arthritis, and other illnesses seem to be ideal candidates for these treatments. Simply stated, cells are taken from a patient, altered by the introduction of normal genes, and then returned to the patient. If all goes well, the imported genes function normally.

Gene therapy was first applied successfully to a human being in 1990, when the technique was used to treat a four-year-old girl suffering from severe combined immunodeficiency disease (SCID), a very serious, and fortunately very rare, condition. Because of a genetic defect, a specific enzyme is not synthesized and its absence leads to the destruction of the white blood cells that protect the body against infection. Children suffering from SCID have essentially no functioning immune system, and the slightest infection can prove fatal. In the past, some children with SCID survived for a few years, but only by living in sterile isolation chambers.

In the procedure used with the first patient, the gene that encodes for the missing enzyme was identified and isolated from other sources. Special viruses were used to introduce copies of this gene into cells that had been removed from the patient. These

Free radicals are also involved in stratospheric ozone depletion (Sections 2.11 and 2.13), in addition polymerization (Section 9.3), and in cooking foods (Section 11.2).

SCID received publicity because of David, the "boy in the plastic bubble." David was 12 years old when he died in 1984 after spending his entire life in isolator containment systems or "bubbles."

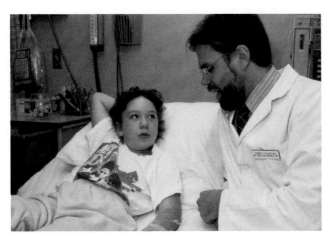

Figure 12.15

Curing disease through genetic engineering. One of two young girls who were the first humans "cured" of a hereditary disorder by transferring into their bodies healthy versions of the gene they lacked. The transfer was successfully carried out in 1990, and the girls remain healthy.

modified cells were then reintroduced into the girl's body. The new genes have continued to function well, producing the previously absent enzyme. As a result, the concentration of white blood cells has increased significantly and the girl's antibody-producing defense mechanism is working (Figure 12.15).

Although gene therapy holds great promise for humans, it cannot be categorized as a general success to date. Some have characterized the efforts of the last decade as having made virtually no progress. The process is perhaps best characterized by a 1998 quote from *The Scientist:* "The gene therapy field resembles a toolbox containing instruments researchers haven't quite mastered, and the number of devices—viral and non-viral vectors—in this toolbox keeps increasing . . . 'People are still trying to figure out what tools to use for what diseases.'" For example, in January 2003, the Food and Drug Administration placed a "clinical hold" on a new SCID approach that uses retroviral vectors to insert genes into blood stem cells after two children in a French clinical trial developed a leukemia-like condition.

Consider This 12.19   **Gene Therapy**

Although gene therapy has had promising developments for treating serious human conditions, success has not been universal.

a.   Why do you think that experimental protocols with gene therapy have shown mixed results? Explain some of the factors that you feel may influence outcomes.

b.   Given the mixed results, under what conditions would you consider gene therapy a valuable tool? Explain the reasons for your criteria.

12.9   Genetic Fingerprinting

We have seen that very small amounts of DNA can be "amplified" or reproduced by the technique of the polymerase chain reaction. This makes DNA fingerprinting a viable and highly accurate method with several possible applications. Because each of us has a unique genetic makeup that we inherited from our biological parents, this technique can be used to establish paternity and maternity. DNA isolated from body fluids like blood, from skin cells, or other genetic evidence left at the scene of a crime can be used to establish guilt or innocence. The "fingerprinting" patterns have been shown useful in establishing the identity of a homicide victim, either from DNA found as evidence or from the body itself. Use in personal identification has not found much application because of the expense, technology, and logistics of keeping such records.

DNA fingerprinting utilizes unique fragments of DNA to identify a specific individual. It is not surprising that the really important genes, those that encode for insulin, hemoglobin, chymotrypsin, and most other proteins, are identical in almost all of us. Here, as we have already noted, mutations are rare. We differ primarily in the "junk" DNA that makes up about 98% of the 3 billion base pairs in each human cell nucleus. Therefore, forensic scientists look to this apparently nonessential DNA when they seek to determine "who dun it."

During fingerprinting, the DNA undergoes reactions with enzymes that cut the DNA into segments or bands. Experimental evidence has shown that in unrelated people, the probability of one band matching is one in four (or 0.25). So the probability of matching two bands is $(0.25)^2$, or 1 in 16 chance. In a fingerprinting case, the goal is to have a large number of bands—10 bands produces $(0.25)^{10}$, or a 1 in over 1 million chance while 20 bands yield $(0.25)^{20}$, or 1 in over 1 trillion chance. Although these are only probabilities, one in a trillion represents only one person on Earth.

Consider a victim who has been assaulted. The victim cannot identify the attacker, who was masked. The police have two suspects. There seems to be little evidence, except for several drops of blood on the victim's clothing. That may be sufficient. Even a very tiny spot contains more than enough DNA for a reliable genetic fingerprint; 1×10^{-9} g of DNA is sufficient. First the DNA is extracted, and then it is multiplied many times over by PCR. These copies are exposed to the action of enzymes that cut the DNA strands into segments.

The segments are then subjected to **electrophoresis,** a method of separating molecules based on their rate of movement in an electric field. In the technique used in DNA fingerprinting, samples are applied to a strip of a polysaccharide gel, and electrophoresis is carried out in this medium. Because the phosphate groups of the DNA are negatively charged, the fragments migrate toward the positive electrode or pole. The speed at which a DNA segment moves depends on the magnitude of its electric charge, and its size or molecular mass. Shorter strands of DNA, consisting of fewer base pairs, move faster than longer strands, which encounter more resistance from the gel.

It is necessary to see and measure how far the DNA fragments have traveled in a fixed period. This is done by using radioactive markers that expose photographic film. The fingerprint thus consists of a film with black bars or bands, each one corresponding to the distance migrated by a particular segment of DNA. The heaviest and longest segments are closest to the point of application, and the lightest and shortest ones are the farthest away.

In most criminal cases, the electrophoresis pattern produced from the evidence collected at the crime scene is compared with the electrophoresis pattern made by DNA obtained from the suspect or suspects. Figure 12.16 shows electrophoretic evidence in

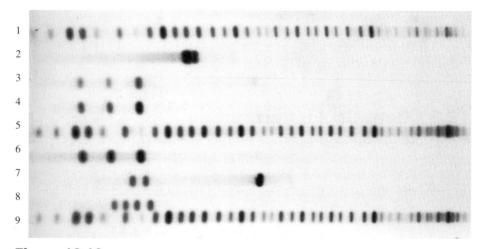

Figure 12.16

DNA fingerprints in an assault case. The various rows represent the electrophoretic migration pattern of DNA segments.

an assault case we are investigating. Each spot indicates how far DNA segments of specific size migrated during the electrophoresis experiment. Rows 1, 5, 8, and 9 are reference markers of DNA exhibiting a known range of molecular masses. Spots at the far left represent the longest, heaviest segments of DNA; spots at the far right represent the lightest, shortest segments. Row 3 is DNA from a blood sample found on the victim's clothing, and row 6 is blood DNA obtained from the sidewalk at the crime scene. Not surprisingly, the positions of the spots in these two rows are identical. To avoid possible misidentification, a sample of the victim's own DNA is included that gives the pattern in row 7. Now look at rows 2 and 4. Row 2 is characteristic of the DNA in a blood sample obtained from suspect A; row 4 is from a blood sample from suspect B. B's genetic fingerprint matches the DNA from the blood samples in rows 3 and 6.

In the case represented by Figure 12.16, the evidence strongly indicates that suspect A is innocent and that suspect B is guilty, but caution is required. A matching DNA fingerprint does not *absolutely* prove the guilt of a suspect. The DNA from two individuals might yield the same electrophoresis patterns, but it is highly unlikely. To increase the odds of accurate identification, comparisons of the sort just described are typically done on DNA from three or more different chromosomes. As we examined earlier when the number of determinations increases, so does the probability differentiating any two individuals based on their DNA.

Such probabilities can be very convincing, especially if the suspect has a motive and can be otherwise placed at the scene of the crime. However, some juries have been skeptical of DNA data. In the O. J. Simpson murder trial, defense attorneys suggested that the blood samples taken from the scene of the crime had been contaminated or planted. It's possible that doubts raised in the jurors' minds helped outweigh the evidence of the DNA fingerprints.

Your Turn 12.20   **Probability from Chromosomal Segments**

Suppose that a DNA fingerprint is based on eight different chromosomal segments. The frequency of a match is determined to be 1 in 4. What is the probability of a match in all eight segments? If the person lived in a city with 50,000 people, could the suspect be ruled out by the statistics alone?

Answer
$(0.25)^8 = 1.52 \times 10^{-5}$, or one in about 65,000

DNA can even be extracted from bone, and the base sequence can be used to identify human remains. Such studies were recently carried out on bones exhumed from a mass grave near Yekaterinburg, Russia, the site of the massacre of Czar Nicholas II and his family in 1918. The DNA was compared with samples from living relatives of the last Russian royal family and found to match sufficiently well to conclude that the remains were indeed those of the Romanovs. In 1998, DNA fingerprinting was used to determine the identity of a soldier who was killed during the Vietnam War and buried in the Tomb of the Unknowns at Arlington National Cemetery. Based on other evidence, there were two possible candidates, and the tests identified the entombed soldier. The profile of DNA taken from the bones of First Lieutenant Michael Blassie matched sufficiently that of a DNA sample taken from his mother, thus allowing proper identification.

DNA analysis is not merely confined to the living and the recently deceased. Researchers have cloned and investigated DNA samples obtained from a 2400-year-old Egyptian mummy and some even older human remains. Scientists have interpreted the results to gain information about the relationship of ancient peoples, their migration routes, and their diseases.

Consider This 12.21   Lincoln's DNA

Some researchers have speculated that Abraham Lincoln suffered from a genetic condition known as Marfan syndrome that causes a person to grow tall and gangly (in addition to having cardiovascular problems). DNA fingerprints could answer the question. Do you support exhuming Lincoln's body from the Oak Ridge Cemetery in Springfield, IL? Give reasons to explain your decision.

The current DNA age record is held by a bee and a termite that lived about 30 million years ago. Since that time, the insects had been entombed and protected in amber, which is solidified plant resin. In 1992, researchers released the perfectly preserved bodies, extracted their DNA, and subjected it to a number of studies. This research yielded important information about the evolutionary connection of these ancient organisms to modern species.

But, could 30-million-year-old DNA yield more? Could it be cloned into a living fossil? That is the premise of *Jurassic Park.* In the science fiction film and the novel, the chief interest is not in the fossilized insects themselves, but in the dinosaur blood they contain. The blood is the source of the DNA that is cloned and introduced into crocodile egg cells, where it replicates until it creates modern copies of long-extinct creatures. "Could this happen in real life?" That question is posed by I. Edward Alcamo in *DNA Technology: The Awesome Skill,* a book that is a useful source of information and illustrations for this chapter. Professor Alcamo's answer may be mildly reassuring for those who would rather not encounter a *Tyrannosaurus rex:* "Possibly. But you would need an entire set of dinosaur chromosomes, and only a minuscule fragment has been obtained up to now. And that's only the first of a thousand problems that must be solved."

Science fiction books and films to the contrary, dinosaurs and humans never coexisted. Dinosaurs were long extinct before the first humans appeared on Earth, a fact confirmed by fossil records.

Consider This 12.22   Science Fiction Success

One reason why science fiction is successful is that it starts with a known scientific principle and extends, elaborates, and sometimes embroiders it. Perhaps you can be as successful as Michael Crichton was with *Jurassic Park.* Start by identifying a scientific principle from this or another chapter, and then writing a one- or two-page outline for a story based on that principle. Be sure to identify the chemical concepts that you plan to include and any pseudoscience that you might employ.

12.10   Mixing Genes from Several Species: Improving on Nature?

One of the more sensational dimensions of genetic engineering are so-called **transgenic organisms,** artificially created and stringently controlled higher plants and animals that share the genes of another species. Inserting foreign DNA becomes progressively more difficult as one moves up the evolutionary ladder from bacteria through plants to animals. Nevertheless, some of the most spectacular successes of recombinant DNA technology have involved modifications of agricultural crops and animals. As a result and in spite of some rather formidable difficulties, altering the genetic makeup of plants by genetic engineering is faster and more reliable than relying on traditional crossbreeding. Moreover, some of the species that have been genetically combined are so dramatically different that interbreeding is impossible.

A report entitled "Transgenic Plants and World Agriculture" was released in July 2000 by the U.S. National Academy of Sciences and the Royal Society of London. The equivalent science academies of Brazil, China, India, Mexico, and a scientific group representing the underdeveloped countries of the world participated in preparing the report. Among other goals it lists recommendations for transgenic crop research and

Table 12.2	Transgenic Plants and Animals
Organism	**Objective of Change**
Transgenic plant	
soybeans, corn, canola, cotton	Resistance to Roundup and Liberty herbicides
corn, cotton, potato	Produce a gene from Bt (*Bacillus thuringiensis*), a soil bacterium that releases a delta-endotoxin in insects
papaya	Resistivity to pests and diseases (ring-spot virus)
grains	Higher-yielding grain on a stronger, shorter stalk
rice	Golden rice capable of synthesizing beta-carotene, the precursor of vitamin A
tomato	Enhanced nutritional factors (e.g., lycopene, a naturally occurring nutrient related to vitamin A)
	Delayed ripening (to enhance flavor and portability)
	Resistance to bacterial speck disease
Plants in development	
canola	Improved nutritional value of canola oil with higher vitamin E content or modified content of fatty acids
sunflower	Disease, pest, and herbicide resistance
turf grass	Herbicide tolerance
	Disease and insect resistance
	Reduced growth rates
	Tolerance to drought, heat, and cold
Animal in Development	
beef cattle	High content of omega-3 fatty acids (nutritional benefit)

development that ". . . should focus on plants that will (i) improve production stability; (ii) give nutritional benefits to the consumer; (iii) reduce the environmental impacts of intensive and extensive agriculture; (iv) increase the availability of pharmaceuticals and vaccines. . . ." Most of the commercially developed crop varieties in the United States and Canada have centered on increasing the shelf life of fruits and vegetables, conferring resistance to pests or viruses, and producing tolerance to specific herbicides. But a host of other variants have been produced or are in development. A list of available commercial or experimental plants and animals is shown in Table 12.2.

Although one might look at delayed ripening as only a commercial issue, that is, a source of increased profit for producers, the issue also addresses nutrition and world hunger. Considering that as much as 50% of agricultural commodities in underdeveloped countries spoil before reaching the consumer, extending the access to fresh food has tremendous repercussions. The same arguments hold for developing transgenic plants that are resistant to specific pests. Some of the crops in Table 12.2 contain insect-resistant genes from a natural source, *Bacillus thuringiensis,* or Bt. A protein produced by these bacteria is deadly to certain caterpillars that otherwise prey on cotton and corn; the caterpillar eats a bit of the plant and dies. A powder that contains the dormant Bt bacteria has been available at garden suppliers to protect home gardens against the ravages of cabbage and broccoli "loopers" for decades. The new corn and cotton varieties now produce the caterpillar toxins on their own without the need for the presence of Bt in the soil.

Many plants including about half of all soybeans grown in the United States have a transgenic gene for resistance to certain herbicides. The DNA segment from *E. coli* enables the use of a minimum amount of herbicide but enough to kill the weeds without affecting the crop. These "internal" protections from insect and plant pests ultimately improve crop yields. Yields of grains have been transferred between different plant species—genes from high-yield semidwarf wheat have been introduced into other species and afford them larger quantities of the grain seeds on a shorter, stronger stalk

Table 12.3	**Millions of Acres of Transgenic Crops Planted in 2000**	
Country	**Area Planted in 2000**	**Crops Grown**
United States	74.8	soybeans, corn, cotton, canola
Argentina	24.7	soybeans, corn, cotton
Canada	7.4	soybeans, corn, canola
China	1.2	cotton
South Africa	0.5	corn, cotton
Australia	0.4	cotton
others[a]	minor area[a]	cotton, corn, soybeans, potatoes[a]

[a] Cotton: Mexico; corn: Bulgaria, Spain, Germany, France; soybeans: Romania, Uruguay; potatoes: Romania

that can better support them. Transgenic rice plants have been developed for use in Africa where yellow mottle virus has destroyed a majority of the annual production of rice. Researchers are investigating ways of incorporating the DNA of nitrogen-fixing bacteria into wheat, rice, and corn. These bacteria are present in the root systems of soybeans, alfalfa, and other legumes. Thanks to the bacteria, these plants can absorb N_2 directly from the atmosphere and use it in biochemical reactions. Genetically modified (GM) plants have been developed with citric acid in their roots to provide better tolerance to aluminum in the soil. A salt-tolerant maize plant can grow at higher concentrations of salt in the soil.

> In Chapter 6 we saw that the removal of N_2 from the air by plants was an important part of the nitrogen cycle.

Issues of nutrition have also been addressed by new plant variants. Golden rice is being grown to produce beta-carotene, the precursor to vitamin A, particularly for countries with blindness related to vitamin A deficiency. Anemia caused by iron deficiency in the diet is being addressed by developing cereal grains that contain essential micronutrients such as iron.

Other benefits of transgenic plants include several aspects related to a reduced environmental impact. Control of weeds or insect pests by a feature of the GM plant often mean that reduced amounts or no pesticides or herbicides are applied. Different tillage systems that reduce erosion can be applied. The reduction of root diseases and a lesser dependence on water are goals being addressed.

During 2000, transgenic plants were grown on approximately 110 million acres worldwide (Table 12.3), an increase of about 11% over the previous year. Figure 12.17

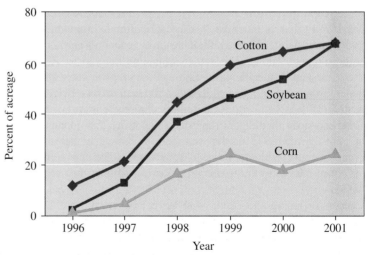

Figure 12.17

The U.S. adoption of transgenic crops as a percentage of the total acreage in the country.

Source: http://www.biotechnol.org/pages/1/index.htm

shows the growth of transgenic crops in the United States over the last several years. For example, cotton and soybeans each were planted on about 50% of available acreage in 1998 and 1999, respectively, but increased to 70% just a few years later. The U.S. corn crop has stabilized at about 25% in transgenic varieties. Most of those plant varieties are either herbicide-tolerant or insect- and pest-resistant. Minor acreage was planted with transgenic potatoes, squash, and papaya.

Consider This 12.23   **Percent of Transgenic Varieties?**

Figure 12.17 shows the change in the percentage of transgenic varieties of cotton, soybeans, and corn grown in the United States over the last several years. Speculate on why cotton and soybean percentages continue to increase while corn appears to have reached a plateau. Find evidence on the Web to support your assertions.

Consider This 12.24   **Turning Over a New Leaf**

The New Leaf Superior potato has been genetically engineered to produce its own insecticide. This gives the potato the ability to resist attack by potato beetles, destructive insects that cause significant damage to potato crops. You may have already eaten some of these potatoes without even knowing it, for they do not have to be labeled as a food produced through biotechnology.

a. Identify some of the benefits and risks associated with genetically engineered potatoes.
b. Would you knowingly eat potato chips made from New Leaf potatoes? Explain the reasons for your opinion.

The grocery store of tomorrow may also contain other products of rearranged genes. For example, it may prove possible to genetically induce cows to give human milk for newborn babies, lactose-free milk for those suffering from lactose intolerance, naturally iron-enriched milk for anemic persons, casein-rich milk for cheese-makers, and skim milk for weight watchers (not all from the same cow, of course). Scientists have developed mice with significantly higher levels of omega-3 fatty acids. The ultimate goal is to transfer this trait into beef cattle and thereby produce meat that has a more healthful form of fat for people's diet.

Plant production of medicines and drug precursors suggest that transgenic plants could be developed to produce a host of pharmaceuticals and vaccines. Most vaccines require refrigeration or other special handling and trained professionals to administer them; some developing countries can't even afford the needles to inoculate their populations. Vaccines against infectious diseases of the intestinal tract have been produced in potatoes and bananas. Anticancer antibodies have been expressed and introduced into wheat recently. The development of transgenic plants that produce therapeutic agents to address problems of disease is just beginning to address its own potential.

So why isn't the whole world enthusiastic about GM or transgenic food? The creature from Mary Shelley's *Frankenstein* evokes images of an entity composed of "parts" from several sources but who collectively produce a being that is out of control. Those opposed to GM food often refer to it as "Frankenfood." In Europe, Japan and other parts of the world legislation has been introduced to ban or control it.

Starlink corn is an example of a bioengineered plant that has achieved notoriety in the last several years. Starlink was never approved for human consumption, in part, because government regulatory agencies were cautious about effects on humans. It has no known deleterious effects on humans, but the concern was whether people who were normally not allergic to corn would develop allergic reactions due to the bacterial protein that was engineered into it. Although Starlink was only approved for animal use,

it accidentally found its way into the human food chain—into several brands of taco shell, for example. No one has suffered any adverse effects to this point, but the mere possibility that it moved from animal feed to human food without sufficient control alarms many individuals. In the summer of 2001, the Centers for Disease Control and Prevention performed some carefully controlled experiments looking for antibodies that would have been produced by a set of individuals in a true allergic reaction who claimed an adverse effect from Starlink corn. Their study showed no link but the report states that ". . . we cannot completely rule out this possibility, in part, because food allergies may occur without detectable [antibodies] to the allergens."

Writing in the May 1999 issue of *Nature,* Cornell researchers reported that Bt-corn, a GM product, is a serious threat to monarch and other butterflies. In laboratory tests, monarchs who were fed milkweed leaves dusted with transgenic corn pollen ate less, grew more slowly, and suffered a higher mortality rate. Nearly half of the larvae who ate leaves dusted with transformed corn pollen died, compared with virtually none who ate leaves dusted with nontransformed corn pollen. Subsequent researchers, however, report that the risk to monarchs is minimal when the leaves are dusted with pollen levels mimicking those actually found in natural settings.

A 1998 report from a research institute in the United Kingdom attacking the safety of genetically modified potatoes may be largely responsible for the "Frankenfood frenzy" in Europe. That document reported that GM potatoes have a specific and negative effect on organ development and the immune system in rats. In June 1999, the Royal Society in Britain conducted a thorough, independent review and concluded that the report and the experiments on which it was based were flawed. They cited poor experimental design, possibly exacerbated by the lack of "blind measurements," uncertainty from the small number of samples, the application of inappropriate statistical techniques, possible dietary differences due to nonsystematic dietary enrichment of protein in the rats, and a lack of consistency of findings within and between experiments. Critics, however, remain undaunted and vocal in spite of few reproducible experiments to support any health concerns.

In the summer of 2003, the European Union (EU) adopted two proposals to help calm the frenzy. The first proposal established a system to trace and label genetically modified products of technology. The second proposal regulated the marketing and labeling of food and feed products derived from GM organisms. The latter is far stricter than previous measures and includes prepared food that might, for example, be baked using oil derived from GM corn or soybeans. The EU argues that people deserve the right to choose or at least know the content of the food they eat. Six EU countries have placed a moratorium on the cultivation of GM crops. However, the United States along with Canada, Argentina, and Egypt filed suit through the World Trade Organization alleging that the moratorium is politically motivated. They claim that in some ways banning GM grain represents the only way EU farmers can compete with foreign ones. The issue remains unresolved.

Consider This 12.25   **EU Moratorium on GM Foods?**

Use the resources of the Web to check the status of the European Union's moratorium on genetically modified foods. Develop a list of arguments supporting the EU and the United States.

Genetic engineering, however, is more than simply a collection of techniques. It has a human face among those who have benefited from it, whether medically or as a result of the GM food items for the underdeveloped world. Before recombinant DNA, there was no hepatitis B vaccine; insufficient erythropoietin, a protein used to stimulate red blood cell growth in dialysis patients; a lack of tissue plasminogen activator (TPA) to dissolve blood clots in cardiac and stroke patients; and limited amounts of insulin and a form that didn't quite match the human molecule.

Using transgenic plants and animals to make compounds currently produced mainly in vats and vials is a far-reaching concept from the standpoint of economics and resource management. In the October 23, 1998, issue of *Science,* DuPont board chair Jack Krol commented about this matter: "In the 20th century, chemical companies made most of their products with nonliving systems. In the next century, we will make many of them with living systems." In support of that premise, DuPont has research underway to use microbes and plants to produce a wide range of compounds, from chiral drugs to plastics. Jerry Caulder, an agricultural genetic engineering entrepreneur, is also "bullish" on this approach. In the same issue of *Science,* he asked: "Organisms are the best chemists in the world. Why not use them to produce the chemical feedstocks you want rather than using petroleum?"

Have the benefits of genetically modified foods outlined in this section been overstated while the problems have been "swept under the table" by those who benefit the most economically? Much of the opposition to Frankenfoods is based on the concern that "natural" plants could be changed forever and perhaps irreversibly on exposure to pollen from altered crops. Opponents also fear that unanticipated health problems could arise in people and animals that consume GM food.

Consider This 12.26   **The Frankenfood Frenzy**

The European Union has enacted stricter controls over the importation of genetically modified foods. Opposition to GM foods was a major theme of American protesters at the 2000 World Trade Organization meeting in Seattle and IMF/World Bank meetings in Washington, D.C.

In 2000 transgenic food opponents filed a class-action suit against Monsanto, a producer of seeds. The Food and Drug Administration announced a new initiative to engage the public about bioengineered products. Search for and describe articles about recent developments in the Frankenfood controversy. Report the reasons for concerns by critics and any evidence refuting them.

12.11   Cloning Mammals and Humans

The announcement of the successful cloning of human embryos in February 2004 sparked fresh debate and raised concern around the world. But this was not a sudden discovery and perhaps not even unexpected. Since first envisioned in 1938 by Hans

(a) Dolly and the lamb she gave birth to by conventional means.

(b) Dolly shortly before her death in 2003.

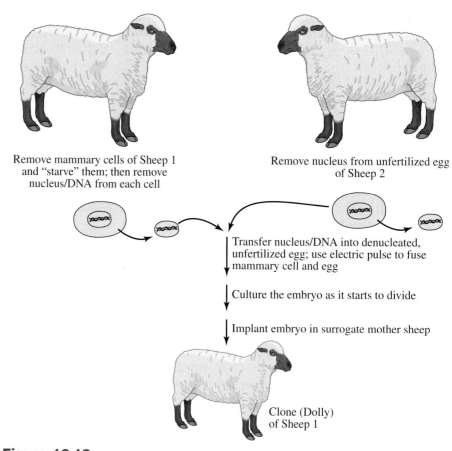

Remove mammary cells of Sheep 1
and "starve" them; then remove
nucleus/DNA from each cell

Remove nucleus from unfertilized egg
of Sheep 2

Transfer nucleus/DNA into denucleated,
unfertilized egg; use electric pulse to fuse
mammary cell and egg

Culture the embryo as it starts to divide

Implant embryo in surrogate mother sheep

Clone (Dolly)
of Sheep 1

Figure 12.18

The cloning of Dolly from a mature animal.

Dolly's cells contain the DNA of only one parent, unlike identical twins who share DNA from both parents.

Mitochondria in cells, but not within cell nuclei, are vital to energy production.

Spemann, slow but steady progress has been made to realize the goal. Through the 1960s and 1970s, British molecular biologist John B. Gurdon made several break-throughs, eventually cloning frogs. The reported cloning of mice, the first mammals, in the early 1980s could not be repeated, but Steen Willadson's cloning of live sheep from immature sheep embryos was replicated for cattle, pigs, goats, rabbits, and rhesus monkeys. In 1996, Ian Wilmut made headlines with the birth of Dolly, which resulted from adult genetic material being fused with another sheep ovum and the resulting embryo being implanted into a surrogate mother.

Dolly was cloned by a technique called nuclear transfer (Figure 12.18). **Nuclear transfer** is a laboratory procedure in which a cell's nucleus is removed and placed into an egg cell that has had its own nucleus removed. The genetic information from the donor nucleus controls the resulting cell, which can be induced to form embryos. Growth was initiated by an electrical "jump start" to create the growing embryo. The egg was then implanted into another sheep's uterus, carried full term, and the lamb Dolly was born, the first mammal produced by cloning. Dolly carried DNA identical to the sheep who had donated the udder cells.

It took 277 such transfers to create Dolly. Until she came along, conventional wisdom had it that adult cells lacked the workable versions of all the genes necessary to create an entire organism. To circumvent this possible difficulty, Wilmut and his coworkers deprived the mammary cells of nutrients for five days before extracting their nuclei, forcing the cells out of their normal growth pattern and into a resting stage. Speculation is that this procedure might have increased the likelihood that once implanted into an egg, the chromosomes could be reprogrammed to establish the growth of an entire organism.

In 1998, Japanese scientists used a more effective nuclear transfer variation employing two kinds of reproductive system cells to clone eight calves from one. In 2000, Tetra (a rhesus monkey) was cloned by a technique called embryo splitting—an embryo formed from in vivo fertilization of monkey sperm and eggs. The embryo was divided at the eight-cell stage—forming four identical two-cell embryos that were implanted in four surrogate mothers, one of which gave rise to Tetra.

The 2004 cloned human embryos discussed at the beginning of the chapter also arose via the nuclear transfer technique—genetic material was added to human donor eggs from which the native nuclear material had been removed. After an electrical jolt, the embryos started growing. However, in four to five days at the stage of about 100 cells, these blastocysts, as they are called, were not implanted. Taking them to the stage of an infant, called reproductive cloning, was never their intention. At the blastocyst stage, the researchers harvested a set of **stem cells,** identical, undifferentiated cells that, by successive divisions, can give rise to specialized ones like blood cells. Stem cells can be cultured or reproduced and then at some point coaxed into forming specific cells like heart or nerve cells. This technique is called therapeutic cloning since the goal is to create a new line of stem cells. In principle a person with a spinal cord injury could donate DNA from her cells, have it converted into stem cells with her genetic material, and then have nerve cells formed to repair the spinal cord injury. Advances in stem cell technology have been touted as holding potential cures for many crippling illnesses and injuries.

But several ethical and moral issues exist. Some people believe that an embryo, even just several minutes old, is a human life and that harvesting stem cells from the blastocyst, thereby destroying its viability, is equivalent to ending that life. In 2003, the U.S. House of Representatives passed a ban on human cloning although the bill failed to get approval in the Senate. President George W. Bush forbade any federally funded research on stem cells lines created after August 2001.

In September 2004, California approved legislation making it legal to derive embryonic stem cells from human embryos and to conduct research using stem cells from "any source."

12.12   The New Prometheus(?)

Nature is an indispensable aid and ally in medicinal and biological chemistry. Much of our success has come from understanding and imitating natural processes. For centuries, animal breeders, agricultural researchers, and observant farmers have brought about genetic transformations in animals and plants by selective breeding. Recent applications of chemical methods to biological systems have greatly increased our capacity to effect such changes. Molecular engineering has made it possible to create nucleic acids, proteins, enzymes, hormones, drugs, and other biologically important molecules that do not exist in nature.

The new molecules can be designed to be more efficient catalysts than their naturally occurring counterparts, more effective and less toxic drugs for treating a wide range of diseases, or modified hormones that actually work better than the original. It is not at all fanciful to imagine a whole range of enzymes, engineered to consume environmentally hazardous wastes that are impervious to naturally occurring enzymes. Likewise, it might be possible to create an enzyme that far surpasses the one that catalyzes photosynthesis, one of the most inefficient of all natural enzymes. As Mark Twain suggested: "Predictions are extremely difficult, especially those about the future." Yet, it is likely that because of genetic engineering, AIDS and at least some forms of cancer may some day become as infrequent as polio or smallpox are now. Even more tantalizing is the possibility of eradicating certain genetic defects. Our growing knowledge of the human genome, coupled with our understanding of the chemistry of genetics, holds the promise of altering our inherited gene pool. Prospects include the elimination of sickle cell anemia, diabetes, hemophilia, phenylketonuria, and dozens of other debilitating hereditary traits.

The next logical step would seem to be the creation of new organisms. Scientists have already cloned "new and improved" animals and vegetables. Mammals have been cloned from adult cells. Human cloning has been dealt with in fiction such as Huxley's *Brave New World*—using eugenics to create different classes of people, and in films like Woody Allen's *Sleeper*—attempts to clone a dead dictator, and *The Boys from Brazil*—cloning Nazis. How should we view the possibility of cloning human beings, perhaps even "new and improved" ones? The question goes straight to our identity as a species and as individuals. One could reason that the best way to treat a genetic disease or disability would be to remove the defective gene from the gene pool by intentionally altering the suspect DNA in the sperm or ova. But one could also argue that *Homo sapiens* was not yet sufficiently wise to assume such god-like power. Our species has had some tragic experiences in the past century with political leaders who used less subtle methods of "genetic cleansing."

Consider This 12.27   A Cloning Clinic?

Plans to develop human cloning clinics have been announced in the last several years. What risks and benefits are associated with such clinics? Has such a clinic been completed at this time? Speculate on why or why not.

Cloning through genetic engineering raises not only the possibility for humans to duplicate themselves by unconventional means but also the chilling potential to design a master race or to subjugate or eliminate "defectives" through genetic manipulation. Hence, there is an intentional irony in the title of this final section. Prometheus was the demigod who stole fire and the flame of learning from the gods and brought these incomparable gifts to humanity. "The New Prometheus" is the subtitle of *Frankenstein,* Mary Shelley's classic study of scientific knowledge run amok. Using our ever-growing knowledge of the chemistry of heredity wisely and well will surely be one of the greatest challenges of the 21st century.

Consider This 12.28   Send in the Clones

The late Isaac Asimov, noted science fiction writer and biochemist, coauthored this verse, called "The Misunderstood Clone."

> *Oh, give me a clone*
> *Of my own flesh and bone*
> *With its Y chromosome changed to an X.*
> *And when it has grown*
> *Then my own little clone*
> *Will be of the opposite sex.*

Source: *The Sun Shines Bright* by Isaac Asimov. Copyright © 1981 Nightfall, Inc. Used by permission of Doubleday, a division of Random House, Inc. and Ralph Vicinanca Agency.

Is the verse describing gene therapy or cloning? Support your answer by explaining the two different processes.

Consider This 12.29   Cloning Humans: For Good or Evil?

James D. Watson said this about manipulating germ cells to create superpersons, "When they are finally attempted, germ-line genetic manipulations will probably be done to change a death sentence into a life verdict—by creating children who are resistant to a deadly virus, for example, much the same way we can already protect plants from viruses

(continued on p. 563)

Consider This 12.29   **Cloning Humans: For Good or Evil?**
                           (*continued*)

by inserting antiviral DNA segments into their genomes." (Source: *Time,* "All for the Good," January 11, 1999, Vol. 153, No. 1.)

Draft a statement supporting or opposing Watson's position on this issue. Your argument should be grounded in the scientific principles learned in this chapter.

CONCLUSION

The last chapter of this book, like almost all that preceded it, ends with a dilemma: How can we balance the great benefits of modern chemical sciences and technology and the risks that seem inevitably to accompany them? Throughout this text, the authors have occasionally looked, with myopic professorial vision, into the cloudy crystal ball of the future. It is in the nature of science that we cannot confidently predict what new discoveries will be made by tomorrow's chemists. Nor can we know the applications of those discoveries, good or bad. Such uncertainty is one of the delights of our discipline. A chemist must learn to live with ambiguity, indeed, to thrive on it, in the search to better understand the nature of atoms and their intricate combinations, in all their various guises.

But all citizens of this planet must at least develop a tolerance for ambiguity and a willingness to take reasonable risks, especially considering that life itself is a biological, intellectual, and emotional risk. Of course, we all seek to maximize benefits, but we must recognize that individual gain must sometimes be sacrificed for the benefit of society. We live in multiple contexts—the context of our families and friends, our towns and cities, our states, our countries, our special planet. We have responsibilities to all. *You,* the readers of this book, will help create the context of the future. We wish you well.

Chapter Summary

Having studied this chapter, you should be able to:

- Understand the chemical composition of deoxyribonucleic acid (DNA), a polymer of nitrogen-containing bases, deoxyribose, and phosphate groups (12.1)

- Recognize the utility of Chargaff's rules: %A = %T, %G = %C (12.1)

- Interpret evidence for the double-helical structure of DNA and its base pairing (12.2)

- Understand DNA replication (12.2)

- Describe the genetic code—codons of three bases corresponding to specific amino acids (12.3)

- Relate to the amount of information encoded in the human genome (12.3)

- Discuss the primary, secondary, and tertiary structure of proteins (12.4)

- Describe the molecular basis of sickle cell anemia (12.4)

- Relate to the Human Genome Project: its aims, significance, and ethical implications (12.5)

- Understand the production of human proteins (insulin and human growth hormone) in bacteria (12.6)

- Recognize the need to establish priorities for the use of limited supplies of drugs produced by recombinant DNA technology (12.6)

- Recognize recombinant DNA techniques—insertion of foreign DNA into bacteria (12.6)

- Relate how new drugs to treat hepatitis, herpes, cancer, heart attack, strokes, AIDS, and other diseases, and new vaccines are developed via genetic engineering (12.7)

- Understand the genetic diagnoses of ALS, Alzheimer's disease, and other conditions (12.8)

- Understand the connection between DNA probes and the polymerase chain reaction and the factors related to appropriate responses to genetically diagnosed hereditary diseases (12.8)

- Describe gene therapy as a treatment for severe combined immunodeficiency disease (SCID), malignant melanoma, and other diseases (12.8)

- Discuss DNA fingerprinting for identification and evolutionary and anthropological studies, and the technical and legal issues associated with DNA fingerprinting (12.8)

- Refute the possibility and wisdom of cloning long-extinct animals (12.9)

- Describe the laboratory techniques in DNA fingerprinting (12.9)

- Describe transgenic organisms and their uses: pest-resistant and nitrogen-fixing plants; animals with genes to produce human proteins and other modified biochemicals (12.10)

- Discuss ethical issues associated with transgenic organisms and the fears about Frankenfood (12.10)

- Understand mammalian cloning using nuclear transfer (12.11)

- Describe the role of nuclear transfer in the cloning of mammalian cells (12.11)

- Debate issues associated with the prudent and ethical applications of mammalian cloning and genetic engineering (12.12)

Questions

Emphasizing Essentials

1. The letters DNA have been called three of the most important letters of the late 20th century. For what is DNA an acronym?

2. **a.** Use Figure 12.1 to determine the percent increase in human life expectancy at birth in the United States between 1940 and 1980 for males and females.

   b. Are these percentages the same as those calculated in Your Turn 12.2? Why or why not?

3. Consider the structural formulas in Figure 12.2.

   a. What functional groups are in adenine?

   b. What functional groups are in deoxyribose?

   c. Does the phosphate group have any functional groups?

4. Consider the structural formula of deoxyribose given in Figure 12.2.

   a. Why is deoxyribose classified as a monosaccharide?

   b. What is the molecular formula for deoxyribose?

   c. Why isn't deoxyribose an acid in aqueous solution?

5. Equation 12.1 shows the general case for a nitrogen-containing base reacting with water. Use the structural formula of thymine in Figure 12.2 to write an equation showing how thymine could react with water to generate hydroxide ions.

6. **a.** What three types of units must be present in a nucleotide?

   b. What type of bonding holds these units together in a nucleotide?

7. Table 12.1 lists the base composition of DNA for various species. The four bases are adenine, cytosine, guanine, and thymine. What relationships exist among these bases, no matter what the species?

8. **a.** What happens experimentally during X-ray diffraction?

   b. The first X-ray diffraction patterns were of simple salts, such as sodium chloride. The X-ray diffraction studies of nucleic acids and proteins did not come until much later. Suggest why.

9. Figure 12.7 shows the pairing of nucleotide bases in DNA.

   a. What type of *intramolecular* bonding occurs within each base?

   b. What type of *intermolecular* bonding holds the base pairs together?

10. Identify the base sequence that is complementary to each of these sequences.

    a. ATGGCAT　　　**b.** TATCTAG

11. Given that the distance between adjacent bases is 0.34 nm, how many base pairs are present in a chromosome that is 3.0 cm long? *Hint:* $1 \text{ m} = 10^2 \text{ cm}$; $1 \text{ m} = 10^9 \text{ nm}$.

12. During cell division, as many as 90,000 nucleotides per minute can be added to the growing DNA chain.

    a. The shortest human chromosome contains 50 million bases. What is the minimum time required to form a strand of this chromosome?

    b. Determine the length of this chromosome (in centimeters) that would be formed in 1 minute if the distance between bases were 0.34 nm. *Hints:* $1 \text{ m} = 10^2 \text{ cm}$; $1 \text{ m} = 10^9 \text{ nm}$.

13. Twenty amino acids commonly occur in proteins.

    a. What is the *general* structural formula for these amino acids?

    b. What functional groups are present in all amino acids?

14. The text states that if you were to use the letters A, T, C, and G in a game of Scrabble, you could generate 64 different three-letter combinations. (Your Scrabble opponent would surely challenge some of these combinations!) Nature pairs the four bases A, T, C, and G, and uses the pairs to encode for amino acids. Write down all the possible paired combinations of A, T, C, and G to find the maximum number of amino acids that can be encoded from these four bases. *Hint:* In nature, unlike Scrabble, a letter can be used more than once in forming a pair.

15. What is a codon and what is its role in the genetic code?

16. Only 61 of the possible 64 triplet codons specify certain amino acids. What is the function of the other three codons?

17. Describe what is meant by the primary, secondary, and tertiary structure of a protein. Is one of these more important than the others? Why or why not?

18. Explain how an error in the primary structure of a protein in hemoglobin causes sickle cell anemia.

19. How can recombinant DNA technology be used to overcome a shortage of insulin for use by diabetics?

20. **a.** What basic molecular building blocks are found in viruses?

    b. Why are viruses referred to as almost living biochemical species?

    c. Why is it not possible to treat a viral infection with antibiotics?

21. The letters DNA have been called three of the most important letters of the late 20th century. A long-shot candidate for this honor might be PCR.

    a. What do those letters stand for?

    b. What problem did PCR technology successfully address?

    c. Name some examples of the success of PCR technology for diagnosis.

22. Give is the minimum number of PCR cycles necessary to convert two DNA molecules to each of these.

    a. 5000 DNA molecules

    b. 50,000 molecules

    c. 500,000 molecules

23. How does electrophoresis separate different molecules?

24. What is meant by the term *molecular pharming?*

25. **a.** What is the Human Genome Project?

    b. Why is it a significant step in understanding the genetic basis of humans?

    c. What progress has been made in mapping the human genome?

26. **a.** What are stem cells?

    b. How are stem cells harvested?

    c. What aspect of stem cell research makes it controversial?

Concentrating on Concepts

27. What is meant by the term *cloning?*

28. Life expectancy in the United States increased dramatically during the 20th century, as shown in Figure 12.1. Is a similar increase possible during the 21st century? Explain your answer.

29. Compare this representation with Figure 12.4.

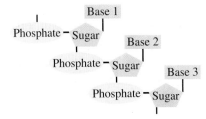

Both show a segment of deoxyribonucleic acid. Discuss the strengths and weakness of each representation.

30. Consider Chargaff's discovery that there are equal percentages of adenine and thymine and of cytosine and guanine in DNA. Was his discovery as important to understanding the nature of DNA as Crick and Watson's discovery of the double helix? Give reasons for your answer.

31. Use Figure 12.7 to help explain why stable base pairing does *not* occur between adenosine and cytosine, thymine and guanine, adenine and guanine, and thymine and cytosine.

32. One of the mechanisms by which DNA sustains damage from UV light is via the formation of covalent bonds between two thymine bases.

    a. What are the potential consequences of cross-linking between two thymines on a single strand of DNA?

    b. What possible problems do you foresee if one thymine from each of two complementary strands undergo a cross-linking reaction?

33. Errors sometimes occur in the base sequence of a strand of DNA. But, not all of these errors result in the incorporation of an incorrect amino acid in a protein for which the DNA codes. Explain how this happens, and why it is advantageous.

34. Human insulin and human growth hormone have both been made through the use of recombinant DNA technology. Which do you believe is a more significant use of this technology and why? Discuss what factors have influenced your opinion.

35. DNA probes are being used for diagnosis. Describe the principle of using DNA probes to identify *Helicobacter pylori,* the cause of gastric ulcers and possibly stomach cancer.

36. **a.** The first successful application of gene therapy to a human being was in 1990. What is meant by gene therapy and what disease was treated in that landmark case?

    b. Has gene therapy been used successfully for treating other diseases as well?

    c. Will it work for all diseases?

37. Lou Gehrig's disease is caused by the alteration of a single amino acid in the enzyme superoxide dismutase. How many base pairs are responsible for specifying this amino acid in the gene that codes for the protein? What is the minimum number of base pairs that would have to be changed to produce the disease?

38. Do you favor the patenting of genes? What are the advantages and disadvantages of this approach?

39. Genetic fingerprinting can be used for personal identification.

    a. List the possible benefits of using this technique.

    b. What are the disadvantages?

40. How widely available is the New Leaf potato? If you wanted to plant this in your garden, would you be able to obtain these potato plants? For what reasons would you want to obtain this plant?

41. Consider the information in Table 12.3. Use a pictorial representation of this information to convey the differences among the countries in using transgenic crops.

42. Consider the idea of mixing genes as an improvement on nature.

    a. What are transgenic organisms?

    b. Why is the alteration of the genetic makeup of plants by genetic engineering preferred to traditional cross-breeding methods?

43. Consider some of the successful transgenic plant experiments given in Table 12.2. What generalizations can be drawn between the source of the genes, the transgenic plant, and the objectives of the experiment?

44. The cloning of Dolly in 1996 and human embryos in 2004 are remarkable scientific events.

    a. In what ways were the techniques that led to each similar?

    b. In what ways were they different?

Exploring Extensions

45. Dolly's birth surprised genetic engineering experts and shocked members of the media responsible for reporting the birth and the method. Locate one or two early reports about the cloning of Dolly. Evaluate these reports for their scientific accuracy and what they reveal about the opinion of experts at that time.

46. Use this information to act as a Sceptical Chymist in checking the correctness of the claim that the DNA in an adult human would stretch from Earth to the Moon and back more than a million times.

    Distance from Earth to the Moon = 3.8×10^5 km

    Number of DNA-nucleated cells in an adult human = 1×10^{13}

    Length of stretched human DNA strand = 2 m

47. Consider the structural formula of deoxyribose shown in Figure 12.2. The prefix "deoxy" means without oxygen; the —OH group is replaced by a hydrogen atom. In the specific case of deoxyribose, the —OH group replaced by a hydrogen was bonded to the only carbon in the ring that bonds to two hydrogen atoms in deoxyribose. The structural formula of ribose appears in Consider This 12.4. Draw another isomer of deoxyribose, one that would still form a nucleoside with a base and with phosphate. Compare your isomer side-by-side with the structural formula of deoxyribose.

48. The text states that the more closely species are related, the more similar the DNA base compositions are in those species. The Sceptical Chymist is having trouble believing this, particularly if it means a close relationship between a fruit fly and a bacterium. Use the information in Table 12.1 to determine whether it supports the generalization about similar base pairs in similar species.

49. Perhaps you have learned some memory aids (mnemonics) when taking music lessons (Every Good Boy Does Fine), memorizing the names of the Great Lakes (HOMES), or learning about oxidation and reduction (OIL RIG). One of the authors learned "All-Together, Go-California" as the mnemonic to remember the correct base pairings in DNA.

    a. What is the relationship in this mnemonic to DNA base pairing?

    b. Design a different mnemonic that will help you remember such base pairings.

50. Of the major players in the discovery of DNA's structure, only Rosalind Franklin had a degree in chemistry. What was her background and experience that enabled her to make significant contributions? Did her contributions receive adequate credit and recognition? Write a short report on the results of your findings, citing your sources.

51. Classify the objectives of the experiments in Table 12.2 into different categories and rate the relative importance of each.

52. Transgenic plants have not been widely accepted in all countries. Give reasons for their rejection in some European markets.

53. A recent focus of participants in the Human Genome Project has been to determine the base sequence for many different microbes (i.e., bacteria and viruses). Use the Web to document recent efforts and progress. Suggest reasons why these sequences would generate so much interest.

54. Genetic diseases are also called inborn errors of metabolism. You may be familiar with some of these diseases, such as hemophilia, PKU, Tay–Sachs disease, or sickle cell anemia. One that does not get much attention is a condition known as Niemann–Pick disease. Find out what inborn metabolic error causes this condition, how many children are born with this disease in the United States each year, what treatments are available, and whether a cure is possible. Write a report to be discussed with your classmates.

55. Remains of a soldier shot down over Vietnam in 1972 and buried in the Tomb of the Unknown in 1984 were identified in 1998 by the Armed Forces DNA Identification Laboratory (see Section 12.9). The confirmatory tests were based on matching mitochondrial DNA (mtDNA) from the soldier's bones with mtDNA from his mother.

    a. Why were these tests not done before the remains were placed in the Tomb?

    b. Find more information about the specificities of test results when using nuclear DNA rather than mtDNA.

c. As of this writing, scientists at the National Institute of Standards and Technology are preparing an mtDNA standard for forensic laboratories to measure the accuracy of their results. Has that standard been issued?

56. Figure 12.17 provides data on the adoption of transgenic crops in the United States. Use the Web to create a timeline with events that help account for the rapid increases and then the leveling of the percentages.

57. The cloning of human embryos to produce infants is a very controversial possibility. Try to find reports of human cloning of this type. What evidence can you find about whether these reports are credible? What are the U.S. regulations about cloning to produce infants?

Measure for Measure
Conversion Factors and Constants

Metric Prefixes

deci (d) $1/10 = 10^{-1}$ deka (da) $10 = 10^{1}$
centi (c) $1/100 = 10^{-2}$ hecto (h) $100 = 10^{2}$
milli (m) $1/1000 = 10^{-3}$ kilo (k) $1000 = 10^{3}$
micro (μ) $1/10^{6} = 10^{-6}$ mega (M) $= 10^{6}$
nano (n) $1/10^{9} = 10^{-9}$ giga (G) $= 10^{9}$

Length

1 centimeter (cm) = 0.394 inch (in.)
1 meter (m) = 39.4 in. = 3.28 feet (ft) = 1.08 yard (yd)
1 kilometer (km) = 0.621 miles (mi)
1 in. = 2.54 cm = 0.0833 ft
1 ft = 30.5 cm = 0.305 m = 12 in.
1 yd = 91.44 cm = 0.9144 m = 3 ft = 36 in.
1 mi = 1.61 km

Volume

1 cubic centimeter (cm^{3}) = 1 milliliter (mL)
1 liter (L) = 1000 mL = 1000 cm^{3} = 1.057 quarts (qt)
1 qt = 0.946 L
1 gallon (gal) = 4 qt = 3.78 L

Mass

1 gram (g) = 0.0352 ounce (oz) = 0.00220 pound (lb)
1 kilogram (kg) = 1000 g = 2.20 lb
1 lb = 454 g = 0.454 kg
1 metric ton (t) = 1 long ton = 1000 kg = 2200 lb = 1.10 ton
1 gigaton (Gt) = 10^{9} t
1 ton = 1 short ton = 2000 lb = 909 kg = 0.909 t

Time

1 year (yr) = 365.24 days (d)
1 day = 24 hours (hr or h)
1 hr = 60 minutes (min)
1 min = 60 seconds (s)

Energy

1 joule (J) = 0.239 calorie (cal)
1 cal = 4.184 joule (J)
1 exajoule (EJ) = 10^{18} J
1 kilocalorie (kcal) = 1 dietary Calorie (Cal)
            = 4184 J = 4.184 kilojoule (kJ)
1 kilowatt-hour (kWh) = 3,600,000 J = 3.60×10^{6} J

Constants

Speed of light (c) = 3.00×10^{8} m/s
Planck's constant (h) = 6.63×10^{-34} J·s
Avogadro's number (N_{A}) = 6.02×10^{23} objects per mole
Atomic mass unit (m) = 1.66×10^{-24} g

The Power of Exponents

Scientific (or exponential) notation provides a compact and convenient way of writing very large and very small numbers. The idea is to use positive and negative powers of 10. Positive exponents are used to represent large numbers. The exponent, written as a superscript, indicates how many times 10 is multiplied by itself. For example,

$$10^1 = 10$$

$$10^2 = 10 \times 10 = 100$$

$$10^3 = 10 \times 10 \times 10 = 1000$$

Note that the positive exponent is equal to the number of zeros between the 1 and the decimal point. Thus, 10^6 corresponds to 1 followed by six zeros or 1,000,000. This same rule applies to 10^0, which equals 1. One billion, 1,000,000,000, can be written as 10^9.

When 10 is raised to a negative exponent, the number being represented is always less than 1. This is because a negative exponent implies a reciprocal, that is, 1 over 10 raised to the corresponding positive exponent. For example,

$$10^{-1} = 1/10^1 = 1/10 = 0.1$$

$$10^{-2} = 1/10^2 = 1/100 = 0.01$$

$$10^{-3} = 1/10^3 = 1/1000 = 0.001$$

It follows that the larger the negative exponent, the smaller the number. The negative exponent is always one more than the number of zeros between the decimal point and the 1. Thus, 1×10^{-4} is equal to 0.0001. Conversely, 0.000001 in scientific notation is 1×10^{-6}.

Of course, most of the quantities and constants used in chemistry are not simple whole-number powers of 10. For example, Avogadro's number is 6.02×10^{23}, or 6.02 multiplied by a number equal to 1 followed by 23 zeros. Written out, this corresponds to $6.02 \times 100,000,000,000,000,000,000,000$, or 602,000,000,000,000,000,000,000. Switching to very small numbers, a wavelength at which carbon dioxide absorbs infrared radiation is 4.257×10^{-6} m. This number is the same as 4.257×0.000001, or 0.000004257 m.

Your Turn Appendix 2.1

Express these numbers in scientific notation.

| a. | 10,000 | b. | 430 | c. | 9876.54 |
| d. | 0.000001 | e. | 0.007 | f. | 0.05339 |

Answers

| a. | 1×10^4 | b. | 4.3×10^2 | c. | 9.87654×10^3 |
| d. | 1×10^{-6} | e. | 7×10^{-3} | f. | 5.339×10^{-2} |

Your Turn Appendix 2.2

Express these numbers in conventional decimal notation.

| a. | 1×10^6 | b. | 3.123×10^6 | c. | 25×10^5 |
| d. | 1×10^{-5} | e. | 6.023×10^{-7} | f. | 1.723×10^{-16} |

Answers

a.	1,000,000	b.	3,123,000
c.	2,500,000	d.	0.00001
e.	0.0000006023	f.	0.0000000000000001723

Clearing the Logjam

You may have encountered logarithms in mathematics courses but wondered if you would ever use them. In fact, logarithms (or "logs" for short) are extremely useful in many areas of science. The essential idea is that they make it much easier to deal with very large *ranges* of numbers, for example, moving by powers of 10 from 0.0001 to 1,000,000.

It is likely that you have met logarithmic scales without necessarily knowing it. The Richter scale for expressing magnitudes of earthquakes is one example. On this scale, an earthquake of magnitude 6 is 10 times more powerful than one of magnitude 5. An earthquake of magnitude 8 would be 100 times more powerful than one of magnitude 6. Another example is the decibel (dB) scale. Each increase of 10 units represents a 10-fold increase in sound level. Therefore, a normal conversation between two people 1 m apart (60 dB) is 10 times louder than quiet music (50 dB) at the same distance. Loud music (70 dB) and extremely loud music (80 dB) are 10 times and 100 times as loud, respectively, as a normal conversation.

A simple exercise using a pocket calculator can be a good way to learn about logs. You will need a calculator that "does" logs and preferably has a "scientific notation" option. Start by finding the logarithm of 10. Simply enter 10 and press the "log" button. The answer should be 1. Next find the log of 100 and then the log of 1000. Write down the answers. What pattern do you see? (The pattern may be more obvious if you recall that 100 can be written as 10^2 and 1000 is the same as 10^3.) Predict the log of 10,000 and then check it out. Then try the log of 0.1 or 10^{-1} and log of 0.01 (10^{-2}). Predict the log of 0.0001 and check it out.

So far so good, but we have only been considering whole-number powers of 10. It would be helpful to be able to obtain the logarithm of any number. Once again, your handy little calculator comes to the rescue. Try calculating the logs of 20 and 200, then 50 (5×10^1) and 500 (5×10^2). Predict the log of 5×10^3, or 5000. Now for something slightly trickier: the log of 0.05. Finally, try the log of 2473 and the log of 0.000404. In each of the three cases, does the answer seem to be in the right ballpark? Remember that your calculator will happily provide you with many more digits than have any meaning, so you will need to do some reasonable rounding.

In Chapter 6, the concept of pH is introduced as a quantitative way to describe the acidity of a substance. A pH value is simply a special case of a logarithmic relationship. It is defined as the negative of the logarithm of the H^+ concentration, expressed in units of molarity (M). Square brackets are used to indicate molar concentrations. The mathematical relationship is given by the equation pH = $-\log [H^+]$. The negative sign indicates an inverse relationship; as the H^+ concentration diminishes, the pH increases. Let us apply the equation by using it to calculate the pH of a beverage with a hydrogen ion concentration of 0.000546 M. We first set up the mathematical equation and substitute the hydrogen ion concentration into it.

$$\text{pH} = -\log [H^+] = -\log (5.46 \times 10^{-4}\,\text{M})$$

Next, we take the negative logarithm of the H^+ concentration by entering it into a calculator and pressing the log button, then the "plus/minus" key to change the sign. This gives 3.26 as the pH of the beverage. (It may display 3.262807357 if you have not preset the number of digits, but common sense prompts you to round the displayed value.) Apply the same procedure to calculate the pH of milk with a hydrogen ion concentration of 2.20×10^{-7} M.

If we can convert hydrogen ion concentration into pH, how do we go in the reverse direction, that is, how to convert pH into a hydrogen ion concentration? Your calculator can do this for you if it has a button labeled "10^x." (Alternatively, it may use two buttons: first "Inv" and then "log." To demonstrate the procedure, suppose you wish to find the hydrogen ion concentration of human blood with a pH of 7.40. Proceed as follows: Enter 7.40, use the "plus/minus" key to change the sign to negative, and then hit 10^x (or follow whatever steps are appropriate for your calculator). The display should give the hydrogen ion concentration as 3.98×10^{-8} M. Now apply the same procedure to calculate the H^+ concentration of an acid rain sample with a pH of 3.6.

Your Turn Appendix 3.1

Find the pH of each sample.

a. tap water, $[H^+] = 1.0 \times 10^{-6}$ M
b. milk of magnesia, $[H^+] = 3.2 \times 10^{-11}$ M
c. lemon juice, $[H^+] = 5.0 \times 10^{-3}$ M
d. saliva, $[H^+] = 2.0 \times 10^{-7}$ M

Answers

a. 6.0   **b.** 10.5
c. 2.3   **d.** 6.7

Your Turn Appendix 3.2

Find the H^+ concentration in each sample.

a. tomato juice, pH = 4.5
b. acid fog, pH = 3.3
c. vinegar, pH = 2.5
d. blood, pH = 7.6

Answers

a. 3.2×10^{-5} M   **b.** 5.0×10^{-4} M
c. 3.2×10^{-3} M   **d.** 2.5×10^{-8} M

Answers to Your Turn Questions
Not Answered in Text

Chapter 1

1.9 Both ozone and sulfur dioxide show the same pattern of higher limits of exposure for shorter time periods. For ozone, the 1-h standard is 0.12 ppm, higher than the 8-h standard of 0.08 ppm. For sulfur dioxide, the 3-h standard of 0.50 ppm is higher than the 24-h standard of 0.14 ppm, which in turn, is higher than the 0.03 ppm standard for annual exposure.

1.10 Although standards for particulate matter are not given in Table 1.5, the given approximate equivalent concentrations are lower for the "fine" particulates for both the 24-h average and annual exposure. This indicates greater health consequences are expected.

1.11 a. $\dfrac{1050\ \mu g\ SO_2}{15\ m^3} = \dfrac{70\ \mu g\ SO_2}{1\ m^3}$

b. The woman's exposure does not exceed either the 24-h standard of 365 $\mu g/m^3$ or the annual standard of 80 $\mu g/m^3$.

1.13 In general, concentrations of pollutants are likely to be higher nearer Earth's surface, where the atmosphere is denser and the pollutants are generated. Temperature inversion layers may trap the pollutants in layers, causing some variation in concentration as a function of altitude. Tall smokestacks may deliver pollutants higher into the atmosphere, helping to distribute them more widely.

1.15 d. element   **e.** compound   **f.** mixture

1.16 b. carbon, chlorine (compound)

c. hydrogen, oxygen (compound)

d. carbon, hydrogen, oxygen (compound)

f. nitrogen, oxygen (compound)

1.18 a. magnesium bromide

d. sodium sulfide

1.19 c. calcium sulfide   **d.** lithium nitride

1.21 b. A molecule of the compound represented by the formula SO_2 consists of one atom of the element sulfur combined with two atoms of the element oxygen.

c. A molecule of the compound represented by the formula N_2O_4 consists of two atoms of the element nitrogen combined with four atoms of the element oxygen.

d. A molecule represented by the formula O_3 consists of three atoms of the element oxygen.

1.22 b. sulfur dioxide

1.23 a.

Balanced equation: $2\ H_2 + O_2 \longrightarrow 2\ H_2O$

b.

Balanced equation: $N_2 + 2\ O_2 \longrightarrow 2\ NO_2$

1.26 b. $2\ C_4H_{10} + 13\ O_2 \longrightarrow 8\ CO_2 + 10\ H_2O$

1.28 Partial list of tail pipe gases: carbon dioxide, carbon monoxide, water vapor, unburned hydrocarbons, nitrogen.

1.29 b. $CuS + O_2 \longrightarrow Cu + SO_2$

1.32 a. Ozone levels are highest from about 1 PM until about 7 PM. Ozone levels drop during the early morning hours and only start to rise again at about 10 AM.

b. Unburned hydrocarbons are emitted as part of the tail pipe gases during morning rush hour.

1.35 Indoor activities that generate pollutants include burning incense, painting, smoking, cooking, having fireplace or woodstove fires, operating a faulty furnace or space heater, using aerosol products for cleaning or personal grooming, installing new carpet, tracking in pollen and mold from yard work, and many others.

1.38 b. No, it wouldn't be more valid. Considering the margin of error expected for a hand-held meter, it is unlikely that the additional digits would provide any real information. Compare this to dividing $20 evenly among seven people. Would each person receive $2.857142857 just because your calculator displayed this result?

Chapter 2

2.2 c. The maximum is 12,000 ozone molecules per billion molecules of all gases that make up the stratosphere.

d. The EPA limit is 0.08 ppm for an 8-h average, equivalent to 80 ppb.

e. The ratio is $\dfrac{12,000}{80} = \dfrac{150}{1}$.

The maximum number of ozone molecules in the stratosphere is 150 times greater than the EPA limit.

2.4 c. 17 protons, 17 electrons

d. 24 protons, 24 electrons

2.5 c. 5 (Group 5A)   **d.** 8 (Group 8A)

2.6 b. Beryllium (Be), magnesium (Mg), calcium (Ca), strontium (Sr), barium (Ba), and radium (Ra) all have two outer electrons and are members of Group 2A.

2.7 c. 53 protons, 53 electrons, 78 neutrons

2.8 b. There are 7 outer electrons per atom of bromine, for a total of 14 outer electrons for Br_2. These are the Lewis structures.

$$:\ddot{Br}:\ddot{Br}: \text{ or } :\ddot{Br}-\ddot{Br}:$$

2.9 b. In dichlorodifluoromethane, CCl_2F_2, the carbon atom has 4 outer electrons, each of the two chlorine atoms has 7 outer electrons, and each of the two fluorine atoms has 7 electrons. The total number of outer electrons is 32. These are the Lewis structures.

2.10 b. In sulfur dioxide, SO_2, the sulfur atom has 6 outer electrons and each of the two oxygen atoms has 6 outer electrons. The total number of electrons is 18. These are the Lewis structures.

2.12 b. Green light has a shorter wavelength and higher frequency than that of red light.

2.14 b. Microwave, infrared, visible, ultraviolet

c. They are arranged in opposite sequence because wavelength and frequency are inversely related; as one increases, the other decreases.

2.19 a. Step 1: photon ($\lambda \leq 242$ nm) $+ O_2 \longrightarrow 2\ O$
Step 2: $O + O_2 \longrightarrow O_3$
Step 3: photon ($\lambda \leq 320$ nm) $+ O_3 \longrightarrow O_2 + O$
Step 4: $O_3 + O \longrightarrow 2\ O_2$

b. Step 2 illustrates formation of O_3. Steps 3 and 4 show its destruction.

2.26 a. There are seven outer electrons in the Lewis structure for the hydroxyl free radical. Here is its Lewis structure.

$$\cdot\ddot{O}:H \text{ or } \cdot\ddot{O}-H$$

b. No. The concentration of water vapor in the stratosphere is too low to cause the observed decrease in ozone concentration.

2.35 b. There are several possibilities. This is the structure for one possible halon with two carbon atoms, halon-2402.

Chapter 3

3.2 c. Higher energy, shorter wavelength UV radiation passes through the windows of the car, and its energy is absorbed. The energy is reradiated as heat, but the lower energy, longer wavelength IR radiation is unable to pass back out through the windows and so it is trapped.

3.3 a. Incoming solar radiation can be reflected back into space or absorbed and scattered by the atmosphere, preventing it all from reaching Earth's surface.

b. Heat energy can also be absorbed and scattered by the atmosphere, preventing it all from escaping back into space.

c. The two colors are a reminder that the incoming and outgoing radiations are not the same mix of wavelengths and energies. The yellow indicates more of the shorter, more energetic wavelengths, and the red indicates more of the longer, less energetic radiation associated with heat.

3.12 b. The total number of outer electrons is $4 + 2(7) + 2(7) = 32$. The central atom is C and there will be four single bonds, one to each halogen atom. The shape is tetrahedral.

c. The total number of outer electrons is $2(1) + 6 = 8$. The central atom is S. There are two bonded pairs of electrons on the central S, and two nonbonded pairs. The shape of the molecule is bent.

3.13 a. bent (only one resonance form shown)

b. triangular planar (only one resonance form shown)

3.15 c. The transmittance minimum (absorbance maximum) at 2350 cm^{-1} corresponds to stretching and 557 cm^{-1} corresponds to bending of CO_2 molecules.

3.18 a. $\dfrac{4000 \text{ Gt}}{5000 \text{ Gt}} \times 100 = 80\%$

b. $\dfrac{5000 \text{ Gt}}{47{,}060 \text{ Gt}} \times 100 = 11\%$

c. $\dfrac{39{,}120 \text{ Gt}}{(47{,}060 - 5000) \text{ Gt}} \times 100 = 93\%$

3.22 a. N-14 has 7 protons, 7 neutrons, and 7 electrons.

b. N-15 has 7 protons, 8 neutrons, and 7 electrons. Only the number of neutrons differs.

c. N-14 is the most abundant natural isotope of nitrogen.

3.24 b. $\left[\dfrac{2.34 \times 10^{-23} \text{g N}}{\text{N atom}}\right] \times \left(5 \times 10^{9} \text{N atoms}\right)$
$$= 1.17 \times 10^{-13} \text{g N}$$

c. $\left[\dfrac{2.34 \times 10^{-23} \text{g N}}{\text{N atom}}\right] \times \left(6 \times 10^{15} \text{N atoms}\right)$
$$= 1.40 \times 10^{-7} \text{g N}$$

3.25 b. 44 g/mol N_2O **c.** 137.5 g/mol CCl_3F

3.26 c. $\dfrac{2(14 \text{ g}) \text{ N}}{44.0 \text{ g } N_2O} = \dfrac{0.636 \text{ g N}}{1.00 \text{ g } N_2O}$

$\dfrac{0.636 \text{ g N}}{1.00 \text{ g } N_2O} \times 100 = 63.6\% \text{ N in } N_2O$

3.27 b. $142 \times 10^{6} \text{ t } SO_2 \times \dfrac{32.1 \times 10^{6} \text{ t S}}{64.1 \times 10^{6} \text{ t } SO_2}$
$$= 71.1 \times 10^{6} \text{ t S}$$

3.29 a. GWP is assigned on the basis of direct and indirect effects on global warming over 100 years. HFC-134a, with a GWP of 1300, is expected to have a far greater effect than CO_2. However, its relatively short lifetime of only 13.8 years and its very low abundance mitigates the effect this gas can have on global warming.

b. Ozone is less abundant than methane in the troposphere, but has a considerably higher potential for contributing to global warming. Ozone has not been assigned a GWP value because it has a relatively short lifetime in the troposphere.

c. Freon-12 can have 25,000 times the effect of CO_2, but its abundance is very low. Likely it was not assigned a GWP, despite its very long lifetime in the troposphere because it has been phased out by the Montreal Protocol, so its abundance will continue to decrease.

3.33 a. The NOAA model predicts a temperature change of approximately 0.8 °C (1.1–0.3 °C).

b. The DOE model predicts a temperature change of approximately 0.3 °C (0.6–0.3 °C).

c. NOAA model: 15.0 °C + 0.8 °C = 15.8 °C
DOE model: 15.0 °C + 0.3 °C = 15.3 °C

3.42 a. The total tons of world CO_2 emissions nearly doubles from 1995 to 2035. In 1995, the total was 6.46×10^{9} t. In 2035, the total is projected to be 11.71×10^{9} t.

b. The percentage for the developed world is expected to drop from 73% to 50% from 1995 to 2035.

c. In 1995, the developed world contributed 73% of 6.46×10^{9} t, or 4.72×10^{9} t. By 2035, the developed world is expected to be 50% of 11.71×10^{9} t, or 5.86×10^{9} t. Thus, although the percentage of emissions from the developed world is dropping, the total amount of CO_2 emissions increases.

Chapter 4

4.2 c. 380 kJ

d. 2.9×10^{3} 22 lb concrete blocks

4.11 Breaking the double bond in O_2 requires 498 kJ/mol. The resonance structures for O_3 include both a double and a single bond. A single O-to-O bond only requires 146 kJ, and the true bond energy between O atoms in O_3 will be intermediate between the single and double bond values. Energy is inversely proportional to wavelength. Therefore, the *higher* bond energy of O_2 requires radiation of *shorter* wavelength to break its bonds.

4.19 a. 18% oxygen in MTBE **b.** 35% oxygen in ethanol

4.28 $770 \text{ lb coal} \times \dfrac{65 \text{ lb C}}{100 \text{ lb coal}} \times \dfrac{44 \text{ lb } CO_2}{12 \text{ lb C}}$
$$= 1.8 \times 10^{3} \text{ lb } CO_2$$

$770 \text{ lb coal} \times \dfrac{2 \text{ lb S}}{100 \text{ lb coal}} \times \dfrac{64 \text{ lb } SO_2}{32 \text{ lb S}}$
$$= 31 \text{ lb } SO_2$$

Chapter 5

5.6 a. soluble   **b.** very soluble   **c.** insoluble

     d. soluble   **e.** insoluble

     f. partially soluble (pure aspirin)

5.7 a. 975 mg/day

     b. No. Calcium requirements vary considerably depending on age and sex, among other factors.

     c. 25 500-L bottles

5.9 a. 16 ppb; 1.6×10^{-2} ppm

     b. No; the lead concentration is over the standard of 15 ppb

5.10 a. 7.5×10^{1} mol NaCl in 500 mL of a 1.5 M NaCl solution.

     7.5×10^{2} mol NaCl in 500 mL of 0.15 M NaCl solution.

     b. To calculate molarity, divide moles of solute by liters of solution. The second solution is more concentrated because the first solution is 2.0 M, less than the 3.0 M concentration of the second solution.

5.11 a. H—F The electrons are more strongly attracted to the F atom.

     b. O—H The electrons are more strongly attracted to the O atom.

     c. S—O The electrons are more strongly attracted to the O atom.

5.13 a. Four hydrogen bonds surround the central H_2O molecule.

     b. Hydrogen bonds are intermolecular forces because they are between, not within, water molecules.

5.17 b. Mg^{2+}, the Lewis structures are •Mg• and Mg^{2+}

     c. O^{2-}, the Lewis structures are

$$\cdot \overset{\cdot\cdot}{\underset{\cdot\cdot}{O}} \cdot \quad \text{and} \quad \left[\overset{\cdot\cdot}{\underset{\cdot\cdot}{:O:}} \right]^{2-}$$

     d. Al^{3+}, the Lewis structures are •$\overset{\cdot}{Al}$• and Al^{3+}

5.18 a. KF, potassium fluoride

     b. Li_2O, lithium oxide

     c. $SrBr_2$, strontium bromide

5.19 c. $Al(C_2H_3O_2)_3$

5.20 c. sodium hydrogen carbonate or sodium bicarbonate

     d. calcium carbonate

     e. magnesium phosphate

5.21 b. Li_2CO_3 **c.** KNO_3 **d.** $BaSO_4$

5.23 b. soluble; all sodium salts are soluble

     c. insoluble; most sulfides are insoluble (except those with Group 1A or NH_4^+ cations)

     d. insoluble; most hydroxides are insoluble (except those with Group 1A or NH_4^+ cations)

5.24

----- indicates a hydrogen bond

5.34 a. 20 ppb = 20 μg/L; higher concentration than 0.003 mg/L = 3 μg/L = 3 ppb

     b. 20 ppb = 15 ppb standard; 3 ppb = 15 ppb standard

5.36 b. Approximately 2–3 ppb Pb^{2+}

     c. Approximately 28–29 ppb Pb^{2+}

Chapter 6

6.2 a. $HI(aq) \longrightarrow H^+(aq) + I^-(aq)$

     b. $HNO_3(aq) \longrightarrow H^+(aq) + NO_3^-(aq)$

     c. $H_2SO_4(aq) \longrightarrow H^+(aq) + HSO_4^-(aq)$

6.4 a. $KOH(s) \longrightarrow K^+(aq) + OH^-(aq)$

     b. $LiOH(s) \longrightarrow Li^+(aq) + OH^-(aq)$

6.5 b.

$H_2SO_4(aq) + 2\,NaOH(aq) \longrightarrow Na_2SO_4(aq) + 2\,H_2O(l)$

$2\,H^+(aq) + SO_4^{2-}(aq) + 2\,Na^+(aq) + 2\,OH^-(aq) \longrightarrow$
$ 2\,Na^+(aq) + SO_4^{2-}(aq) + 2\,H_2O(l)$

$2\,H^+(aq) + 2\,OH^-(aq) \longrightarrow 2\,H_2O(l)$
$H^+(aq) + OH^-(aq) \longrightarrow H_2O(l)$

     c.

$2\,H_3PO_4(aq) + 3\,Mg(OH)_2(aq) \longrightarrow$
$ Mg_3(PO_4)_2(aq) + 6\,H_2O(l)$

$6\,H^+(aq) + 2\,PO_4^{3-}(aq) + 3\,Mg^{2+}(aq) + 6\,OH^-(aq) \longrightarrow$
$ 3\,Mg^{2+}(aq) + 2\,PO_4^{3-}(aq) + 6\,H_2O(l)$

$6\,H^+(aq) + 6\,OH^-(aq) \longrightarrow 6\,H_2O(l)$
$H^+(aq) + OH^-(aq) \longrightarrow H_2O(l)$

6.6 b. The solution is basic because $[OH^-] > [H^+]$.

$$[H^+] = \frac{1 \times 10^{-14}}{1 \times 10^{-6}} = 1 \times 10^{-8}$$

     c. The solution is basic because $[OH^-] > [H^+]$.

6.7 a. $OH^-(aq) = K^+(aq) > H^+(aq)$; basic

     b. $H^+(aq) = NO_2^-(aq) > OH^-(aq)$; acidic

     c. $H^+(aq) = HSO_3^- > SO_3^{2-}(aq) = OH^-(aq)$; acidic

6.9 a. The lake water with pH = 4 is 10 times more acidic and has 10 times more H^+ than the rainwater with pH = 5.

6.12 $H_2SO_3(aq) \longrightarrow H^+(aq) + HSO_3^-(aq)$
$HSO_3^-(aq) \longrightarrow H^+(aq) + SO_3^{2-}(aq)$
$H_2SO_3(aq) \longrightarrow 2\,H^+(aq) + SO_3^{2-}(aq)$

6.13 c. 3.36×10^4 tons SO_2

   d. The SO_2 is carried by the wind into the surrounding areas. It eventually dissolves in the water droplets of clouds, fog, and rain to form acidic precipitation.

6.14 Between 1940 and 2002, the NO_x emissions from industry did not change much. In contrast, the emissions from combustion engines used for transportation and from fossil fuel combustion increased substantially. For SO_2, industrial emissions led to the large increase in the 1970s.

6.15 a. metal, Fe   **b.** metal, Al

   c. nonmetal, F   **d.** metal, Ca

   e. metal, Zn   **f.** nonmetal, O

6.16 Add equations 6.22 and 6.23 together, and cancel like terms.

$4\,Fe(s) + 2\,O_2(g) + \cancel{8\,H^+(aq)} \longrightarrow \cancel{4\,Fe^{2+}(aq)} + \cancel{4\,H_2O(l)}$
$\cancel{4\,Fe^{2+}(aq)} + O_2(g) + \cancel{4\,H_2O(l)} \longrightarrow 2\,Fe_2O_3(s) + \cancel{8\,H^+(aq)}$
$\overline{4\,Fe(s) + 3\,O_2(g) \longrightarrow 2\,Fe_2O_3(s)}$

6.17 Fe has 26 electrons, Fe^{2+} has 24 electrons, and Fe^{3+} has 23 electrons.

6.18 $MgCO_3(s) + 2\,H^+(aq) \longrightarrow$
$$Mg^{2+}(aq) + CO_2(g) + H_2O(l)$$

6.19 $CaCO_3(s) + H_2SO_4(aq) \longrightarrow$
$$CaSO_4(s) + CO_2(g) + H_2O(l)$$

6.22 $S(s) + O_2(g) \longrightarrow SO_2(g)$
$SO_2(g) + \frac{1}{2}\,O_2(g) \longrightarrow SO_3(g)$
$SO_3(g) + H_2O(l) \longrightarrow H_2SO_4(aq)$

6.23 a. $H_2SO_4(aq) + NH_4OH(aq) \longrightarrow$
$$NH_4HSO_4(aq) + H_2O(l)$$

   b. $H_2SO_4(aq) + 2\,NH_4OH(aq) \longrightarrow$
$$(NH_4)_2SO_4(aq) + 2\,H_2O(l)$$

6.25 a. 78%   **b.** $:N{\equiv}N:$

   c. The N-to-N triple bond is one of the most difficult bonds to break, requiring 946 kJ/mol. For comparison, the C-to-C triple bond requires 813 kJ/mol. See Table 4.2 for other values.

6.26 a. Using NH_4^+ as an example:
$H_2SO_4(aq) + 2\,NH_4OH(aq) \longrightarrow$
$$(NH_4)_2SO_4(aq) + 2\,H_2O(l)$$

   b. Lewis structure for NH_4^+

$$\left[\begin{array}{c} H \\ | \\ H-N-H \\ | \\ H \end{array} \right]^+$$

c. Both NO and NO_2 have unpaired electrons, making them reactive.

6.33 a. $N_2 + O_2 \xrightarrow{\text{high temperature}} 2\,NO$

   $2\,NO + O_2 \longrightarrow 2\,NO_2$ (net reaction)

   $NO_2 \xrightarrow{\text{sunlight}} NO + O$

   $O + O_2 \longrightarrow O_3$

   b. Goals for NO_x emission still have not been reached and nitrogen saturation is observed in many lakes.

Chapter 7

7.4 a. U-234 has 92 protons and $234 - 92 = 142$ neutrons.

   b. There are 92 protons in all uranium atoms.

7.5 b. $^{1}_{0}n + ^{235}_{92}U \longrightarrow ^{137}_{52}Te + ^{97}_{40}Zr + 2\,^{1}_{0}n$

7.6 $^{1}_{0}n + ^{235}_{92}U \longrightarrow ^{143}_{54}Xe + ^{90}_{38}Sr + 3\,^{1}_{0}n$

7.7 1.00×10^{14} J/day; 1.11 g

7.8 $^{241}_{95}Am \longrightarrow ^{237}_{93}Np + ^{4}_{2}He$

   $^{9}_{4}Be + ^{4}_{2}He \longrightarrow ^{12}_{6}C + ^{1}_{0}n + ^{0}_{0}\gamma$

7.9 The cloud is condensed water vapor. People sometimes call this "steam," although technically this is not correct, as steam is water vapor, which is invisible until it condenses. The cloud does not contain any nuclear fission products.

7.14 a. U-238; not fissionable in a nuclear reactor

   b. Isotopes of Sr, Ba, Kr, and I may be products of the fission reactions.

7.17 b. Gamma radiation refers to gamma rays, a part of the electromagnetic spectrum of radiant energy.

   c. UV rays refers to UV radiation, a part of the electromagnetic spectrum. The word *rays* is used to connect UV radiation with the Sun's rays.

   d. Radiation in this case refers to the particles and rays spontaneously emitted by unstable uranium atoms.

7.18 b. $^{239}_{94}Pu \longrightarrow ^{235}_{92}U + ^{4}_{2}He$

   c. $^{131}_{53}I \longrightarrow ^{131}_{54}Xe + ^{0}_{-1}e$

7.21 Answers will vary with individuals based on their exposure to background radiation and medical procedures.

7.23 After 36.9 years (3 half-lives), only 12.5% of the original tritium will remain.

7.24 a. The radon was produced from uranium, a radioactive atom that decays through several steps to form radon.

   b. Four half-lives (4×3.8 days = 15.2 days) are required for the radiation to drop from 16 pCi to 1 pCi.

   c. If uranium is still present in the rock formation, further decay will take place.

7.25 $:\!\overset{\cdot\cdot}{\underset{\cdot\cdot}{I}}\!\cdot$ $:\!\overset{\cdot\cdot}{\underset{\cdot\cdot}{I}}\!\!:\!\overset{\cdot\cdot}{\underset{\cdot\cdot}{I}}\!\!:$ $\left[:\!\overset{\cdot\cdot}{\underset{\cdot\cdot}{I}}\!\!:\right]^-$

   atom   molecule   ion

   The iodide ion is used for medical imaging.

7.31 a. U-238, the most abundant isotope.

b. Thorium is found with uranium because uranium decays to produce thorium. See Figure 7.17.

Chapter 8

8.2 Equations **a**, **c**, and **e** are reduction half-reactions because electrons are gained, appearing on the reactant side. Equations **b** and **d** are oxidation half-reactions because electrons are lost, appearing on the product side.

8.3 a. Each equation is balanced from the standpoint of number and type of atoms.

b. Each equation is balanced from the standpoint of charge, but the total charge does not have to be zero on each side, just the same.

c. Electrons do not appear in the overall cell reaction, but are in half-reactions.

8.3 a. $Zn \longrightarrow Zn^{2+} + 2\,e^-$

b. $Cu^{2+} + 2\,e^- \longrightarrow Cu$

c. In this galvanic cell, zinc(*s*) is oxidized to Zn^{2+} (*aq*). The mobile zinc ions cannot be conveniently reduced to reform a solid zinc electrode, which would be essential during a recharging step.

8.7 a. Pb(*s*)   **b.** PbO_2(*s*)

c. $PbSO_4(s) + 2\,H_2O(l) \longrightarrow$
        $Pb(s) + PbO_2(s) + 2\,H_2SO_4(aq)$

$PbSO_4$(*s*) is both oxidized and reduced during recharging.

8.8 a. A fuel cell converts the chemical energy of a fuel directly into electricity. Usually the fuel is a gas, unlike the solids used in all batteries discussed previously in this chapter.

b. A PEM fuel cell cannot be recharged as was the case for the storage battery. Instead, a fuel cell is a "flow battery," one that operates so long as fuel and oxidizing gases are supplied.

c. Heat or light are not produced, requirements for combustion reactions.

8.21 a. Bonds broken: 4 mol(C—H) + 4 mol (O—H)
        = 4 mol (416kJ/mol) + 4 mol (467 kJ/mol)
        = 1664 kJ + 1868 kJ
        = 3532 kJ (endothermic step)
    Bonds formed: 4 mol(H—H) + 2 mol (C=O)
        = 4 mol (436 kJ/mol) + 2 mol (803 kJ/mol)
        = 1744 kJ + 1606 kJ
        = 3350 kJ (exothermic step)
    Overall reaction: (+3532 kJ) + (−3350 kJ)
        = +182 kJ

b. The reaction is endothermic. More energy is required to break bonds than in released in bond formation.

c. Although there is general agreement (+182 kJ vs. 165 kJ), remember that Table 4.2 gives *average* bond energies, not specific energies associated with the bonds in these compounds (with the exception of CO_2).

8.22 a. $Na(s) + \tfrac{1}{2}\,H_2(g) \longrightarrow NaH(s)$

b. NaOH(*aq*)

c. Nickel metal hydride batteries do not release hydrogen. Only active metals such as Li and Na form hydrides capable of releasing hydrogen when combined with water.

8.24 a. $E = \dfrac{hc}{\lambda} = \dfrac{(6.63 \times 10^{-34}\text{J} \cdot \text{s})(3.00 \times 10^8 \text{ m/s})}{500 \text{ nm} \times 1 \text{ m}/10^9 \text{ nm}}$

$E = 4.0 \times 10^{-19}\text{J per photon}$

and

$E = \dfrac{4.0 \times 10^{-19}\text{J}}{1 \text{ photon}} \times \dfrac{6.02 \times 10^{23}\text{photons}}{1 \text{ mol}}$

$E = \dfrac{2.4 \times 10^5\text{J}}{1 \text{ mol}} \times \dfrac{1\text{kJ}}{10^3\text{J}} = \dfrac{240 \text{ kJ}}{1\text{mol}}$

Both numbers agree with text values.

b.

Wavelength	Energy, J per photon
400 nm	5.0×10^{-19}
700 nm	2.8×10^{-19}

Both energies are above the minimum 1.8×10^{-19} J per photon required for Si to release and electron.

8.25 a. P forms a *n*-type semiconductor. Both P and As are in Group VA.

b. B forms a *p*-type semiconductor. Both B and Ga are in Group IIIA.

8.26 Ga is in Group 3A and has 3 outer electrons. As is in Group 5A and has 5 outer electrons. When they bond in a similar array to Ge or Si, the Group 4A element shown in Figure 8.20, each will have a share in a stable octet of electrons.
Similarly, Cd is in Group IIB (a Group that behaves much like the IIA Group) and has 2 outer electrons. Se is in Group VIA and has 6 outer electrons, allowing each to have a share in a stable octet of electrons when bonded.

8.28 a. U.S. Sunrayce 2001 (3700 km) was longer than the Australian Sunrace (2300 km).

b. Ultimate Solar Car Challenge (4000 km) was longer than the U.S. Sunrayce (3700 km).

Chapter 9

9.8

$$\begin{array}{c}\text{H H H H H H H H H H H H H H H H H H}\\ \text{| | | | | | | | | | | | | | | | | |}\\ \text{—C—C—C—C—C—C—C—C—C—C—C—C—C—C—C—C—C—C—}\\ \text{| | | | | | | | | | | | | | | | | |}\\ \text{H H H H H H H H H H H H H H H H H H}\end{array}$$

9.10

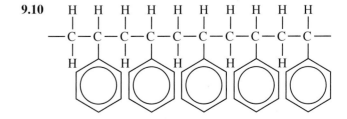

This arrangement is favored over the head-to-head arrangement because it minimizes electrostatic repulsion between the rings. In the head-to-head arrangement, the rings would be closer together, leading to more electrostatic repulsion.

9.11 a.

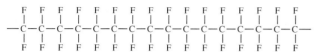

b. Head-to-tail arrangements are not distinguishable from head-to-head or tail-to-tail arrangements because all the atoms attached to the carbon chain are the same.

9.19

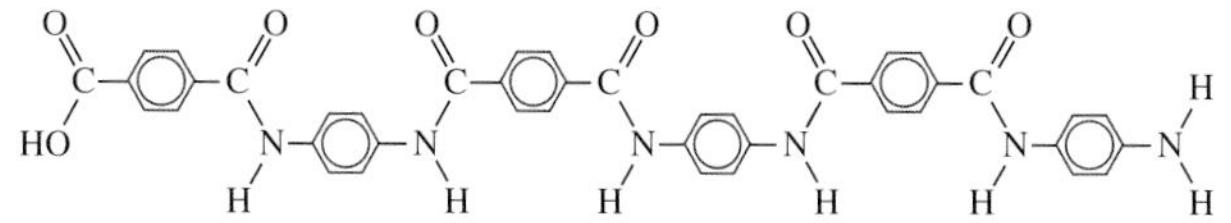

9.24 Propylene is $H_2C=CHCH_3$ or C_3H_6.

$$2500\ C_3H_6 + 11250\ O_2 \longrightarrow 7500\ CO_2 + 75000\ H_2O$$

9.29 Procedures will vary. Identification is based on relative densities. If the density of the plastic is greater than the density of the liquid, the plastic will sink. If the density of the plastic is less than the density of the liquid, the plastic will float. For example, if a plastic sinks in every one of the test liquids except a saturated solution of $ZnCl_2$, you have found polyethylene terephthalate.

Chapter 10

10.4 Each carbon atom in Figure 10.2 is surrounded by 8 electrons (4 bonds), so the octet rule is followed.

10.5 *Note:* The hydrogen atoms have been omitted to make the linkage of the carbon atoms clear.

Structural formulas:

(a)

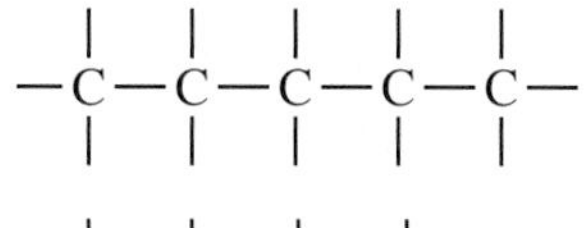

(b)

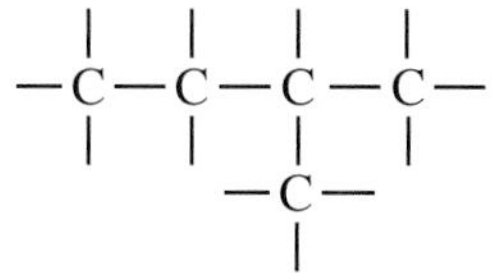

(c)

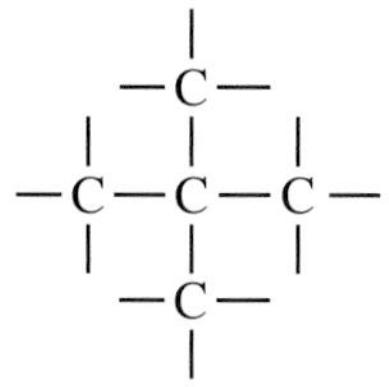

Condensed formulas:
(a) $CH_3CH_2CH_2CH_2CH_3$
(b) $CH_3CH_2CH(CH_3)CH_3$
(c) $C(CH_3)_4$

10.6 a. Yes, *n*-butane and isobutane are isomers. They have the same formula, C_4H_{10}, but different structures.

b. No, *n*-hexane (C_6H_{14}) and cyclohexane (C_6H_{12}) are not isomers. They have different formulas as well as different structures.

10.7 c. NH$_2$ (an amine)

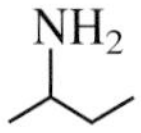

d. O (an ester)

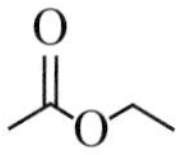

e. O (an aldehyde)

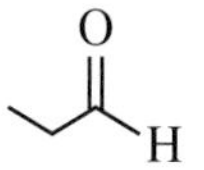

10.8 a.

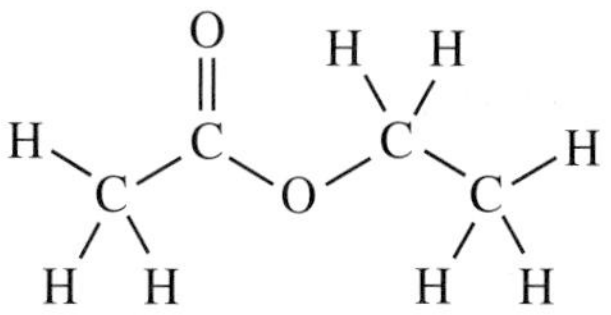

b.

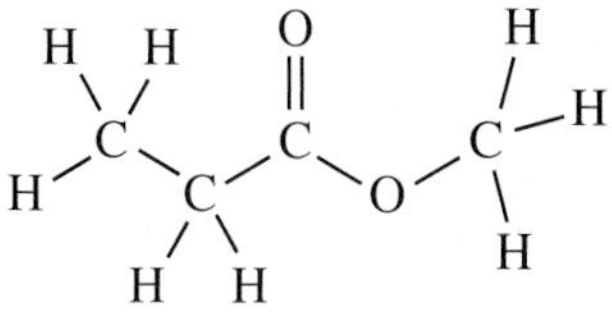

10.9 All three molecules have a benzene ring with functional groups attached that can form hydrogen bonds with water. Both aspirin and ibuprofen have a carboxylic acid group attached.

10.13 a. D-ibuprofen   **c.** L-ibuprofen.

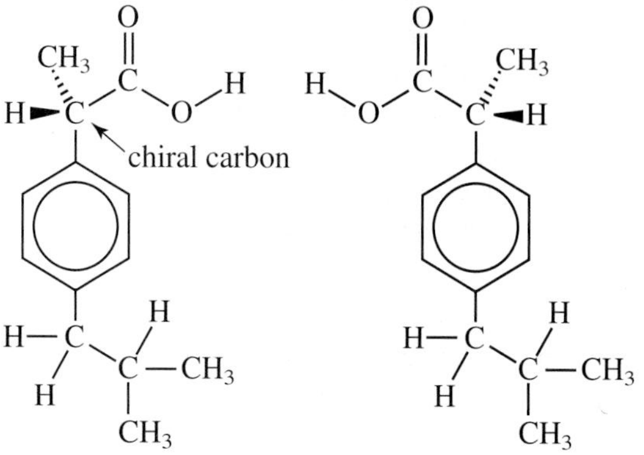

D-ibuprofen   L-ibuprofen

b. carboxylic acid functional group

10.14 a.

1. estradiol	testosterone
—OH group on the D ring	—OH group on the D ring
—CH$_3$ group on C-D ring intersection	—CH$_3$ group on C-D ring intersection
2. estradiol	progesterone
—OH group on the D ring	C=O group on the D ring
—CH$_3$ group on C-D ring intersection	—CH$_3$ group on C-D ring intersection

3. cholic acid   cholesterol

no double bonds in any ring

no double bonds in A, C, or D rings

—OH group on the A ring

—OH group on the A ring

—CH_3 group on C-D ring intersection

—CH_3 group on C-D ring intersection

b. estradiol $C_{18}H_{24}O_2$ progesterone $C_{21}H_{30}O_2$

testosterone $C_{19}H_{28}O_2$ cholic acid $C_{24}H_{40}O_5$

cortisone $C_{21}H_{28}O_5$ cholesterol $C_{27}H_{46}O$

10.16 The molecular formula for RU-486 is $C_{29}H_{35}NO_2$. Its structure differs from that of progesterone in several ways. The "B" ring in RU-486 has a double bond but not in progesterone. The side groups attached to the "D" ring are different for the two steroids, and a complex group is attached to the "C" ring of RU-486 that is not present in progesterone.

10.24 **a.** $Al(OH)_3$ **b.** $Mg(OH)_2$ **c.** $Ca(OH)_2$

10.29 Hydrochloric acid, HCl(aq), is used to form the salt of oxycodone. This is the structure for the salt. The N atom no longer has a lone pair of electrons, which is what characterizes the freebase form.

Chapter 11

11.3 **a.** The mass percent (in the second column of Table 11.2) is ratio of the mass of an element in grams to 100 g of body weight. The relative abundance (column 3 of Table 11.2) accounts for the *number* of atoms of an element. For example, 12 g of hydrogen (atomic mass = 1) represents 12 times as many atoms as 12 grams of carbon (atomic mass = 12).

b. Even though oxygen is about 2.5 times more abundant, oxygen has a smaller number of moles present per 100 g of body weight, because oxygen has a higher atomic mass than does hydrogen.

11.5 **a.** Answers will vary.

b. The BMI is defined as a person's weight in kilograms divided by the square of the person's height in meters. A BMI of less than 25 is considered to be healthy.

11.8 Salivary enzymes break down the complex carbohydrates in unsweetened crackers into simple sugars, which are responsible for the sweet taste.

11.9 Enzymes catalyze breaking bonds during digestion. There must be a close "lock-and-key" fit between the enzyme and the substrate molecule upon which it acts. The molecular architecture of human enzymes apparently is not compatible with the beta linkage between glucose units in cellulose, but the enzymes fit the alpha linkage in starches.

11.11 **a.** Canola oil is 6% saturated (likely capric, lauric, myristic, or palmitic acids), 36% polyunsaturated (linoleic or linolenic acids), and 58% monounsaturated fat (oleic acid).

b. Olive oil is 14% saturated (likely capric, lauric, myristic, or palmitic acids), 9% polyunsaturated (linoleic or linolenic acids), and 77% monounsaturated fat (oleic acid).

c. Lard is 41% saturated (likely stearic acid), 12% polyunsaturated (linoleic or linolenic acids), and 47% monounsaturated fat (oleic acid).

11.16 No, it does not meet these guidelines. Only 8-10% of Calories should come from saturated fat.

$$16.5 \text{ g saturated fat} \times \frac{9 \text{ Cal}}{1 \text{ g fat}} = 148.5 \text{ Cal from saturated fat}$$

$$\frac{148.5 \text{ Cal}}{1230 \text{ total Cal}} \times 100 = 12\% \text{ Calories from saturated fat}$$

11.20 GlyGlyGly, GlyGlyAla, GlyAlaAla, GlyAlaGly, AlaAlaAla, AlaAlaGly, AlaGlyGly, AlaGlyAla

11.24 Vitamin A has only one —OH that can form hydrogen bonds with water. The rest of the molecule is composed of nonpolar C-to-C and C-to-H bonds, making this vitamin soluble in nonpolar fats. Vitamin C has several —OH groups, making it a polar molecule that is soluble in water through hydrogen bonding. Both vitamins are examples of the general principle that "like dissolves like."

11.26 Answers will vary.

11.28 **a.** $\frac{130 \text{ Calories}}{2000 \text{ Calories}} \times 100 = 6.5\%$

b. Assuming that the serving size is the same for both the bagel chips and the potato chips, the potato chips are slightly higher in Calories per serving. Remembering that fats release 9 Cal/g when digested, a serving of bagel chips releases 4 g $\times$ 9 Cal/g = 36 Cal. The percentage of daily caloric intake from this fat is only 1.8% based on a 2,000-Calorie diet. A serving of potato chips releases 10 g $\times$ 9 Cal/g = 90 Cal, which is 4.5% of the caloric intake based on a 2000-Calorie diet. It appears that these bagel chips are slightly lower in fat and Calories per serving than these particular potato chips. Both types of chips are snack food and not meant to be a major component of the diet.

11.31 Answers will vary.

Chapter 12

12.2 **a.** Between 1900 and 1950 for males (from 47 to 66), a 40% increase and for females (from 49 to 72), a 47% increase. Between 1950 and 2001, for males (from 66 to 74), a 12% increase and for females (from 72 to 80), also a 12% increase.

b. Factors that have been proposed are that women have better diets and have hormonal differences that offer some protection against heart attacks before menopause. Another suggested factor is that the percentage of women who smoke is lower. Job-related stress was once suggested as a factor in lowering life expectancy for men more than for women, but that may not be making much difference in today's society.

12.7 3×10^9 base pairs $\times \dfrac{0.34 \text{ nm}}{\text{base pair}} \times \dfrac{1 \text{ m}}{1 \times 10^9 \text{ nm}} = 1 \text{ m}$

Section 12.1 notes that the length is 2 m, but this is for the combined length of both DNA strands.

12.9 Chargaff's rules pointed to the complementarity of the base pairs, leading to the conclusion that adenine and thymine must occur together, as do guanine and cytosine.

12.12 Redundancy in the code means that a mistake in the amino acid produced by the codon and introduced into a protein may not necessarily result in a mistake in the amino acid introduced into the protein. Multiple codons for certain amino acids may also speed up protein synthesis.

12.18 **b.** 8 cycles; $2^8 = 256$ DNA molecules

c. 20 cycles; $2^{20} = 1.0 \times 10^6$ DNA molecules

5

Answers to Selected End-of-Chapter Questions Indicated in Color in the Text

Chapter 1

1. $\dfrac{1\ L}{1\ \text{breath}} \times \dfrac{15\ \text{breaths}}{1\ \text{min}} \times \dfrac{60\ \text{min}}{1\ \text{hr}} \times \dfrac{8\ \text{hr}}{1\ \text{working day}}$
$= 7200\ L$

5. a. $N_2 > O_2 > Ar > CO_2 > CO > Rn$

b. The concentrations of CO_2 and CO can conveniently be expressed in ppm. Using percent is more reasonable for N_2, O_2, and Ar. The concentration of Rn in the troposphere is on the order of 1 part in 10^{21}, which is too small to be expressed conveniently in any manner.

6. a. $9000\ \text{ppm} \times \dfrac{100\ \text{pph}}{1{,}000{,}000\ \text{ppm}} = 0.9\ \text{pph or } 0.9\%$

10. a. 85,000 g

b. 10,000,000 gallons

11. a. 7.2×10^7 cigarettes

b. $1.5 \times 10^4\ {}^\circ C$

13. a. Group 1A, Group 7A

b. Group 1A: hydrogen, lithium, sodium, potassium, rubidium, cesium, and francium

Group 7A: fluorine, chlorine, bromine, iodine, and astatine

16. a. potassium oxide

b. aluminum chloride

18. a. One atom of the element carbon, two atoms of the element hydrogen, and one atom of the element oxygen are combined to form one molecule of formaldehyde.

19. a. $N_2 + O_2 \longrightarrow 2\ NO$

b. $O_3 \longrightarrow O_2 + O$

24. a. Platinum = Pt, palladium = Pd, rhodium = Rh

b. All three metals are in Group 8B on the periodic table. Platinum is directly under palladium, and rhodium is just to the left of palladium.

c. These metals are solids at the temperature of the exhaust gases, so they must have relatively high melting points. Also, they do not undergo permanent chemical change in catalyzing the reaction of CO to CO_2 in the exhaust stream.

29. CO is termed the "silent killer" because your senses cannot detect this colorless and odorless gas. The same term cannot be applied to either O_3 or SO_2. Each has a distinctive odor that can be detected at concentrations below the level of toxicity.

32. a. $5.0 \times 10^{-3}\ m < 1\ m < 3.0 \times 10^2\ m$

36. Sample **a** represents a compound since two different atoms are joined. Sample **b** represents a mixture since two different types of atoms are shown. Sample **c** represents a mixture since two different types of atoms and a compound are shown. Sample **d** represents an element since only one type of atom is shown.

38. a. Jogging outdoors, as opposed to sitting outdoors, increases your exposure to air pollutants because you will be breathing harder and exchanging more air during your exercise.

42. Formaldehyde can be released from cigarette smoke, and from synthetic materials such as foam insulation, and from the adhesives used in dying and gluing carpet pads, carpets, and laminated building materials. The air indoors is often not well circulated, leading to an accumulation of formaldehyde and other pollutants. Efforts to make homes airtight, leading to greater energy efficiency, have in some cases, made indoor air pollution worse, rather than better.

Chapter 2

1. a. Yes; 0.118 ppm is equivalent to 118 ppb, well above the detection limit of 10 ppb.

b. Yes; 25 ppm is equivalent to 25,000 ppb, well above the detection limit of 10 ppb.

4. a. Diamond and graphite are allotropes. They are two different forms of the same element, carbon.

b. Water and hydrogen peroxide are not allotropes. They are not different elements, but different compounds.

c. White phosphorus and red phosphorus are allotropes because they are two different forms of the same element.

7. a. A neutral atom of oxygen has 8 protons and 8 electrons.

   b. A neutral atom of nitrogen has 7 protons and 7 electrons.

9. a. He, helium

   b. K, potassium

   c. Cu, copper

10. c. U-238 has 92 protons, 146 neutrons, and 92 electrons if neutral.

   f. Ra-226 has 88 protons, 138 neutrons, and 88 electrons if neutral.

11. c. The element is radon, Rn-222. The symbol is $^{222}_{86}Rn$.

12. a. $\cdot\overset{\cdot\cdot}{Ca}\cdot$ **b.** $\cdot\overset{\cdot\cdot}{N}\cdot$ **c.** $:\overset{\cdot\cdot}{\underset{\cdot\cdot}{Cl}}\cdot$

13. b. There are $2(1) + 2(6) = 14$ outer electrons. The Lewis structures are:

$$H:\overset{\cdot\cdot}{\underset{\cdot\cdot}{O}}:\overset{\cdot\cdot}{\underset{\cdot\cdot}{O}}:H \quad \text{and} \quad H-\overset{\cdot\cdot}{\underset{\cdot\cdot}{O}}-\overset{\cdot\cdot}{\underset{\cdot\cdot}{O}}-H$$

   c. There are $2(1) + 6 = 8$ outer electrons. The Lewis structures are:

$$H:\overset{\cdot\cdot}{\underset{\cdot\cdot}{S}}:H \quad \text{and} \quad H-\overset{\cdot\cdot}{\underset{\cdot\cdot}{S}}-H$$

15. a. Wave 1 has longer wavelength than wave 2.

   b. Wave 1 has lower frequency than wave 2.

   c. Both waves have the same forward speed because they are both part of the electromagnetic spectrum.

19. In order of increasing energy per photon: radiowaves < infrared < visible < gamma rays

21. a. In order of increasing wavelength: UV-C < UV-B < UV-A

   b. In order of increasing energy: UV-A < UV-B < UV-C

   c. In order of increasing potential for biological damage: UV-A < UV-B < UV-C

23. a. Cl has 7 outer electrons. Its Lewis structure is:

$$:\overset{\cdot\cdot}{\underset{\cdot\cdot}{Cl}}\cdot$$

NO_2 has $5 + 2(6) = 17$ outer electrons. Its Lewis structure is:

$$\overset{\cdot\cdot}{\underset{\cdot\cdot}{O}}::\overset{\cdot}{N}:\overset{\cdot\cdot}{\underset{\cdot\cdot}{O}}: \quad \text{or} \quad \overset{\cdot\cdot}{\underset{\cdot\cdot}{O}}=\overset{\cdot}{N}-\overset{\cdot\cdot}{\underset{\cdot\cdot}{O}}:$$

ClO has $7 + 6 = 13$ outer electrons. Its Lewis structure is:

$$:\overset{\cdot\cdot}{\underset{\cdot\cdot}{Cl}}:\overset{\cdot\cdot}{\underset{\cdot\cdot}{O}}\cdot \quad \text{or} \quad :\overset{\cdot\cdot}{\underset{\cdot\cdot}{Cl}}-\overset{\cdot\cdot}{\underset{\cdot\cdot}{O}}\cdot$$

OH has $6 + 1 = 7$ outer electrons. Its Lewis structure is:

$$\cdot\overset{\cdot\cdot}{\underset{\cdot\cdot}{O}}:H \quad \text{or} \quad \cdot\overset{\cdot\cdot}{\underset{\cdot\cdot}{O}}-H$$

   b. Each of these free radicals has an atom with a single electron.

27. a. Methane, CH_4, has $4 + 4(1) = 8$ outer electrons.

$$H-\overset{\displaystyle H}{\underset{\displaystyle H}{C}}-H$$

Ethane, C_2H_6, has $2(4) + 6(1) = 14$ outer electrons.

$$H-\overset{\displaystyle H}{\underset{\displaystyle H}{C}}-\overset{\displaystyle H}{\underset{\displaystyle H}{C}}-H$$

   b. Fourteen different CFCs can be formed from methane. They are:

CH_3F, CH_2F_2, CHF_3, CF_4, CH_3Cl, CH_2Cl_2, $CHCl_3$, CCl_4, CH_2FCl, CHF_2Cl, CF_3Cl, $CHCl_2F$, CCl_3F, and CCl_2F_2.

30. The message is that ground-level ozone is a harmful air pollutant. Ozone in the stratosphere, on the other hand, has a beneficial effect in protecting Earth's surface from damaging UV rays.

31. Allotropes of oxygen, O_2 and O_3, are different forms of the same element. Isotopes of oxygen, such as O-16 and O-18, differ in the number of neutrons found in each nucleus.

35. Energy is directly proportional to frequency but inversely proportional to wavelength. Frequency and wavelength are inversely proportional to each other.

39. UV-C radiation is extremely dangerous, but it is completely absorbed by normal oxygen, O_2, as well as by ozone, O_3, before it can reach Earth's surface.

42. Step 2 of the Chapman cycle forms ozone as a produce. Ozone is a reactant in steps 3 and 4.

45. The structural formula of CF_3CH_2F is:

$$:\overset{\cdot\cdot}{\underset{\cdot\cdot}{F}}-\overset{\displaystyle :\overset{\cdot\cdot}{\underset{\cdot\cdot}{F}}:}{\underset{\displaystyle :\overset{\cdot\cdot}{\underset{\cdot\cdot}{F}}:}{C}}-\overset{\displaystyle :\overset{\cdot\cdot}{\underset{\cdot\cdot}{F}}:}{\underset{\displaystyle H}{C}}-H$$

53. O_2, O_3, and N_2 all have an even number of electrons to place in their Lewis structures. N_3 would have 15 electrons. Molecules with odd numbers of electrons are generally more reactive than those in which all electrons are present in pairs, and often would not be stable enough to even form.

Chapter 3

1. a. Yes. Earth is warmer than predicted based on distance from the Sun and the amount of radiation reaching Earth. Without the greenhouse effect, our planet would be too cold to be hospitable to life.

   b. Yes. Most scientists now conclude that observed increases in Earth's average temperature are evidence that an enhanced greenhouse effect, or global warming, is taking place.

4. a. The number of atoms of each element on either side is the same. $C = 6$, $O = 18$, $H = 12$.

   b. The number of molecules is not the same on either side of the equation. There are 12 on the left, but only 7 on the right. The large molecule glucose has formed on the right-hand side of the equation, using 24 atoms per molecule.

6. a. $\dfrac{343\ W}{1\ m^2} \times \dfrac{30\ parts}{100\ parts} = \dfrac{103\ W}{m^2}$

   b. Under steady-state conditions, 103 W/m^2 leaves our atmosphere.

7. a. The concentration of CO_2 in the atmosphere at the present time is about 376 ppm; 20,000 years ago, the concentration was about 190 ppm, far lower. However, 120,000 years ago the CO_2 concentration was about 370 ppm, closer to today's levels.

   b. The mean temperature at present is higher than the 1950–1980 mean temperature of the atmosphere. 20,000 years ago, the mean temperature was lower by about 9 °C. However, 120,000 years ago, the mean temperature was lower by only about 1 °C.

   c. Although there appears to be a *correlation* between mean temperature and CO_2 concentration, this figure does not prove *causation* of either factor by the other.

10. These are the two Lewis structures.

$$H{-}H \quad \text{and} \quad H{-}\overset{..}{\underset{..}{O}}{-}H$$

For H_2 with only two atoms, the atoms can only be arranged in a straight line. For H_2O, even though the Lewis structure has been *shown* as a straight line, this does not mean that the molecule is linear. In fact, the bent structure of water is so well known that the Lewis structure is often written in this manner.

$$H{-}\overset{..}{\underset{|}{O}}:$$
$$H$$

13. a. $4 + 4(1)\ 6 = 14$ outer electrons; The Lewis structure is:

$$\begin{array}{c} H \\ | \\ H{-}C{-}\overset{..}{\underset{..}{O}}{-}H \\ | \\ H \end{array}$$

   b. There are four regions of electrons around the C atom, making the H-to-C-to H bond angle 109.5°.

   c. There are 2 bonding and 2 nonbonding regions of electrons around the O atom, for a total of four regions. The H-to-O-to-C bond angle is predicted to be 109.5°.

16. All three modes of vibration contribute to the greenhouse effect. In each case, there is a change in polarity of the molecule during the vibration.

17. a. $E = h\nu = \dfrac{hc}{\lambda}$

   For 4.26 μm (or 2350 cm^{-1})

   $$E = \dfrac{\left(6.63 \times 10^{-34}\ J \cdot s\right)\left(3.00 \times 10^8\ m/s\right)}{4.26 \times 10^{-6}\ m}$$
   $$= 4.67 \times 10^{-20}\ J$$

   For 15.00 μm (or 667 cm^{-1})

   $$E = \dfrac{\left(6.63 \times 10^{-34}\ J \cdot s\right)\left(3.00 \times 10^8\ m/s\right)}{15.00 \times 10^{-6}\ m}$$
   $$= 1.33 \times 10^{-20}\ J$$

20. $C_6H_{12}O_6(aq) \xrightarrow{\text{yeast}} 2\ C_2H_5OH(aq) + 2\ CO_2(g)$

22. a. A neutral atom of Ag-107 has 47 protons, 60 neutrons, and 47 electrons.

   b. A neutral atom of Ag-109 has 47 protons, 62 neutrons, and 47 electrons. The only difference is the number of neutrons per atom.

25. a. $2(1) + 16 = 18$ g/mol

   b. $12\ 2(19) \cdot 2(35.5) = 121$ g/mol

26. a. Mass percent Cl in CCl_3F (Freon-11) is:

   $$\dfrac{3(35.5)}{12.0 + 3(35.5) + 19.0} \times 100 = 77.5\%$$

29. a. CO_2

   b. Although CO_2 accounts for the largest percentage of greenhouse gas contributions, the case could be made that N_2O has the greatest impact. It is almost 300 times more effective than CO_2, but its concentration is far lower than that of either CO_2 or CH_4.

Gas	% Contribution	GWP	Net Effectiveness
	(graph)	(Table 3.5)	(Product)
CO_2	55	1	0.6
CH_4	15	23	3.5
N_2O	5	296	14.8

Although CO_2 accounts for the largest percentage of greenhouse gas emissions, the net effectiveness of N_2O is the largest of these three gases.

33. Drilled ocean cores can be analyzed for the number of types of microorganisms present. Another correlating piece of evidence is the changing alignment of the magnetic field in particles in the sediment over time. Still another possibility is to analyze the H-2 to H-1 ratio in ice cores.

36. Lewis structures *show* linkages and can be used to *predict* shape by counting bonding and nonbonding regions around a central atom. For example, the Lewis structure for H_2O can be written in either of these ways.

$$H-\overset{\cdot\cdot}{\underset{\cdot\cdot}{O}}-H \quad or \quad H-\overset{\cdot\cdot}{\underset{\overset{|}{H}}{O}}\text{:}$$

Both show the same linkages. Either one can be used to predict the H-to-O-to-H bond angle by noting that there are two bonded pairs and two nonbonded pairs around the central O atom. The bond angle is predicted to be 109.5°, not shown explicitly in either Lewis structure. Compare H_2O to H_2S, which also has a bent shape.

40. a. $C_2H_5OH(l) + 3 O_2(g) \longrightarrow 3 H_2O(l) + 2 CO_2(g)$

b. 2 mol CO_2

c. 30 mol O_2

45. $73 \times 10^6 \text{ t } CH_4 \times \dfrac{12 \text{ t C}}{16 \text{ t } CH_4} = 55 \times 10^6 \text{ t C}$

Chapter 4

2. The temperature is the same, 70 °C, in each container. Temperature is commonly measured in degrees Celsius (°C) (for scientific work, or in degrees Fahrenheit (°F) for household applications. The heat content of the water is not the same. Heat depends on both the temperature and the mass of the water sample. The water in container 1 has twice the heat of the water in container 2 because twice the mass of water is present in container 1.

7. a. $2 C_2H_6 + 7 O_2 \longrightarrow 4 CO_2 + 6 H_2O$

8. $\dfrac{52.0 \text{ kJ}}{1 \text{ g } C_2H_6} \times \dfrac{30.1 \text{ g } C_2H_6}{1 \text{ mol } C_2H_6} = \dfrac{1560 \text{ kJ}}{\text{mol } C_2H_6}$

12. a. Exothermic; a charcoal briquette releases heat as it burns.

b. Endothermic; liquid water gains the necessary heat for evaporation from your skin, and your skin feels cool.

13. Bonds broken in the reactants
2 mol C-to-O triple bonds   = 2(1073 kJ) = 2146 kJ
1 mol O-to-O double bond   = 1(498 kJ) = 498 kJ
Total energy *absorbed* in breaking bonds   = 2644 kJ
Bonds formed in the products
4 mol C-to-O double bonds   = 4(803 kJ) = 3212 kJ
Total energy *released* in forming bonds   = 3212 kJ
Net energy change is $(+2644 \text{ kJ}) + (-3212 \text{ kJ})$
    $= -568 \text{ kJ}$
Notice that the overall energy change has a negative sign, characteristic of an exothermic reaction.

16. a. $2 C_5H_{12}(g) + 11 O_2(g) \rightarrow 10 CO(g) + 12 H_2O(l)$

$8(356 \text{ kJ}) + 24(416 \text{ kJ}) + 11(498 \text{ kJ}) \rightarrow$
$10(1073 \text{ kJ}) + 24(467 \text{ kJ})$

The net energy change is 3628 kJ, and the reaction is highly exothermic.

21. a. Estimating values from the graph, domestic oil production accounted for approximately 80% in 1970

(12 million barrels out of 15 million), 65% in 1980 (11 million out of 17 million), 59% in 1990 (10 out of 17 million), 48% in 2000 (9.5 out of 20 million), and is predicted to be about 36% in 2005 (8 out of 22 million).

22. $\dfrac{650,000 \text{ kcal}}{1 \text{ day}} \times \dfrac{365 \text{ days}}{1 \text{ yr}} = \dfrac{2.4 \times 10^8 \text{ kcal}}{\text{yr}}$

This value can be related to each of the energy sources.

a. $\dfrac{2.4 \times 10^8 \text{ kcal}}{1 \text{ yr}} \times \dfrac{1 \text{ yr}}{65 \text{ barrels}} = \dfrac{3.7 \times 10^6 \text{ kcal}}{\text{barrel}}$

27. There are two different isomers.

30. All of these refer to the idea that energy is not consumed during a chemical reaction. Energy can be transformed, but the total energy is constant during a chemical reaction. For example, when energy is transformed, it can be converted from the potential energy stored in chemical bonds to heat energy.

33. The entropy values increase with increasing complexity and randomness. Diamond is a highly ordered crystalline structure and therefore has low entropy. Methanol is a liquid and its molecules have greater complexity and more opportunity for random movement.

35. a. The C—F bond requires 485 kJ/mol, the C—Cl bond requires 327 kJ/mol, and the C—Br bond requires 285 kJ/mol to break the bond. The C—Br bond is the least energetic, and the bromine free radical can interact with ozone even more effectively than the chlorine free radical.

b.

bond most
easily broken

Because the C-to-Cl bond is broken most easily, chlorine free radicals can be released. Such free radicals can catalyze the destruction of ozone.

Chapter 5

1. a. 75–98 lb

b. 9–12 gal

4. a. Partially soluble. Orange juice concentrate needs to be stirred to prevent some of the suspended pulp from settling out of the mixture.

b. Very soluble. The detergent and water form a solution that spreads evenly throughout the clothes being cleaned.

c. Very soluble. Ammonia and water form a clear solution useful for many purposes.

6. a. 55 mg of Ca^{2+} per liter of bottled water is the same as 55 ppm.

$$\frac{55 \text{ mg Ca}^{2+}}{1 \text{ L H}_2\text{O}} \times \frac{1 \text{ L}}{10^3 \text{ mL}} \times \frac{1 \text{ mL H}_2\text{O}}{1 \text{ g H}_2\text{O}} \times \frac{1 \text{g}}{10^3 \text{mg}}$$

$$= \frac{55 \text{ mg Ca}^{2+}}{10^6 \text{ mg H}_2\text{O}} \text{ or 55 ppm Ca}^{2+}$$

b. Evian, with 78 mg Ca^{2+}/L, has a higher Ca^{2+} concentration than this bottled water in which the Ca^{2+} concentration is 55 ppm, or 55 mg Ca^{2+}/L.

9. a. The number of moles of H_2SO_4 in 100 mL of 12 M H_2SO_4 is the same as the number of moles of HCl in 100 mL of 12 M HCl.

$$\frac{12 \text{ mol H}_2\text{SO}_4}{1 \text{ L solution}} \times \frac{1 \text{ L solution}}{10^3 \text{ mL solution}} \times 100 \text{ mL solution}$$

$$= 1.2 \text{ mol H}_2\text{SO}_4$$

$$\frac{12 \text{ mol HCl}}{1 \text{ L solution}} \times \frac{1 \text{ L solution}}{10^3 \text{mL solution}} \times 100 \text{ mL solution}$$

$$= 1.2 \text{ mol HCl}$$

b. The number of grams is not the same because the molar mass of H_2SO_4 is greater than that of HCl.

$$1.2 \text{ mol H}_2\text{SO}_4 \times \frac{98.1 \text{ g H}_2\text{SO}_4}{1 \text{ mol H}_2\text{SO}_4} = 120 \text{ g H}_2\text{SO}_4$$

$$1.2 \text{ mol HCl} \times \frac{36.5 \text{ g HCl}}{1 \text{ mol HCl}} = 44 \text{ g HCl}$$

11. Although there is slight polarity to each C-to-H bond in CH_4, overall the molecule is symmetrical and nonpolar. The forces among nonpolar molecules are relatively weak, allowing the molecules to escape from the liquid to the gas phase at room temperature and pressure. In contrast to CH_4, the O-to-H bonds in H_2O are very polar and so is the molecule, thanks to its bent shape. The forces among polar water molecules, the hydrogen bonds, are strong. Far more energy is required to change liquid water to a gas than is the case with CH_4.

13. a. N and C; $3.0 - 2.5 = 0.5$
S and O; $3.5 - 2.5 = 1.0$
N and H; $3.0 - 2.1 = 0.9$
S and F; $4.0 - 2.5 = 1.5$

b. N will attract the bonded electron pair more strongly than C.
O will attract the bonded electron pair more strongly than S.
N will attract the bonded electron pair more strongly than H.
F will attract the bonded electron pair more strongly than S.

15. a. This is the Lewis structure for ammonia, NH_3.

$$H-\overset{\displaystyle ..}{N}-H$$
$$|$$
$$H$$

b. Each N-to-H bond is polar. The difference in electronegativity is $3.0 - 2.1$, or 0.9.

c. The molecule is polar. The molecular shape is triangular pyramidal.

21. a. A chlorine atom satisfies the octet rule by gaining one electron to form a chloride ion, with a charge of 1−.

$$:\!\overset{..}{\underset{..}{Cl}}\!\cdot \quad \text{and} \quad \left[:\!\overset{..}{\underset{..}{Cl}}\!:\right]^-$$

A barium atom satisfies the octet rule by losing two electrons to form a barium ion, with a charge of 2+.

$$\cdot Ba\cdot \quad \text{and} \quad \left[Ba\right]^{2+}$$

22. a. Na_2S sodium sulfide

b. Al_2O_3 aluminum oxide

23. a. $Ca(HCO_3)_2$

b. $CaCO_3$

24. a. potassium acetate

b. calcium hypochlorite

25. a. The lightbulb will shine. $CaCl_2$ is an electrolyte. It dissolves in water to form a solution that conducts electricity.

b. The lightbulb will not shine. C, H, and O atoms are covalently bonded in C_2H_5OH. Although C_2H_5OH dissolves in water to form a solution, the solution will not conduct electricity.

26. a. Ca^{2+} and OCl^- ions are present.

b. No ions are present. C_2H_5OH (ethanol) is not an electrolyte.

28. a. 55 ppm $CaCO_3$ is within the range classified as slightly hard water.

b. 225 mg $CaCO_3$/L is within the range classified as very hard water.

35. a. The electronegativities of the elements generally increase from left to right across a period (until the 8A Group is reached) and from bottom to top within any group. This means that of the positions indicated, the element in position 2 is predicted to have the highest electronegativity.

b. Ranking the other elements is not straightforward. Element 1 is expected to be more electronegative than element 3, based on their relative positions in the same group. Element 4 will likely be more electronegative than the 1 and 3 and less electronegative than 2. However, because element 4 is not in the same period with any other element, this prediction cannot be made with certainty. Here are the values found in references; they do not appear in Table 5.4. EN 1 = 0.8; EN 2 = 2.4; EN 3 = 0.7; EN 4 = 1.9. These values confirm the relative order: 3 < 1 < 4 < 2.

38. NH_3, like water, is a polar molecule. Therefore, despite its low molar mass, considerable energy must be added to liquid NH_3 to break the intermolecular forces among NH_3 molecules.

39. a. A single covalent bond holds two hydrogen atoms together in H_2. It is an example of an *intra*molecular force.

b. Hydrogen bonding is a type of *inter*molecular force, a force between molecules, not within the molecule.

43. With the exception of contaminants that are known carcinogens, MCLG and MCL values are usually very close to being the same.

Chapter 6

1. a. Possibilities include nitric acid (HNO_3), hydrochloric acid (HCl), sulfuric acid (H_2SO_4), phosphoric acid (H_3PO_4), and hydroiodic acid (HI).

b. In general, acids taste sour, turn litmus paper red (and have characteristic color changes with other indicators), are corrosive to metals such as iron and aluminum, and release carbon dioxide ("fizz") from a carbonate. These properties may not be observed if the acid is not in sufficient concentration.

3. a. $HBr(aq) \longrightarrow H^+(aq) + Br^-(aq)$

b. $H_2SO_3(aq) \longrightarrow H^+(aq) + HSO_3^-(aq)$

4. a. Possibilities include sodium hydroxide (NaOH), ammonium hydroxide (NH_4OH), magnesium hydroxide ($Mg(OH)_2$), and calcium hydroxide ($Ca(OH)_2$).

b. In general, bases taste bitter, turn litmus paper blue (and have characteristic color changes with other indicators), have a slippery feel in water, and are caustic to your skin.

6. a. $RbOH(s) \longrightarrow Rb^+(aq) + OH^-(aq)$

8. For the sulfate ion, one possibility is sulfuric acid neutralizing sodium hydroxide.

$$2\,H^+(aq) + SO_4^{2-}(aq) + 2\,Na^+(aq) + 2\,OH^-(aq) \longrightarrow$$
$$2\,Na^+(aq) + SO_4^{2-}(aq) + 2\,H_2O(l)$$

For the nitrate ion, one possibility is nitric acid neutralizing sodium hydroxide.

$$H^+(aq) + NO_3^-(aq) + Na^+(aq) + OH^-(aq) \longrightarrow$$
$$Na^+(aq) + NO_3^-(aq) + H_2O(l)$$

For the ammonium and carbonate ions, one possibility is carbonic acid neutralizing ammonium hydroxide.

$$H_2CO_3(aq) + 2\,NH_4OH(aq) \longrightarrow$$
$$2\,NH_4^+(aq) + CO_3^{2-}(aq) + H_2O(l)$$

Note: Ammonium hydroxide is written in its undissociated form, as explained in the text; see equation [6.4c]. Similarly, carbonic acid is written in an undissociated form.

9. a. The acid–base reaction in aqueous solution.

$$HNO_3(aq) + KOH(aq) \longrightarrow KNO_3(aq) + H_2O(l)$$

Showing the ions involved:

$$H^+(aq) + NO_3^-(aq) + K^+(aq) + OH^-(aq) \longrightarrow$$
$$K^+(aq) + NO_3^-(aq) + H_2O(l)$$

And now canceling the ions that appear on both sides:

$$H^+(aq) + OH^-(aq) \longrightarrow H_2O(l)$$

13. a. The solution of pH = 6 has 100 times more $[H^+]$ than the solution of pH = 8.

d. The solution with $[OH^-] = 1 \times 10^{-2}$ M has 10 times more $[OH^-]$ than the solution with $[OH^-] = 1 \times 10^{-3}$ M.

16. a. Some soft drinks have a pH of about 2.6. The pH of acid rain is in the range of 4.3 to 6.0. So picking 4.6 for acid rain, soft drinks are about 100 times more acidic.

17. $S(s) + O_2(g) \longrightarrow SO_2(g)$

24. a. Combustion engines, such as those associated with jet aircraft, directly emit CO, CO_2, and NO. If small amounts of sulfur are present in the fuel, SO_2, and SO_3 will be emitted in small amounts as well.

25. No, the relative importance differs. Fossil fuel combustion is the main source of SO_2 and transportation is the largest source of NO_x, with fuel combustion second. Sulfur dioxide has only one major source, the burning of coal. In contrast, nitrogen monoxide is produced from N_2 and O_2 in the air wherever there is a high temperature. Thus, both car engines and power plants are big contributors of NO_x.

28. To do this calculation, you need the molar masses of both $CaCO_3$ and SO_2. It is not necessary to change tons to grams, because the ratio of the number of grams per mole is the same as the ratio of the number of kilograms per kilomole, or the number of tons to the number of "ton moles".

$$1.00\ \text{ton } SO_2 \times \frac{1\ \text{ton} \cdot \text{mol } SO_2}{64.1\ \text{ton } SO_2} \times \frac{2\ \text{ton} \cdot \text{mol } CaCO_3}{2\ \text{ton} \cdot \text{mol } SO_2}$$
$$\times \frac{100\ \text{ton } CaCO_3}{1\ \text{ton} \cdot \text{mol } CaCO_3} = 1.56\ \text{ton } CaCO_3$$

33. c. Since the H appears first in the chemical formula $HC_2H_3O_2$, this representation stresses that one H^+ can dissociate. It also is easier to use when writing a chemical equation that contains acetic acid. The structural formula for acetic acid shows how the all atoms are bonded together, which is not evident from $HC_2H_3O_2$. However, it is not convenient to use in a chemical equation.

35. a. In aqueous NaOH, four species are present: $H^+(aq)$, $OH^-(aq)$, $H_2O(l)$, and $Na^+(aq)$. Water is present in the greatest amount, and hydrogen ion in the least.

The concentrations of sodium ion and hydroxide ion are equal, since they are formed when sodium hydroxide dissociates. Thus:

$$H_2O(l) > Na^+(aq) = OH^-(aq) > H^+(aq)$$

Chapter 7

1. E represents energy, m represents the mass (in kg) lost in a nuclear transformation, and c represents the speed of light (in m/s).

3. **a.** 94 protons

   b. Neptunium contains one more proton than uranium, and plutonium contains two more.

4. **a.** C-14 has 8 neutrons.

6. **a.** Because there are only two naturally occurring isotopes of boron, and because 10.81 is closer to 11 than to 10, there must be a higher percentage of B-11. In fact, the natural abundances of B-10 and B-11 are 19.9% and 80.1%, respectively.

10. **a.** $^2_1H + ^2_1H \longrightarrow ^3_2He + ^1_0n$

    b. $^{238}_{92}U + ^{14}_7N \longrightarrow ^{247}_{99}Es + 5\,^1_0n$

12. **a.** The sum of the masses of the reactants is 5.02838 g and the sum for the products is 5.00878 g. Thus, the mass difference is 0.0196 g.

    b. Use $E = mc^2$.

    $$E = 0.0196 \text{ g} \times \frac{1 \text{ kg}}{10^3 \text{ g}} \times \left[\frac{3.00 \times 10^8 \text{m}}{\text{s}}\right]^2$$

    $$E = 1.76 \times 10^{12} \text{ J}$$

13. **A** = control rod assembly, **B** = cooling water out of the core, **C** = control rods, **D** = cooling water into the core, **E** = fuel rods

16. There are two advantages of using water rather than graphite as the moderating material in U.S. reactors: (1) Water has a higher heat capacity than graphite, and (2) water does not burn, as was the case with the graphite in the Chernobyl reactor.

20. These all represent the alpha particle, but differ in the amount of information they provide. The last, showing the charge of the alpha particle, gives the most information.

22. **a.** $^{131}_{53}I \longrightarrow ^{131}_{54}Xe + ^0_{-1}e$

25. A table can help answer these questions.

# of half-lives	% Remains	% Decayed
1	50	50
2	25	75
3	12.5	87.5
4	6.25	93.75
5	3.12	96.88
6	1.56	98.44
7	0.78	99.22

The table shows that after two half-lives, 25% remains and 75% has decayed. After four half-lives, 6.25% remains and 93.75% has decayed. After six half-lives, 1.56% remains and 98.44% has decayed.

26. To a first approximation, the person is correct in that after 7 half-lives (see the previous question), less than 1% of the radioisotope remains. However, if you have a large sample to begin with, 1% still might be a substantial amount. Also, having a small percent does not mean zero percent.

Chapter 8

1. **a.** Oxidation is a process in which an atom, ion, or molecule *loses* one or more electrons.

   b. Reduction is a process in which an atom, ion, or molecule *gains* one or more electrons.

   c. Electrons must be transferred from the species losing electrons to the species gaining electrons.

3. **a.** Anode is $Zn(s)$.

   b. $Zn(s) \longrightarrow Zn^{2+}(aq) + 2\text{ e}^-$

   c. Cathode is $Ag(s)$.

   d. $2Ag^+(aq) + 2\text{ e}^- \longrightarrow 2\,Ag(s)$

5. A galvanic cell is a device that converts the energy released in a chemical reaction into electrical energy. A collection of several galvanic cells wired together constitutes a battery, such as the battery in your automobile. In an electrolytic cell, electrical energy is converted to chemical energy. During recharging of your automobile's battery, the reaction that usually takes place in the galvanic cell is reversed and your battery is acting an electrolytic cell.

8. In every electrochemical process described in this chapter, energy is produced through electron transfer. Such transfer takes place because of a chemical reaction, such as takes place in galvanic cells and batteries. The transfer may be because light strikes a photovoltaic cell, resulting in the movement of electrons.

11. In both 1970 and 1980, the major use for mercury was for batteries. By 1990, the major use was in the chlor-alkali process. By 1990, awareness of the dangers of mercury in urban trash had grown. Safer batteries and the need to recycle batteries led to the passage of the Mercury-Containing and Rechargeable Battery Management Act (The Battery Act) in 1996.

13. **a.** The electrolyte provides a medium for transfer of both electrons and ions.

    b. KOH paste

    c. $H_2SO_4(aq)$

14. A storage battery converts chemical energy into electrical energy by means of a reversible reaction. No reactants or products leave the "storage" battery and the reactants can be reformed during the recharging cycle. A fuel cell also

converts chemical energy into electrical energy but the reaction is not reversible. A fuel cell continues to operate only if fuel is continuously added, which is why it is classed as a "flow" battery.

18. Oxidation half-reaction:
$$CH_4 + 8\,OH^- \longrightarrow CO_2 + 6\,H_2O + 8\,e^-$$

Reduction half-reaction:
$$2\,O_2 + 4\,H_2O + 8\,e^- \longrightarrow 8\,OH^-$$

Overall reaction:
$$CH_4 + 2\,O_2 \longrightarrow CO_2 + 2\,H_2O$$

20. In current usage, the term *hybrid car* refers to the combination of a gasoline engine together with a nickel-metal hydride battery, an electric motor and an electric generator. Other hybrids using fuel cells are being developed.

24. $370\ kg\ H_2 \times \dfrac{1000\ g}{1\ kg} \times \dfrac{1\ mol\ H_2}{2.0\ g\ H_2} \times \dfrac{286\ kJ}{1\ mol\ H_2}$
$$= 5.3 \times 10^7\ kJ$$

25. a. First write the chemical equation for the reaction.
$$H_2(g) + \tfrac{1}{2}\,O_2(g) \longrightarrow H_2O(l) + energy$$

Then consider the Lewis structures for the reaction.

$$H{-}H + \tfrac{1}{2}\,\overset{..}{\underset{..}{O}}{=}\overset{..}{\underset{..}{O}} \longrightarrow H{-}\overset{..}{\underset{..}{O}}{\diagdown}_H$$

Energy needed to break bonds:

$436\ kJ + 1/2\ (498\ kJ) = 685\ kJ$

Energy released as new bonds form:

$2(467\ kJ) = -934\ kJ$

Overall, according to this calculation, the reaction releases 249 kJ of energy.

b. Average bond energies are based on bonds within molecules in the gaseous state. In the given chemical equation, the H_2O formed is present as a liquid rather than as a gas. Additional energy is released when gaseous water condenses to the liquid state, so the stated value of 286 kJ is greater than the 249 kJ calculated in part **a**.

29. Note that there are eight electrons around each silicon atom, but there are nine electrons around the central atom. Each carbon atom has four outer electrons, so the central atom in the figure must have five outer electrons. This is consistent with arsenic, which is in Group 5A. The additional electron forms an *n*-type silicon semiconductor.

31. Equations 1 and 3 represent redox reactions. Electrons must be transferred to change an element into its combined form. Equation 2 is a neutralization reaction in which ions combine to form a soluble salt as well as covalently bonded water molecules. Electron transfer does not occur in this case.

35. a. Conversion of fuel:
$$C_8H_{18}(l) + 4\,O_2(g) \longrightarrow 9\,H_2(g) + 8\,CO(g)$$

Fuel cell reaction:
$$CO(g) + H_2O(g) \xrightarrow{\text{catalyst}} CO_2(g) + H_2(g)$$

45. a. When water boils, intermolecular hydrogen bonds must be broken.

b. When water is electrolyzed, a chemical reaction takes place and O_2 and H_2 are formed. Intramolecular covalent bonds within H_2O must be broken for these changes to take place.

Chapter 9

1. Plastics are synthetic polymers. This means that plastics are polymers, but not all polymers are called plastics.

8. a. Pounds of plastic in 2003 = 107 billion lb (107×10^9 lb)

b. Pounds of plastic per person in 2003
$$= \frac{107 \times 10^9\ lb}{290 \times 10^6\ people} = \frac{369\ lb}{1\ person}$$

c. % change in production per person, 1977 to 2003 =
$$\frac{369\ lb/person\ -\ 155\ lb/person}{155\ lb/person} \times 100 = 138\%\ change$$

d. % change in production per person, 1997 to 2003 =
$$\frac{369\ lb/person\ -\ 331\ lb/person}{331\ lb/person} \times 100 = 115\%\ change$$

11.

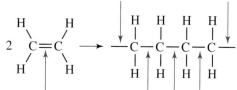

365 kJ *released* in forming 2 half-bonds; equivalent in energy to 1 covalent C-to-C bond

598 kJ *added* in breaking each double bond; 1196 kJ for 2 bonds

365 kJ *released* in forming each single bond; 1068 kJ for 3 bonds

One C-to-C double bond is broken in each of the two ethylene molecules. That is an energy *input* of 2×598 kJ/mol = 1196 kJ. In forming the dimer, three new carbon-to-carbon single bonds are formed and 2 half-bonds are formed. That is an energy *release* of 4×356 kJ/mol or 1424 kJ. The net change is that (1424 kJ – 1196 kJ) = 228 kJ/mol of energy are released from the system, so the reaction is exothermic.

13. a. $n = 40{,}000\ g\ polymer \times \dfrac{1\ monomer\ unit}{28.0\ g\ monomer}$
$$= 1430\ monomer\ units$$

b. $1430\ monomer\ units \times \dfrac{2\ carbon\ atoms}{1\ monomer\ unit}$
$$= 2860\ carbon\ atoms$$

17. There are *three* regions of electrons around each carbon in the monomer, making the geometry around the carbon trigonal planar and the Cl-to-C-to-H bond angle 120°. Each carbon in the polymer has *four* regions around the carbon, making the geometry tetrahedral and the bond angle 109.5° in the polymer.

19. The bottle on the left is likely made of flexible, low-density branched polyethylene. The one on the right is likely made of rigid, high-density, linear polyethylene. The structures of LDPE and HDPE, found in Figure 9.7, can be used to help explain this difference in properties at a molecular level. The low-density polyethylene is highly branched, preventing close interactions between the chains and allowing the plastic to be softer and more easily deformed. The high-density polyethylene chains are linear. The chains can more closely approach each other, creating opportunity for interactions. The bottle on the right is less flexible and more rigid than the one on the left. The bottle on the left may also have had plasticizers added to it, which help to increase flexibility.

21. All except PETE share a common structural feature. The other five of the Big Six polymers have the same basic structure as the ethylene molecule, but one of the hydrogen atoms has been replaced with a different atom or group of atoms. PETE is different because it is a combination of two different monomers, neither of which has the carbon-to-carbon double bond seen in the others.

24. This is the head-to-head, tail-to-tail arrangement of PVC forming from three monomer units.

26. The CO_2 most likely replaces CFCs formerly used for this process. Most of the CO_2 used comes from existing commercial and natural sources, so no additional CO_2 is contributed to global warming. CFCs are implicated in the depletion of the ozone layer and can no longer be used for blowing Styrofoam. For several years, pentane, C_5H_{12}, has been used to replace CFCs, but pentane is flammable and CO_2 is not.

28. **a.** This is the Lewis structure.

b. When Acrilan fibers burn, one of the products is the poisonous gas hydrogen cyanide, HCN.

34. **a.** The approximate increase in plastic production for any five-year period is between 8 and 12 billion pounds.

b. In 1977 production was 34 billion pounds. By about 1992, 15 years later, this figure had doubled.

35. Several other features of polymer chains influence the properties of the polymer formed. These include:

length of the chain (the number of monomer units)
three-dimensional arrangement of the chains
branching of the chain
bonding between chains
orientation of monomer units within the chain
additives (such as plasticizers and coloring agents)

42. The Big Six polymers are almost completely nonpolar molecules and therefore do not dissolve in polar water molecules. The generalization, developed in Chapter 5, is that "like dissolves like." Some of the Big Six dissolve or soften in hydrocarbons or chlorinated hydrocarbons because these nonpolar solvents interact with the nonpolar polymeric chains.

45. A monomer must have a C-to-C double bond in its structure. Although C-to-C double bonds in rings may be present, the double bond used for addition polymerization must be along the chain. That C-to-C double bond must be accessible to attack by a free radical, resulting in a single carbon-to-carbon bond that is left along the chain. As that reaction repeats itself, the addition polymer grows. An example is the formation of PVC.

Chapter 10

1. **a.** An antipyretic drug is intended to reduce fever.

b. An analgesic drug is intended to reduce pain.

c. An antiinflammatory drug is intended to reduce inflammation, which is redness, heat, swelling, and pain caused by irritation, injury, or infection.

d. Yes. Aspirin is an example of a drug with all three properties.

3. The condensed formulas are $CH_3CH_2CH_2CH_2CH_3$ (or $CH_3(CH_2)_3CH_3$), $CH_3CH_2CH(CH_3)CH_3$, and $CH_3C(CH_3)_2CH_3$. These are the line-angle drawings.

5. Four possible isomers have the formula C_4H_9OH. Here is the structural formula for each isomer. The hydrogen atoms have been omitted for clarity.

7. There are one-carbon examples of alcohols, aldehydes, and acids. The other classes of compounds require more than one carbon atom by the nature of their functional groups.

a. Alcohol. The only example is methanol, CH_3OH.

b. Aldehyde. The only example is methanal (commonly called formaldehyde), CH_2O.

c. Carboxylic acid. The only example is methanoic acid (commonly called formic acid), HCO_2H.

d. Ester. There is no example of a one-carbon ester. To be an ester, a compound must have a carbon-containing group bonded to the oxygen atom that is singly bonded to the carbonyl carbon atom.

e. Ether. There is no example of a one-carbon ether. To be an ether, the central oxygen atom must have a carbon-containing group on both sides.

f. Ketone. There is no example of a one-carbon ketone. To be a ketone a compound must have a carbon-containing group on both sides of the carbon that is double bonded to an oxygen atom. If hydrogen atoms occupy one or both these positions, the compound is an aldehyde.

11. a. There are two amide groups.

b. There are two amide groups ( ), one carboxylic acid group (H), and one sulfide group ( ).

c. There is one ester group (C), and one amine in which the two methyl (—CH$_3$) groups have replaced two hydrogens.

13. This molecule is predominantly nonpolar, so it is likely to be very soluble in the lipids that form the cell membranes. The amide group is somewhat polar, helping the molecule to dissolve in the aqueous fluids inside and outside the cells.

18. "Superaspirins" are a new class of medicines that preferentially block the COX-2 enzyme that makes the prostaglandins associated with inflammation, pain, and fever. These superaspirins have no effect on COX-1 enzymes, which means there should be fewer side effects such as stomach irritation and liver or kidney dysfunction.

21. A pharmacophore is the three-dimensional arrangement of atoms, or groups of atoms, responsible for the biological activity of a drug molecule.

23. a. This compound cannot exist in chiral forms. The central carbon atom is bonded to two equivalent —CH$_3$ groups.

b. This compound can exist in chiral forms, as the four groups attached to the central carbon are all different.

c. This compound can exist in chiral forms, as the four groups attached to the central carbon are all different.

d. This compound cannot exist in chiral forms. The central carbon atom is bonded to two equivalent —CH$_3$ groups

25. A freebase is a nitrogen-containing molecule in which the nitrogen is in possession of its lone pair of electrons. Treating methamphetamine hydrochloride with a base (like hydroxide ions, OH$^-$) will strip off one of the hydrogen atoms attached to the nitrogen atom, freeing up its lone pair.

31. There are just three distinct isomers shown here, because some of the structures the student drew represent duplicates.

Numbers 1 and 5 are different paper-and-pencil representations of the *same* isomer.

Numbers 2, 3, and 4 are all different paper-and-pencil representations of the *same* isomer.

Number 6 is an isomer different from numbers 1 and 5, and from numbers 2–4.

35. a. Aspirin produces a physiological response in the body.

b. Superaspirins produce a physiological response in the body.

c. Antibiotics kill or inhibit the growth of bacteria that cause infections.

d. Hormones produce a physiological response in the body.

e. Amphetamine produces a physiological response in the body.

f. Penicillin kills or inhibits the growth of bacteria that cause infections.

37. L-dopa is effective because the molecule fits in the receptor site, but the nonsuperimposable mirror image form, D-dopa, does not. This is the structure of L-dopa, with the chiral carbon atom marked in red. Note that there are four different groups attached to the starred carbon atom.

Chapter 11

1. The four fundamental types of materials provided by food are water, energy sources, raw materials, and metabolic regulators.

4. There is too much carbohydrate for steak, and too much protein for chocolate chip cookies. The pie chart is likely a representation of peanut butter. (See Table 11.1 for confirmation.)

9. Although hydrogen atoms are far more abundant than oxygen or carbon atoms, hydrogen atoms have far smaller masses than those of either oxygen or carbon atoms.

12. A. 6–11 servings of bread, cereal, grain, and pasta

    B. 3–5 servings of vegetables

    C. 2–4 servings of fruits

    D. 2–3 servings of milk, yogurt, and cheese

    E. 2–3 servings of meat, poultry, fish, dry beans, eggs, and nuts

    F. fats, oils, and sweets used sparingly

15. The structure of glucose is based on a six-member ring of five carbon atoms and one oxygen atom. The structure of fructose is based on a five-membered ring of four carbon atoms and one oxygen atom. Glucose has one —CH_2OH side chain, and fructose has two.

20. a. Fats and oils are both composed of nonpolar hydrocarbon chains. Edible fats and oils both contain some oxygen. Most fats and oils are triglycerides, which are esters of three fatty acid molecules and one glycerol molecule. Both oils and fats feel greasy and are insoluble in water.

    b. Oils tend to contain more highly unsaturated fatty acids and smaller fatty acids than fats. If the triglycerides are solid at room temperature, the material is usually termed a fat. If the triglycerides are liquid at room temperature, the material is usually termed an oil.

28. a. Among these are the countries of sub-Saharan Africa, many former Soviet Union countries, Afghanistan, and Cuba.

    b. The countries of North America—U.S., Canada, and Mexico—grow or import sufficient food.

30. According to Table 11.1, peanut butter is 25% protein, which makes it a very good source of protein indeed. However, it is also 50% fat, which is quite high if one needs to limit fat in the diet. Peanut butter is relatively low in saturated fat if it has not been hydrogenated.

33. It would be unlikely to meet the guidelines. Only 2–3 servings per day are recommended for milk, yogurt, and cheese; 6–11 servings of bread, cereal, grains, and pasta are required, 3–5 servings of vegetables, 2–4 servings of fruit, and 2–3 servings of meat, poultry, fish, dry beans, eggs, and nuts. If 40% came from milk or dairy products, there is not enough room in the diet for the other food groups to be well represented.

Chapter 12

1. DNA stands for **d**eoxyribo**n**ucleic **a**cid.

3. a. The base adenine has the amine group, —NH_2.

   b. The sugar deoxyribose has several —OH groups.

   c. There are no functional groups *in* the phosphate, but the phosphate itself is a functional group. Do not mistake the doubly bonded oxygen atom for a ketone, for there is no bond to a carbon atom.

6. a. A nucleotide must contain a base, a deoxyribose molecule, and a phosphate group linked together.

   b. Covalent bonding holds the units together.

8. a. A beam of X-rays, which have relatively high energy and relatively short wavelengths, is directed at a target. The X-rays are then diffracted at certain angles, which are related to the distance between atoms. The process of diffraction makes it appear like the X-ray waves are deflected or bent by the atoms in the structure.

   b. Ions in a salt like sodium chloride have a very regular structure that is easily determined by X-ray studies. Atoms in nucleic acids and proteins do not show the same well-known patterns of crystalline regularity, making the interpretation of the X-ray diffraction pattern more difficult.

13. a. This is the general formula for an amino acid, where R represents a side chain that is different in each of the 20 amino acids.

   b. The functional groups are —COOH, which is the carboxylic acid group, and the —NH_2 group, the amino group.

15. A codon is a grouping of three RNA bases. An appropriate RNA molecule transfers the order of bases in DNA into a specific amino acid that should appear in a protein sequence. Thus, codons are used to signal that a molecule of a certain amino acid should be incorporated into a protein.

18. Only a minor change in the amino acid composition of human hemoglobin leads to sickle cell anemia. In hemoglobin S, two of the residues that should be glutamic acid are replaced with valine. This seemingly innocuous change has rather drastic results for the person with this genetic disease.

21. a. PCR stands for **p**olymerase **c**hain **r**eaction.

    b. Using DNA for diagnostic probes was not effective if the concentration of the invading DNA is too low. PCR technology made it possible to start with a single segment of DNA and make millions or billions of copies of it in a relatively short time period.

c. Amplification of DNA has made possible early diagnosis of HIV infection, for example, as well as several genetic diseases.

24. This is the term coined to describe the practice of using domesticated animals to produce drugs and other medically significant substances.

30. The discovery that %A = %T and that %C = %G provided the basis for asking *why* this pattern was observed. Chargaff's contribution was in finding that the bases were paired. Crick and Watson took this information a step further to discover both *how* and *why* they were paired, and the influence the pairing had on the structure of DNA.

46. The statement is false. Calculations show that the DNA in an adult would stretch just about 26,000 times to the moon and back.

$$\frac{2 \text{ m}}{1 \text{ DNA thread}} \times \frac{1 \times 10^{13} \text{ cells}}{1 \text{ adult}}$$
$$= 2 \times 10^{13} \text{ m of DNA in an adult}$$

$$3.8 \times 10^5 \text{ km} \times \frac{10^3 \text{ m}}{1 \text{ km}} \times 2 = 7.6 \times 10^8 \text{ m},$$
$$\text{the distance to the moon and back}$$

The ratio of these two distances is $\dfrac{2 \times 10^{13} \text{ m}}{7.6 \times 10^8 \text{ m}} = \dfrac{26,000}{1}$.

Glossary

The numbers at the end of each term indicate the page(s) on which the term is defined and explained in the text.

A

acid a substance that releases hydrogen ions, H^+, in aqueous solution *269*

acid anhydride literally, "an acid without water." For example, the nonmetallic oxide SO_3 forms an acid when reacted with water. *278*

acid deposition the deposition of either wet or dry forms of acidic substances from the atmosphere to the Earth. The wet forms include rain, snow, fog, and cloud-like suspensions of microscopic water droplets and may be more acidic and damaging than acid rain. *275*

acid-neutralizing capacity (ANC) the capacity of a lake or other body of water to resist a decrease in pH *295*

acid rain rain that is more acidic and has a lower pH value than "normal" rain *274*

activation energy the energy necessary to initiate a reaction *185*

active (receptor) site the region of an enzyme molecule in which its catalytic effect occurs *541*

addition polymerization a process in which monomers simply add to a growing polymer chain in such a way that the product contains all the atoms of the starting material *404*

aerosols particles, both liquid and solid, that stay suspended in the air rather than settling out *39*

alkane a hydrocarbon with only single bonds between carbons *193*

allotropes two or more forms of the same element that differ in their molecular or crystal structure, and therefore in their properties *62*

alpha particle (α) a positively charged ($+2$) particle that consists of the nucleus of a helium atom—two protons and two neutrons *327*

amino acid a molecule that contains an amine group ($-NH_2$) and a carboxylic acid group ($-COOH$) *413*

amino acid residue an amino acid that was once incorporated into a peptide or protein chain *504*

amniocentesis a medical procedure in which a sample of the amniotic fluid is withdrawn from the mother's uterus. This fluid contains fetal cells that are then analyzed for their genetic makeup. *550*

amorphous substance a compound in which the molecules are found in a random arrangement *403*

anabolic steroid a synthetic steroidal hormone used to stimulate muscle and bone growth *459*

anaerobic bacteria bacteria that thrive without the use of molecular oxygen *144*

androgens male sex hormones *454*

anion a negatively charged ion *234*

anode the electrode where oxidation takes place. The anode delivers electrons to the cathode via an external circuit. *359*

antagonist a drug that fits into a receptor site but does not have the customary drug effect *457*

antibody a protein "marker" that remains in the body and offer protection against subsequent infection by the virus or bacterium *547*

antioxidant a compound that prevents packaged, processed foods from becoming rancid due to oxidation of oil or fats *520*

aqueous solution a solution in which water is the solvent *224, 233*

aquifer a great pool of water trapped in sand and gravel 50–500 ft below Earth's surface *222*

atmospheric pressure the force with which the atmosphere presses down on a given area *23*

atom the smallest unit of an element that can exist as a stable, independent entity *28*

atomic mass the mass of an atom expressed relative to a value of exactly 12 for an atom of C-12 *138*

atomic mass unit the mass one atom of C-12, 1.66×10^{-24} g *138*

atomic number the number of protons in an element *65*

Avogadro's number the number of atoms in exactly 12 g of C-12 (6.02×10^{23}); also the number of particles per mole of substance such as a covalently bonded molecule *139*

B

background radiation the average daily amount of radiation to which we each are exposed. The amount of background radiation depends on location. *331*

basal metabolism rate (BMR) the minimum amount of energy required daily to support basic body functions *514*

base any compound that produces hydroxide ions, OH^-, in aqueous solution *270*

battery a system for the direct conversion of chemical energy to electrical energy *358*

beta particle (β) a high-speed electron emitted from a nucleus *327*

biomass materials produced by biological processes *201*

bond energy amount of energy that must be absorbed to break a specific chemical bond *181*

breeder reactor a type of nuclear reactor that creates new fissionable fuel (e.g., Pu-239) while the current fissionable fuel of the reactor (U-235) undergoes fission *326*

C

calibration graph a graph made with carefully measured absorbencies (or other properties) of several solutions of known concentration of the species being analyzed *253*

calorie the amount of heat necessary to raise the temperature of exactly one gram of water by one degree Celsius (exactly 4.184 J) *173*

Calorie the unit used in nutrition. 1 Cal = 1 kcal = 4.184 kJ *486*

calorimeter a device for experimentally determining the quantity of heat energy released in a combustion reaction *179*

carbohydrate a compound containing carbon, hydrogen, and oxygen. The hydrogen and oxygen are in the same 2:1 atom ratio found in water. *491*

carcinogen a compound capable of causing cancer *242*

carcinogenic capable of causing cancer *48*

case–control studies experimental methods that compare a group of people with an outcome (i.e., the health problem) with a group who do not have the outcome *516*

catalyst a chemical substance that participates in a chemical reaction and influences its speed without itself undergoing a permanent change *40*

catalytic converter a device installed in the exhaust stream of a vehicle in order to reduce emissions *40*

catalytic cracking a process by which catalysts are used to promote molecular breakdown at lower temperatures than those used in thermal cracking *197*

cathode the electrode where reduction takes place. The cathode receives the electrons sent from the anode through the external circuit. *359*

cation a positively charged ion *234*

chain reaction a reaction that becomes self-sustaining. For example, nuclear fission becomes a chain reaction under certain conditions. *315*

Chapman cycle the set of chemical reactions that determines the steady state of stratospheric ozone *81*

Chargaff's rules the observation that in every species, the percent of adenine almost exactly equals the percent of thymine and the percent of guanine matches the percent of cytosine. Put more simply: %A = %T and %G = %C. *534*

chemical equation a representation of a chemical reaction using chemical formulas *33*

chemical formula a symbolic way to represent the elementary composition of a chemical compound that indicates the kinds and numbers of atoms present *29, 435*

chemical reaction the process whereby reactants are transformed into products (e.g., combustion) *33*

chemical symbol one- or two-letter abbreviation for an element *25*

chiral (optical) isomers compounds that have the same chemical formula but which differ in their three-dimensional molecular structure and their interaction with polarized light *449*

chlorofluorocarbon a compound composed of the elements chlorine, fluorine, and carbon *92*

chromosomes 46 compact structures of intertwined molecules of DNA in a human cell nucleus that contain all of the genetic material *530*

clone a collection of cells or molecules identical to an original cell or molecule *545*

codon the genetic code written in groupings of three DNA bases. The code is transferred from DNA to RNA and ultimately to the amino acid sequence in proteins. *539*

coenzyme a molecule that works in conjunction with an enzyme to enhance the enzyme's activity *509*

cohort studies experimental methods that follow large groups of people over long periods of time *516*

combustion the rapid combination of oxygen with a substance (e.g., fuel) to form products *33, 178*

complementarity the process of combining foods that complement each others' essential amino acid content so that the total diet provides a complete supply of amino acids *506*

complementary bases bases capable of forming a hydrogen-bonded base pair *534*

compound a pure substance made up of two or more elements in a fixed, characteristic chemical combination *28*

concentration the ratio of amount of ingredient to amount of water (solvent or solution) *226*

condensation polymerization a process in which monomers join by eliminating (splitting out) a small molecule, often water *411*

condensed structural formula a representation in which C-to-H bonds are not drawn out explicitly, but simply understood to be single bonds *435*

conductivity meter an apparatus that produces a signal to indicate that electricity is being conducted *234*

copolymer a combination of two or more different monomers *411*

covalent bond a chemical bond in which two electrons are shared by the atoms involved *68*

cracking a chemical process by which large molecules are broken into smaller ones suitable to be used in gasoline *197*

criteria air pollutants the six major air pollutants for which the EPA has set permissible levels in the air based on their effects on human health and the environment *11*

critical mass the amount of fissionable fuel required to sustain a chain reaction *315*

crystalline region a region in which the molecules are arranged neatly and tightly in a regular pattern *403*

curie (Ci) a measure of radioactivity equivalent to the number of decays per second from one gram of radium *332*

current the rate of electron flow *360*

D

denitrification the process of converting the nitrate ion back to nitrogen gas, thus removing reactive nitrogen from the environment *293*

density the ratio of mass per unit volume *232*

deoxyribonucleic acid (DNA) the molecule that carries genetic information in all species *530*

depleted uranium (DU) uranium (U-238) that has been depleted of the small amount of U-235 that it once naturally contained *325*

desalination any process that removes ions from salty water *259*

diatomic molecule a molecule such as N_2 or CO that contains two atoms *29*

dietary supplements vitamins, minerals, amino acids, enzymes, and herbs and other botanicals *471*

disaccharide a "double sugar" formed by joining two monosaccharide units *492*

dispersion forces attractions between molecules that result from a distortion of the electron cloud that causes an uneven distribution of the negative charge *408*

distillation a purification or separation process in which a solution is heated to its boiling point and the vapors are condensed and collected *193, 259*

distributed generation the method of placing power-generating modules of 30 megawatts or less near the end user *368*

DNA fingerprinting a technique that utilizes unique fragments of DNA to identify a specific individual *552*

DNA probes DNA segments selected or engineered so that they are complementary to some segment of the infecting viral or bacterial DNA (the target) *548*

doping a process of intentionally adding small amounts of other elements to pure silicon *382*

double bond a covalent bond consisting of two pairs of shared electrons *71*

double helix description of the molecular structure of DNA *535*

E

effective stratospheric chlorine a term reflecting both chlorine- and bromine-containing gases in the stratosphere *101*

electricity the flow of electrons from one region to another that is driven by a difference in potential energy *358*

electrode an electrical conductor placed in the cell as a site for chemical reaction *359*

electrolysis the process of passing a direct current of electricity of sufficient voltage through water to decompose it into H_2 and O_2 *377*

electrolyte a conducting solute in aqueous solution *234*

electrolytic cell a device in which electrical energy is used to bring about a nonspontaneous chemical change *358*

electromagnetic spectrum a continuum of waves ranging from very low energy radio waves to very high energy X-rays and gamma rays *75*

electron a subatomic particle of much smaller mass than a proton or neutron (approximately 1/2000th the mass) with a negative charge equal in magnitude to that of the proton, but opposite in sign *65*

electronegativity (EN) a measure of an atom's attraction for the electrons it shares in a covalent bond *229*

electrophoresis a method of separating molecules based on their rate of movement in an electric field *552*

element a pure substance that cannot be broken down into simpler ones by chemical means *25*

endothermic describes any chemical or physical change that absorbs heat *180*

enhanced greenhouse effect a return of greater than 81% of radiated energy from Earth *118*

enriched uranium uranium that has a higher percent of U-235 than 0.7%, the natural abundance *324*

entropy randomness or disorder in position or energy *177*

enzyme a protein that acts as a biochemical catalyst, influencing the rate of chemical reactions *443*

estrogens female sex hormones *454*

exothermic describes any chemical or physical change that releases heat *179*

exposure the amount of the substance encountered, generally referring in chemistry to human contact with toxic substances or disease-causing organisms *18*

F

fats triglycerides that are solid at room temperature *495*

fatty acid a molecule with two structural features: a long hydrocarbon chain generally containing an even number of carbon atoms (typically 12–24) with a carboxylic acid group (—COOH) at the end of the chain *495*

first law of thermodynamics the principle that energy is neither created nor destroyed in a chemical reaction; also called the law of conservation of energy *174*

free radical an atom or molecule with an unpaired electron *90, 521*

freebase a nitrogen-containing molecule in which the nitrogen is in possession of its lone pair of electrons *441*

frequency the number of waves passing a fixed point in one second *73*

fuel cell a galvanic cell that produces electricity by converting the chemical energy of a fuel directly into electricity without burning the fuel *364*

fullerenes a class of compounds based on C_{60} *380*

functional groups distinctive arrangements of atoms that impart characteristic chemical properties to the molecules that contain them *401, 438*

G

galvanic cell a device that converts the energy released in a spontaneous chemical reaction into electrical energy *358*

galvanized iron iron coated with zinc to minimize corrosion *287*

gamma rays (γ) high-energy, short-wavelength photons that may be emitted from the nucleus during the process of nuclear decay *328*

gaseous diffusion   a process in which a gas is forced through a series of permeable membranes to separate molecules of different masses   *325*

gene   a biological component that conveys one or more hereditary traits   *530*

gene therapy   a medical technique that supplies cells with normal copies of missing or flawed genes   *550*

generic drug   a drug that is chemically equivalent to the pioneer drug, but that cannot be marketed until the patent protection on the pioneer drug has run out after 20 years   *465*

genetic engineering   the manipulation and alteration of genetic material (DNA) for a wide variety of purposes   *528*

global warming (enhanced greenhouse effect)   the result if concentrations of greenhouse gases increase, returning more than the usual portion of the heat (infrared radiation) radiated by Earth   *118*

global warming potential (GWP)   a number that represents the relative contribution of a molecule of the indicated substance to global warming   *145*

green chemistry   the process of designing chemical products and processes to reduce or eliminate the use or generation of hazardous substances   *43*

greenhouse effect   the process by which atmospheric gases trap and return a major portion of the heat (infrared radiation) radiated by Earth   *118*

greenhouse gas   a gas capable of absorbing and reemitting infrared radiation   *117*

groundwater   water pumped from wells that have been drilled into underground aquifers   *223*

group   a vertical column in the periodic table designated by a number and letter   *26*

H

half-life   for a particular radioisotope, the time required for the level of radioactivity to fall to one-half of its value   *335*

half-reaction   a type of chemical equation that shows the electrons either lost or gained   *358*

halons   compounds in which bromine or fluorine atoms replace some or all of the chlorine in CFCs   *93, 106*

hard water   water that contains high concentrations of dissolved calcium and magnesium ions   *247*

heat   energy that flows from a hotter to a colder object   *172*

heat of combustion   the quantity of heat energy given off when a specified amount of a substance burns in oxygen   *179*

high-level radioactive waste (HLW)   nuclear waste with high levels of radioactivity. HLW requires essentially permanent isolation from the biosphere because of the long half-lives of the radioisotopes involved. It consists of the radioactive materials that result from the reprocessing of spent nuclear fuel.   *338*

hormone   a substance produced by the body's endocrine glands that can have a wide range of physiological functions, including serving as "chemical messengers"   *443*

human genome   the totality of human hereditary information in molecular form   *530*

Human Genome Project   the effort to map all the genes in the human organism   *543*

hybrid car   a vehicle that combine conventional gasoline engines with battery technology   *374*

hydrocarbon   a compound of hydrogen and carbon   *36, 193*

hydrochlorofluorocarbon (HCFC)   a compound of hydrogen, chlorine, fluorine, and carbon   *100*

hydrofluorocarbon (HFC)   a compound of hydrogen, fluorine, and carbon   *104*

hydrogen bond   the attraction between a hydrogen atom attached to O, N or F and another electronegative atom due to partial charges on the atoms   *231*

hydronium ion (H_3O^+)   formed from a proton interacting with a solvent water molecule; responsible for acidic properties in aqueous solution   *366*

hygroscopic   describes a substance that seeks water and readily absorbs it   *289*

I

infrared radiation   heat radiation; the region of the electromagnetic spectrum adjacent to the red end of the visible spectrum and characterized by wavelengths longer than red light   *75*

intermolecular force   a force that exists *between* molecules. For very large molecules an intermolecular force can exist between different regions of the same molecule.   *231, 408*

intramolecular force   a force that exists *within* a molecule   *230*

ion   an electrically charged species that carries current when dissolved in aqueous solution   *234*

ion exchange   a process in which ions are interchanged, usually between a solution and a solid   *249*

ionic bond   a chemical bond formed by the electrostatic attraction between oppositely charged ions   *234*

ionic compound   a compound consisting of oppositely charged ions   *236*

irradiation   a process of using short-wavelength, high-energy gamma radiation to kill microorganisms   *521*

isomers   molecules with the same chemical formula (same number and kinds of atoms) but different molecular structures and properties   *197, 435*

isotope   an atom of the same element (same number of protons) in which the nucleus has a different number of neutrons   *67*

J

joule   a unit of energy approximately equal to the energy required for one beat of a human heart   *78*

K

kinetic energy   the energy of motion   *177*

L

law of conservation of matter and mass   in a chemical reaction, atoms are neither created nor destroyed, and the elements present do not change when converted from reactants to products. Matter and mass are conserved.   *34*

lead compound a drug (or a modified version of that drug) that shows high promise for becoming an approved drug *448*

Lewis structure a representation of a molecule that uses dots to show the outer (valence) electrons *69*

line-angle drawing a simplified version of a structural formula that is most useful when representing larger molecules *437*

linear, nonthreshold model a model that assumes that the adverse effects of radiation increase linearly with dose, with radiation being harmful at all doses, even low ones *334*

lipids a class of compounds that includes cholesterol and other steroids and molecules of some complex compounds, such as lipoproteins, that contain fatty segments *495*

liter (L) the volume occupied by 1000 g of water at 4 °C *227*

low-level radioactive waste (LLW) waste such as clothing, shoes, filters, or medical equipment that is contaminated with smaller quantities of radioactive materials than HLW contains. This category specifically excludes spent nuclear fuel. *344*

M

macromolecule a molecule that contains thousands of atoms and that have molecular masses that can reach over a million *398*

malnutrition a condition caused by a diet lacking in the proper mix of nutrients, even though the energy content of the food eaten may be adequate *486*

mass numbers the sum of the number of protons and the number of neutrons in an atom *67*

maximum contaminant level (MCL) the legal limit for the concentration of a contaminant, expressed in ppm or ppb *243*

maximum contaminant level goal (MCLG) the level, expressed in ppm or ppb, at which a person weighing 70 kg (154 lb) could drink 2 L (about 2 qt) of water containing a contaminant every day for 70 years without suffering any ill effects *243*

mesosphere the region of the atmosphere at an altitude above 50 km *22*

metal an element that is shiny and conducts electricity and heat well, such as iron, gold, and copper. Metals tend to form cations by losing electrons. *26, 235*

metallic bonding chemical bonding in which outer (valence) electrons are shared among all the atoms in the substance *381*

microcell a very small fuel cell *369*

microgram (μg) 10^{-6}, or a millionth of a gram *18*

micrometer (μm) 10^{-6} of a meter (m); sometimes referred to as a micron *11*

mixture a physical combination of two or more substances present in variable amounts *12*

moderator a material that slows the speed of the neutrons in nuclear reactor, making them more effective in causing fission *319*

molar mass the mass of one mole, or Avogadro's number, of whatever particles are being specified *141*

molarity (M) the number of moles of solute present in one liter of solution *227*

mole (mol) Avogadro's number of particles being specified *140*

molecule a combination of a fixed number of atoms such as H_2O or Cl_2 held together in a particular spatial arrangement *29*

monomer a small molecule used to synthesize a polymeric chain *398*

monosaccharide a single sugar molecule *492*

N

nanometer (nm) one one-billionth of a meter, 1 nanometer (nm) = 1×10^{-9} m *73*

nanotechnology work at the atomic and molecular (nanometer) scale *29*

neutral solution a solution that is neither acidic nor basic, that is, it contains equal concentrations of H^+ and OH^- ions *272*

neutralization a chemical reaction in which H^+ from an acid combines with the OH^- from a base to form H_2O molecules *271*

neutron a subatomic particle that is electrically neutral with almost exactly the same mass as a proton *65*

nitrification the process of converting ammonia in the soil to the nitrate ion *293*

nitrogen cycle a set of chemical pathways whereby nitrogen moves in different chemical forms such as N_2, NO, and nitrates through the biosphere *293*

nitrogen-fixing bacteria bacteria that remove nitrogen from the air and convert it to ammonia *293*

nitrogen saturation the process in which an area is overloaded with nitrogen, that is, when the reactive forms of nitrogen entering an ecosystem exceed the system's capacity to absorb the nitrogen *296*

noble gases elements that are inert and do not readily undergo chemical reactions, found in Group 8A (helium family) of the periodic table *26*

nonelectrolyte nonconducting solute in aqueous solution *234*

nonmetal an element that does not conduct electricity or heat well, such as sulfur, chlorine and oxygen. Nonmetals tend to gain electrons to form anions. *26, 235*

nonspontaneous events that will not occur by themselves; they require that work be done by someone or something *178*

***n*-type semiconductor** a semiconductor in which there are freely moving negative charges, the electrons *383*

nuclear fission the splitting of a large nucleus such as U-235 into smaller ones with the release of energy *313*

nuclear transfer a laboratory procedure in which a cell's nucleus is removed and placed into an egg cell that has had its own nucleus removed. The genetic information from the donor nucleus controls the resulting cell, which can be induced to form embryos. *560*

nucleotide the combination of a base, a deoxyribose molecule, and a phosphate group *532*

nucleus the minuscule but highly dense center of every atom that contains the protons and neutrons *64*

O

octet rule the generalization that states that electrons in many molecules are arranged so that every

atom (except hydrogen) has eight bonded or nonbonded outer electrons *69*

oils triglycerides that are liquid at room temperature *495*

optical (or chiral) isomers compounds that have the same chemical formula but which differ in their three-dimensional molecular structure and their interaction with polarized light *449*

organic chemistry the branch of chemistry devoted to the study of carbon compounds *434*

organic compound a compound containing carbon, often bonded with hydrogen *40–41*

osmosis the natural tendency for a solvent (e.g., water) to move through a membrane from a region of higher solvent concentration to a region of lower solvent concentration *260*

outer (valence) electrons the electrons that account for many of the chemical and physical properties of the corresponding elements *66*

oxidation half-reaction a type of chemical equation that shows the reactant that loses electrons *358*

oxygenated gasolines blends of petroleum-derived hydrocarbons with oxygen-containing compounds such as MTBE, ethanol, or methanol *199*

ozone layer the region of the stratosphere with the maximum ozone concentration *63*

P

parts per billion (ppb) one part out of one billion, or 1000 times less concentrated than 1 part per million *21*

parts per million (ppm) one part out of a million; 10,000 times less concentrated than 1 part per hundred *14*

peptide bond a bond that joins a —C=O and a —NH fragment and thereby attaches the remaining parts of two amino acids *413*

percent parts per hundred, sometimes abbreviated as pph *12*

periodic table an orderly arrangement of all the elements based on similarities in their properties *26*

periodicity of properties the regular recurrence of certain chemical aspects of atoms that is chiefly the consequence of the number and distribution of electrons in the atoms of the elements *66*

pH a number, usually between 0 and 14, that indicates the acidity of a solution *273*

pharmacophore the specific part of the molecule that gives the compound its biological activity *447*

photon an individual bundle or "particle" of light energy *77*

photovoltaic cell (solar cell) a device that converts radiant energy directly to electrical energy *381*

plasmids rings of DNA *545*

plasticizer a substance that improves the flexibility of a polymer *404*

PM_{10} particulate matter with an average diameter of 10 μm, or less *11*

$PM_{2.5}$ particulate matter with an average diameter less than 2.5 μm, also called fine particles *11*

polar covalent bond a covalent bond in which the electrons are not equally shared, but rather displaced toward the more electronegative atom *230*

polar stratospheric cloud (PSC) a thin stratospheric cloud made up of frozen water vapor *96*

polyamide a polymer of amino acids *413*

polyatomic ion a group of covalently bonded atoms bearing a positive or negative electrical charge *237*

polyatomic molecule a compound whose molecules contain more than two bonded atoms *70*

polymer a large molecule made up of long chains of atoms covalently bonded together *398*

polymerase chain reaction (PCR) a technology that uses a single segment of DNA to make millions or even billions of copies of itself in a few hours *548*

polysaccharide a polymer made up of thousands of sugar units *492*

potable water water that is fit for human consumption *220*

potential energy the form of energy related to the positions of atoms and molecular structure and stored in the chemical bonds *175*

precursor a molecule that can be converted directly to a different molecule *455*

primary coolant a liquid that comes in direct contact with a nuclear reactor to carry away heat *319*

primary structure the identity and sequence of the amino acids present in a sample *541*

primers synthetic, single-stranded nucleotides that bracket and identify the section of the DNA to be copied *548*

products the substances formed from reactants as a result of a chemical reaction *33*

protein a polyamide (polypeptide) polymer made from amino acid monomers; essential components of the body and diet *503, 539*

proton a positively charged subatomic particle *64*

p-type semiconductor a semiconductor in which there are freely moving positive charges, or "holes" *383*

Q

quantized a noncontinuous energy distribution that consists of many individual, or discrete, steps *77, 133*

R

racemic mixture a mixture that consists of equal amounts of each chiral (optical) isomer of a compound *451*

rad (radiation absorbed dose) a unit of radiation that indicates the absorption of 0.01 J of radiant energy per kilogram of tissue *332*

radiant energy the entire collection of different wavelengths, each with its own energy *75*

radiation sickness the illness produced by exposure to large amounts of radiation, characterized by early symptoms of anemia, malaise, and susceptibility to infection *330*

radioactive decay series a characteristic pathway of radioactive decay that begins with a heavier isotope such as U-238 and ends with a stable isotope such as Pb-206 *330*

radioactivity the spontaneous emission of radiation by certain radioisotopes, such as C-14 or I-131 *327*

radioisotope a radioactive isotope of an element *330*

reactants the starting materials in a chemical reaction that are transformed into products during the reactions *33*

reactive nitrogen compounds of nitrogen that are biologically active, chemically active, or active with light in the atmosphere *292*

recombinant DNA DNA that has incorporated DNA from another organism *529*

reduction half-reaction a type of chemical equation that shows the reactant that gains electrons *358*

refinery gases the most volatile components of the fractionating tower that boil far below room temperature *195*

reforming process the process by which the hydrocarbons in petroleum are converted to highly branched alkanes to eliminate engine "knocking" *367*

reformulated gasoline (RFG) an oxygenated gasoline that contains a lower percentage of certain more volatile hydrocarbons such as benzene found in conventional gasoline *199*

rem (roentgen equivalent man) a unit of equivalent dose that indicates the damage done by a particular dose of radiation. A rem is the number of rads multiplied by the quality factor Q *332*

replication the process by which copies of DNA are made *537*

resonance forms forms that are hypothetical extremes of electron arrangements that do not exist exactly as represented by any one Lewis structure *72*

reverse osmosis a purification process that reverses the natural tendency for a solvent (e.g., water) to move through a membrane from a region of higher solvent concentration to a region of lower solvent concentration; often used to remove ions and other contaminants from seawater *260*

risk assessment the process of evaluating scientific data and making predictions in an organized manner about the probabilities of an occurrence *18*

S

saturated hydrocarbon a hydrocarbon chain containing only C-to-C single bonds between the carbon atoms and no C-to-C double bonds *496*

scientific notation a system for writing a number as the product of one number and 10 raised to the appropriate power or exponent *19*

second law of thermodynamics the principle that states that it is impossible to completely convert heat into work without making some other changes in the universe *177*

secondary coolant the water in the steam generators that does not come in contact with the reactor *319*

secondary pollutant a pollutant produced from chemical reactions among two or more other pollutants *44*

secondary structure the intermediate level of molecular organization in proteins or DNA *541*

semiconductor a material that does not normally conduct electricity well, but that can do so under certain conditions *381*

sequestration literally means keeping something apart. Chemically this is accomplished by forming stable bonds between the sequestering agent and the substance "trapped." *156*

sievert (Sv) a unit of equivalent dose equal to 100 rem *333*

significant figure a number that correctly represents the accuracy with which an experimental quantity is known *51*

single bond a covalent bond in which only one pair of electrons is shared between two atoms *69*

soft water water that contains few dissolved calcium or magnesium ions *247*

solutes those substances that dissolve in a solvent *224*

solution a homogeneous mixture of uniform composition *224*

solvent a substance capable of dissolving other substances *224*

specific heat the quantity of heat energy that must be absorbed to increase the temperature of 1 g of a substance by 1 °C *233*

spent nuclear fuel (SNF) the radioactive material remaining in fuel rods after they have been used to generate power in a nuclear reactor *326*

steady state a condition in which a dynamic system is in balance so that there is no net change in concentration of the major species involved *81*

stem cells identical, undifferentiated cells that, by successive divisions, can give rise to specialized ones like blood cells *561*

storage battery a battery capable of storing electrical energy *363*

stratosphere the region of the atmosphere above the troposphere that includes the ozone layer *22*

structural formula a representation that shows the atoms and their arrangement with respect to one another in a molecule *435*

substituents an atom or functional group substituted for a hydrogen atom in an organic compound *441*

substrate the molecule or molecules whose reaction is catalyzed by an enzyme *447, 541*

surface water water from lakes, rivers, and reservoirs *223*

T

temperature a property that determines the direction of heat flow *172*

tertiary structure the overall shape or conformation of a molecule *541*

tetrahedron a four-cornered figure with four equal triangular sides *125*

thermal cracking the heating of starting materials to a high temperature *197*

thermal energy the random motion of molecules *177*

thermoplastic polymer a polymer that can be melted and shaped and tend to be flexible *403*

toxicity the intrinsic health hazard of a substance *18*

transgenic organisms artificially created and stringently controlled higher plants and animals that share the genes of another species *554*

triglyceride an ester of three fatty acid molecules and one glycerol molecule *495*

trihalomethanes (THMs) compounds in which any three halogen atoms replace three of the four H atoms in methane, CH_4 *254*

triple bond a covalent linkage made up of three pairs of shared electrons *72*

troposphere the region of the atmosphere that lies directly above Earth's surface *22*

U

ultraviolet radiation the region of the electromagnetic spectrum adjacent to the violet end of the visible spectrum and characterized by wavelengths shorter than violet light *76*

undernourishment a condition in which the daily caloric intake is insufficient to meet the metabolic needs of a person *486*

unsaturated hydrocarbon a hydrocarbon molecule containing one or more C-to-C double bonds between carbon atoms *496*

V

vaccine a preparation of killed microorganism, attenuated living organisms, or fully virulent organisms that produce or increase immunity to a particular disease *547*

valence electrons the number of electrons in the outer energy level of an atom; helps account for observed trends in physical and chemical properties *66*

vector a modified plasmid used to carry DNA back into a bacterial "host" *545*

virus a simple, infectious, almost living biochemical species *547*

vitamins nutrients essential to the human diet that are required in very small amounts. They are organic molecules with a wide range of physiological functions that are generally are not used as a source of energy, although some of them help break down macronutrients *508*

VOCs, or volatile organic compounds incompletely burned gasoline molecules or fragments of these or other small molecules that served as solvents *41*

volatile refers to a substance that evaporates readily *40*

voltage the difference in electrochemical potential between the two electrodes *359*

volumetric flask a type of glassware that contains a precise amount of solution when filled to the mark on its neck *228*

W

wavelength the distance between successive peaks of waves *73*

wavenumber a number, inversely proportional to wavelength, that is

used to describe a characteristic of a wave *131*

work the form of energy describing movement against a restraining force; equal to the force multiplied by the distance over which the motion occurs *172*

X

X-ray high-energy, short-wavelength photons produced from an electronic source (such as an X-ray tube) *76*

X-ray diffraction a process in which a beam of X-rays is directed at a crystal. The X-rays strike the atoms in the crystal, interact with their electrons, and bounce off the atoms *534*

Z

ZEV zero emission vehicle *370*

Credits

Photographs

CHAPTER 0

OPENER: © R. Ian Lloyd/Masterfile;
PAGE 4: © R.W. Jones/CORBIS;
PAGE 6: Lucy Pryde Eubanks.

CHAPTER 1

OPENER: Image by Reto Stöckli,
NASA/Goddard Space Flight Center.
Enhancements by Robert Simmon;
FIGURE 1.1: Image provided by
ORBIMAGE. © Orbital Imaging
Corporation and processing by NASA
Goddard Space Flight Center;
FIGURE 1.2: © Shelley Gazin/Image
Works; **FIGURE 1.3:** Cathy Middlecamp;
FIGURE 1.5: © Lon C. Diehl/Photo Edit;
FIGURE 1.6: EPA; **FIGURE 1.8:**
© Galen Rowell/CORBIS; **FIGURE 1.9:**
Cathy Middlecamp; **FIGURE 1.13:**
Courtesy IBM, Almaden Research Center;
FIGURE 1.14: © The McGraw-Hill
Companies, Inc./Bob Coyle, photographer;
FIGURE 1.16a: Adapted from How Stuff
Works, Convex Group; **FIGURE 1.16b:**
© Courtesy Corning, Incorporated;
Figure 1.17a: © The McGraw-Hill
Companies, Inc./Photo by Eric Misko,
Elite Images Photography;
FIGURE 1.17b: © Chinch Gryniewicz,
Ecoscene/CORBIS; **FIGURE 1.19a:**
© Image Source/CORBIS RF website;
FIGURE 1.19b: © Digital Vision Vol.
DV384, /Getty; **FIGURE 1.20:** © The
McGraw-Hill Companies, Inc./Photo by
Ken Karp; **FIGURE 1.21:** © David M.
Grossman/Photo Researchers;
FIGURE 1.22: © Sheila Terry/Photo
Researchers.

CHAPTER 2

OPENER (1–3): Ozone Processing Team
at NASA's Goddard Space Flight Center;
FIGURE 2.2 (all): © The McGraw-Hill
Companies, Inc./Stephen Frisch,
photographer; **FIGURE 2.5:** © Philip
Schermeister/National Geographic Image
Collection; **FIGURE 2.16:** Courtesy Blue
Lizard Products; **FIGURE 2.19 (all):**
© The McGraw-Hill Companies,
Inc./Stephen Frisch, photographer;
PAGE 96: Courtesy Dr. Susan Solomon.
Photo by Carlye Calvin; **FIGURE 2.22:**
United Nations Environment Program/World
Meteorological Organization. Image taken
by Dr. Ross Salawitch, Jet Propulsion
Laboratory; **FIGURE 2.23:** NOAA-
National Oceanic and Atmospheric
Administration; **FIGURE 2.28:** Courtesy
Pyrocool Technologies, Inc.

CHAPTER 3

OPENER: © John Dominis/Time Life
Pictures/Getty; **FIGURE 3.1:** NASA;
FIGURE 3.3: © Maria Stenzel/National
Geographic Image Collection;
FIGURE 3.7: © David Young-Wolff/
PhotoEdit; **FIGURE 3.21:** Conrad
Stanitski; **FIGURE 3.22 (both):** Ocean
Drilling Program; **FIGURE 3.23:**
© Marko Riikonen; **PAGE 152:**
www.wisconsinbutterflies.org;
FIGURE 3.26 (top): © Ira Kirschenbaum/
Stock Boston; **FIGURE 3.26 (bottom):**
© Michele Burgess/Stock Boston.

CHAPTER 4

OPENER (8 images): clockwise from
upper left: © Digital Vision/Getty R-F
website; © Photodisc Vol. 44/Getty;
© Brand X Vol. X195/Getty; © Photodisc
Vol. 88/Getty; © AP Wide World Photos;
© Photodisc Vol. 31/Getty; © Photodisc
Vol. 44/Getty; © Photodisc Vol. 39/Getty;
FIGURE 4.1: © Digital Vision Vol.
DV418/Getty; **FIGURE 4.7:** © Brand X
Vol. X147/Getty; **FIGURE 4.14:**
© Charles E. Rotkin/CORBIS;
FIGURE 4.17: © Michael Newman/
PhotoEdit; **FIGURE 4.18:** © David
Young-Wolff/PhotoEdit; **FIGURE 4.19:**
Courtesy Wilmer Stratton; **FIGURE 4.20:**
© Bob Daemmrich/Stock Boston;
FIGURE 4.21 (both): © AP Wide World
Photos; **FIGURE 4.22:** Courtesy, A.
Truman Schwartz; **FIGURE 4.25:**
© Argus Fotoarchiv/Peter Arnold;
FIGURE 4.27: © AP/Toyota/Wide World
Photos.

CHAPTER 5

OPENER: © Network Productions/Image
Works; **PAGE 219:** City of San Diego
Water Department; **FIGURE 5.1a:**
© David Young-Wolff/PhotoEdit;
FIGURE 5.1b: © David Raymer/CORBIS;
FIGURE 5.4: © D. Cavagnaro/Visuals
Unlimited; **FIGURE 5.12 (all):** © Tom
Pantages; **FIGURE 5.19:** © Jerry Mason/
SPL/Photo Researchers; **CONSIDER
THIS 5.30:** © R.W. Jones//CORBIS;
FIGURE 5.20: Courtesy of GE
Infrastructure Water and Process
Technologies; **FIGURE 5.22:** Knapper's
Water Store, LLC; **FIGURE 5.31:** Image
supplied courtesy of Katadyn.

CHAPTER 6

OPENER: © PressNet/Topham/The
Image Works; **FIGURE 6.1:** © Ted
Spiegel/CORBIS; **FIGURE 6.2:**

© PhotoDisc Vol. 77/Getty; **FIGURES 6.3, 6.4:** © The McGraw-Hill Companies, Inc./Photo By Eric Misko, Elite Images Photography; **FIGURE 6.5:** © The McGraw-Hill Companies, Inc./Photo by C. P. Hammond; **FIGURE 6.8:** © Charles D. Winters/Photo Researchers; **FIGURE 6.9:** National Atmospheric Deposition Program (NRSP-3)/National Trends Network. (2004). NADP Program Office, Illinois State Water Survey, 2204 Griffith Dr., Champaign, IL 61820; **FIGURE 6.12:** © E. R. Degginger/Color-Pic; **FIGURE 6.14:** © Kennon Cooke/Valan Photos; **FIGURE 6.17 (left):** © NYC Parks Photo Archive/Fundamental Photographs; **FIGURE 6.17 (right):** © Kristen Brochmann/Fundamental Photographs; **FIGURE 6.18:** © A. J. Copley/Visuals Unlimited; **FIGURE 6.19 (both):** U.S. National Park Service; **FIGURE 6.20b:** © Pittsburgh Post Gazette Archives, 2002. All rights reserved. Reprinted with permission; **FIGURE 6.21:** © Custom Medical Stock Photo; **FIGURE 6.25:** © Phil McCarten/PhotoEdit Inc.; **FIGURE 6.26:** Clean School Bus USA, USEPA Office of Transportation and Air Quality; **FIGURE 6.27:** National Energy Technology Laboratory/U.S. Department of Energy; **PAGE 305 (left):** ©Stock Portfolio/Stock Connection/PictureQuest; **PAGE 305 (right):** Cathy Middlecamp.

CHAPTER 7

OPENER: U.S. Department of Energy; **FIGURE 7.1:** © The McGraw-Hill Companies, Inc./Photo by C. P. Hammond; **FIGURE 7.2:** © The McGraw-Hill Companies, Inc./Photo by Eric Misko, Elite Images Photography; **FIGURE 7.3:** Courtesy, Nuclear Energy Institute; **FIGURE 7.4:** © Rob Crandall/Image Works; **FIGURE 7.5:** Photo by Heka Davis, courtesy AIP Emilio Segre Visual Archives, Physics Today Collection; **FIGURE 7.9:** © Joe Sohm/Image Works; **FIGURE 7.11:** © Igor Kostin/Corbis-Sygma; **FIGURE 7.12 (both):** EC/IGCE, Roshydromet (Russia)/Minchernobyl (Ukraine)/Belhydromet (Belarus); **FIGURE 7.13:** © Bettmann/CORBIS; **FIGURE 7.14:** Courtesy USEC, Inc.; **FIGURE 7.15:** © AP/Wide World Photos; **FIGURE 7.16:** © AIP Emilio Segre Visual Archives/W. F. Meggers Collection; **FIGURE 7.21:** © Southern Illinois University/Photo Researchers; **FIGURE 7.22:** © Chuck Nacke/Time Life Pictures/Getty; **FIGURE 7.23:** © Visuals Unlimited; **FIGURE 7.25:** © Science Source/Photo Researchers; **FIGURE 7.26b:** U.S. Dept. of Energy; **FIGURE 7.27:** © U.S. Dept. of Energy/SPL/Photo Researchers.

CHAPTER 8

OPENER (top left): © Digital Vision Vol. DV673/Getty; **(top middle):** Image courtesy Segway LLC. www.Segway.com; **(top right):** Courtesy, Ballard Power Systems; **(bottom):** Copyright 2003 Michelin North America, Inc. Used by permission; **FIGURE 8.1:** Courtesy BLACK & DECKER; **FIGURE 8.3:** © The McGraw-Hill Companies, Inc./Photo by Eric Misko, Elite Images Photography; **FIGURE 8.9:** Proton Exchange Membrane Fuel Cell powered Segway HT. Fuel cell integration performed at the DoD Fuel Cell Test and Evaluation Center (FCTec) operated by Concurrent Technologies Corporation (CTC) Johnstown, PA.; **FIGURE 8.11:** Courtesy of Advanced Vehicle Development, Georgetown University; **FIGURE 8.12:** © Reuters/Wolfgang Rattay/CORBIS; **FIGURE 8.13:** © Brett Dewey; **FIGURE 8.14:** Image courtesy Global Electric Motorcars, a company of DiamlerCrysler, Fargo, N.D.; **FIGURE 8.15 a, b:** © AP/Wide World Photos; **FIGURE 8.18:** Reprinted with permission from C & EN Vol. 125 #24, pp. 7152–7153, © 2003 American Chemical Society. Image courtesy Koichi Komatsu, Institute for Chemical Research, Kyoto University; **FIGURE 8.19:** Warren Gretz/DOE/NREL; **FIGURE 8.23:** © Westinghouse/Visuals Unlimited; **FIGURE 8.26:** © Ken Lucas/Visuals Unlimited; **FIGURE 8.27:** Image courtesy of GEOSOL; **FIGURE 8.28:** Image courtesy of www.cleanenergy.org; **FIGURE 8.29:** © AP/Wide World Photos; **FIGURE 8.30:** © AP/Tribune-Star/Wide World Photos.

CHAPTER 9

OPENER: © John Shafer/Photo John; **FIGURE 9.1:** © Hank deVre/Mountain Stock; **FIGURE 9.6a:** © Bill Aron/PhotoEdit; **FIGURES 9.9, 9.10:** Courtesy DuPont; **FIGURE 9.13:** © The Garbage Project, University of Arizona; **FIGURE 9.14:** © Gayna Hoffman/Stock Boston; **FIGURE 9.15:** Image courtesy Andy Wyeth; **PAGE 426:** © The McGraw-Hill Companies, Inc./Photo by Eric Misko, Elite Images Photography.

CHAPTER 10

OPENER (top left): © Bushnell/Soifer/Getty; **(top right):** © Michael Newman/PhotoEdit; **(bottom left):** © Quill/Getty; **(bottom right):** © Comstock Images/Getty R-F website; **FIGURE 10.1:** © Terry Wild Studio; **FIGURE 10.7b 1,2,3:** © The McGraw-Hill Companies, Inc./Photo by Eric Misko, Elite Images Photography; **FIGURE 10.10:** Courtesy of The Alexander Fleming Laboratory Museum, St. Mary's Hospital, Paddington, London; **PAGE 448:** John M. Rimoldi, University of Mississippi; **FIGURE 10.14:** PDB ID: 1CQE. D. Icot, P.J. Loll, A.M. Mulichak, R.M. Garavito. The X-Ray Crystal Structure of the Membrane Protein Prostaglandin H2 Synthase-1 *Nature* 367 pp. 243 (1994); **PAGE 455:** © The McGraw-Hill Companies, Inc./Jill Braaten, photographer; **FIGURE 10.28:** © Michael Newman/PhotoEdit; **FIGURE 10.29 (both):** © Michael P. Gadomski/Photo Researchers; **FIGURE 10.30 (left):** © Gerald & Buff Corsi/Visuals Unlimited; **FIGURE 10.30 (right):** © James Leynse/CORBIS; **FIGURE 10.35 (right):** © 2004, Publishers Group. www.streetdrugs.org.

CHAPTER 11

OPENER: © 2004 Newsweek, Inc. All rights reserved. Reprinted by permission; **FIGURE 11.1:** Federal Citizen Information Center; **FIGURE 11.9 (top):** © Lund/Custom Medical Stock Photo; **FIGURE 11.9 (bottom):** © Roseman/Custom Medical Stock Photo; **FIGURE 11.14:** © Len Lessin/Peter Arnold; **FIGURE 11.19:** © Tony Freeman/PhotoEdit.

CHAPTER 12

OPENER: © AP/Wide World Photos; **FIGURE 12.5:** © Oesper Collection in the History of Chemistry; **FIGURE 12.6:** © Bettmann/CORBIS; **FIGURE 12.12 (both):** © Bill Longcore/Photo Researchers; **PAGE 545:** © Michael Nicholson/CORBIS; **FIGURE 12.15:** Courtesy Dr. Ken Culver, photo by John Crawford, National Institutes of Health; **FIGURE 12.16:** Courtesy Lifecodes Corporation; **PAGE 557:** © Bettmann/CORBIS; **PAGE 559 (left):** © APTV via Media/AP/Wide World Photos; **PAGE 559 (right):** © AP/Wide World Photos.

Index